Lecture Notes in Networks and Systems 723

The series "Lecture Notes in Networks and Systems" publishes the latest developments in Networks and Systems—quickly, informally and with high quality. Original research reported in proceedings and post-proceedings represents the core of LNNS.

Volumes published in LNNS embrace all aspects and subfields of, as well as new challenges in, Networks and Systems.

The series contains proceedings and edited volumes in systems and networks, spanning the areas of Cyber-Physical Systems, Autonomous Systems, Sensor Networks, Control Systems, Energy Systems, Automotive Systems, Biological Systems, Vehicular Networking and Connected Vehicles, Aerospace Systems, Automation, Manufacturing, Smart Grids, Nonlinear Systems, Power Systems, Robotics, Social Systems, Economic Systems and other. Of particular value to both the contributors and the readership are the short publication timeframe and the world-wide distribution and exposure which enable both a wide and rapid dissemination of research output.

The series covers the theory, applications, and perspectives on the state of the art and future developments relevant to systems and networks, decision making, control, complex processes and related areas, as embedded in the fields of interdisciplinary and applied sciences, engineering, computer science, physics, economics, social, and life sciences, as well as the paradigms and methodologies behind them.

Indexed by SCOPUS, INSPEC, WTI Frankfurt eG, zbMATH, SCImago.

All books published in the series are submitted for consideration in Web of Science.

For proposals from Asia please contact Aninda Bose (aninda.bose@springer.com).

Radek Silhavy · Petr Silhavy
Editors

Networks and Systems in Cybernetics

Proceedings of 12th Computer Science On-line Conference 2023, Volume 2

Springer

Editors
Radek Silhavy
Faculty of Applied Informatics
Tomas Bata University in Zlin
Zlin, Czech Republic

Petr Silhavy
Faculty of Applied Informatics
Tomas Bata University in Zlin
Zlin, Czech Republic

ISSN 2367-3370 ISSN 2367-3389 (electronic)
Lecture Notes in Networks and Systems
ISBN 978-3-031-35316-1 ISBN 978-3-031-35317-8 (eBook)
https://doi.org/10.1007/978-3-031-35317-8

This Springer imprint is published by the registered company Springer Nature Switzerland AG
The registered company address is: Gewerbestrasse 11, 6330 Cham, Switzerland

Preface

We are honored to present the refereed proceedings of the 12th Computer Science Online Conference 2023 (CSOC 2023), composed of three volumes: Software Engineering Perspectives, Artificial Intelligence Trends, and Cybernetics Perspectives in Systems. This Preface is intended to introduce and provide context for the three volumes of the proceedings.

CSOC 2023 is a prominent international forum designed to facilitate the exchange of ideas and knowledge on various topics related to computer science. The conference was held online in April 2023, using modern communication technologies to provide researchers worldwide with equal participation opportunities.

The first volume, Software Engineering Research in System Science, encompasses papers that discuss software engineering topics related to software analysis, design, and the application of intelligent algorithms, machine, and statistical learning in software engineering research. These papers provide valuable insights into the latest advances and innovative approaches in software engineering research.

The second volume, Networks and Systems in Cybernetics, presents papers that examine theoretical and practical aspects of cybernetics and control theory in systems or software. These papers provide a deeper understanding of cybernetics and control theory and demonstrate how they can be applied in the design and development of software systems.

The third volume, Artificial Intelligence Application in Networks and Systems, is dedicated to presenting the latest trends in artificial intelligence in the scope of systems, systems engineering, and software engineering domains. The papers in this volume cover various aspects of artificial intelligence, including machine learning, natural language processing, and computer vision.

In summary, the proceedings of CSOC 2023 represents a significant contribution to the field of computer science, and they will be an excellent resource for researchers and practitioners alike. The papers included in these volumes will inspire new ideas, encourage further research, and lead to the development of novel and innovative approaches in computer science.

April 2023

Radek Silhavy
Petr Silhavy

Organization

Program Committee

Program Committee Chairs

Petr Silhavy	Tomas Bata University in Zlin, Faculty of Applied Informatics
Radek Silhavy	Tomas Bata University in Zlin, Faculty of Applied Informatics
Zdenka Prokopova	Tomas Bata University in Zlin, Faculty of Applied Informatics
Roman Senkerik	Tomas Bata University in Zlin, Faculty of Applied Informatics
Roman Prokop	Tomas Bata University in Zlin, Faculty of Applied Informatics
Viacheslav Zelentsov	Doctor of Engineering Sciences, Chief Researcher of St. Petersburg Institute for Informatics and Automation of Russian Academy of Sciences (SPIIRAS)
Roman Tsarev	Department of Information Technology, International Academy of Science and Technologies, Moscow, Russia
Stefano Cirillo	Department of Computer Science, University of Salerno, Fisciano (SA), Italy

Program Committee Members

Juraj Dudak	Faculty of Materials Science and Technology in Trnava, Slovak University of Technology, Bratislava, Slovak Republic
Gabriel Gaspar	Research Centre, University of Zilina, Zilina, Slovak Republic
Boguslaw Cyganek	Department of Computer Science, University of Science and Technology, Krakow, Poland
Krzysztof Okarma	Faculty of Electrical Engineering, West Pomeranian University of Technology, Szczecin, Poland

Monika Bakosova	Institute of Information Engineering, Automation and Mathematics, Slovak University of Technology, Bratislava, Slovak Republic
Pavel Vaclavek	Faculty of Electrical Engineering and Communication, Brno University of Technology, Brno, Czech Republic
Miroslaw Ochodek	Faculty of Computing, Poznan University of Technology, Poznan, Poland
Olga Brovkina	Global Change Research Centre Academy of Science of the Czech Republic, Brno, Czech Republic and Mendel University of Brno, Czech Republic
Elarbi Badidi	College of Information Technology, United Arab Emirates University, Al Ain, United Arab Emirates
Luis Alberto Morales Rosales	Head of the Master Program in Computer Science, Superior Technological Institute of Misantla, Mexico
Mariana Lobato Baes	Research-Professor, Superior Technological of Libres, Mexico
Abdessattar Chaâri	Laboratory of Sciences and Techniques of Automatic control & Computer engineering, University of Sfax, Tunisian Republic
Gopal Sakarkar	Shri. Ramdeobaba College of Engineering and Management, Republic of India
V. V. Krishna Maddinala	GD Rungta College of Engineering & Technology, Republic of India
Anand N Khobragade (Scientist)	Maharashtra Remote Sensing Applications Centre, Republic of India
Abdallah Handoura	Computer and Communication Laboratory, Telecom Bretagne, France
Almaz Mobil Mehdiyeva	Department of Electronics and Automation, Azerbaijan State Oil and Industry University, Azerbaijan

Technical Program Committee Members

Ivo Bukovsky, Czech Republic
Maciej Majewski, Poland
Miroslaw Ochodek, Poland
Bronislav Chramcov, Czech Republic
Eric Afful Dazie, Ghana
Michal Bliznak, Czech Republic

Organizing Committee Chair

Radek Silhavy — Tomas Bata University in Zlin, Faculty of Applied Informatics
radek@silhavy.cz

Conference Organizer (Production)

Silhavy s.r.o.
Website: https://www.openpublish.eu
Email: csoc@openpublish.eu

Conference Website, Call for Papers

https://www.openpublish.eu

Contents

Analysis of Changes in the Pricing of Renewable Energy Sources

K. V. Selivanov(✉) and A. D. Zaharova

Bauman Moscow State Technical University, Moscow, Russia
selivanov_kv@mail.ru

Abstract. The article systematizes the observed changes in the cost of generating renewable energy sources in the period from 2005 to 2021 around the world. A number of scientific papers and articles on this topic have been studied and a rapid decrease in the cost of renewable energy production and, as a consequence, the electricity generated by them has been revealed. The main criteria by which the price of renewable energy generation is formed are outlined. As a result of the analysis, dependencies were obtained that allow us to make a conclusion about the profitability of using renewable energy for electricity generation. #CSOC1120.

Keywords: Analysis · Renewable Energy · Pricing

1 Introduction

The actualization of the climate agenda reveals a trend towards a gradual increase in the role of renewable energy sources in the world [1]. The reason for this is the rapid change of the global climate to a state that threatens the established way of life, caused by anthropogenic emissions of greenhouse gases into the atmosphere, man-made disasters, unsustainable use of natural resources, etc. [2].

At the same time, renewable energy sources as the most efficient, accessible and safe for the environment began to act as an alternative to traditional resources [3].

The purpose of alternative energy is to use renewable or virtually inexhaustible resources to generate energy. The main types of renewable resources include:

- solar energy;
- wind energy;
- energy of water (including waste water energy), except for use of such energy in pumped-storage power plants;
- geothermal energy;
- biomass, except for waste derived from the use of hydrocarbon raw materials and fuels.

At the same time, the key issue for the discussion of the transition to alternative energy was the economic component of power generation, namely the production costs of electricity, pricing scheme and profitability [3, 4].

R. Silhavy and P. Silhavy (Eds.): CSOC 2023, LNNS 723, pp. 1–7, 2023.
https://doi.org/10.1007/978-3-031-35317-8_1

In this article will establish trends in the price of kWh of generated energy from renewable energy sources, made a diagram of the cost of capacity installation and electricity generation from various renewable energy sources. The impact of the cost of traditional resources on the cost of power generation is evaluated and a conclusion about the profitability of renewable energy sources at the moment is made.

2 The Problem of Energy Pricing

Today, conventional hydrocarbon resources account for about 80% of the world's energy consumption. Despite the gradual decrease in the price advantage of energy from fossil sources, the slow transition to alternative energy is primarily due to the hesitancy of many countries to follow the goals of reducing fossil fuel consumption at the expense of the current stability of energy supply and price stability [2, 3]. The share of energy production based on various fossil fuels, biomass and direct electricity generation is presented in Fig. 1.

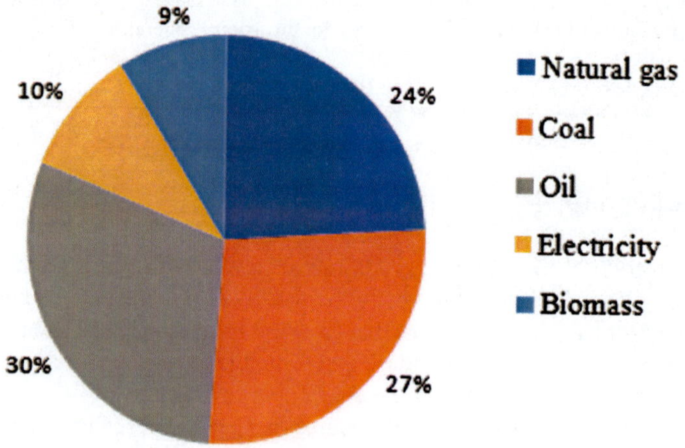

Fig. 1. Proportion of energy produced by different methods [5].

In order for renewable energy sources to become more competitive for electricity generation, the cost of electricity generated from them must be less than the wholesale electricity price for industrial consumers.

Below is a diagram (Fig. 2) illustrating the development of prices for power generation from renewable and non-renewable sources without taking into account subsidies or other state tariffs.

As can be seen, over the past 12 years, the cost of energy from renewable energy sources has decreased significantly, while hydrocarbon resources are practically unchanged in their pricing, and in some cases are very expensive in the short term.

To assess the profitability of renewable energy sources and fossil fuels, it is necessary to consider how the cost of electricity is calculated. A common method is to calculate the average estimated cost of electricity generation over the entire life cycle of a generating

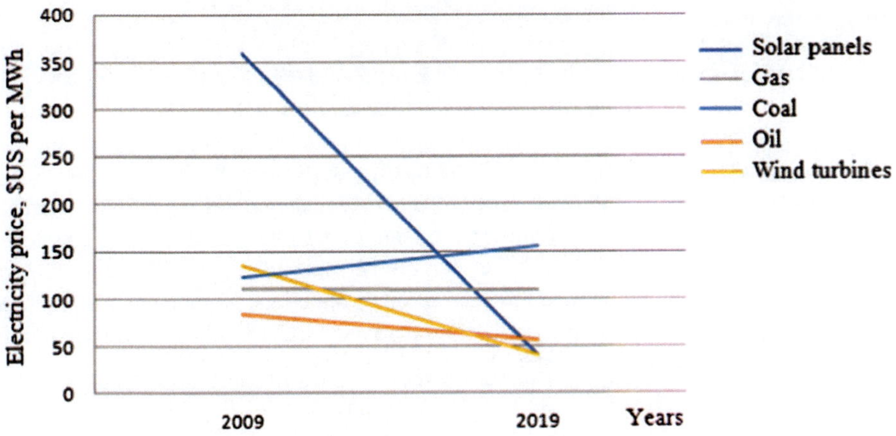

Fig. 2. The dynamics of electricity prices from different types of sources.

facility, referred to as LCOE (the levelized cost of electricity) - the cost per kWh over the life of a power plant, which equates the present value of revenues from generating and selling electricity to the present value of the cost of constructing and operating a power plant.

This indicator is calculated using the following formula and includes several parameters:

$$LCOE = \frac{\sum \frac{Cap_t + O\&M_t + F_t + Carb_t + D_t}{(1+r)^t}}{\sum \frac{MWh_t}{(1+r)^t}} \quad (1)$$

where MWh_t is the amount of energy produced in year t;
$(1 + r)^t$ - discount factor for year t;
Cap_t - total capital costs in year t;
$O\&M_t$ - operating costs in year t;
F_t - fuel costs in year t;
$Carb_t$ - greenhouse gas costs in year t;
D_t - waste management and decommissioning costs in year t;

In this paper, a simplified form was used and only measures of electricity price dynamics in terms of capacity installation and maintenance costs were considered.

3 Peculiarities of the Pricing of Energy from Renewable Energy Sources

The main criteria for assessing the relative cost of renewable energy:

- Cost of Service;
- Price of installations per kWh - shows the price of equipment per hour of energy it produces;
- Volume of installed capacity.

3.1 Solar Panels

In the case of solar panels, the cost of maintenance depends on the equipment manufacturer itself, the number of panels, the technological state of the panels, and the territorial location of the installation. Examples of such costs can be problems with monitoring, maintenance of wires or individual modules, installation of snow screens, lack of communication with the inverter. A properly installed system requires virtually no maintenance costs [6]. Firstly, most manufacturers of solar installations guarantee on average 25–30 years of trouble-free service life [7]; secondly, growth of technological progress allows reducing the rate of degradation of materials, so such costs do not contribute significantly to the final cost of solar energy and LCOE indicator and account for only 5–10% of the final cost of energy.

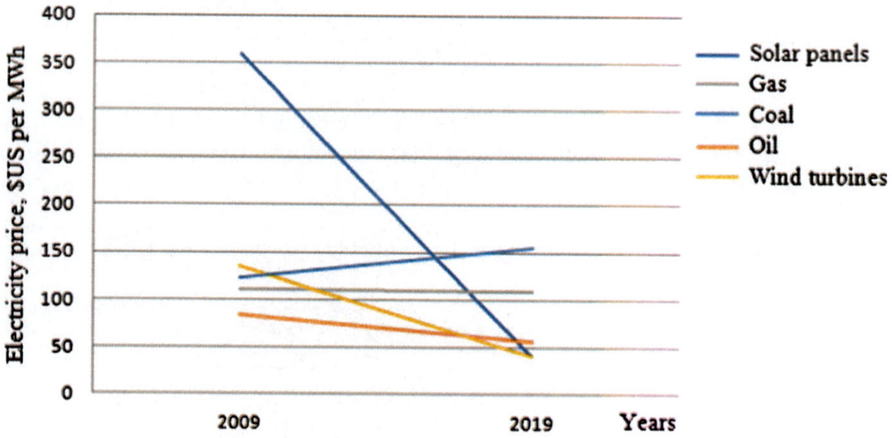

Fig. 3. Cost per kWh and installed power volumes for wind turbines and solar panels from 2005 to 2021 [9, 11].

A chart illustrating the change in solar panel price and installed capacity from 2005 to 2022 is shown in Fig. 3. It clearly demonstrates that, at the global level, the reduction in equipment costs was about 80% from 2005 to 2021, mainly due to the multiple reductions in module prices, as well as the reduction in costs of components required to convert PV module output into usable electrical energy.

The relative cheapness of solar panels is also due to the simplicity of the designs themselves, guaranteeing their reliability and mobility compared to other sources of inexhaustible energy [7].

Note that with the exponential increase in the number of installed capacity, the price for solar modules decreased exponentially. That is, for each doubling of installed capacity, the price of the equipment decreased by 20.2%. This is the peculiarity of renewable energy cost formation: as the amount of energy produced increases, the demand for equipment increases, and, as a consequence, its price decreases, which leads to an even greater increase in production growth [9, 10].

3.2 Wind Turbines

In contrast to solar panels, the calculation of the cost of wind energy seems more complex. An important characteristic of wind energy production is the absence of its fixed price. There is no linear relationship mathematically, because the annual production of electricity will vary greatly depending on the amount of wind available at a given wind turbine site. Thus, there is no single price for wind power, but rather a price range depending on wind speed.

The LCOE of wind turbines is determined by several parameters: total installed cost; lifetime capacity factor; electricity costs; economic life of the project; and cost of capital.

Approximately 75% of the LCOE for a wind turbine is related to initial costs, such as the cost of the turbine, its installation, electrical equipment, connection to the grid, and so on. Obviously, fluctuations in fuel prices have no effect on the cost of wind turbine power generation. However, wind generator is more capital-intensive compared to traditional technologies of power generation based on fossil fuels, on the other hand, when generating electricity from classic fuels, their price in power generation is somewhere between 30–55%. Figure 3 illustrates the amount of installed power capacity for wind turbines in the period from 2005 to 2022.

Similar to solar panels, the cost of wind power is also decreasing, although not exponentially. As production capacity increases, wind generator prices peaked between 2007 and 2010, but have since fallen 44% to 78% by the end of 2021, with recent prices ranging from $1.3 to $1.8 per kWh.

O&M costs for wind power often account for a significant portion (up to 30%) of LCOE, which is much higher than for solar panels. However, improvements in technology, increased competition among service providers, and increased experience of service providers are driving down plant maintenance prices.

3.3 Hydroelectric Power Plants

Hydroelectric power plants are unique in comparison to other types of renewable power generation. They require a significant capital investment, but offer extremely low operating costs and a long service life of 40–50 years, which can often be extended to 100 years. The result is extremely competitive energy production costs [12, 13].

However, investment costs are highly dependent on geographic location, which explains the wide range of equipment installation costs and the amount of installed power capacity. It is also important to note that hydropower projects can be designed to vary greatly in efficiency, making it difficult to estimate LCOE.

Figure 4 shows the change in installed power from 2005 to 2021.

The slow growth of installed power can be explained by the fact that the predominant part of the green energy balance is accounted for by hydropower plants. Operation of hydropower plants is associated with a number of difficulties, but at the moment we are most interested in the problem of "seasonality" of power generation volumes. The flow of the rivers during the year is uneven, which entails varying levels of load on the power plant's power units. This leads to the fact that in most cases, many power plants are forced to stand idle, which entails a low level of installed capacity.

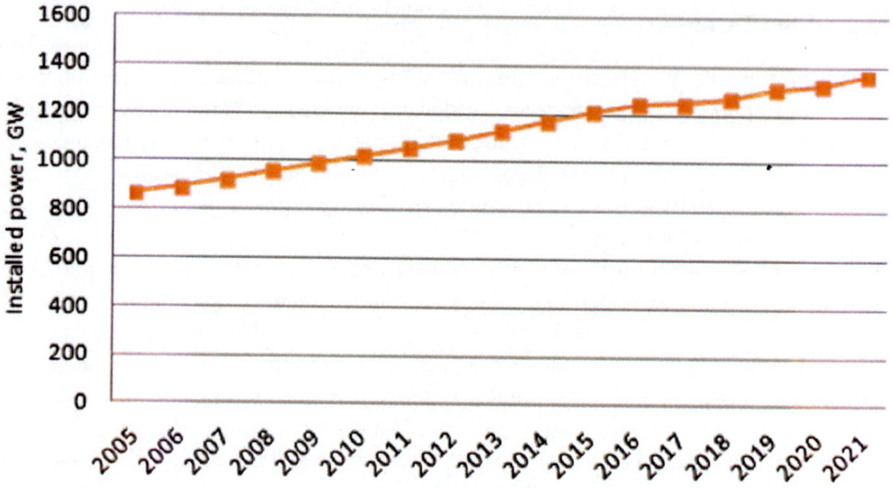

Fig. 4. Changes in the volume of installed power generation of hydroelectric power plants from 2005 to 2021.

Maintenance costs account for 20% to 61% of the LCOE, while materials account for about 4%. Such costs typically include periodic mechanical and electrical equipment upgrades such as turbine overhauls, generator rewinds, and reinvestment in communication and control systems, but exclude overhauls of electromechanical equipment or repairs to plugs, backstops, etc. [14]. They are replaced infrequently, with a design life of 30 or more years for electromechanical equipment, and 50 or more years for manual and backstroke drives. However, hydropower is distinguished from other renewable energy-based power sources by minimal stochasticity of power coming, which favorably distinguishes it from other green energy units.

4 Conclusion

The main feature that favorably distinguishes renewable energy sources from fossil fuels is the potential cost of operating plants of this type of raw materials, as they use inexhaustible fuel sources and the main costs associated with this area of energy are due to the maintenance of power plants. We can conclude that to achieve a minimum cost of widespread use of power plants based on renewable energy sources, it is necessary to develop technological methods of production of power units, which would entail a qualitative increase in the profitability of renewable energy. All areas of green energy are subject to the same principle: every doubling of power leads to a corresponding decrease in costs. The more widespread hydroelectric, wind and photovoltaic generators become, the faster their cost drops.

In the long term, green energy has every chance of gaining a significant share in the global energy balance. In this case, we can assume that in general its cost will be lower than that of classical energy, which will undoubtedly entail large monetary investments and will cause an increase in the electricity produced. Greater energy availability, especially in developing countries, will lead to global economic growth and higher overall living standards.

Acknowledgements. The results were partially obtained within the project under the Development Program of Bauman Moscow State Technical University as part of the Priority-2030 federal academic leadership program.

References

1. Selivanov, K.V.: Development trend of electrification and small-scale power generation sector in Russia. In: Radionov, A.A., Karandaev, A.S. (eds.) RusAutoCon 2019. LNEE, vol. 641, pp. 409–416. Springer, Cham (2020). https://doi.org/10.1007/978-3-030-39225-3_44
2. Yergin, D.: The Quest: Energy, Security and the Remaking of the Modern World. Penguin Books, New York (2012)
3. Todorov, G.N., Volkova, E.E., Vlasov, A.I., Nikitina, N.I.: Modeling energy-efficient consumption at industrial enterprises. Int. J. Energy Econ. Policy **9**(2), 10–18 (2019)
4. Vlasov, A.I., Artemiev, B.V., Selivanov, K.V., Mironov, K.S., Isroilov, J.O.: Predictive control algorithm for a variable load hybrid power system on the basis of power output forecast. Int. J. Energy Econ. Policy **12**(3), 1–7 (2022). https://doi.org/10.32479/ijeep.12912
5. The price of electricity from the long-standing sources: fossil fuels and nuclear power. https://ourworldindata.org/cheap-renewables-growth. Accessed 23 Aug 2022
6. Vlasov, A.I., Grigoriev, P.V., Krivoshein, A.I., Shakhnov, V.A., Filin, S.S., Migalin, V.S.: Smart management of technologies: predictive maintenance of industrial equipment using wireless sensor networks. Entrepreneursh. Sustain. Iss. **6**(2), 489–502 (2018)
7. Selivanov, K.V., Shakhnov, V.A.: Analysis of development trends of power industry market for power consumption purposes. In: Proceedings of DEFIN-2021: IV International Scientific and Practical Conference, vol. 38, pp. 1–6 (2021). https://doi.org/10.1145/3487757.3490878
8. Vlasov, A.I., Shakhnov, V.A., Filin, S.S., Krivoshein, A.I.: Sustainable energy systems in the digital economy: concept of smart machines. Entrepreneursh. Sustain. Iss. **6**(4), 1975–1986 (2019)
9. Todorov, G.N., Vlasov, A.I., Volkova, E.E., Osintseva, M.A.: Sustainability in local power supply systems of production facilities where there is the compensatory use of renewable energy sources. Int. J. Energy Econ. Policy **10**(3), 14–23 (2020)
10. Lantz, E., Hand, M., Wiser, R.: The Past and Future Cost of Wind Energy. U.S. Department of Energy. https://www.nrel.gov/docs/fy12osti/54526.pdf. Accessed 23 Aug 2022
11. The share of production of renewable energy sources in the global energy balance. https://yearbook.enerdata.ru/renewables/wind-solar-share-electricity-production.html. Accessed 23 Aug 2022
12. Sioshansi, R., Short, W.: Evaluating the impacts of real-time pricing on the usage of wind generation. IEEE Trans. Power Syst. **24**(2), 516–524 (2009)
13. Vlasov, A.I., Echeistov, V.V., Krivoshein, A.I., Shakhnov, V.A., Filin, S.S., Migalin, V.S.: An information system of predictive maintenance analytical support of industrial equipment. J. Appl. Eng. Sci. **16**(4), 515–522 (2018)
14. Akberdina, V., Kalinina, A., Vlasov, A.: Transformation stages of the Russian industrial complex in the context of economy digitization. Probl. Perspect. Manage. **16**(4), 201–211 (2018)

Management of a Replacement Policy of Learning-Based Software System Based on a Mathematical Model

Eze Nicholas, Okanazu Oliver(✉), Ifeoma Onodugo, Madu Maureen, Ifeoma Nwakoby, Ifediora Chuka(✉), Eze Emmanuel, Onyemachi Chinedu, and Onyemachi Chinmma

University of Nigeria, Nsukka, Nigeria
{nicholas.eze,okanazu.oliver,ifeoma.onodugo,madu.maureen, ifeoma.nwakoby,ifediora.chuka,ezec.emmanuel,chinedu.onyemachi, chidinma.onyemachi}@unn.edu.ng

Abstract. Although school's addition of web-based services to teaching and learning has increased, many effective quality systems to achieve replacement of software product due to software expansion and ageing has still not been well established. The primary objective of this study was therefore to develop a mathematical model for software developers advanced as potential breakthroughs in determining a replacement of ageing educational software products. This was a developmental case with retrospective review of software on maintenance for many years. The primary endpoint of such software was incidences of adverse breakdowns and malfunctions caused by system size, age, structure, incorrect, out of date documentation, complexity of the system and insufficient knowledge of the maintainer of the system. For comparison, data was generated from 2 software development firms to determine the rate of inflation, depreciation, salvage value and varying maintenance cost with the age of the software. The quality loss function was obtained to ascertain if software products are to continue undergoing maintenance process or a replacement. 84 software was developed within the 13years period of study. Their age mix ascertained. Findings revealed a sharp rise in year-wise development cost and a sharp reduction in year-wise cost of the salvage value of the software as they aged. Cost due to maintenance and repairs also vary with age and indicated a sharp rise in the fixed annual payment per year on the 10th year, an indication that software may be considered for replacement at the end of the 10th year based on the model. However, some common adverse effects of software failures due to age were noted. The requirement for this model for software developers therefore warrants further investigation into its applicability to achieve optimum economic life of software. Larger, comparative studies need to be performed with this model before its use can be routinely recommended to software development firms.

Keywords: Learning-Based Software · Model · Quality Lost Function · Software maintenance

R. Silhavy and P. Silhavy (Eds.): CSOC 2023, LNNS 723, pp. 8–22, 2023.
https://doi.org/10.1007/978-3-031-35317-8_2

1 Introduction

It is often considered unreasonable to ask a civil engineer to move an old bridge 200 miles or rotate it through 90° but it is perfectly acceptable to tell a software engineer to rewrite half an aged operating system over a 5-year period. This is because civil engineers know that redesigning half a bridge is so expensive and dangerous but is cheaper and safer to rebuild it from scratch. On the contrary, software engineers are equally well aware that, in the long run, extensive maintenance of software is unwise [1] and that replacing the software by writing the code from scratch will sometimes prove to be less expensive. Nevertheless, clients had frequently demanded major changes that most times require writing the code from scratch and subsequent replacements to software [2, 3]. Brooks points out that there will always be pressures from clients to replace ageing software. It has been observed that the problems caused by frequent and drastic replacement of software are merely problems caused by system size, age, structure, incorrect, out of date documentation, complexity of the system and the insufficient knowledge of the maintainer of the system [4] and if the public at large were better educated with regard to the nature of software, then demands for software replacement would not occur [5, 6]. But brooks still maintained that changeability is a property of the essence of software, an inherent problem that cannot be surmounted. That is to say, the very nature of software is such that, no matter how the public is educated, there will always be pressures for replacements of software, and often these replacements will be drastic.

However, there are seven principal reasons why learning software should undergo replacement. First, the learning institutions knows that software is a model of reality, and as the reality changes, so the software must adapt or be replaced [7]. Second, if a learning software is found to be useful, then there are pressures, chiefly from the institution or the satisfied learners, to extend the functionality of the product beyond what is feasible in terms of the original design [4, 6, 7]. Third, one of the greatest strengths of software is that it is so much easier to change than hardware. In other words, successful learning software survives well beyond the lifetime of the hardware for which it was written. In part this is due to the fact that, after 4 or 5 years, hardware often does not function as well as it did [6]. But more significant is the fact that technological change is so rapid that more appropriate hardware components, such as larger disks, faster CPU'S, or powerful monitors, become available while the software is still viable. In general, the software will have to be modified to some extent in order to run on the new hardware [8]. Fourth, true retirement of software is a somewhat rare event that occurs when a software product has outgrown its usefulness. Most educational institutions may no longer require the functionality provided by the product, and may wish to replace it. This is because, after many years of service, a stage is reached when further maintenance of software is no longer cost- effective [9]. Sometimes the proposed changes are so drastic that the design as a whole would have to be replaced [1, 3]. In such a case it is less expensive to redesign and recode the entire product [1]. Fifth, so many changes may have been made to the original design that interdependencies have inadvertently been built into the product, and there is a real danger that even a small change to one minor module might have a drastic effect on the functionality of the product as a whole [10]. Sixth, the documentation may not have been adequately maintained [11, 12], thus increasing the risk of a regression fault [13]. Seventh possibility is that if the hardware (and operating system) on which the product runs is to be replaced; it may be more economical to rewrite than to modify [1].

However, in each of these factors the current version of the software after undergoing extensive maintenance is advised to be replaced by a new version, and the software process continues. This is because part of the essence of software is that it has to be maintained and replaced when deemed old [2] and this inexorable and continual change has a deleterious effect on the quality of software. However, since software deteriorate with time because of so many activities involved during maintenance [14] every concern has to take on building a sustainable framework for a replacement of aging software at a certain point in time. This significant and dynamic research zone has always been neglected by researchers. Instead what researchers are only interested on is developing models based on different assumptions ranging from software testing strategies and reliabilities [15]; to reliability models for software within the framework of perfect debugging and imperfect debugging, [16, 17] as well as a fault-correction models [18, 19]. This present study to the best of our knowledge appears to be the very first to develop a model within the framework of determining a replacement pattern of software system. This is therefore how our study contributes to the existing knowledge.

2 Classification Matrix of Existing Methodologies on Related Work

This section reviews previous research from different scholars on issues of maintenance and replacement of software systems and the optimum economic level. The aim is to enable the readers gain more insight on software maintenance and replacement using the model so developed.

2.1 State of the Art

Overview of Maintenance Issues in Software System

Software products are developed and afterwards are handed down to the client for use. However, the moment the client accepts the product, whatever changes that comes after constitute maintenance, [20]. This is because software product is deemed as a model of the real world, which no doubt perpetually changes. For example, new information introduced in most of our institutional websites may necessitate that the website has to be maintained constantly so as to reflect an exact representation of the real world. What we mean here is that if the variable say, school fee rate, for instance, increases from say, 8 to 10 percent, then almost every software product that deals with paying school fees or invoice generation has to be changed. Let assume the product contains the C++ statement like below;

Const Float School FeeRate 8.0;

or the equivalent Java statement

Public Static Final Float School FeeRate (Float) 8.0;

and we declare school feeRate as a floating-point constant and initialize the value to 8.0, then it makes maintenance to be relatively easy by simply changing a line of code using text editor to replace the value of 8.0 by 10.0 and thereafter the code is recompiled and relinked. But, if say in a situation where the name school feeRate is not used for instance in a product and the actual value of the school feeRate of 8.0 has been used in a product wherever the value of the school feeRate is invoked, then it becomes extremely difficult to modify such software product. An instance is, there could be areas where the value 8.0 occurred in the source code that needed to be changed to 10.0 but are overlooked, or instances where the value 8.0 that has nothing to do with school feeRate are incorrectly changed to 10.0.

With these instances, it becomes extremely difficult to find these faults almost always. To deal with this problem requires that Testing and proofing of correctness of software must be adequately carried out during development in other to discover faults earlier to achieve software products that are reliable, [21]. Reliability here will mean that every phase in software development process, (the requirement; the specification documents; the design; e.t.c.) must be tested, [22]; and thereafter is certified error free by the software quality assurance team, [23] before handing it to the client. See Fig. 1.

Fig. 1 Testing in software development

However, often times, it is considered more expensive to continue to make modifications or determine which of the many constants that needs to be entirely scraped if a software product is badly damaged. In fact, in real sense it is more advisable at this stage to discard the software product; recode and redesign the entire software, [1] than maintaining software products that are bad. But in real sense, software developers do the opposite: they tend to discard bad products, and repair good ones for 10, 15, or even 20 years; leading most educational institutions into heavy loss in terms of increased maintenance cost, time, less profit, delay in delivery of jobs, learners' dissatisfaction and project deviating from earlier budget, [24, 10, 25]. According to Shaw, a major limitation in maintaining software products is that it consumes more money than all other software development activities combined. Research carried out independently for over two decades, 2006 and 2019 respectively to investigate the proportional software maintenance cost, in other words, the cost ratio of new development versus maintenance points to the fact that almost two-third of total software cost went into maintenance, [26, 11]. See the pie chart in Fig. 2.

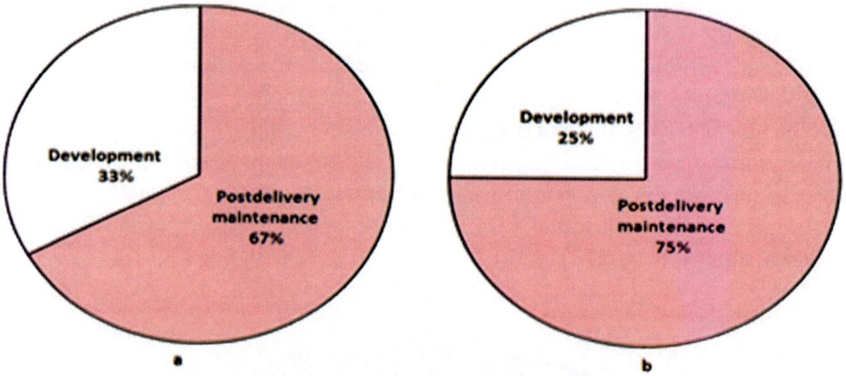

Fig. 2 Approximate average cost percentage of development and post delivery maintenance. (a) Between 1996 and 2002 and (b) Between 2012 and 2016. (Source Schach 2017)

Further research evidence from the National Institute of Standards and Technology (NIST) revealed that United States loses a whopping $59.5 billion every year from software maintenance, [27]. Between 1990 and 1994, annual software maintenance cost alone in the United States was estimated to be more than $70 billion for ten billion lines of existing code [28]. At the company level, Nokia Inc. Used about $90 million for preventive Y2K-bug corrections, [29]. The Cutter Consortium in 2002 also reported that about 45% of software development cases they studied were full of bugs that could not be managed hence could not be used leading to almost 78% of software development firms to settle their disputes in court and the financial implications of these crises were too horrendous on both the firms and the customer, [30].

The duo of researchers, [31, 32] maintained same position but this time on other limitations to software maintenance. Both are of the view that a common limitation peculiar to software maintenance is documentation, or rather a lack of it. They stated that project managers are always overanxious in placing targets on task duration times, and in trying to develop and deliver software against this time deadline, the original specification and design documents are frequently not updated and are consequently almost useless to the maintenance team. Most often documentation like the database manual or the operating manual may not have been written, because project managers decided that delivering the product to the client on time was more important than developing the documentation in parallel with the software leaving only the source code in many cases as the only documentation available to the maintainer. Corroborating this assertion, [7] maintained that over 100 billion lines of code in software production in the world, 80% of it is unstructured, patched, and badly documented. It is therefore necessary to keep these software systems operational and source code well documented. Figure 3 below clearly illustrates the concept.

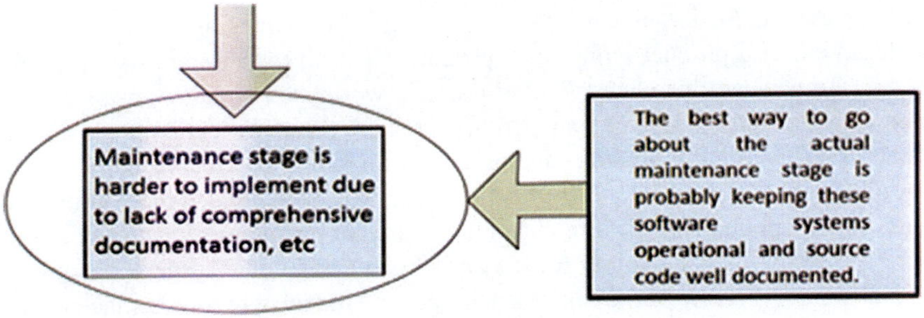

Fig. 3 The success of a postdelvery maintenance process. (Source: Eze 2017)

However, [33] have argued that even when the source code is available in most cases, personnel turnover in the software industry as a result of occupational mobility constitute a huge problem during maintenance. For example, if the original developers of a particular software firm migrate to other firms, maintenance of the software product become almost impossible in that none of the original developers may be working for the organization at the time when maintenance is needed to perform. Consider the case of Toronto in 2000 where they lost a whopping $700,000 in pet fees simply because nearly half of Toronto's dog and cat owners were never billed because the original developers of the computerized billing system who knew how to run the system were laid off; some left due to better opportunity while some due to downsizing. However, none of the city employees understood the system well enough to debug it when problems arose leaving no one to get things going again when the system ran into trouble and collapsed. By so doing the required maintenance may not have been done correctly and the best organization can do is to discard the software and go for a new one.

2.2 Replacement of Software at Optimum Economic Level

To maintain software operations at optimum economic levels, replacement of aged software becomes necessary through an application of this replacement model based on revolutionary quality concepts for arriving at a policy of replacement of ageing software products. This quality concept is often misunderstood especially when used within the software context. This is because many interpret quality as excellence of some sort, [34] but this unfortunately is not the meaning intended by the researcher. Quality level does not also mean the level of fraction, but the level that indicates the magnitude of loss. To this end, we define quality as the losses a product imparts to the society from the time the product is shipped, other than the losses caused during the development of the product, [35]. We also recognize the vitality of producing an outcome based on target without which could lead to losses through excessive variation in performance over time between the units or deviation from target and these may affect the quality level of a product leading to great losses to the society, [36]. These losses to the society does not include only the inevitable losses through series of maintenance activities, but the cost of production, the cost to the user during the life time of the product, the cost to the environment, and the difficulties in integrating or interfacing with other parts of a product, etc

However, these losses can only be minimized by changing the process through replacement of a product with a new product especially when the existing product is of age. To achieve this change in the process, we looked on adoption of a proactive technique based on measurement, analysis, prediction and prevention while developing an entirely new quality products and processes instead of continuous maintenance and spending into bad products, [37]. We have successfully integrated these strong mathematical and applied statistical methods into software processes so as to achieve higher product stability and capability for the aim of achieving both organization and client's satisfaction, [38]. However, we first sketch a general model of product replacement based on quality loss function which we believe can illuminate learning and understanding of product developers as well.

2.3 Theoretical Framework and Model Formulation

Our model is developed based on the "Great Replacement" theory propounded by a French author Renaud Camus in 2010. The theory states that the easiest thing to do for a government (institutions) was to replace or change anything that has forfeited its confidence [39]. In other words, in a situation where the gains from using software outweigh the cost, the easiest thing to do is to change or replace it. However, based on the literature so far reviewed, we have seen that a software product is said to be bad when the output of such product fails to meet the specification and customers' satisfaction, [40]. On the contrary if the product is one that meet the output specifications and satisfies the customers' needs, such product is said to be of best quality. Knowing this, we never rate the quality of any software product based on whether the products meets its specifications or not. Rather, we based our facts on functional deviation observed in the product's response from the target. We maintained that this response is a software quality characteristic that need to be quantified to ascertain whether the products is still at its best quality or whether the product has deviated from planned target and need to be replaced, [41]. However, for any software product to deviate from its planned target, some attribute of such product may either have been observed in the form of product failure before its expected lifespan or its response becoming poorer with time, leading to high cost due to post-delivery maintenance, cost of warranty, energy, money and time spent by clients, clients' complaints in the form of dissatisfaction, eventual loss of market share and company's reputation. All these together is known as quality loss and it needs to be quantified in this study to ascertain if a software product is to continue undergoing maintenance process or how much time is reasonable for such software products to be discarded or rather replaced. To achieve this, we generated a quality loss function based on the standard deviation (σ) and variation of software product from the target (μ-μ0) as given below:

$$Q = K'[(\mu - \mu 0)2 + \sigma 2]$$

From the above equation, the quality loss is seen to reduce only if the product variation from the mean is reduced leading to an adjustment of the mean nearer to the target through a scaling factor. Thus, Quality loss function if adjusted is given below:

$$Qa' = \eta = 10Log10[\mu 2/\sigma 2]$$

3 Research Method

The most commonly used methods, either to replace any product or to make changes in existing products, are given below;

1. Payback period method.
2. Rate of return method.
3. Mathematical model with present worth method.
4. M.A.P.I. method.

In this research, the mathematical model with present worth method were considered and used to determine the time that is reasonable for aged software products to be replaced so as to avoid wastes and achieve optimum economic life. To achieve this, the following procedures were adopted in working out the economic life of a product.

3.1 Research Procedures Adopted

1. Collection and processing of the data, such as number of software developed and their age mix, development cost, operating and maintenance cost, etc., year-wise.
2. Construction of suitable mathematical model.
3. Deriving solution from the model by feeding the data

3.2 Collection and Processing of the Data

The values of the development cost, salvage value of the software, discounting factor and the running cost of maintaining a software product was collected and computed. Two software development firms were selected one each from Nigeria and Japan. There are 84 software developed within the 13 years period of study. (38 nos. From Tenecee (Nigeria) and 46 nos. From Microtelesoft (Japan)). The age mix of this software is given below in Table 1.

Development Cost

The accepted tender price was taken as the development cost of software. Year-wise price of the software is indicated below in Table 2.

Meanwhile Rate of Price inflation of the software so developed has been calculated using the following formular:

$$\mathrm{Cn} = \mathrm{C}(1 + \mathrm{j})^{\mathrm{n}-1},$$

where Cn = development cost of the software in the nth year.

C = Price of the software in the base year, (2007–08).

j = Annual rate of inflation of new software (average)

n = Number of years, the age of software as taken from the beginning of base year of purchase and commissioning

Substituting, we have

$3{,}12{,}578 = 1{,}20{,}694\ (1 + j)^{11-1}$ or $j = 10\%$.

Table 1. The age mix of software developed between 2007–2019.

		Software developers		
S/N	Year of developed software	Tenecee (Nigeria)	Microtelesoft (Japan)	Total
1	2007	5	–	5
2	2008	4	–	4
3	2009	1	3	4
4	2010	–	–	–
5	2011	–	1	1
6	2012	–	14	14
7	2013	–	3	3
8	2014	–	–	–
9	2015	–	6	6
10	2016	8	4	12
11	2017	14	–	14
12	2018	6	6	12
13	2019	–	9	9
	Total	38	46	84

Table 2. Purchase price of software

Year	Purchase price of software in (Dollars)
2007–08	1,20,694
2008–09	1,32,763
2009–10	1,46,040
2010–11	1,60,523
2011–12	1,76,213
2012–13	1,94,317
2013–14	2,13,628
2014–15	2,35,353
2015–16	2,58,285
2016–17	2,84,838
2017–18	2,96,452
2018–19	3,12,597

Table 3. Salvage value of the software

Year	Ages of the software in years	Salvage values of software in (Dollars)
2007–08	1	1,14,808
2008–09	2	1,09,209
2009–10	3	1,03,883
2010–11	4	98,851
2011–12	5	93,996
2012–13	6	89,413
2013–14	7	85,052
2014–15	8	80,904
2015–16	9	76,958
2016–17	19	73,205
2017–18	11	69,444
2018–19	12	66,232

Table 4. Maintenance cost of software

Age of the software	Cost of maintenance and repair
1	1781
2	2493
3	2904
4	4,694
5	4,113
6	10,660
7	9,521
8	11,430
9	19,230
19	22,954
11	44096
12	76,407

Salvage Value of the Software

This is the amount new software will fetch while replacing the old software. It has been calculated with following formula:

$$S_n = C^* e^{-0.05n}$$

where Sn = Purchase price of the software in the base year, year of commissioning, 2007.

n = Age of the software in No. of years as taken from the beginning of base year of purchase and commissioning.

0.05 = is a constant factor, ie., rate at which old software value depreciate annually in book of account. Depreciation if the difference between the purchase price of the software and the salvage value, which has been taken as 20% per year in the case of software.

e = Exponential constant, i.e., 2.718.

Salvation values of the software computed based on the above formula are given at Table 3.

Discounting Factor

All future expenditures are to be discounted by multiplying with a discounting factor **'V'** and the present worth of money spent years on maintenance of the software hence it to be computed. For this purpose the following formula has been adopted:

$$V^{n} = \{1/(1 + i)\}^{n} \quad (1)$$

where, V^n = Discounting factor for the nth year.

i = Cost of capital or interest rate charged by the bank on capital borrowing in case the client borrowed from the bank, i.e., 16%.

Running Cost

Running cost of the software includes:

1. Maintenance and repair of software
2. Making changes and updates to the software

Only cost due to maintenance and repair, which vary with the age of the software was considered in this model.

Based on the above concept, maintenance and repair cost of the software has been compiled as given in Table 4.

Development of Mathematical Model

Let,

C = Purchase price of the software Rn = Running cost in year '*n*'

i = Rate of interest,

$V = \{1/(1 + i\}$ is the present worth of a dollar to be spent a year hence.

'*V*' is also known as the discounting factor.

Suppose the software is to be replaced after *n* year. Then assuming that expenditure can be considered to take place at the beginning of each year, the present worth of all the years expenditure associated with replacing the software after *n* years, Pn is given by:

$$Pn = C + V^{0}R1 + V^{1}R2 + \ldots + V^{n-1}Rn$$

Pn increases as *n* increases; thus, if one decides to replace software in the $(n + 1)^{th}$ year, instead of nth year, an expenditure is incurred whose present worth is *P(n + 1)* instead of *Pn*. Against this additional expenditure of { *P(n + 1)* - *Pn* } one is getting an additional year's service. The moment one feels it is costlier for the service, one replaces it, see [1]. To determine when it becomes costlier to take service, the present value *Pn* is converted into the sum of *n* fixed annual payment, each equal to *X* at the start of each year for *n* years such that,

$$Pn = V^0X + V^1X + V^2X + \ldots V^{(n-1)}X$$
$$Pn = X(1 - V^n)/(1 - V) \ldots\ldots\ldots.[\text{Geometricseries}] \quad (2)$$
$$X = Pn(1 - V)/(1 - V^n)$$

If the salvage value of the software is also taken into account, the capital component of the software will be determined as given below:

Capital Component of software = purchase price – Discounted salvage value after *n* years

The salvage value of the software will be realized after *n* years, hence, the present worth of discounted salvage value

$= SnV^n$, where,

$Sn =$ Salvage value after *n* years

$V^n =$ Discounted factor for n^{th} year

SO the difference of these two is the capital component = (C - SnV^n).

Now, considering the salvage value of the software the present worth, *Pn* of all the expenditures associated with replacing the software after *n* years may be modified as follows:

$$Pn = C - SnV^n + VR2 + V^2R3$$
$$\ldots\ldots + V^{(n-1)}Rn$$
$$= C - SnV^n + \text{Summation}(V^{(n-1)}Rn)$$
$$forn = 1, 2\ldots.., n.$$

Now, the fixed annual payment '*X*' can be written by substituting the above value of *Pn* as:

$$X = Pn(1 - V)/(1 - V^n)$$
$$= [C--SnV^n + Summation(V^{(n--1}Rn)](1 - V)/(1--V^n)$$
$$Forn = 1, 2, \ldots., n.$$

We have now to determine the value of *n* for which the equivalent fixed annual payment or, [*Pn (1 - V)] / (1 - V^n)* is the minimum. The value for the *n* gives the optimum economic life of the software.

Deriving Solution from the Model

Relevant cost data based on the above model are presented in Annexure 1 below.

Year	Age of Software	Discount Factor	Acquisition Cost	Salvage Value	Present worth of salvage value	Capital Expenditure	Maint Cost	Present worth of maint. cost	Cumulative maint. cost	Tot expene
7-08	1	0.86	12094	114808	98735	21959	1781	1781	1781	237
8-09	2	0.74		109208	80814	39880	2493	2144	3925	438
9-10	3	0.64		103882	66485	54209	2904	2148	6073	602
0-11	4	0.55		98816	54349	66345	4694	3004	9077	754
1-12	5	0.48		93997	45118	75576	4113	2262	11339	869
2-13	6	0.41		89412	36659	84035	10660	5117	16456	1004
3-14	7	0.35		85052	29768	90926	9521	3904	20359	1112
4-15	8	0.31		80904	25080	95614	11430	4001	24360	1199
5-16	9	0.26		76958	20009	100685	19230	5961	30321	1310
6-17	10	0.23		73205	16837	103857	22954	5968	36289	1401
7-18	11	0.2		69634	13927	106767	44096	10142	46431	1531
8-19	12	0.17		66238	11261	109433	76407	15281	61713	1711

4 Discussion

From Annexure 1, it is observed that the software may be replaced at the end of the 5th year of its operation to derive maximum economy. However, considering that the rate of maintenance done on the software, the replacement at the end of the 5th year may not be desirable because the fixed annual payment 'X', between the 6th and the 10th year is more or less constant, i.e., within 23845 to 25481per year. After the 10th year, there is a sharp rise in the fixed annual payment per year. Therefore, the software may be considered for replacement at the end of the 10th year, i.e., during the 11th year of its operation, before 3rd overall or major repair of R4 / Rv type is taken.

5 Conclusion

It is concluded that the optimum economic life of the software under the above working conditions is 10 years. This is useful as a guide while planning for development of any software.

Acknowledgment. I would like to thank all respondents who have participated in this study.

References

1. Chikofsky, E.J., Cross, J.H.: Reverse engineering and design recovery: a taxonomy. IEEE Softw. **7**(1), 13–17 (1990) . Copyright IEEE
2. Brooks, F.: No silver bullet. In: Kugler, (ed.), Information Processing. Elsevier, New York (1986). Reprinted in IEEE Computer, pp. 10–19, April 1987

3. Schach, S.R.: Object Oriented and Classical Software Engineering, 10th edn. McGraw-Hill, New York, pp. 18–23 (2010)
4. Martin, J., Mcclure, C.: Software Maintenance: The Problem and Its Solutions. Prentice Hall, Englewood Cliffs (1983)
5. Biggerstaff, T.: An assessment and analysis of software reuse. In: Yovits, M., ed., Advances in Computers. Academic Press, New York (1992)
6. Takang, A.A., Grubb, P.A.: Software Maintenance Concepts and Practic. Thompson Computer Press, London (1996)
7. Van,V.H.: Software Engineering: Principles and Practices, 2nd Edition. John Wiley & Sons, West Sussex (2000)
8. Coenen, F.P., Bench Capon, T.J.M.: Maintenance of Knowledge Based System Theory, Tools and Techniques. Academic Press, London (1993)
9. Chapin, N., Cimitile, A.: Announcement. J. Softw. Mainten. Evol. Res. Pract. **13**(1) (2001)
10. Mcconnell, S.: The nine deadly sins of project planning. IEEE Softw. **18**(1), 5–7 (2001)
11. Boehm, B.W.: Software engineering: R & D trends and defense needs. In: ICSE 1979 Proceedings of the 4th International Conference on Software Engineering, Germany, pp 11–21 (1979)
12. Kelly, J.C., Sherif, J.S., Hops, J.: An analysis of defect densities found during software inspections. J. Syst. Softw. **17**(2), 111–117 (1992)
13. Myers, G., Glenford, J.: The Art of Software Testing, 2nd edn. John Wiley and Sons, New York (1979)
14. Vigder, M.R.: Building Maintainable Component-Based Systems, Carnegie Mellon Software Engineering Institute (1999). http://www.sei.cmu.edu/icse99/papers/38/38.htm
15. Eze, N.U., Obichukwu, P.U., Ibezim, N.E.: Testing the correctness of educational software system based on Testmatica model to explore its impact on productivity gains. Int. J. Eng. Res. Technol. **12**(3), 321–332 (2019)
16. Saraf, I., Iqbal, J.: Generalized multi-release modelling of software reliability growth models from the perspective of two types of imperfect debugging and change point. Qual. Reliab. Eng. Int. **35**, 2358–2370 (2019). https://doi.org/10.1002/qre.2516
17. Pandey, S.K., Mishra, R.B., Tripathi, A.K.: Machine learning based methods for software fault prediction: a survey. Expert Syst. Appl. **172**, 114595, 15 June 2021. https://doi.org/10.1016/j.eswa.2021.114595
18. Li, Q., Pham, H.: Modeling software fault-detection and fault-correction processes by considering the dependencies between fault amounts. Appl. Sci. **11**, 6998 (2021). https://doi.org/10.3390/app11156998
19. Eze, N., et al.: Fault detection model for software correctness and reliability. In: Silhavy, R., et al. (eds.), Software Engineering Application in Informatics, CoMeSySo 2021. LNNS 232, pp. 1–20 (2021). https://doi.org/10.1007/978-3-030-90318-3_79
20. IEEE: IEEE Standard for Software Maintenance. IEEE Std 1219-1993, Institute of Electrical and Electronics Engineers, inc., New York (1993)
21. Welman, C., Kruger, F., Mitchell, B.: Research Methodology. 3rd edn. Oxford University Press, London (2005)
22. Smith, R.K., Hale, J.E., Parrish, A.S.: An empirical study using task assignment patterns to improve the accuracy of software effort estimation. IEEE Trans. Softw. Eng. **27**(3), 264–271 (2001)
23. Gregory, J., Crispin, L.: More Agile Testing. 1st edn. Addison-Wesley Professional, pp. 23–39 (2014)
24. Johnson, R.A.: The ups and downs of object-oriented system development. Commun. ACM **43**(10), 69–73, October 2000

25. De Lenos, R., Giese, H., Muller, H., Shaw, M., Anderson, J., Baress, I.: Software engineering for self - adaptative system: a second research roadmap. In De Lemon, R., Giese, H., Muller, H., Shaw, M., (eds.) Software Engineering for self Adaptative system, Schloss Dagstuhl, Germany. No. 10431 in Seminar Proceeding (2011)
26. Zelkowitz,Shaw, Gannon.: Principles of Software Engineering and Design. Prentice Hall Englewood Cliff, NJ (1979)
27. The Economic Impacts of Inadequate Infrastructure for Software Testing. National Institute of Standards and Technology, May 2002. Accessed 19 Dec 2017
28. Sutherland, J.: Business objects in corporate information systems. ACM Comput. Surv. **27**(2), 274–276(1995)
29. Koskinen, J.: Software Maintenance Cost (2003). http://www.cs.jyu.fi/~koskinen/smcosts.htm
30. Highsmith, J.: Cutter Consortium Reports: Agile Project Management: Principles and Tools 4. 2nd edn. Cutter Consortium, Arlington (2003)
31. Mathis: The last 10 percent. IEEE Trans. Softw. Eng. SE-12, pp. 705–712, June 2010
32. Eze, N.U.: Development process for controllable software. Int. J. Educ. Res. **5**(7), 37–52 (2017)
33. Bowker, P.: Lost pet fees cost Toronto $700,000 (2001).http://catless.ncl.ac.uk/Risks/subj2.1
34. Chung, L., Nixon, B.A., Yu, E., Mylopoulos, J.: Non-Functional Requirements in Software Engineering. Kluwer, Boston (2000)
35. Rosa, J.L., Robin, A., Silva, M.B., Baldan, C.A., Peres, M.P.: Electrodeposition of copper on titanium wires: Taguchi experimental design approach. J. Mater. Process. Technol. **209**(3), 1181–1188 (2009). https://doi.org/10.1016/j.jmatprotec.2008.03.021
36. Davis, A.M.: Software Requirements: Objects, Functions and States. Prentice-Hall, Upper Saddle River (1993)
37. Ross, P.J.: Taguchi Techniques for Quality Engineering, 2nd edn. McGraw-Hill, New York (1996)
38. Samruddhi, R., Pragati, S., Athira, K., Reshma, S.: An overview of Taguchi method: evolution, concept and interdisciplinary applications. Int. J. Sci. Eng. Res. **4**(10), 621, October-2013. ISSN 2229-5518
39. Camus, J.-Y., Mathieu, A.: "D'où vient l'expression 'remigration'?". Le Soleil, 19 August 2017. Archived from the original on 24 May 2019
40. The, D.E., Programmer, H.: Commun. ACM **15**(10), 859–866 (1972)
41. Goodenough, J.B.: A survey of program testing issues. In: Wegner, P. (ed.) Research Directions in Software Technology, pp. 316–340. The MIT Press, Cambridge, MA (1979)

Elements of Analytical Data Processing for a Factual Database: Statistical Processing of Historical Facts on the Example of a Database 'for Christ Suffered'

Alexander Soloviev[1,2], Anna Bogacheva[2], and Vladimir Tishchenko[1,2](✉)

[1] Federal Research Center 'Computer Science and Control' of Russian Academy of Sciences, Moscow, Russia
vtischenko@isa.ru
[2] St. Tikhon's Orthodox University, Moscow, Russia

Abstract. The article presents method of statistical data processing of a database of those repressed for their Orthodox faith by the Soviet authorities in the period 1917–1952. The task of statistical data processing for an object-oriented database containing some historical facts, structured according to selected fields, is posed. An example of statistical data processing for a factual database 'for Christ suffered' is given. Different categories of victims exist: 'bishops', 'clerics', 'laity'. For the 'Year of arrest' field, a graph of the number of arrests by year for persons of different categories is plotted. Against the background of the general schedule of arrests for the category of 'bishops', a trend is revealed associated with an attempt by the Soviet authorities to 'decapitate' the Russian Church already during the seizure of church valuables in 1922. *The graph of arrests of 'bishops' has the form of a trend that differs from the general dependence.* Researchers noted that the nature of the general dependence is manifested most of all for the category of 'clerics'. Researchers concluded that such a database is not only a historical source, but also a means of statistical processing of historical data, which can serve as a way to confirm events in history, as well as identify trends. #CSOC1120.

Keywords: statistical processing of historical facts · OLAP data mining · NIKA DBMS · graph of percecution · wave of percecution

1 Introduction

Currently, the database 'for Christ suffered' or the database on those repressed for their Orthodox faith in the USSR in the XX-th century contains information about more than 36 thousand people. Information is structured according to approximately 100 fields of a hierarchical database with the possibility make links to any vertex. To each section of information in the hierarchy a text comment of arbitrary length is possible. Criteria for entering the database 'for Christ suffered' are the presence of repressions, that is, information about a person must contain at least one arrest and/or violent death. In this case, among all the repressions that occurred for various reasons, only repressions

R. Silhavy and P. Silhavy (Eds.): CSOC 2023, LNNS 723, pp. 23–39, 2023.
https://doi.org/10.1007/978-3-031-35317-8_3

against Orthodox Christians are considered. This criterion is not completely formal, because the criminal article under which the victim was convicted, for example, 'anti-Soviet agitation' does not speak of this. Almost all the victims demonstrating a holy dignity were repressed for their faith. This confirms also the fact that some of them were tried in an attempt to force them to renounce their priesthood. Similarly, one can consider repressions against those who were members of the church or clergy. For the rest of the suffered application the criteria for entering the database become much less obvious. Here additional information about the person is needed, which is far from always preserved. Sometimes, indirect evidence of repression for the Orthodox faith can be information about kinship with the victim, who demonstrates a holy dignity. However, even this is not enough; he could be in another investigation. Therefore, the number of victims and the distribution of the number of victims should be studied by category and year.

2 Methods

2.1 The Number of Victims and Some Remarks

The Number of Victims for the Orthodox Faith

There are different estimates of the number of victims. The literature (Zemskov 1994) gives a figure of the total number of victims of repression by the Soviet authorities. It is approximately four million people. We will be interested in how many of them suffered for the Orthodox faith. Publications (Emelyanov 1999; Emelyanov 2005a) give an estimate based on the number of Orthodox Christians before the 1917 revolution. Professor N.E. Emelyanov believes that the number of Orthodox who suffered during the persecution ranges from half million to one million people. At the same time, he clarifies that information can only be collected about one hundred thousand of them. In other words, information about most of the victims is missing or lost. PhD N.V. Somin proposes to consider the possible number of people who suffered about whom information can be collected (Somin 2019), i.e. hundred thousand people.

English Literature on Repressions for the Orthodox Faith in the USSR

Among contemporary English-language books is a book (Ware 1997) by Bishop Kallistos Ware, who from 1966–2001 was a Lecturer in Eastern Orthodox Studies at the University of Oxford. Here, indicating a link to a resource on the Internet, is also necessary, which tells about the saints on the example of the icon of the New Martyrs and Confessors of Russia of the 20th century (ALLSAINTSOFAMERICA.ORG 2008; OCA.ORG 1996). A book has just been released (Christensen 2019). Best known source in English about the Russian martyrs of the 20th century remains the book of Archpriest Michael Polsky (Polsky 2000), the first edition of which was published in 1949 in Jordanville and in 1957, the second volume of the book was published. A modern English-language author who has written numerous books on the Russian New Martyrs and Confessors is Dr Vladimir Moss, who studied at Oxford University from 1967 to 1970 (Moss 2021).

Remarks on the Statistics from the Database 'for Christ Suffered'
Before studying the statistical data obtained on the basis of queries to the database 'for Christ suffered', emphasizing the following provisions is necessary, taking into account which of the results obtained should be considered.

1. The database 'for Christ suffered' means the database of those repressed in the USSR for the Orthodox faith in the 20th century.[1]
2. All statistics calculated on the basis of the database are relative, since the database is constantly *corrected* and *replenished based on new incoming data.*
3. On the basis of statistics from the database, building various statistical models that can reconstruct historical reality in different ways is possible (Emelyanov 1999; Somin 2019). See Sect. 2.
4. Finally, the very historical interpretation of the figures obtained may be different.

The fourth point needs clarification. According to Emelyanov, the number of victims for the Orthodox faith in the USSR in the 20th century is about half a million victims (lower limit), and about one hundred thousand people can collect information about them based on archival documents, oral testimonies of relatives, and publications. Documents about the rest of the victims either did not exist (in 1918 they were deprived of their lives without trial or investigation), or they did not survive. Somin considers the resulting figure of one hundred thousand victims to be the total number of victims for the Orthodox faith.

From his point of view, the rest of the believers simply 'dressed up in Soviet clothes', and became ordinary Soviet people, 'forgetting' under the threat of death that they were Orthodox Christians.

To the last statement, one can object to him "weren't Orthodox churches overflowing with believers during the persecution?!" For example, see the photograph in Fig. 1, taken during the service of the future Archbishop Ermogen (Golubev), in 1948.[2] Later, in the article, researches believed that Somin clarified the total number of victims, and its estimate (Somin 2019) will be a refinement in the model of Emelyanov.

Description of Data for the Database 'for Christ Suffered'
The printed edition of the database "for Christ suffered" is the biographical reference book of the same name edited by Archpriest Vladimir Vorobyov (Vorobyev 2015). The database 'for Christ suffered' is built on the basis of the NIKA OODBMS (Emelyanov et al. 1996), the model of which is a network with distinguished vertex hierarchies. Vertices can be nonterminal (structure, array, template reference, value reference) and terminal (string, integer, etc.). Indexes are built on the terminal fields with links to the main branch of the database. The conceptual schema of the subject area, is a data description schema for the database. An example of biographical information can be found at martyrs.pstbi.ru. In the article an overview of the work of Professor Emelyanov in English on creating a database in the field of history is given (Solovyev 2020a).

Fig. 1. Archimandrite Hermogen (Golubev). Astrakhan, Intercession Cathedral, 1948.

Graph of Persecution and Groups of the Suffered

Prof. Emelyanov N.E. (Emelyanov 1999) built a graph of persecution based on statistical data on the database 'for Christ suffered' and the model he proposed. The model is as follows: to obtain estimates of the number of victims, all statistical data on repressions registered in the database must be multiplied by the coefficient:

$$K = N^0_{real}/N_{db},$$

where, $N^0{}_{real}$ is the lower estimate of the total number of victims for the Orthodox faith for the range $[N^0{}_{real}; N^1{}_{real}]$, and N_{db} is the total number of victims registered in the database 'for Christ suffered'. As mentioned above, the estimate of Somin N.V. is taken as an estimate of the total number of victims. $N^0{}_{real} = 10^5$ (Somin 2019). The graph was tested on different data volumes and kept all the main trends and appearance.

The graph in Fig. 2 and its description, Professor Emelyanov called "History of the Church, written by a computer", published an article with that title in an Italian newspaper.

The model data shown in Fig. 2 fit into the real statistics on all repressions published by the historian Zemskov V.N. (Zemskov 1994), shown as a graph in Fig. 3. The data were obtained from the statistical reporting of the Soviet state security agencies. Model data are statistical estimates and are not precise. In accordance with the data in Fig. 3, no peak of repression occurs in 1922, but a decline is observed. However, this can be explained by the fact that in 1922 the peak of repression was only among those who suffered for the Orthodox faith. Conversely, the peak of repression in 1933 is the maximum only in the general statistics of Zemskov and it is not on the graph of the persecution of the Orthodox Church.

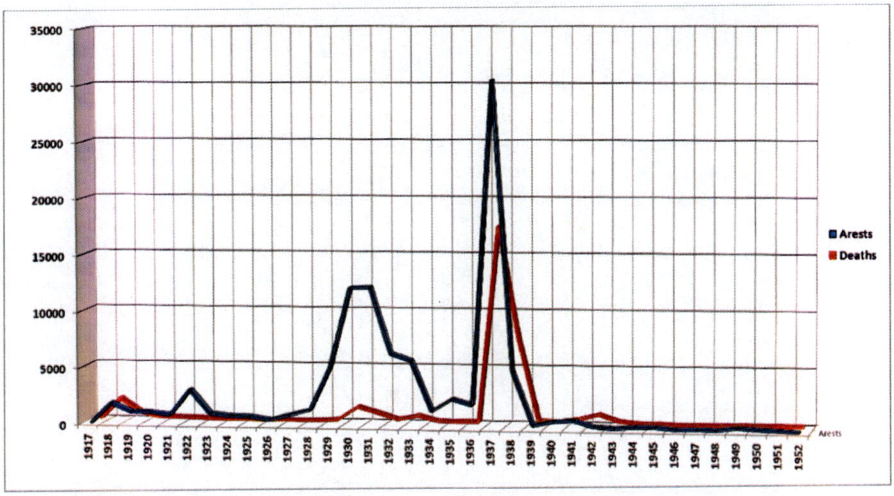

Fig. 2. "History of the Church, written by a computer" (Emelianov 1998; Emelyanov 2005b)

The same model and graph (Fig. 2) remain relevant for the current version of the database. Prof. Emelyanov gave a historical interpretation and a detailed description for all the 'peaks' on the graph, or 'waves of persecution' as he called them. For example, 1918: the seizure of power by the Bolsheviks in 1917; 1922: seizure of church valuables; 1930: collectivization; 1937: 'great terror'. These 'waves of persecution' affected all classes of Orthodox believers in all regions of the country. Below, we will consider graphs of persecution, similar to those given, for three nonoverlapping main groups or categories of Orthodox Christians—bishops, clergy, and laity. All three groups, together, make up the total number of victims (see Fig. 2). The group of bishops includes all those who exhibit the episcopal dignity, that is, patriarchs, (schema-) metropolitans, (schema-)archbishops, (schema-)bishops. The group of clerics includes all those who are not included in the first group and demonstrate a completed field Dignity-ChurchService in the database. The group of laity includes everyone who left the field Dignity-ChurchService blank. The smallest group of bishops is represented in the database in an exhaustive way, since information about them is most complete. The group of clerics is the most numerous, because information about clerics is much easier to find than about laity. The reason for this is the presence of a dignity or church service, because in the investigative case, as a rule, a political accusation of anti-Soviet activity was indicated in order to show the absence of persecution of Orthodox Christianity. For the laity, an indirect sign of persecution could be the fact that they went through the same investigative case with bishops and clerics.

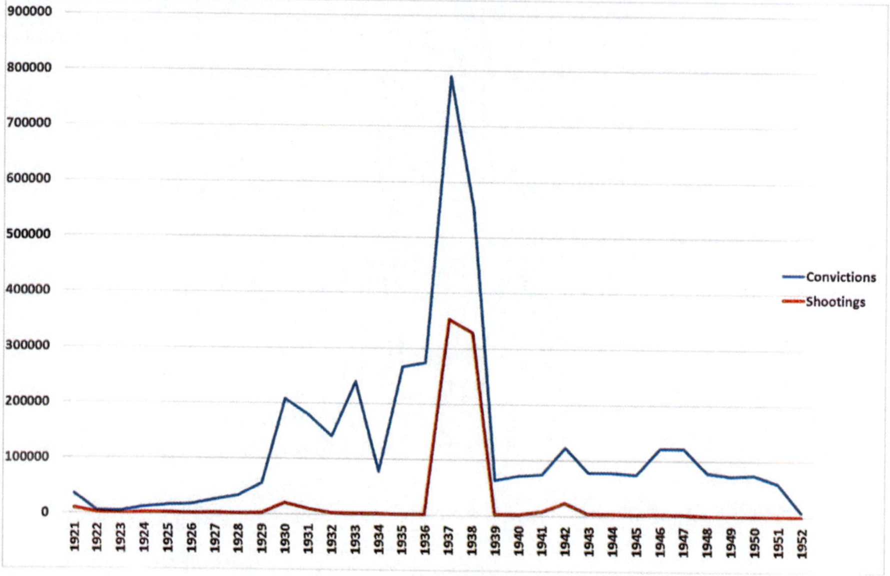

Fig. 3. The number of those convicted and the number of those executed for counter-revolutionary and other especially dangerous state crimes in 1921–1952 according to Zemskov V.N.

2.2 Statistical Processing of OODB Data of Historical Facts

OLAP Data Mining

The technology of online analytical processing of multidimensional data OLAP is used to select multidimensional data, and the search for patterns is carried out using data mining (Data Mining). The integration of these two technologies in the form of 'OLAP Data Mining' as introduced in (Parsaye 1998), enables full decision support. With regard to the historical sphere, such integration allows you to confirm historical events based on historical facts contained in the database, as well as to find trends in statistical data obtained on the basis of the original database. The 'OLAP Data Mining' technology is implemented in a hypertext system based on the OODBMS NIKA NKWSystem (Emelyanov et al. 1996). The NKWSystem provides the ability to record statistical data in the form of spreadsheets stored in electronic documents (Solovyev 2020b, Lecture Notes). An example of statistical processing of data obtained in the NKWSystem is described below.

Distribution of Victims Based on the Year of Birth

Consider the YearOfBirth field. By dividing all frequencies by the total number of births, you can obtain a probability function for the YearOfBirth field. The form of this function is shown in Fig. 4.

The graph in Fig. 4 (a fragment from 1865 to 1900 is shown) covers the period from 1830 to 1939. To the boundaries of the period, the probability function approaches zero. Such a probability function can be approximated by a continuous normal probability density function.

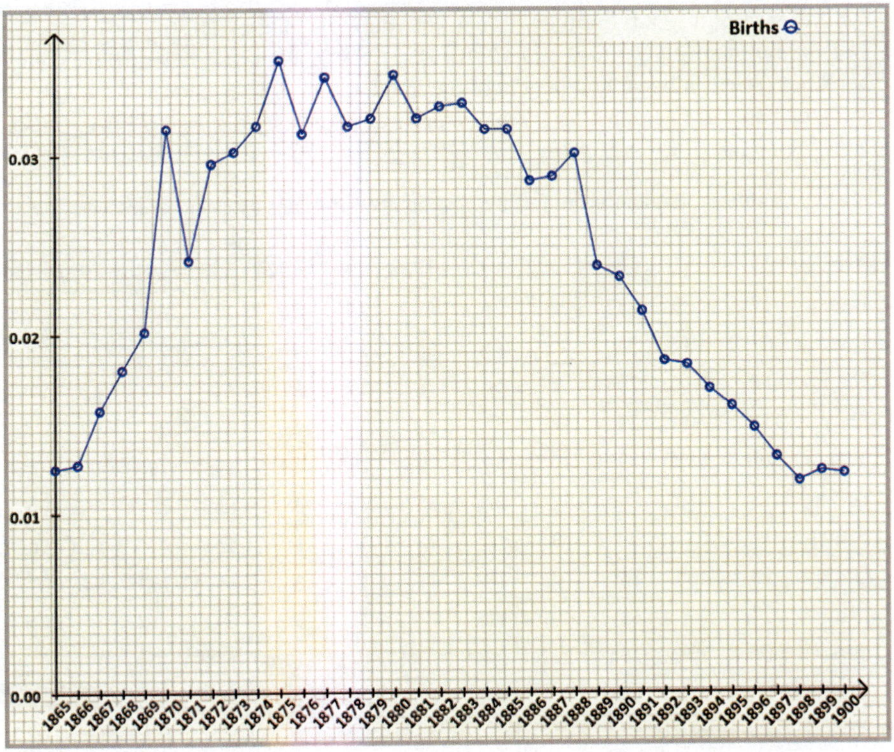

Fig. 4. Fragment of the probability function for the YearOfBirth field

By the form of the probability function in Fig. 4, we can hypothesize that no trend is found in the time series. Assuming that this hypothesis is true, the probability function for the YearOfDeath field in the absence of a trend will demonstrate the same form, but it will be shifted along the time axis by 70–80 years: "The days of our lives are seventy years; And if by reason of strength they are eighty years" (Ps. 89.10). According to the database 'for Christ suffered', the average age of victims is ~ *57 years and 8 months*, which confirms the deviation from the usual course of history.

Distribution of Victims Based on the Year of Death

Table 1 shows part of the time series (1915–2018) for the YearOfDeath field from 1917 to 1952, according to the three categories of the suffered. As can be seen from Table 1, the sum of the three lower rows of bishops, clerics, and laity gives the second row, the total number. In the period under review, the global maximum of the total number of deaths and deaths by category of the suffered corresponds to 1937.

In all of the time series shown in Table 1, a trend is detected; in other words, these data are not random (see also Fig. 2). The article (Lemeshko et al. 2015) examines the power of criteria for the presence of a trend in mathematical expectations. In the Russian-language monograph by the same authors, when studying the trend, models of the form $x_i = a * t_i + b * \sin(2k\pi t_i) + \xi_i$ are used. Here ξ_i are independent normally distributed random variables. Researchers emphasized that these models are associated only with the presence of a trend, but they do not determine its nature. This is a separate issue. *The nonparametric test of the cumulative sum* CST in all cases except for the category of clerics reveals a linear component of the trend at a significance level of $\alpha = 0.1$: CST ≤ 2 (see Table 2, Column CST). This criterion is mainly used to detect linear trends.

Table 1. Total number of deaths and number of deaths by category and year

Year	1917	1918	1919	1920	1921	1922	1923	1924	1925	1926	1927	1928
Deaths	17	669	235	88	75	63	20	26	14	6	15	28
Bishops	0	13	5	3	4	5	2	4	3	2	2	8
Clerics	14	548	208	67	66	46	15	21	11	4	12	17
Laity	3	108	22	18	5	12	3	1	0	0	1	3
Year	1929	1930	1931	1932	1933	1934	1935	1936	**1937**	1938	1939	1940
Deaths	65	482	296	97	223	55	43	48	**6357**	2792	98	94
Bishops	5	6	7	10	18	7	3	9	**128**	41	4	6
Clerics	54	406	246	79	184	41	37	34	**5536**	2433	84	73
Laity	6	70	43	8	21	7	3	5	**693**	318	10	15
Year	1941	1942	1943	1944	1945	1946	1947	1948	1949	1950	1951	1952
Deaths	178	335	118	45	43	16	21	23	28	28	26	25
Bishops	5	8	4	3	5	2	2	1	0	0	2	5
Clerics	142	256	92	35	32	10	12	14	24	26	21	15
Laity	31	71	22	7	6	4	7	8	4	2	3	5

Notes to the Table. 1. In the statistics by years, only those suffered persons for whom the field YearOfDeath is known are taken into account. Their number is approximately 2.6 times less than the total number of suffered persons recorded in the database. 2. Table 2 uses the frequency probabilities from Table 1 to calculate the statistics: $P_{ki} = n_{ki} / n_k$, $i = 1,\ldots,n$, $k = 0,\ldots,3$. Here, k denotes the category: 0 is the total number of deaths (Row 2 of Table 1), 1, 2, 3 is the number of deaths in the category of bishops, clergy, laity correspond to 3, 4, 5 rows of Table 1 respectively. N_{ki} is the frequency in category k with index i. n_k is the sum of all frequencies for category k. All probabilities for categories $k = 1, 2, 3$ are conditional. This is not further specified. Similarly, statistics are calculated for the YearOfArrest field (Tables 3 and 4)

The statistics corresponding to the time series of Table 1 are given in Table 2. Columns—1: categories, 2: mathematical expectation (EXP), 3: standard deviation (STD), 4: cumulative sum statistics (CST), 5: Abbé statistics (ABL), 6: statistics of the inversion criterion I, 7: normalized statistics of the inversion criterion I*, 8: normalized statistics of the Savage rank test Sr*, 9: sum of squared differences from the least squares method (SUM), 10: multiple coefficients of determination R^2. Time period under consideration: 1915–2018. The amount of datum in the time series $n = 94$.

Table 2. Statistics on the YearOfDeath field by three categories

	EXP	STD	CST	ABL	I	I*	S_r^*	SUM	R^2%
Deaths	3574.9805	2729.8889	2	0.6426	3418	8.0514	– 7.6856	0.0000	100.00
Bishops	46.3286	54.8436	0	0.6750	2904	4.6936	– 6.0502	0.0402	83.81
Clerics	3188.7571	2362.8154	3	0.6427	3356	7.6463	– 7.5979	0.0002	99.92
Laity	364.4325	294.7378	0	0.6378	2878	4.5238	– 5.9201	0.0031	98.74

To confirm the trend for all categories, the Abbé test was used, which is resistant to deviations in the distribution of the initial random variable from the normal one and detects nonlinear trends (Lemeshko 2006). The test rejects the hypothesis of randomness of the time series data, if the test statistic is less than the critical value $q < q_\alpha$. At the significance level $\alpha = 0.01$, $q = \text{ABL} < q_\alpha = 0.7626$ (see Table 2, Column ABL). Here q_α is the critical value of the statistic calculated for $n = 94$ (number of data in the time series) and $\alpha = 0.99$ (significance level) using the formula for $n > 60$:

$$q_\alpha \approx 1 + \psi_\alpha((n-2)/(n^2-1))^{1/2},$$

where Ψ_α is the α-quantile of the normal distribution (Hart 1942).

Among the *nonparametric* tests for determining the trend in mathematical expectation, the most powerful test is the *test of inversions* (Lemeshko et al. 2015). The test statistic I must lie within the given limits in order to accept the randomness hypothesis. Normalized statistics $I^* = (I-\mu_I)/\sigma_I$, where $\mu_I = n(n-1)/4$, and $\sigma_I = ((2n^3 + 3n^2-5n)/72)^{1/2}$ are distribution parameters of statistics I, approximately described by the standard normal law. The hypothesis of a trend is accepted if $|I^*| \geq \Phi_{1-\alpha/2}$. At a significance level of $\alpha = 0{,}01$—$\Phi_{0.995} = 2.586$ and from Table 2, Column 7, it can be seen that the statistics I^* exceeds $\Phi_{1-\alpha/2}$ for all rows of Table 2. This means that the hypothesis of randomness is also rejected by the inversion test, and an alternative hypothesis is accepted that a trend is found in the expectation, both for the general case of the number of deaths and for categories.

When analyzing the power of tests for the presence of a trend in the variance in the article (Lemeshko et al. 2016), models of the form $x_i = \xi_i(1 + ct_i + d\sin(2k\pi t_i))$ were used. In accordance with the conclusions of the authors of this article, the most powerful test is the Hsu test. However, it is parametric, i.e., the distribution of the test statistic is affected by the distribution of the random variable. In addition, a large number of trials (more than 900) are required to distinguish between hypotheses in the case of some types of trends. To detect a trend in the dispersion for the distribution of the number of deaths of the suffered by years, using Savage rank test is acceptable. Normalized statistics of the test $S_r^* = (S_r-\mu_r)/\sigma_r$, where $\mu_r = n(n + 1)/2$; $\sigma_r = (n(n + 1)/12(n-\Sigma(1/i)))^{1/2}$, $i = 1,\ldots,n$, is approximately described by the normal distribution for $n > 30$. When $|S_r^*| > \Phi_{1-\alpha/2}$, the hypothesis of no trend is rejected. At the significance level $\alpha = 0.01$ and $n = 94$, the critical value is equal to $\Phi_{0.995} = 2.586$ and $|S_r^*| > \Phi_{0.995}$ for all rows of Table 2 in Column 8.

The conclusion regarding the distribution of the number of deaths over the years is that various tests reveal a trend visually observed on the graph of probability functions (Fig. 5), both in the mathematical expectation and in the variance. In particular, the trend in variance manifests itself in the form of a growing amplitude of oscillations of 'persecution waves'. This can also be seen on the graph for the YearOfArrest field (Fig. 6).

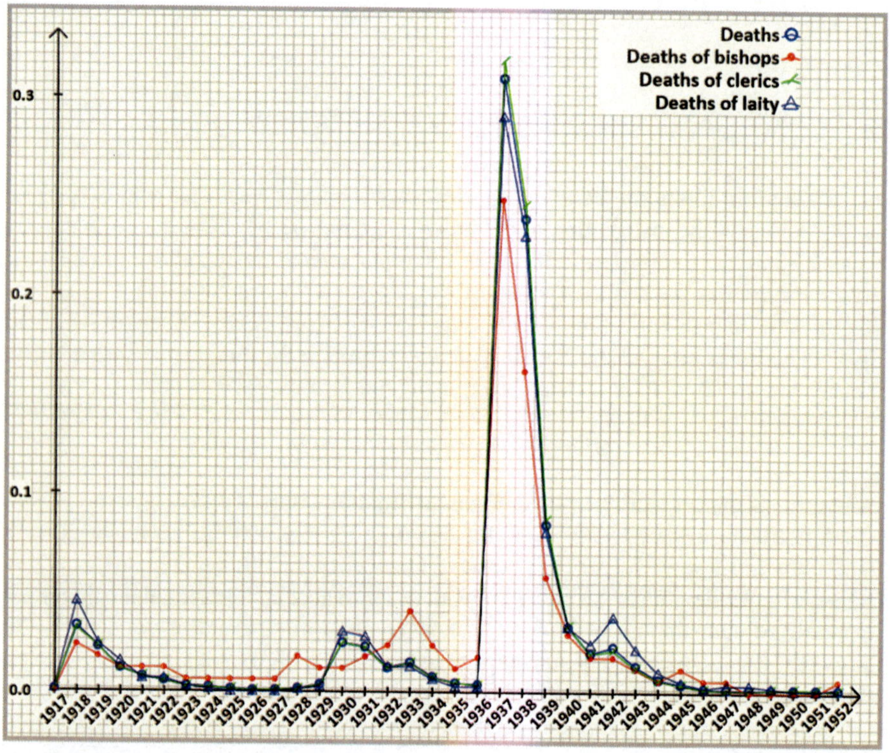

Fig. 5. Fragments of probabilistic functions for the YearOfDeath field by category.

Figure 5 shows the smoothed probability functions for the YearOfDeath field for the total number and for three categories. For a qualitative comparison of graphs by categories and with the general graph of deaths, frequency probabilities are taken instead of frequencies. Each frequency is divided by the total sum of frequencies for each line.

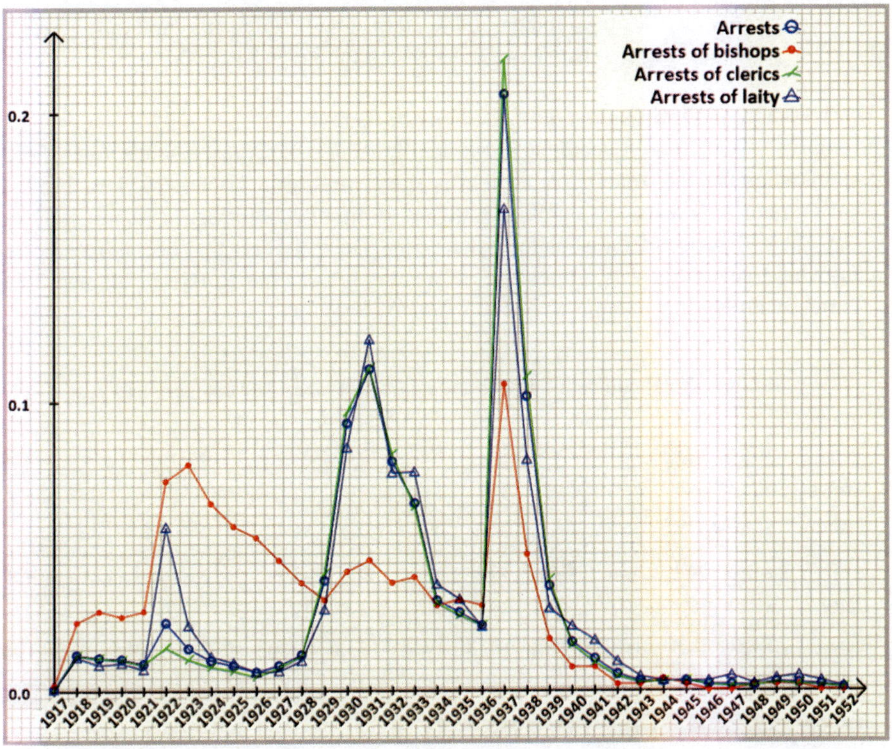

Fig. 6. Fragments of probabilistic functions for the YearOfArrest field by category.

Previously, the time series were smoothed by the simple moving average method with a small interval of 3 years (to preserve the nature of the dependencies). The basis of this method is the linear regression of the original random data u_t on the time axis within a given interval. Let an interval of an odd number of members of the time series equal to 2*m + 1 be considered: $u_{-m}, u_{-(m-1)}, \ldots, u_0, \ldots, u_{m-1}, u_m$. Based on these terms, a linear dependence $u_r = a_0 + a_1 t$ is constructed, the parameters of which are determined by the least SUM. The value of this dependency is in the middle of its domain of definition. After that, a shift exists to the right by one member of the series, and the procedure is repeated for the next group of members of the series, and so on, until the entire series is completed (Kendall et al. 1976). In accordance with least squares method (LSM), derivatives with respect to parameters are equated to zero.

$$\frac{\partial}{\partial a_j} \sum_{t=-m}^{m} (u_t - a_0 - a_1 t) = 0; j = 0, 1$$

From the equation for j = 0 it turns out that

$$a_0 = \sum_{t=-m}^{m} u_t/(2m + 1)$$

and for $j = 1$—$a_1 = 0$. Thus, at the point u_0, the value for the smoothed time series will be equal to the arithmetic mean of all points u_t in the given interval.

Distribution of Victims Based on the Year of Arrest

Similarly to the YearOfDeath field, we can consider the YearOfArrest field. Part of the statistics for the YearOfArrest field is shown in Table 3 from 1917 to 1952: the total number of arrests, the number of arrests in three categories: bishops, clerics and laity. The difference in statistics from the YearOfDeath field is that one suffered person can exhibit several arrests. Otherwise, all results are similar to those obtained for the YearOfDeath field.

Table 3. Total number of arrests and number of arrests by category and year

Year	1917	1918	1919	1920	1921	1922	1923	1924	1925	1926	1927	1928
Arrests	40	684	404	397	317	1143	379	311	293	198	367	530
Bishops	5	37	32	26	30	97	84	59	55	54	44	35
Clerics	27	539	322	308	248	536	262	210	184	114	278	418
Laity	8	108	50	63	39	510	33	42	54	30	45	77
Year	1929	1930	1931	1932	1933	1934	1935	1936	**1937**	1938	1939	1940
Arrests	1905	4467	4496	2371	2147	542	938	761	**11114**	1854	129	258
Bishops	29	48	48	34	42	25	35	31	**148**	18	3	4
Clerics	1644	3718	3571	2009	1630	408	721	621	**9472**	1607	107	132
Laity	232	701	877	328	475	109	182	109	**1494**	229	19	122
Year	1941	1942	1943	1944	1945	1946	1947	1948	1949	1950	1951	1952
Arrests	301	111	69	119	124	58	74	56	132	87	64	17
Bishops	10	1	3	7	2	1	2	5	2	2	1	1
Clerics	198	70	50	97	102	36	34	46	97	49	47	13
Laity	93	40	16	15	20	21	38	5	33	36	16	3

Table 4 summarizes the statistics for the YearOfArrest field, which correspond to the statistics calculated for the YearOfDeath field (see Table 2 and the description of the columns to it). Time period under consideration: 1887–1983. The number of datum samples in the time series: $n = 57$. All tests for detecting a trend in mathematical expectation confirm the trend. The CST cumulative sum test in all cases detects a linear trend component at a significance level of $\alpha = 0.1$: CST ≤ 1 (see Table 4, Column CST). The Abbé test rejects the hypothesis of the randomness of the time series data at the significance level $\alpha = 0.01$, since $q = \text{ABL} < q_\alpha = 0.6999$ (see Table 4, Column ABL). The hypothesis about the presence of a trend is accepted by inversion test at the significance level $\alpha = 0.05$, since $|I^*| \geq \Phi_{0.975} = 1.96$ for all values in Column I^* (see Table 4, Column I^*). At a significance level of $\alpha = 0.01$ and $n = 57$, the critical value is equal to $\Phi_{0.995} = 2.586$ and $|S_r^*| > \Phi_{0.995}$ for all rows of Table 4 in Column 8 (Savage rank test), which means that there is a trend in the dispersion for the YearOfArrest field.

Table 4. Statistics on the field YearOfArrest by three categories

	EXP	STD	CST	ABL	I	I*	S_r^*	SUM	R^2%
Arrests	4985.2456	4225.0708	1	0.6611	1004	2.8361	– 4.2763	0.0000	100.00
Bishops	62.6131	41.2487	0	0.4133	1003	2.8224	– 4.0959	0.0569	50.98
Clerics	4363.2144	3648.100 8	1	0.6655	1012	2.9463	– 4.4124	0.0006	99.49
Laity	686.5565	520.9265	1	0.6611	977	2.4644	– 4.7242	0.0078	93.31

3 Results

Now, as a result, a comparison of data on the fields of YearOfDeath and YearOfArrest by category can be made. The proximity of data for three categories—bishops, clergy, laity—to the general graphs for the YearOfDeath and YearOfArrest fields can be determined through the sum of the squared differences between the corresponding probabilities for the general graph and for this category:

$$RSS = \sum_{i=1}^{n} (P_{0i} - P_{ki})^2, k = 0, \ldots, 3$$

where, n is the number of points, and k is the category number (0 is general data, 1,2,3 are data on the category of bishops, clerics, and laity, respectively). General data should be understood as Row 2 in Tables 1–4. Such a sum of squares is used in *the least SUM* when approximating data by some function (Draper and Smith 1998). In this case, the result for the YearOfDeath field is presented in Table 2 and for the YearOfArrest field in Table 4 in the SUM column. From Table 2, it turns out that $0.0002 < 0.0031 < 0.0402$, that is, for clerics the sum of squares is less than for laity and for laity less than for bishops. It is visually seen in Fig. 5 that the probability function for clerics is closest to the overall graph. This is because data on clerics is a large part of all data collected. Similarly, from Table 4 for the YearOfArrest field, it turns out that $0.0006 < 0.0078 < 0.0569$ (clerics, laity, bishops). Qualitatively, this is the same result as for the YearOfDeath field. A R^2 multiple coefficient of determination characterizes the accuracy of the constructed function that describes the data. When calculating R^2, the formula is used:

$$R^2 = 1 - \sum_{i=1}^{n} (P_{0i} - P_{ki})^2 / \sum_{i=1}^{n} (P_{0i} - \overline{P}_0)^2, k = 0, \ldots, 3$$

where $\overline{P}_0$ is the mean value for P_{0i}. Tables 2 and 4 of Column R^2 show that the most accurate general data are 'described' by data on the category of clerics (in percent): 99.9% for the YearOfDeath field and 99.5% for the YearOfArrest field. Most of all, probabilistic functions differ from general probability functions in the category of bishops. In the case of arrests for data on the category of bishops, the coefficient of determination is only about 51.0%. According to *the graph* of the probability function for *the category of bishops*, it can be seen that it demonstrates a slightly *different dependence* than *the general probability function*. This difference *can be explained* by the desire of the Soviet authorities *to behead the Russian Orthodox Church*. On the graph of arrests, this is manifested in the fact that in the period from 1917 to 1928 the probability of arrest of bishops *demonstrates a characteristic peak in 1922*, in which the seizure of church property took place, and then, it gradually decreases until 1928. Since 1928 to 1936 the level of arrests of bishops exhibits small bursts, and in 1937 a wave of 'great terror' occurs. Further persecution was 'stopped' by the Great Patriotic War of 1941–1945. In the case of the Year of Death field, the difference between the graph for the category of bishops and the general one is not so great. The coefficient of determination in this case is 83.8%. This can be explained by the fact that the Bolsheviks, first of all, tried to deprive the episcopal sees of their heads through the exile and arrest of bishops. Many bishops have been arrested several times. For example, clergyman Bishop Athanasius (Sakharov) spent a total of about 30 years in prisons and camps.

2022 marks the 100th anniversary of the persecution that took place in 1922. In accordance with the corrected model of prof. Emelyanov, according to data from the database 'for Christ suffered', it turns out that in 1922, during the seizure of church valuables, about 3.1 thousand people were arrested by the Soviet authorities for the Orthodox faith. At the same time, about 173 of them were martyred, that is, about every 18th.

4 Discussions

The paper (Somin 2019) gives a more accurate estimate of the total number of victims than 100 thousand. It is about 109.8 thousand. In a subsequent oral report, Somin refines the estimate with a figure of 85 thousand. Since the work (Somin 2019) indicates a spread of values of $\pm$ 47%, it is quite possible to consider 100 thousand people as an estimate of the total number of victims. When building statistics on the database, the actual issue is the total number of victims, because distribution of the number of casualties by category and year use this estimate. The LSM method allows us to separately estimate how close the probability functions for various groups of victims are to the general graph of persecution in terms of deaths and arrests. In the general case, when fitting a curve, you can use the minimization of the loss function $\Sigma\varphi(\varepsilon) \rightarrow \min$ Instead of LSM, other methods such as the method of least absolute deviations (LAD) can be considered. Then $\varphi(\varepsilon) = |\varepsilon|$. Statistical analysis methods are not available for the general case therefore, the main statistical properties have been studied using the LSM method.

5 Conclusion

Key takeaway: the graph of bishops' persecution is qualitatively different from the general one (general persecution graph). Figures 5 and 6 show trends in executions and arrests as probability functions. In the absence of a trend, all graphs in Fig. 5 (the YearOfDeath field) would be graphs of approximately normally distributed random variables. An example of such a distribution is shown in Fig. 4 (field YearOfBirth). The graph in Fig. 6 (field YearOfArrest) would represent a fluctuation of a random variable around zero (as it was before the revolution). It can be seen from the statistical data on the arrests of bishops that the probability function demonstrates its own distribution ($R^2 \approx 51\%$), which is different from the general probability function of arrests. Most of the victims in the database are clerics, so the probability functions in this category are closest to the general probability functions in terms of the sum of squared differences and the multiple coefficient of determination (Columns SUM and R^2 in Tables 2 and 4). The third category, the laity, is much smaller than the second, and information on it is more difficult to identify, because documentary evidence is needed that the repressed suffered for their faith. In terms of proximity to the general probability function, this category is located between the categories of clerics and bishops, but in terms of the value of R^2, it is much closer to the probability functions for the general category and for the category of clerics.

As a result of the construction of probability functions, we can draw the following conclusion. Along with the fact that the graphs of repressions for the selected categories demonstrate similar trends with the general graphs for the fields YearOfDeath and YearOfArrest, some features were found in the probability functions for the category of bishops: in Fig. 5: small peaks for 1928, for 1933, Fig. 6 shows that the likelihood of bishops being arrested increases by 1922, and then, after 1922, it gradually decreases until 1928, because many bishops were already in exile. *Historically, this is explained by the Bolsheviks' attempt to deprive the Orthodox Church of its governance.*

The statistics for the different categories are different. Data on the category of bishops are almost exhaustive, for example, it is known that in 1922, ninety-three arrests of bishops occurred, and five of them died: Hieromartyr Metropolitan Veniamin (Kazan) was shot on August 13, 1922, Archbishop Filaret (Nikolsky) froze to death in exile in 1922, Metropolitan Jacob (Pyatnitsky) was arrested in 1922 and died from an operation in the same year, Bishop Nikolai (Orlov) was shot in 1922, and Bishop Evgeny (Berezhnov) was arrested in 1918 and died in 1922.

A further direction in the development of research can be the allocation of various categories of the suffered according to the CauseOfDeath field, as well as a more detailed consideration of the categories in the field of Dignity-ChurchService.

Notes

1. Short title. Database "Suffered for Christ". Full title: "New Martyrs, Confessors Who Suffered for Christ During the Years of Persecution of the Russian Orthodox Church in the 20th Century." (c) St. Tikhon Orthodox University for the Humanities (Institute until 2004) pstbi@ccas.ru (c) Brotherhood in the Name of the All-Merciful Savior spasbr@ccas.ru.
   http://martyrs.pstbi.ru
2. See the biographical information of Archbishop Ermogen (Golubev Alexei Stepanovich) in the database "Suffered for Christ".
   http://martyrs.pstbi.ru/bin/db.exe/spc_1_prn/ans/nm1/?HYZ9EJxGHoxITYZCF2JMTdG6Xbu3s8micuKh60W1fejXcOKW660fdOfVdOYUYS8Zfu0hfe8ctmY*

References

Allsaints of America.ORG.: New Martyrs, Confessors, and Passion-Bearers of Russia (2008). https://web.archive.org/web/20080505101425/, http://www.allsaintsofamerica.org/martyrs/nmruss.html. Accessed 14 April 2022

Draper, N.R., Smith, H.: Applied Regression Analysis, p. 716. John Wiley Sons, USA (1998)

Emelianov, N.E.: Una storia dei martiri scritta a computer. 'La Nuova Europa', 3, 23–34 (1998)

Emelyanov, N.E.: Evaluation of the statistics of persecution of the Russian Orthodox Church from 1917 to 1952. Theological Collect. PSTBI, Moscow, **3**, 258–274 (1999)

Emelyanov, N.E.: On the number of new martyrs and confessors of the Russian orthodox church in the 20th Century. In: Proceedings of the 15th Annual Theological Conference of St. Tikhon's Orthodox University for the Humanities, vol. 1, pp. 265–271 (2005a)

Emelyanov, N.E.: Studies in history based on the database In: 'The 20th Century New Martyrs and confessor of the Russian Orthodox Church'. The Second International Symposium 'Christianity in Our Life: Past, Present, Future'. Tbilisi 24–26 Nov 2005, pp. 33–34 (2005b)

Emelyanov, N.E., Muhanov, I.V., Tishchenko, V.A.: Web server on the basis of NIKA DBMS. In: Proceedings of the Third International Workshop on "Advances in Databases and Information Systems", ACM SIGMOD, Moscow, vol. 2, pp. 58–59, 10–13 September 1996

Hart, B.I.: Tabulation of the probabilities for the ratio for the mean square successive difference to the variance — AMS, vol. 18, pp. 207–214 (1942)

Christensen, K.H.: The Making of the New Martyrs of Russia. Soviet Repression in Orthodox Memory ISBN 9780367886141 Published by Routledge, p. 246 (2019)

Kendall, M.G., Stuart, A.: The Advanced Theory of Statistics, vol. 3, Design and Analysis, and Time-Series. 3rd edn., p. 552. Charles Griffin & Co. Ltd., London and High Wycombe (1976)

Moss, V.E.: A Century of Russian Martyrdom. A Selection of the Lives of the Holy New Martyrs and Confessors of Russia, vol. 1, p. 482 (2021). ISBN 978-1-304-60041-7. https://orthodoxchristianbooks.com/downloads/894_A_CENTURY_OF_RUSSIAN_MARTYRDOM_VOLUME_1.pdf. Accessed 14 April 2022

Lemeshko, B.Yu., Veretelnikova, I.V.: Tests for an absence of trend. In: Proceedings of the International Workshop "Applied Methods of Statistical Analysis. Nonparametric Approach" – AMSA'2015, Novosibirsk–Belokuricha, Russia, pp. 80–91, 14–19 September 2015. https://ami.nstu.ru/~headrd/seminar/publik_html/Veretelnikova_Lemeshko_AMSA_2015.pdf. Accessed 14 April 2022

Lemeshko, B.Yu., Veretelnikova, I.V.: Criteria of test against absence of trend in dispersion characteristics. In: Proceedings 2016 11th International Forum on Strategic Technology (IFOST), Novosibirsk, Russia, Part 1, pp. 333–337, 1–3 June 2016. https://ami.nstu.ru/~headrd/seminar/publik_html/IFOST_Veretelnikova_Lemeshko.pdf. Accessed 14 April 2022

Lemeshko, S.B.: The Abbé independence test with deviations from normality. Meas Tech **49**, 962–969 (2006)https://doi.org/10.1007/s11018-006-0220-7 Accessed 14 April 2022

OCA.ORG.: The Orthodox Church in America (1996). https://www.oca.org/saints/lives/2019/01/27/205310-new-martyrs-and-confessors-of-russia. Accessed 14 April 2022

Parsaye, K.A.: Characterization of data mining technologies and processes. J. Data Warehouse. **1**, 12–24 (1998). http://citeseerx.ist.psu.edu/viewdoc/download;jsessionid=B3AF12AA760A022A2A3EBA79F2C3A836?doi=10.1.1.194.7676&rep=rep1&type=pdf. Accessed 14 April 2022

Polsky, M.A.: The New Martyrs of Russia. This edition was published by Monastery Press, p. 353 (2000)

Solovyev, A.V.: Information Technology and the Experience of Creating a Database on the History of the 20th Century in the Personalities. ISTORIYA 11(3) (89) (2020a). https://history.jes.su/s207987840005930-2-1/?sl=en. Accessed 14 April 2022

Solovyev, A.V.: Long-term digital documents storage technology. Lect. Notes Elec. Eng. **641**, 901–911 (2020b). https://www.researchgate.net/publication/339347717_Long-Term_Digital_Documents_Storage_Technology. Accessed 14 April 2022

Somin, N.V.: Using a representative sample to estimate the number of those who suffered for their faith in Russia in the 20th century. St Tikhon's University Review, Series II: History. History of the Russian Orthodox Church, vol. 87, pp. 98–107 (2019)

Vorobyev, V.N.: (archpriest Vladimir Vorobyev): "Suffering for Christ": Persecution of the Russian Orthodox Church 1917–1956. Biographical reference book. Book 1 (A). - M.: Publishing house. PSTGU, p. 736 (2015)

Ware, T.R.: (Bishop Kallistos): The Orthodox Church, New Edition. Penguin Books, London, p. 164 (1997)

Zemskov, V.N.: Political repressions in the USSR (1917-1990)."Russia XXI" 1–2, 107–125 (1994)

Analysis of Facial Expressions of an Individual's Face in the System for Monitoring the Working Capacity of Equipment Operators

Maxim Khisamutdinov[1], Iakov Korovin[2], and Donat Ivanov[2](✉)

[1] LLC "Research Institute of Multiprocessor Computing and Control Systems", 150-g Socialisticheskaya St., 3479305 Taganrog, Russia
[2] Southern Federal University, Chehova St., 347928 Taganrog, Russia
donat.ivanov@gmail.com

Abstract. The monitoring of the performance and health status of operators of critical equipment must be carried out in real time. Analysis of video images from cameras installed at the workplace can be used. Modern algorithms allow you to select faces in images and determine the coordinates of key points of faces in images. This paper proposes a method and algorithm for selecting an image fragment suitable for analysis and detecting on it some specific facial features of the operator's face in order to further analyze the operator's state and performance. The results of the practical application of the proposed approach are presented.

Keywords: Pose Estimation · Body Parts Recognition · Pose Recognition · Video Recognition

1 Introduction

Decreased reaction speed and deterioration in the health of the operator of critical equipment can lead to errors in operation. These errors in the management of equipment can lead to serious failures and even to the occurrence of man-made disasters. In this regard, monitoring the health and performance of operators of such equipment is necessary. For example, operators of nuclear power plant equipment undergo a pre-shift medical examination before each shift.

Despite regular medical check-ups, an operator may become incapacitated during a work shift. To prevent dangerous consequences, it is necessary to monitor the health and performance of operators of critical equipment. For this, systems for monitoring the state of the operator are being developed.

Some of these systems contain contact sensors that are located on the operator's body. However, the use of such sensors creates discomfort for workers. In this regard, there is a need to use non-contact monitoring tools.

One of the directions for solving this problem is the automated analysis of video from a camera located at the workplace. Video analysis allows you to identify postures and mimic signs specific to those states in which the operator cannot fully perform his work.

R. Silhavy and P. Silhavy (Eds.): CSOC 2023, LNNS 723, pp. 40–48, 2023.
https://doi.org/10.1007/978-3-031-35317-8_4

Intellectual analysis of the combination of facial features and specific postures of the operator allows the system to issue warning signals in cases where there is a possibility of loss of efficiency by the operator.

Modern algorithms make it possible to select the face area [3] in the image and highlight the key points of the face [7, 8]. This article proposes a method and algorithm for automatic detection of such facial features as:

- downcast eyes,
- eyes open,
- upper eyelid lifted,
- frequent blinking,
- raised eyebrows,
- furrowed eyebrows,
- open mouth.

The results of the practical application of the proposed method and algorithm made it possible to determine these facial features in real time.

2 Detection of an Individual's Face Area

2.1 Proposed Method for Detecting an Individual's Face Area

The algorithm for detecting a human face area is designed to select a fragment containing a human face for further analysis of the eye area and facial expressions. Preliminary selection of the face area in the input image can significantly reduce the computational costs of further procedures for analyzing a person's face. The algorithm for detecting a human face area is based on the mathematical relationships described below.

At the first stage of the algorithm, the input image I is processed using convolutional neural networks [1], and specifically the implementation of OpenCV cv::dnn [2], the YOLOv4 neural network architecture [3], in order to obtain a set of bounding boxes R_M and corresponding image fragments $R_M(I)$ faces of a person or several faces of people:

$$DNN(I, W) \to R_M(I) \tag{1}$$

Also, the input is the argument W - the weights of the YOLOv4 neural network, they are obtained by extracting a class of a person's face in open image databases for machine vision [4, 5] using the procedure for training an artificial neural network on a GPU.

Further, the operation of the algorithm is based on choosing the largest pixel size of the face $R_{max}(I)$ from the set R_M, but not less than the minimum size (150 pixels on one of the sides) required for further algorithms for recognizing facial expressions, gaze direction, etc.:

$$R_M(I) \to R_{max}(I) \tag{2}$$

$R_{max}(I)$ is the result of the algorithm for detecting the area of a person's face, if the face is not detected, the bounding box R_{max} contains zero vertex coordinates.

2.2 Algorithm for Detecting the Face Area of an Individual

An algorithm for detecting an individual's face based on the proposed method has been developed. At the first step of the algorithm, the input is a three-color image *I*. The OpenCV cv::Mat container is used to provide the original image. Then the weight coefficients W for the initialization of the artificial neural network YOLOv4 are input.

The next step is to search for all fragments $R_M(I)$ faces in image *I* (1).

Next, the condition is checked that $R_M(I)$ is an empty set (no face was found in the image *I*). In case of successful fulfillment of the R_{max} condition, zero values of the vertices of the bounding box are assigned and the transition to the next step of the algorithm is performed.

If the condition is negative, the transition to the step of selecting the largest pixel size of the face $R_{max}(I)$ from the set R_M takes place and the transition to the minimum size condition check (150 pixels on one of the sides) is performed.

In case of successful fulfillment of the condition, a transition to the next step of the algorithm is made, in case of a negative result of the fulfillment of the condition, zero values are assigned to the vertices $R_{max}(0)$ and the transition to the final step of the algorithm is made.

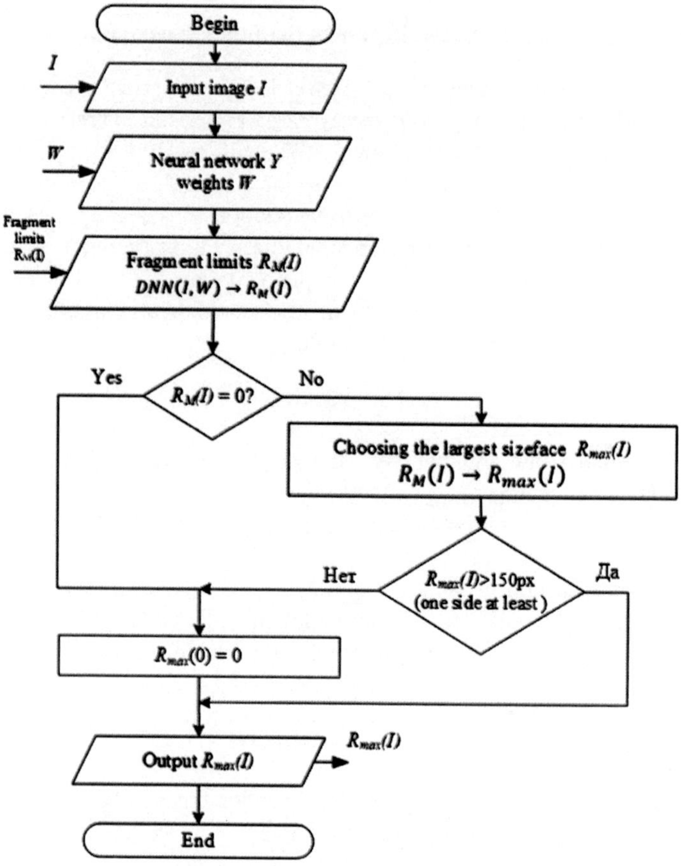

Fig. 1. Block diagram of the human face area detection algorithm

At the final step of the algorithm, $R_{max}(I)$ is output, which is the result of the human face area detection algorithm: if the face is not detected, the R_{max} bounding box contains zero vertex coordinates. Figure 1 shows a block diagram of the algorithm for detecting a region of a human face.

3 Analysis of Mimic Signs of the Operator's Face

3.1 The Proposed Method for the Analysis of Mimic Features of the Operator's Face

The analysis of human facial expressions is intended to highlight the key points of a person's face: based on them (obtained key points), various facial expressions are detected, such as - eyes closed, eyes open, upper eyelid raised, frequent blinking of the eyes, raised eyebrows, shifted eyebrows, open mouth, it is possible to adapt the algorithm to work with other additional key facial features. The algorithm for analyzing human facial expressions is based on the following mathematical relationships.

At the first step, the input image I is processed. We assume that the image contains only a person's face. For processing, you can use the DLib libraries (implementation of the Face Detection Landmark algorithm) [6] or OpenPose [7, 8] in order to obtain a tuple of key points F (each element contains pixel's coordinates (X, Y) of the key point), describing the key points of a person's face.

$$F_d(I) \to F \quad (3)$$

Further, the operation of the algorithm is based on the heuristic ratios of the parts of a person's face, Fig. 2 schematically shows a person's face as a set of 68 key points.

Fig. 2. Key points of a person's face

Let us introduce notation to describe the ratios of facial fragments when highlighting the key features of the desired mimic states. The distance between two key points A and B will be denoted as $L_{A,B}$, note that $L_{A,B} = L_{B,A}$, the visibility of point A is V_A, if point A is not visible $\neg V_A$. Point A is above point B $U_{A,B}$. Consider the logical formulas of the desired facial features of a person's face.

Eyes closed (C_F):

$$\begin{aligned} C_F = {} & V_{38} \& V_{39} \& V_{40} \& V_{41} \& V_{42} \& V_{43} \& V_{44} \& V_{45} \& V_{46} \& \\ & \& \left(\left(\frac{L_{38,42} + L_{39,41}}{2} < \frac{L_{37,40}}{4} \right) \& \left(\frac{L_{44,48} + L_{45,47}}{2} < \frac{L_{43,46}}{4} \right) \right) \end{aligned} \tag{4}$$

Eyes opened (A_F):

$$\begin{aligned} A_F = {} & V_{38} \& V_{39} \& V_{40} \& V_{41} \& V_{42} \& V_{43} \& V_{44} \& V_{45} \& V_{46} \& \\ & \& \left(\left(\frac{L_{38,42} + L_{39,41}}{2} > \frac{L_{37,40}}{4} \right) \& \left(\frac{L_{44,48} + L_{45,47}}{2} < \frac{L_{43,46}}{3} \right) \right) \end{aligned} \tag{5}$$

Upper eyelid lifted (Z_F):

$$\begin{aligned} Z_F = {} & V_{38} \& V_{39} \& V_{40} \& V_{41} \& V_{42} \& V_{43} \& V_{44} \& V_{45} \& V_{46} \& \\ & \& \left(\left(\frac{L_{38,42} + L_{39,41}}{2} > \frac{L_{37,40}}{3} \right) \& \left(\frac{L_{44,48} + L_{45,47}}{2} > \frac{L_{43,46}}{3} \right) \right) \end{aligned} \tag{6}$$

Frequent blinking of the eyes (M_F). An analysis is made of the retrospective of the frequency V_b of blinking (transition from the state of the eyes closed to the state of the eyes open), when the threshold of 0.4 Hz is exceeded, it is considered that the person blinks frequently:

$$M_F = V_b > 0.4 \tag{7}$$

Raised eyebrows (P_F):

$$\begin{aligned} P_F = {} & V_{20} \& V_{25} \& V_{37} \& V_{40} \& V_{43} \& V_{46} \& V_{38} \& V_{45} \& \\ & \& (L_{20,38} > L_{37,40}) \& (L_{25,45} > L_{43,46}) \end{aligned} \tag{8}$$

Shifted eyebrows (S_F):

$$\begin{aligned} S_F = {} & V_{18} \& V_{22} \& V_{23} \& V_{27} \& \\ & \& \left(L_{22,23} < \frac{L_{23,27}}{2} \right) \& \left(L_{22,23} < \frac{L_{18,22}}{2} \right) \end{aligned} \tag{9}$$

Open mouth (R_F):

$$\begin{aligned} R_F = {} & V_{52} \& V_{58} \& V_{63} \& V_{67} \& \\ & \& (L_{63,67} > L_{52,63}) \& (L_{63,67} > L_{67,68}) \end{aligned} \tag{10}$$

The algorithm can be supplemented with other key signs / desired mimic states.

The result of the algorithm is an OF tuple of seven Boolean values of functions (4–10):

$$O_F = (C_F, A_F, Z_F, M_F, P_F, S_F, R_F) \tag{11}$$

3.2 The Proposed Algorithm for the Analysis of Mimic Features of the Operator's Face

The beginning of the implementation of the algorithm for analyzing the configuration of the human skeleton consists in entering the input three-color image I, which is represented by the container OpenCV cv::Mat. It is worth noting that the algorithm assumes a sequence of images I for retrospective analysis and blink rate calculation. Then, at the next step, key points of a person's face are detected using the OpenPose library [7, 8] – the result is a tuple of 68 key points F (3). Next, the output resulting tuple O_F is initialized to seven (according to the number of mimic features sought) with logical zeros.

At the next step, the execution of the "eyes closed" function (4) is checked. In the case of a negative result, a transition is made to the next stage of the algorithm. If condition (4) is successfully met, the first element of the tuple O_F is assigned a logical unit and the transition to the next step of the algorithm occurs.

Checking the performance of the "eyes open" function (5). In the case of a negative result, the transition to the next step of the algorithm (5) is performed, in the case of a positive result, the second element of the tuple O_F is assigned a logical unit and the transition to the next step of the algorithm.

Checking the fulfillment of the condition "frequent blinking of the eyes" (7). In the case of a negative result, the transition to the next step of the algorithm (6) is performed, in the case of a positive result, the fourth element of the tuple O_F is assigned a logical unit and the transition to the next step of the algorithm.

Checking the performance of the "upper eyelid lifted" function (6). In the case of a negative result, the transition to the next step of the algorithm is performed, in the case of a positive result, the third element of the tuple O_F is assigned a logical unit and the transition to the next step of the algorithm.

Checking the performance of the "raised eyebrows" function (8). In the case of a negative result, the transition to the next step of the algorithm is performed, in the case of a positive result, the fifth element of the tuple O_F is assigned a logical unit and the transition to the next step of the algorithm.

Checking the "shifted eyebrows" function (9). In the case of a negative result, the transition to the next step of the algorithm is performed, in the case of a positive result, the sixth element of the tuple O_F is assigned a logical unit and the transition to the next stage of the algorithm.

Checking the "open mouth" function (10). In the case of a negative result, the transition to the next step of the algorithm is performed, in the case of a positive result, the seventh element of the tuple O_F is assigned a logical unit and the transition to the next step of the algorithm.

The output of the resulting tuple O_F (11), and then the completion of the algorithm for analyzing human facial expressions.

Figure 3 shows a block diagram of the algorithm for analyzing human facial expressions.

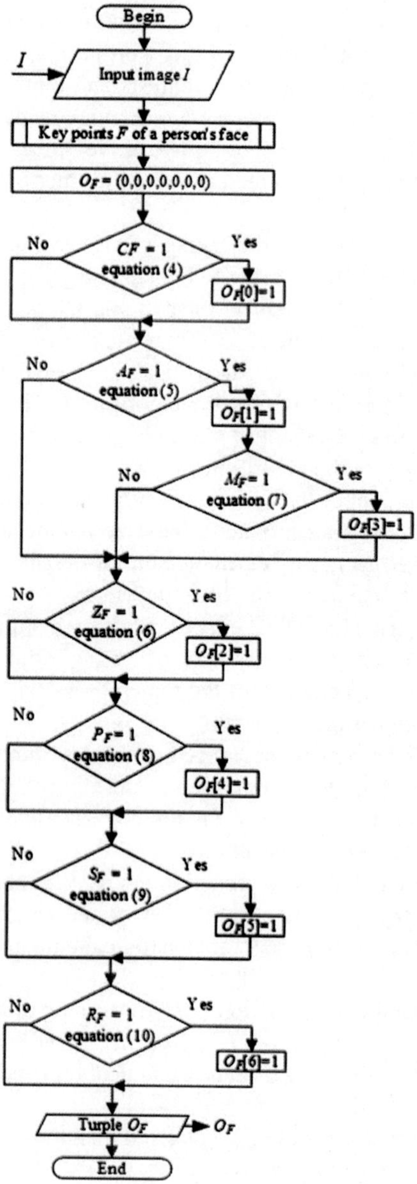

Fig. 3. Block diagram of the algorithm for the analysis of facial features

4 Practical Implementation of the Proposed Method

The method and algorithm for detecting the face area on video frames proposed in the article were used in the development of software that makes it possible to determine the poses of equipment operators. This is necessary for the timely detection of a possible loss of performance by the operator. For example, if the operator lost consciousness or had a heart attack.

Figure 4 shows a fragment of a screenshot of the software with visualization of recognition of the characteristic points of the face, the position of the pupils and the percentage of visibility of the face in the frame.

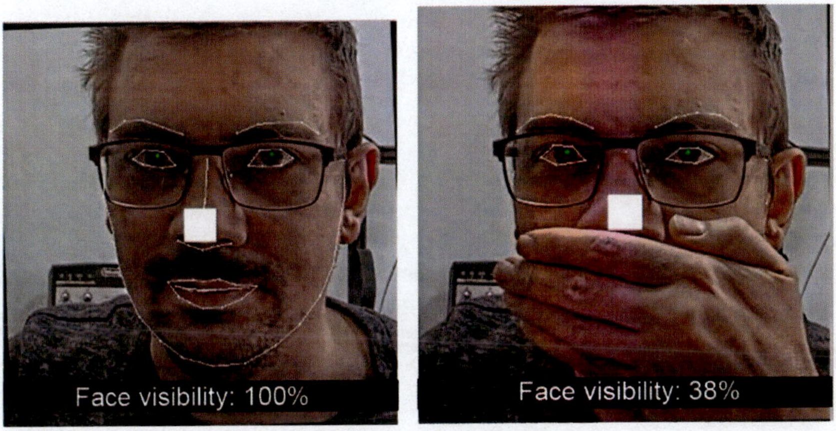

Fig. 4. Фрагменты снимков экрана с визуализацией распознавания характерных точек лица, положения зрачков и процентом видимости лица в кадре

A fragment of a screenshot of the graphical interface while the program is running: visualization of the registered facial features is shown of Fig. 5.

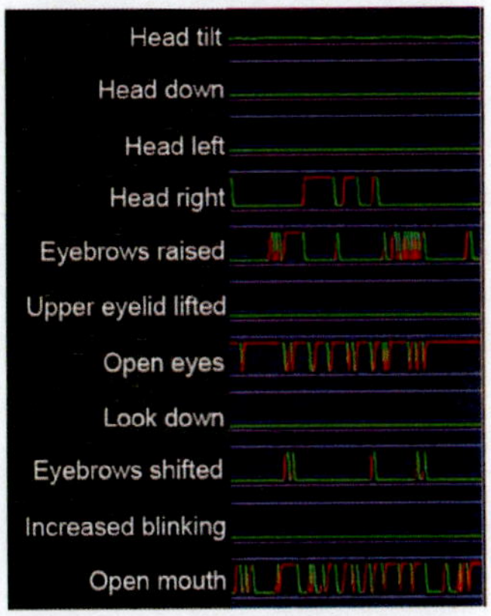

Fig. 5. Fragment of a screenshot of the graphical interface while the program is running: visualization of registered facial features

The results of the practical application of the proposed method and algorithm made it possible to identify mimic signs in real time, which are necessary for analyzing the health status and performance of operators of critical equipment.

5 Conclusions and Future Work

The article proposes a method and algorithm that allow detecting facial areas on video frames, as well as identifying a number of facial features of a person's face: eyes lowered, eyes open, upper eyelid raised, frequent blinking, raised eyebrows, shifted eyebrows, open mouth. The algorithm was applied in the development of software for a system for analyzing the state of operators of critical equipment.

The results of the practical application of the proposed approach showed its efficiency and low computational complexity. Due to this, this algorithm can be used in real-time monitoring systems.

Further work will be aimed at improving the algorithm by expanding the list of detected facial features and by adding the ability to work with frames obtained from different angles.

Acknowledgment. The study was carried out within the framework of the scientific program of the National Center for Physics and Mathematics (direction No. 9 "Artificial intelligence and big data in technical, industrial, natural and social systems").

References

1. Romanuke, V.: Appropriate number and allocation of ReLUs in convolutional neural networks. Res. Bull. Natl. Tech. Univ. Ukr. Kyiv Politech. Institute, № 1, pp. 69–78 (2017)
2. Deep Neural Network module [Electronic resource]. https://docs.opencv.org/3.4/d6/d0f/group__dnn.html
3. yolov4_darknet [Electronic resource]. https://github.com/kiyoshiiriemon/yolov4_darknet
4. Kaggle: Your Machine Learning and Data Science Community [Electronic resource]. https://www.kaggle.com
5. COCO - Common Objects in Context [Electronic resource]. https://cocodataset.org
6. Kazemi, V., Sullivan, J.: One millisecond face alignment with an ensemble of regression trees. In: Proceedings of the IEEE Conference on Computer Vision and Pattern Recognition, pp. 1867–1874 (2014)
7. Qiao, S., Wang, Y., Li, J.: Real-time human gesture grading based on OpenPose. In: 10th International Congress on Image and Signal Processing, BioMedical Engineering and Informatics (CISP-BMEI), pp. 1–6 (2017)
8. Cao, Z., et al.: OpenPose: Realtime Multi-Person 2D Pose Estimation using Part Affinity Fields. sarXiv. 2018

Using Design Science Research to Iteratively Enhance Information Security Research Artefacts

S. G. Govender(✉), M. Loock, E. Kritzinger, and S. Singh

University of South Africa (UNISA), Pretoria 0001, Gauteng, South Africa
32393113@mylife.unisa.ac.za, {loockm,kritze,singhs}@unisa.ac.za

Abstract. A report published in 2020 by the FBI on Internet Crime revealed that organizations lost USD$4.2 billion as a result of cybercrime. This was a 200% increase since last analyzed three years earlier. Cyber vulnerability is one of the most critical risk areas. Inaction to review, understand and react to cyber threat is simply too great for any organisation to ignore. Companies need to be confident that information security is managed effectively so that the organizations can be protected against data breaches and malicious cyber-attacks. Breaches negatively affect organizations in that time-to-recover is long and expensive. The reputational impact is also detrimental to organizations especially when strategic or personal information is stolen and/or disseminated. Design Science Research (DSR) is a research paradigm that allows for the development and iterative enhancement of practical applicative information security artefacts that can help organizations improve their security position.

Keywords: Design Science Research · Information Security Culture · Information Security Assessment · Information Security Cost · Information Security Evaluation · Information Security · The ARCS Security Framework · Artefacts

1 Introduction

Design Science Research (DSR) is a set of process driven, analytical methods to conduct research in information systems [1]. These techniques complement positivist, interpretive, and critical methods of research. DSR generally involves creating an artefacts or design theories to enhance the current state of application and existing research knowledge [2].

Two key activities in DSR that assist in improving and understanding the behavioural characteristics of Information Systems are [1]:

- Developing of new knowledge through the creation of inventive artefacts.
- Analyzing the artefacts use or performance through evaluation and iterative enhancement.

DSR lends itself well to the information systems and information security field as new concepts and ideas in the technology field leads to researchers being able to posit

R. Silhavy and P. Silhavy (Eds.): CSOC 2023, LNNS 723, pp. 49–61, 2023.
https://doi.org/10.1007/978-3-031-35317-8_5

new ideas or concepts to be tested [3]. This is an exploration type approach to research in IS/IT [3], which DSR greatly supports.

2 Design Science Research Methodology

In DSR, the research paradigm may be fully or partially created instead of naturally occurring [4, 5]. The research conducted using DSR methodology must be of value and interest to the research community in which it is conducted in order for it to be accepted and valued. Design science supports the creation and evaluation of IT artefacts developed to solve organizational problems that have been articulated [6]. DSR follows a strict and iterative process to create artefacts so that these can be evaluated to aid in contributing knowledge to the relevant research community. Any designed object that contains a possible provided solution can be the output of DSR. Research contributions include evaluating the output and understanding it, and communicating it through publications, articles and books [7]. The artefacts created in the DSR process include but are not limited to processes, frameworks or methods. Theories developed through DSR lead to knowledge contribution by developing a set of rules or concepts with a set of possible specific outcomes based on theory.

The definition of design is to create or bring something (an artefact) into being that did not exist previously. There are various types of artefacts, according to DSR, which generally introduces new ideas and concepts to be tested, that can create a state change in a researcher's understanding [8].

The DSR process model selected was by Pfeffers et al. [7] as described in Table 1. In this model artefacts must be created to address the problem defined. The artefact must have some method to be quantitatively or qualitatively evaluated.

Table 1. DSR Process Model adapted from Peffers, Tuunanen, Rothenberger and Chatterjee 2007

Activity	Description
Activity 1: Problem Identification and Motivation	In this phase, the researcher will define the research problem. The researcher should break the problem down into the smallest possible concepts to define its complexity. In this phase, the value of the solution is also motivated. The significance of doing so lies in motivating the researcher's efforts and motivating the research community to pursue the proposed developed outcome
Activity 2: Define the Objectives for a Solution	In this phase, the purpose for the development of an artefact is described to address each of these smaller conceptual problems defined in phase 1. The purpose can be quantitative in that the proposed solution may be superior to any similar artefact currently available, or qualitative in that the proposed solution describes a solution to a problem that that has nit previously been considered

(continued)

Table 1. *(continued)*

Activity	Description
Activity 3: Design and Development	In this phase, the artefact which is potentially a construct, model, method or re-design of properties of an existing construct is developed. A DSR artefact may be any artefact that contributes to research through its design. In designing this artefact, the researcher considers its functionality and architecture. Once designs are considers the actual artefact is created
Activity 4: Demonstration	In this phase, the artefact must be demonstrated. It must be able to solve in part or fully, the problem hypothesised. The demonstration of an artefact may be conducted through its use in testing, simulation, case study and verification
Activity 5: Evaluation	In this phase, the researcher examines and quantifies how well the artefact supports the objectives of addressing the problem. Proposed outcomes n must have been developed, and these must be compared to the tangible results of executing the usage of the artefact. The artefact can be quantitatively evaluated through output results or satisfaction surveys or qualitatively through response times or availability metrics. Based on the evaluated responses, researchers may choose to re-conduct phase 3 to enhance the artefact or just move on to the next phase and communicate the outputs with improvements left for additional research
Activity 6. Communication	In this phase, the researcher must convey the problem and its importance. The artefact and its applicability to contribute to a solution, its uniqueness and its rigour of design must also be communicated. Lastly its effectiveness to the pertinent researchers and other pertinent audiences such as practical applications in organisation must be conveyed

Artefacts developed could be created as the following:

- Constructs
- Models
- Frameworks
- Architectures,
- Design principles

- Methods
- Instantiations and design theories.

Instantiations are generally considered as material artefacts, while the rest of artefacts described can be considered abstract artefacts. Abstract artefacts and material artefacts are generally combined to form a design theory. In the context of possible outputs of artefact development, the following can be defined:

- Constructs are defined when problems are conceptualized and are refined throughout the DSR process.
- A model is a set of plans defining relationships among constructs. Models are generally associated with the problem and solution statements [9]. A model is presented in terms of what it does and may be part of a construct that describes the model's relationship with a theory.
- A framework is a supporting construct around which a concept can be developed. A framework defines a classification of rules, concepts or principles that is used to develop or determine something. A framework can be an abstract depiction of a set of ideas or concepts that aid solving a problem.
- A method is a set of steps such as a process or a policy that is used to perform a task.

3 Information Security Artefacts Developed

A study was conducted that developed three models and a framework that were used as a support or guide for information security research. The models, framework and tool created were discussed in detail in previously published papers [10–12] and are briefly described as follows:

- Model 1- Social and Technical Cost Reduction Factors which describes factors that influence information security cost.
- Model 2 - Human Intervention in Information Security Capability which describes information security assessment methodologies.
- Model 3 - Five Pillars of Information Security Culture Enhancement Model which describes factors that improve information security culture
- ARCS Security Framework – Combines components of Model 1 to 3 to form which key features which is then expanded into evaluations areas.

Another artefact from the noted study was the ARCS evaluation tool which is formulated from the ARCS Security Framework.

Figure 1 separates information security cost-reducing factors into social and technical factors. When isolated six of these factors are socially (people, managerial or structurally) influenced and depicted in orange to the left. The other six factors are technically influenced and depicted in the blue color to the right. Therefore, concentrating the information security management effort on these twelve factors will provide the best information protection at the lowest cost. For each of the six technical factors, some human intervention and response for these factors are required to be successful. For example, the technical factors may include management, participation, configuration, administration, continuous monitoring and evaluation and periodic ad hoc processes [13, 14]. Since this human interaction is social in nature, people's behaviour and values within an organisation directly influence whether these actions, supporting information security management, are successful.

Improving the values and behaviour of technical resources (e.g. server and network administrators, application developers, desktop support specialists and email and file-server administrators) that support information security remediation requirements will also assist in reducing the risk of information security breach incidents. In effect, developing and enhancing the socially relevant factors creates a stronger foundation for the success of the technical factors. This is depicted in Fig. 1 by the green circle, representing the dependence on human culture and behaviour to succeed in the social and technical cost-saving factors.

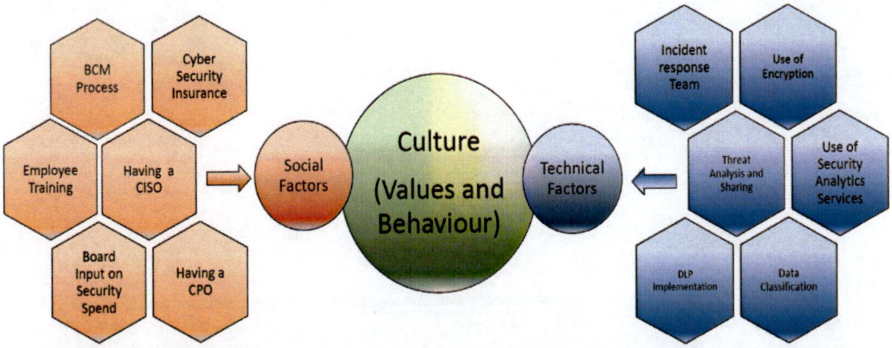

Fig. 1. Cost Reducing Remediation Factors [24]

Figure 2 describes the relationship between generalized common information security evaluation methods and the reliance on human resources to run, manage, monitor and maintain information security systems that are identified through these methods. Security tools are not always managed by the security function within the organisation, and staff that do manage these solutions are from alternative functional areas within the IT department, i.e. application development, infrastructure, end-user computing or networks. The motivation and behaviour for these IT staff members to consider security first is generally incongruent with their motivation for their primary job responsibilities. The effect of what staff consider additional work to their primary job responsibility is lower motivation to consider their information security responsibilities as important.

The model in Fig. 3 proposes five pillars of cultural change that are applicable in redefining staff members' values and behaviors, which will develop and enhance information security culture and behaviour. The theory describing motivation, organizational behaviour, rewards systems and information security culture covered in Chapter 2 supports each pillar described. Pillars 1, 3, 4 and 5 are related to, intrinsic rewards which relates to intangible rewards but enhances organizational culture, and introjected motivation which is developed based on human want or requirement to finish an activity to prove an accomplishment to themselves. Pillar 2 is based on extrinsic reward which is a tangible reward system that enhances organizational behaviour. Each of the five pillars supports organizational behaviour theory and provides and external and internal stimulus to motivation. As such this motivation supports sustained improvement in organizational culture and in the context of information security supports sustained improvement in information security culture.

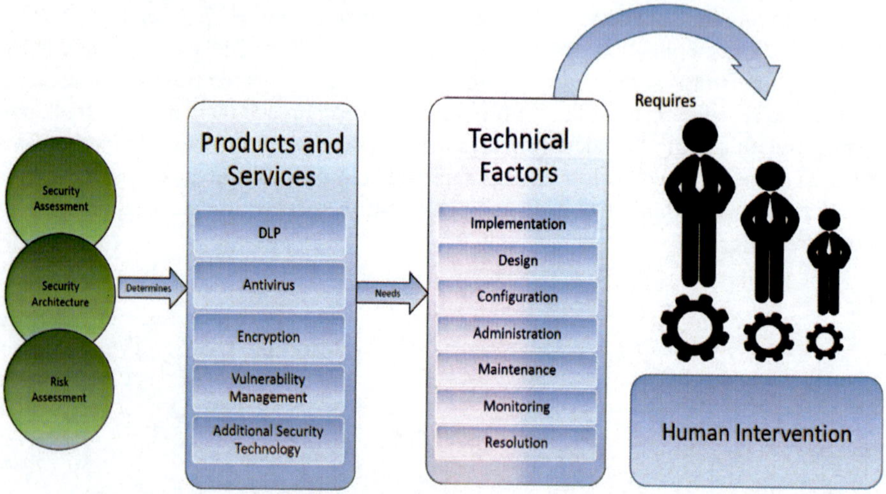

Fig. 2. Information Security Assessment and Human Intervention [24]

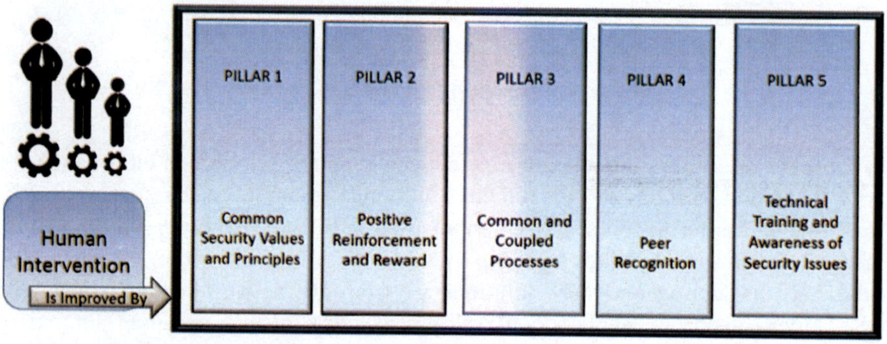

Fig. 3. Five Pillar Model to enhance Information Security Culture [24]

The effect of improving IT employees' information security culture is two-fold: firstly staff will be motivated within their job functions to consider information security a priority and secondly the enhancement of information security cultural aspects will allow for long-term value for the organisation and create the foundation for information security practices to become a prioritized norm. The model proposes five practical streams of activity that can be applied to enhance the information security culture of IT staff. The model is not interdependent, and an organisation may execute each pillar independently or select to execute the necessary pillar that may be relevant to that organisation.

The framework discussed below combines the three concepts of Information Security Risk Assessment, Information Security Cost and Information Security Culture to establish a structured evaluation method, which will assist an organisation in addressing and improving its level of information security maturity.

The framework extracted from the models described in Figs. 1, 2 and 3 is structured into three features (F1, F2 and F3) called Assessment of Security Risk (F1), Reduction of Information Security Cost (F2) and Sustainability of Culture (F3) and is referred to as the ARCS Security Framework as described in Fig. 4.

To implement the ARCS Security Framework, an evaluation tool has been developed in Microsoft Excel. The tool contains tabs related to the Features described in the Framework, which each contain the Evaluation Areas split into questions. The response tables are embedded in the tool, and the scoring logic is calculated based on the given responses. Weighting and output charts are also embedded in the tool.

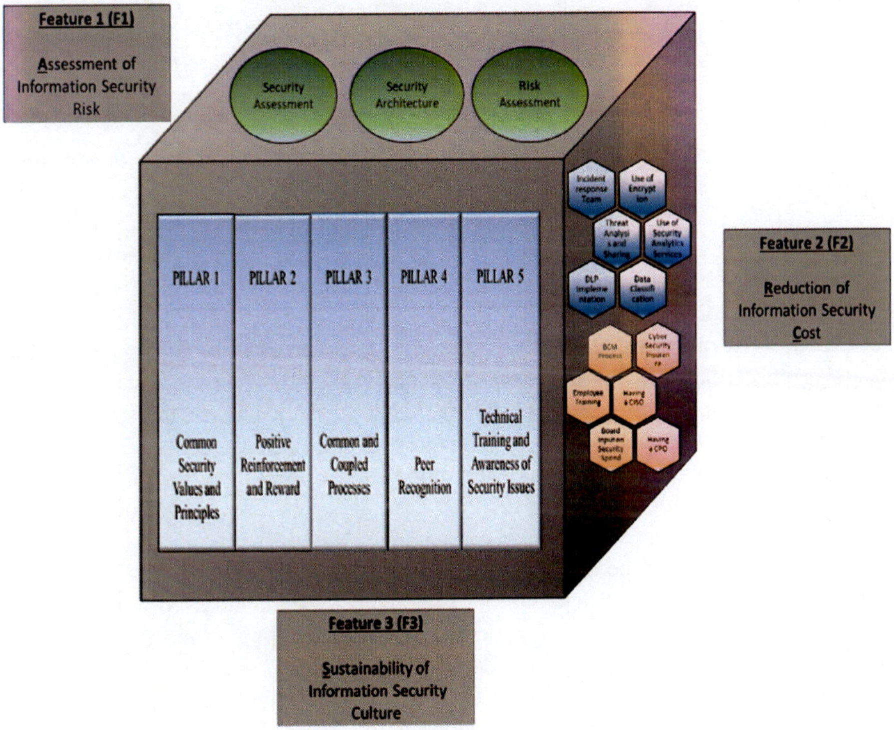

Fig. 4. ARCS Security Framework

4 Evaluation of Artefacts

In order to conduct the study, five steps were completed to demonstrate the artefacts developed and elicit results for improvement in iterations and communication of the study.

The five steps are described as follows:

Step 1 - Presentation of the framework and evaluation tool – The details and function of the Security Framework and Evaluation Tool was presented and explained to the participants. The underlying concepts and models used to generate the Security Framework were explained to the participants.

Step 2 – Conduct the evaluation, and present results of evaluation – The evaluation tool developed was directly aligned to the framework developed and used to evaluate the implementation of the framework. This tool contains scoring logic that will analyze responses and output a quantitative score for the organisation evaluated. The scoring methodology was developed based on the value of the relationship of questions developed for the security framework. Where areas of importance are linked, scores were deemed to be weighted higher than non-related questions.

After the scoring was concluded, the results of the evaluation generated by the evaluation tool was presented to participants in the form of the output scoring charts generated by the tool. The scoring charts and the outputs depicted therein was explained to the participants.

Step 3 - Review of evaluation results - Participants were given an opportunity to review the results. The researcher then collected information on the participants' views of the outcomes and whether the outcomes were an accurate reflection of the information security position of the organisation in its current state. This is considered to be an expert review as the participants chosen, display significant years of experience and capability in the information security field.

Step 4 - Review of the Security Framework – The researcher then conducted interviews with participants on their views of the components of the Security Framework, the value and applicability of the components to their organisation, the structure of the framework, the components that they feel were not covered, their views of improvements or changes and their general views on the framework. A semi-structured interview questionnaire was used for this part of the interview, along with researcher observations in regard to the participant's responses.

Step 5 - Review of the Security Evaluation Tool – The researcher conducted interviews with participants on their views of the Security Evaluation Tool, the value and applicability of the tool, the scoring mechanism and weighting of questions, the quality and value of the output charts generated, their views on improvements and changes and their general views on the evaluation tool. A semi-structured interview questionnaire was used for this part of the interview along with researcher observations in regard to the participant's responses.

Further to the first pass evaluation, input regarding the framework from the expert review was considered to improve and redevelop parts of the evaluation questions and tool.

5 Iterative Updates of Artefacts

In interviews with expert reviewers and based on observations by the researcher while demonstrating the artefacts, it was established that enhancements were required. These enhancements related to the expansion of the framework and evaluation tool in order to improve ease of use, provide more detailed information and address areas of information security that the expert reviewers considered to be of importance. The enhancements discussed in the following sections forms the second iteration of the framework and evaluation tool as expected as part of the iterative improvement process described by the DSR process selected.

The response tables, in the evaluation tool, that were originally created in Microsoft Excel were embedded into an enhanced evaluation tool which was developed as a web application that contains information about the framework, an evaluation jump page, as well as links to an online forms system that collects the data. Expert reviewers felt that Excel based tools needed to be modernized. The framework's structure remained the same with three feature areas, twenty-one evaluation areas and eighty-three questions. The questions remained weighted and scored to give each evaluation area an output score. However, the scoring logic was updated to reflect partial execution or achievement of an action or structure within the expert reviewer's environment. This was requested as the original scoring was binary and did not give reviewers an opportunity to note that action was partially implemented in their organisation leading to an inaccurate representation of their security position.

A third iteration of the framework and tool was not concluded but an expansion of the framework and question model was proposed, based on inputs by expert reviewers. In the original framework model Fig. 5, each feature (F1, F2 and F3) is broken down into an evaluation area. Each evaluation area (E1, E2, En) is given a shortened tagged description and is broken down further into questions (Q n.1 to Q n.m) related to that evaluation area. The questions were then developed in line with the evaluation aim of each of the evaluation areas.

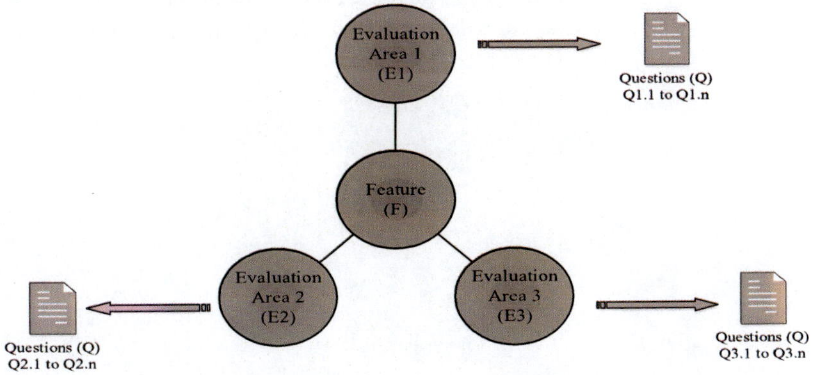

Fig. 5. Original ARCS Security Framework Relationship Model

The framework comprised of three features established from the models developed in Sect. 2. Twenty-one evaluation areas were created for the framework, and eighty-three questions were developed for the framework that is aligned to the evaluation areas. In discussion with expert reviewers, it was established that the framework was positioned at a management reporting level and did not interrogate more technical aspects of information security. Furthermore, one expert reviewer advised that the evaluation areas did not cover the compliance realm as much as the information security realm. The researcher therefore redeveloped the framework relationship model with expansion in mind to evolve over time based on newer information security concepts, technology, and processes. The researcher would consider future studies to reconfigure the framework to expand the relationship between evaluation areas and questions to a relationship between evaluation areas and sub-evaluation areas, as noted in Fig. 6.

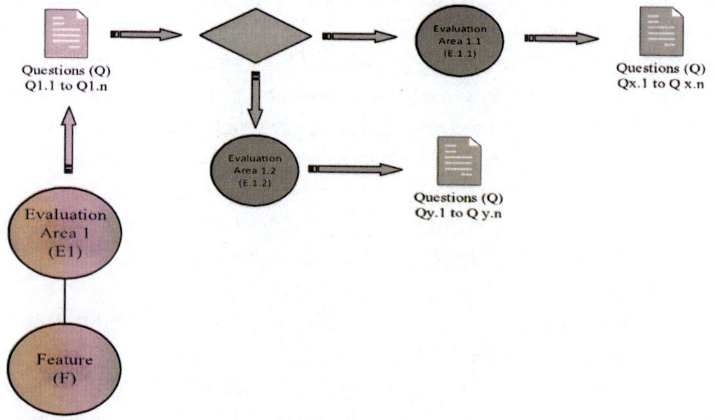

Fig. 6. Updated Relationship Model

The questions in each section or sub-section will be contextual, thus leading a user of the framework to more defined and detailed evaluations of their environments. The questions would be related as per a defined tree model, as depicted in Fig. 7.

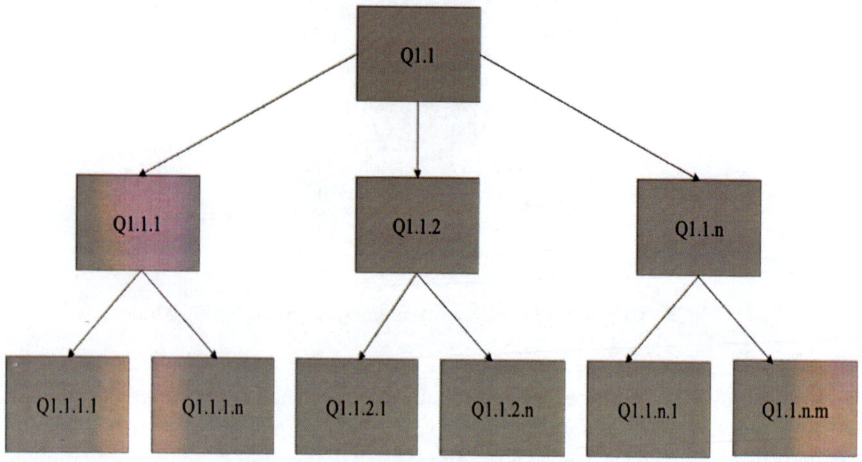

Fig. 7. Updated Questionnaire Tree Model.

Expanding the models addresses the concerns of expert reviewers in regard to bringing in more technical aspects of information security management and moving the reporting output away from just high-level management reporting to more actionable technical responses to the evaluated gaps.

Also based on expert reviewer input two additional Features were added. These were IT Compliance and Data Privacy. IT Compliance refers to the regulatory, legislative or standardised rules defined to ensure that components of computing systems and data meet acceptable standards based on industry or statutory body requirements. This alignment to standards allows for the reduction risk in terms of threats, or malicious intent [15]. External bodies such as auditors, customers, regulators or legislators define, enforce and test an organization adherence to the expected compliance regime. There are well known general cyber standards within the IT industry, such as ISO 27000 and the NIST Cyber Security Framework, but as cyber-threats become more prevalent there is also a significant increase in standards being developed and enforced by regulatory bodies and government organizations [16].

Compliance like the information security discipline focuses on practicing due diligence in order to protect computing assets. Compliance however is positioned on the requirements of a third party, such as a government, security framework, or regulatory bodies requirement. Compliance frameworks define the controls and the parameters of control an organisation needs to adhere to in order to conduct business [17]. Generally it is expected that if an organisation wants to conducts business in a specific country or a specific industry that there will be strong regulations that must be adhered to. The healthcare, finance or any industry where customer privacy is of importance are such an example. For example, regulations like HIPAA [18] and Sarbanes Oxley [19] or standards like PCI-DSS [20] or ISO:27001 [21], outline very specific security criteria that a business must meet to be deemed compliant. Adherence to standards are key as non-adherence could lead to a loss of business, loss of customer trust or illegal business operations.

When an organization is compliant it implies all defined standards are adhered to and evidence is available to support such a claim [15]. IT Compliance can be considered to be the process of adhering to externally defined digital security requirements in order to enable business operations in a specific industry or with specific customers [18].

Data privacy or digital ethics is focused on the processing and management of data in respect of consent, notice and regulatory requirements. Data privacy differs from data security in that the latter focuses more on malicious data compromise while the former is concerned with how data is collected, shared and used [22]. Data privacy concerns often revolve around the processing and sharing of data with third parties, the legal collection or storage of data and legislative obligations such as GDPR, HIPAA, or POPIA [23].

Data is an important company asset as it defines the data economy. Organizations find vast value in accumulating, distributing, and utilizing data. Regulatory frameworks define how organizations must gain consent to access, store and process user data [24]. Furthermore, the number of countries that enacted data privacy legislation or are currently in the process of doing so has increased significantly in the last three to five years. Data privacy as a feature of the security framework is important as the increase in data regulation imply that organisation need to evaluate the efficiency, effectiveness and maturity of their controls in the data privacy space [20]. The new set of features are depicted in Fig. 8.

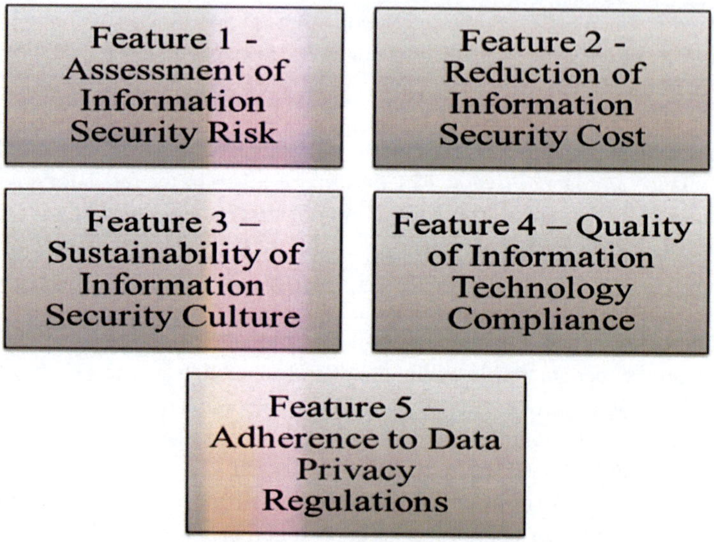

Fig. 8. Updated Framework Feature Model

6 Conclusion

Cyber security breaches are becoming more prevalent. Awareness of such events are mainstream and therefore companies are very much more aware of the impact and threats that are faced. Therefore there is a need to rapidly provide newer tools, techniques and technology that practically assist organizations in improving their security position. Artefacts that can be tested and improved with organisation through a iterative improvement process will bring quicker more focused value.

References

1. Vaishnavi, V., Kuechler, B.: A framework for theory development in design science research: multiple perspectives. J. Assoc. Inf. Syst. **13**(6), 3 (2015)
2. Baskerville, R.: What design science is not. Eur. J. Inf. Syst. **17**(5), 441–443 (2008). https://doi.org/10.1057/ejis.2008.45
3. Orlikowski, W.J., Iacono, C.S.: Research commentary: desperately seeking the "IT" in IT research - a call to theorizing the IT artifact. Inf. Syst. Res. **12**(2), 121–134 (2001). https://doi.org/10.1287/isre.12.2.121.9700
4. Lakatos, I.: Falsification and the methodology of scientific research programmes. In: Can Theories be Refuted? Springer Netherlands, Dordrecht, pp. 205–259 (1976). https://doi.org/10.1007/978-94-010-1863-0_14
5. Kuhn, T.: The Structure of Scientific Revolutions, vol. 111. University of Chicago Press, Chicago (2012)
6. Hevner, A.R., Chatterjee, S.: Design research in information systems, intergrated series. Des. Res. Inf. Syst. Intergr. Ser. Inf. Syst. **22**(1), 9–22 (2010). https://doi.org/10.1007/978-1-4419-5653-8_2

7. Peffers, K., Tuunanen, T., Rothenberger, M.A., Chatterjee, S.: A design science research methodology for information systems research. J. Manag. Inf. Syst. **24**(3), 45–77 (2007). https://doi.org/10.2753/MIS0742-1222240302
8. Gregor, S., Hevner, A.R.: Positioning and presenting design science research for maximum impact. MIS Q. Manag. Inf. Syst. **37**(2), 337–355 (2013). https://doi.org/10.25300/MISQ/2013/37.2.01
9. March, S.T., Smith, G.F.: Design and natural science research on information technology. Decis. Support Syst. **15**(4), 251–266 (1995). https://doi.org/10.1016/0167-9236(94)00041-2
10. Govender, S.G., Loock, M., Kritzinger, E.: Enhancing information security culture to reduce information security cost: a proposed framework. In: Castiglione, A., Pop, F., Ficco, M., Palmieri, F. (eds.) CSS 2018. LNCS, vol. 11161, pp. 281–290. Springer, Cham (2018). https://doi.org/10.1007/978-3-030-01689-0_22
11. Govender, S.G., Kritzinger, E., Loock, M.: Information security cost reduction through social means. In: Venter, H., Loock, M., Coetzee, M., Eloff, M., Eloff, J. (eds.) ISSA 2019. CCIS, vol. 1166, pp. 1–14. Springer, Cham (2020). https://doi.org/10.1007/978-3-030-43276-8_1
12. Govender, S.G., Kritzinger, E., Loock, M.: A framework and tool for the assessment of information security risk, the reduction of information security cost and the sustainability of information security culture. Pers. Ubiquit. Comput. 1–14 (2021)
13. Takemura, T., Komatsu, A.: An empirical study on information security behaviors and awareness. In: Böhme, R. (ed.) The Economics of Information Security and Privacy, pp. 95–114. Springer, Heidelberg (2013). https://doi.org/10.1007/978-3-642-39498-0_5
14. Bojanc, R., Jerman-Blažič, B., Tekavčič, M.: Managing the investment in information security technology by use of a quantitative modeling. Inf. Process. Manage. **48**(6), 1031–1052 (2012). https://doi.org/10.1016/j.ipm.2012.01.001
15. Chatterjee, C., Sokol, D.: Data security, data breaches, and compliance. In: Cambridge Handbook on Compliance, 1st ed. Cambridge University Press, pp. 1–17 (2019)
16. Edwards, B., Jacobs, J., Forrest, S.: Risky Business: Assessing Security with External Measurements (2019). arXiv preprint arXiv:1904.11052. Accessed 6 October 2019
17. Haqaf, H., Koyuncu, M.: Understanding key skills for information security managers. Int. J. Inf. Manage. **43**, 165–172 (2018). https://doi.org/10.1016/j.ijinfomgt.2018.07.013
18. Herold, R., Beaver, K.: Security rule requirements overview. In: The Practical Guide to HIPAA Privacy and Security Compliance, pp. 236–259, 20 October 2014. https://doi.org/10.1201/b17548
19. Kim, N.Y., Robles, R.J., Cho, S.E., Lee, Y.S., Kim, T.H.: SOX act and IT security governance. In: Proceedings - 2008 International Symposium on Ubiquitous Multimedia Computing, UMC 2008, pp. 218–221 (2008). https://doi.org/10.1109/UMC.2008.51
20. Wu, S.M., Guo, D., Wu, Y.J., Wu, Y.C.: Future development of Taiwan's smart cities from an information security perspective. Sustainability. **10**(12), 4520 (2018). https://doi.org/10.3390/su10124520
21. Prislan, K., Bernik, I.: Risk management with ISO 27000 standards in information security. Inf. Secur. 58–63, December 2010
22. Mehmood, A., Natgunanathan, I., Xiang, Y., Hua, G., Guo, S.: Protection of big data privacy. IEEE Access. **4**, 1821–1834 (2016). https://doi.org/10.1109/ACCESS.2016.2558446
23. Torra, V.: Data Privacy: Foundations, New Developments and the Big Data Challenge. Springer International Publishing, vol. 28 (2017). https://doi.org/10.1007/978-3-319-57358-8
24. Martin, K.D., Borah, A., Palmatier, R.W.: Data privacy: effects on customer and firm performance. J. Mark. **81**(1), 36–58 (2017). https://doi.org/10.1509/jm.15.0497

Improving Test Quality in E-Learning Systems

Roman Tsarev[1,2](✉), Abhishek Bhuva[3], Dipen Bhuva[4], Irina Gogoleva[5], Irina Nikolaeva[6], Natalia Bystrova[7], Ivetta Varyan[8,9], and Svetlana Shamina[10]

[1] MIREA - Russian Technological University (RTU MIREA), Moscow, Russia
rtsarev@acadst.org
[2] International Academy of Science and Technologies, Moscow, Russia
[3] University of Massachusetts Boston, 100 Morrissey Blvd, Boston, MA 02125, USA
[4] Cleveland State University, 2121 Euclid Avenue, Cleveland, OH 44115, USA
[5] FGBOU VO «Arctic State Agrotechnological University», Building 3, Sergelyakhskoe Highway, 3 Kilometer, Yakutsk 677007, Russia
[6] North Eastern Federal University, 57, Dzerzhinsky Street, Yakutsk 677009, Russia
[7] Minin State Pedagogical University of Nizhny Novgorod, Nizhny Novgorod, Russia
[8] N.M. Emanuel Institute of Biochemical Physics, Russian Academy of Sciences, Moscow, Russia
[9] Plekhanov Russian University of Economics, Moscow, Russia
[10] South Ural State Agrarian University, Troitsk 457100, Russia

Abstract. Electronic learning systems play an important role in the educational process. They typically provide access to lecture material and practical tasks and are used for testing. A significant advantage of e-learning systems is that they provide instant verification of test results and tools to analyze them. This article presents a number of indicators that allow assessing the ability of test questions to grade students by their levels of knowledge. Test questions not meeting the threshold requirements are proposed to be excluded from the test, thus improving the quality of the test. #CSOC1120.

Keywords: Automated Testing · Difficulty · Standard Deviation · Grading Factor · Iterative Procedure · Moodle

1 Introduction

The development of information technology has inevitably led to its integration into the educational process. Electronic learning systems and computer-based educational support tools have become an integral part of the educational environment [1–4]. E-learning systems increase the efficiency of the educational process and expand its capabilities with new computer-related functions [5–7]. In critical situations such as the COVID-19 pandemic, e-learning systems have become the only way to receive education remotely and fully [8].

A widely-known learning management system that is used in universities around the world is LMS Moodle [9, 10]. The system has a full set of tools that instructors can use to give their students theoretical material and practical assignments. It also

R. Silhavy and P. Silhavy (Eds.): CSOC 2023, LNNS 723, pp. 62–68, 2023.
https://doi.org/10.1007/978-3-031-35317-8_6

offers testing facilities, allowing students to take tests not only in classrooms, but also remotely. Moodle allows creating tests that include both open and closed questions, including the choice of one or more options from several proposed options, true/false questions, comprehension questions, numeric answers, answers calculated by formulas, matching, etc. In addition, Moodle has a rich analytical arsenal of analysis of test results and answers to individual questions.

In terms of the application of information technology in e-learning systems, we can distinguish such tasks as the implementation of student learning on individual educational trajectories [11, 12], gamification in e-learning [13, 14], and automated testing [15]. Moreover, while the first two areas are optional and depend on the objectives of the course and the preferences and competencies of the e-course creator, testing is an integral part of the vast majority of courses implemented in the Moodle system. The advantage of Moodle is that you not only can create tests and conduct testing, but also have the opportunity to perform a comprehensive analysis of the test results.

LMS Moodle allows instantly verifying answers to questions automatically and carrying out automatic statistical analysis of the test and its elements. If the former is of equal interest to both the e-course creator and the student, the latter is the most important for the test developer, as it allows understanding the effectiveness of the test and taking the necessary steps to improve it.

The bank of test questions that is further used in testing allows adequately assessing students' knowledge levels, and, therefore, correctly grading them based on their preparation levels [16, 17]. The test results obtained can be considered as feedback on the basis of which the quality of test questions, their difficulty, and the ability to grade students based on their knowledge levels can be assessed. Test questions not meeting the testing objectives can then be excluded or improved based on this assessment.

This article presents indicators that allow assessing the grading ability of test questions. Test quality can be improved by removing test questions that fall below the given thresholds.

2 Grading Ability Indicators of Test Questions

The difficulty assessment allows identifying questions that are inappropriate for the training levels of students taking the test. Analytically, the difficulty assessment of a question can be written down as follows:

$$e_i = \frac{\sum_{j=1}^{m} x_{ij}}{m}; i \in \overline{1, n}, \quad (1)$$

where n is the number of questions in the test, m is the number of students who completed the test, x_{ij} is the assessment of the result of the answer to the ith question by the jth student.

The lower the value of the difficulty assessment, the smaller the number of students who cope with the ith question, and vice versa, the higher its value, the greater the number of students who successfully answer the ith test question.

Naturally, the difficulty of the test as a whole should correspond to the knowledge levels of the students taking the test. A test can include questions of varying difficulty from easy to difficult. However, questions that are too easy, for which e_i approximately equals 1, and questions that are too difficult, for which e_i approximately equals 0, are recommended to be excluded from the test, since they do not allow grading students by their knowledge levels.

The contribution that each question makes to the grading ability of the test can be estimated by the value of the standard deviation of the results, which is calculated by the formula:

$$\sigma_i = \sqrt{\frac{\sum_{j=1}^{m} (x_{ij} - \overline{x_j})}{m}}; i \in \overline{1, n}, \tag{2}$$

Modern pedagogical theory determines that the value of the standard deviation of a test question assessment should be no more than 0.3 [18, 19]. Otherwise, the test question is assumed to not have any grading ability and is excluded from the test.

Finally, an important indicator reflecting the grading ability of test questions is the grading factor, which reflects the correlation between the assessment of the answer to the ith question and the assessment of the test as a whole. The grading factor is calculated by the formula:

$$d_i = \frac{C(x_i, R)}{C_{\max}(x_i, R)} \cdot 100; i \in \overline{1, n}, \tag{3}$$

$$C(x_i, R) = \frac{1}{Q} \sum_{q \in Q} (x_i(q) - \overline{x_i}) \cdot (R_q - \overline{R_q}),$$

where R_q is the average assessment of the answer to the question, $\overline{R_q}$ is the average assessment of the test as a whole, $\overline{x_i}$ is the arithmetic mean of the assessments received by all students of the answer to the ith question.

The grading factor takes values in the interval from –1 to + 1. This indicator shows how well a particular question is able to grade students between those who are well prepared and those who are poorly prepared.

Positive values allow determining the test questions with a good grading ability, and can grade students between those who are well prepared and those who are poorly prepared. Negative values of the grading factor show that poorly prepared students answer this task better on average than well prepared students. In this case, the test questions do not reflect the objective situation with the students' preparation for the tests and are unable to adequately grade the students. Test questions with negative grading factor values are excluded from the test.

In modern pedagogical theory, a test question is accepted as having sufficient grading ability if its grading factor is equal to or more than 0.3 [18, 19].

3 Results and Discussion

The Moodle learning management system allows analyzing test results by providing all the necessary information [20, 21]. Below are the results of calculating the indicators of the grading ability of test questions received from students passing the test. The total number of test questions is 200 ($n = 200$). The figures in this section show the calculations for the first twenty questions to clearly demonstrate the calculation results.

We will use the test question results to calculate the difficulty of assessment of questions according to Eq. (1). The closer e_i is to 1, the easier the question is. The closer e_i is to zero, the harder the question is.

In practice, thresholds for assessing test difficulty are defined at the 0.9 and 0.2 levels (see Fig. 1). Questions with difficulty assessments greater than 0.9 are too easy, and questions with difficulty assessments less than 0.2 are too difficult. Since both do not allow grading students by their knowledge levels, they should be excluded. Note that the questions are removed after calculating the other indicators to be able to maintain the calculated indicator values.

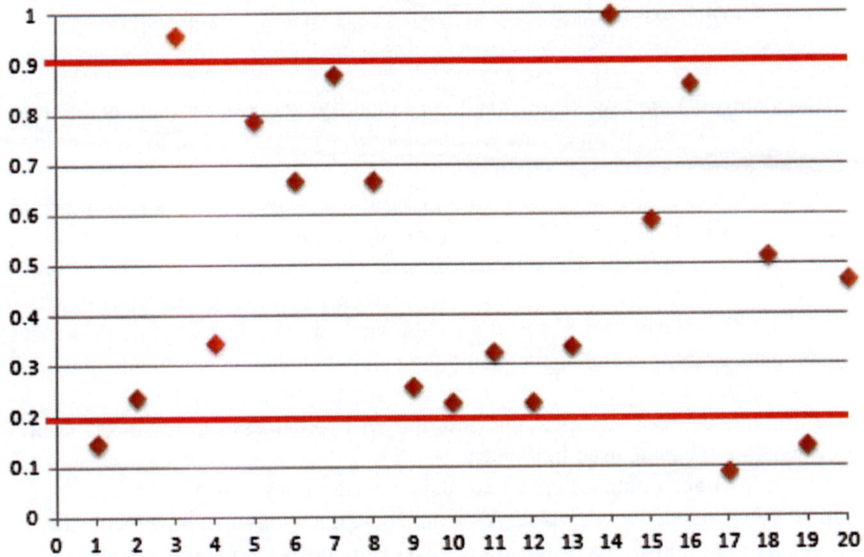

Fig. 1. Results of calculating the difficulty assessment of test questions.

Questions 1, 17, 19 (too difficult) and 3, 14 (too easy) will be excluded from the test (see Fig. 1).

Figure 2 shows the results of calculating the standard deviation of the results obtained by the formula (2). The threshold value is equal to 0.3. Questions with values below the threshold have no grading ability and should be excluded from the test. These are questions 1, 3, 14, 17, and 19. We excluded these questions after calculating the next indicator – the grading factor (Fig. 3).

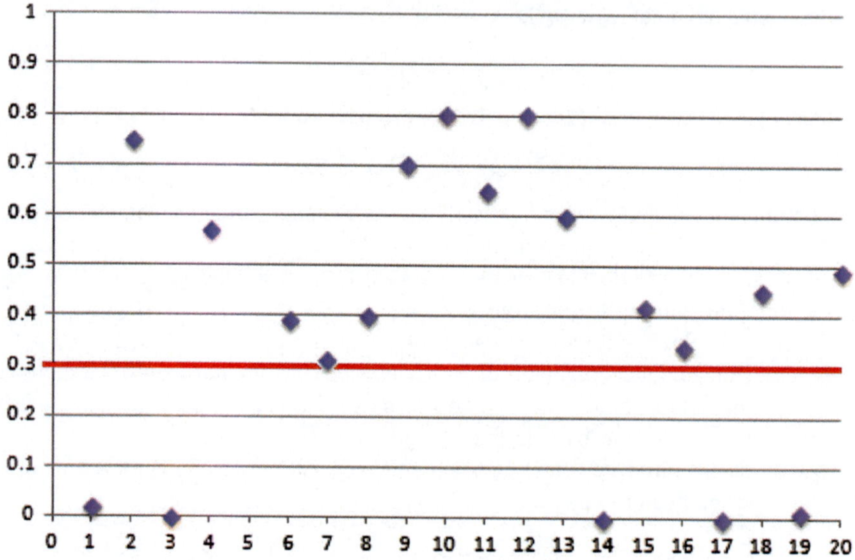

Fig. 2. Results of calculation of the standard deviation of the results.

Calculating the grading factor (3) allowed identifying questions lacking the required grading ability. These are questions 1, 3, 14, 17, 19, which are below the threshold value of 0.3 (see Fig. 3).

After calculating the indicator values and removing the questions falling below the specified thresholds, the indicator values are recalculated. This procedure is iterative, and is performed until all the test questions not meeting the requirements have been excluded. If the exclusion leaves too few questions remaining, the question bank is updated and the questions in the new question bank are checked whether they meet the requirements again.

It is important to note that you can modify the questions instead of excluding them. Since this procedure is iterative, the question will be assessed at the next iteration to check whether it can be used in the test.

Also of particular note is that we excluded the easy questions. However, these can be retained to keep students motivated. Although these questions do not meaningfully grade students by their knowledge levels, they give even poorly-prepared students the opportunity to feel confident that they can answer the tests successfully and stay interested during testing.

The easy questions that the vast majority of students can answer successfully can be evenly distributed at intervals over a number of questions or randomly distributed across the test. The impact of such easy questions on the test results is a subject of further research.

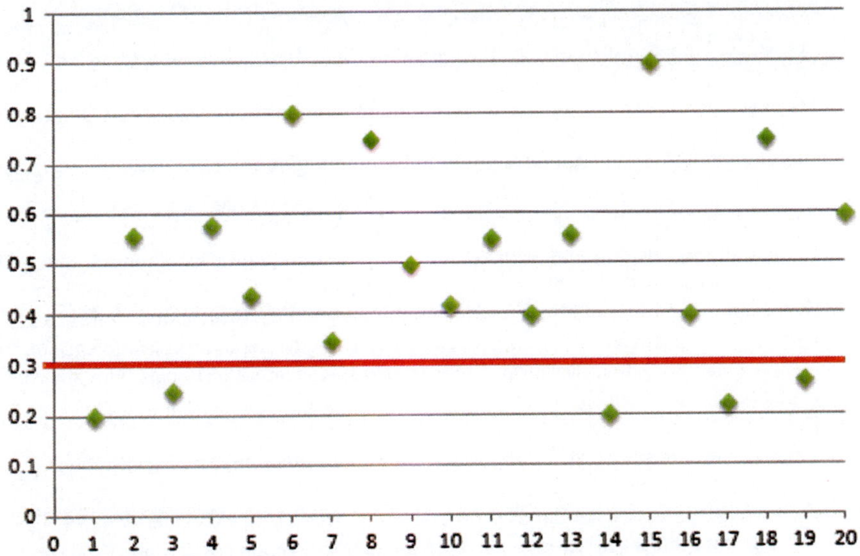

Fig. 3. Results of calculation of the grading factor.

4 Conclusion

Testing consists in providing students with a system of questions to assess their knowledge levels and structures effectively. Questions provided to the student obviously must meet certain requirements. The Moodle e-learning system not only allows creating a bank of test questions and conducting testing, but also assessing the test quality. This article discussed indicators reflecting how test questions are able to grade students based on their knowledge levels. Questions not having this ability should be excluded from the test or reformulated. The test quality is significantly improved through iterative calculation of the grading ability indicators of the test questions and excluding questions falling below the set thresholds. This procedure can be used to assess test questions in various learning management systems such as Moodle, eLearning Server 4G, eFront, and others.

References

1. Aldiab, A., Chowdhury, H., Kootsookos, A., Alam, F., Allhibi, H.: Utilization of Learning Management Systems (LMSs) in higher education system: a case review for Saudi Arabia. Energy Proc. **160**, 731–737 (2019)
2. Cavus, N., Zabadi, T.: A comparison of open source learning management systems. Proc. Soc. Behav. Sci. **143**, 521–526 (2014)
3. Akhmetjanov, M., Ruziev, P.: Fundamentals of modeling fire safety education. Inform. Econ. Manag. **1**(2), 0301–0308 (2022). https://doi.org/10.47813/2782-5280-2022-1-2-0301-0308
4. Zenyutkin, N., Kovalev, D., Tuev E., Tueva, E.: On the ways of forming information structures for modeling objects, environments and processes. Mod. Innov. Syst. Technol. **1**(1), 10–22 (2021). https://doi.org/10.47813/2782-2818-2021-1-1-10-22

5. Dobre, I.: Learning management systems for higher education-an overview of available options for higher education organizations. Proc. Soc. Behav. Sci. **180**, 313–320 (2015)
6. Hassouni, B.E., et al.: Realization of an educational tool dedicated to teaching the fundamental principles of photovoltaic systems. J. Phys: Conf. Ser. **1399**(2), 022044 (2019). https://doi.org/10.1088/1742-6596/1399/2/022044
7. Lonn, S., Teasley, S.D.: Saving time or innovating practice: investigating perceptions and uses of Learn. Manag. Syst. Comput. Educ. **53**(3), 686–694 (2009)
8. Alzahrani, L., Seth, K.P.: Factors influencing students' satisfaction with continuous use of learning management systems during the COVID-19 pandemic: an empirical study. Educ. Inf. Technol. **26**(6), 6787–6805 (2021). https://doi.org/10.1007/s10639-021-10492-5
9. Kakasevski, G., Mihajlov, M., Arsenovski, S., Chungurski, S.: Evaluating usability in learning management system Moodle. In: Proceedings of the 30th International Conference on Information Technology Interfaces, pp. 613–618. IEEE, Cavtat (2008). https://doi.org/10.1109/ITI.2008.4588480
10. Zabolotniaia, M., Cheng, Z., Dorozhkin, E., Lyzhin, A.: Use of the LMS Moodle for an effective implementation of an innovative policy in higher educational institutions. Int. J. Emerg. Technol. Learn. (iJET) **15**(13), 172–189 (2020)
11. García Peñalvo, F.J., Conde García, M.Á., Alier Forment, M., Casany Guerrero, M.J.: Opening learning management systems to personal learning environments. J. Univ. Comput. Sci. J. UCS **17**(9), 1222–1240 (2011)
12. Tsarev, R.Y., et al.: An approach to developing adaptive electronic educational course. In: Silhavy, R. (ed.) CSOC 2019. AISC, vol. 986, pp. 332–341. Springer, Cham (2019). https://doi.org/10.1007/978-3-030-19813-8_34
13. Strmecki, D., Bernik, A., Radosevic, D.: Gamification in e-learning: introducing gamified design elements into e-learning systems. J. Comput. Sci. **11**(12), 1108–1117 (2015)
14. Rebelo, S., Isaías, P.: Gamification as an engagement tool in e-learning websites. J. Inform. Technol. Educ. Res. **19**, 833 (2020)
15. Wielicki, T.: Integrity of online testing in e-learning: empirical study. In: Proceedings of the Fourth Annual IEEE International Conference on Pervasive Computing and Communications Workshops (PERCOMW 2006), pp. 5–210. IEEE, Pisa (2006). https://doi.org/10.1109/PERCOMW.2006.69
16. Aljarbouh, A., et al.: Application of the K-medians clustering algorithm for test analysis in e-learning. Lect. Notes Netw. Syst. **596**, 249–256 (2023). https://doi.org/10.1007/978-3-031-21435-6_21
17. Nikolaeva, I., Sleptsov, Y., Gogoleva, I., Mirzagitova, A., Bystrova, N., Tsarev, R.: Statistical hypothesis testing as an instrument of pedagogical experiment. AIP Conf. Proc. **2647**, 020037 (2022). https://doi.org/10.1063/5.0104059
18. Glass, G.V., Hopkins, K.D.: Statistical Methods in Education and Psychology. 3rd edn. Pearson, London (2008)
19. Protasova, I.V., Tolstobrov, A.P., Korzhik, I.A.: Method of test quality analysis and improvement in Moodle e-learning system. In: Proceedings of Voronezh State University. Series: Systems Analysis and Information Technologies, vol. 3, pp. 61–72 (2014)
20. Koretska, V.O., Shlianchak, S.O.: Use of Moodle-based informational technologies for test tasks analysis. Inform. Technol. Learn. Tools **62**(6), 130–139 (2017). https://doi.org/10.33407/itlt.v62i6.1859
21. Vaganova, O.I., Smirnova, Z.V., Vezetiu, E.V., Kutepov, M.M., Chelnokova, E.A.: Assessment tools in e-learning Moodle. Int. J. Adv. Trends Comput. Sci. Eng. **9**(3), 2488–2492 (2020). https://doi.org/10.30534/ijatcse/2020/01932020

Models and Algorithms for Process Management of Enterprises Equipment Repair in the Oil and Gas Industry

A. A. Dnekeshev[1], V. A. Kushnikov[2,4], A. D. Selyutin[3], V. A. Ivashchenko[3], A. S. Bogomolov[3,4](✉), E. V. Berdnova[5], J. V. Lazhauninkas[5], T. V. Pakhomova[5], and L. G. Romanova[5]

[1] Zhangir Khan West Kazakhstan Agrarian Technical University, 51 Zhangir Khan Street, Uralsk 090009, Kazakhstan

[2] Saratov Federal Scientific Centre of the Russian Academy of Sciences, 24 Rabochaya Street, Saratov 410028, Russia

[3] Institute of Precision Mechanics and Control, Saratov Federal Scientific Centre of the Russian Academy of Sciences, 24 Rabochaya Street, Saratov 410028, Russia

alexbogomolov@ya.ru

[4] Saratov State University, 83 Astrakhanskaya Street, Saratov 410012, Russia

[5] Saratov State University of Genetics, Biotechnology and Engineering named after N.I. Vavilov, 1 Teatral'naya pl. Street, Saratov 410012, Russia

Abstract. The article describes a method for the operational identification of production situations and the search for information about them in a distributed database of an enterprise specialized in repairing equipment in the oil and gas sector. The method we propose is based on new models and algorithms for the operational management of an enterprise in complex production situations based on the use of the precedent method. We prove a new method of operational identification of production situations and information retrieval in a distributed database of an enterprise specializing in the repair of equipment in the oil and gas sector. The method is based on forming a metric space of production situations and determining the distance between its points according to formulas traditionally used in image recognition systems when assessing the similarity between compared objects. A heuristic algorithm for solving the complex tasks of operational management of an industrial enterprise for repairing equipment of the oil and gas sector in complex production situations based on the use of proximity functions between information objects is proposed and justified.

Keywords: Situations Identification Process · Situational Management · Production Situations · Management · Oil Refining · Industrial Enterprise · Information Technology

R. Silhavy and P. Silhavy (Eds.): CSOC 2023, LNNS 723, pp. 69–77, 2023.
https://doi.org/10.1007/978-3-031-35317-8_7

1 Introduction

The integration of modern information technologies that contribute to the increase of gross national product (GNP), increase the level of living, and reliable provision of the country's defence cannot be carried out without improving the competitiveness and quality of industry product, including enterprises that repair equipment in the oil and gas sector. One way to solve the problem associated with the development of the computerized integrated production concept, a feature of which is an integrated approach to the automation of the entire production process, which makes it possible to combine individual information systems of an industrial enterprise as part of a single integrated management system. In the opinion of experts in the gas and oil sector, this allows an average of 15–35% reduction in the cost of implementing the primary production, 50–60% speed up the execution of technological processes, 50% reduction in losses from defects and 50–60% reduction in the irregular performance of production tasks and increase the duration of trouble-free operation of the equipment used (Fig. 1).

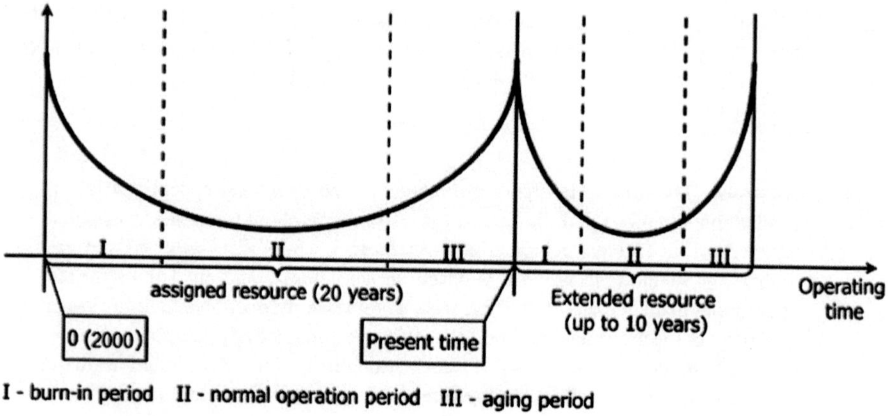

Fig. 1. The main operation periods of the oil and gas sector equipment

Since the late 80s, Russia has been developing the methodology, technical and software tools necessary for creating and implementing the industry's first two stages of the integrated production concept. At the same time, the main emphasis is on creating hybrid systems combining formalized models and methods of traditional automated control systems and automated process control systems with situational management systems.

The theoretical substantiation of the functioning principles of production process control systems using precedents has been carried out in the research of many scientists [1–16]. As a result of the practical application of this theory, effective hardware and software management tools for production complexes have now been created.

Meanwhile, as the experience of combining existing automation systems into a single information complex shows, for the more successful creation of a computerized integrated production, it is necessary to develop new models, methods, algorithms and software that will significantly expand the system functionality of production processes operational management. At the same time, the primary attention should carry out the tasks of improving the mathematical support of industrial enterprise management systems in complex production situations.

The above considerations determine the relevance and practical significance of the research devoted to improving the mathematical methods of the enterprise management system for repairing equipment in the oil and gas sector by creating new models, methods, algorithms and software packages for situational management.

2 The Statement of the Problem

Let us assume that in the process of functioning of an enterprise for the repair of equipment in the oil and gas sector, difficult production situations periodically arise $S(\vec{a}(t), \vec{u}(t)) \in \{S(\vec{a}(t), \vec{u}(t))\}$, requiring decisions based on the analysis of documents and data stored in a distributed database ($\{S(\vec{a}(t), \vec{u}(t))\}$ – set of different emergency situations that arrived at an oil and gas equipment repair enterprise, $\vec{a}(t)$, $\vec{u}(t)$ – vectors of environmental parameters and control actions characterizing the production situation).

We will assume that each complex production situation $S(\vec{a}(t), \vec{u}(t)) \in \{S(\vec{a}(t), \vec{u}(t))\}$ is uniquely characterized by characteristics:

$$\left\{ \begin{matrix} Name, Reason, Consequence, Time, Division, Action, \\ A_1, A_2, A_3, \ldots, A_n, Documents, Data \end{matrix} \right\} \tag{1}$$

Name, *Reason*, *Consequence*, *Time*, – name, cause, consequences, start and end time of the production situation; *Division* – departments affected by production situation; *Action* – measures necessary to resolve the production situation; A_1, A_2, A_3,..., A_n – parameters of the control object, control system and environment taken into account by the decision-maker in the current situation; *Documents, Data*–data and documents used by the decision-maker).

Let us also assume that there is an information repository at the enterprise, in which many documents are placed in $\{D\}$ dataset. $\{D\}$, is used by the decision-maker when managing the production process in production situations $\{S(\vec{a}(t), \vec{u}(t))\}$.

Taking into account the assumptions made, the formalized statement of the problem being solved has the following formulation:

For the management systems of an oil and gas equipment repair enterprise, it is necessary to develop methods and algorithms for identifying production situations that allow for a time interval $[t_h, t_k]$ with known environment parameters $\vec{a}(t) \in \vec{A}(t)$ and control actions $\vec{u}(t) \in \vec{U}(t)$, characterizing a complex production situation, during the time allotted for solving the problem, perform the following actions:

- to establish the degree of compliance with the difficult production situation that has arisen $S(\vec{a}(t), \vec{u}(t)) \in \{S(\vec{a}(t), \vec{u}(t))\}$ with others from the set $\{S(\vec{a}(t), \vec{u}(t))\}$;

- arrange the set $\{S(\vec{a}(t), \vec{u}(t))\}$ according to the degree of proximity of production situations to the arisen production situation $S(\vec{a}(t), \vec{u}(t)) \in \{S(\vec{a}(t), \vec{u}(t))\}$ in the metric space of production situations;
- recommend for execution those measures taken in the event of a production situation closest to the one that has arisen.

3 Mathematical Models

To determine the distance between two production situations $S_1(t)$,$S_2(t) \in \{S(t)\}$ (Table 1) it is necessary to define the functions that establish the degree of similarity between the characteristics of these production situations. To make this, we divide the characteristics of set (1) into five subsets, the first of which contains textual information, the second – the time of occurrence and end of the production situation, the third contains an oriented graph used by the decision-maker in the process of preparation. The fourth subset contains names of production situations stored in the dataset. The fifth subsetcontains quantitative indicators of the production process taken into account by the decision-maker. Let us consider the procedure for forming similarity functions for each subset. Then, based on these functions, we will form a metric that determines the distance between the compared emergencies $S_1(t)$, $S_2(t) \in \{S(t)\}$.

Table 1. Classification of typical production situations characteristics

Characteristics of typical production situations	Notation	Data type	Classification Group
Name of the situation	*Name* (t)	Text	D
Occurrence causes	*Reason* (t)	Text	A
Departments affected by the production situation	*Division* (t)	Text	A
Consequences of the production situation	*Consequence* (t)	Text	A
The start and end times of the production situation	*Time* (t)	Number	B
Decision-maker documents	*Documents* (t)	Text	A
Qualitative indicators of the enterprise that affect the production situation	$A_1, A_2, A_3,\ldots,A_n$	Text	A
Actions that eliminate the production situation	*Action* (t)	Oriented graph	C
Quantitative indicators of the enterprise	*Data* (t)	Number	E

As similarity functions for the characteristics of the first subset (*Reason, Consequence, Division, Documents*), one of the indicators that used in information search engines to compare the degree of coincidence of text documents can be used: the similarity function based on the cosine coefficient, Dice, overlap and Jacquard. Determination of the distance between production situations in metric space on their basis is carried out according to known formulas [13].

In order to determine the similarity between the characteristics of the second subset (*Time*), the next function was selected: $\psi_2(S_1, S_2) = 1 - \frac{\Delta t}{\Delta t_{\max}}$. This function has the following properties: monotonically increases with increasing degree of similarity between the compared characteristics;$\psi_2 = 0$ in the complete absence of similarity (when $\Delta t = \Delta t_{\max}$); $\psi_2 = 1$ with the greatest similarity (when $\Delta t = 0$).

Characteristics of the third subset(*Action*)it is oriented graphs. To determine the similarity function $\psi_2(S_1, S_2)$ quantitative analysis methods are used based on the assessment of the degree of coincidence of sets of vertices, arcs and weights of arcs in the compared graphs.

The characteristics of the fourth group (*Name*) represent the names of situations $S_1(t), S_2(t) \in \{S(t)\}$, stored in the memory of the dataset in the text form. The similarity function between the compared characteristics of production situations has the following form:

$$\psi_4(S_1, S_2) = \begin{cases} 1, \text{by coincidence}' \, Name(S_1(t))'' \, u' Name(S_2(t))' \\ 0, \text{otherwise} \end{cases} \tag{2}$$

$\psi_4(S_1, S_2)$ takes the value 1 if there is a complete match between the names of the characteristics being compared; otherwise, 0. The distance between the characteristics of the fifth group is determined by the formula:

$$\psi_5((Data(S_1(t)), Data(S_1(t))) = \sqrt{\sum_{k=1}^{m} (x_{S_1} - x_{S_2})_k^2 \eta_k} \tag{3}$$

where $\eta_k, k = \overline{1, m}$ – weight coefficients characterize the degree of influence of quantitative characteristics of situations $S_1(t), S_2(t)$ – on the decision-making process).

Similarity functions of various characteristics of the compared production situations $\psi_i, \overline{i = 1, 9}$ are part of the metric ρ_S, determining the distance between $S_1(t), S_2(t) \in \{S(t)\}$, in the space of production situations. When forming a metric, it was taken into account that $\rho_S(S_1(t), S_2(t))$ there should be a real numerical function for which the known axioms of the metric are fulfilled. As a function that has these properties, the distance of the Riemannian space was chosen, determined by the formula $\rho_S(S_1(t), S_2(t)) = \left[\sum_{k=1}^{9} \psi_{4k}^2 \mu_k\right]^{1/2}$ ($\mu_k, k = \overline{1, 9}$ – weight coefficients determined by experts).

The developed methodology for determining the distance between various production situations that arise during the operation of the control object allowed the formation of a number of new algorithms used in identifying production situations and searching for data in the information systems of an industrial enterprise for the repair of equipment in the oil and gas sector.

4 The General Algorithm for Solving the Problem

The leading specialists of the enterprise compile the production situations controlled using computing equipment in such a way that it includes all production situations that significantly affect the process of executing the production program, associated with the operational processing of significant information amounts, as well as time-consuming search for data and documents in the distributed database of the enterprise. When analyzing the production situation that has arisen, it is compared with known situations recorded in the computer memory. If, in the metric space of production situations, formed by the methodology developed above, its complete coincidence with known situations is found, then the situation is considered known, and the decision-maker is issued a list of documents, data, and recommendations necessary for making adequate management decisions. Otherwise, in the metric space of situations, the point closest to the situation that has arisen is determined, and management personnel are given documents, data, and recommendations directly related to the situation that this point characterizes. If the management personnel deem the information received insufficient to make a decision, and then the corresponding point is excluded from consideration, the management system automatically determines the next point located closest to the initial situation and issues related documents, data and recommendations to the management personnel. This process continues until the decision-maker considers the information received sufficient and makes decisions adequate to the situation that has arisen.

After the end of the production situation, the updated information on the data, documents and recommendations used by the management personnel in the decision-making process is entered by experts into the computer memory, and the situation itself is included in the database used by the management system.

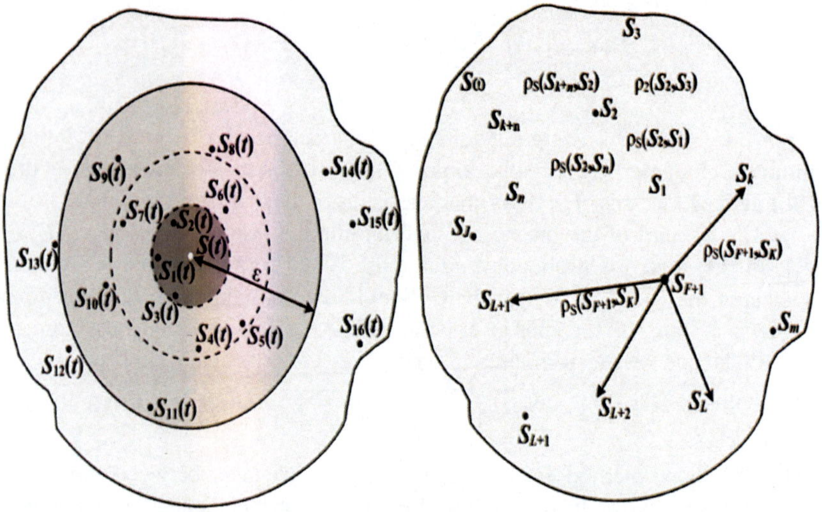

Fig. 2. Ordering production situations $S_1(t)$, ..., $S_{16}(t)$, stored in the database of the production situation being solved, according to the degree of their proximity to the new production situation $S(t)$

Illustration of the procedure for ordering situations in a Riemannian space with a metric ρ_S according to the degree of their proximity to the new production situation $S(t)$ present in Fig. 2.

The organizational and technical structure of the operational management system of the production processes of the enterprise for the repair of equipment in the oil and gas sector with the allocation of a subsystem developed based on the mathematical support proposed in the article is shown in Fig. 3.

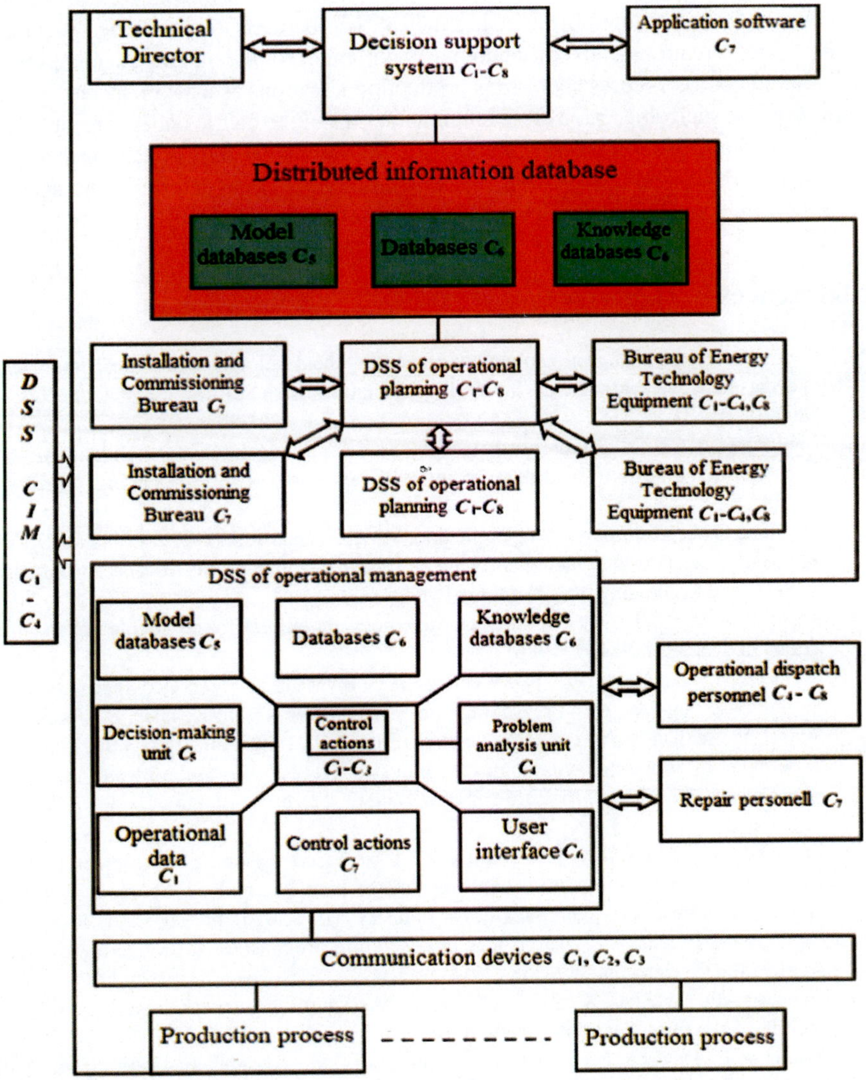

Fig. 3. The production process management system of the oil and gas sector equipment repair enterprise

The following designations are used in Fig. 3: DSS – DecisionSupportSystem; CIM – computer integrated production; C_1 - collecting information; C_2 - transmission of information; C_3- accumulation of information; C_4 - analysis, forecasting; C_5 - decision making; C_6–database insert; C_7- implementation of control actions; C_8- control of decisions made.

5 Discussion

The method of assessing objects using metrics can be used in cases when it comes to complicated situations in the production of equipment for the oil and gas complex. The information objects represent various production situations as a set of specific parameters. Suppose such objects are close enough in terms of the metric under consideration in some situations. In that case, we can also talk about the proximity of actions that should be taken in these similar situations. This makes it possible to determine the necessary control actions in situations of incomplete certainty.

References

1. Bogomolov, A.S., et al.: The problem of preventing the development of critical combinations of events in large-scale systems. In: Intelligent Algorithms in Software Engineering, pp. 274–280 (2020)
2. Dnekeshev, A.A., et al.: Models and algorithms for improving the safety of oil refineries of the Republic of Kazakhstan. In: Advances in Intelligent Systems and Computing, pp. 230–239 (2020)
3. Dolinina, O.N., Kushnikov, V.A., Pechenkin, V.V., Rezchikov, A.F.: The way of quality management of the decision making software systems development. In: Advances in Intelligent Systems and Computing, pp. 90–98 (2018)
4. Engelhardt, M., Bain, L.J.: On the mean time between failures for repairable systems. IEEE Trans. Reliab. **35**, 419–422 (1986)
5. Eremeev, A., Varshavskii, P., Kozhevnikov, A., Polyakov, S.: Integrated approach for data mining based on case-based reasoning and neural networks. In: Kovalev, S., Tarassov, V., Snasel, V., Sukhanov, A. (eds.) Proceedings of the Fifth International Scientific Conference "Intelligent Information Technologies for Industry" (IITI 2021). Lecture Notes in Networks and Systems, vol. 330, pp. 15–23. Springer, Cham (2022). https://doi.org/10.1007/978-3-030-87178-9_2
6. Eremeev, A., Varshavskiy, P., Alekhin, R.: Case-based reasoning module for Intelligent Decision Support Systems. In: Proceedings of the First International Scientific Conference "Intelligent Information Technologies for Industry" (IITI 2016), pp. 207–216 (2016)
7. Ferraz, I.N., Garcia, A.C.: Turbo machinery failure prognostics. In: Modern Advances in Applied Intelligence, pp. 349–358 (2014)
8. Gong, H., Shi, L., Zhai, X., Du, Y., Zhang, Z.: Assembly process case matching based on a multilevel Assembly Ontology Method. Assem. Autom. **42**, 80–98 (2021)
9. Khatir, M.E., Davison, E.J.: Cooperative control of Large Systems. In: Cooperative Control, pp. 119–136 (2004)
10. Lytvyn, V., Vysotska, V., Dosyn, D., Lozynska, O., Oborska, O.: Methods of building intelligent decision support systems based on Adaptive Ontology. In: 2018 IEEE Second International Conference on Data Stream Mining & Processing (DSMP) (2018)

11. Ragab, A., Ghezzaz, H., Amazouz, M.: Decision fusion for reliable fault classification in energy-intensive process industries. Comput. Ind. **138**, 103640 (2022)
12. Saffiudeen, M.F., Mohammed, F.T., Syed, A.: A case study on procedure standardization of heat exchanger retubing in KSA oil and Gas Industries. J. Fail. Anal. Prev. **20**, 1451–1455 (2020)
13. Yamamoto, E., et al.: Guidelines for repair welding of pressure equipment in refineries and Chemical plants. J. Press. Vess. Technol. **135** (2013)
14. Yan, H., Wang, F., Yan, G., He, D.: Hybrid approach integrating case-based reasoning and Bayesian network for Operational Adjustment in industrial flotation process. J. Process Control **103**, 34–47 (2021)
15. Zhmud, V., Dimitrov, L.: Using the fractional differential equation for the control of objects with delay. Symmetry. **14**, 635 (2022)
16. Zhmud, V., Liapidevskiy, A., Avrmachuk, V., Sayapin, V., Sedinkin, J.N., Hardt, W.: Critical technologies in the cluster of virtual and augmented reality. In: IOP Conference Series: Materials Science and Engineering, **1019**, 012065 (2021)

Check for updates

Novel and Simplified Scheduling Approach for Optimized Routing Performance in Internet-of-Things

Gauri Sameer Rapate[1](✉) and N. C. Naveen[2]

[1] Department of Computer Science and Engineering, PES University, Bengaluru, India
gauri.rapate@gmail.com

[2] Department of Computer Science and Engineering, JSS Academy of Technical Education, Bengaluru, India

Abstract. Effective planning of routing protocol plays a critical role in controlling communication for large and complex form of network like Internet-of-Things (IoT). Review of existing routing strategy formulation states that there is a large scope of improving its performance. Therefore, the proposed scheme introduces a novel computational framework of scheduling in order to optimize the performance of routing in resource constrained IoT nodes. The core idea of this implementation is to accomplish a good balance between reduced energy consumption and reduced delay using a unique time-slot management considering newly construct of carrier message. The simulation study of the work exhibits that proposed scheme exhibits higher retention of residual energy, lower energy consumption, and lower delay in contrast to existing scheme of IoT and hence proves its effectiveness.

Keywords: Internet-of-Things · Routing Protocol · Scheduling · Energy · Time Slot Management · delay

1 Introduction

Internet-of-Things (IoT) consists of multiple number of physical heterogeneous devices called as 'things' connected with each other via internet [1]. The physical devices called as IoT nodes are characterized by resource constraints and hence sophisticated protocols are challenging to be executed [2]. One primary demand to maintain a seamless connectivity among all the IoT devices is to ensure higher level of resource utilization, especially energy [3, 4]. At present, there are evolving number of routing protocols in IoT which has emphasized on energy efficiency [5], traffic load management [6], and security [7]. Although, these studies have claimed of its effectiveness in its routing scheme, but still there are manifold shortcomings associated with all these routing schemes in IoT [8]. The primary challenge is associated with formulating deployment strategy of IoT nodes that are highly heterogeneous in nature. The secondary challenge is associated with availability of highly wide and varied ranges of networking standards for which reason

R. Silhavy and P. Silhavy (Eds.): CSOC 2023, LNNS 723, pp. 78–88, 2023.
https://doi.org/10.1007/978-3-031-35317-8_8

it is computationally challenging task to develop a uniform and generalized architecture. Apart from this, it is also noted that there are various categories of research work where deep learning has been used towards improving and optimizing the routing performance in IoT [9]; however, they too have shortcoming associated with resource dependency and higher processing time. Hence, a new form of scheduling approach is highly demanded which can consider simplified mechanism towards controlling routing management in resource constrained IoT nodes. Existing scheduling studies are still associated with limitation towards imbalance between energy and time [10]. Therefore, the proposed study presents a novel and simplified computational architecture of scheduling where a unique time-slot management is carried out in order to optimize the routing performance in IoT. The organization of the paper is as follows: Sect. 2 discusses about the existing research work followed by problem identification in Sect. 3. Section 4 discusses about methodology followed by algorithm implementation in Sect. 5. Comparative analysis of result is discussed in Sect. 6 and conclusion in Sect. 7.

2 Related Work

At present, there are evolution of various routing protocols in IoT environment where the highlights are towards addressing multiple communication-based problems. The work carried out by Ding et al. [11] have implemented a dynamic routing scheme using software defined network. Adoption of software-defined network was also noted in work of Xu et al. [12] where optimization of routing operation is carried out targeting cloud services. Ekler et al. [13] have developed a multi-hop scheme towards routing data for extending the lifespan of node. Adoption of fuzzy logic was witnessed in work of Chithaluru et al. [14] where a scheduling scheme is designed based on ranks computed from fuzzy inference. Mohammadsalehi et al. [15] have enhanced the conventional RPL protocol customized with mobility factor for IoT environment. Similar form of study towards adoption of RPL protocol was also reported in work of Tsai et al. [16], Vaezian and Darmani [17], and Mahyoub et al. [18]. Li et al. [19] have used blockchain based learning framework for mechanizing lightweight routing scheme using reinforcement learning technique. Patel et al. [20] have presented a design of routing scheme which addresses problems of collision and energy using link quality information. Work towards conversion of routing protocol has been presented by Narayanan and Murthy [21] where the scheme computes the path with least overhead. Apart from the above-mentioned studies, there are also studies carried out emphasizing on different set of unique problems viz. Routing scheme for congestion mitigation (Chanak and Banerjee [22]), energy efficiency using nature-inspired routing scheme (Khan et al. [23]), cognitive approach in routing (Ghosh et al. [24]), energy-based data transmission for unmanned vehicle (Yang et al. [25]), collaborative routing scheme (Zhu et al. [26]), intelligent routing using reinforcement learning (Kaur et al. [27]), handover mechanism for reducing data redundancy (Selem et al. [28]), routing scheme based on layering (Han et al. [29]), energy harvesting based approach for scheduling (Jiang et al. [30]).

Hence, it can be noted that there are various recent studies using multiple approaches focusing towards improving routing efficiency of IoT networks. The next section highlights the core research problem associated with the existing studies that are found to be unaddressed.

3 Problem Description

The core identified research problems are i) existing routing techniques are not much investigated with respect to optimization, ii) although scheduling approaches are reported but they are highly sophisticated and lacks balance between energy and delay, iii) majority of the existing schemes of routing are highly complex with less emphasis towards achieving cost effective computation, and iv) exact scenario of IoT is less found to be implemented in the context of fulfilling intrinsic communication demands of resource constraint IoT nodes. Apart from the above stated problems, existing schemes also suffers from non-inclusion of effective temporal factors associated with scheduling and hence there is no consistency of routing performance for critical and non-critical applications in IoT. The next section presents solution to overcome this problem.

4 Proposed Methodology

The proposed scheme presents a simplified and novel routing scheme with simplified optimization in IoT environment using a unique scheduling approach. The prime idea of the proposed scheme is to adopt a unique time-slot based management which can keep a balance of prioritized and normal communication over peak traffic condition in IoT. Figure 1 highlights the architecture of proposed scheme.

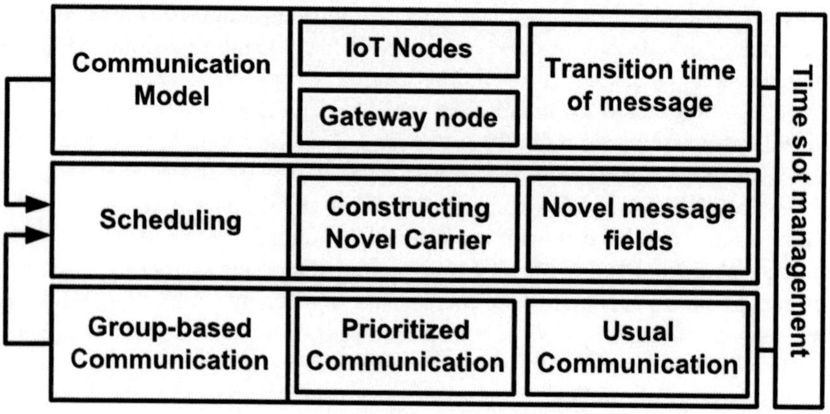

Fig. 1. Propose Scheduling Architecture

According to Fig. 1, the primary block of operation associates with building a communication module considering IoT nodes, gateway nodes, and transition time of exchange of messages among the devices in IoT. The second block of architecture relates to developing a scheduling operation where a novel carrier is designed considering novel message fields to emphasize over controlling energy consumption. The third block of architecture associated with management of prioritized and normal message communication, while all the blocks are linked with a unique time slot management for facilitating scheduling in IoT.

5 System Design

The proposed scheme introduces an optimal routing strategy by using a scheduling approach considering the use case of IoT. This section discusses about all the operation being carried out towards developing this novel routing strategy.

i) Constructing a Novel Carrier

Prior to discussion of the carrier message, it is to be noted that proposed scheme considers sensory application to represent the IoT device which are characterized by minimal computational capabilities and restricted resources. Hence, the novel beacon is constructed towards addressing the demands of scheduling as well as resource management while performing routing. The proposed scheme offers a specific representation of the carrier as depicted in Fig. 2.

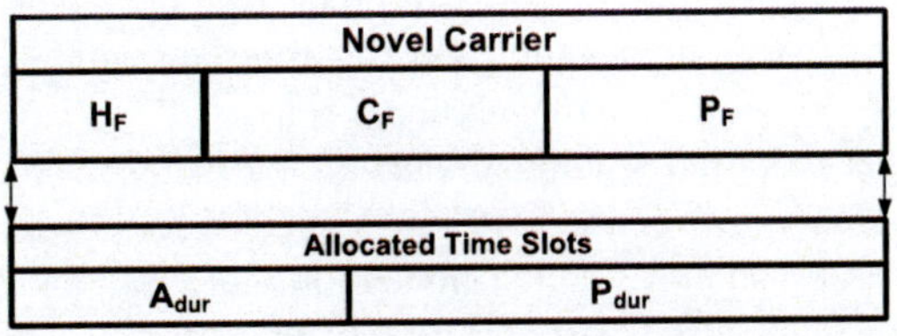

Fig. 2. Construction of Novel Carrier

Figure 2 represent the considered design of the novel carrier and its allocated time slots. The core fields of novel carrier are header field H_F, content field C_F, and priority field P_F. The complete carrier is further classified with respect to allocated time slots of active duration A_{dur} and passive duration P_{dur}. The active duration A_{dur} and passive duration P_{dur} is allocated with an arbitrary access and scheduled access respectively.

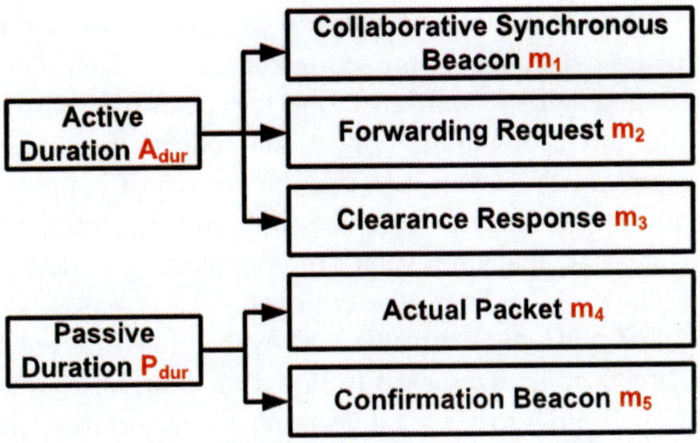

Fig. 3. Allocation of sub-fields in carrier

Figure 3 highlights 5 different forms of allocations in respective fields of time slots. While performing an arbitrary access in A_{dur}, there are three types of allocations i.e., i) *collaborative synchronous beacon* m_1 which performs syncing operation with different device in IoT prior and after performing routing thereby assisting in updated routing information, ii) *forwarding request* m_2 that is responsible for forwarding a request message from transmitting IoT node for participating in data forwarding process, and iii) *clearance response* m_3 is the response message for m_2. On the other hand, while performing scheduled access in P_{dur}, there are two types of allocation i.e., i) *actual content* m_4 which is the actual data to be forwarded and ii) confirmation beacon m_4 that is responsible for forwarding a confirmation message from the recipient IoT device to the transmitting IoT device.

All the IoT nodes with similar schedules formulate a group while a hierarchy of IoT nodes with single hop connection. The similar configuration is used for formulating mutlihop connections too. The system also allocated a discreet index number to all the IoT nodes in such a way that node with multihop connection will bear unique index number. The scheme also introduces an *operational time* for an IoT node which represents the exact duration when the IoT node is busy in processing data/message. The selection of operational time is carried out in such a way that time slot with P_{dur} should be higher in order to sufficiently forward an actual content m_4 as well as confirmation beacon m_4. It should be noted that proposed scheme renders all IoT nodes to consider any time slot to carry out transmission; however, the field P_F is used only for prioritized transmission. On the basis of the operational time, the IoT nodes configure A_{dur} and P_{dur} time. Apart from this, the proposed scheme also performs an exchange of the state information among the communicating IoT nodes.

According to proposed scheme, the IoT nodes will undergo a state of P_{dur} and save its energy and then based on operational time, it will transit to a state of A_{dur}. Similar schedule is followed by the IoT nodes belong to same group of communication in proposed scheme. The moment the IoT node transit towards the state of P_{dur}, it will assess if it has still any actual content m4 to be forwarded or if there is a priority set to m_4. In case of prioritized data indexed by P_F, the IoT device extends its schedule to forward the priority message otherwise it confirms its transition towards the state of P_{dur}. Another interesting contribution of proposed scheme is that it can perform concurrent transmission of data to each other as there is a separate hop table maintained for the adjacent nodes, which makes the routing task easier. Hence, higher throughput is expected. It is to be noted that forwarding request m_2 bears the information about the link of both transmitting and receiving IoT nodes and it reduces the quantity of the candidate IoT nodes for expecting the access towards the system of communication will be of reduced dependency so that it can cater up the concurrent transmission. For this purpose, it is necessary to alter the allocation vector of communication network in IoT use-case which is actually meant to resisting the collision. The proposed system also offers a novelty where the IoT node will not be blocked by the time for vector of communication channel when it receives the overheard information from other IoT node. Rather, the proposed scheme will analyze all the feasibility to perform concurrent transmission prior to blocking of IoT nodes. However, if there are presence of certain IoT nodes which have been queued for longer duration exceeding the threshold time limit (that can be set by the user), than these slots will be automatically set to priority slot. In order to let other nodes about the priority slot, the IoT nodes forwards separately an advertisement message bearing information of total number of such slots. This information when sensed by neighboring IoT devices, checks their buffer and checks their synchronization of slots with other nodes and fine tune the individual slots automatically. Hence, this process over a long number of links progressively updates link information which assists in faster routing performance.

Another significant contribution of the proposed scheme is that it is capable of finetuning the power required during the operation of IoT nodes with an agenda is to offer a significant control over energy consumption. Hence, a simplified mathematical expression is formed for this purpose as follows:

$$E_{anti} = \Delta E.A_1 \tag{1}$$

In the above mathematical expression (1), the variable E_{anti} will represent anticipated energy which is required to be computed while performing the transmission among IoT devices. The difference it makes with existing scheme is that it doesn't allocate a static energy value which could be either over-utilized or under-utilized owing to the fluctuating form of data transmission over dynamic IoT network. Further, the second variable ΔE will represent ratio of maximum energy to obtained level of energy. The third variable A_1 will represent product of strength of signal that is minimally required and network coefficient. It should be noted that proposed scheme performs transmission of forward request message m_2 and clearance response message m_3 with highest possible energy $E_{\max}$. Upon receiving m_2, the receiving IoT nodes provides m_3 message at same $E_{\max}$. Upon receiving m_3 message, the transmitting IoT nodes computes E_{anti} on the basis of obtained level of energy i.e., E_{rec}. Therefore,

$$\Delta E = \frac{E_{\max}}{E_{rec}} \tag{2}$$

Hence, on the basis of mathematical expression (1) and (2), it can be stated that E_{anti} energy is utilized by the transmitting IoT node in order to perform data propagation. On the other hand, the signal energy is utilized by the receiving IoT node in order to obtain m_2 message which is further utilized for computing the level of energy to be adopted for computing E_{anti}. The study model also considers that level of any form of artifacts (scattering, fading, interference, noise, etc.) are retained below specific cut-off score. The next section discusses about the outcome obtained from proposed scheme.

6 Results Discussion

The implementation of the proposed scheme towards optimized routing is carried out by constructing a specific environment of an IoT consisting of 100 IoT devices randomly distributed over 1000x1000m^2 simulation area with a presence of gateway node. The analysis is carried out considering operational time of 0.15s, channel capacity of 15 kbps, with length of actual content to be 15 bytes. The transmit and receive power is configured to 36x10^6 joules per second and 15x10^6 joules per second. The outcome of the proposed study are compared with existing system Exist1 which is motivated from core approach implemented by Zeb et al. [31], Khalifa et al. [32], Shah and Sharma [33]. A closer look into the core technique used for optimizing data transmission in above-mentioned work is to minimize the consumption of energy caused due to multiple reasons viz. Collision, idle listening, or overhearing. The typical performance analysis is carried out with respect to energy and delay.

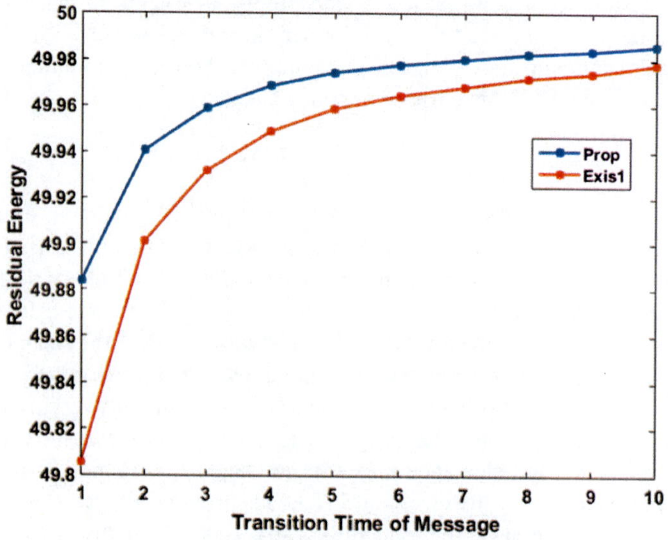

Fig. 4. Comparative Analysis for Residual Energy

A closer look into Fig. 4 and Fig. 5 showcase that proposed scheme *Prop* offers better energy conservation performance in contrast to existing scheme *Exist1*. The analysis is carried out considering increasing values of transition time of messages. Figure 4 exhibits higher energy retention while Fig. 5 showcase lower dissipated energy for proposed scheme. The prime reason behind this is proposed scheme offers a dynamic scheduling process where the slot allocation among the IoT device is carried out with respect to dynamic topology. Further, unlike *Exist1* scheme, proposed scheme doesn't allocate a static energy, rather it computes the necessary energy and allocates them which significantly saves allocation of higher states of energy values unnecessarily.

The next performance parameter assessed was delay. From Fig. 6, it can be seen that proposed scheme offers slightly better delay reduction in comparison to existing scheme. The justification behind this outcome is as follows: The scheduling approach implemented in existing scheme has a static activation time, which will mean that there is a higher possibility of IoT nodes to ignore some messages when it goes to P_{dur} state. On the contrary, the carrier allocations of field of proposed scheme are designed in such a way that if there is any prioritized message than such message will be indexed in P_F field of carrier and will be prioritized for transmission compared to usual messages. At the same time, the normal messages are allocated over multiple carriers and using collaborative synchronous beacon i.e., m_1, the information of queued messages are updated to adjacent nodes. Upon availability, the adjacent nodes assist in forwarding the queued

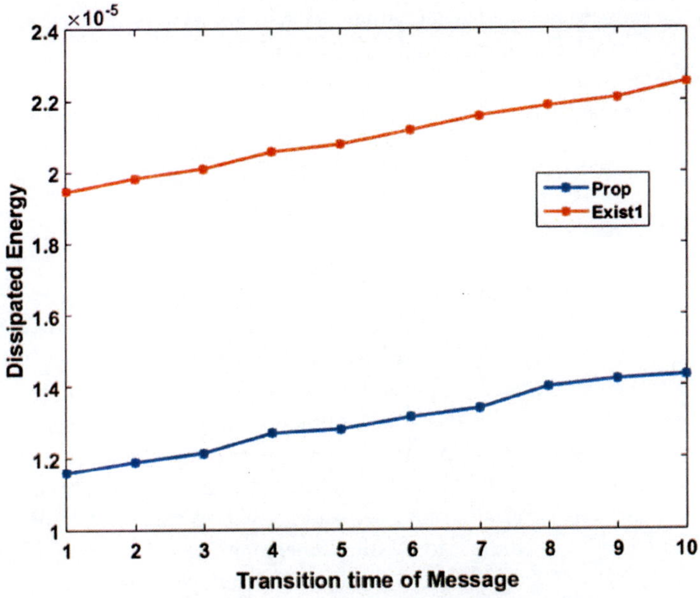

Fig. 5. Comparative Analysis of Dissipated Energy

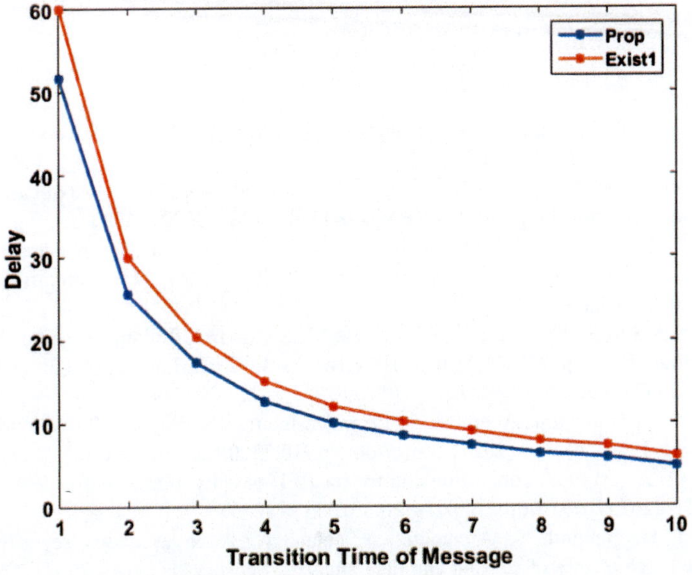

Fig. 6. Comparative Analysis of Delay

message. It will mean that proposed scheme offers two forms of communication channel where the primary channel is meant for propagating priority message and secondary channel is meant for usual message propagation. Hence, a balance is maintained which significantly reduces delay. Further, a network will not have prioritized message every time and in such case the prioritized channel will be allocated with normal messages which further increases the rate of data delivery and hence delay is significantly reduced with increasing transition time of message. Therefore, proposed scheme offers better routing performance in IoT.

7 Conclusion

This paper has presented a solution towards optimizing the routing performance with an emphasis towards energy efficiency using a novel scheduling approach for IoT use case. The proposed study model introduces three novel contributions viz. i) A unique scheduling approach has been presented which is based on propagation (or exchange of message) and path-information based scheduling, ii) the model also assists in facilitating concurrent transmission for both prioritized and non-prioritized communication channel, iii) the proposed scheme introduces a simplified mathematical expression which can adjust the energy based on the traffic demand so that no excessive transmit energy is underutilized, and iv) in contrast to existing scheme, proposed scheme is found to offer approximately 35% of retention of energy, 40% of reduced energy consumption, and 27% of reduced delay. Our future work will be continued towards further improving the routing performance and further improvement towards the energy modelling will be carried out to accomplish more optimized routing performance in IoT.

References

1. Misra, S., Mukherjee, A., Roy, A.: Introduction to IoT. Cambridge University Press (2021). ISBN: 9781108842952, 110884295X
2. Chatterjee, J.M., Mohanty, S.N., Satpath, S.: Internet of Things and Its Applications. Springer International Publishing (2021). ISBN: 9783030775285, 3030775283
3. Kafle, V.P., Muktadir, A.H.A.: Intelligent and agile control of edge resources for latency-sensitive IoT services. IEEE Access **8**, 207991–208002 (2020). https://doi.org/10.1109/ACCESS.2020.3038439
4. Hosen, A.S.M.S., Sharma, P.K., Cho, G.H.: MSRM-IoT: A Reliable Resource Management for Cloud, Fog, and Mist-Assisted IoT Networks. In: IEEE Internet of Things Journal **9**(4), 2527–2537 (2022). https://doi.org/10.1109/JIOT.2021.3090779
5. Farhan, L., et al.: Energy efficiency for green internet of things (IoT) networks: a survey. Network **1**(3), 279–314 (2021). https://doi.org/10.3390/network1030017. Nov.
6. Lim, C.: A survey on congestion control for RPL-based wireless sensor networks. Sensors **19**(11), 2567 (2019). https://doi.org/10.3390/s19112567
7. Mrabet, H., Belguith, S., Alhomoud, A., Jemai, A.: A survey of IoT security based on a layered architecture of sensing and data analysis. Sensors **20**(13), 3625 (2020). https://doi.org/10.3390/s20133625
8. Shah, Z., Levula, A., Khurshid, K., Ahmed, J., Ullah, I., Singh, S.: Routing protocols for mobile internet of things (IoT): A survey on challenges and solutions. Electronics **10**(19), 2320 (2021). https://doi.org/10.3390/electronics10192320
9. Lakshmanna, K., et al.: A review on deep learning techniques for IoT data. Electronics **11**(10), 1604 (2022). https://doi.org/10.3390/electronics11101604

10. Urke, A.R., Kure, Ø., Øvsthus, K.: A survey of 802.15.4 TSCH schedulers for a standardized industrial internet of things. Sensors **22**(1), 15 (Dec. 2021). https://doi.org/10.3390/s22010015
11. Ding, Z., Shen, L., Chen, H., Yan, F., Ansari, N.: Energy-efficient relay-selection-based dynamic routing algorithm for IoT-oriented software-defined WSNs. IEEE Internet Things J. **7**(9), 9050–9065 (2020). https://doi.org/10.1109/JIOT.2020.3002233
12. Xu, S., Wang, X., Yang, G., Ren, J., Wang, S.: Routing optimization for cloud services in SDN-based Internet of Things with TCAM capacity constraint. Journal of Communications and Networks **22**(2), 145–158 (2020). https://doi.org/10.1109/JCN.2020.000006
13. Ekler, P., Levendovszky, J., Pasztor, D.: Energy aware IoT routing algorithms in smart city environment. IEEE Access **10**, 87733–87744 (2022). https://doi.org/10.1109/ACCESS.2022.3199757
14. Chithaluru, P., Kumar, S., Singh, A., Benslimane, A., Jangir, S.K.: An energy-efficient routing scheduling based on fuzzy ranking scheme for internet of things. In: IEEE Internet of Things Journal **9**(10), 7251–7260 (2022). https://doi.org/10.1109/JIOT.2021.3098430
15. Mohammadsalehi, A., Safaei, B., Monazzah, A.M.H., Bauer, L., Henkel, J., Ejlali, A.: ARMOR: a reliable and mobility-aware RPL for mobile internet of things infrastructures. IEEE Internet of Things Journal **9**(2), 1503–1516 (2022). https://doi.org/10.1109/JIOT.2021.3088346
16. Tsai, R.-G., Tsai, P.-H., Shih, G.-R., Tu, J.: RPL based emergency routing protocol for smart buildings. IEEE Access **10**, 18445–18455 (2022). https://doi.org/10.1109/ACCESS.2022.3150928
17. Vaezian, A., Darmani, Y.: MSE-RPL: mobility support enhancement in RPL for IoT mobile applications. IEEE Access **10**, 80816–80832 (2022). https://doi.org/10.1109/ACCESS.2022.3194273
18. Mahyoub, M., Hasan Mahmoud, A.S., Abu-Amara, M., Sheltami, T.R.: An efficient RPL-based mechanism for node-to-node communications in IoT. IEEE Internet of Things Journal **8**(9), 7152–7169 (2021). https://doi.org/10.1109/JIOT.2020.3038696
19. Li, Z., Su, W., Xu, M., Yu, R., Niyato, D., Xie, S.: Compact learning model for dynamic off-chain routing in blockchain-based IoT. IEEE J. Sel. Areas Commun. **40**(12), 3615–3630 (2022). https://doi.org/10.1109/JSAC.2022.3213283
20. Patel, N.R., Kumar, S., Singh, S.K.: Energy and collision aware WSN routing protocol for sustainable and intelligent IoT applications. IEEE Sensors Journal **21**(22), 25282–25292 (15 Nov. 2021). https://doi.org/10.1109/JSEN.2021.3076192
21. Narayanan, R., Murthy, C.S.R.: A routing framework with protocol conversions across multiradio IoT platforms. IEEE Internet of Things Journal **8**(6), 4417–4432 (2021). https://doi.org/10.1109/JIOT.2020.3028239
22. Chanak, P., Banerjee, I.: Congestion free routing mechanism for IoT-enabled wireless sensor networks for smart healthcare applications. IEEE Trans. Consum. Electron. **66**(3), 223–232 (2020). https://doi.org/10.1109/TCE.2020.2987433
23. Khan, I.U., Qureshi, I.M., Aziz, M.A., Cheema, T.A., Shah, S.B.H.: Smart IoT control-based nature inspired energy efficient routing protocol for flying Ad Hoc network (FANET). IEEE Access **8**, 56371–56378 (2020). https://doi.org/10.1109/ACCESS.2020.2981531
24. Ghosh, S., Dagiuklas, T., Iqbal, M., Wang, X.: A cognitive routing framework for reliable communication in IoT for industry 5.0. IEEE Trans. Industr. Inf. **18**(8), 5446–5457 (2022). https://doi.org/10.1109/TII.2022.3141403
25. Yang, Z., Liu, H., Chen, Y., Zhu, X., Ning, Y., Zhu, W.: UEE-RPL: A UAV-based energy efficient routing for internet of things. IEEE Transactions on Green Communications and Networking **5**(3), 1333–1344 (2021). https://doi.org/10.1109/TGCN.2021.3085897
26. Zhu, M., Chang, L., Wang, N., You, I.: A smart collaborative routing protocol for delay sensitive applications in industrial IoT. IEEE Access **8**, 20413–20427 (2020). https://doi.org/10.1109/ACCESS.2019.2963723

27. Kaur, G., Chanak, P., Bhattacharya, M.: Energy-efficient intelligent routing scheme for IoT-enabled WSNs. IEEE Internet of Things Journal **8**(14), 11440–11449 (15 July, 2021). https://doi.org/10.1109/JIOT.2021.3051768
28. Selem, E., Fatehy, M., El-Kader, S.M.A.: mobTHE (Mobile Temperature Heterogeneity Energy) Aware Routing Protocol for WBAN IoT Health Application. IEEE Access **9**, 18692–18705 (2021). https://doi.org/10.1109/ACCESS.2021.3054367
29. Han, D., Du, X., Liu, X.: CELR: connectivity and energy aware layering routing protocol for UANs. IEEE Sensors Journal **21**(5), 7046–7057 (1 March, 2021). https://doi.org/10.1109/JSEN.2020.3039808
30. Jiang, M., Chin, K.-W., He, T., Soh, S., Wang, L.: Joint link scheduling and routing in two-tier RF-energy-harvesting IoT networks. IEEE Internet of Things Journal **9**(1), 800–812 (1 Jan., 2022). https://doi.org/10.1109/JIOT.2021.3085862
31. Zeb, A., Wakeel, S., Rahman, T., Khan, I., Uddin, M.I., Niazi, B.: Energy-efficient cluster formation in IoT-enabled wireless body area network. Hindawi-Computational Intelligence and Neuroscience Volume 2022, Article ID 2558590, p. 11. https://doi.org/10.1155/2022/2558590
32. Khalifa, A.B.: Medium Access Control Layer for Dedicated IoT Networks, Doctorial Thesis, NSALyon (2020)
33. Shah, M.K., Sharma, L.K.: Study on 6LoWPAN Routing Protocols with SD aspects in IoT. In: 2018 2nd International Conference on I-SMAC (IoT in Social, Mobile, Analytics and Cloud) (I-SMAC)I-SMAC (IoT in Social, Mobile, Analytics and Cloud) (I-SMAC), 2018 2nd International Conference on, pp. 60–65 (2018). https://doi.org/10.1109/I-SMAC.2018.8653660

Check for updates

Data Forecasting Models and Decision-Making Methods in the Management System of a Flooded Object or Territory

M. V. Khamutova[1], V. A. Kushnikov[1,2], A. D. Selyutin[3], V. A. Ivashchenko[3], A. S. Bogomolov[1,3](✉), E. V. Berdnova[4], J. V. Lazhauninkas[4], T. V. Pakhomova[4], and L. G. Romanova[4]

[1] Saratov State University, 83 Astrakhanskaya Street, Saratov 410012, Russia
mariuka7d@rambler.ru

[2] Saratov Federal Scientific Centre of the Russian Academy of Sciences, 24 Rabochaya Street, Saratov 410028, Russia

[3] Institute of Precision Mechanics and Control, Saratov Federal Scientific Centre of the Russian Academy of Sciences, 24 Rabochaya Street, Saratov 410028, Russia

[4] Biotechnology and Engineering Named after N.I. Vavilov, Saratov State University of Genetics, 1 Teatral'naya pl. Street, Saratov 410012, Russia

Abstract. The paper considers the processes of generating data necessary for making decisions in the management system of a flooded object or territory. We present an overview of the main models and methods for forecasting hydrological parameters. A system dynamics model has been developed to determine the characteristics of the flood impact. Based on the analysis of the influence of flood impact characteristics on the amount of damage, the problem of managing a flooded object or territory according to the criterion of minimizing flood damage is formulated. The proposed algorithm for solving the problem is based on a set of mathematical models and methods that make it possible to predict the dynamics of hydrological parameters, form digital relief models, and determine the characteristics of the flood impact. The complex of mathematical models includes cause-and-effect matrices and graphs, as well as differential equations of system dynamics. These equations form a control model for solving the main problem. The optimal solution is selected from the set of considered action plans.

Keywords: Management System · Flood Effects Characteristics · Mathematical Models and Methods

1 Introduction

Flood is the water level rise in rivers and other water bodies, resulting in an overflow of water. There are basic types of floods: fluvial floods (river floods), pluvial floods (flash floods), snowmelt floods, ice-jam floods, estuarine floods, coastal floods. The causes of flood may be different, but the effects is socio-economic consequences. Floods are one of the most common natural disasters and their number is only increasing, which is associated with global warming and anthropogenic impact [10, 15].

R. Silhavy and P. Silhavy (Eds.): CSOC 2023, LNNS 723, pp. 89–95, 2023.
https://doi.org/10.1007/978-3-031-35317-8_9

There is a need to develop a new or improve the efficiency of an existing management system a flood-prone territory or object, considering the increase in floods around the world. The development of a set of actions aimed at preventing and eliminating flood rely on the analysis of mathematical support, which includes models that allow predicting hydrological parameters, the flood area and possible flood effects, and models that allow optimizing available resources [9].

2 The Main Actions of the Stages of Functioning of the Management System for Flood-Prone Object or Territory

The functional structure of the management system should take into account forecasting, prevention and preparation for flood, as well as the elimination of flood effects. The following stages of the functioning of the management system for a flood-prone object or territory should be noted [1, 9, 14].

Stage 1. There are no prerequisites for flood. During this period, information is accumulated and analyzed, forecasting of various flood scenarios, and development of response measures aimed at ensuring the safety of the object or territory, taking into account the possible flood effects.

Stage 2. There are prerequisites for flooding. Typical for this period are actions to prevent flooding and mitigate possible effects during the flood.

Stage 3. The object or territory is flooded. During this period, various emergency rescue actions and actions aimed at eliminating the flood effects are conducted.

The implementation of any action affects the quality of management of an object or territory during a flood. The lack of preventive actions entails an increase in the negative flood effects.

The following actions are typical for the first stage of the functioning of the management system:

- Monitoring of hydrological phenomena is an actions associated with constant supervision of the water bodies state, measuring the parameters necessary for analysis and forecasting (water levels, flow rate, flow velocity, ice and snow cover thickness, precipitation, air temperature, etc.).
- Forecasting the flood timing, its scale and effects is an actions associated with the analysis of the data (including meteorological and hydrological) obtained using various forecasting methods and models.
- Flood prevention is the action taken in advance to reduce the risk of flooding, as well as to preserve human health, reduce environmental damage and material losses. They include preventive flood control actions aimed at preventing or reducing the negative flood effects.
- Creation of a reserve of fuel and lubricants, food, medicines, necessities and financial resources for flood relief.

At the second stage of the functioning of the management system, the implementation of actions aimed at preventing flooding and reducing possible damage continues. The data obtained with forecasting models and methods update depending on changes in the monitoring data. According to the forecast data on the possible development of the situation, the necessary forces and means are determined for flood elimination.

Unlike the first and second stages, the third stage is characterized by tactical rather than strategic actions. The main actions of this stage are operational actions to protect object or territories from the damaging factors of flood, carrying out emergency rescue and other urgent actions.

The choice of a set of actions for each stage depends on the flood type. For forecasting various flood types certain hydrological, meteorological, topographic and other data, and various mathematical methods and models are required to predict specific flood parameters.

3 Statement of the Management Problem

To analyze the impact of a flood on an object or territory, we considered the flood effects characteristics [5–8]. There are basic flood effects characteristics:

- The population in the flood zone (human loss, number of injured, the number of people left homeless, etc.).
- The number of settlements in the flood zone.
- The number of objects of the economy various sectors in the flood zone.
- The length of railways and roads, power lines and communication lines in the flood zone.
- The number of bridges and tunnels flooded, destroyed and damaged by flood.
- The number of residential buildings flooded, destroyed and damaged by flood.
- The area of agricultural land covered by flooding.
- The number of dead farm animals.

These characteristics affect the amount of flood damage, so their minimization is the main problem of managing a flood-prone object or territory. The choice of a set of flood effects characteristics depends on the specifics of the flood-prone object or area.

Thus, for information management systems of EMERCOM needto develop models, methods and algorithms that allow determining controlfunction p(t) that minimize the cost function [5–8]

$$\sum_{i=1}^{n} (X_i(t, p(t)) - X_i^*)^2 \gamma_i \rightarrow \min \qquad (1)$$

where $X_i{}^*$ are the preferable values of the flood effects characteristics, $X_i(t)$ are the flood effects characteristics, γ_i are the weight coefficients of the characteristics $X_i(t)$, $i = 1,\ldots,n$.

The sets of action implemented at Stages 1–3 are control functions. It is necessary to develop and implement models, methods and algorithms that take into account the flood type and specifics of the flood-prone object or territory subject to create a set of actions (hereinafter - action plan) that minimizes function (1).

4 Models, Methods and Algorithms of the Management System for Flood-Prone Object or Territory

Flood forecasting models and methods are a necessary part of management systems, given that no preventive or protective measures can be completely effective. In order to form actions from Stage 1, 2, in particular, to allocate and distribute resources aimed at flood elimination, it is necessary, based on hydrological, meteorological, and others monitoring data, to obtain a forecast of possible water levels, flow rate, flooding area, etc.

When selecting models (methods) for predicting hydrological parameters, it is necessary to take into account the specifics of the object or territory. Thus, the following main forecasting hydrological parameters models and methods should be considered [3, 11, 13]:

1. Rain flood forecasting methods.The methods rely on the regularities of runoff processes occurring in the catchment. One way to implement the method is to establish empirical dependences of runoff on rainfall by regression analysis.It should be noted that some methods based on water balance equations and Saint-Venant equations.
2. Snowmelt flood forecasting methods. The main idea of the methods is that the runoff during the flood period defined bythe amount of snow accumulated during the winter in the river basin, the precipitation amount that fell during the snowmelt flood period, and the water absorption capacity of the river basin.The water balance equation for the river basin during the snowmelt flood period can be represented as follows:$Y = S + P–(I + Z)$, where Y is runoff, S is the water supply in snow and ice crust on the soil surface, P is liquid precipitation, I is infiltration, Z is evaporation.
3. Dam-break flood, estuarine flood and coastal flood forecasting methods.The main idea of the methods is hydrodynamic modeling based on the application of Saint-Venant equations (shallow-water equations).

Because most hydrological problems are of a spatial nature, geographic information systems (GIS) are an effective tool for the comprehensive analysis of spatial data. In particular, the reconstruction of a digital elevation model (DEM), which, taking into account the objective monitoring data of the water level or its predicted values, will allow to determine the flooding area, or can be used to simulate runoff in conjunction with a hydrological model. The use of GIS technologies to analyze and predict the scale of the flood provides more information for decision-makers. However, for completeness of information, models are needed that allow calculating possible flood effects characteristics, based on data obtained from hydrological models, methods, and the DEM.

Thus, let us consider a model for determining the flood effects characteristics based on the mathematical apparatus of system dynamics. The mathematical model is a system of the first order nonlinear differential equations [4, 12].

$$\frac{dX_i(t)}{dt} = f_i^+ - f_i^-, i = \overline{1,n} \tag{2}$$

where$f_i{}^+, f_i{}^-$, $i = 1,\ldots,n$ is continuous or piecewise continuous functions that determine the positive and negative rate of change in system variable (flood effects characteristics)$X_i(t)$, i = 1,…,n. In turn,$f_i{}^+F_1,\ldots,F_m), f_i{}^-F_1,\ldots,F_m)$ is functions of the factorsF_j, $j = 1,\ldots,m$ affecting the rate of change of the variable $X_i(t)$, while they can be system variables and environmental parameters (flooding area, water level, flow velocity, water temperature, population density, etc.). Let us assume that the functions of the right part (2) have the form [2].

$$f_i^{+/-}(F_1, \ldots, F_n) = \sum_{l=1}^{n} k_{i,l}^{+/-} \prod_{j=1}^{n} f_{i,l}^{F_j}(F_j) \tag{3}$$

where the coefficients $k_{i,l}{}^{+/-}$ are determined at the stage of adaptation of the model to the object of study. The functions of the right part (2) determines the dependence of the variable $X_i(t)$ on the factor F_j, and are defined by approximating the statistical data. Thus, the system of Eqs. (2) with $t > 0$ and initial conditions is a Cauchy problem for a system of ordinary first order differential equations, and is solved numerically [5–8].

Forecasting the possible flood effects will allow decision-makers to form the necessary forces and means aimed at eliminating flooding, to allocate resources, to develop the basic action plan, and as a result, improve the quality of management of an object or territory.

In order for the presented models to maintain their accuracy and relevance, it is necessary to calibrate and adapt them to changing conditions.

The optimal allocation of resources for the organization of emergency rescue and other urgent actions in the flood zone solved by a linear programming problem. The solution of the problem will allow, for example, rescue teams to evacuate the affected population to special evacuation points or medical centers faster and with minimal costs. Thus, management systems need models to optimally allocate available resources in flood response.

The volume of data, including monitoring data and data obtained from the presented models and methods, will allow the development and adjustment of an effective action plan aimed at improving the quality of management of a flood-prone object or territory (see **Ошибка! Источник ссылки не найден.**).

Consider the algorithm for solving the problem (1).

Step 1. Creation of a set of actions aimed at flood prevention (Stage 1)$p_1,\ldots,p_k$.

Step 2. If there is a flood threat, then identify the flood type otherwise Step 1.

Step 3. Taking into account the flood type, and the features of the object or territory, select a hydrological, hydrometeorological or hydraulic model, forecasting hydrological parameters and create a DEM of the flooded area using monitoring data and meteorological data.

Step 4. To determine the possible flood effects characteristics using the model (1).

Step 5. Calculation of emergency rescue and other actions using resource allocation optimization models.

Step 6. Creation of a set of actions aimed at preparing for flood and eliminating flood effects (Stage 2) $p_{k+1},\ldots,p_m$ based on presented models. Formation of the basic action plan $p = (p_1,\ldots,p_k, p_{k+1},\ldots, p_m)$.

Step 7. If there is a change in monitoring and forecasting data, then the basic action plan is adjusted.

The resulting action plan will improve the efficiency of the management system and reduce the negative impact of the flood effects (Fig. 1).

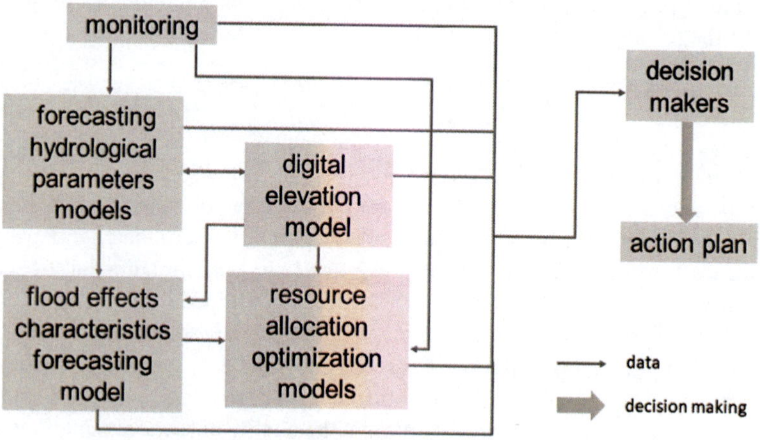

Fig. 1. The process of generating data for decision-making

Based on the complex of presented models, an automated management system can be developed that allows visualizing the calculated data and forming a set of actions aimed at preventing and eliminating floods. Automation of the management system will make the management process less dependent on the competence of decision-makers.

5 Conclusion

To improve the quality and efficiency of the management systems for flood-prone objects and territories, taking into account the stages of its functioning, it is necessary to create a set of models and methods that will allow forecasting hydrological parameters, the flooding area, flood effects characteristics and flooding timing, as well as optimizing resources. Based on the analysis of data obtained from a complex of forecasting models and methods, an action plan is formed, the implementation of which will allow minimizing the flood effects for a flood-prone object or territory.

References

1. Adams, T.E., Pagano, T.C.: Flood Forecasting – A Global Perspective, p. 480. Academic Press, Cambridge (2016)
2. Brodsky, Yu.I.: Lectures on Mathematical and Simulation Modeling, p. 240. M.- Berlin: Direkt-Media, Moscow (2015). [in Russian]
3. Felder, G., Zischg, A., Weingartner, R.: The effect of coupling hydrologic and hydrodynamic models on probable maximum flood estimation. J. Hydrol. **550**, 157–165 (2017)

4. Forrester, J.W.: World Dynamics, 2nd ed., p. 144 Productivity Press, Portland (1973)
5. Khamutova, M., Rezchikov, A., Kushnikov, V., et al.: Forecasting characteristics of flood effects. J. Phys: Conf. Ser. **1015**, 052012 (2018)
6. Khamutova, M., Rezchikov, A., Kushnikov, V., et al.: Algorithms for the management of liquidation process of floods consequences. Recent Research in Control Engineering and Decision Making **199**, 540–551 (2019)
7. Khamutova, M., Kushnikov, V.: Models and algorithms for managing the process of eliminating the flood effects at flood-prone objects and areas. In: Proceedings of 2021 14th International Conference Management of Large-Scale System Development, MLSD 2021. 14. C. 9600228 (2021)
8. Khamutova, M.V., Kushnikov, V.A., Dranko, O.I.: A mathematical model for choosing an action plan for the prevention and elimination of flood effects. IFAC-Papers on Line **55**(3), 113–118 (2022)
9. Kosyachenko, S.A., Kuznetsov, N.A., Kulba, V.V., Shelkov, A.B.: Models, methods and automation of control in emergency situations. Automation and Remote Control, issue **6**, 3–66 (1998). [in Russian]
10. Munich Re. Natural Catastrophes 2015: Analyses, assessments, positions. Topics Geo, Munich Re, Munich, p. 82 (2016)
11. Popov, E.G.: Hydrological forecasts. L.: Hydrometeoizdat, 256 (1979). [in Russian]
12. Sadovnichiy, V., Akayev, A., Korotayev, A., Malkov, S.: Modelling and Forecasting World Dynamics. Scientific Council for Economics and Sociology of Knowledge Fundamental Research Programme of the Presidium of the RAS. RAS ISPR, Moscow (2012). (in Russian)
13. Sitterson, J., Knightes, C., Parmar, R., Wolfe, K., Avant, B., Muche, M.: An overview of rainfall-runoff model types. International Congress on Environmental Modelling and Software **41** (2018)
14. Sene, K.: Flood Warning, Forecasting and Emergency Response, p. 303. Springer, Berlin (2008)
15. World Meteorological Organization (WMO). Manual on flood forecasting and warning, WMO-No. 1072, p. 120. WMO, Geneva (2011)

Check for updates

VSS Cyber Security Oriented on IP Cameras

Kristýna Knotková(✉)

Faculty of Applied Informatics, Univerzita Tomáše Bati Ve Zlíně, The Czech Republic, Nad Stráněmi 4511, 760 05 Zlín, Czech Republic
k2_knotkova@utb.cz

Abstract. This article discusses the cyber security of camera systems, especially IP cameras, and their vulnerabilities. It focuses on warning about camera system cyber security risks, threats, and openness. It includes a description of the risks and hazards associ-ated with the cyber security of IP cameras, scenarios of possible attacks, their coun-termeasures, and the course of the testing.

Keywords: VSS · camera systems · cyber security · vulnerability of IP cameras · cyber security of IP cameras

1 Introduction

Nowadays, cameras are almost everywhere around us. They are primarily meant to provide protection, but there is another side to the coin, so it is necessary to look closely at webcams to see how secure they are and how easy it is to hack and access these systems. Cameras are widely used, and thus there is a high percentage of insuf-ficiently secured cameras and access points. Since these are IP cameras, i.e., cameras connected to the Internet (or to a network that is connected to the Internet), many threats are associated with them. In this article, the author tries to capture the current security of webcams, both in terms of hardware and software, and then write down scenarios of possible attacks on the vulnerabilities of webcams and their security measures.

Risks in cyber security can be understood as the sum of careless actions of the user, the installer and, in many cases, the developer of the camera interface, which leads to greater attention of the attacker. These risks and attacks can take place as follows:

- attacks on websites, obtaining sensitive data,
- imposing unwanted e-mails, which can subsequently cause damage to the given device,
- illegall downloading of files that may be infected, leading to malware running on the device.

Cyber Security Threats

A threat can be understood as the possibility of exploiting an asset's weakness, which negatively impacts the protected asset. In terms of cyber security, vulnerabilities and their exploits can be described as: [1].

- default login data leading to easy access to the given device,

R. Silhavy and P. Silhavy (Eds.): CSOC 2023, LNNS 723, pp. 96–107, 2023.
https://doi.org/10.1007/978-3-031-35317-8_10

- weak credentials leading to fast decryption and access to the given device,
- negligence of security or ignorance from the user's point of view, leading to an attack on the device, e.g., via phishing messages or spam,
- downloading and installing files without verifying their security, leading to malware infection of the device,
- illegal downloads of files that are usually infected with malware,
- user carelessness in securing devices connected to the Internet leads to different at-tacks (depending on the weakness and the device involved),
- insufficiently secured devices connected to the Internet, leading to various attacks (depending on the weakness and the device),
- inadequate network security leads to easy access to the network and devices connected.

Vulnerabilities in cyber security:

- HW error of the server where the camera recordings are backed up,
- insecure server where camera recordings are backed up,
- insecure web camera interface,
- weak password,
- implemented bug,
- default configuration settings,
- default login information,
- spying options.

Webcam Vulnerabilities:

From a programmer's point of view, this is an error found in the device's hardware or software and thus causes a security problem. Most of the time, these errors are used to take over the home computer network or to monitor and find out sensitive infor-mation, which is subsequently used.

Examples of web cameras using: [2, 3].

- traffic monitoring,
- construction progress,
- news regarding winter resorts,
- weather monitoring in different countries,
- tourist events,
- babysitters,
- tracking of mobile devices,
- monitor shops and working places,
- compound security.

2 Configuration Settings

Setting up the configuration, i.e., changing the username and password, is essential to add a new camera to a household. If the design is careless, the criminal can connect to the camera that monitors the home and the entire network.

2.1 Default Configuration Settings

The camera in operation, which is set in default mode (login data from the manu-facturer, etc.), publishes the captured scene publicly on the Internet. The first security principle is to change these default login credentials to completely different ones, preferably so the password is considered vital. Several software packages on the In-ternet work on searching for devices with a default configuration. These devices must be connected to the Internet. An example of such software is the Shodan server. It later adds the mentioned device to its database. In the case of web cameras, Shodan users can watch the recorded scene from the given camera [4, 5]. Another search for cameras and devices connected to the Internet with a default configuration is a search on the google.com server. Users can find other cameras if they know the basic parameters of inurl, intitle, and the web interface of various manufacturers. An ex-ample of a search link looks like this: inurl:view/viewer_index.shtml [4, 6]. Where view/viewer_index.shtml is the web interface of the webcam manufacturer. When inserting this link into the Google search engine, various camera devices that can be viewed will pop up. The mere fact that these captured scenes from the camera can be displayed is a vulnerability that must be avoided [4, 6]. Another server working simi-larly to Shodan is called Insecam. These servers are created to alert owners that their devices are not secure, and they should do something about it. Insecam works on the principle of registering in its database low-security private webcams that are allowed for anonymous access from the Internet [4, 7].

Insufficient Password. Many programs only serve to connect to a poorly secured device, like a webcam. These programs can operate using a brute force attack meth-od, an analytical test to try all possible key settings and evaluate if the correct one has been found. some programs can work on the principle that they try all passwords that have already been cracked and located in a published database.

WiFi. WiFi is one of the options to connect to the IoT in any compound. It is neces-sary to pay attention to the security of the configuration related to the Internet and the web camera. If an attacker cracks the router's password, they can see other de-vices on the network.

- if the webcam does not contain a password, the attacker will take control immediate-ly after breaking access to the Wi-Fi network,
- if the webcam contains a default password, searching for it in existing password da-tabases on the Internet is enough,
- if there is an implemented bug in the camera, it will create the possibility of abuse for easier access to the webcam,
- if an attacker gets to the camera via Wi-Fi, he can monitor the camera's operation, but he will have limited access to control it. This is a scenario where the attacker did not crack the webcam password, only the Wi-Fi network.

Wi-Fi network security measures.

- placing AP in the center of the building,
- make sure the AP is not near windows,

- Wi-Fi network coverage should not extend further than the wall and windows of the building,
- regular inspections are also important
  - checking the signal outside the building,
  - checking whether an unapproved AP is on the network,
- limit the connection of MAC addresses
  - perform an update,
- limit or disable DHCP (Dynamic Host Configuration Protocol) = automatic assignment of static IP addresses
  - When limiting DHCP, limit the allocation of IP addresses only to the number of devices that use Wi-Fi [8].

Protection against access to the local network.

- do not share the Wi-Fi password with an unauthorized person,
- secure the place where it is possible to connect with a cable against unauthorized persons,
- secure the router,
- have a good password to access devices on this network,
- have a good password to access the router.

3 Description of the Experiment

The author chose the following tools for testing webcams:

- Shodan
- SSL Labs
- AbuseIPDB
- Security headers

3.1 Testing the Webcam Itself

The camera for the test equipment is called Day/Night Surveillance Camera. Its description is as follows:

- Model: TL-SC3171.
- Power: 12 V DC 1 A.
- Default settings: User: admin Password: admin
- SN: 112CD101484
- MAC: 940C6DB0796E

Procedure for Putting the Camera into Operation. A switch was used to connect the camera, which was connected to the router. Another network cable was connected from

Fig. 1. IP Camera

the switch to the camera and another to the PC. The power cable from the switch and the camera were plugged into the power source.

For the GUI interface, it was necessary to download the installation file from the Internet.

The access data was changed because the camera was previously used in laboratory work. Therefore, it was necessary to reset the camera, which can be seen in Fig. 1. in the left rear part of the camera. Subsequently, after the reset, the default login data worked, i.e., user: admin, password: admin. The camera was connected to a local IP address, so it was only possible to connect to the web interface from the given local network—IP address: 192.168.100.238. List of properties from a safety perspective.

Pros

- The camera has no connection via a VPN.

Cons

- The HTTPS protocol does not secure the web interface of the page,
- the website does not have a valid certificate,
- the camera is using outdated software/firmware,
- most camera settings do not work with browsers other than Internet Explorer,
- the camera interface does not support any two-step verification.

3.2 Possible Attacker Scenarios and Their Countermeasures

The IP camera was connected to the local network in the test area, which means it is impossible to connect directly to the camera from another network. It would be pos-sible if the attacker had remote access to a device in this local network. Examples of abuse can be, e.g., remote desktop or VPN. As the camera does not support a VPN connection, there is no threat of connecting to the camera through its server. In this case, the VPN would only be in danger if the attacker got through the VPN (of anoth-er device) to the local network to which the camera is connected. If an attacker has access to a local network using remote access, he also has access to a camera located on this network. In this case, the computer can be protected by requiring us always to be careful when downloading files and folders. There is a need for both passive and active security of devices connected to the Internet. If devices allow two-step verifi-cation, then apply it. What plays a significant role here is, i.e., human foolishness and ignorance. Threats can come from any social network, or the user's actions can cause them. Any malware download may contain remote access capabilities.

3.3 Prevention Against the Emergence of a Threat

- Read the package manuals and activate any security measures that can be activated on the device (two-step verification),
- change the original login data immediately after connecting the device to start up,
- do not download attachments or click on links from emails from unknown recipients (prevention of PC viruses),
- protecting the network and the devices within it through which the camera could be connected,
- do not download illegal files from the Internet (do not visit suspicious websites),
- have an anti-virus software activated,
- set a sufficiently secure and robust password - apply this to any device connected to the Internet that allows you to enter login information,
- when using websites, check the validity of certificates, and check connections.

3.4 Threats Associated with Other Devices

The case when the camera is connected to a local network. Many threats are associated with devices that are connected to the Internet. This can be a desktop computer, laptop, mobile phone, tablet, etc. If an attacker connects to one of these devices, he can also connect to the local network in which the camera is connected. Therefore, there is the same threat as with a camera connected to a public network. In this case, it would be advisable to set up the network so that it is possible to connect to the camera from only one point (e.g. a laptop). Not every camera includes this setting, but if it does, it is advisable to set a whitelist packet that can pass to the given device either in the camera's firewall (if it allows it) and at the same time in the network settings, or just set this restriction in the router. Another appropriate measure against this threat is to secure the device and avoid threats such as phishing, spam/ham, malware - computer virus, spyware, keylogger, Trojan horse.

As with devices connected to the Internet, it is necessary to protect the network itself, the correct settings of the router and the inaccessibility of the Wi-Fi network outside the building. If the local network is not secured, the threats are the same as if it is a public network. Easy access to the network - easy access to the device connected to it - web cam.

3.5 Threats Associated with the Camera Itself

The camera itself uses outdated software, which means there are no updates from the manufacturer to help keeping this webcam's active protection up-to-date. Outdated software can contain many unsolved security bugs, which then help an attacker. After a possible attack on this system, the attacker can have control over the device forever due to the non-updated software.

Another threat and quite important vulnerability these days is that the webcam is not connected via the HTTPS protocol, i.e. it does not have a secure connection. Attackers then have options to attack the given web address. These types of attacks can look like the following: An attacker can eavesdrop on the communication between the browser (user) and the server. In the case of a web camera application, an attacker finds out the credentials of the web camera and gains access to it. Thus, the attacker can watch the video transmission of the webcam without the owner being aware of it, until he reveals himself or until the owner disconnects the Internet connection from the webcam. Another danger from not using the https protocol is the possibility of falsifying messages. This means that DDoS attacks are possible with an unsecured http protocol, which would cause the webcam to be inoperable for a longer period of time.

3.6 Threats Associated with a Camera Connected to a Public Network

Although the tested webcam is not located in a public network, it is also worth mentioning the threats related to a camera that is connected to a public network or uses a VPN for its transmission. The first threat found with webcams connected to a public network is that anyone can access the camera from any device on the network. In such a situation, it is necessary to make firewall settings either in the interface of the camera itself, or settings in the network. The best way would be to limit access to the camera so that only one computer on the network can access it. Next, secure this computer with a strong password and restrict access to this computer.

If the camera is connected to a publicly accessible data socket, it is possible to disconnect the camera from this socket and the transmission will thus continue to be non-functional. Therefore, it is also necessary from a physical point of view to connect the camera in such a way that the power cables will not be easily disconnected. For example, insert the cables leading to the socket into the wall in armored tubes, then plug the end into the socket, which cannot be accessed from the outside. In the best case, connect the cables in a closed room to which a limited number of people have access.

3.7 List of Tested Addresses and Their Results

For this step, two online web applications were used to determine their vulnerabilities and possible attack history. These web applications are as follows: https://securityheaders.com/ and https://www.abuseipdb.com/ [9].

The AbuseIPDB site works on the basis of a database in which IP addresses that have been attacked in history or are considered dangerous or suspicious are located. The Securityheaders web application is used to display vulnerabilities and then describe possible header deficiencies for the specified IP address. This page ranks addresses from A to F, where A is the best grade and F is the worst.

The list of tested addresses and their information collection was carried out in May 2020 on the shodan.io server under the "web cams" category, and IP addresses of IP cameras were randomly selected from this list and subsequently tested. Currently, this information can change according to security updates, changes in camera IP addresses, etc. The shodan.io tool was chosen because it is a type of OSINT tool, which means that it is publicly available information.

For example, three tested IP addresses of publicly available web cameras will be listed here [5].

Address 1 - 101.132.145.56. When using AbuseIPDB, the outcome is that this address is not in the given database. The Web application Securityheaders rated this address as D, where XCTO and XFO prevent XXE, Injection, or Phishing attacks. This page does not have CSP, RP and FP security headers. Where CSP can be the cause of XXE attacks on this IP address ie. The insertion of images, scripts, fonts, media, etc., is not blocked here. The RP header would offer more security, which could be used against XSS attacks. As for FP, It's feature-policy, which is a new header that takes care of controlling which features and APIs can be used in the browser. Another thing is that address 1 does not use the HTTPS protocol, which means that there is no secure communication between the browser (user) and the server.

Address 2 - 82.228.230.28. When using AbuseIPDB, the outcome is that this address is not in the given database. The Securityheaders web application rated this address an F. This is the worst possible rating for this web application. It is justified by the fact that the given address does not contain any security headers and does not have a connection using the https protocol. And type of attacks like XSS, XXP, clickjacking, XXE, etc. are possible here.

Address 3 - 47.252.23.1. AbuseIPDB has no information about this page in its database. Securityheaders rated this page as D, just like address 1. The address has 2 headers: XCTO and XFO, these headers prevent XXE, Injection, or Phishing attacks. This page does not have CSP, RP and FP security headers, where failure to implement the CSP header can be the cause of XXE attacks, i.e. the insertion of images, scripts, fonts, media, etc. is not blocked here. The RP header would offer more security, which could be used against XSS attacks. As for FP, it is a Feature-Policy header that takes care of controlling which features and APIs can be used in a given browser. And the last important thing is that both addresses 1 and 2 and address 3 do not use the https protocol, which means that there is no secure communication between the browser (user) and the server.

4 Experiment Results

Table 1. Address list security evaluation

Address no	Abuse IPDB	Evaluation	XCTO	XFO	CSP	RP	FP	HTTPS
1	NO	D	YES	YES	NO	NO	NO	NO
2	NO	F	NO	NO	NO	NO	NO	NO
3	NO	D	YES	YES	NO	NO	NO	NO
4	NO	D	YES	YES	NO	NO	NO	NO
5	NO	D	YES	YES	YES	NO	NO	NO
6	YES	-	-	-	-	-	-	-
7	NO	D	YES	YES	NO	NO	NO	NO
8	NO	D	YES	YES	NO	NO	NO	NO
9	NO	D	YES	YES	NO	NO	NO	NO
10	NO	D	YES	YES	NO	NO	NO	NO
11	NO	F	NO	NO	NO	NO	NO	NO
12	NO	D	YES	YES	NO	NO	NO	NO
13	NO	D	YES	YES	NO	NO	NO	NO
14	NO	D	YES	YES	NO	NO	NO	NO
15	NO	F	NO	NO	NO	NO	NO	NO

From Table 1 – Address list security evaluation, it follows that the AbuseIPDB page does not contain any of the selected addresses in its database except for address number 6. The assessment of the addresses was, in most cases, D, and 3 cases were F. In addition to the mentioned 3 cases, XCTO – X-Content-Type-Options and XFO – X Frame Options headers are recorded for all addresses. The security header CSP - Content Security Policy was recorded only for address 5; the others no longer contained it. Furthermore, the RP-Refferer-Policy and FP - Feature Policy headers were not found in any address lists. As the last result of the testing, the HTTPS protocol was not detected at any of the tested addresses, which is used for secure communication between the browser and the server.

4.1 How to Provide Individual Security Headers in a Web Application

Each header can be secured with the settings explained below. If possible, the full measure for the tested cameras will not be at the server level but would have to change and set the camera firmware, which shows that most cameras are fundamentally insecure against this. (This can be secured by network and router settings). These syntaxes and codes are explained for use with Apache and Nginx software web servers [10, 11].

X-Content-Type-Options Syntax

```
X-Content-Type-Options: nosniff
```

Settings in Nginx

```
add_header X-Content-Type-Options "nosniff" always;
```

Settings in Apache

```
header always set X-Content-Type-Options "nosniff"
```

X Frame Options Syntax

```
X-Frame-Options: SAMEORIGIN
```

Settings in Nginx

```
add_header X-Frame-Options "SAMEORIGIN" always;
```

Settings in Apache

```
header always set X-Frame-Options "SAMEORIGIN"
```

Content Security Policy Syntax

```
Content-Security-Policy: <policy-directive>; <policy-di-
rective>
```

Refferer-Policy Syntax

```
enum ReferrerPolicy { "", "no-referrer", "no-referrer-
when-downgrade", "same-origin", "origin", "strict-
```

```
origin", "origin-when-cross-origin", "strict-origin-when-
cross-origin", "unsafe-url" };
```

Feature Policy

Syntax for Feature-Policy

```
Feature-Policy: autoplay 'none'; camera 'none'
```

Settings in Nginx

```
add_header Feature-Policy "autoplay 'none'; camera
'none'" always;
```

Settings in Apache

```
header always set Feature-Policy "autoplay 'none'; camera
'none'"
```

Syntax for Expect-CT
This header enables certificate checking for websites.

```
Expect-CT: max-age=604800, enforce, report-
uri=https://www.example.com/report
```

Settings in Nginx

```
add_header Expect-CT "max-age=604800, enforce, repor-
turi='https://www.example.com/report' always;
```

Settings in Apache

```
header always set Expect-CT "max-age=604800, enforce, re-
porturi=https://www.example.com/report
```

5 Conclusion

This article aims to inform about the issues related to the cyber security of web cameras. It describes and tests selected webcams from the point of view of cyber security and presents the results. During testing, vulnerabilities such as missing website headers, default configuration settings, weak passwords, etc., were detected. Shodan, abuseIPDB, and security header tools were chosen for testing public web cameras.

The physical camera was tested in a local network and scenarios of possible attacks were created. The research results from the mentioned tools are listed in Table 1. At the end of the article there are syntax and codes that are explained to use for Apache and Nginx software web servers.

In the future, it would be possible to design an application for testing web cameras, or more in-depth testing of one webcam IP address and testing this camera from different ones from a cyber security point of view. It would also be a good idea to buy twenty to thirty of the most popular cameras on the market and test them for all the vulnerabilities mentioned in this article. Another possible topic from the point of view of this article is creating enough secure passwords to devices connected to the Internet, or to cameras.

References

1. Threat: general terms. The Ministry of the Interior of the Czech Republic [online]. Praha: © 2019 The Ministry of the Interior of the Czech Republic (in Czech), 2019 (2003). [viewed 2022-11-03]. Available from: https://www.mvcr.cz/clanek/hrozba.aspx
2. Vulnerability: (in Czech) In: ManagementMania.com [online]. Wilmington (DE) 2011-2020, 11.04.2016 [viewed 2022-11-03]. Available from: https://managementmania.com/cs/zranitelnost-vulnerability
3. Vulnerability: Vulnerability. (in Czech) In: Wikipedia: the free encyclopedia [online]. San Francisco (CA): Wikimedia Foundation (2019). 27. 4. 2019 [viewed 2022-11-03]. Available from: cs.wikipedia.org
4. Cizek, J.: Live: IoT is leaky. There are thousands of private cameras and network boxes on the web. (in Czech) Live: Computers [online]. Czech News Center: Czech News Center, 2016, 21 (january 2016). [viewed 2022-11-03]. Available from: https://www.zive.cz/clanky
5. Shodan [online]. © 2013-2020, All Rights Reserved - Shodan®, ©2013-2020 [viewed 2022-11-03]. Available from: https://www.shodan.io/
6. Google [online]. Menlo Park, California, Alphabet, USA (1998). [viewed 2022-11-03]. Available from: https://www.google.com/
7. Insecam [online]. [viewed 2022-11-03]. Available from: www.insecam.org
8. Malanik, D.: Security in a corporate environment, corporate FW, Phishing, Pharming, Social engineering, Cross Site Scripting, DoS, DDoS, …: Cross site scripting. (in Czech) UTB Zlin. Zlin
9. Security Headers: Everything you ever wanted to know about Security Headers (but were afraid to ask) in one place. Security Headers (in Czech) [online]. Nove sady 988/2, 602 00 Brnostred: Copyright © 2018 Web security (2018). [viewed 2022-11-03]. Available from: https://securityheaders.cz/
10. NGINX. NGINX [online]. Copyright © F5 [viewed 2022-11-03]. Available from: https://www.nginx.com/
11. APACHE: APACHE [online]. Copyright © 1997-2020 The Apache Software Foundation (1997). [viewed 2022-11-03]. Available from: https://httpd.apache.org/

Using Patent Analytics in Additive Manufacturing Evaluation for Monitoring and Forecasting Business Niches

V. V. Somonov[1,2], A. S. Nikolaev[1(✉)], S. V. Murashova[1], and E. Y. Gordeeva[3]

[1] IP Management, Faculty of Technological Management and Innovations, ITMO University, St. Petersburg, Russian Federation
ip@itmo.ru

[2] World-Class Research Center, St. Petersburg State Marine Technical University, St. Petersburg, Russian Federation

[3] St. Petersburg State University of Aerospace Instrumentation, St. Petersburg, Russian Federation

Abstract. This article presents a study of the applying patent analytics to predict the state and manage complex projects in the field of additive manufacturing market carried out with the help of big patent data and recommendations for seeking promising fields of activity. In the course of patent data research, the leading countries were defined that possess engineering centers concerned with developing and implementing technical solutions in the field of additive manufacturing; and leading countries where such solutions are implemented, being protected by respective patents. The most demanded fields of activity were determined for enterprises developing additive manufacturing technology, equipment, materials, and software. Using the example of the best-in-class organizations of top patent strength nations, fields of activities outlined where new-to-market enterprises could see consumer interest. Unoccupied niches revealed for certain countries whose leaders have not secured a stable market ground protected by patent documents.

Keywords: patent analytics · intellectual property analysis · patent landscape · additive manufacturing market · technology intelligence

1 Introduction

The modern economy has an increased information capacity of strategical decisions. It is a rule of thumb: The larger the company, the more it is concerned about risks associated with a wrong market conduct strategy. This is especially relevant to next-gen breakthrough technology that significantly affects the economy [1]. Among such disruptors is additive production, which comes with huge growth potential [2]. Growing at a high gear, today's technology always entails change and progress. One of the most definitive information sources—helping assess the current status and determine innovation growth directions in a particular market—is patent information [3]. Intellectual property assets mainly define innovations. They require proper treatment, both in legal and managerial

R. Silhavy and P. Silhavy (Eds.): CSOC 2023, LNNS 723, pp. 108–121, 2023.
https://doi.org/10.1007/978-3-031-35317-8_11

aspects [4]. Most companies rely on patent data and patent analytics tools in their development and management strategies [5]. Patent analytics allows a company to discover what its rivals are up to, contemplate trends in a particular market, determine the key developers in a particular field, find the most powerful and sought-after technical solutions (which are protected by patents that every market entity must observe). Effective use of patent analytics tools contributes to successful market activity. It helps reveal promising niches and directions, favorable for growth and bearing strategic benefits [6]. Using patent analytics tools facilitates and improves the quality of decision-making. A common solution in this field is technology intelligence methods. They employ data visualization to analyze intellectual property [7]. Recently, a wide scope of research has emerged, where scientists utilize various patent analytics tools to discover opportunities induced by emerging technology across various economic sectors. In their work, researchers rely on patent documentation analysis and invest much effort in visualizing and processing the data collected [8, 9]. This paper analyzes the opportunity for using a patent analytics tool in the context of additive manufacturing—one of the most innovative fields of the last 7–8 years. Various consulting firms issue annual patent landscapes [10–15], the most renowned of which is Wohlers Associates. However, each of those materials is more of a general review: it does not investigate either patent strategies of market leaders or trends in implementing the developed solutions across the globe. It's often hard for managers and investors to handle patent data. Such documents provide incomplete information about the owner, miss details of a patent controlling organization; the applicant may have a name that differs from those of the patent holder, or be a subsidiary or an inventor. These factors don't let researchers form a holistic picture of a company's patent strategy. Another factor is the genuine information on the reasons why a particular patent became void. (It may have expired or deemed invalid, or become void due to no duties properly and timely paid). What also affects the completeness of patent status information is timely tracking of licensing and alienation of rights and correctness of translating data acquired from Asian patent authorities. The latter is significant, since Asian patent authorities hold the largest share of all patent documents registered worldwide [16]. All this requires special instruments that could curb risks and save time processing patent documentation before making managerial or investment decision.

The original purposes of this paper were to demonstrate the capabilities of patent analytics in handling the said tasks—by means of visualization, grouping, and use of analytics tools and data acquired; assess the current technology development status in the additive manufacturing market; reveal the trends in leading companies' patent strategies; and define recommendations for commercializing solutions in this market.

2 Discussion

This research employed the patent analytics tools [17]. It grants access to a large volume of patent documents managed by patent authorities in various countries. For the purposes of additive manufacturing market study, the descriptive and aggregated approaches to analyzing patent information utilized. The said software defined additive manufacturing as a technology for manufacturing three-dimensional objects with the help of a pre-designed physical 3D model and a system of additional or multilevel development frameworks. In this model, layers are stacked one after another—to eventually design a three-dimensional object. With that, segmented/layered graphic data of a model sent to a 3D printer. The latter uses a particular volume of welding material to reproduce the object layer after layer, until it is finished in conformity with the design criteria. During the study, a strategy for searching patent documentation, adjusted for the multitasking capabilities of such software developed.

Three-dimensional printing patents, issued during the last 20 years in various countries, taken for analysis. To facilitate the process, documentation grouped into patent families, each consisting several patents pertaining to a particular technical solution. Information in patent documentation assessed in terms of various indicators, including grouping by IPC code and keyword found in patent descriptions, by technology cluster, by patent portfolio strength (significance of a technical solution, titled Patent Asset Index). The Patent Asset Index calculated for a company as a whole or for some of its patents on a particular subject [18].

A particular patent family's Patent Asset Index depends on the relevance of the field of technology in which the related solution is currently protected. The Patent Asset Index is a ratio between: (a) the number of a parent family's citations in the latest patent documents across the globe (adjusted for the age of patent, patent authority expertise, and maturity of the field of technology to which the protected technical solution is related); and (b) the parameter designating size of the market in which the technical solution is protected. This parameter is determined by the number of active patents and those being reviewed. The said ratio can reflect how a particular patent influences the market or the value of the company that uses it. This influence called Competitive Impact. The sum of all Competitive Impact values defines the Patent Asset Index for a particular patent family or company in a particular field.

During the study, major engineering centers developing technical solutions, as well as countries where they patented, discovered. The researchers also compared the leading companies protecting the most significant technical solutions. That was done to find out the market trends and methods leading entities employ to maintain their market positions.

3 Processing

3.1 The Current State of the Face Recognition Technology Market

Government and defense sector has the highest market share due to the growing demand for recognition solutions in law enforcement and security services (Fig. 1). These solutions used to identify and verify suspected criminals, as well as cross-border monitoring, among other things, in real time, while providing a secure environment. For example, in April 2018, the Indian Police in New Delhi implemented facial recognition systems to identify lost or abducted children. It applied to over 45,000 children, of which about 3,000 children have found throughout the city.

Having set the patent document search criteria and selected the most significant patents by Patent Asset Index, the researchers have formed a collection of patent families for further analysis of the global additive manufacturing market.

From the patent documents found, it was revealed that globally there are no more than 4 major regional centers concerned with developing significant technical solutions in the field of additive manufacturing (see Fig. 1). Those include North America (United States), European Union (Germany), Israel, Asia (China, Japan). It was found out that original technical solutions are protected not only in the said countries but also in France, Italy, Switzerland, Great Britain, South Korea, and Canada (see Fig. 1). Israel, on the other hand, prefers to protect its home-invented products not with the domestic authorities. Instead, engineers file requests with PCT authorities (submitting requests to several countries including the US) or European patent authorities. Such a strategy indicates Israel's motivation to introduce its solutions in the leading international markets (e.g., the States, China, and Europe) (Table 1).

Table 1. Top Patent Regions for Additive Manufacturing

Country of origin of R&D centers	Patent portfolio size, 1000 patent families	Patent asset index for R&D centers	Countries implement patented solutions in accordance with ISO	Patent portfolio size, 1000 patent families	Patent asset index for enterprises implementation in production according to ISO patented solutions
Uinted Kingdom	>1	3,7	Uinted Kingdom	>4	5,3
France	>1	2,4	France	>4	5,1
Germany	>6	2,6	Germany	>8	4,0
Israel	<1	5,6	Israel	>1	–
Switzerland	<1	3,9	Switzerland	>1	–
South Korea	>1	0,6	South Korea	>4	3,0
Japan	>7	1,3	Japan	>9	3,3
USA	>10	3,5	USA	>17	3,6
China	>20	0,8	China	>20	2,0
Canada	<1	3,6	Canada	>2	5,7

By analyzing Fig. 1a, one may conclude that the most significant inventions in the field of additive manufacturing—which are more than 5.5 times greater than average in terms of competitive impact—are created in Israel. In terms of the number of patent families affecting the global market, the United States, Germany, and China (see Fig. 1a) are among the top performers, being leading sales markets for technical solutions (see Fig. 1b). For further analysis of the additive manufacturing market, we need to review development and patenting processes in the said countries (see Fig. 2, 3, 4 and 5).

The pie chart (Fig. 2) demonstrates that, in terms of the number of patent families, the most important clusters for Israel are fabrication (mainly additive manufacturing or plastic 3D printing), healthcare (dentistry), chemistry (organic chemistry and coatings), information (computing and imagery), and electronics (electric circuits). The largest of

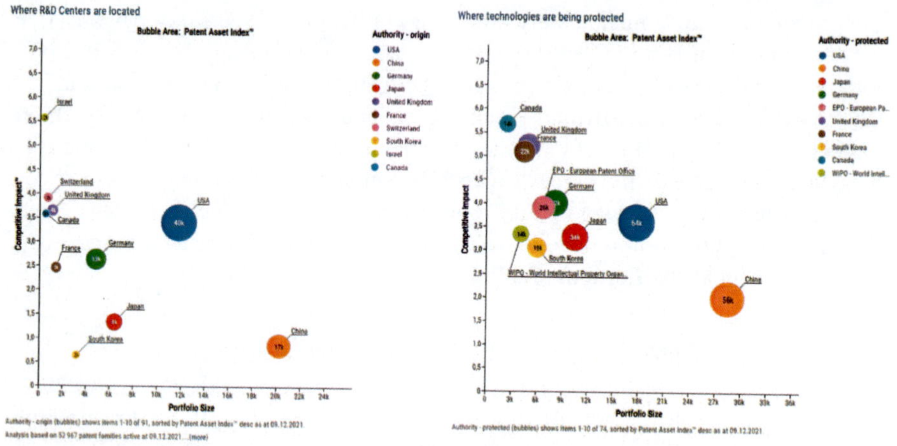

a b

Fig. 1. Geographic location: a) Major engineering centers where the most significant additive manufacturing solutions are developed. b) Regions where the most significant additive manufacturing solutions are protected.

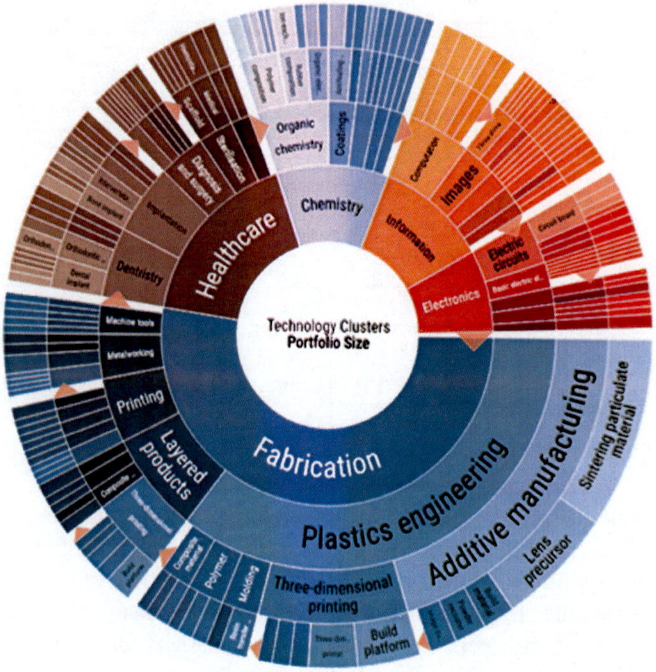

Fig. 2. Distribution of Israel's additive manufacturing-related inventions between technology clusters

these is the additive manufacturing cluster, comprising over 60% of patent families and seeing plastics engineering as the leading manufacturing method. All this indicates that Israel is focused on additive technology that could be applied in light manufacturing. It's apparent from other clusters that Israel also seeks to develop additive manufacturing methods potentially interesting to other countries in such traditionally powerful economic sectors as dentistry and organic chemistry.

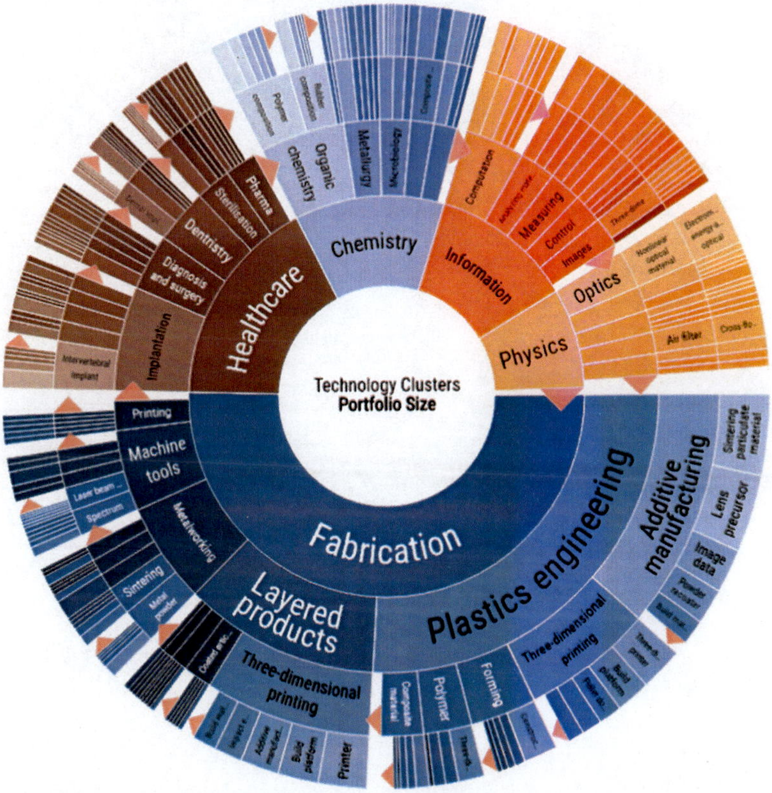

Fig. 3. Distribution of the US additive manufacturing-related inventions between technology clusters

The pie chart (Fig. 3) clearly shows that in terms of the number of patent families, the most important clusters for the US—the global market's leader according to various consulting firms [11, 12, 15]—are fabrication (mainly additive manufacturing or plastic 3D printing; over 50% of all patent families), healthcare (implantation), chemistry (organic chemistry), information (computing, measurements, control), and physics (optics). That said, the global significance of solutions invented in the US is somewhat lower than that of those created in Israel. Although plastic additive manufacturing recognized as the larger cluster, there is a considerable number of patents related to 3D printing equipment. This proves the US concern for trading not only additive technology but also equipment, including optical elements therefor. Technical solutions created in China, the world's leader in terms of additive manufacturing-related patents (see Fig. 4), have a several times lower market impact than those from Israel, the States, and Germany. But since Chinese solutions are superior in number, they still maintain a significant share of the global additive manufacturing market.

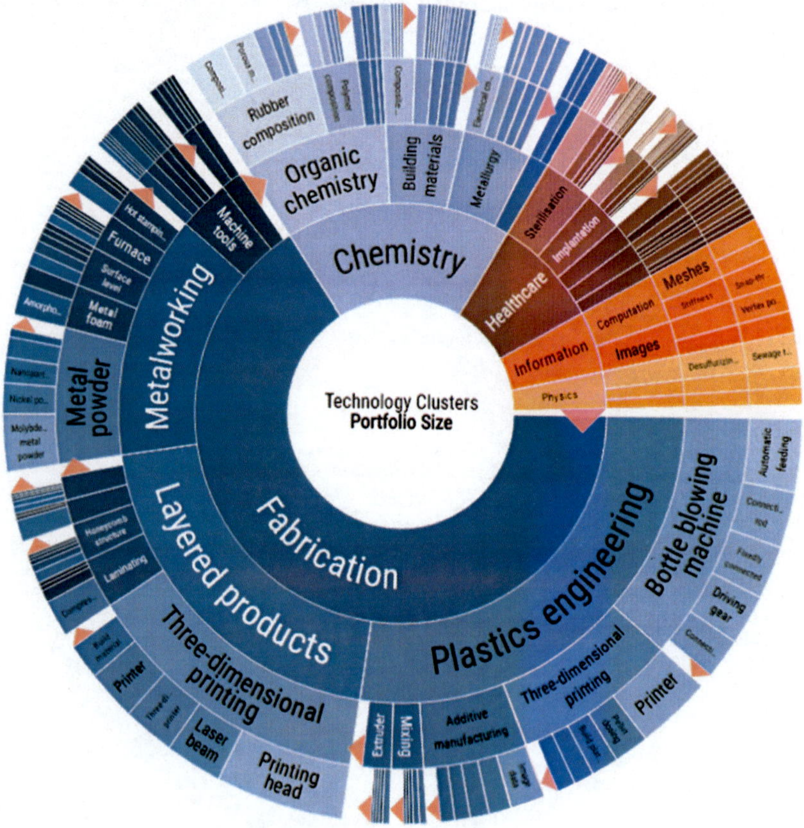

Fig. 4. Distribution of Chinese additive manufacturing-related inventions between technology clusters

China focused on real production (over 70% of all patent families). Alongside plastics, engineering, other major fields are development of 3D printers and additive technology equipment, as well as additive manufacturing from metal powders. This indicates China's scientific and technological preparedness for rivaling the US and interest in introducing metal additive manufacturing around the globe. As for distribution of inventions from Germany (Europe's additive manufacturing leader), the situation is very similar to China, except for one difference. Since additive manufacturing emerged in Germany earlier, the country was first of these two to embrace the importance of metal 3D printing and managed to make some significant advancements in this field. The share of patent families associated with metalworking is higher in Germany than in China. Figure 6 helps visualize how American companies influence the Chinese market.

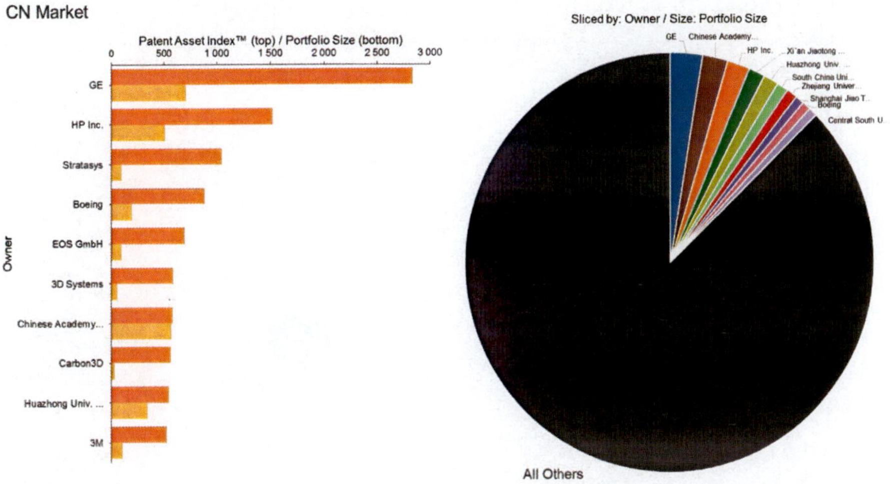

Fig. 5. Competitive situation around additive manufacturing patenting in the Chinese market

On the left is a bar chart of China's top 10 patent holders. It demonstrates the ratio of the average strength of a company's technical solution to the number of registered patents. Apparently, none of the Chinese enterprises is on the squad of top five patent holders possessing significant technical solutions. Meanwhile, the squad comprises four US-based companies (GE, HP Inc., Stratasys, and Boeing) and one German enterprise (EOS GmbH). China's leading additive manufacturing patent holder is the Chinese Academy of Sciences, which sits on the sixth line in the rating. There are two conclusions from these findings. First, German companies and US are interested in the Chinese market. Second, Chinese companies have not yet conquered this territory with any significant technology—otherwise they would have invested much in patenting advanced solutions. That said, the right part of Fig. 7 reveals that all other Chinese patentees leave top-10 companies far behind by the number of patents registered. Obviously, there is no monopolist that would hold the patents securing the entire market control. For the purposes of analyzing how top-10 companies influence the global market, Fig. 7 was plotted with the use of the patent analytics tools. The Fig. 6 illustrates the quality-quantity ratio of patent families of particular patentees.

This Figure suggests that US-based company Carbon 3D, the most successful 3D printing startup of the recent years, is leading the market in terms of the number of technology advancements. Besides, there are no Chinese patentees among the leaders. Stratasys (US), EOS (Germany), 3DSystems (US), which were titled global market leaders by some consulting firms, also made it to the top-5 owners of the most technically significant solutions. Such giants as GE and HP Inc. Devote much attention to additive manufacturing and are ahead of the pack in terms of the number of patent families. This implies their interest in using such technology in their products and introducing it to other enterprises. A key to success in the additive manufacturing market lies in staying on the ball about what is happening around significant patented solutions. Since fabrication is the leading cluster for additive manufacturing top performers, we will analyze the

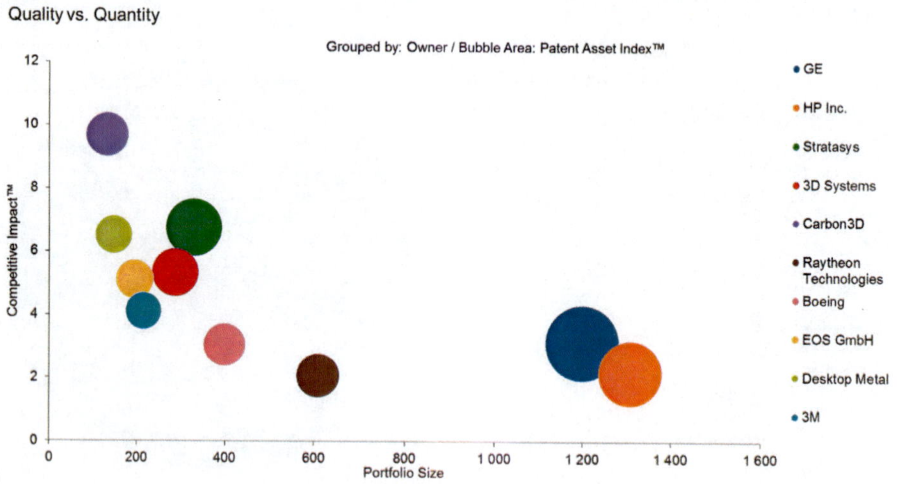

Fig. 6. Top-10 companies compared by the quality vs. quantity ratio of additive manufacturing-related patent families

competitive situation in this very field to evaluate the trends and market status. For that, we will plot a chart for the ten leading companies that will reflect the quality vs. quantity ratio of patent families of a particular patentee (Fig. 7).

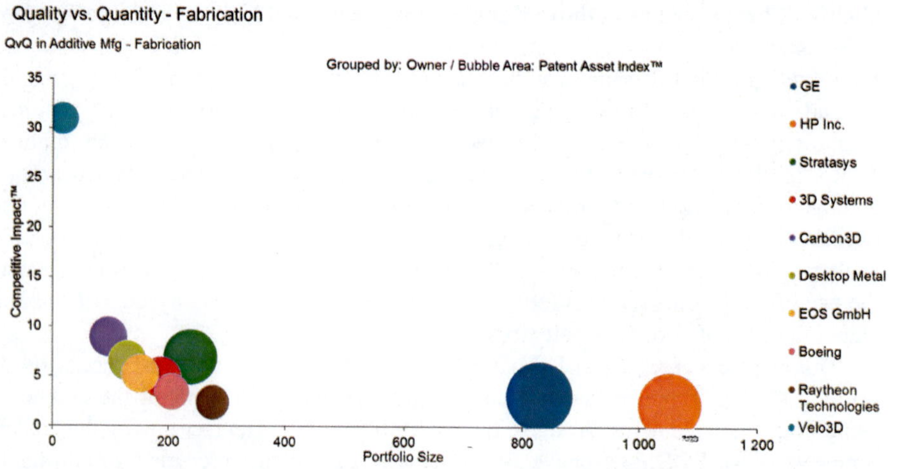

Fig. 7. Top-10 additive manufacturing companies compared by the quality vs. quantity ratio of additive manufacturing-related patent families

It's seen from Fig. 8 that Carbon 3D yielded the leadership to its subdivision Velo3D, which holds patents for over 4 times more advanced solutions—prepared for introduction into final production—than Carbon 3D as a whole. Other top-10 companies distributed almost the same way, except a whale of a difference in the number of patents between HP

Inc. And other companies in this cluster, including GE. This indicates HP Inc.'s readiness to focus on manufacturing a diverse line of products and render services mainly with the use of additive manufacturing. The latter requires protection of the technical solutions created and used in this sphere and zero legal barriers hindering business growth. All this indicates that major additive manufacturing entities employ all-ways patenting, while small and disruptive companies prefer selective patenting. To reveal the direction in which additive manufacturing is developing, we analyzed the patent portfolio against the Patent Asset Index of Velo3D, the market leader in terms of advanced technical solutions. The patent families found in PatentSight sorted by significance.

As part of the study, a number of company's patent families sorted by significance and relevance to the external market in relation to the patent holder country. It's seen from the numbers of patent documents that the company protected the most significant technical solutions concurrently in several countries, and European countries (EP) were among them more often than others. This suggests that the US is interested in introducing its advancements in Europe. An innovative startup, having a solution that may have a powerful impact on the global market, tries to conquer the European market first—since the competition is lower there than in the US. Among the latest significant technical solutions is patent family EP3263316 (selected by date of initial application; first document filed on June 27, 2017). It concerns 3D printing and the structure of the 3D printer. This points to the fact that development of new 3D equipment structures and 3D printing technology remains popular. Patenting processes are consistent with the policy of the companies from this country toward selling non-metal fabrication 3D printers and rendering 3D printing services. One can assess the significance of Velo3D's technical solutions for the market and its rivals by the share of direct citations to the company's patent documents in patents of other companies in the fabrication cluster, additive manufacturing group (see Table 2).

HP Inc. And GE (patenting leaders in this field), other US market leaders, and German enterprise EOS mainly cite to the company's patents. Apparently, they need to have similar advancements. To find out what a strategically significant solution for the global market is, let us analyze patent portfolios of two other technology leaders of the global market, Stratasys and EOS.

Obviously, the company is actively patenting both in the domestic and foreign markets, being one of the additive manufacturing pioneers. For this company, the foreign markets are European countries and Russia. European patent family EP1631439 has the highest Patent Asset Index so far. This family dates back to 2004 by the US patent document, devoted to the 3D printer structure. As seen from further patenting of this technical solution, the company started protecting it in the Chinese market (after Europe). This proven by CN patent documents dating back to 2008, which are in this family. If this family still maintained, it means the company is building its business around this direction and aspires to grow it worldwide. This also proven by consulting firms' reports that suggested growth of the Chinese additive manufacturing market over the recent years [12–15]. This is why the global technology leader had to protect its advancements with patents in this major emerging market—to protect its position.

Analyzing the patent portfolio of EOS GmbH, the world's third technology leader, the authors found out that the most significant and cited solution of the company was

Table 2. Direct citations to Velo3D's patent documents in the field of additive manufacturing.

Owner of citing active patents	Owner Patent Asset Index™	Share in total portfolio strength (...		Portfolio Size Share in total portfolio size of ow...	Avg. number of citations received	
Velo3D	Velo3D	431	77.4%	15	83.3%	74.8
HP Inc.	Velo3D	408	73.2%	11	61.1%	100.7
GE	Velo3D	407	73.0%	10	55.6%	109.9
Divergent 3D	Velo3D	374	67.2%	8	44.4%	129.0
DAIFU GENTE TECH	Velo3D	350	62.8%	7	38.9%	138.9
EOS GmbH	Velo3D	339	60.8%	7	38.9%	133.1
Raytheon Technologies	Velo3D	323	58.0%	7	38.9%	122.1
Kinpo Electronics	Velo3D	313	56.1%	7	38.9%	119.9
XYZprinting	Velo3D	313	56.1%	7	38.9%	119.9
Collins Aerospace (in: Raytheon Te...	Velo3D	294	52.7%	5	27.8%	154.8

the one related to the method for laser-sintering fabrication of three-dimensional objects (RU2009136179). The original German application dates back to 2007. Later, the company introduced this solution in the Russian, Austrian, and Chinese markets. This proves that the company's primary field of business is laser sintering and related matters. Since it also tries to capture leadership in the selected metal 3D printing in the Chinese market, this field is also considered promising.

To discover free niches in additive manufacturing markets of various countries, a technology matrix was designed for top 10 technology companies in their primary cluster (fabrication—additive manufacturing) (see Fig. 8).

The Fig. Illustrates that the leaders have concentrated their efforts on patenting the main solutions regarding creating three-dimensional polymer objects using additive manufacturing and 3D printing. Such enterprises as Velo3D, Stratasys, and 3D Systems are not interested in patenting unique metal powders used in 3D printing. The other leaders have few significant solutions in this field. Advancements in this sector may entail considerable investments and help capture a certain niche in the Chinese and US markets. Additive manufacturing from composite materials is not too popular among German companies. Original solutions protected in this market can help capture a solid position in the market. Since the study revealed that the most significant solutions—protected by the market leaders— related to 3D printers, we are going to determine which technology leader plays a pivotal role in the Russian market in terms of solution

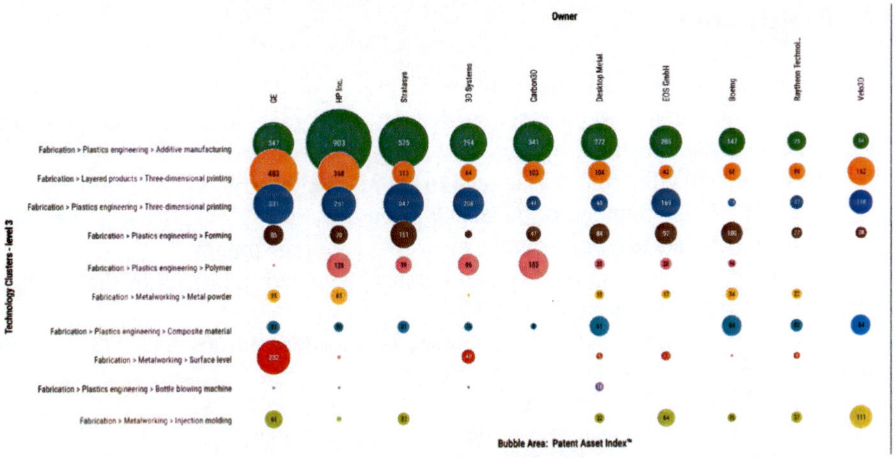

Fig. 8. Additive manufacturing technology matrix based on patent documentation of the global market's leading technology companies

significance. For that, we will analyze the ratio of the Patent Asset Index to the patent portfolio size. You can see the results in Fig. 9.

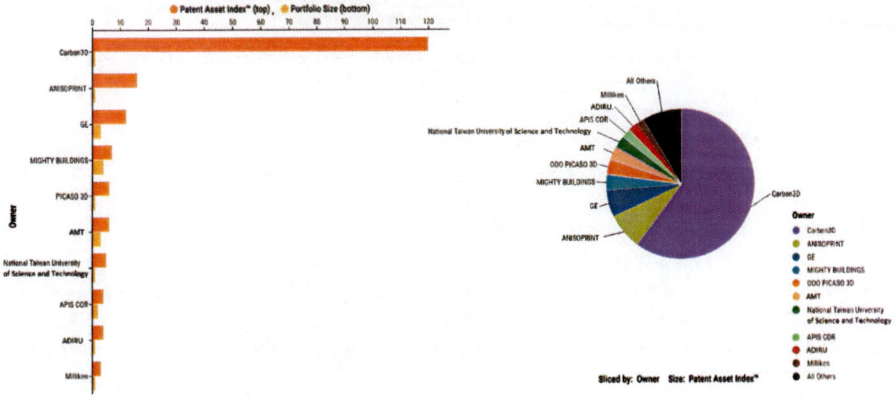

Fig. 9. Comparison of technology leaders patenting 3D printer solutions in the Russian market by solution impact on the equipment market competition

Figure 14 provides more evidence of Carbon3D's leading position (also in the Russian market) in terms of its impact on 3D printer structure advancements. In terms of patent strength, Carbon3D significantly outperforms its rivals protecting their solutions in the Russian market. All of their solutions combined are inferior in strength and quantity to those of Carbon3D, indicating that the latter controls advanced development—both globally and in the Russian market. That said, Carbon3D also focuses on the emerging Chinese market in the field of plastics engineering 3D printers.

4 Conclusion

The technology leaders in the field of 3D printing are the US, China, and Germany.

Solutions of the highest patent strength created in Israel, while the country protects them mainly in Europe and the US.

The most innovative solutions related to the 3D printer structure.

In terms of the number of patented inventions, the global leaders are major international companies employing the all-ways patenting strategy.

Technology leaders aspire to introduce their patented solutions in the emerging markets of China and Russia.

Among the promising patenting niches are composite materials in the US and China and metal powder in Germany.

Acknowledgments. The research described in this paper is partially supported by the Ministry of Science and Higher Education of the Russian Federation within the framework of the «World-class Science Center» program: Advanced digital technologies (Grant Agreement No. 075-15-2022-312, 20.04.2022) and financially supported by the ITMO University, NIR No. 622150 «Development of approaches to system design of integration of university science and business (pilot study)» and also by the ITMO University Research program (№622150).

References

1. Harrison, M.: Small innovative company growth: barriers, best practices and big ideas. Lessons from the 3D printing industry. Entrepreneur in residence U.S. Small business administration office of advocacy, p. 56 (2015)
2. Syed, A.M., Tofail, E.P., Koumoulos, A.B., Susmita. B., Lisa O', D., Costas, C.: Additive manufacturing: scientific and technological challenges, market uptake and opportunities. Mater. Today **21**(1), 22–37 (2018)
3. Yena, O., Popov, N.: Methodology for developing patent landscapes of the Federal Institute for Industrial Property project office. Stankoinstrument **41**(1), 28–35 (2019)
4. OECD: The Measurement of Scientific and Technological Activities: Guidelines for Collecting and Interpreting Innovation Data: Oslo, M., Third Edition, prepared by the Working Party of National Experts on Scientific and Technology Indicators, OECD, Paris, para. 146 (2005). Accessed 22 Dec 2014
5. Qi, H., Florian, H., Joan, C-F., Steffen, L., Leo, W., Thomas, E.: Visual patent trend analysis for informed decision making in technology management. World Patent Inf. **49**, 34–42 (2017)
6. Zuykov & Partners: What is patent analytics and what is its purpose? https://zakon.ru/blog/2020/02/11/chto_takoe_patentnaya_analitika_i_dlya_chego_ona_nuzhna Accessed 1 Jan 2022
7. Sevim, S.-M.: The supply side of IP management: understanding firms' choices regarding IP intermediaries. World Patent Inf. **50**, 55–63 (2017)
8. Kim, C., Lee, H.: A patent-based approach for the identification of technology-based service opportunities. Comput. Ind. Eng. **144**, 106–464 (2020)
9. Leonidas, A., Frank, T.: The state-of-the-art on Intellectual Property Analytics (IPA): a literature review on artificial intelligence, machine learning and deep learning methods for analysing intellectual property (IP) data. World Patent Inf. **55**, 37–51 (2018)

10. Has 3D Printing Reached the Plateau of Productivity? https://amfg.ai/2021/04/16/has-3d-printing-reached-the-plateau-of-productivity/. Accessed 7 May 2021
11. AMFG: The Additive Manufacturing Landscape 2020. https://amfg.ai/whitepapers/the-additive-manufacturing-landscape-2020-report/. Accessed 7 May 2021
12. Wohlers Associates: Wohlers reports 2020. 3D Printing and Additive Manufacturing Global State of the Industry, p. 380 (2020)
13. 3D printing trends 2020. Industry highlights and market trends. https://downloads.3dhubs.com/3D_printing_trends_report_2020.pdf. Accessed 17 Dec 2020
14. Gridlogics Technologies Pvt. Ltd. 3D Printing Technology Insight Report. An analysis of patenting activity around 3D Printing from 1990-Current, p. 44 (2014)
15. Sculpteo: The State of 3D Printing Report (2020). https://www.sculpteo.com/en/ebooks/state-of-3d-printing-report-2020/. Accessed 17 Dec 2020
16. LexisNexisIP: PatentSight. Big Data Innovation Analytics for Investors (2018). https://service.bloomberg.com/track_download/assets/content/examples/alt-data-content/PatentSight/PatentSight_Finance_Presentation_Bloomberg_long.pdf
17. LexisNexis: Products & Services. PatentSight Business Intelligence (2022). https://www.patentsight.com/patent-analytics-software
18. Holger, E., Nils, O.: The patent asset index – a new approach to benchmark patent portfolios. World Patent Inf. **33**(1), 34–41 (2011)

Check for updates

A Computational Model of Biotechnology

Raditya Macy Widyatamaka Nasution[1] and Mahyuddin K. M. Nasution[2(✉)]

[1] XII US A, Syafiyyatul Amaliyyah, Medan, Sumatera Utara, Indonesia
[2] Data Science and Computational Intelligence Research Group, Universitas Sumatera Utara, Medan 20155, Sumatera Utara, Indonesia
mahyuddin@usu.ac.id

Abstract. One of the human endeavors to understand nature is to decipher natural objects, such as biology, to form novelties, such as in biotechnology. The descriptions then have a model according to the characteristic importance of the biological study. The model reveals that decomposition characteristics are necessary for biology, technology, and product for biotechnology to be suitable as either a concept or a technology where mathematics plays a role as a foundation. With several methodological considerations, this paper reveals the existence of a computational model for biotechnology to develop this scientific field, that is, a consideration of a brief review of some of the literature for the basis of biotechnology mathematics.

Keywords: Biology · mathematics · computation · technology · statistics · artificial

1 Introduction

Science or technology is the achievement of human efforts to organize/continue its existence on the earth [1]. First, science is to uncover natural laws to protect human life [2], while second, technology is to use these natural laws to elevate human life [3]. Both are knowledge that increases human dignity [4,5]. One of the sciences of this knowledge is (from) natural science, which is known as biology [6,7]. Biology is a science that tells stories about organisms, plants, animals, humans, and their environment with the interactions between them [8,9]. A principle was underlying the life sciences. However, science doesn't just record stories about the object of its discussion. Science requires a framework capable of expressing deep insight into that life, where it needs the completeness of infrastructure and facilities in the form of what is known as technology [10–15]. Human culture derives scientific principles. The principle collaboration based on the word "logy" and "bio" has the name "biotechnology" [16–18].

Biotechnology - a scientific framework involving technology - is to express profound ideas from biology as an innovation, both for scientific development and its utilization for humans and their environment [7,19]. Unfortunately, it's not easy to present any technology to support the development of any science

R. Silhavy and P. Silhavy (Eds.): CSOC 2023, LNNS 723, pp. 122–133, 2023.
https://doi.org/10.1007/978-3-031-35317-8_12

without the spirit of computing from computers. Of course, it is related to a framework where the principles of mathematics play an important role. However, it is not easy also to involve any mathematical principles in a field of science whose framework is poor in computing [20] without revealing the characteristics of the object of study [21]. Therefore, this paper expresses a computational side of biotechnology based on modeling. Mathematical concepts that form the biotechnology mathematics foundation.

2 Some Ideas with Problems

Apart from the use of meta-scientific terms, the term logy comes from Greek the word $\lambda \acute{o} \gamma o \zeta$ (read *logos*) means "*account, explanation, and narrative*" provides a logical reason for language psychologically to represent different kinds of knowledge. How could it not be natural science deals with natural objects from the highest to the lowest level of abstraction: physics, earth science, space, chemistry, and biology - with its framework, where there is a small degree of similarity between them [22]. Therefore, biology as a science grows and stands alone in starting with a natural framework that deals with all the facts about life whiles technology appears from a framework of innovation [23]. Under the output of each study/research, an innovation has a level of achievement from scientific work to a product. The product may be a technology or be something that becomes clothing, and food [24–26].

Biology has characteristics following the objects of discussion in the form of chemical compounds, and it depends on interactions to form large structures in an organism [27]. Therefore, the universe of discussion from biology revolves around discovery, even if it is related to the language of life as genes or physical language such as interactions between organisms, or leads to individual interactions known as social networks [28]. Each organism carries its modality in the composition of genes to interact with other organisms. That is an output of the respective modality slices [29–31]. So, any similarity between them or interests between one and another brings up innovation differently [32]. However, all modalities used to interact will require and produce tools to understand it. These tools, generally, are called technology. For example, in a way that involves the concept of similarity in genetics [33].

Technology, according to the Demand Readiness Level (DRL) at the second level [34], is the specific need of the innovation that already has been identified, namely innovation that is already available according to level 9 (nine) of the Technology Readiness Level (TRL) [35]. Therefore, technology as an innovation is always present as something new from what already exists [36,37]. Technology - the applied science with a framework to answer life challenges with better infrastructure or facilities [36,38,39]. Therefore, biotechnology is a science that discusses breakthroughs that are more beneficial to interested parties, whether from the organism's point of view, whether it is from the environmental or natural perspective, or from both at the same time [40,41]. Likewise, to produce a technology that is useful for supporting the development of biology, each object of the technology requires a description that explains the purpose of using the technology. That description characterizes each thing with characteristics, with which technology also elevates those characteristics to become part of it based on mathematics principles [13,42], as a technological way of speaking to biology toward developing biotechnology. It requires a framework such as the following methodological discourse.

3 Some Methodological Discourses

Conceptually, a framework for a scientific field is the first proposal in a definition that covers the study and describes its characteristics to develop alternatives [43]. The mentioned definition is the foundation of the development of the science of biology [44]. Therefore, that framework states biotechnology (biotech) as follows [45]:

D1 "*Biotechnology is the use of biology to develop new products, methods and organisms intended to improve human and societal health.*"[1]

A definition that expresses the existence of an object of study, in the form of biology [46], is written as $\mathcal{B}$, or

D2 "*Biology is the study of life and living organisms. That is a science including their structure, function, growth, evolution, distribution, and taxonomy.*"[2]

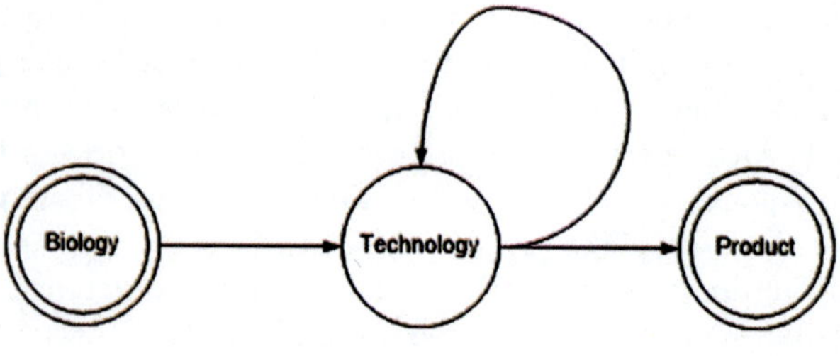

Fig. 1. Relation between biology, technology, and products in a model

The definition is a limitation of the scope of the discussion [47]. The word "use" in the D1 above reveals the existence of technology that stretches to the D2, a framework, and scientific activity with various studies that lead to the emergence of novelties [48,49]. Adaptation to the interests of biology to produce innovation indirectly states that there is an interface between biology and technology, namely a new technology or novelty itself, which explains itself through a definition [50].

[1] https://www.techtarget.com/whatis/definition/biotechnology.
[2] https://www.livescience.com/44549-what-is-biology.html.

D3 "*Technology is the product of scientific knowledge for practical purposes or applications, whether in industry or in our everyday lives.*"[3]

Therefore, the terminology "use of biology" is a transformation τ to produce a set of outputs $\mathcal{L}$, whether it is a new something of product, method, organization, or technology. In a formulation

$$\tau : \mathcal{B} \to \mathcal{L}, \tag{1}$$

or D1 = D2 + D3. So the transformation of τ is not one way outward but also towards the technology itself. In this case, τ is recursive Fig. 1 formally describes it.

Figure 1 and the formulation of the Eq. (1) is a model that forms a framework for developing the science of biotechnology, where studies that describe the characteristics of biology, technology, and outcomes of τ provide challenges to each about the framework of the biology, technology, or product demanding one of [51–57]:

1. Approach [58,59]: A scientific framework consists of a series of patterns or organizations of actions based on principles directed systematically toward achieving goals. Usually, it involves philosophy.
2. Adaptation [60–62]: A scientific framework aims to overcome challenges to carrying out an activity scientifically by generating an interpretation of the characteristics of an object. Generally, it involves learning like artificial intelligence or computational intelligence.
3. Modeling [63,64]: A scientific framework aims to create a model of the system. Usually, it involves math, like optimization, simulation, statistics, or mining.
4. Formalization [65,66]: A scientific framework conceptually arranged in the field of knowledge inquiry. In general, it involves formal theory [67].

All frameworks are to drive teaching or research mechanisms in theory-based or experimental laboratories. In general, to obtain descriptions ranging from the characteristics in the axioms to the propositions.

4 A Model of Computation and Discussion

Today, discussing any object of study in biology means that discuss the object of study in biotechnology. This interpretation is to get a stepping stone for developing it or at least a model for computation. Philosophically, it intends to describe the characteristics of the object. Each object has properties, features, details, and so on, in general, can refer to as characteristics. That is, in a set as something that humans express from their thoughts to their coding as an example, cases, or scenario for supporting the modeling of computation [68,69].

[3] https://study.com/academy/lesson/what-is-technology-definition-types.html.

4.1 Example as a Case in a Review

For example, in biology, or the case: the food, there is a plant object called rice, and each type of rice has its features [24,25]. Rice seeds are seeds fulfilling the quantity and quality of food must be complete and have a measurement of the weight of each seed in the form of milligrams, or the number of seeds in 100 g on average (a case of computation). The method and duration of germination have the potential to affect subsequent growth. Germination method - the initial technology of growing rice - or germination time depends on the different ways or technology directly or indirectly affect it. That is how or how long to cultivate rice [70]. The during (a technology) of growing rice will affect how long it takes for rice to produce (harvest) where effectiveness and efficiency are the target sizes [71]. The size of the rice weevil per x rice seed grains may require a balance with root length and rice height after planting until harvest [72,73].

4.2 General Modeling as an Abstraction in Review

General modeling of the description above is possible. Any plant object O of a different kind, say $k = 1, 2, \ldots, K$, forms a characteristic description of $l = 1, 2, \ldots, L$, which arrange in a two-dimensional array in a table or matrix, where t_{ij} are any measurement appropriate to the type and characteristics [74,75],

$$O = \begin{pmatrix} t_{11} \; t_{12} \cdots t_{1k} \\ t_{21} \; t_{22} \cdots t_{2j} \\ \cdots\cdots\cdots\cdots \\ t_{l1} \; t_{l2} \cdots t_{lk} \end{pmatrix} \tag{2}$$

Although each technology is for all types, technological development continuously carries out, where each technology has fixed characteristics that are not the same as one another, each technology that differs in the arrangement of columns, $t = 1, 2, \ldots, T$, has features, $p = 1, \ldots, P$, are parsed row by row from a table or matrix.

4.3 A Scenario as a Review for Generating the Model

Each plant species will grow in a suitable environment [59,71]. Likewise, technological capabilities provide adaptation services for each environmental characteristic of any species. Therefore, the natural environment has properties, for example, seasons, rainfall, sunlight, wind, nutrients (soil structure), and others. The environment has the potential to provide a technology that seeks to adapt any plant to either different species or different habitats. For example, fertilizer treats different plants differently in places, so does temperature control technology affect the appearance of flowers and fruit for plants. Nature and plants provide their challenges for any changes that occur. Therefore, to recondition the challenge or better known as simulating natural conditions for plant objects, it is necessary to have technology that provides treatment [76]. In addition, plants also need protection from enemies who are always lurking. Enemies from roots to fruits of plants, such as rice, have many enemies, both from other plants and animals. Rice will not grow well when there are many weeds in the fields, as animals, such as mollusks, can kill rice, while birds will attack when the rice begins to bear fruit. Therefore, to prevent rice enemies from becoming rampant, technology is needed. Technology may be toxic, which can kill or repel plant enemies. It may also be natural conditioning that interferes with the enemy's life, such as involving frequencies or using predatory animals such as ducks against mollusks or tobacco plants to kill mollusks naturally [77]. All of these plant prevention and protection tools are the results of studies in the form of products, technology, or other organisms, that are ready to be applied when needed. This technology is an outcome of research in biotechnology that will affect production costs according to the appropriate choice [78]. Likewise, technology does not always succeed in providing treatment if it does not change the nature of the plant. Genetic engineering has been widely carried out on plants ranging from changes in the gene structure of seeds to the fruit they produce, including adaptation to the growing environment [79].

4.4 A Formula as Brief Review Toward the Steps for Computing

As a review [80]: With a wide selection of plant objects that produce harvested fruit to meet human food needs, based on consideration of the amount of crop production, production costs, storage costs, transportation costs, market absorption, and others. If the selling price of U for each type of plant varies based on need and taste, formulation

$$u_1 o_1 + u_2 o_2 + \cdots + u_k o_k \ \{<, =, >\} \ w \tag{3}$$

gives the meaning of the number of sales obtained, where $U = \{u_1, u_2, \ldots, u_K\}$, $O = \{o_1, o_2, \ldots, o_K\}$, and w as the market demand [81]. Production formulations do not always stand alone but have boundaries by constraints, such as land boundaries, where the amount of crop production depends on the square unit of land. For example, for each type of plant, there is an amount of production of $A = \{a_1, a_2, \ldots, a_K\}$, and there are constraint formulations $a_1 o_1, a_2 o_2, \ldots, a_k o_k$. If the computation considering the profit other than the formula (3) is given a maximum target, then the additional constraints on production costs, from planting to harvesting, must be added $B = \{b_1, b_2, \ldots, b_k\}$, where the constraint has formulation as follows $\sum_{k=1}^{K} b_k o_k$. Other constraints are possible depending on the plant object in this discussion, and this C constraint forms a two-dimensional array by involving in calculations with Eq. (3),

$$C = \begin{pmatrix} c_{11} & c_{12} & \cdots & c_{1k} \\ c_{21} & c_{22} & \cdots & c_{2k} \\ \cdots & \cdots & \cdots & \cdots \\ c_{l1} & c_{l2} & \cdots & c_{lk} \end{pmatrix} \tag{4}$$

The final formalization is a conclusion from what has already been. A way to model computations used in matters relating to biology. This way is one of what is talked about in biotechnology according to D1 for generating computational model [82–84], i.e.

1. Select plant objects based needed by humans according to possible alternative criteria, choice of criteria, and multi-criteria considerations.
2. Choose technological objects that increase the quantity and quality of the results of the plant under the potential alternative technologies and multi-technology considerations.
3. Selecting useful outcome products according to potential alternative objectives and multi-objective considerations.

5 Conclusion

The computational model of biotechnology is to develop a scientific field called biotechnology. The model has based on the description of biological and technological objects into characteristics that become choices, turn into a collection of symbols, and are linked to one another through formulations. The computational model also has an implementation in three options: biology, technology, and product. Likewise, further studies are needed to determine formulations that have the potential to reveal the computational side of biotechnology.

References

1. Wong, P.K.: Basic needs of human beings as the purposes and foundations for the existence of governing institutions and the advancement of science and technology. In: American Society of Mechanical Engineers, Technology and Society Division (Publication) TS, vol. 6, pp. 35–40 (2001)
2. Lange, M.: What would natural laws in the life sciences be? In: Kampourakis, K. (eds.) The Philosophy of Biology. History, Philosophy and Theory of the Life Sciences, vol. 1, pp. 67–85. Springer, Dordrecht (2013). https://doi.org/10.1007/978-94-007-6537-5_4
3. Turczynski, C., Dodd, A., Urlakis, M.A.: Assisted reproductive technology and natural law: how seven years as an embryologist revealed IVF's disordered approach to patient care. Linacre Q. (2022, in press). https://doi.org/10.1177/00243639221128393
4. Riley, S.: Subsidiarity and human dignity: democracy, authority and knowledge. In: Human Dignity and Democracy in Europe: Synergies, Tensions and Crises, pp. 60–76 (2022)
5. Nasution, M.K.M.: World on data perspective. World **3**(3), 736–752 (2022). https://doi.org/10.3390/world3030041
6. Bolyen, E., et al.: Reproducible, interactive, scalable and extensible microbiome data science using QIIME 2. Nat. Biotechnol. **37**(8), 852–857 (2019). https://doi.org/10.1038/s41587-019-0209-9
7. Nasution, M.K.M., Syah, R., Elveny, M.: What is data science. Data science with semantic technologies: a new trend and future development, pp. 1–16. https://doi.org/10.1201/9781003310785-1

8. Nasution, M.K.M.: Ontology. J. Phys. Conf. Ser. **1116**(2), 022030 (2018). https://doi.org/10.1088/1742-6596/1116/2/022030
9. Nijhout, H.F., Emlen, D.J.: Developmental biology, evolution competition among body parts in the development and evolution of insect morphology. Proc. Natl. Acad. Sci. U.S.A. **95**(7), 3685–3689 (1998). https://doi.org/10.1073/pnas.95.7.3685
10. Butt, U.M.: A deep insight into signature verification using deep neural network. In: Advances in Intelligent Systems and Computing 1252 AISC, pp. 128–138. https://doi.org/10.1007/978-3-030-55190-2_10
11. Razeghian, E.: A deep insight into CRISPR/Cas9 application in CAR-T cell-based tumor immunotherapies. Stem Cell Res. Ther. **12**(1) (2021). https://doi.org/10.1186/s13287-021-02510-7
12. Joseph, J., Iftekhar, S., Srivastava, V., Fallah, Z., Zare, E.N., Sillanp, M.: Iron-based metal-organic framework: synthesis, structure and current technologies for water reclamation with deep insight into framework integrity. Chemosphere **284**, 131171 (2021). https://doi.org/10.1016/j.chemosphere.2021.131171
13. Chupradit, S.: Various types of electrochemical biosensors for leukemia detection and therapeutic approaches. Anal. Biochem. **654**, 114736 (2022). https://doi.org/10.1016/j.ab.2022.114736
14. Lubis, A.R., Nasution, M.K.M., Sitompul, O.S., Zamzami, E.M.: spelling checking with deep learning model in analysis of Tweet data for word classification process. In: International Conference on Electrical Engineering, Computer Science and Informatics (EECSI), pp. 343–348 (2022). https://doi.org/10.23919/EECSI56542.2022.9946476
15. Hasan, M.Z., Hanapi, Z.M., Hussain, M.Z., Hussin, M., Sarwar, N., Akhlaqi, M.Y.: Deep insight into IoT-enabled agriculture and network protocols. Wirel. Commun. Mob. Comput. (2022). https://doi.org/10.1155/2022/5617903
16. Zurbriggen, M.D., Moor, A., Weber, W.: Plant and bacterial systems biology as platform for plant synthetic bio(techno)logy. J. Biotechnol. **160**(1–2), 80–90 (2012). https://doi.org/10.1016/j.jbiotec.2012.01.014
17. Meyer, V., Nai, C., Idnurm, A.: Openness and visibility of fungal bio(Techno)logy. Fungal Biol. Biotechnol. **4**(1), 1–3 (2017). https://doi.org/10.1186/s40694-017-0038-x
18. Tebeje, A., Tadesse, H., Mengesha, Y.: Synthetic bio/techno/logy and its application. Biotechnol. Biotechnol. Equip. **35**(1), 1156–1162 (2021). https://doi.org/10.1080/13102818.2021.1960189
19. Ramachandran, K.B., Hussain, M.A.: Special theme issue: bioprocess and environmental biotechnology research. Dev. Chem. Eng. Min. Proc. **13**(5–6), 517–518 (2005). https://doi.org/10.1002/apj.5500130501
20. Reardon, S.: The geometry of life: when mathematics meets synthetic biology. Nature (2022, in press). https://doi.org/10.1038/d41586-022-02176-y
21. Womack, R. P.: Research data in core journals in biology, chemistry, mathematics, and physics. PLoS ONE **10**(12), 0143460 (2015). https://doi.org/10.1371/journal.pone.0143460
22. Cooper, G., Berry, A.: Demographic predictors of senior secondary participation in biology, physics, chemistry and earth/space sciences: students' access to cultural, social and science capital. Int. J. Sci. Educ. **42**(1), 151–166 (2020). https://doi.org/10.1080/09500693.2019.1708510

23. Kennedy, B., Buikema, A., James, J.K.S.N.: Integrating biology, design, and engineering for sustainable innovation. In: ISEC 2015–5th IEEE Integrated STEM Education Conference, pp. 88–93 (2015). https://doi.org/10.1109/ISECon.2015.7119952
24. Abu Bakar, U.K., Pillai, V., Hashim, M., Daud, H.M.: Sharing Malaysian experience with the development of biotechnology-derived food crops. Food Nutr. Bull. **26**(4), 432–435 (2005)
25. Amin, L., et al.: Ethical perception of modern biotechnology with special focus on genetically modified food among Muslims in Malaysia. Asia Pac. J. Mol. Biol. Biotechnol. **18**(3), 359–367 (2013)
26. Show, P.L.: Special issue on biotechnology for sustainability and social well being. Processes **9**(2), 1–3 (2021). https://doi.org/10.3390/pr9020216
27. Saha, R.: Systems and synthetic biology of photosynthetic organisms for biorenewable chemicals. In: Education Division 2015 - Core Programming Area at the 2015 AIChE Annual Meeting, **93** (2015)
28. Cahalan, S.M.: Chemical and genetic tools to explore s1p biology. Curr. Top. Microbiol. Immunol. **378**, 55–83 (2014). https://doi.org/10.1007/978-3-319-05879-5_3
29. Hamzah, A., Ibrahim, N.: Microbiology students performance at school of biosciences and biotechnology. Procedia Soc. Behav. Sci. **18**, 301–306 (2011). https://doi.org/10.1016/j.sbspro.2011.05.043
30. Amin, L.: Social aspects of biotechnology. Afr. J. Biotechnol. **10**(58) (2011)
31. Nasution, M.K.M.: Social Network Mining (SNM): a definition of relation between the resources and SNA. Int. J. Adv. Sci. Eng. Inf. Technol. **6**(6), 975–981 (2016). https://doi.org/10.18517/ijaseit.6.6.1390
32. Nasution, M.K.M.: A method to prove the existence of a similarity. In: Advances in Intelligent Systems and Computing (AISC), vol. **1295**, pp. 243–252 (2020). https://doi.org/10.1007/978-3-030-63319-6_21
33. Hedrick, P.W.: A new approach to measuring genetic similarity. Evolution **25**(2), 276–280 (1971). https://doi.org/10.2307/2406918
34. Paun, F.: The demand readiness level scale as new proposed tool to hybridise market pull with technology push approaches in technology transfer practices. Technology Transfer in a Global Economy, 353–366 (2012). https://doi.org/10.1007/978-1-4614-6102-9_18
35. Moorhouse, D.J.: Detailed definitions and guidance for application of technology readiness levels. J. Aircr. **39**(1), 190–192 (2002)
36. Janardhan, B.S.: Promising achievements and new challenges in agriculture biotechnology. Curr. Sci. **93**(8), 1052–1054 (2007)
37. Nasution, M.K.M.: Research serendipity: a challenge for changing. J. Phys. Conf. Ser. **1542**, 012072 (2019). https://doi.org/10.1088/1742-6596/1542/1/012072
38. Nasution, M.K.M., Nuradi, T.E., Syah, R.: SumutSiana: a framework for applying ICT to preserve the cultural heritage of Sumatera Utara Indonesia. J. Telecommun. Electron. Comput. Eng. (JTEC) **9**(2–4), 123128 (2017)
39. Abuduxike, G., Aljunid, S.M.: Health biotechnology in Malaysia: issues and challenges faced by the innovative biotechnology firms. Innovation in Developing and Transition Countries, pp. 131–155 (2017)
40. Smith, M.K., Ko, H.L., Hamill, S.D., Sanewski, G.M., Graham, M.W.: Biotechnology, he pineapple: botany, production and uses (2003). https://doi.org/10.1079/9780851995038.00
41. Clark, D.P., Pazdernik, N.J.: Biotechnology, 2nd edn. Elsevier, Amsterdam (2016)

42. Bahri, N.M., Suryawati, E., Osman, K.: Students' biotechnology literacy: the pillars of STEM education in Malaysia. Eurasia J. Math. Sci. Technol. Educ. **10**(3), 195–207 (2014). https://doi.org/10.12973/eurasia.2014.1074a
43. Fidel, R.: Qualitative methods in informati data science: a reviewon retrieval researach. LISR **15**, 219–247 (1993)
44. Lusch, R.F., Vargo, S.L., Wessels, G.: Toward a conceptual foundation for service science: contributions from service-dominant logic. IBM Syst. J. **47**(1), 5–14 (2008). https://doi.org/10.1147/sj.471.0005
45. Roco, M.C., Bainbridge, W.W.: overview converging technologies for improving human performance. converging technologies for improving human performance, 1–27 (2003). https://doi.org/10.1007/978-94-017-0359-8_1
46. Green, J.L., Bohannan, B.J.M., Whitaker, R.J.: Microbial Biogeography: from taxonomy to traits¿ Science **320**(5879), 1039–1043 (2008). https://doi.org/10.1126/science.1153
47. Nasution, M.K.M., Sawaluddin: Redefining the magic square on numerical characters, IOP Conf. Ser. Materi. Sci. Eng. **308**, 012033 (2017). https://doi.org/10.1088/1757-899X/308/1/012033
48. Wahab, R.A., et al.: Article; food biotechnology enzymatic production of a solvent-free menthyl butyrate via response surface methodology catalyzed by a novel thermostable lipase from Geobacillus Zalihae. Biotechnol. Biotechnol. Equip. **28**(6), 1065–1072. https://doi.org/10.1080/13102818.2014.978220
49. Nasution, M.K.M.: Methodolgy. J. Phys.: Conf. Ser. **1566**, 012031 (2019). https://doi.org/10.1088/1742-6596/1566/1/012031
50. Meyer, G.G., Främling, K., Holmström, J.: Intelligent products: a survey. Comput. Ind. **60**(3), 137–148 (2009). https://doi.org/10.1016/j.compind.2008.12.005
51. Kossmann, J.: Grand challenges in plant biotechnology. Frontiers in Plant Science 3(APR) (2012). https://doi.org/10.3389/fpls.2012.00061
52. Khaldari, M., Ghiasei, H.: Challenges and opportunities of biotechnology in animal production of developing countries. Genetics Third Millennium **12**(4), 3884–3899 (2015). ISSN 24237159
53. Bernardo-Álvarez, M.Á.: Biotechnology and communication in democratic societies old challenges for a new era. Metode **2019**(9), 17–21 (2019). https://doi.org/10.7203/metode.9.10388
54. Toscano, B.M.: The Use of Biotechnology in Food: Challenges, Problems and New Perspectives. Food Secur. Iss. Challenges 331–353 (2021). ISBN 978-168507131-8
55. Al-Delaimy, A.K.: Perspective of ethical issues and challenges of biotechnology in developing country. Malays. J. Public Health Med. **22**(1), 242–251 (2022). https://doi.org/10.37268/MJPHM/VOL.22/NO.1/ART.1449
56. Melzer, R., McCabe, P.F., Schilling, S.: Evolution, genetics and biochemistry of plant cannabinoid synthesis: a challenge for biotechnology in the years ahead. Curr. Opin. Biotechnol. **75**, 102684 (2022). https://doi.org/10.1016/j.copbio.2022.102684
57. Sharma, A., et al.: Prospects of chitinase in sustainable farming and modern biotechnology: an update on recent progress and challenges. Biotechnol. Genet. Eng. Rev. Article in Press (2023). https://doi.org/10.1080/02648725.2023.2183593
58. Syah, R.: A laboratory approach to measure enhanced gas recovery from a tight gas reservoir during supercritical carbon dioxide injection. Sustainability **13**(21) (2021). https://doi.org/10.3390/su132111606
59. Tan, Y.C., Kumar, A.U., Wong, Y.P., Ling, A.P.K.: Bioinformatics approaches and applications in plant biotechnology. J. Genet. Eng. Biotechnol. **20**(1) (2022). https://doi.org/10.1186/s43141-022-00394-5

60. Elveny, M., Syah, R., Elfida, M., Nasution, M.K.M.: Information retrieval on social network: an adaptive proof. IOP Conf. Seri. Mater. Sci.Eng. **300**, 012055 (2018). https://doi.org/10.1088/1757-899X/300/1/012055
61. Musa, H., Kasim, F.H., Nagoor Gunny, A.A., Gopinath, S.C.B.: Salt-adapted moulds and yeasts: potentials in industrial and environmental biotechnology. Process Biochem. **69**, 33–44 (2018). https://doi.org/10.1016/j.procbio.2018.03.026
62. Nasution, M.K.M.: Industry 4.0. IOP Conf. Ser. Mater. Sci. Eng. **1003**, 012145 (2020). https://doi.org/10.1088/1757-899X/1003/1/012145
63. Nasution, M.K.M.: Modelling and simulation of search engine. J. Phys. Conf. Ser. **801**, 012078 (2016). https://doi.org/10.1088/1742-6596/801/1/012078
64. Yoo, D.-I., Chavas, J.-P.: Dynamic modeling of biotechnology adoption with individual versus social learning: an application to US corn farmers. Agribusiness **39**(1), 148–166 (2022). https://doi.org/10.1002/agr.21772
65. McKelvey, M., Alm, H., Riccaboni, M.: Does co-location matter for formal knowledge collaboration in the Swedish biotechnology-pharmaceutical sector? Res. Policy **32**(3), 483–501 (2003). https://doi.org/10.1016/S0048-7333(02)00020-3
66. Nasution, M.K.M.: The uncertainty: a history in Mathematics. J. Phys. Conf. Ser. **1116**(2), 022031 (2018). https://doi.org/10.1088/1742-6596/1116/2/022031
67. Friedkin, N.E.: A formal theory of social power. J. Math. Soc. **12**(2), 103–126 (1986). https://doi.org/10.1080/0022250X.1986.9990008
68. Kemp, M., Walker, S., Scott, P.: Coding of thoughts, words and things. HIM J. **34**(2), 54–56 (2005). https://doi.org/10.1177/183335830503400207
69. Bérard, S., Chateau, A., Chauve, C., Paul, C., Tannier, E.: Computation of perfect DCJ rearrangement scenarios with linear and circular chromosomes. J. Comput. Biol. **16**(10), 1287–1309 (2009). https://doi.org/10.1089/cmb.2009.0088
70. Jalil, M.T.M., Ibrahim, D.: Effect ethyl acetate extract of lasiodiplodia pseudotheobromae industrial biotechnology research laboratory OS-64 on growth and cell morphology of a foodborne bacterium, Bacillus cereus. Thai J. Pharm. Sci. **45**(6), 545–551 (2021)
71. Awasthi, M.K.: Microbial biotechnology approaches for conversion of pineapple waste in to emerging source of healthy food for sustainable environment. Int. J. Food Microbiol. **373**, 109714 (2022). https://doi.org/10.1016/j.ijfoodmicro.2022.109714
72. Kadhimi, A.A., et al.: Use of biotechnology to improve the tolerance in rice (Oryza sativa) to drought stress. J. Pure Appl. Microbiol. **8**(5), 4001–4010 (2014)
73. Somasundram, C., Sagran, A., Razali, Z., San, R.O.P.: From low technology to biotechnology: practical solutions to indigenous communities' postharvest needs. Acta Horticulturae **1088**, 273–276 (2015). https://doi.org/10.17660/ActaHortic.2015.1088.44
74. Stephenson, F.H.: Calculations for Molecular Biology and Biotechnology: A Guide to Mathematics in the Laboratory. Calculations for Molecular Biology and Biotechnology: A Guide to Mathematics in the Laboratory, 1–458 (2010). https://doi.org/10.1016/C2009-0-01985-0
75. Abayeva, N.F., Yegorov, V.V., Golovachyova, V.N., Mustafina, L.M., Yerakhtina, I.I., Mustafina, B.M.: About professional orientation of the mathematics as a discipline for students majoring in biotechnology. Indian J. Sci. Technol. **9**(19), 93891 (2016). https://doi.org/10.17485/ijst/2016/v9i19/93891

76. Teoh, T.P., Ong, S.A., Wong, Y.S., Ho, L.N., Noor, N.M., Matei, M.: Constructed wetland-microbial fuel: biotechnology for simultaneous wastewater treatment and electricity generation. In: Mohamed Noor, N., Sam, S.T., Abdul Kadir, A. (eds) Proceedings of the 3rd International Conference on Green Environmental Engineering and Technology. Lecture Notes in Civil Engineering, vol. 214, pp. 379–384 (2022). https://doi.org/10.1007/978-981-16-7920-9_45
77. Nasution, R.M.W., et al.: NIMO - Natural Molluscicide. Poster - Shafiyyatul Amaliyyah(2021)
78. Loan, L.W., Kassim, M.A., Muda, K., Kheang, L.S., Abdullah, N.: Performance evaluation of Industrial Effluent Treatment Systems (IETSs) - An insight for biotechnology advances in agro-based wastewater treatment. World Appl. Sci. J. **30**(30 A), 316–325 (2014). https://doi.org/10.5829/idosi.wasj.2014.30.icmrp.45
79. Maran, S., Kannan, T.P., Sasongko, T.H.: Bioinformatics and biotechnology in human genetic research: a current scenario. Biotechnology and Bioinformatics: Advances and Applications for Bioenergy, Bioremediation and Biopharmaceutical Research, 459–488 (2014). https://doi.org/10.1201/b17104
80. Kamyab, H., Lee, C.T., Chelliapan, S., Khademi, T., Talaiekhozani, A., Rezania, S.: Role of microalgal biotechnology in environmental sustainability-a mini review. Chem. Eng. Trans. **72**, 451–456 (2019). https://doi.org/10.3303/CET197207
81. Loy, A.C.M., Lim, J.Y., How, B.S., Yoo, C.K.: Blockchain as a frontier in biotechnology and bioenergy applications. Trends Biotechnol. **40**(3), 255–258 (2022). https://doi.org/10.1016/j.tibtech.2021.09.006
82. Nasution, M.K.M.: Multi-criteria as decisions. IOP Conf. Ser. Mater. Sci. Eng. **1003**, 012118 (2020). https://doi.org/10.1088/1757-899X/1003/1/012118
83. asution, M.K.M., Nasution, R.M.W., Syah, R., Elveny, M.: Biotechnology among computer science and data science: a review of scientific development. In: Silhavy, R., Silhavy, P., Prokopova, Z. (eds.) Data Science and Algorithms in Systems. CoMeSySo 2022. Lecture Notes in Networks and Systems, vol 597. Springer, Cham (2023). https://doi.org/10.1007/978-3-031-21438-7_76
84. Hartmann, F.S.F., Udugama, I.A., Seibold, G.M., Sugiyama, H., Gernaey, K.V.: Digital models in biotechnology: towards multi-scale integration and implementation. Biotechnol. Adv. **60**, 108015 (2022). https://doi.org/10.1016/j.biotechadv.2022.108015

An Empirical Analysis of the Switching and Continued Use of Mobile Computing Applications: A Structural Equation Model

Alfred Thaga Kgopa(✉), Raymond Kekwaletswe, and Agnieta Pretorius

Department of Informatics, Tshwane University of Technology, Pretoria, South Africa
{KgopaAT,PretoriusAB1}@tut.ac.za

Abstract. It has been determined that there is insufficient explanation for people switching and continuously using mobile computing applications. Knowing and being able to explain this behavior was regarded as essential and might be useful to both application developers and researchers, particularly in forecasting future behavior. The objective of this study was to model the determinants influencing switching and continued use of mobile computing applications using Cronbach's alpha and confirmatory factor analysis. Data were collected from academics in South African universities through a survey using structured questionnaires. Cronbach's alpha and confirmatory factor analysis were used for reliability and validity testing [19]. Five decision variables were supported as significant with Cronbach's alpha and confirmatory factor analysis. Model significance and strength were assessed using SEM to estimate the model's coefficients [8]. The variance of the dependent variable, "continue use," was 38%, and switching behavior was 32% as a contribution to this study.

Keywords: Academics · continued use · switching behavior · mobile computing applications

1 Introduction

The current study was prompted by a lack of reasons for people's switching and continuous usage of mobile computing applications and services. The current study was influenced by the lack of research models that can be used to predict switching and continued usage of mobile computing applications.

The insufficiency or absence of frameworks or models that explain academics' switching and continuous usage of mobile computing applications and services is the theoretical knowledge gap. The continued usage and switching behavior of mobile computing applications are not fully described for practitioners and developers of mobile computing apps. Application developers who want to forecast future switching or continued usage must understand individual switching and continuing usage behavior [4, 23]. This is the first study to look into the switching of mobile computing applications, with academics from South Africa as participants.

R. Silhavy and P. Silhavy (Eds.): CSOC 2023, LNNS 723, pp. 134–148, 2023.
https://doi.org/10.1007/978-3-031-35317-8_13

Switching behaviour and continued use can estimate the projected life cycle of existing mobile computing applications [24]. Companies that create and sell mobile computing apps, for example, must maintain strong consumer loyalty by favourable continuing usage. Most previous technology usage studies focused on factors that influenced early adoption and often did little to study the factors that encouraged continued use. This phenomenon prompted several Information Systems (IS) researchers to investigate a deeper understanding of post-adoption intention and technical conduct [3, 4]. The key component of post adoption studies argues that customer understanding of a product may be switch to another product once prospective consumers first use the product. Several authors [14] write to clarify this position, stating that the various characteristics of a technology in use may be opposed, ignored, used infrequently, or routinized.

According to [10], most related research does not use the combination of constructs that drive the current investigation. The previous related study concentrated on hypothesis testing [10]. This current study was vital because it modeled the relevant factors that influence continued use and switching behavior through the use of Cronbach's alpha, confirmatory factor analysis, and the structural equation model (SEM) [8]. In addition, the variance explained for the contribution of this study was explained based on the dependent or latent variables "continue use" and "switching behavior."

2 Literature Review

According to Tran [25], mobile applications are "those mobile applications developed by companies in order to build the brand." The brand name and logo are frequently displayed in these mobile applications. The goal of mobile applications is to connect with present and potential customers [25]. The amount of 14 personalisation of mobile advertising (such as mobile applications) has a beneficial impact on customer sentiments regarding mobile advertising [25]. Consequently, the interaction of the mobile application can influence the consumer's attitude toward the use of mobile applications [17]. According to Tran [25], mobile applications provide a valuable utility to the consumer while also creating an emotional connection with the consumer. However, one concern has been raised about the negative impact of privacy concerns on consumers' attitudes toward mobile applications [17]. The rise in smartphone sales has contributed to the rise in popularity of mobile applications.

There several of studies conducted to investigate switching or continued use of mobile computing services. Some of those studies have demonstrated a positive link between the functional value of mobile computing services and user adoption of such services. The existing protocol is required for the study of such a phenomenon in order to perform an empirical assessment of mobile computing application switching and continued use. In the study of [20], the systematic approach to the adoption and use of technology is followed in two phases: the first focuses on acceptance and intention to use, and the second on post-adoption: continuation behavior and outcome. The first phase of acceptance examines the consumer's ability or intent to use a product during an opening period [20]. The second stage of post-adoption (continuous purpose, conduct, and performance effects) examines aggregate expectations and experiences over time over initial use.

Users of the mobile money transfer application called M-Pesa are more likely to stick with it if certain criteria are taken into account [6]. The research of Osah and Kyobe [16] set out to develop a theoretically grounded conceptual framework, ii) recognize factors from the literature that are most likely to affect user performance expectancy towards M-pesa, and iii) test the model within different sampling contexts in order to determine the dimensions and factors of user performance expectancy towards M-pesa in Kenya To create the questionnaire survey used to evaluate the study's variables, literature research, an expert who was before, pilot testing, and statistical verification were utilized. Users' persistent intentions are influenced by elements such as object, control, attitude, and behavioral beliefs, according to the study. As a result of the study's surprising findings, new research pathways are now available for future research [16].

[34] Developed and tested a theoretical switching intention model with a data collected on switching intentions. Switching intention behaviour investigated which included relationship of alternative attractiveness, perceived worth, and the switching costs. The actual switching behaviour data was then comparable to the theoretical switching intentions, and results were discussed. Both contexts of switching are examined to determine the role of interpersonal traits. [34] Used online questionnaires for obtaining primary data that was collected from a cross-sectional forum. The participants were people who had or have a mobile device contracts with a mobile network operator companies. In AMOS, feature estimates were obtained using maximum likelihood, and confidence intervals were calculated using bootstrapping. Researchers found that switching intent and switching behaviour are significantly different.

Reviewing mobile application development literature, just a limited number of studies concentrate on the switching or continued use of mobile applications [4, 21–24, 27]. Most existing studies, especially those in developing countries have been focused on adoption, readiness, security and implementations of mobile applications [2, 7, 11, 14, 17, 26, 28]. Therefore, it was very important for the present study to incorporate the theory of switching and continued use behaviour.

3 Data Collection Methods

The data gathering method is used to provide evidence for investigations (Leedy & Ormrod, 2005). A quantitative sample was gathered to help with the prediction and endorse the results obtained from the sample of academic consumers. A questionnaire was used as a source of data to enable an effective understanding of the sustainable use of mobile computing applications in the socio-cultural field of academics. The questionnaire was designed using Survey Monkey, and the link was distributed via the workgroup's email addresses, WhatsApp, LinkedIn and Instagram. The study first shared the survey link with known academics, who in turn shared the link with their colleagues. This method of referential selection, called "snowballing," would enable the study to get as many participants as necessary for statistical purposes.

To analyse the collected data, the study focused on the data received via the questionnaires and used descriptive and inferential statistical analysis methods, using SPSS. The data was analysed for validity based on exploratory factor analysis to discover the principal component matrix factors using varimax methods. Cronbach's alpha test for the constructs for reliability was carried out in this study [18]. Cronbach's alpha assessed the internal consistency of the constructs [18]. To assess causality, a Structural Equation Modeling (SEM) and multiple regression analysis test were used [18].

4 Demography Profile Data

The demographic data from the 220 eventual academic participants. The participants were asked to accept or decline taking part in the survey in the first section of the questionnaire. A total of 216 academics responded to the survey and accepted it, while 4 others did not select accept but nonetheless took part. The check box was on the initial page that featured user consent information, so it's possible that the four participants who took part without hitting the accept button did not see it.

As regards to gender, there were 114 males who participated, which is representing (51.8%), and 106 were females, (48.2%). The age group with most participants was 36–45 years represented by 79 (35.9%) participants, followed by age 25–35 years (22.7%), participants of age below 25 years were 39 (17.7%) and the fourth place was age category 46–55 years with 33 (15%) participants being the older academic who are above 55 years were 19 (8.6%).

Frequency of academic specialisation indicate that majority of the participants (63) were from the Information and Communication Technology (28.6%), followed by 47 participants from Humanities (21.4%), 27 participants are from Education (12.3%) while 18 specialized in Commerce and Management (18.2%). There are also 16 participants who specialize in Management Sciences (7.3%), 14 Engineering (6.4%), 12 Social Sciences (5.5%). 9 participants were from Agriculture and natural resources (4.1%), 6 from Health and Medical Sciences (2.7%), 5 from Law (2.3%), the least was 2 from Fine and Performing Arts (0.9%). 1 participant select "others" which is only (0.5%) of the sample.

5 Formulation of Hypothesis

The following hypotheses were developed to guide the study and data analysis for the groups:

5.1 Technological Features

The research is about providing technical and administrative assistance to academic end users who want to adopt mobile computing apps. According to [3], the most critical variables influencing user's mobile application adoption intentions are a lack of administrative assistance and a lack of mobile computing technologists. If a group of academics from various disciplines wanted to collaborate online on a project, or a health practitioner wanted to consult an expert through video conference to operate on a patient, they would require a lot of technological help [9]. Academics will choose to adopt alternative dependable solutions if the technology parts of their organization are not reliable [25]. This leads to the following hypotheses:

H1: *Technological features have positive influence towards mobile computing applications switching behavior.*

H2: *Technological features have positive influence towards continued use of mobile computing applications.*

5.2 Task Features

The study focuses on academics' capacity to accomplish tasks on a mobile computing application on their own. The task characteristics state that a person's usage of technology is contingent on the technology being customized to the needs of a specific task and its capacity to aid the consumer in completing the job [9]. A misalignment between the mobile application's features and the user's ability to complete a job will diminish task fit. The mismatch has a detrimental impact on the person's ability to do the activity at hand, as well as poor technology views [9].

H3: *Task features have influence towards mobile computing applications switching behaviour.*

H4: *Task features have influence towards continued use of mobile computing applications.*

5.3 Individual Characteristics

The research focuses on academic end users' capacity to use mobile computing applications. Individuals have played a significant part in the research of IT use and adoption. In a critical literature analysis, [25] identifies four key features of IT usage: IT use, IT use behaviour, IT use procedures, and time. This IT usage element includes the consumer, IT equipment, and job activities [25].

H5: *Individual features have influence towards of mobile computing applications switching behaviour.*

H6: *Individual features have influence towards continued use of mobile computing applications.*

5.4 Satisfaction

The research focuses on whether academics' expectations or demands for new mobile computing applications are met, as well as the enjoyment obtained from utilizing existing mobile computing apps [20].

H7: *Satisfaction have influence towards mobile computing applications switching behaviour.*

H8: *Satisfaction have influence towards continued use of mobile computing applications.*

5.5 Perceived Usefulness

According to academics' perspectives, perceived usefulness may be defined as the benefits consumers want while using mobile computing apps, and this includes rewarding, trust, application quality, and service quality [16, 27].

***H9:** Perceived Usefulness have influence towards mobile computing applications switching behaviour.*

***H10:** Perceived Usefulness have influence towards continued use of mobile computing applications.*

5.6 Switching Cost

According to the study, there are both financial and noneconomic barriers preventing instructors from using smart phone apps. The charges spent by a client while switching apps or service providers are known as switching costs. Although monetary value is the most common cost of moving, there are other psychological costs based on work and time [21, 23].

***H11:** Switching costs have influence towards continued use of mobile computing applications.*

***H12:** Switching costs have influence towards mobile computing applications switching behaviour.*

5.7 Social Norm

Students' opinions on whether or not their significant others would prefer that they transfer systems or stay with the one they have. They are described using the social-norms-construct, which is defined as the degree to which these individuals in their life believe it is acceptable for them to engage in a certain behavior [12, 16].

***H13:** Social norm have influence towards mobile computing applications switching behaviour.*

***H14:** Social norm have influence towards continued use of mobile computing applications.*

6 Data Presentation and Analysis

The degree of correlation between the constructs and other measures that have been expected in theory to correlate with them was determined using discriminant and convergent validity in this study [18]. Furthermore, to see if these constructs are relating with other variables that are supposed to be unrelated to them [2]. Cronbach's alpha was used to assess the reliability of variables.

6.1 Reliability of the Constructs

The constructs were assessed for reliability before being tested for discriminant convergent validity and correlation. Cronbach's alpha was used to assess the construct internal consistency [18]. As indicated in Table 1, the remaining values of the constructs were kept were above 0.7, indicating that their adjusted item-total correlation is significant [18]. Cronbach's alpha values for all items vary from 0.809 to 0.888, with an overall internal consistency reliability of 0.850 for all Likert scale items, according to the data.

6.2 Convergent and Discriminant Validity and Correlation

The proportion of variance for each construct that is linked is defined as convergent validity [28]. Convergent validity is determined by computing the composite or construct reliability (CR) and variance extracted (VE) for each construct indicator. Composite reliability and the average VE (AVE) were employed to test the convergent validities [28]. While discriminant validity was assessed by looking at whether the squared root of AVE surpasses the correlations across constructs, reliability was assessed by looking at the internal consistency reliability.

All factors in the measurement model had acceptable composite reliability and convergent validity because all factor loadings (the indicators' standardised loadings [λ]) are significant ($p < 0.001$), the composite or construct reliabilities exceed acceptable criteria of 0.7, and the AVEs in all cases were greater than the threshold value of 0.5 [10]. The diagonal items that represent the square roots of AVE and off-diagonal elements that represent correlations between constructs were deemed significant. All diagonal elements were higher than off-diagonal elements in the respective columns and rows, and internal consistency reliability was above 0.7, indicating discriminant validity and reliability.

6.3 Structural Equation Modelling

There are five logical steps in Structural Equation Modelling (SEM): model specification, identification, parameter estimation, model evaluation, and model modification [8]. Model specification defines the hypothesized relationships among the variables in an SEM based on one's knowledge. Model identification is to check if the model is over-identified, just identified, or under-identified. Figure 1 show the developed structural equation modelling followed by steps followed for validity and reliability processes.

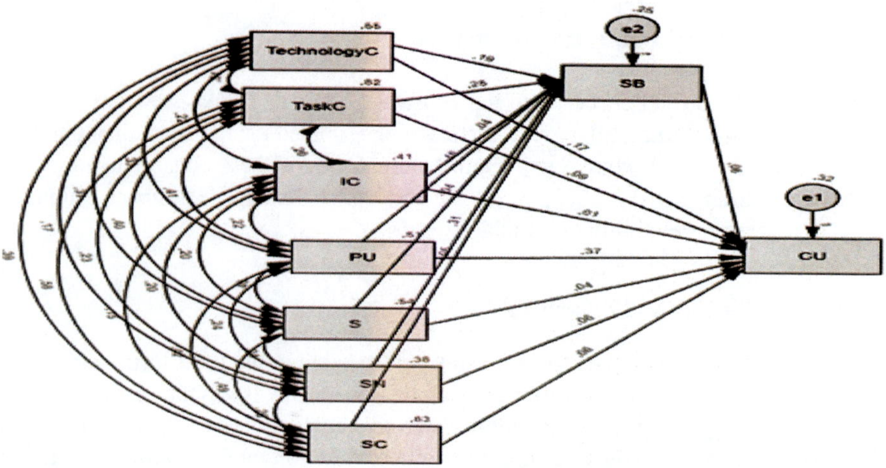

Fig. 1. Structural equation modelling.

The critical ratio values of the causality between the hypothesized constructs prompt that their causality are significant except that of hypothesis H5, H7, H11, H15, H3, H6, H8, H13 and H12. This implies that the anticipated hypothesis H5, H7, H11, H15, H3, H6, H8, H13 and H12 are not supported. The following hypotheses were supported H9, H2, H4, H14, H1 and H10 with regards to this study as shown in Table 1.

Table 1. Extracted standardised significance levels of the structural model.

Relationship			Hypotheses	Estimate	S.E.	C.R.	P	Results
SB	<---	PU	H9	0,484	0,081	5,952	***	Supported
SB	<---	Technology	H2	0,19	0,077	2,477	0,013	Supported
SB	<---	Task	H4	0,254	0,067	3,805	***	Supported
SB	<---	IC	H5	0,043	0,067	0,635	0,526	Not Supported
SB	<---	S	H7	0,041	0,082	0,494	0,622	Not Supported
SB	<---	SN	H14	0,308	0,071	4,331	***	Supported
SB	<---	SC	H11	0,049	0,063	0,774	0,439	Not Supported
CU	<---	Technology	H1	0,172	0,089	1,935	0,053	Not Supported
CU	<---	Task	H3	0,092	0,079	1,169	0,243	Not Supported
CU	<---	IC	H6	0,007	0,076	0,097	0,922	Not Supported
CU	<---	PU	H10	0,369	0,1	3,693	***	Supported
CU	<---	S	H8	0,044	0,094	0,47	0,638	Not Supported
CU	<---	SN	H13	0,062	0,084	0,732	0,464	Not Supported
CU	<---	SC	H12	0,06	0,072	0,833	0,405	Not Supported

Table 2. Squared multiple correlations

Estimate	
SB	0.30
CU	0.39

Multiple correlation coefficients squared for the two multiple regressions are shown in Table 2. Continued use and switching behaviour both have a squared multiple correlation coefficient of 0.30 and 0.39. This indicates that the variable has a strong relationship with latent constructs discovered by the other variables [19]. For the latent variables switching behaviour and continued use, squared multiple correlation coefficients suggest that roughly 39% of the variability is captured in this model.

The covariance of the independent variables in this study was all significant. This implies that multiple regression analysis might have bias with regards to covariance of the decision variables. This study when evaluating the causality of the dependent variables with independent variable covariance issues were considered as shown in Table 3.

Table 3. Covariance

Relationship			Estimate	S.E	C.R	P
PU	<-->	IC	0,223	0,035	6,458	***
IC	<-->	S	0,198	0,035	5,719	***
IC	<-->	SN	0,197	0,03	6,587	***
IC	<-->	SC	0,155	0,041	3,792	***
Task	<-->	IC	0,197	0,042	4,735	***
Technology	<-->	IC	0,221	0,035	6,242	***
PU	<-->	SN	0,24	0,034	7,056	***
PU	<-->	SC	0,419	0,052	8,014	***
PU	<-->	Task	0,406	0,052	7,832	***
PU	<-->	Technology	0,33	0,042	7,81	***
S	<-->	SC	0,487	0,056	8,711	***
Task	<-->	S	0,404	0,053	7,647	***
Technology	<-->	S	0,311	0,042	7,323	***
Task	<-->	SN	0,233	0,041	5,691	***
Technology	<-->	SN	0,174	0,033	5,265	***
Task	<-->	SC	0,58	0,068	8,515	***
Technology	<-->	SC	0,39	0,053	7,418	***
Technology	<-->	Task	0,515	0,057	8,997	***
PU	<-->	S	0,404	0,045	8,998	***
S	<-->	SN	0,244	0,035	7,007	***
SN	<-->	SC	0,264	0,042	6,313	***

6.4 Interpretation of Model Fit Indices

The initial run of the model gave the chi-square value of 36.234 with degrees of freedom d.f = 15 giving a ratio of chi-square / degree of freedom of 2.4156 / 15 = 2.4156 at p = 0. 000 significant at $p \leq 0.01$. Much as the ratio chi-square / degree of freedom was within the acceptable range as shown in Table 2, though, the probability value of less than 0.05 recommends that, rejecting the null hypothesis that the model has a good fit mighty give a minimum error [19]. As debated by several experts, these measurements must be compared with other fit indices before the model would be used. The output was; GFI = 1.000; CFI = 1.000 and RMSEA = 0.040. The CFI fitting index was below the threshold and the RMSEA was above the threshold as illustrated in Table 4.

Table 4. Interpretation of model fit indices

Fit indices	Measurement model	Threshold	Recommendations for the measurement model	Values of the model
X^2	36.234	Ratio $2.1 \leq (X^2/d.f) \leq 3.1$	X^2 /d.f is within the range of threshold	36.234
D.f	15			15
X^2 /d.f	15.943			2.4156
RMSEA	0.189	$0.05 \leq (RMSEA) \leq 0.080$	Less than the threshold, shows the model is good	0.04
CFI	0.882	≥ 0.950	More than the threshold, suggests model is good	1
GFI	0.926	≥ 0.90	More than the threshold, shows the model is good	1

7 Discussion of Results

7.1 Technological Features and Continued Use Behavior of Academics

H1: *Technological features have positive influence towards mobile computing applications switching behavior (MCA).*

The current study found that the technological qualities of mobile computing applications had a beneficial impact on South African academics' switching behavior. The theory was shown to be correct. Technological factors had a favorable impact on MCA switching behavior, as predicted by the research model. Users' switching behavior is impacted by the application's technical features [5]. The findings of this study agree with [15], who discovered that a specialist using video conferencing to operate a patient requires a significant amount of technological support to achieve their goal if they are not technologically oriented, and that academics require a similar amount of support. The findings of this study show that technical factors have a favorable and significant impact on South African academics' MCA switching behavior.

H2: *Technological features have positive influence towards continued use of mobile computing applications.*

However, the findings of this study did not show that technological qualities had a favorable impact on academics' continuous usage of MCA in South Africa. This theory was found to be false. According to the study model, technological qualities may have a favorable impact on South African academics' continuous usage of MCA. However, technical factors influencing the continuous usage of MCA in South Africa have a negative impact on academics. This conclusion, on the other hand, contradicts Task Technology Fit hypothesis.

7.2 Task Features and Continued Use Behavior of Academics

H3: *Task features have positive influence towards mobile computing applications switching behaviour.*

The findings revealed that task factors influence MCA switching behavior in a favorable way. This theory received a lot of support. The findings support the study model's forecast, which is in line with the Task Technology Fit model's proposal. The findings of hypothesis H3 are in line with those of [12], who found that task qualities are essential in deciding whether or not technology is acceptable. Similarly, the switching behavior of South African academics is influenced by task factors.

H4: *Task features have pôsitive influence towards continued use of mobile computing applications.*

The findings did not support the hypothesis that task factors had a beneficial impact on MCA usage. This hypothesis was not supported, contradicting the study model's prediction.

The findings of hypothesis H4 contradict [12], who believe that task qualities are crucial in deciding whether or not technology is appropriate. Similarly, academic activities and qualities have little bearing on whether or not South African academics continue to use a computing program.

7.3 Individual Features and Continued Use Behavior of Academics

H5: *Individual features have influence towards of mobile computing applications switching behaviour.*

The study found no evidence that an individual's attributes influence MCA switching behavior in a good way. The theory was shown to be false. The Task Technology Fit hypothesis does not explain this finding. Users' adoption and usage of technology is not affected by an individual's traits, according to [23]· The ability and efficacy (individual qualities) of a person to use mobile computing apps is unimportant, and hence has little bearing on the academic's switching behavior.

H6: *Individual features have influence towards continued use of mobile computing applications.*

Individual qualities do not appear to have a beneficial impact on the sustained use of MCA among South African academics, according to the findings of this study. This theory was found to be false. This finding contradicts the Task Technology Fit paradigm, which claims that an individual's qualities play a role in deciding whether or not a technology is appropriate for them.

7.4 Satisfaction Features and Continued Use Behavior of Academics

H7: *Satisfaction have influence towards mobile computing applications switching behaviour.*

The study found little evidence that contentment had a favorable impact on academics' MCA switching behavior in South Africa. The theory was shown to be false. Satisfaction traits do not have a beneficial impact on academic switching behavior, as indicated by the research model. Users' usage of technology is determined by satisfaction factors, according to [22]. As a result, neither academic expectations, requirements for new mobile computing apps, nor the enjoyment obtained from utilizing existing mobile computing applications influenced mobile application switching behavior.

H8: *Satisfaction have influence towards continued use of mobile computing applications.*

Satisfaction qualities do not have a favorable relationship on sustained usage behavior among South African academics, contrary to what the study model expected. The Information System Continuance model does not support this finding. The qualities of user happiness affect whether or not they continue to utilize technology [25, 26]. This means that even if academics are happy with the app's features, they are unlikely to continue using it.

7.5 Perceived Usefulness Features and Continued Use Behavior of Academics

H9: *Perceived Usefulness have influence towards mobile computing applications switching behaviour.*

The study found that perceived utility qualities had a beneficial impact on South African academics' MCA switching behavior. The theory was shown to be correct. That is, perceived utility qualities have a positive impact on MCA switching behavior, as anticipated by the research model. This outcome is in line with the Switching Intention model's predictions. Users' switching behavior is determined by perceived utility qualities, according to [29]. As a result, it can be deduced that academics' perceptions of the utility of new mobile computing apps are influenced by trust, application quality, and service quality [30].

H10: *Perceived Usefulness have influence towards continued use of mobile computing applications.*

The findings of this study show that perceived utility qualities have a beneficial impact on academics' continuous usage of MCA in South Africa. This theory was found to be correct. Users' continuing usage of a technology is determined by perceived utility qualities, according to [20]. Academics' perceptions of the use of existing mobile computing applications are another important factor that influences their continuing use.

7.6 Switching Costs Features and Continued Use Behavior of Academics

H11: *Switching costs have influence towards continued use of mobile computing applications.*

The study found no evidence that switching fees had a favorable impact on South African academics' MCA switching behavior. This theory was found to be false. This finding contradicts the continuing use and switching intention theories for IS. Users' technology switching behavior is typically affected by cost features, according to [29]. Because the switching process may entail complicated psychological, emotional, and behavioral concerns, as well as the need to overcome some switching costs, the consumer switching behavior process may necessitate a significant lot of work and deliberation on the part of [34].

H12: *Switching costs have influence towards mobile computing applications switching behaviour.*

The study found no evidence that switching fees had a favourable impact on South African academics' ongoing usage of MCA. This theory was found to be false. This finding contradicts the IS ongoing usage and switching intention models once again. Academics' continued usage behaviour was not influenced by the expenditures connected with the use. This means that academics do not consider the cost of using an application in comparison to the cost of utilizing a new application. It is obvious that the bulk of them can afford and learn quickly, therefore wasting time on an application that does not suit them is a waste of time.

7.7 Social Norm Features and Continued Use Behavior of MCA

H13: *Social norm have influence towards mobile computing applications switching behaviour.*

The study found that social norm factors had a beneficial impact on South African academics' MCA switching behavior. The research model projected that the hypothesis will be supported. Users' acceptance of technology may be influenced by social norms, according to [3]. This suggests that social norms may have an impact on academics' decision to use a new mobile computing app. Academics are autonomous in the sense that they can make their own decisions regarding which mobile application to use. They have their own culture and conventions that they follow while making professional decisions.

H14: *Social norm have influence towards continued use of mobile computing applications.*

The study found no evidence that social norm factors had a favorable impact on MCA usage among South African academics. As expected by the study model, the hypothesis was not validated. This finding contradicts [1], who argue that social norms may influence how people utilize technology. This suggests that social norms have little effect on academics' decision to continue using a mobile computing app. As indicated in the introduction, the rationale for selecting academics for this study is because their behavior of choice may differ as significant elements influencing the decision to continue utilizing various mobile computing apps.

8 Conclusion

Individual academics' switching to and sustained usage of mobile computing applications is explained in this study. Although there has been a surge in the usage of mobile computing applications, research on what drives end-users to switch and continued usage is still in its infancy. Switching behaviour is a key aspect in the post-adoption of newly produced mobile computing apps, and the desire to use the application in the future is a determining factor in actual usage. Exploratory factor analysis results discover the principal component matrix factors that influenced switching and continued behavior. The study used Cronbach's alpha to assess the internal consistency of the constructs. Structural equation model and multiple regression analysis test were also used to evaluate the causality. The contribution of this study towards theory is that it merged the appropriate information systems literature in order to enhance what is known regarding the use of mobile computing applications. This means that the study brought out the factors that explain why individuals switch between mobile applications and why they may continue using mobile computing applications. By triangulating three theoretical frameworks, the study showed how the theories of Task Technology Fit, IS Continuance and Switching Intention model could best explain the behaviour of individuals towards switching and continued use of mobile computing applications.

References

1. Abbas, H.A., Hamdy, H.I.: Determinants of continuance intention factor in Kuwait communication market: case study of Zain-Kuwait. Comput. Hum. Behav. **49**, 648–657 (2015)
2. Ajzen, I., Fishbein, M.: A Bayesian analysis of attribution processes. Psychol. Bull. **82**(2), 261 (1975)
3. Andrade, A.D., Doolin, B.: Information and communication technology and the social inclusion of refugees. MIS Q. **40**(2), 405–416 (2016)
4. Bhattacherjee, A., Lin, C.P.: A unified model of IT continuance: three complementary perspectives and crossover effects. Eur. J. Inf. Syst. (2014)
5. Chen, W., Hirschheim, R.: A paradigmatic and methodological examination of information systems research from 1991 to 2001. Inf. Syst. J. **14**(3), 197–235 (2004)
6. Creswell, J.W.: Qualitative Inquiry & Research Design: Choosing Among Five Approaches. 3rd edn. Sage, Los Angeles (2013)
7. Ebner, M.: Mobile applications for math education–how should they be done. In: Mobile Learning and Mathematics. Foundations, Design, and Case Studies, pp. 20–32 (2015)
8. Fan, Y., et al.: Applications of structural equation modelling (SEM) in ecological studies: an updated review. Ecol. Process. **5**(1), 1–12 (2016)
9. Goodhue, D.L.: Understanding user evaluations of information systems. Manage. Sci. **41**(12), 1827–1844 (1995)
10. Kgopa, A.T., Kekwaletswe, R.M., Pretorius, A.: A model for switching and continued use of mobile computing applications amongst South African academics'. South African J. Inf. Manage. **24**(1), a1470 (2022). https://doi.org/10.4102/sajim.v24i1.1470
11. Kim, H.J., Lee, J.M., Rha, J.Y.: Understanding the role of user resistance on mobile learning usage among university students. Comput. Educ. **113**, 108–118 (2017)
12. Middleton, C., Scheepers, R., Tuunainen, V.K.: When mobile is the norm: researching mobile information systems and mobility as post-adoption phenomena. Eur. J. Inf. Syst. **23**(5), 503–512 (2014)
13. National Communications Authority: Quarterly Statistical Bulletin on Communications in Ghana **2**(3), 211 (2017)
14. Ndayizigamiye, P., Kante, M., Shingwenyana, S.: An adoption model of mHealth applications that promote physical activity. Cogent Psychol. **7**(1), 1764703 (2020)
15. Nel, J., Boshoff, C.: Cell phone banking adoption and the continuance of use in an internet-banking context: a study of consumer cognitive evaluations, PHD, Department of Business Management, Marketing Management: University of Pretoria (2013)
16. Osah, O., Kyobe, M.: User continuance intention towards mobile technology-enabled services in developing regions, PHD, Department of Information System: University of Cape Town (2015)
17. Osakwe, J.O., Iyawa, M.U.G., Florich, K.: Enabling quality education in Namibia through mobile learning technologies. The high school teachers' perspective. In: Proceedings IST-Africa Week Conference (IST-Africa), pp. 1–9) (2018)

18. Pallant, J.: SPSS Survival Manual: A Step by Step Guide to Data Analysis using IBM SPSS. Routledge (2020)
19. Pomykalski, J.J., Dion, P., Brock, J.L.: A structural equation model for predicting business student performance. J. Educ. Bus. **83**(3), 159–164 (2008)
20. Purnami, L.D., Agus, A.A.: The effect of perceived value and mobile game loyalty on in-app purchase intention in mobile game in Indonesia (case study: mobile legend and love nikki). ASEAN Mark. J. **12**(1), 9–19 (2020)
21. Ray, S., Kim, S.S., Morris, J.G.: Research note-online users' switching costs: their nature and formation. Inf. Syst. Res. **23**, 197–213 (2012)
22. Rouibah, K.: Switching factors in mobile service providers: qualitative study from an Arab country. In: International Conference on Transformations and Innovations in Management (ICTIM-17) (2017)
23. Salo, M., Makkonen, M.: Why do users switch mobile applications? Trialing behavior as a predecessor of switching behavior. Commun. Assoc. Inf. Syst. **42**, 386–407 (2018)
24. Subhadin, R.: App adoption and switching behavior: applying the extended TAM in smart phone App adoption and switching behaviour. J. Inf. Syst. Technol. Manage. **14**, 239–261 (2017). https://www.scielo.br/pdf/jistm/v14n2/1807-1775-jistm-14-02-00239.pdf. Accessed 2 Jun 2019
25. Tran, V.D.: Using mobile food delivery applications during the COVID-19 pandemic: Applying the theory of planned behavior to examine continuance behavior. Sustainability **13**(21), 12066 (2021)
26. Tutubala, N., Mathonsi, T.E.: A hybrid framework to improve data security in cloud computing. In: 2021 Big Data, Knowledge and Control Systems Engineering (BdKCSE), pp. 1–5. IEEE (2021)
27. Van der Merwe, M.C.: A comparison between switching intension and switching behavior in the South African mobile telecommunications industry, DCom Marketing Management, University of Pretoria (2015)
28. Yavuz, M., Çorbacioğlu, E., Başoğlu, A.N., Daim, T.U., Shaygan, A.: Augmented reality technology adoption: case of a mobile application in Turkey. Technol. Soc. **66**, 101598 (2021)

The Effect of Internet Threat on Small Businesses in South Africa

Awosejo Oluwaseun Johnson(✉), Agnieta Pretorius, and Tendani Lavhengwa

Department of Informatics, Soshanguve South Campus, Tshwane University of Technology Soshanguve, Gauteng Pretoria, South Africa
awosejooluwaseun@gmail.com

Abstract. This study cultivates security threats in small businesses as well as operations to improve business strategies, cyber-attacks, and rapidly changing digital performance and is still widely known. However, the study investigates socioeconomic factors, security breaches, technical factors that could cause failure, and factors that affect internet-based security in business and proposes a way out. The study investigates 387 small businesses, and a set of 253 (65.4%) of the respondents reported that their business has been in operation between 1 and 10 years, 46 (11.9%) said it has been in existence between 11 and 20 years, and 2.6%) confirmed that their business has been in operation between 21 and 30 years and 31 and 40 years, respectively. Data analysis was done utilizing techniques such as binary logistic regression analysis and factor analysis as part of the study. The study used small businesses (SMEs) operating in South Africa and developed a quantitative approach. The results revealed that the use of the internet adequately supports small businesses and significantly supports the electronic sustainable growth of SMEs, which is correlated with completing tasks quickly, accomplishing the goals and objectives of the business, and implementing sustainable computer software and network architecture. The cloud computing services that cloud computing provides could handle large storage of data in the electronics business.

Keywords: Electronics Business · Internet of Things · Security and Threat · Small Business · South Africa

1 Introduction

Rapid worldwide company change is progressively transforming it into a unique industry that requires network providers to offer consumers a utility network [18]. The foundational elements of Internet commerce worldwide are to be determined by network providers and virtual businesses [17]. However, this is primarily related to point-of-sale processing and security, which are better methods of managing digital businesses to provide suppliers and customers with the quickest access possible via the Internet, due to its speed and adaptability [24]. The latest advancement in business motivates organizations to keep responding to information systems and business applications, and global information technology investment is necessary for South Africa [8]. It is very important to

R. Silhavy and P. Silhavy (Eds.): CSOC 2023, LNNS 723, pp. 149–168, 2023.
https://doi.org/10.1007/978-3-031-35317-8_14

protect user identity, sensitive data sets, and business information in a safe and effective manner [15]. In recent years, the adoption of business innovation has increased competition among businesses. The amount of time small businesses operating in the various areas of the City of Tshwane spend online has dramatically grown over the past several years [20]. Most small electronics businesses frequently do business online. They are not sufficiently shielded from internet-based risks, though [2, 3]. Although internet usage has increased dramatically over the past several decades, online dangers have also increased during that time [9]. The internet has substantially improved the climate for SMME growth [2, 3]. The internet has made it possible for regular people to do commerce and freely share information [18]. People are drawn to web apps, and online communication and online companies appreciate the usefulness of the Internet [22]. People are now able to make quick judgments and carry out profitable commercial transactions thanks to it. E-commerce may now readily supply services and products to clients and engage with them thanks to the Internet [25]. According to [28], approximately half of all newly founded small businesses and SMMEs in South Africa fail within the first 3.5 years of operation due to unexpected network topology [20], and [29], both reported findings of a similar kind [9]. According to [30] and [20] research, newly established SMMEs in and around Tshwane are characterized by a lack of entrepreneurial skills and difficulty obtaining from open systems interconnection. To support online transactions, SMMEs must develop their capabilities [19]. Newly created SMMEs in Tshwane have a low survival rate; thus, there must be a greater understanding of how to utilize the Internet wisely among these businesses [25]. In newly founded SMMEs around the city of Tshwane, there are significant internet-based security vulnerabilities [9].

2 Literature Review

Malware is the most typical means by which computers are harmed [2]. It enables internet users to view inappropriate files and access inappropriate websites [3]. However, viruses, worms, keyloggers, and spyware Trojans are examples of harmful software [4]. Small businesses may be penetrated and attacked by threats because of their inability to recognize and address threats to internet security development [4]. In South Africa, electronic business conducted over a computer network sometimes exposes users to internet threats, include ing phishing and spear phishing, online fraud, corrupted websites, and denial of service attacks [5]. In today's business, many insurance companies, courier companies, restaurants, clothing stores, and hospitality businesses are primarily commercials navigating through social networks [17] and able to attract consumers and make payments through electronic systems for cash machines that are vulnerable to wireless internet vulnerabilities [6]. Although, at the time, keeping customers' payment methods on the network regardless of their location [2, 3] may have made them vulnerable to internet fraud [3]. Due to the threat of the worst form of criminality known as cybercrime, the internet and computer technology make life easy and quick but also put people in danger [10]. However, this is the reason that most small businesses fail. South Africa has witnessed an increase in usage of the internet over the past 10 years, and the capacity of usage has enabled small businesses to survive and gain a competitive advantage in recent years. But the importance of trust is seen as a key factor when conducting business online [31]. Nevertheless, another key aspect to be considered under internet capacity and threat in small business is "phishing" and "spear phishing," which relate to the fraudulent use of login user information, such as passwords or personal information, while impersonating a trustworthy organization, such as a bank [13]. The purpose of spear phishing is to use fictitious identities to obtain computer data on a company computer [17]. By appearing to originate from a service provider that the company works with, for example, it increases perceived credibility. To a smaller group of recipients as opposed to a large one, spear phishing emails are sent. The major challenge facing small businesses in South Africa today is the hacking of internet protocol (IP) addresses and email compromise [11]. Any user who uses the internet for work is suffering from a digital phobia. There are several reasons for security issues [12]. However, the IP address on the system has been compromised, and the sender is secured using the above methodology [13]. Security modules might include network security, data security, IP security, and more [14]. The work that has been provided can be expanded in this area because it is possible to measure the effectiveness and complexity of IP addresses [15].

According to Awosejo and Edward Lovin, executive directors of the ANCOM, the Internet's transition from a local area network to a wide area network is increasing interest among small businesses in South Africa in users using their mobile phones to access the Internet [6]. However, nearly all SMME personnel in South Africa are familiar with business software aimed at managing computer system operations and ICT's capacity to increase enterprises' and, more specifically, SMEs', competitiveness [31]. The aim influences modern society, which now routinely makes use of new technology including the Internet, digital television, mobile phones, home appliances, and "smart" electronics but is vulnerable to economic factors, environmental factors, organizational factors, individual factors, and technological factors [10]. Compromised websites should be considered, and a website may be compromised and used to host illegal information without the site owners' awareness [7]. If a company's website cannot be trusted, it will inevitably be affected by a hacked website and suffer a loss of reputation [16]. However, sometimes in electronic business, attacks unknowingly undertaken as denial of service are undertaken against companies in the name of extortion, rivalry, protest, or retaliation [13]. Online extortion enables threats against customers' financial gain, denies them access to their information, and privately demands their gain [8]. However, most of the time, a business's reputation will suffer if its website cannot be trusted, and having a hacked website has an immediate negative impact on the business. In South Africa, many business owners are now familiar with Internet connectivity and can access computer networks remotely, which allows them to share knowledge more quickly [15]. Business owners and internet users have been warned to protect their information via the internet and demand privacy and honesty when they use the internet [1]. Internet capacity threats in digital business areas are increasing, whereas security breaches are a common phenomenon that can collapse a business because of significant data and financial losses [33]. Meanwhile, a security awareness campaign is essential to educate people about the risks and threats to internet capacity, as well as threats that may affect the electronics industry, to identify problems with the local area network and wide area network [2]. For instance, instructing your staff not to open emails from unknown senders may prevent malware from relying on victims to run an executed file to propagate and infect a computer or network [17]. They must be made aware that doing so jeopardizes all their business activities. However, small businesses benefit from policies on security-based applications. Such regulations might be utilized to provide electronics business owners with training opportunities about online security risks [33]. For the protection of corporate earnings and operational information, security threat information is essential. Policies outline the proper course of action for the firm as well as the method to be followed, all of which are supported by management information systems [8]. They must inform employees of the necessity to regulate and enforce a firm through domain policies. Government measures must be able to appropriately safeguard the regional economy and enterprises.

3 Literature Review

A review of academic research on the effect of internet threats on small businesses in South Africa has been developed, and some of the issues relate to online business. According to a study by [14], a percentage of SMEs experience failure in their operations. The study looks at firms in South Africa and how much exogenous macroenvironmental, variability, crime, unemployment, low product demand, incorrect pricing strategy implementation, and business location have an impact on South African SMEs Furthermore, the failure of SMEs was also influenced by inflation and lending rates. To provide small businesses in South Africa's economy with suitable security measures, the task is to develop the skills and experiences of SMEs. While the results are positive, a case study was not used to identify the problems faced by SMEs in South Africa when it comes to internet threats. [3] suggests small and medium suggests small and medium-sized businesses should take data protection seriously. A problem with small and medium-sized businesses' data protection was identified because of the study's investigation of the variables that have increased data protection risk [33]. Afterward, counsel them on scenarios where data causes decision-making to be delayed. The business operation may be temporarily halted as a result. To minimize the impact of this issue on small and medium-sized firms, the study decided to present some scientific recommendations and implement them. The study demonstrates that the concept is sound but that no quantitative approach to identifying the issue was used and that binary logistic regression analysis and factor analysis were used as data analyses. An earlier study [2], for example) looked at the expanding breadth of information security vulnerabilities in SMEs. The study addressed security measures to protect small businesses and revealed some of the consequences of security breaches on consumer information. According to the study's recommendations, SMEs need to know where to go for the right technology to safeguard their enterprises. To defend their firms, the study's findings suggest that SMEs train their employees in security. Although the study's findings are positive, they do not signal a significant change in the security or risks facing SMEs. According to research by [9], access to the Internet influences American households with small farms as well as their household income and expenses. According to the study, having access to the Internet has a significant favorable impact on off-farm income as well as total. Household income. According to the study's findings, an internet connection helps small farms by increasing business income, off-farm income, and overall household income.

It also looks at how internet users can lower some input costs and living expenses. The study found that households in the farming industry had mainly investigated the factors influencing Internet adoption, as well as patterns of Internet use and applications. In some cases, they also realized the importance of discussing the effects of communication and information technologies, such as the Internet, on large-scale farms. As a result, little is known about how Internet connectivity affects small farm enterprises and how it affects the welfare of small farm business households. The outcome demonstrates that the results are good but that the framework, binary logistic regression analysis, and factor analysis were unable to identify internet households.

4 A Proposed Theory for Managing Internet-Based Threats in Small Business

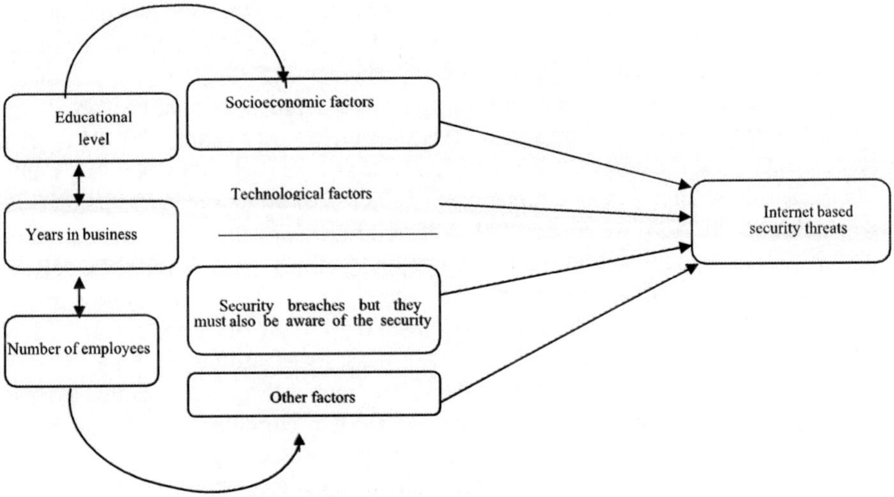

Fig. 1. A Conceptual Information System Framework

According to the conceptual framework, which was based on the philosophy of theoretical risk and threat in small business, the study suggests that respondents follow strategic plans that may advance sustainable security and threat. However, few of them disagreed with their counterparts over such plans. A greater percentage of respondents, 378 (94.8%), explained that security policy (documented information security policy, liaising with service providers, registering users on the systems, planning against a site disaster, information security policy, email, and internet conduct, for which someone is responsible for backup) can sustain security measures in the electronics business, while only a small number, 20 (5.2), of them, thought this plan was impractical. Organizational security, according to 364 (94.1) of the respondents, may significantly reduce internet-based threats in SMMEs (senior managers must approve access to our computer systems, and information security expertise is accessible inside). However, this idea was opposed by just 23 (5.9) of their co-workers 361 respondents (93.3%) concurred that asset classification and control (including software, hardware, controlling local and remote access to our information assets, and backup) were important, whereas 26 respondents (6.7%) disagreed. 362 (93.5%) of respondents suggested personnel security as a strategy to maintain security in SMMEs, while 25 (6.5%) disapproved of this proposal. In addition, a larger percentage of respondents (353 (91.2) recommended that physical and environmental security (theft, server maintenance, business premises, and liability through compromised data) be used to address and sustain security in small businesses, though 34 (8.8) of them chastised their co-workers for taking this approach to security threats in business. As a strategic strategy for long-term security in business, 345 (89.1) of the respondents selected communication and operation management (equipment failure, backup, theft, or site disaster), whereas 42 (10.9) of them reprimanded their colleagues for making this choice. Furthermore, more respondents (353 (91.2) recommended physical and environmental security to combat security threats in small businesses, even though 34 (8.8%) of them reprimanded their co-workers for making this recommendation. Environmental security, which includes theft, server maintenance, business premises, and liability from compromised data, was also recommended. Additionally, 42 (10.9) of the respondents reprimanded their colleagues for this choice, whereas 345 (89.1) of the respondents listed communication and operation management (equipment failure, backup, theft, or site disaster) as a strategic strategy for SME security. Once more, 329 (85.0) of the respondents offered Access Control (changing passwords every 2 weeks) as a solution to internet-based threats in electronics, even as 58 (15.0) of their business partners criticized this proposition. Again, 350 (90.4) of respondents assured their contemporaries that Systems Development and Maintenance (outsourcing from developers, software, server monitoring, and) can maintain e-commerce security, but 37 (9.6) disagreed. The assumption that business continuity management (in the case of a disaster) may serve as a means of sustainability security in small businesses was still put forth by 356 (92.0) of the entrepreneurs, while 31 (8.0) of the collaborators disagreed with this assertion. In conclusion, 346 (89.4) of the respondents said that compliance with international information security standards might have an impact on SMMEs' security, whereas 41 (10.6) of their partners disagreed with them on this point (Fig. 1).

5 Research Methodology

Data from a stratified random sample of 387 small enterprises in South Africa were collected for the study. A standardized questionnaire and in-depth interviews were used to collect the data. The sample size for the study was n = 387, and the sampling method utilized was stratified random sampling. The study was conducted using a quantitative method of data collection. The unit of analysis is a group of people who work in small businesses (SMEs) and use the Internet support system. To conduct the data analysis for the study's quantitative components, techniques such as crosstab analysis, binary logistic regression, and factor analysis were used. Based on the results of the statistical analysis, it can be deduced that the use the of internet adequately supports small businesses globally and suggest that internet security software will protect small business from a range of internet security threats which includes, hacking, virus, and malware.

6 Results of Data Analysis

6.1 Socio-demographic Variables

Table 1. Socio-demographic

Question	Frequency	Percentage
20 years or less	56	14.5
21 to 40 years	207	53.5
41 to 60 years	105	27.1
61 years or more	19	4.9

The socio-demographic characteristics table above indicates the respondents as follows: 56 (14.5%) fell between 20 years or below; 207 (53.5%) of the respondents fell between 21 to 40 years; 105 (27.1%) the respondents fell between the ages of 41 to 60 years, while 19 (4.9%) of the respondents were of the age category 61yrs or more (Table 1).

6.2 The Highest Level of Education

Table 2. Socio-demographic

Question	Frequency	Percentage
Primary level or less	5	1.3
Matric level	67	17.3
Certificate level	95	24.5
Diploma level	145	37.5
Bachelor's or Honours degree	56	14.5
Master's degree	10	2.6
Doctoral degree or above	9	2.3

The table above also shows that 5 (1.3%) of the total respondents have primary or less primary education, 67 (17.3%) have a primary education, 71 (30.5%) have matric level, 95(24.5%) have certificate level, 145(37.5%) possessed diploma certificate, 56(14.5%) of the respondentsm obtained bachelor's or Honors degree, 10(2.6%) acquired master's degree in their academic pursuit and 9 (2.3%) had Doctoral degree or above (Table 2).

Table 3. Socio-demographic

Years in operation	Frequency	Percent
1 to 10 years	253	65.4
11 to 20 years	46	11.9
21 to 30 years	10	2.6
31 to 40 years	10	2.6
41 to 50 years	7	1.8
51 to 60 years	5	1.3
61 to 70 years	1	0.3
No response	55	14.2
Total	387	100.0

6.3 The Years the Business

From the table above, research shows that the majority 253(65.4%) of the respondents reported that their business has been in operation between 1 to 10 years, 46(11.9%) said it has been in existence between 11 to 20 years, 10(2.6%) of the respondents affirmed that their business has been in operation between 21 to 30 years and 31 to 40 years respectively, 7(1.8%) between 41 to 50 years, 5(1.3%) between 51 to 60 years while just 1(0.3%) indicated 61 to 70 years that their business has been in use, 55(14.2%) could not specify the operational year of their business (Table 3).

6.4 How Many Employees Does Your Business Have at the Moment?

Table 4. Socio-demographic

Question	Frequency	Percentage
5 or fewer	178	46.0
6–10	100	25.8
11–25	39	10.1
26–35	32	8.3
36–50	6	1.6
51 or more	32	8.3

The table above shows the respondents' workforce as follows: 178 (46.0%) 5 or fewer, 100 (25.8%) 6–10 staff strength, 39(10.1%) 11–25 workers, 32(8.3%) 26–35 and 51 or more workforce, 6(1.6%) 36–50 staff members. The Report shows that a larger proportion of 178 (46.0%) of the respondents have 5 or fewer staff (Table 4).

6.5 How Large is Your Current Turnover in Millions of Rand?

Table 5. Socio-demographic

Question	Frequency	Percentage
50,000 Rand or less	77	19.9
51,000–80,000	37	9.6
81,000–90,000	24	6.2
91,000–100,000	17	4.4
101,000–200,000	36	9.3
201,000–500,000	112	28.9
500, 001 Rand or more	84	21.7

Respondents revealed their total annual turnover in millions of Rand as follows: 77(19.9%) 50,000 Rand or less, 37(9.6%) 51,000–80,000, 24(6.2%) 81,000–90,000, 17(4.4%) 91,000–100,000, 36(9.3%) 101,000–200,000, 112(28.9%) 201,000 500,000 and 84(21.7%) 500, 001 Rand or more. Results show that most of the respondents 112(28.9%) Earn 201,000–500,000 turnover in millions of Rand annually (Table 5).

6.6 How Many Computers Do You Use in Your Business?

Table 6. Socio-demographic

Computers	Frequency	Percent
1 to 20 computers	312	80.6
21 to 40 computers	20	5.2
41 to 60 computers	6	1.6
81 to 100 computers	4	1.0
101 to 200 computers	2	0.5
201 to 520 computers	2	0.5
No response	41	10.6
Total	387	100.0

Research exhibits the respondents' response regarding the number(s) of computers they have in their organization as follows: A good number 312 (80.6%) of the respondents have 1 to 20 computers, 20(5.2%) 21 to 40 computers, 6(1.6%) 41 to 60 computers, 4(1.0%) 81 to 100 computers, 2(0.5%) 101 to 200 computers, and 201 to 520 computers respectively, on the other hand, 41(10.6%) could not signify the number of computers their organization purchased. What are the most common internet-based security threats to electronics business operations conducted by small businesses in Tshwane of South Africa? (Table 6)

6.7 Socioeconomic Factors that Could Cause Failure in Small Businesses

Table 7. Fieldwork on Socioeconomic factor

Socioeconomic Factors	Yes	No
1. Environmental factors	356 (92.0)	31 (8.0)
2. Lack of business capital	354 (91.5)	33 (8.5)
3. Innovation, infrastructural upgrade, advertisement or procedures	332 (85.8)	55 (14.2)
4. Mismanagement of resources	363 (93.8)	24 (6.2)
5. Corruption	350 (90.4)	37 (9.6)
6. Global economic challenges such as inflation and recession	354 (91.5)	33 (8.5)
7. Wrong business policies	351 (90.7)	36 (9.3)
8. Adverse social or cultural factors	324 (83.7)	63 (16.3)
9. Lack of knowledge	358 (92.5)	29 (7.5)
10. Indigenous factors	330 (85.3)	57 (14.7)
11. Bad intention from role players and stakeholders	344 (88.9)	43 (11.1)
12. Wrong business location	335 (86.6)	52 (13.4)
13. Properties procedures and agreement	334 (86.3)	53 (13.7)
14. Government policies	319 (82.4)	68 (17.6)
15 Theft of data, information, hackers, or sabotage	342 (88.4)	45 (11.6)

(continued)

Table 7. (*continued*)

Socioeconomic Factors	Yes	No
16. Gender differences	268 (69.3)	119 (30.7)
17. Racial differences	272 (70.3)	115 (29.3)
18. Poor infrastructure	307 (79.3)	80 (20.7)
19. Natural disasters and other uncertainties	318 (82.2)	69 (17.8)

This table revealed the most common internet-based security threats to business operations conducted by SMMEs in Tshwane as was reported by the respondents; 356 (92.0) of the respondents specified Environmental factors as the major factor while 31 (8.0) of them count it to be untrue, again, the majority 354 (91.5) of the respondents regard Lack of business capital as the major threat to their business, on the other hand, 33 (8.5) of them viewed this state ment to be false. Once more, a larger proportion of 332 (85.8) of them affirmed that Innovation, infrastructural upgrades, and advertisement procedures posed a major threat to their business while 55 (14.2) proved this declaration to be wrong. More so, 363 (93.8) of the people under study said Mismanagement of resources has been the major source of threat to their business conversely, 24 (6.2) of the respondents considered it to be fallacious. Still, 350 (90.4) declared that Corruption stands as the most common internet-based security threat to business operations whereas, 37 (9.6) of them disprove this report to be incorrect. Almost all the respondents 354 (91.5) concurred those Global economic challenges such as inflation and recession posed major and common internet-based security threats to their firms, on the contrary, 33 (8.5), of them counted it not to be true. While 351 (90.7) of the respondents approved Wrong business policies to be the most common threat to their business, 36 (9.3) of the respondents fault this proclamation. Results show that a larger population of 324 (83.7) respondents admitted that Adverse social or cultural factors stand as the most common threat to their organization, and 63 (16.3) of them disagreed with this reality. Findings affirmed that almost all the respondents 358 (92.5) admitted that Lack of knowledge was pictured as the internet-based security threat to their corporation but 29 (7.5) of them viewed it as mere speculation. Information made available revealed that 330 (85.3) of the respondents consented that Indigenous factors serve as a common major threat to their business while 57 (14.7) of them were not in support of this proclamation. Findings affirmed that 344 (88.9) of the respondents consented that Bad intention from role players and stakeholders has been the most common threat to their corporations while 43 (11.1) of them contradicted the statement. Research also shows that 52 (13.4) of the respondents did not concur that the wrong business location is the most common threat to their establishment while on the other hand, while 335 (86.6) of the respondents accepted the statement. As regards Property procedures and agreements being the major threat to the respondents' business, results show that 334 (86.3) of them accepted the expression while 53 (13.7) of the respondents did not see it as the most common threat to their business. Furthermore, 319 (82.4) of the respondents recognized Government policies being the main internet-based threat to their establishment, and 68 (17.6) of them reported the statement to be false. The table further reported that 342 (88.4) of the respondents upheld that Theft of data, information, hackers, or sabotage are among the threat hampering the organization, and 45 (11.6) of the respondents did not see it as threatening tools aiding their business setback. Research affirmed that 268 (69.3) of the businessmen regarded Gender differences as one of the major common threats but an appreciable number 119 (30.7) of them rejected the utterance. Still, the report shows that 272 (70.3) of the respondents affirmed that Racial differences a pose major internet-based threat to their business while 115 (29.3) of them did not see it as a threat a, 307 (79.3) of the respondents regarded Poor infrastructure as the most common internet-based threat to their corporation whereas 80 (20.7) of them did not view it that way. While 318 (82.2) of the respondents admitted that Natural disasters and other uncertainties have been the major internet-based threat they were battling with, 69 (17.8) of them did not regard this as a major threat to their business (Table 7).

6.8 Technical Factors that Could Cause Failure in Electronics Small Businesses

Table 8. Fieldwork technical factors

	Yes	No
1. Accidental entry of bad data	356 (92.0)	31 (8.0)
2. Intentional entry of bad data	343 (88.6)	44 (11.4)
3. Accidental destruction of data	345 (89.1)	42 (10.9)
4. Intentional destruction of data	340 (87.9)	47 (12.1)
5. Unauthorized access to the data system	358 (92.5)	29 (7.5)
6. Unauthorized access to the data and or systems by outsiders (hackers)	358 (92.5)	29 (7.5)
7. Sharing of passwords by employees	348 (89.9)	39 (10.1)
8. Fire accidents, floods, or loss of power	345 (89.1)	42 (10.9)
9. Introduction of computer viruses to the system	346 (89.4)	41 (10.6)
10. Suppression or destruction of output	327 (84.5)	60 (15.5)
11. Creation of fictitious or incorrect output	333 (86.0)	54 (14.0)
12. Theft of data or information	347 (89.7)	40 (10.3)
13. Unauthorized copying of output	338 (87.3)	49 (12.7)
14. Unauthorized document visibility of output	344 (88.9)	43 (11.1)
15. Printing and distribution of information by an unauthorized person	340 (87.9)	47 (12.1)
16. Directing print and distributing information to un-entitled people	335 (86.6)	52 (13.4)
17. Shredding of sensitive documents by non-security people	331 (85.5)	56 (14.5)
18. Transmission of data from remote locations	321 (82.9)	66 (17.1)

The table above discusses some technical factors that could cause failure in Business Information Systems (BIS). Research divulged that almost all the respondents 356 (92.0) admitted that Accidental entry of bad data could cause the above-mentioned nemesis in BIS while only 31 (8.0) failed to support this fact. 343 (88.6) of the respondents agreed that the Intentional entry of bad data stands as the technical factor that could cause failure in Business Information Systems, on the other hand, 44 (11.4) of them disagreed with this statement. Still, 345 (89.1) of the respondents accepted that part of the technical factors that could cause failure in business Information systems is the Accidental destruction of data whereas 42 (10.9) of them counted this declaration as untrue. Also, 340 (87.9) of the respondents proclaimed that the Intentional destruction of data served as the technical factor that could cause failure in Business Information Systems while 47 (12.1) of them considered the utterance false. Larger proportion of the businessmen under investigation 358 (92.5) reported Unauthorized access to the data system as technical factors that could cause failure in Business Information Systems, in contrast 29 (7.5) of them disapproved this assertion. Still, 358 (92.5) of the respondents be of the same opinion that Unauthorized access to the data and or systems by outsiders (hackers) consisted parts of the technical factors that could cause failure in Business Information Systems con versely, 29 (7.5) of the respondents objected to this fact Again, 348 (89.9) of the respondents established that the technical factors that could cause failure in Business Information Systems is Sharing of passwords by employees but 39 (10.1) of them counted this statement baseless. Results show that a larger population 345 (89.1) of the business owners under investigation indicated Fire accidents, flood or loss of power serve as technical factors that could cause failure in Business Information Systems, on the contrary, 42 (10.9) of them were not of the support of the statement While 41 (10.6) of the respondents disagreed over the statement "Introduction of computer viruses to the system" being the technical factors that could cause failure in Business Information Systems, the majority of them 346 (89.4) affirmed the proclamation to be true. Again, research revealed that 327 (84.5) of the businessmen investigated established that Suppression or destruction of output stands as the technical factor that could cause failure in Business Information Systems (BIS) in contrast, 60 (15.5) of them argued out the statement to be correct. Discoveries asserted that 333 (86.0) of the respondents conceded that the Creation of fictitious or incorrect output remains a specialized element that could cause failure in Business Information Systems 54 (14.0) of them were not in support of the announcement. Bigger populace 347 (89.7) of the entrepreneurs under scrutiny showed the Theft of data or information filled in as specialized elements that could cause disappointment in Business Information Systems, despite what might be expected, 40 (10.3) of them were not of the support of the declaration. From the examination done, 338 (87.3) of the respondents approved that Unauthorized copying of output could fill in as stands as technical factors that could cause failure in Business Information Systems, then again, 49 (12.7) of them couldn't acknowledge the announcement to be valid. From the above table, a larger proportion of the respondents 344 (88.9) expressed their opinion towards Unauthorized document visibility of output stands as part of technical factors that could cause failure in Business Information Systems while 43 (11.1) of the respondents did not accept the utterance truthfully. Quite a higher pro-portion 340 (87.9) of the respondents of the same opinion that Printing and distribution of information by unauthorized persons stands as technical components that could cause disappointment in Business Information Systems while 47 (12.1) of the respondents could not consent to that proclamation. Furthermore, from the above table, 335 (86.6) of the respondents affirmed that Directing print and distributing information to un-entitled people remains as specialized parts that could cause disappointment in Business Information Sys tems though, 52 (13.4) of the respondents be of contrary opinion on the subject matter. Once more, the majority 331 (85.5) of the respondents believed that the Shredding of sensitive documents by non-security people will remain technical factors that could cause failure in Business Information Systems (BIS) in contrast, 56 (14.5) of them concurred that the avowal was not true. In conclusion, 66 (17.1) of the respondents disputed that the transmission of data from remote locations may not serve as a technical factor that could cause failure in electronics small businesses, quite many of them 321 (82.9) confirmed this statement to be true (Table 8).

6.9 My Business has Suffered a Loss Due to the Following Security Breaches

Table 9. Fieldwork security breaches

Question Items	Yes	No
1. User accidentally deleted a data file or change the d computer configuration	287 (74.2)	100 (25.8)
2. Deliberate attack by the hacker or a disgruntled staff gained access and deleted or stole data	273 (70.5)	114 (29.5)
3. Theft or misplacement of software caused delay or re-installation of the program	293 (75.7)	94 (24.3)
4. Equipment breakdown (e.g., hardware crashed causing data loss and business disruption)	312 (80.6)	75 (19.4)
5. Lack of a backup (e.g., system restore failure due to corrupt or inadequate backups)	296 (76.5)	91 (23.5)
6. Theft of data (e.g., espionage which resulted in data loss and possible legal problems)	280 (72.4)	107 (27.6)
7. Site disaster (e.g., fire or flood causing damage to systems and business disruption)	264 (68.2)	123 (31.8)
8. Copyright violation (e.g., staff loading pirated software, legally exposing my business)	268 (69.3)	119 (30.7)
9. Compliance (e.g., passing on confidential information, legally exposing my business)	271 (70.0)	116 (30.0)

From the table above respondents gave various reasons responsible for the loss they have acquired in their businesses while some did not count those security breaches as causation factors to their loss. 287 (74.2) of them said they have suffered a loss in their business because of a User accidentally deleting a data file or changing computer configuration but 100 (25.8) of them did not accept this factor as a cause of their loss in business. A good number of the respondents 273 (70.5) declared that a Deliberate attack by the hacker or a disgruntled staff gained access and deleted or stole data has seriously aided their loss in business, although, few of them 114 (29.5) disagreed with their counterparts over this statement. Most of the respondents 293 (75.7) consented that the Theft or misplacement of software caused delay or re-installation of the program has fostered their misfortune in business whereas, 94 (24.3) of the businessmen disapproved of this avowal. The larger proportion of the respondents 312 (80.6) reported that they experienced loss in the business as a result of Equipment failure (e.g., hardware crashed causing data loss and business disruption) albeit 75 (19.4) of them regarded it not as the cause of their loss in business. Also, a larger proportion of the respondents 296 (76.5) strongly agree with the statement "Backup failure (e.g., system restore failure due to corrupt or inadequate backups)" constituted the loss they acquired in their business, on the other hand, 91 (23.5) entrepreneurs proscribed this assertion. Moreover, 280 (72.4) of the respondents gave an account of loss in their businesses because of Data theft (e.g., espionage which resulted in data loss and possible legal problems) while 107 (27.6) of the businessmen investigated rejected this verdict. In addition, 264 (68.2) businesspeople explored explained that Site dis asters (e.g., fire or flood causing damage to systems and business disruption) have caused them to suffer loss in their business, even though 123 (31.8) of them were on the contrary opinion to this fact. Furthermore, 268 (69.3) of the respondents narrated their ordeal in the business because of Copyright infringement (e.g., staff loading pirated software, legally exposing my business), even as 119 (30.7) of their colleagues censured the proclamation. False. Finally, 271 (70.0) of the respondents affirmed that Compliance (e.g., passing on confidential information, legally exposing my business) has effects loss in their organizational set-up, when116 (30.0) of their associates deprecated the edict (Table 9).

6.10 Factors that Affect Internet-Based Security in Electronic Small Business

Table 10. Fieldwork on factors that could affect internet-based security in small business

Question Items	Yes	No
1. Business environment not being conducive	346 (89.4)	41 (10.6)
2. Internal factors of business not being conducive	349 (90.2)	38 (9.8)
3. Legal and regulatory frameworks not being good	336 (86.8)	51 (13.2)
4. Lack of access to finance	334 (86.3)	53 (13.7)
5. Lack of human resources	310 (80.1)	77 (19.9)
6. Lack of entrepreneurial skills	322 (83.2)	65 (16.8)
7. Lack of managerial capacity	317 (81.9)	70 (18.1)
8. Lack of marketing skills	319 (82.4)	68 (17.6)
9. Lack of technological capacity	344 (88.9)	43 (11.1)
10. Corruption	325 (4.0)	42 (16.0)
11 Lack of Internet connectivity and Component	325(84.9)	65 (3.3)
12 Lack of ideas to Protect the Internet and firewall	326 (98.9)	88 (9.9)

Respondents identified the following as factors that affect internet-based security in small businesses 346 (89.4) of them said the Business environment not being conducive has formed a major factor affecting internet-based security in SMMEs, 41 (10.6) of their counterparts did not see it as influencing components to the web-based security in SMMEs. 349 (90.2) of the respondents regarded Internal factors of business not being conducive as a causative factor affecting internet-based security in SMMEs, while 38 (9.8) of them disagreed with this statement. Again, a larger proportion of the respondents 336 (86.8) accepted Legal and regulatory frameworks as not being good a threat to internet-based security in SMMEs, whereas 51 (13.2) regarded the declaration as false. Also, 334 (86.3) of the businessmen be aware of the Lack of access to finance as a factor that affects internet-based security in small businesses and 53 (13.7) of their colleagues are against the statement. Research shows that 310 (80.1) of the entrepreneurs reported a Lack of human resources as a major element influencing internet-based security in SMMEs, and 77 (19.9) of them declared the statement untruth. 322 (83.2) respondents believed that Lack of entrepreneurial skills stands as a factor affecting internet-based security in SMMEs while disapproving of the statement. Equally, 317 (81.9) of the businessmen conceded that Lack of managerial capacity serves as a major factor constituting internet-based security in SMMEs, whereas 70 (18.1) of the respondents refuted the statements as reality. Moreover, 319 (82.4) of the respondents reported that Lack of marketing skills functions as a causative factor to internet-based security in SMMEs, whilst 68 (17.6) of them objected to the declaration. Again, 344 (88.9) of the respondents accepted that Lack of technological capacity remains a renowned element affecting internet-based security SMMEs though 43 (11.1) of them rejected this avowal. In conclusion, 325 (84.0) of the people investigated asserted Corruption as a prominent factor hindering internet-based security in SMMEs, however, 62(16.0%) of them failed to agree to the report (Table 10). HYPERLINK "SPS:refid::tbl10"

6.11 Which of the Following Strategic Plans Can Promote Sustainable Security in Small Businesses?

Table 11. Fieldwork on plans can promote sustainable security in small business.

	Yes	No
1. Security policy (Documented information security policy, liaising with the service provider, and registered users on the systems, planning against a site disaster, information security policy, Email, and internet conduct someone is responsible for backup)	367 (94.8)	20 (5.2)
2. Organizational security (Expertise in information security is available internally, access to our information systems re-quires approval by a senior manager)	364 (94.1)	23 (5.9)
3. Asset classification and control (Including Software, hardware, control local and remote access to our information assets adequately and backup)	361 (93.3)	26 (6.7)

(*continued*)

Table 11. (*continued*)

	Yes	No
4. Personnel security (Violated our security policy, staff training to secure computers all time, and security incidental must be reported to management immediately)	362 (93.5)	25 (6.5)
5. Physical and Environmental Security (Theft, server maintenance, business premises, and liability through compromised data)	353 (91.2)	34 (8.8)
6. Communication and Operation Management (Equipment failure, backup, theft, or site disaster)	345 (89.1)	42 (10.9)
7. Access Control (Changing passwords every 2 weeks)	329 (85.0)	58 (15.0)
8. Systems Development and Maintenance (Outsource from developers, software, server monitoring, and)	350 (90.4)	37 (9.6)
9. Business Continuity Management (In the event of a disaster)	356 (92.0)	31 (8.0)
10. Compliance (International information security standards)	346 (89.4)	41 (10.6)

Respondents proffer following strategic plans that can promote sustainable security in SMMEs although few numbers among them disagreed with their counterparts over such plans A larger proportion of the respondents 367 (94.8) explain that Security policy (Documented information security policy, liaising with a service provider, registering users on the systems, planning against a site disaster, information security policy, Email, and internet conduct someone is responsible for backup) can sustain security measure in SMMEs, just few 20(5.2) them counted this plan impossible. 364 (94.1) of the respondents suggested that Organizational security (Expertise in information security is available internally, access to our information systems requires approval by a senior manager) can go a long way to curb internet-based threats in SMMEs on the other hand, only 23 (5.9) of their colleagues disagreed over this suggestion The majority of the respondents 361 (93.3) agreed that Asset classification and control (Including Software, hardware, control local, and remote access to our information assets adequately and backup) just as 26 (6.7) of them discarded this opinion. Also, 362 (93.5) of the respondents recommended Personnel security (Violated our security policy, staff training to secure computers all time, and security incidentals must be reported to management immediately) as a strategy that can sustain security in SMMEs, at the same time as, 25 (6.5) of the show disapproval of this proposal. Furthermore, a greater number of the respondents 353 (91.2) advised that Physical and Environmental Security (Theft, maintenance, business premises, and liability through compromised data) can solve and sustain security in small businesses, although, 34 (8.8) of them reprimanded their colleagues on this way out to security threats in SMMEs. Besides, 345 (89.1) of the respondents gave Communication and Operation Management (Equipment failure, backup, theft, or site disaster) as a strategic plan for sustainable security in SMMEs, while 42 (10.9) of them chide their associate over this option. Once more, 329 (85.0) of the respondents offered Access Control (Changing passwords every 2 weeks) as a solution to internet-based threats in small businesses, even as 58 (15.0) of their partners in business criticized this proposition. Once more, 350 (90.4) of the respondents assured their contemporaries that Systems Development and Maintenance (Outsourcing from developers software, server monitoring, and) can sustain security in SMMEs, but 37 (9.6) of them opposed the statement Still, 356 (92.0) of the entrepreneur gave the idea that Business Continuity Management (In the event of a disaster) can stand as a means of sustainability security in small businesses, on the other hand, 31 (8.0) of collaborators argued over this statement. Finally, 346 (89.4) of the respondents stated that Compliance (International information security standards) could affect the security in small businesses while 41 (10.6) of their partners disagreed with them on this point (Table 11).

7 Conclusion

Small business enterprises, commonly known as SMEs, are challenged both by the ability and the desire to safeguard themselves against cyberattack, making them particularly susceptible to threat given the capacity of the business. However, small businesses are unaware of an attack until it is too late. Every year, small businesses incur higher costs because of cybercrime-related expenses. Cybercriminals can easily exploit their targets for financial gain with the help of the Internet, which offers a clear path for their activities. Because of this, it is crucial to perform an in-depth study on how cyberattacks affect SMEs' electronic businesses, as well as their causes, costs, and driving factors, especially in South Africa.

8 Suggestions for Future Works

The study suggests that business owners and internet users should be very sensitive and observant during the time they spend on the internet, and they should be familiar with various cyber threats and attacks. E.g., business email compromise continues to be a high-risk problem since there are various ways in which the threat can penetrate anything online, and actors can control everything from the email communication system itself to the user by duping the user into inadvertently initiating a destructive activity.

References

1. Apau, R., Nti Koranteng, F., Adu Gyamfi, S.: Cyber-crime and its effects on e-commerce technologies. J. Inform. **5**(1), 39–59 (2019)
2. PEARSON, 2011: Telecommunications, the Internet, and Wireless Tech nology Essentials of Management Information Systems, Telecommunication, and networking in today's business corporate network infra-structure, vol. 6, pp. 82–96 (2011)
3. Omar, H.M.S.: The important of data protection process in small and medium enterprises. Int. J. Comput. Secur. **6**, 193–195 (2011)
4. Cant, M.C., Johannes, A.: Establishing the challenges affecting South African SMEs. Int. Bus. Econ. Res. J. **12**, 30–36 (2013)

5. Smith, A., Anderson, M.: Online Shopping and E-commerce. Review Research Centre Internet & Technology, vol. 201 (2016). http://www.pewintenet.org/2016/12/19/online-shopping-and-e-commerce/
6. Awosejo, O.J.: Adoption of accounting information systems in an organization. Afr. J. Comput. & ICTs **7**, 127–136 (2014)
7. Atzori, L., Iera, A., Morabito, G.: The internet of things: a survey. Comput. Netw. **54**, 52787–52805 (2010)
8. Awosejo, O.J.,: Factors influencing accounting information systems us age in an accounting organization, vol. 1, pp. 54–55 (2014)
9. Awosejo, O.J., Kekwaletswe, R.M., Pretorius, P.: Recommendation of information systems to motivate accounting firm in South Africa. Lect. Notes Inform. Theory **2**, 328–334 (2014)
10. Castro, D., Misra, J.: The internet of things. http://www2.datainnovation.org/2013-internet-of-things.pdf (2013). Accessed 21 May 2015
11. Consoli, D.: Literature analysis on determinant factors and the impact of ICT in SMEs. Procedia – Soc. Behav. Sci. **62**, 93–97 (2012)
12. Chen, T.: A compositive education scenario for network security protocols. In: International Workshop on Education Technology and Computer Science (2010)
13. Pawan, K.R., Chaurasia, P.: Proposed model for IP address security. Int. J. Adv. Res. Comput. Sci. Software Eng. **4**, 363–365 (2014)
14. Lin, L., Xiao, D.-B.: Research on network security management based on data mining. Int. Conf. Adv. Comput. Theory Eng. **23**, 75–82 (2010)
15. Mathew, T.J.: A new packet cipher to secure IP based communications. IEEE **16**, 24–35 (2009)
16. Hong, S., Lim, S.: Analysis of attack models via UML in WSN: a survey study. IEEE **10**, 17–25 (2010)
17. Xu, L.D., He, W., Li, S.: Internet of things in industries: a survey. IEEE Trans. Ind. Inf. **10**, 2233–2243 (2014)
18. Hadaya, P., Pellerin, R.: Determinants and performance outcome of manufacturing SMEs use of internet-based IOISs to share inventory information. Int. J. Electron. Bus. **8**, 504 (2010)
19. Veiga, A.D., Eloff, J.H.P.: An information security governance framework. Inform. Syst. Manag. **24**, 367–368 (2015)
20. White, C.C., Cheong, T.: In-transit perishable product inspection. Transport. Res. Part E: Logistics Transport. Rev. **48**, 310–330 (2012)
21. Worku, Z.: Predictors of incubation of small, micro, and medium-sized businesses in Gauteng province. South Africa. Prob. Perspect. Manage. **13**, 7–17 (2015)
22. Wymer, S., Regan, E.: Factors influencing e-commerce adoption and use by small and medium businesses. Elec. Markets **15**, 438–453 (2005)
23. Zheng, C., O'Neill, G., Morrison, M.: Enhancing Chinese SME performance through innovative HR practices. Pers. Rev. **38**, 175–194 (2009)
24. Harris, M.A., Patten, K.P.: Mobile device security considerations for small-and medium-sized enterprise business mobilit. Inform. Manag. Comput. Secur. **22**, 97–114 (2014)
25. Harris, L., Rae, A.: The online connection: transforming marketing strategy for small businesses. J. Bus. Strategy **31**, 4–12 (2010)
26. Hajji, A.M., Lewis, M.P.: Development of a productivity-based estimating tool for fuel use and emissions from earthwork construction activities. J. Construct. Eng. Project Manage. **3**, 58–65 (2013)
27. Anah, T.: Managing information security in small and medium-sized enterprises. J. Manag. Inform. Secur. **2**, 23–31 (2010)
28. Szabo, K.H.B.: The penetration of business information systems in small and medium-sized enterprises. Int. J. Manag. Sci. Inform. Technol. **1**, 32–39 (2016)

29. Marivate, S.P.: The impact of entrepreneurial skills on the viability and long-term survival of small businesses: a case of the city of Tshwane, South Africa. Eur. J. Bus. Econ. Accountancy **2**, 53–72 (2015)
30. Khale, S., Worku, Z.: Factors that affect municipal service delivery in Gauteng and north west provinces of South Africa. Afr. J. Sci. Technol., Innovation Dev. **5**(1), 61–70 (2013)
31. Gundu, T., Flowerday, S.V.: Ignorance of awareness of information security. Proc. Conf. Inform. Secur. South Africa **32**, 15–17 (2013)
32. Abu-Musa, A.A.: Investigating the security policies of computerized accounting information systems in the banking industry of an emerging economy: the case of Egypt. Rev. Bus. Inform. Syst. (RBIS) **8**, 87–102 (2004)
33. Awosejo, O.J., Hans, R.: Support of electronic systems in small business in South Africa, pp. 2767–9640. International Business, Information Management Association (IBIMA) (2021)

The Conceptual Design of a Data Collection System for Predictive Maintenance

Lenka Halenarova(✉), Igor Halenar, and Pavol Tanuska

Faculty of Materials Science and Technology, Trnava Slovak University of Technology, Trnava, Bratislava, Slovakia
{lenka.halenarova,igor.halenar,pavol.tanuska}@stuba.sk

Abstract. The core of this work is the design of a conceptual data collection system for predictive maintenance. As a physical device for data collection is the automatic production line – Festo AFB Factory. Via the designed data collection architecture, which consists of several layers, it is possible to obtain complex data from the physical device. The designed system considers the momentary state of the device, it has the ability of a prediction of a future state and that all on the basis of the collected data. Simultaneously thanks to decision-making processes within the Intelligent Devices System (IDS), it is possible to achieve a wider overview about the state of the devices and eventually determine a percentual probability of a malfunction occurrence or a collapse of the given device. The advantage of this kind of solution making is the instant interaction with the induced failure in the production process and a prospective machine shutdown.

Keywords: Maintenance · Predictive Maintenance · Prognostics · Industry 4.0 · Innovations

1 Introduction

Industry automation, the usability of digitalization, and virtualization of systems is consequential in the massive improvement and evolution in the area of information technologies. New progressive approaches to fast changing market requirements are not always a feasible goal for a common institution. Multiple firms fight with the implementation of systems to original devices, another with the evaluation of data, and other have trouble ensuring the reliability and safety of the introduction of new technology. Ultimately, all of this has a great impact on the course of maintenance. Inefficient operation and unplanned shutdowns in the production process cost companies a lot of money, so it is not surprising that they are looking for a solution to minimize their downtime.

Historically, maintenance was considered only as activities that had the task of carrying out repairs only after the necessary period or until the occurrence of a malfunction. As a result, maintenance was minimized to cleaning and lubrication only, and in many cases it was performed very haphazardly.

Maintenance is a complex process of fault identification based on several diagnostic systems. Its task is to identify and remove damaged units or parts of equipment, in

R. Silhavy and P. Silhavy (Eds.): CSOC 2023, LNNS 723, pp. 169–178, 2023.
https://doi.org/10.1007/978-3-031-35317-8_15

the shortest time possible, and to reestablish the machine production. The trend and necessity of today's production is to eliminate equipment malfunctions before they occur. Predictive maintenance deals with this routing.

The core of this paper is to design a data collection system for predictive maintenance. As a physical device, the automatic production line - AFB Factory from Festo will be used and it is located in the Laboratory of control systems. The device offers a high degree of variability and real-time operation data obtained through intelligent sensors.

2 Related Work

Thanks to the development of Industry 4.0 technology [1–4], nowadays much attention is paid to the integration of innovative elements into industrial production. Industry 4.0 enables the connection of people, machines, and objects, within the Internet and in real time, and has a great impact on the introduction of information and communication means for decision-making for higher management processes. On the one hand, it provides companies with the possibility of decentralization and the incorporation of human-machine cooperation in production [5], and on the other hand, it exposes them to the challenging course of changes and innovations [6]. In the mentioned articles [5–7], the authors highlight and recommend the decentralization and implementation of Cyber-physical systems (CPS) in enterprises from the point of view of their usability in managing, planning, and monitoring work and logistics processes. In its ideology, Industry 4.0 considers and does not neglect developments in the field of maintenance and its activities. Therefore it is necessary to include the term Maintenance 4.0 in this category [8–10].

In the work of Dolatabadi and Budinská [11], there is a constructive and targeted inquiry into research and studies on the implementation of maintenance in small and medium-sized enterprises around the world. And since the result of this inquiry is the need for prediction and the development of this technology only on a theoretical level, further research is needed to assess the implementation of digitization and prediction in maintenance at a practical level.

Current trends offer companies a new direction in the field of maintenance, using a predictive approach, exemplified by Prognostics and Health Management (PHM) [8, 12] and Condition-Based Maintenance (CBM) [8, 13]. We understand the sustainability of production in a state that we can predict and prevent based on the acquired anomalies of the machines as one of the relevant strategies for achieving the company's goals. For the successful introduction of maintenance in the company, it is necessary to achieve openness and standardization of processes, both at the company level and beyond, and thus obtain a higher degree of flexibility and independence in the automation of production processes [14, 15].

The primary goal of Predictive Maintenance (PdM) [8, 16] is to detect an ongoing equipment failure and then to further predict the time or state of a potential failure. When PdM is put into practice and compared to other maintenance strategies used, such as Preventive Maintenance (PvM) or Reactive Maintenance (RtM), the results show a significant reduction in the appearance of unplanned equipment failures [17, 18]. And therefore we can claim that this maintenance strategy is able to achieve the desired prediction effect. Here, however, we have to realize that within this description, maintenance is still taking place, determined on the basis of the given course and the final state of the equipment [9, 19].

Predictive maintenance can be taken to a higher level using the collected data. With advanced data analysis and their appropriately chosen application, we can to significantly increase maintenance support by providing a controlled and intelligent form of decision making [8, 20]. Some companies are relatively far in the field of data collection [21–24] – they have a defined data collection concept and the technology used, others are just at the beginning and still collect data in paper form through their employees. This does not appear to be the most suitable solution when there is an increased need to implement process automation.

By using intelligent diagnostics, monitoring, and early detection of abnormalities, we are able to collect data to the extent necessary and desired for maintenance optimization. PdM is based on data obtained from sensors in real time and during continuous operation [19, 25, 26]. By combining various data sources, we can select ongoing anomalies and historically analyze the scenario of the course of operation. We can use the information obtained for maintenance planning based on preventive replacement of components that do not require service at that particular time [19].

Data obtained from machine diagnostics can be very complex and therefore need to be processed and analyzed using a powerful tool [27, 28]. The PdM module is made up of two systems: a diagnostic system and prognostic system. Through PdM analysis, we are able to directly identify defects on the device or its component [28]. We can usually implement the subsequent evaluation and decision making of the process using Machine Learning (ML) algorithms. However, there are companies that focus on implementing maintenance using key machine performance parameters. The primary task of ML is to process large amounts of data and find correlations between them [27]. Through Big Data, it is possible to create more accurate procedures for data acquisition from the proposed algorithms and thus better decisions for the maintenance plan. In this way, we can gain greater oversight of critical equipment and a Just In Time (JIT) maintenance plan [29].

3 Conceptual System Design

For the better identification of anomalies and defects in the monitored production process and the setting of predictive maintenance, it is necessary to obtain a comprehensive picture of the equipment, correct, high-quality, and usable data. Through the proposed architecture, we can to use the collected data to set a maintenance strategy to avoid unexpected equipment failures.

For each machine/entity that we want to monitor, we propose a separate layered architecture scheme for data collection into the predictive system. The proposed system will take into account the current state of the device and predict its future state, given the data mined from various sources. Through the sensors, we recognize the unusual behavior of the device in the flow of measured data. The proposed architecture of the predictive maintenance system is shown in Fig. 1.

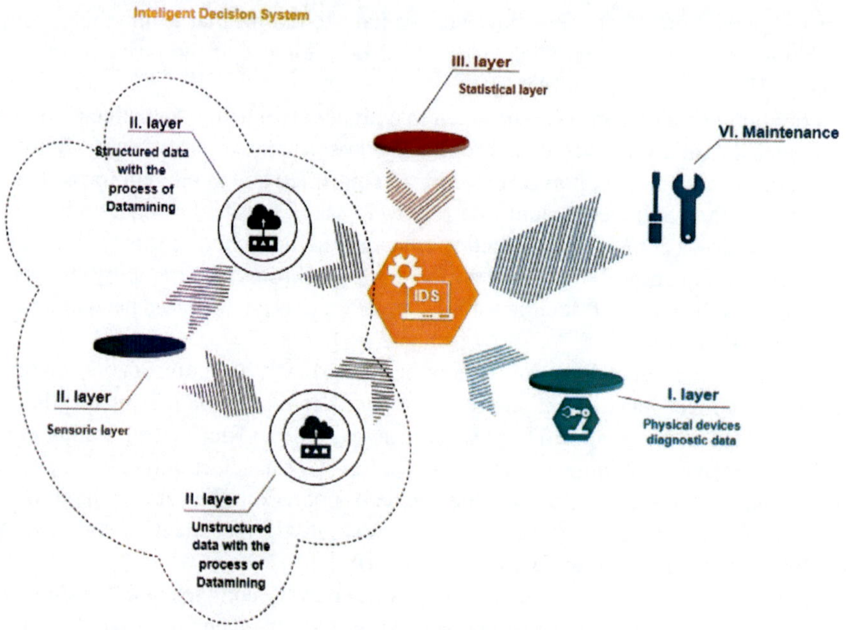

Fig. 1. Design of predictive maintenance system

In the first step, we are able to obtain data directly from the physical device, and in the second step, through IoT sensors, we implement a sensory collection layer, from which we obtain data that we are not able to identify directly. Depending on the type of data, these data will be processed using Data Mining. This will allow us to acquire identifiers based on which we can evaluate and work with the data. In the third step, through statistical data from the Historian Server, which is part of the MES, we can compare the data, statistically process it, and use it in decision-making processes. In this architecture, it is essential that we also include a classic maintenance plan. It is necessary to supplement the completeness of the strategy in the decision-making process.

The overall data collection process is concluded in the Intelligent Decision System (IDS). After obtaining all the necessary data, the IDS takes over the decision-making authority. It compares all obtained data with historical data and the set maintenance plan. It then determines the percentage probability of a malfunction and evaluates the situation as critical or negligible. In case of high criticality, it will suggest machine shutdown.

Description of proposed layers:

I. Layer **Physical devices diagnostic data** – direct data. This category includes diagnostic data about the physical device. They are immediate alarms, triggers – data/symptoms that we can read directly from physical devices. We will evaluate data from this group using direct measurement devices.

II. Layer **Sensory layer** – indirect data. In this layer, we will include physical data from the environment and additional data of physical devices that native devices do not offer us. These data contain physical quantities such as temperature, pressure, or complex data such as vibrations, audio, video, etc.

- Structured data – data that we can directly identify, quantify, and use their value directly. They are physical quantities provided by the IoT layer, and we can measure them, e.g. motor current load, bearing temperatures…
- Unstructured data – data that we need to process, e.g. vibrations, sound, video, photo. It is necessary to use a suitable form of artificial intelligence (AI) on this type of data (e.g. neural network, machine learning, regression methods, etc.). Through data mining methods, we can obtain relevant and usable information from unstructured data.

III. Layer **Statistical layer** – completely indirect data. The data is based on the device's operating history. All data is recorded in the Wonderware Historian Server. The data from this group, obtained by statistical analysis, will directly form one of the inputs to the evaluation process.

IV. Layer **Maintenance** – the defined maintenance plan is an accompanying element in the evaluation. Criteria for the maintenance plan are important in decision-making processes.

If we were to use the proposed architecture for all the stations of the presented equipment, the output would be a diagram where each entity/machine would be displayed separately and data would be available on the expected downtime of the equipment under standard working conditions. The resulting predictive maintenance system will consist of a set of autonomous IDS (it will be necessary to create a separate scheme containing all data collection layers for each monitored device).

3.1 Technical Backround

The automatic production line - Festo AFB Factory is located in the Laboratory of control systems, which is under the ownership of The Institute of Applied Informatics, Automation and Mechatronics of Faculty of Materials Science and Technology in Trnava – Fig. 2.

The modular hybrid automatic production system consists of the basic manufacturing systems stated:

- production,
- manipulation with seals,
- packing line,
- filling line,
- system of conveyors,
- automatic shelf stacker/storage,
- inputs/outputs.

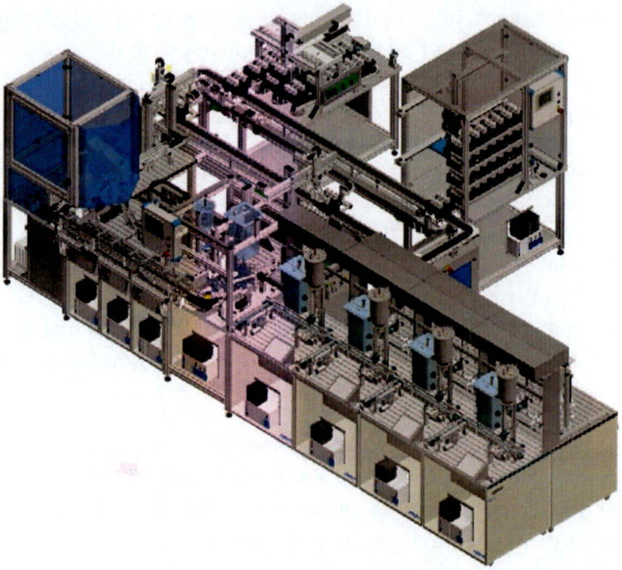

Fig. 2. Automatic production line – Festo AFB Factory [30]

AFB Factory – acts as a complete automatised production line. The system also contains procedural and factory automatization. The whole device is able to work as a complex entity but can also be separated to individual systems. This model of production line offers a high level of flexibility, and due to the PLC control system it abble to solve complicated problems e.g. logistics, manipulation, etc., as well as partial problems like e.g. optimization of control loops. The control system of separate modules contains industrial PLC Siemens S7 mounted on a bar connected to regulated technology via IEEE 488/24 pin connectors and to a visualisation through an external interface. Operator panels (HMI) are mounted on a plate with outsourced connectors for communication interfaces.

3.2 Sensory Layer Practical Proposal

An example of the implementation of the sensory layer in laboratory conditions is a part of the automated AFB Factory line from Festo, which is part of the recycling station – Unpacking Station. The station ensures the unpacking of bottles from the crates. The bottles are then transported using a conveyor to the recycling station. Using a two-axis industrial manipulator with a clamp, for three bottles at a time, the unloading of bottles is ensured.

Basic sensors are currently incorporated into the actuators of the mentioned device, which monitor the end positions of the servomotor and the current state of the manipulator arm (open/close). For a more adequate evaluation of the data, we suggest incorporating IoT sensors within the proposed sensor layer, which are shown in Fig. 3.

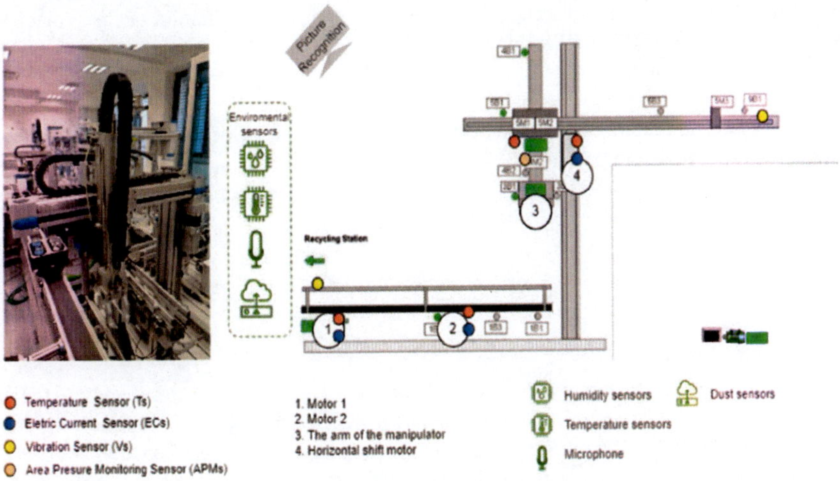

Fig. 3. Sensory layer for Unpacking Station

The sensory layer is formed by a set of sensors of basic physical variables, e.g. temperature, electric current, and additional variables, e.g. vibration, area pressure monitoring, and environment sensors (humidity, temperature, dust, and noises), thanks to which we get a more detailed overview of the state of the device.

4 Conclusion

It is obvious that data collection for the purpose of predicting the state of equipment and planning maintenance is still in its infancy. The aim of this article was to design a conceptual system architecture for predictive maintenance in laboratory conditions. The given solution can also be used in ordinary industry, but we must take into account the uniqueness of each company and condition the design of the equipment digitization system and maintenance plan settings. The application of the latest technologies and the integration of the Internet of Things elements into the collection and evaluation of data is of practical importance for businesses. Truly real and usable data is a support for many activities of the company. On the basis of them, we are able to optimize equipment performance, reduce the risk of outages, reduce maintenance costs, increase production, and extend the life of machines. All of which have an enormous impact on the operation of the entire company. This work represents only a part of the overall solution. The main ideology in this direction is for companies to receive more usable and inexpensive solutions for data collection and digitization of their devices. The proposed system represents a complex solution containing the technologies of machine learning, statistical data processing, Big Data, and the overall implementation exceeds the publishing possibilities. The goal of the final phase is to create an interface for visualize the condition of individual parts (entities) of the production process.

Acknowledgment. This publication is the result of the implementation of the project VEGA 1/0193/22: "Proposal of identification and monitoring of production equipment parameters for the needs of predictive maintenance in accordance with the concept of Industry 4.0 using Industrial IoT technologies." supported by the VEGA.

This contribution was published with the support of the Operational Program Integrated Infrastructure within the project of "Výskum v sieti SANET a možnosti jej ďalšieho využitia a rozvoja, code ITMS 313011W988 (Research into SANET and options of its further utilisation and development), co-financed by the ERDF".

References

1. Hassan Reza, M.N., Agamudai Nambi Malarvizhi, C., Jayashree, S. Mohiuddin, M.: Industry 4.0–technological revolution and sustainable firm performance. In: 2021 Emerging Trends in Industry 4.0 (ETI 4.0), pp. 1–6 (2021). https://doi.org/10.1109/ETI4.051663.2021.9619363
2. Büchi, G., Cugno, M., Castagnoli, R.: Smart factory performance and Industry 4.0, Technol. Forecast. Soc. Chang. **150,** 119790 (2020) ISSN 0040-1625, https://doi.org/10.1016/j.techfore.2019.119790.
3. Dixon, P.: The Industry 4.0 Lexicon. In: IEEE IAS Pulp and Paper Industry Conference (PPIC), pp. 191–195 (2022). https://doi.org/10.1109/PPIC52995.2022.9888880
4. Oluwaseun, A.A., Chaubey, M.S. Numbu, L.P.: Industry 4.0: the fourth industrial revolution and how it relates to the application of internet of things (IoT). J. Multi. Eng. Sci. Stud. (JMESS), **5** (2019)
5. Schuhmacher, J., Hummel, V.: Decentralized control of logistic processes in cyber-physical production systems at the example of ESB logistics learning factory. Procedia CIRP **54**, 19–24 (2016). https://doi.org/10.1016/j.procir.2016.04.095
6. Sakurada, L., Leitão, P.: Multi-Agent systems to implement Industry 4.0 components. In: 2020 IEEE Conference on Industrial Cyberphysical Systems (ICPS), pp. 21–26 (2020). https://doi.org/10.1109/ICPS48405.2020.9274745
7. Müller, J.M., Kiel, D., Voigt, K.-I.: What drives the implementation of Industry 4.0? The role of opportunities and challenges in the context of sustainability. Sustainability **10**(1), 247 (2018). https://doi.org/10.3390/su10010247
8. Cachada, A., et al.: Maintenance 4.0: intelligent and predictive maintenance system architecture. In: 2018 IEEE 23rd International Conference on Emerging Technologies and Factory Automation (ETFA), pp. 139–146 (2018). https://doi.org/10.1109/ETFA.2018.8502489
9. Poór, P., Basl, J., Zenisek, D.: Predictive Maintenance 4.0 as next evolution step in industrial maintenance development. In: 2019 International Research Conference on Smart Computing and Systems Engineering (SCSE), pp. 245–253 (2019). https://doi.org/10.23919/SCSE.2019.8842659
10. Câmara, R.A., Mamede, H.S. Santos, V.D.d.: Predictive industrial maintenance with a viable systems model and maintenance 4.0. In: 2019 8th International Conference On Software Process Improvement (CIMPS), pp. 1–8 (2019). https://doi.org/10.1109/CIMPS49236.2019.9082435
11. Hassankhani Dolatabadi, S., Budinska, I.: Systematic literature review predictive maintenance solutions for SMEs from the last decade. Machines **9**, 191 (2021). https://doi.org/10.3390/machines9090191
12. Zhang, T., Zhang, W., Du, G., Wang, J.: PHM of Rail vehicle based on digital twin. In: 2021 Global Reliability and Prognostics and Health Management (PHM-Nanjing), pp. 1–5 (2021). https://doi.org/10.1109/PHM-Nanjing52125.2021.9613068

13. Li, Q., Peng, X.: Application of large-data-driven PHM technology in satellite test and on-orbit management. In: 2017 Prognostics and System Health Management Conference (PHM-Harbin), pp. 1–5 (2017). https://doi.org/10.1109/PHM.2017.8079214
14. Weyer, S., Schmitt, M., Ohmer, M., Gorecky, D.: Towards Industry 4.0-Standardization as the crucial challenge for highly modular, multi-vendor production systems. In: IFAC PapersOnLine, vol. 48, pp. 579–584 (2015). https://doi.org/10.1016/j.ifacol.2015.06.143
15. Zhang, Y., Ye, Z.: Data maintenance and control strategy of group management. In: 2011 Second International Conference on Innovations in Bio-inspired Computing and Applications, pp. 141–144 (2011). https://doi.org/10.1109/IBICA.2011.39
16. Motaghare, O., Pillai, A. S. Ramachandran, K.I.: Predictive maintenance architecture. In: 2018 IEEE International Conference on Computational Intelligence and Computing Research (ICCIC), pp. 1–4 (2018). https://doi.org/10.1109/ICCIC.2018.8782406
17. Jezzini, A., Ayache, M., Elkhansa, L., Makki, B. Zein, M.: Effects of predictive maintenance (PdM), Proactive maintenace (PoM) & Preventive maintenance(PM) on minimizing the faults in medical instruments. In: 2013 2nd International Conference on Advances in Biomedical Engineering, pp. 53–56 (2013). https://doi.org/10.1109/ICABME.2013.6648845
18. Ma, L., Sun, Y., Mathew, J.: Effects of Preventive Maintenance on the reliability of production lines. In: 2007 IEEE International Conference on Industrial Engineering and Engineering Management, pp. 631–635 (2007). https://doi.org/10.1109/IEEM.2007.4419266
19. Abidi, M.H., Mohammed, M.K., Alkhalefah, H.: Predictive maintenance planning for Industry 4.0 using machine learning for sustainable manufacturing. Sustainability **14**(6), 3387 (2022). https://doi.org/10.3390/su14063387
20. Yuanyuan, L. Jiang, S.: Research on equipment predictive maintenance strategy based on big data technology. In: 2015 International Conference on Intelligent Transportation, Big Data and Smart City, pp. 641–644 (2015). https://doi.org/10.1109/ICITBS.2015.163
21. Gogliano Sobrinho, O., et al.: Big data analytics in support of the under-rail maintenance management at Vitória – Minas Railway. In: 2021 IEEE International Conference on Big Data (Big Data), pp. 6026–6028 (2021). https://doi.org/10.1109/BigData52589.2021.9671739
22. Kong, Q., Lu, R., Yin, F., Cui, S.: Privacy-preserving continuous data collection for predictive maintenance in vehicular Fog-Cloud. IEEE Trans. Intell. Transp. Syst. **22**(8), 5060–5070 (2021). https://doi.org/10.1109/TITS.2020.3011931
23. Cachada, A. et al.: Using internet of things technologies for an efficient data collection in maintenance 4.0. In: 2019 IEEE International Conference on Industrial Cyber Physical Systems (ICPS), pp. 113–118 (2019). https://doi.org/10.1109/ICPHYS.2019.8780217
24. Yanming, T., Ping, Y., Zongjun, G.: Predictive maintenance strategy based upon management information system. In: Proceedings of 2001 International Symposium on Electrical Insulating Materials (ISEIM 2001). 2001 Asian Conference on Electrical Insulating Diagnosis (ACEID 2001). 33rd Symposium on Electrical and Ele, pp. 225–228 (2001). https://doi.org/10.1109/ISEIM.2001.973622
25. Huynh, K.T., Grall, A., Bérenguer, C.: A parametric predictive maintenance decision-making framework considering improved system health prognosis precision. IEEE Trans. Reliab. **68**(1), 375–396 (2019). https://doi.org/10.1109/TR.2018.2829771
26. Lin, C.-Y., Hsieh, Y.-M., Cheng, F.-T., Huang, H.-C., Adnan, M.: Time series prediction algorithm for intelligent predictive maintenance. IEEE Robot. Autom. Lett. **4**(3), 2807–2814 (2019). https://doi.org/10.1109/LRA.2019.2918684
27. Cinar, E., Kalay, S., Saricicek, I.: A predictive maintenance system design and implementation for intelligent manufacturing. Machines **10**, 1006 (2022). https://doi.org/10.3390/machines10111006
28. Samatas, G.G., Moumgiakmas, S.S. Papakostas, G.: Predictive maintenance - bridging artificial intelligence and IoT. In: 2021 IEEE World AI IoT Congress (AIIoT), pp. 0413–0419 (2021)

29. Lee, C., Cao, Y., Ng, K.H.: Big Data analytics for predictive maintenance strategies (2017). https://doi.org/10.4018/978-1-5225-0956-1.ch004
30. The Festo documentation. https://www.festo-didactic.com/ov3/media/customers/1100/afb_en_monitor_056961.pdf
31. The Festo Stack magazine module documentation. https://www.festo-didactic.com/int-en/learning-systems/mps-the-modular-production-system/stack-magazine-module.htm?fbid=aW50LmVuLjU1Ny4xNy4xOC41ODUuNDA5Ng

Optimized Framework for Spectrum Resource Management in 5G-CRN Ecosystem of IoT

B. Kursheed[1] (✉) and Vijayashree R. Budyal[2]

[1] Sri Venkateshwara College of Engineering, Bengaluru, Karnataka, India
bkursheedofficial@gmail.com

[2] Sai Vidya Institute of Technology, Bengaluru, Karnataka, India

Abstract. An effective utilization of spectrum and allocation of radio resources is one of the potential concerns as well as factor towards ensuring better communication performance in 5G network. Reviewing existing multiple access scheme, spectrum utilization scheme, as well as resource allocation scheme exhibited open scope of further carrying out investigation. Therefore, the proposed scheme contributes towards a novel model harnessing the potential of Cognitive Radio Network (CRN), Orthogonal Frequency Division Multiple Access (OFDMA), and Code Division Multiple Access (CDMA) to construct a hybrid model that is capable of empirically computing a novel parameter of spectrum quality mark targeting towards improvement of radio resource allocation problem considering large scale distributed environment of Internet-of-Things (IoT). The simulation outcome of the study model exhibited better throughput, delay, and fairness index for proposed scheme in contrast to conventional multiple access scheme.

Keywords: 5G · Cognitive Radio Network · Spectrum Utilization · Radio Resource Allocation · Internet-of-Things

1 Introduction

With the wider usage of 4G network in present time, there has been exponential increment towards usage of radio spectrum [1]. Unfortunately, such abundant amount of radio spectrum is not utilized to its appropriate level despite knowing the fact that it is very scarce resources [2]. This problem multiplies to higher degree when 4G will be completely replaced by 5G services globally. 5G services has multiple beneficial points towards assisting in faster rate of data transmission suitable especially for large form of network like Internet-of-Things (IoT) [3]. However, there are many studies to exhibit the open-end challenges in 5G too [4, 5]. As a part of solution towards such challenges, Cognitive Radio Network (CRN) has been evolved towards ensuring an effective utilization of spectrum [6]. Sensing of spectrum is periodically carried out by CRN nodes in order to resist challenges associated with transmission of secondary and primary users concurrently [7]. In this process, the mechanism of spectrum sharing has evolved which deals with parallel transmission over equivalent band of frequency [8]. However, this is only permissible for tolerable limit of interference by primary users in

R. Silhavy and P. Silhavy (Eds.): CSOC 2023, LNNS 723, pp. 179–197, 2023.
https://doi.org/10.1007/978-3-031-35317-8_16

deployment zone populated by secondary users. Such challenges can be cumulatively deal with by investigating improvements in conventional multiple access of 5G i.e., Orthogonal Frequency-Division Multiple Access (OFDMA), Non-Orthogonal Division Multiple Access (NOMA), Sparse Coding Multiple Access (SCMA), and Code Division Multiple Access (CDMA). At present, there are various studies being carried out more potentially on SCMA [9] and NOMA schemes [10]; however, less emphasis is given towards OFDMA and CDMA in the context of cost-effective modelling towards spectrum utilization and resource allocation. By optimizing the performance factor associated with the secondary OFDMA along with potential control towards CDMA in context of interference, it is possible to improve the communication performance of 5G with effective spectrum resource management. It is to be noted that gain in channel between the secondary receiver and transmitter in multiple access scheme is highly related to throughput termed as primary channel gain. However, gain of cross channel can be obtained from identifying the signal quality between primary receiver and secondary transmitter, especially in OFDMA termed as secondary channel gain. Existing schemes towards 5G performance has been carried out with predefined knowledge about the above stated gain information of channel. Although, it is quite possible to compute primary channel gain, but computation of secondary channel gain is near to impossible practically. The reason behind this problem is that signal of secondary transmitter is required to be decoded by PR that is developed only for estimating primary channel gain.

Therefore, this paper introduces a parameter termed as spectrum quality mark which acts as an alternative solution towards the challenges in gain computation in multiple access over 5G network. The proposed paper makes following contribution: i) a computational analytical model is presented for 5G based CRN in order to introduce a novel hybrid multiple access scheme, ii) the proposed model introduces a simplified mechanism to optimize empirically calculated spectrum quality mark which in turns yield optimal allocation mechanism, iii) the simulated and benchmarked outcome exhibited proposed scheme of better performance in contrast to existing multiple access scheme.

The organization of the paper is as follows: Sect. 2 discusses about the existing approaches towards improving 5G performance while its associated issues are identified in Sect. 3. Briefing of adopted research methodology is carried out in Sect. 4 while algorithm illustration is carried out in Sect. 5. Result analysis is carried out in Sect. 6 while conclusive remarks of paper are stated in Sect. 7.

2 Existing Techniques

At present, there are evolution of various methodologies towards strengthening the communication performance in 5G based CRN. However, the studies towards this direction are carried out emphasizing towards overall performance with respect to spectrum utilization and resource allocation schemes in this section as follows:

2.1 Multiple Access Scheme

There are different variants of multiple access scheme presented in existing system towards improving the communication system in 5G. The study presented by Mohammad karimi et al. [11] have presented a multiple access scheme for uplink transmission in presence of sporadic traffic in large network of 5G. The work carried out by Mhedhbi et al. [12] have constructed a pattern matrix for evaluation of shared resources while using multiple Multiplexing approaches. Similar direction of study using pattern-based multiple access is carried out by Zeng et al. [13] targeting towards optimizing power allocation. The scheme uses an iteration approach and has a dependency towards channel state information. Luo et al. [14] have carried out investigation using Non-Orthogonal Multiple Access (NOMA) and Sparse Code Multiple Access (SCMA) with respect to error computation. The study presented optimization scheme using probability of pair-wise error and used decoding for uplink communication. The study towards use case of Radio Access Network (RAN) in fog computing is analyzed by Papanikolaou et al. [15] with respect to Hierarchical Multiple Access. The optimization is carried out using rate fairness metric with max-min entity for carrying out this optimization towards better transmission rate. The study also contributes to showcase that NOMA doesn't perform well all the times. Use-case considering RAN is further investigated in work of Ahmad et al. [16] where multiple access of splitting rate is introduced towards achieving scalable performance. The optimization of rate is carried out using beaming technique. The work towards rate splitting is also carried out by Gamal et al. [17] improving the multiple access scheme in CRN. Chen et al. [18] have presented another multiple access scheme using reserved resources. The study controls the complexity using centralized coordination while performing frame transmission. Shi et al. [19] have used learning-based method in NOMA using multiple agents in order to facilitate communication for large number of users in 5G. A seamless interaction is facilitated between smaller base station and environment that perform selection of an optimal state.

2.2 Spectrum Utilization Scheme

Spectrum utilization is another significant area of research work focusing towards improving the communication performance in 5G based CRN system. Saha [20] have developed a scheme towards improving dual spectrum utilization on millimeter wave. Sharing of spectrum has been identified towards enhancing 5G performance using CRN as noted in work of Ahmad et al. [21]. Allocation of spectrum is another essential area of research work towards generating a seamless communication using 5G millimeter waves as discussed by Saha [22, 23]. Researchers have also implemented a spectrum slicing mechanism in order to cater up dynamic 5G demands. The slicing is carried out by base station dynamically while the allocated spectrum is shared to the services as witnessed in work of Srinivas et al. [24]. Approaches towards slicing is also carried out by Munoz et al. [25] for the purpose of spectrum planning in next generation network of 5G. Study considering satellite-based network of 5G has been carried out by Tang et al. [26] targeting towards improving the throughput for cognitive users using greedy algorithm. Adoption of learning-based algorithms are reported in work of Chang et al. [27] towards sharing of spectrum dynamically. The author has used deep learning approach for maximizing the spectrum utilization. A unique study carried out by Abubakar et al. [28] where a concept of spectrum leasing is discussed. The analysis of the study is carried out over delay tolerant network further followed up by introducing cell switching concept considering constraints of quality of service in 5G network. Qian et al. [29] have presented discussion on spectrum sharing for large wireless network where a game theory has been implemented towards developing framework for minimizing cost of deployment.

2.3 Resource Allocation Scheme

By allocating proper resources, the devices involved in performing communication perform efficiently to cater up the communication demands of users in 5G. Ahmed et al. [30] have developed a resource allocation technique using Stackelberg game dynamically. The idea of this work is towards resource sharing among mobile edge computing server. Madi et al. [31] have developed a framework for ensuring better fairness considering heterogeneous network in 5G. Resource blocks are prioritized using greedy approach. The work carried out by Yu et al. [32] have used deep reinforcement learning approach for energy efficiency. Khan and Jacob [33] have developed a model over RAN system towards addressing issues of fronthaul capacity. The work carried out by Jankovic et al. [34] have developed a load balancing scheme towards addressing the radio resource allocation problem in 5G. Wu [35] have introduced an auxiliary variable to signify the degree of criticality associated with bits of information. Slapak et al. [36] has considered the case of mobile network in order to formulate multitier edge computing using Bayesian optimizer. The outcome of the study is found to excel better than pseudorandom allocation of resources. Yun et al. [37] have used deep leraning algorithm which targets towards increasing the system utility and reduce energy consumption. Ibrahim et al. [38] have developed a mechanism for placing the controller in the environment of software defined network integrated with network function virtualization. The study has used multi-objective optimization using heuristics over distributed environment of 5G. Pratyusha and Sheu [39] have investigated schedulers in order to maximize the expected rate to be achieved over NOMA. The study outcome shows effective solution towards addressing resource allocation problem.

Hence, there are various approaches implemented towards improving the communication performance of 5G. The next section outlines the identified research problems associated with existing approaches discussed in this section.

3 Research Problem

The identified problems associated with the existing schemes discussed in prior section are as follows:

- *Adoption of Sophisticated Multiple Access Approach:* A closer look at the existing approaches of multiple access e.g. [11–19] showcase the consideration of complex process of developing the access constructs. Apart from this, studies have been carried out explicitly towards considering the transmission flow and not much towards considering the possible constraints associated with user equipment. Therefore, the outcome of such approaches has been evaluated with respect to traffic flow; however, their impact on the performance on the devices are yet not presented in concrete way. An effective multiple access formulation should offer better performance along with consistency irrespective of the fluctuating load of traffic that is not found to be reported in existing approaches.

- *Issues with Existing Spectrum Utilization*: There is no denying the fact that existing schemes of spectrum utilization has introduced some novel solutions towards next generation network e.g. [20–29]. However, they could not address the issues relative to small cells which are highly essential to offer minimal delay and maximized data delivery performance in 5G. In order to deploy a 5G spectrum of maximized frequency, cell density is demanded to be higher owing to an event of frequent path loss. Hence, it is quite challenging to improve the spectrum utilization by deploying a greater number of small cells, which is not only expensive but also demands more power. Adoption of existing approaches are also witnessed with demand of training data or certain heuristic or adopt iterative methods in order to make a decision of an effective utilization of spectrum. Unfortunately, these processes are not suitable for distributed large-scale implementation with massive number of heterogeneous devices effectively.
- *Issues with Existing Resource Allocation*: Irrespective of various methods of resource allocation schemes [30–39], it is noted that the computation of resources to be allocated doesn't map with the dynamic scenario of network especially in case of mobility. Adoption of Cognitive Radio Network (CRN) offers a better solution by allowing the licensed channel to be dynamically accessed by the user equipment. Although the spectrum utilization is improved in this manner but there is no effective mitigation towards various other artifacts present in secondary and primary user (e.g., interference). The implementation of spectrum sharing approach is endowed with fairness problem still. Apart from this, it is reported in existing studies that cloud and fog-based radio access network could offer a better performance in perspective of spectrum utilization and resource management; however, the studies doesn't address bottleneck issues of cloud-based RAN due to fronthaul bandwidth. Such work doesn't address allocation of resources between remote and baseband unit. Hence, the problems associated with resource allocation is 5G and CRN still demands to be explored more.

4 Research Methodology

The prime aim of the proposed scheme is to optimize the first objective in order to offer full fledge spectrum utilization and thereby enhance the resource allocation to higher degree. The primary work has been carried out using OFDMA as a multiple access policy in 5G-based CRN system where one user is allocated with a carrier or a set of carriers. The enhancement carried out on it is by adding a spreading code or sparse code. This technique can offer an efficient mechanism where large number of users can be added in order to retain better spectrum efficiency. The research problem in this case is to formulate a technique to do so. The proposed scheme implements an analytical approach in order to simplify the integration and progressive approach of implementing both the multiple access schemes. The solution offered by proposed system is to integrate OFDMA with CDMA and to formulate a novel multiple access of SCMA using a novel mathematical modeling. Figure 1 highlights the proposed scheme.

According to Fig. 1, proposed scheme uses a set of input viz. User equipment, resource blocks, time slots, system bandwidth, and transmit power. Two discrete and revised multiple access schemes using OFDMA and CDMA is presented where the

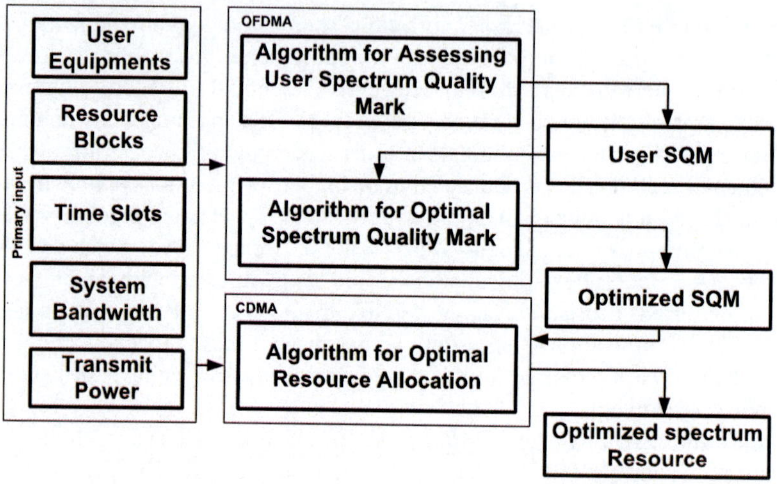

Fig. 1. Proposed Scheme of Spectrum Resource Optimization

evaluation of a novel attribute called as Spectrum Quality Mark (SQM) is carried out in order to empirically represent the degree of spectrum resources. The SQM value is spontaneously subjected to sequences of proposed algorithm in order to obtained its optimized score that finally results in optimized spectrum resource management. The core idea of the proposed scheme is towards developing a simplified analytical model to improve the performance of 5G-based CRN communication system. This part of the system will construct a novel mathematical model that will perform dual optimization on resources of OFDMA as well as receive power of CDMA as a novel SCMA algorithm. The first phase of the study will focus on optimizing assignment of sub-carriers offered to received power of CDMA followed by optimizing transmit power of OFDMA. The network model will consist of OFDMA system as well as CDMA system with their discrete receive antenna but common site of cell. The novel SCMA algorithm will also emphasize on maintaining a balance between optimal throughput and minimal interference using a newly constructed objective function. Further the optimization performance can be further enhanced incorporating a novel search optimization algorithm in order to explore an elite state of resource allocation.

5 Algorithm Implementation

The proposed algorithm targets towards accomplishing higher utilization of the spectrum resource management considering two multiple access scheme OFDMA and CDMA. The core idea of this algorithm implementation is to incorporate both these revised and hybrid multiple access scheme in network devices within 5G CRN ecosystem. The scenario of proposed algorithm implementation is showcased in Fig. 2

Figure 2 showcase three different forms of layers. The first layer is used for core network that manages radio resources and knowledge base. This is the layer where CRN ecosystem is considered to be existing where cognition and observation is carried out

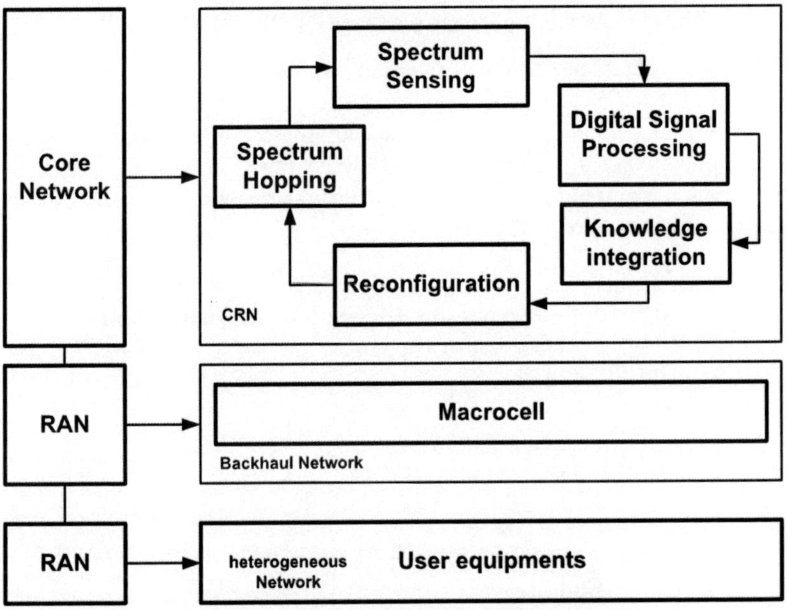

Fig. 2. Scenario of Algorithm Implementation

towards spectrum hopping leading to processing the information for spectrum sensing using digital signal processing. Further knowledge extraction is carried out using reasoning and learning followed by performing reconfiguration using decision process. The second middle layer is for RAN which is connected with macrocell in backhaul network while the last layer is used for RAN connected with heterogeneous network system with varied applications over microcell and picocell. The prime input for the proposed algorithm design is mainly number of user equipment, time slots, resource block, and transmit power. The study model considers individual time-slots to be invariant with respect to all the wireless channel. The proposed study algorithm also assumes that equivalent distribution for the antennas in base station for the channel coefficient for same for both OFDMA and CDMA scheme. The algorithm also constructs a logic for path loss distance between specific number of user and base station where further complex Gaussian random variable is utilized for constructing the channel coefficient of Rayleigh fading. The prime motive of this algorithm is to construct a novel empirical attribute called as spectrum quality mark followed by optimizing it in order to obtain better quality of signal under different ranges of signal availability on 5G traffic system. A simplified algorithmic approach is used by integrating both OFDMA and CDMA scheme mainly focusing on allocation of resources and maximum utilization of spectrum. The complete detailing of proposed algorithm towards optimization is discussed as follows:

5.1 Algorithm for Assessing User Spectrum Quality Mark

The proposed scheme introduces a simplified parameter in order to assess the quality of a spectrum called as Spectrum Quality Mark (SQM). This parameter is utilized by the UE in order to understand the quality of spectrum associated with its connection with eNB. Identification of this outcome of SQM will let the UE know the degree of coding and modulation required for assigning optimal sub-carriers in 5G-based CRN network. The algorithmic steps are as follows:

Algorithm for Assessing User Spectrum Quality Mark

```
Input: α_low, α_high, sq_low, sq_high
Output: user_α
Start
1. init α_low, α_high, sq_low, sq_high
2. For k=1: TS_n
3.    For i=1: Rb
4.       For j=1:K
5.          sq=f1(j). arb
6.       End
7.    End
8. For i=1: size(sq, 1)
9.    For j=1: size(sq, 2)
10.      user_α(i, j)=(A1.A2)/A3
11.      If user_α(i, j)> α_high
12.         user_α(i, j)= α_high
13       Else If user_α(i, j)< α_low
14.         user_α(i, j)= α_low-1
15.      End
16.   End
17. End
End
```

The algorithm takes the input of low spectrum quality metric α_{low}, high spectrum quality metric α_{high}, low signal quality sq_{low}, and high signal quality sq_{high}(Line-1) which after processing yields an outcome of user spectrum quality user_α. Considering the user-defined entries of number of UE i.e., *K* (Line-4), resource block *Rb* (Line-3), and Time slots TS_n (Line-2). In the next process of computation, the proposed scheme implements a discrete function $f_1(\mathrm{x})$ in order to compute signal quality *sq* (Line-5). The computation of this process is shown as follows:

$$\mathrm{f_1(x) = abs(P_T + C_g - N_p)} \tag{1}$$

In the above expression (1), the formulation of the function f1(x) is dependent on transmit power at base station PT, channel gain for kth user Cg, and SNR of kth user Np. The transmit power of the base station is initialized to 0 while the computation of other two dependable variables are as follows:

$$\mathrm{C_g = -P_l + 10.\log_{10}(\gamma)} \tag{2}$$

$$N_p = \sigma + 10.\log_{10}(\beta.\,B) \tag{3}$$

In the above expression (2) and (3), the dependable attribute Pl represents an array for retaining path loss distance while variable γ represents channel coefficient for kth user in case of fading effect. The attribute σ, β, and B represents noise figure, noise spectral density, and bandwidth of system respectively. The signal quality metric for kth used allocated with Rb resource block is finally obtained by multiplying the SNR value using function f1(x) with an arbitrary value (Line-5). The algorithm further considers the range of SNR of kth user with respect to signal quality sq (Line-8 and Line-9) follow computation of user spectrum quality i.e. user_α(i, j) as exhibited in Line-10. The empirical expression of the dependable variable A1, A2, and A3 are represented as follows:

$$A_1 = sq(i,j) - sq_{low} \tag{4}$$

$$A_2 = \alpha_{high} - \alpha_{low} \tag{5}$$

$$A_3 = (sq_{high} - sq_{low}) + \alpha_{low} \tag{6}$$

From the above expressions (4)–(6), it can be noted that computation of user spectrum quality mark is quite simplified empirical form with lower dependencies of variable. Once the outcome of user_α is obtained than it is compared with higher value of spectrum quality mark αhigh (Line-11). If positive condition is found, the higher value αhigh is assigned to α (Line-12) otherwise (αlow-1) is assigned. Therefore, the final computation of spectrum quality mark of user i.e., user_α is computed.

5.2 Algorithm for Optimal Spectrum Quality Mark

This algorithm is responsible for computing spectrum quality mark optimally by optimizing the transmit power. However, transmit power cannot be directly optimized in OFDMA and it requires more control over varying number of users. One effective mechanism is to develop a temporary matrix for allocation of transmit power based on certain quality indicator of transmission. Such mechanism can be used for balancing the computational problem for adoption OFDMA in IoT based resource constraint application in 5G. The algorithmic steps are as follows:

Algorithm for Optimal Spectrum Quality Mark
Input: user_α
Output: osqm
Start
1. **For** i=1: Rb
2. $\alpha_{max}(i) = \arg_{min}(user\_\alpha(1))$;
3. $user\_\alpha = f_2(user\_\alpha)$
4. $ind = [\alpha_{max}(i) == user\_\alpha(i)]$
5. $a_{mat}(i) = ind$;
6. **If** $\Sigma ind > 1$
7. $r_n = arb(\Sigma ind)$
8. $s_n = f_3(r_n)$
9. $i_n = f_4(ind = 1)$
10. $a_{mat}(i) = i_n(s_n(1))$
11. **Else If** $\Sigma ind == 1$
12. $a_{mat}(i) = f_4(ind == 1)$
13. **For** $i = 1:size(a_{mat})$
14. $a_t(i) = arg_{max}(f_4(a_{mat}(i) == 1))$
15. **End**
16. $osqm(k) = a_t$
17. **End**

End

The algorithm takes the input of user_α (user Spectrum Quality Mark) which after processing yields to osqm (Optimal Spectrum Quality Mark). The initial step of this algorithm is to consider all the resource block Rb for ensuring better allocation process (Line-1). The algorithm initially accesses the elements with the newly obtained matrix for user_α obtained from prior algorithm with respect to its minimal value and retains that outcome for unit resource block of αmax (Line-2). In parallel, the algorithm also applies a discrete function f2(x) in order to eliminate any form of singleton dimension of the matrix user_α (Line-3). The resulting value is stored in original matrix user_α (Line-3). The next part of implementation is to extract the value of indices ind (Line-4). All the values of amax equivalent to newly obtained user_α from prior algorithm with respect to resource blocks is considered as indices and stored in a matrix ind (Line-4). The newly obtained indices matrix ind is stored in another matrix amat which is meant to reserve the allocation information termed and thereby termed as allocation matrix amat (Line-5). A conditional construct is written to find if the sum of all these indices matrix ind is more than 1 (Line-6). In case of positive scenario, the algorithm obtains the arbitrary number from the summation of all the indices ind and store in another matrix of random number i.e., rn (Line-7). The random number matrix rn is further subjected to function f3(x) which carries out sorting operation and further the outcome is stored in sorted indices matrix i.e., sn (Line-8). The number of indices i.e., in plays a bigger role here for controlling transmit power of OFDMA and hence a function f4(x) is developed which is responsible for searching a condition when the value of indices matrix ind is equivalent to 1 (Line-9) and the obtained outcome with respect to sorted indices sn is stored back in allocation matrix amat (Line-10). On the contrary to conditional logic stated in Line-6, if the summation of indices ind is found to be equivalent to 1 (Line-11) than only the indices ind equivalent to one is stored back in allocation matrix amat (Line-12) and it doesn't perform sorting operation. In the next step, the maximum argument is obtained for all the allocation matrix amat which are equivalent to 1 using prior function f4(x) (Line-14) considering all sizes of allocation matrix amat (Line-13). The outcome of this process i.e., at for each resource block Rb is finally considered as outcome of optimized spectrum quality mark osqm (Line-16). Hence, proposed algorithm assists in significantly controlling of information over multiple sub-carriers which has positive effect on transmit power.

It should be noted that the proposed implementation is compliant of standard 802.11ax that is capable of processing data for concurrent clients. As the conventional deployment of OFDMA usually perform resource allocation of complete communication channel for a single user in order to facilitate transmission towards multiple clients, therefore, the proposed scheme mechanizes a way to control this transmission degree with respect to computed spectrum quality mark in prior algorithm. By adoption this technique, the mechanism of mission of data by OFDMA can be significantly control thereby optimizing the transmit power. Hence, this algorithm targets towards its applicability on applications that are anticipated to run on low/restricted bandwidth. By this mechanism, proposed scheme of OFDMA can be used for both uplink and downlink transmission whereas conventional OFDMA is used only for downlink transmission. The next algorithm is the final step of optimization of data transmission in 5G-CRN.

5.3 Algorithm for Optimal Resource Allocation

This is the final part of algorithm implementation that mainly emphasize towards optimal resource allocation. It is to be noted that proposed algorithm is designed explicitly towards optimizing both OFDMA as well as CDMA in order to enhance the performance of multiple access in 5G based CRN application. As the proposed approach is expected to be executed in the devices with resource constraint therefore, it is necessary to ensure a two essential task i.e., utilization of maximal spectrum and allocation of proposed amount of resources. Hence, this algorithm contributes towards addressing the issues related to management of spectrum resources. The steps of proposed algorithm are as follows:

Algorithm for Optimal Resource Allocation
Input: K, osqm
Output: p_s
Start

```
1. For l=1:K
2.    For i=size(o_c, 1)
3.       For j=1:size(o_c, 2)
4.          u=u+1
5.          o_c(i, j)=u
6.          If u=K
7.             u=0
8.       End
9.    End
10. For i=1:size(p_s)
11.      If mod(i, 2)= = 1
12.         itr_1=itr_1+1
13.          p_s (i)=osqm(itr_1)
14.      Else
15.         itr_2=itr_2+1
16.          p_s(i)=o_c(itr_2)
17.     End
18. End
19. End
End
```

The inputs to this algorithm are K (number of UE) and osqm (optimized spectrum quality mark) while upon processing it gives the outcome of p_s(optimal resource allocation). The algorithm considers all the number of the UE i.e., K (Line-1) followed by constructing a matrix to store spectral information of CDMA termed as variable o_c with respect to time slots, resource blocks, and transmit power. Considering the two number of input argument for constructing o_c matrix, the algorithm evaluates its sizes (Line-2 and Line-3) followed by incrementing the count of user u (Line-4). The users$u$ is then allocated to o_c matrix (Line-5). Upon this allocation process, the beneficial point is better accomplishment of time synchronization which is one core issues in conventional CDMA. Apart from this beneficial factor, if the number of user equipment K is found to be equivalent to user u (Line-6), the users u is assigned with 0 value (Line-7). The next line of implementation of proposed scheme is towards further optimizing the spectrum resources ps which is a matrix constructed equivalent to size of time slot, resource block, and transmit power. The proposed scheme considers the size of all such spectrum resources p_s (Line-10) followed by constructing a condition to assess if the modulus of p_s is equivalent to 1 (Line-11). Upon positive condition, the algorithm increases the first iterations itr_1 followed by allocation individual optimal spectrum quality mark to p_s (Line-13). Otherwise, it increases the second iteration variable itr_2 followed by allocating updated value of o_c to p_s (Line-16). Therefore, the proposed scheme computes an effective value of p_s which is capable of ensuring maximal spectrum utilization and efficient resource allocation.

The prime novelty of this algorithm is its mechanism of multiple access by balancing the demand of spectrum and resources. It is to be noted that majority of the algorithm implementation is carried out by OFDMA itself. On the other hand, the receiving power is being fine-tuned on the basis of transmission by CDMA under the supervision of second algorithm constructed. This form of communication system for multiple access using CDMA and OFDMA doesn't demand any form of signaling efforts of inter-system instantaneously. However, from the practical implementation perspective of device in IoT system with 5G services, it is anticipated for the OFDMA system to have a proper knowledge about highest possible receive power, target signal quality, and load of system. As the availability of such forms of information are quite infrequent therefore, the system doesn't have any potential dependency of signaling efforts for inter-system.

6 Result Analysis

The implementation of the proposed scheme is carried out in MATLAB considering 64-bit Windows environment with 16 GB RAM. The initial part of the deployment of test bed consists of 5G-based CRN system considering simulation parameters presented in Table 1.

Table 1. Simulation Parameters

Parameters	Values
Width of Operating environment	500
Length of Operating environment	500
Number of antennas at base station	2
Number of antennas at UE	1
System bandwidth	1.6
System noise figure	10 dB
Noise spectral density	174 dBm

The observation is performed considering 25 number of UE, 9 resource blocks, 6 time slots, 1.6 MHz of system bandwidth, and 10 dBm of transmit power of base station. Further, the discussion of the simulation outcome has been carried out on the observed initialized values of following variables e.g., $\alpha_{low} = 1$, $\alpha_{high} = 15$, $sq_{low} = -6$, and $sq_{high} = 20$. The outcome of the proposed scheme is compared with conventional multiple access i.e., OFDMA and CDMA with respect to standard performance metric viz. Network throughput, network fairness index, and delay over multiple test cases of Signal-to-Noise Ratio (SNR). The algorithms discussed in prior section are sequentially implemented on above test environment and comparative analysis is made to arrive to specific conclusion.

6.1 Analysis of Throughput

Network throughput is the first parameter that is evaluated to assess the successful deployment of proposed multiple access scheme. This computation is carried out by evaluation quantity of successfully transmitted data from one to another node in 5G environment for a specific period of time. The analysis of network throughput is carried out with respect to increasing values of base station transmit power and average SNR values. From Figs. 3 and 4, it is noted that proposed scheme that hybridizes both CDMA and OFDMA performs better than individually assessed conventional CDMA and OFDMA. The prime reason behind this is the adoption of optimized spectrum quality mark osqm which is capable of generation of allocation matrix efficiently with highly structured information of sorted index associated with nodes with its higher values of sqm. Hence, a simpler and faster process of data delivery is carried out over allocated optimized sub-carriers that cannot be witnessed in either OFDMA and CDMA.

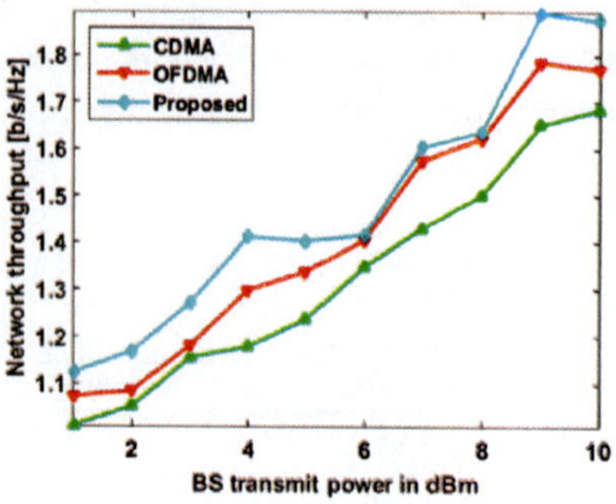

Fig. 3. Throughput Vs Transmit power

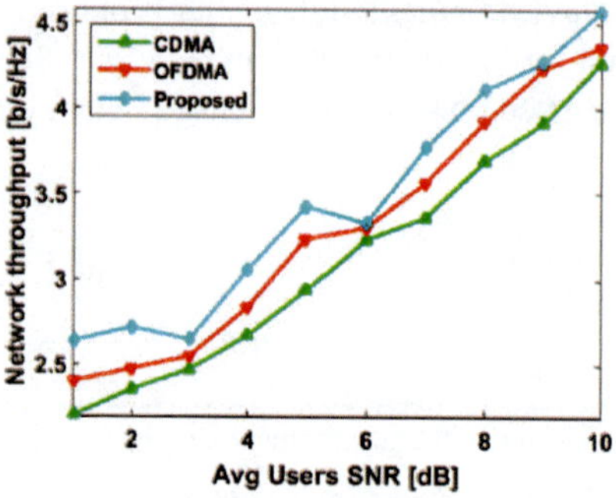

Fig. 4. Throughput Vs Average user

6.2 Analysis of Fairness Index

The second important parameter for assessing performance of proposed scheme is the fairness index which is responsible for computing the proximity or appropriateness of allocated resources from the anticipated goal. The computation of fairness index offers an information that determines if there are justified amount of fair share of system resource allocation for all the users in 5G test environment.

The simulation outcome in Fig. 5 of network fairness index for proposed scheme is found to be better compared to existing multiple access scheme over increasing average user SNR values. The prime reason behind this outcome is as follows – It is to be noted that core purpose of the proposed resource allocation mechanism calls for equal sharing of resources of e-Node B in 5G-based CRN present in macro cell. However, there is a possibility of higher traffic fluctuation which may be impediment to obtained fair share. Hence, the proposed scheme obtains the optimized value of *sqm* in simpler steps which is indirectly responsible for allocating the throughput via the 5G-CRN based gateway system to optimize the network performance. On the other hand, OFDMA introduces complexity due to conditional frequency diversity for assigning the subcarriers to the user. Nearly similar problems are there for CDMA where the subscribers are presented

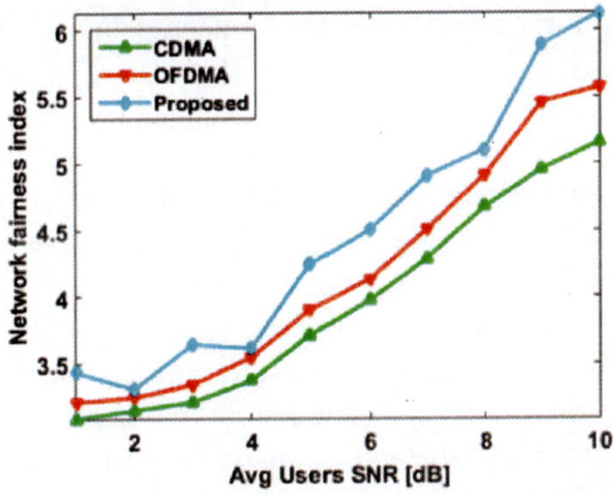

Fig. 5. Analysis of fairness index vs average user SNR

with surplus amounts of signals originating from multiple cell sites. These cause unequal allocation of resources over fluctuating traffic causing imbalanced fairness index for existing multiple access scheme. The proposed scheme has a unique mechanism of constructing an allocation matrix for various condition on the basis of summation of indices (referring to maximized *sqm* value) which is capable of properly justifying the resource blocks resulting in on-demand resource allocation. Hence, the proposed technique computes exact number of resources, assigns optimal sub-carriers, optimize transmit power followed by updating all the variables in order to deal with dynamically changing ecosystem of 5G environment. Hence, fairness index is found with better consistency and improved compared to existing multiple access scheme.

6.3 Analysis of Delay

An effective and efficient multiple access scheme should offer minimal delay irrespective of any traffic condition. The proposed assessment evaluates the delay performance with respect to increased transmit power of base station and average user SNR. The computation of delay is carried out by duration of lag associated with data transmission from one to another node.

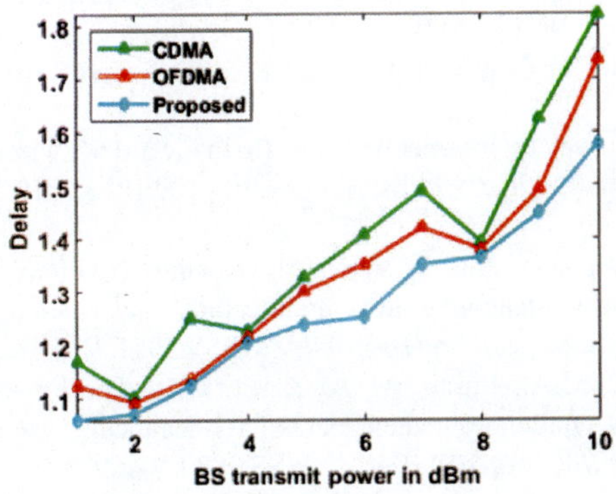

Fig. 6. Delay Vs Transmit power

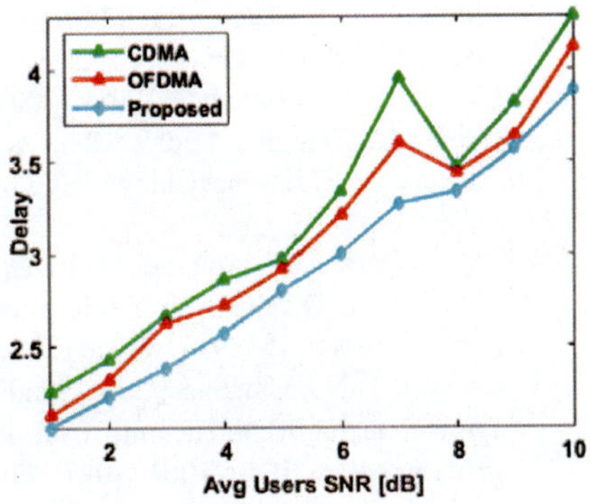

Fig. 7. Delay Vs Average user SNR

The outcomes in Figs. 6 and 7 exhibit reduced delay for proposed system in contrast to existing one. A closer look into Fig. 6 showcase that increase of transmit power of base station doesn't have much significant impact on delay of proposed scheme. It will eventually mean the reduced dependency of transmit power of base station referring to cost effectiveness during resource allocation process. Figure 7 highlights that delay increases for proposed scheme with increase of average user SNR although it is lesser than existing scheme. The prime reason behind this is proposed scheme adopts a progressive scheme without any form of iterative operation involved in process of allocation. At the same time, as proposed scheme hybridizes both OFDMA and CDMA, the probabilities towards mitigating any forms of impediment towards seamless communication is potentially controlled due to its management of path loss parameters. Hence, a better control of data transmission is noted in proposed scheme as compared to existing multiple access scheme.

7 Conclusion

This paper has presented a novel modelling targeting towards improving the spectrum utilization to maximum level as well as accomplishing better resource allocation. The contributions of the proposed scheme are as follows: i) the proposed scheme present a simplified architecture of 5G based CRN considering minimal number of inputs, ii) the proposed system harnesses OFDMA for optimizing transmit power and uses CDMA for optimizing receive power and assignment of sub-carriers, iii) the scheme presents a novel parameter called as spectrum quality mark which empirically calculates the quality of spectrum on e-Node B considering assigned resource blocks and signal quality, iv) the outcome of proposed scheme exhibited approximated 35% of increased throughput, approximately 41% of increased network fairness index, and approximately 45% of reduced delay compared to existing multiple access approaches of OFDMA and CDMA. The next direction of future work will be carried out further towards optimizing the model performance on variable traffic environment of 5G based CRN ecosystem.

References

1. Hunukumbure, M., Coon, J.P., Allen, B., Vernon, T. (eds.): The Technology and Business of Mobile Communications: An Introduction. Wiley (2021)
2. Thakur, P., Singh, G.: Spectrum Sharing in Cognitive Radio Networks: Towards Highly Connected Environments. Wiley (2021)
3. Chettri, L., Bera, R.: A comprehensive survey on internet of things (IoT) toward 5G wireless systems. IEEE Internet Things J. **7**(1), 16–32 (2020). https://doi.org/10.1109/JIOT.2019.2948888
4. Prados-Garzon, J., Ameigeiras, P., Ordonez-Lucena, J., Muñoz, P., Adamuz-Hinojosa, O., Camps-Mur, D.: 5G Non-public networks: standardization, architectures and challenges. IEEE Access **9**, 153893–153908 (2021). https://doi.org/10.1109/ACCESS.2021.3127482
5. Qureshi, H.N., Manalastas, M., Zaidi, S.M.A., Imran, A., Al Kalaa, M.O.: Service level agreements for 5G and beyond: overview, challenges and enablers of 5G-healthcare systems. IEEE Access **9**, 1044–1061 (2021). https://doi.org/10.1109/ACCESS.2020.3046927

6. Diab, R.A., Bastaki, N., Abdrabou, A.: A survey on routing protocols for delay and energy-constrained cognitive radio networks. IEEE Access **8**, 198779–198800 (2020). https://doi.org/10.1109/ACCESS.2020.3035325
7. Zheng, K., Liu, X., Zhu, Y., Chi, K., Li, Y.: Impact of battery charging on spectrum sensing of CRN with energy harvesting. IEEE Trans. Veh. Technol. **69**(7), 7545–7557 (2020). https://doi.org/10.1109/TVT.2020.2994005
8. Hu, F., Chen, B., Zhu, K.: Full spectrum sharing in cognitive radio networks toward 5G: a survey. IEEE Access **6**, 15754–15776 (2018). https://doi.org/10.1109/ACCESS.2018.2802450
9. Vameghestahbanati, M., Marsland, I.D., Gohary, R.H., Yanikomeroglu, H.: Multidimensional constellations for uplink SCMA systems—a comparative study. IEEE Commun. Surv. Tutorials **21**(3), 2169–2194 (2019). https://doi.org/10.1109/COMST.2019.2910569
10. Maraqa, O., Rajasekaran, A.S., Al-Ahmadi, S., Yanikomeroglu, H., Sait, S.M.: A survey of rate-optimal power domain NOMA with enabling technologies of future wireless networks. IEEE Commun. Surv. Tutorials **22**(4), 2192–2235 (2020). https://doi.org/10.1109/COMST.2020.3013514
11. Mohammadkarimi, M., Dobre, O.A., Win, M.Z.: Massive uncoordinated multiple access for beyond 5G. IEEE Trans. Wireless Commun. **21**(5), 2969–2986 (2022). https://doi.org/10.1109/TWC.2021.3117256
12. Mhedhbi, M., Boukour, F.E.: Analysis and evaluation of pattern division multiple access scheme jointed with 5G waveforms. IEEE Access **7**, 21826–21833 (2019). https://doi.org/10.1109/ACCESS.2019.2893362
13. Zeng, J., Fan, X., Zhong, Y., Peng, M., Lv, T., Jiang, T.: Downlink power allocation optimization in pattern division multiple access. IEEE Access **9**, 14649–14659 (2021). https://doi.org/10.1109/ACCESS.2021.3049493
14. Luo, Q., et al.: An error rate comparison of power domain non-orthogonal multiple access and sparse code multiple access. IEEE Open J. Commun. Soc. **2**, 500–511 (2021). https://doi.org/10.1109/OJCOMS.2021.3064504
15. Papanikolaou, V.K., Mitsiou, N.A., Diamantoulakis, P.D., Ding, Z., Karagiannidis, G.K.: Hierarchical multiple access (HiMA) for Fog-RAN: protocol design and resource allocation. IEEE Trans. Wireless Commun. **21**(2), 960–975 (2022). https://doi.org/10.1109/TWC.2021.3100538
16. Ahmad, A.A., Mao, Y., Sezgin, A., Clerckx, B.: Rate splitting multiple access in C-RAN: a scalable and robust design. IEEE Trans. Commun. **69**(9), 5727–5743 (2021). https://doi.org/10.1109/TCOMM.2021.3085343
17. Gamal, S., Rihan, M., Hussin, S., Zaghloul, A., Salem, A.A.: Multiple access in cognitive radio networks: from orthogonal and non-orthogonal to rate-splitting. IEEE Access **9**, 95569–95584 (2021). https://doi.org/10.1109/ACCESS.2021.3095142
18. Chen, Q., et al.: Design and demonstration of a TDD-based central-coordinated resource-reserved multiple access (CRMA) scheme for bidirectional VLC networking. IEEE Access **9**, 7856–7868 (2021). https://doi.org/10.1109/ACCESS.2020.3049004
19. Shi, Z., Liu, J., Zhang, S., Kato, N.: Multi-agent deep reinforcement learning for massive access in 5G and beyond ultra-dense NOMA system. IEEE Trans. Wireless Commun. **21**(5), 3057–3070 (2022). https://doi.org/10.1109/TWC.2021.3117859
20. Saha, R.K.: Approaches to improve millimeter-wave spectrum utilization using indoor small cells in multi-operator environments toward 6G. IEEE Access **8**, 207643–207658 (2020). https://doi.org/10.1109/ACCESS.2020.3037684
21. Ahmad, W.S.H.M.W., et al.: 5G technology: towards dynamic spectrum sharing using cognitive radio networks. IEEE Access **8**, 14460–14488 (2020). https://doi.org/10.1109/ACCESS.2020.2966271

22. Saha, R.K.: Licensed countrywide full-spectrum allocation: a new paradigm for millimeter-wave mobile systems in 5G/6G era. IEEE Access **8**, 166612–166629 (2020). https://doi.org/10.1109/ACCESS.2020.3023342
23. Saha, R.K.: Millimeter-wave spectrum utilization improvement in multi-operator networks: a framework using the equal likelihood criterion. IEEE Access **9**, 72980–72999 (2021). https://doi.org/10.1109/ACCESS.2021.3080326
24. Srinivasan, M., Murthy, C.S.R.: Efficient spectrum slicing in 5G networks: an overlapping coalition formation approach. IEEE Trans. Mobile Comput. **19**(6), 1299–1316 (2020). https://doi.org/10.1109/TMC.2019.2908903
25. Munoz, P., Adamuz-Hinojosa, O., Navarro-Ortiz, J., Sallent, O., Perez-Romero, J.: Radio access network slicing strategies at spectrum planning level in 5G and beyond. IEEE Access **8**, 79604–79618 (2020). https://doi.org/10.1109/ACCESS.2020.2990802
26. Tang, F., Chen, L., Li, X., Yang, L.T., Fu, L.: Intelligent Spectrum Assignment Based on Dynamical Cooperation for 5G-Satellite Integrated Networks. IEEE Transactions on Cognitive Communications and Networking **6**(2), 523–533 (2020). https://doi.org/10.1109/TCCN.2020.2988480
27. Chang, H.-H., Liu, L., Yi, Y.: Deep echo state q-network (DEQN) and its application in dynamic spectrum sharing for 5G and beyond. IEEE Trans. Neural Netw. Learning Syst. **33**(3), 929–939 (2022). https://doi.org/10.1109/TNNLS.2020.3029711
28. Abubakar, A.I., et al.: Revenue maximization through cell switching and spectrum leasing in 5G HetNets. IEEE Access **10**, 48301–48317 (2022). https://doi.org/10.1109/ACCESS.2022.3172280
29. Qian, B., Zhou, H., Ma, T., Yu, K., Yu, Q., Shen, X.: Multi-operator spectrum sharing for massive IoT coexisting in 5G/B5G wireless networks. IEEE J. Sel. Areas Commun. **39**(3), 881–895 (2021). https://doi.org/10.1109/JSAC.2020.3018803
30. Ahmed, J., Razzaque, M.A., Rahman, M.M., Alqahtani, S.A., Hassan, M.M.: A stackelberg game-based dynamic resource allocation in edge federated 5G network. IEEE Access **10**, 10460–10471 (2022). https://doi.org/10.1109/ACCESS.2022.3144960
31. Madi, N.K.M., Nasralla, M.M., Hanapi, Z.M.: Delay-based resource allocation with fairness guarantee and minimal loss for eMBB in 5G heterogeneous networks. IEEE Access **10**, 75619–75636 (2022). https://doi.org/10.1109/ACCESS.2022.3192450
32. Yu, P., et al.: Intelligent-driven green resource allocation for industrial internet of things in 5G heterogeneous networks. IEEE Trans. Industr. Inf. **18**(1), 520–530 (2022). https://doi.org/10.1109/TII.2020.3041159
33. Khan, J., Jacob, L.: Resource allocation for CoMP enabled URLLC in 5G C-RAN Architecture. IEEE Syst. J. **15**(4), 4864–4875 (2021). https://doi.org/10.1109/JSYST.2020.3018308
34. Jankovic, J., Ilic, Z., Oracevic, A., Ahsan Kazmi, S.M., Hussain, R.: Effects of differentiated 5g services on computational and radio resource allocation performance. IEEE Trans. Netw. Serv. Manage. **18**(2), 2226–2241 (2021). https://doi.org/10.1109/TNSM.2021.3060865
35. Wu, H.: Efficient allocation of the amount of radio resources in 5G NR to efficient allocation of radio resources in 5G NR. IEEE Trans. Wireless Commun. **21**(5), 3321–3332 (2022). https://doi.org/10.1109/TWC.2021.3120271
36. Šlapak, E., Gazda, J., Guo, W., Maksymyuk, T., Dohler, M.: Cost-effective resource allocation for multitier mobile edge computing in 5G mobile networks. IEEE Access **9**, 28658–28672 (2021). https://doi.org/10.1109/ACCESS.2021.3059029
37. Yun, J., Goh, Y., Yoo, W., Chung, J.-M.: 5G multi-RAT URLLC and eMBB dynamic task offloading with MEC resource allocation using distributed deep reinforcement learning. IEEE Internet Things J. **9**(20), 20733–20749 (2022). https://doi.org/10.1109/JIOT.2022.3177425

38. Ibrahim, A.A.Z., Hashim, F., Noordin, N.K., Sali, A., Navaie, K., Fadul, S.M.E.: Heuristic resource allocation algorithm for controller placement in multi-control 5G based on SDN/NFV architecture. IEEE Access **9**, 2602–2617 (2021). https://doi.org/10.1109/ACCESS.2020.3047210
39. Prathyusha, Y., Sheu, T.-L.: Coordinated resource allocations for eMBB and URLLC in 5G communication networks. IEEE Trans. Veh. Technol. **71**(8), 8717–8728 (2022). https://doi.org/10.1109/TVT.2022.3176018

An Efficient Network Slicing Navigation Scheme in a Network Virtualization Environment

Ducsun Lim and Inwhee Joe(✉)

The Department of Computer and Software, Hanyang University, Seoul 04763, Republic of Korea
{imcoms,iwjoe}@hanyang.ac.kr

Abstract. This study analyzes network virtualization slicing and proposes an efficient network slice setting scheme. Topology and address virtualization based on software-defined networking (SDN) has many advantages for cloud computing by providing control, monitoring, and flexible resource provisioning in virtual network environments. If the process of navigating nodes in the network virtualization service is not guaranteed, the network service quality will greatly degrade. To address this problem, we propose an improved navigation scheme to increase the efficiency of navigating nodes in a virtual network environment. Compared to the existing scheme, the proposed scheme reduces the search time for network slices and minimizes unnecessary overhead, according to the simulation results.

Keywords: network slicing · network virtualization · SDN · NFV · 5G

1 Introduction

With the increase in the number of applications and the demand for high-quality services owing to the recent development of smart devices, many changes to the existing mobile communication network have been required. In response to the need for such change, 5G mobile communication technology provides flexible and reliable network access with higher bandwidth and connectivity of various devices while providing low-latency end-to-end communication [1]. However, in this development process, there are several disadvantages in terms of existing networks.

Existing networks have not evolved significantly from the structure of processing simple data at end-to-end nodes, and the current Internet environment structure is limited in accommodating all the growing services [2, 3]. In addition, the network complexity of current network equipment is increasing to meet technical requirements, network operators' capabilities are limited, and supplier dependence has increased because vendors produce network equipment with differing interface settings or usage methods. As a result, there is a need for scalability based on ongoing network utilization, innovative features that consider virtualization, and technologies that can support resources [4]. To address this problem, OpenFlow, an open networking technology, has emerged, and OpenFlow-based software-defined networking (SDN) is spotlighted as the most suitable model to design a new structure instead of relying on the existing network structure.

R. Silhavy and P. Silhavy (Eds.): CSOC 2023, LNNS 723, pp. 198–208, 2023.
https://doi.org/10.1007/978-3-031-35317-8_17

Unlike traditional network environments, SDN technology can be centrally controlled through a software platform [5, 6]. Therefore, far from the fixed method of the existing network, it is possible to change management according to the purpose of use, so it can be used flexibly. Research and development related to SDN, a future Internet technology, is actively carried out in academia and industry, and many studies have been conducted in terms of its own network virtualization functions while supporting OpenFlow technology [7–9]. In addition, studies have been conducted to address the problem by separating network routing functions and hardware-based switches through network virtualization technology [10]. This means dividing the network into logical elements. In this manner, several physical resources can be used as one logical resource, and one physical resource can be divided into several logical resources and used.

OpenFlow-based SDN classifies traffic in the control plane and performs network virtualization through software platforms, such as Flowvisor [2], rather than directly dividing the resources of physical devices into software platforms. The Flowvisor is located in between the controller and switch to virtualize the network, divides the virtualized network into slices, a logical unit, and analyzes and converts all control information prior to the delivery to the switch. There is significant delay because FlowVisor navigates each slice when processing flow entries. It continues to explore the entire entry even if it finds a suitable slice of flow space, so as the number of entries increases, the latency continues to increase. This paper proposes a scheme to reduce network slice navigation.

In this paper, we introduce SDN and NFV technologies, analyze network slicing technologies in 5G mobile networks, and examine the configuration management problems. Furthermore, we propose algorithms for reducing navigation time of efficient network slicing and compare their performances. Finally, we consider the direction of improvement for the optimized network slicing search scheme, which considers scenarios involving 5G communication network in the future.

2 Related Work

In this paper, we examine SDN and NFV to propose a scheme for efficient network slicing navigation. We also describe network virtualization technologies and network slicing.

2.1 SDN and NFV

Software Defined Networking (SDN) is a network architecture that can centrally control the functions of network equipment, such as switches and routers, by separating into control planes and data planes by abstracting the equipment consisting of existing hardware and software [3]. The control plane is in charge of controlling numerous devices, and the data plane is in charge of forwarding. In this manner, it provides centralized control of the entire network and improves the scalability of the entire network. This refers to a technology that can realize a scalable 5G network that can control and coordinate various applications and services from a network perspective. As shown in Fig. 1, the SDN controller controls network slices based on network policies and maintains and manages network slice information. Resource allocation and control are performed based on this.

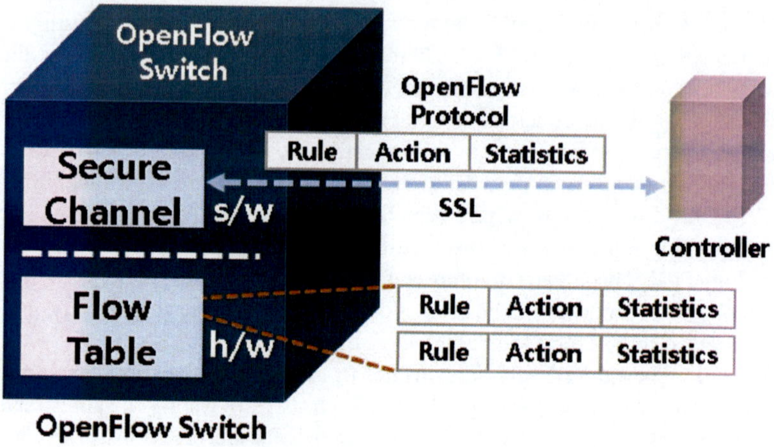

Fig. 1. The architecture of SDN

Network function virtualization (NFV) refers to a technology that implements the functions of network hardware devices to enable software management and control through virtualization [7]. This signifies a different approach to using hardware and software together with devices provided by existing network equipment manufacturers. Network functions can be installed relatively easily. In addition, via NFV technology, network resources may be dynamically allocated to virtual network functions (VNFs) more efficiently. In conclusion, NFV can provide scalability to meet the needs of users, reduce CAPEX and OPEX, and reduce the time it takes to deliver new services to the market.

SDN and NFV have many commonalities, but they require a new network platform to separate the control plane and data plane. However, because NFs operate on general-purpose servers, NFV has the capability to operate on existing networks.

2.2 Network Virtualization

Early network equipment was provided as independent devices. It was built to combine various network equipment and connect with each other, and because it was added or changed whenever conditions on the system were added or changed, the network configuration could become complex. Resource allocation techniques using network virtualization emerged as a solution to these problems (Fig. 2).

Network virtualization technology means that one network resource is logically divided and used as multiple network resources. Virtualization can raise the utilization rate of network resources to a high level [8]. Among network virtualization services, virtual networking is an attractive field for telecommunication businesses. By introducing network virtualization, it is possible to freely provide independent networks according to requirements without building network devices, such as routers and switches, separately. Tele-communications operators are trying to solve this problem by introducing network virtualization technology rather than simply expanding network equipment.

Recently, cloud computing technology and network virtualization have been combined to improve services [9]. This technology allows operators to access and manage network equipment through a central control method without having to operate the network only for specific vendors. OpenFlow technology uses a method of dividing traffic in the control plane to implement a virtual network.

2.3 Flow Visor

Flowvisor enables network virtualization of OpenFlow-based networks via virtualization technology that intercepts and modifies information about all controls between the OpenFlow controller and switch [2]. Flowvisor virtualizes the network by modifying messages exchanged between the controller and switch. It then passes information to the controller that matches the virtualization policy. An OpenFlow network defines only the traffic serviced through a virtual network as slices.

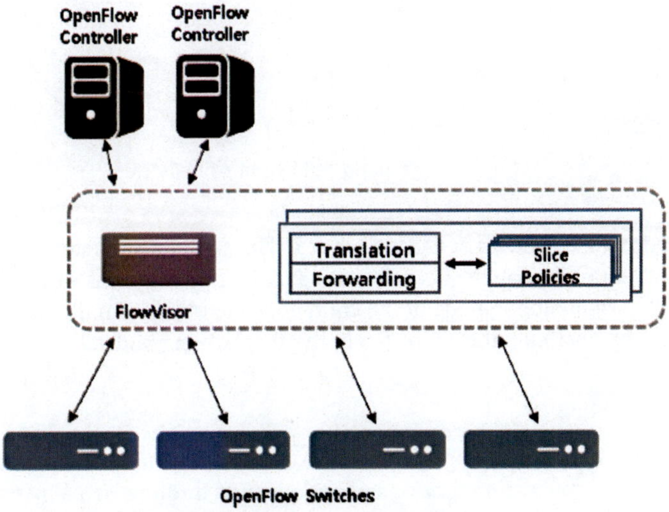

Fig. 2. The architecture of Flow visor

- When a client transmits a packet, it checks the flow entry information to the switch, and if there is no flow entry in the switch, the switch requests the path information from the controller.
- The Flowvisor intercepts the Openflow message between the switch and controller, checks the slice policy of the packet, and sends the message to the corresponding controller.
- When the controller sends a flow entry to the switch, the Flowvisor checks the slicing policy and sends it to the switch.

Each slice is controlled by the Flowvisor so that they do not interfere with each other, and each slice serves to connect the controller and switch. At this time, the controller and switch do not detect the Flowvisor.

2.4 Light Child-Right Sibling Binary Tree

Light child-right sibling binary tree (LC-RS) [11] can solve the problem of N-Link, which is difficult to apply because the tree changes from node to node. This is because the tree may have zero order or hundreds to thousands. Using linked lists by allocating dynamic memory can solve this problem. However, the disadvantage is the possibility of creating a more complex tree. In LC-RS, nodes with N orders can be expressed only by left child-right sibling. For one node to get all child nodes in a tree using this method, only pointing for the left right node is needed.

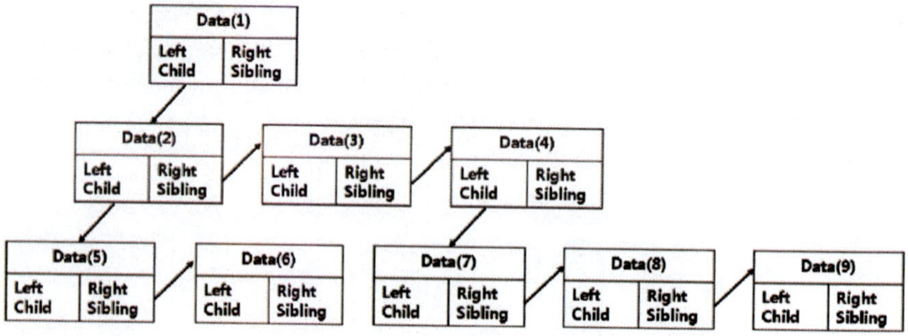

Fig. 3. Light child-right sibling binary tree

Figure 3 is a tree constructed using LC-RS. This is a node structure that has pointers to left-light and right sibling. By using this pointing, the address of the left child and address of the right-sibling node of the child node are obtained until all the child nodes are obtained by obtaining the addresses of the right sibling nodes.

2.5 Network Slicing

Network slicing means dividing a physical network into multiple virtual networks. It can be operated as an application service and optimization through virtual network division. With virtualization technology, cloud computing, and edge computing, physical network resources shared according to user requirements can be more effectively scheduled to logical network slices [14].

According to the definition of NGMN, network slicing consists of three layers: Network Slice Instance Layer (NSIL), Service Instance Layer (SIL), and Resource Layer (RL). NSIL provides several network characteristics required for a service instance, SIL includes network slice instances, RL provides physical or virtual resources and network functions needed to create a network slice instance [15].

3 The Proposed Scheme

In this paper, we propose a tag slice navigation scheme to improve the process of navigating nodes to minimize delays caused by existing sequential navigation searches in network slicing. The existing scheme performs flow space search when a flow entry message is transmitted from a controller to a switch during a message exchange process and when a flow request message is transmitted from a switch to a controller. In the process, the controller is implemented to sequentially search all flow space entries registered for network virtualization when a message containing path information is delivered. At this time, even if the flow space of the suitable slice is found, the delay time increases as the number of entries increases because the entire entry is searched.

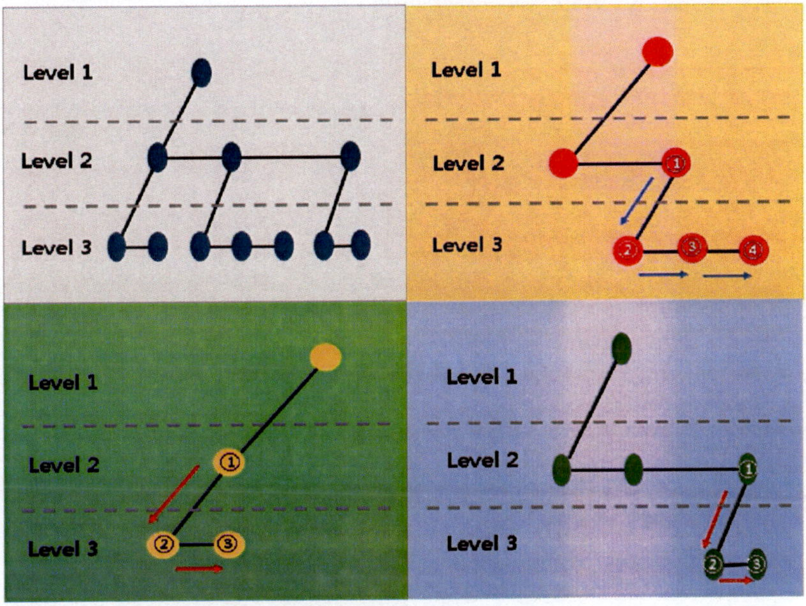

Fig. 4. The proposed network slicing exploration scheme

Each network slice configures the TCP port, MAC address, and IP address information present in the packet in the flow table to distinguish the traffic flow. In this scheme, nodes with child nodes are tagged and have related information. If there is a node you want to find in the flow space entry, the node with the tag attached is found, and the group to which the node belongs is more efficiently browsed to find the node. Thus, the tag-attached slice scheme can improve the search efficiency by quickly searching resources in the corresponding network slice and reducing the time taken to search for duplicate entries. Table 1 displays the details of the tags attached to the node (Fig. 4).

Table 1. The tag information attached to the node

Node	Left-child	Right-Sibling	Tag
Alice	Netxt.Alice	Next.Bob	Area(Alice)-1
Bob	Next. Bob	Next.Cindy	Area(Bob)-2
Cindy	Netxt. Cindy	Next.Dan	Area(Cindy)-3

Algorithm 1 shows a part of the code that searches for network slicing information by adding a tag. The lookRight function finds TAG in arhear_tag, similar to the method of finding a sibling in the LC-RS algorithm. The first if statement checks for the existence of a tag referring to slice, and the second switch statement searches for a node until the tag is equal to the area of Entry X. The node's investigation is done through recursive work. If the target node is not found in the second process, the search starts with the next tag value through the third switch statement.

```
BOOLEAN lookRight(Entry X, Slicer_tag T){
      if(T == NULL)
            return false;
      else if (X > T > element)
            return lookDown(x, T > LeftChild);
      else  /* x must be > T > element */
            return lookRight (X, T > rightSibling);
}
void lookdown(int A[], int i, int n){
      int child;
      child = 2 * i;
      if(child < n && A[child + 1] > A[child])
              ++child;

      if (child <= n && A[i] < A[child])
                swap(A, i, child);
                lookdown(A, child, n);
}
```

The lookRight function looks for tags in arhear_tag, similar to finding a sibling in the LC-RS scheme. In the first if statement, we check the existence of a TAG referring to Slicer, and in the second switch statement, we search for nodes only until the TAG is equal to the band of Entry X. At this time, the search of the node is performed through recursive work. If the target node is not found in the second area, the search is started with the next tag through the third switch statement [12].

$$P(n) = 1 + \frac{2}{n^2}\sum_{k=1}^{n-1} kP(k) \tag{1}$$

$$\sum_{i=1}^{\log_2 n} in/2^{i+1} \tag{2}$$

LookDown is a function that performs a search between nodes that have the same area, i.e., the same tag, and only searches for the first odd number of nodes. This shows that Eq. 2 achieves better performance in terms of efficiency than searching sequentially like in Eq. 1. The first switch statement searches for an odd-numbered node as above, and the second switch statement searches for an even-numbered node.

4 Simulation Results

In this section, we discuss the results obtained using computer simulations to demonstrate the performance of the proposed scheme. The simulation settings are as follows. We performed simulations on computers with Intel i7 CPU 3.7 GHz 6 core and 16 GB RAM. The software created a virtual topology using Mininet [16], linked Flowvisor and Opendaylight [17], and implemented a network topology that comprises multiple nodes on the OpenFlow switc h. The network bandwidth was 1 Gbps, and two controllers were registered in Flowvisor to insert 5,000 flow space entries and 10,000 flow space entries, respectively. Simulation tests were conducted on the existing network slicing technique and the proposed scheme. The performance index measured the delay for the packet of each slice.

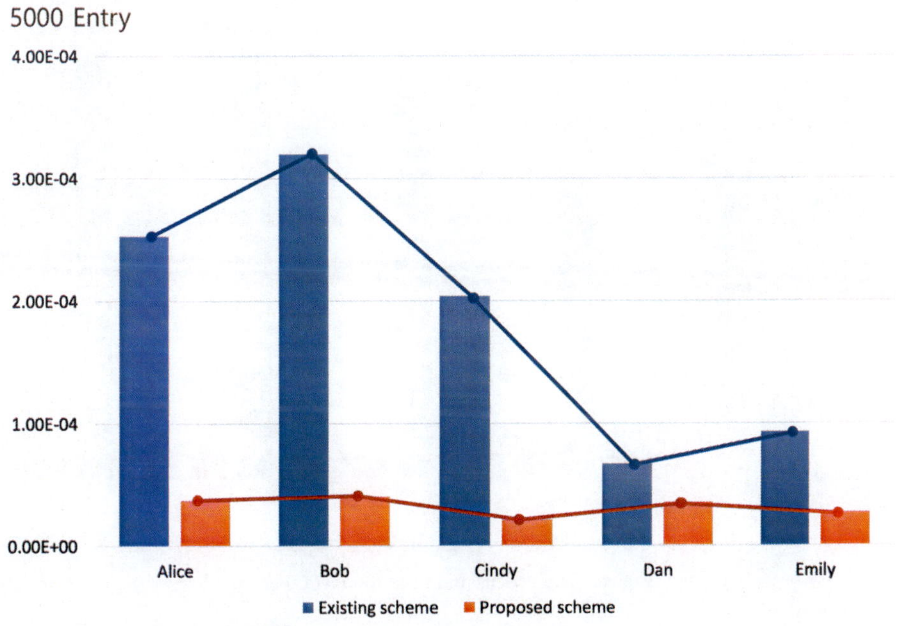

Fig. 5. The processing time for 5,000 entries

Figure 5 shows the results of a simulation test conducted by configuring 5000 flow space entries. Existing schemes find nodes through sequential search algorithm. As a result, many unnecessary delays occur. However, it was confirmed that TSS can reduce delay in navigating nodes. In addition, in the case of 5,000 entries summarized in Table 2, the delay of approximately 80% or more on average could be reduced compared to that of the existing scheme.

Table 2. The result of processing time for 5,000 entries

Node	Existing scheme	Proposed scheme	Reduction Ratio
Alice	2.16E–04	0.37E–04	85.3%
Bob	2.80E–04	0.30E–04	87.5%
Cindy	1.77E–04	0.21E–04	86.7%
Dan	0.32E–04	0.35E–04	47.7%
Emily	0.66E–04	0.27E–04	70.9%

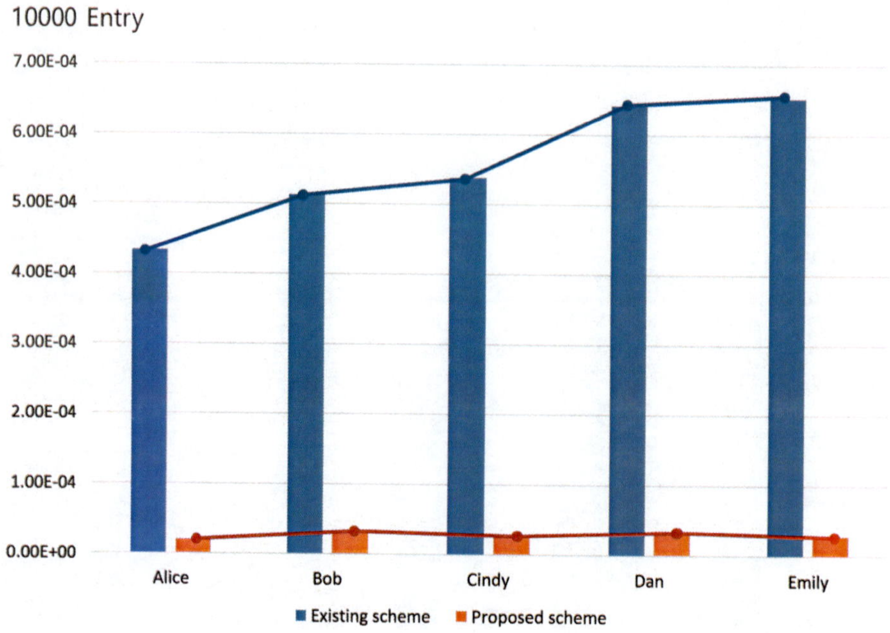

Fig. 6. The processing time for 10,000 entries

In addition, the results of the test with flow space entries set to 10,000 can be seen in Fig. 6. The existing scheme shows a high search time. However, the proposed scheme shows a low search time, and it can be confirmed that all nodes are constant. Table 3 summarizes the execution speed and processing time of the simulation test with the entry set at 10000. A network slice requires an independent physical network and performance from a network provider. Reducing the search time of the network slice can provide more efficient and smooth network performance.

Table 3. The result of processing time for 5,000 entries

Node	Table column subhead	Subhead	Subhead
Alice	4.14E–04	0.20E–04	95.3%
Bob	4.83E–04	0.40E–04	94.1%
Cindy	5.12E–04	0.21E–04	95.1%
Dan	6.07E–04	0.35E–04	94.5%
Emily	6.25E–04	0.27E–04	95.7%

5 Conclusion

In this paper, we proposed a network slicing navigation scheme on a basis of network virtualization for efficient network slice navigation so as to not explore unnecessary items. The proposed technique efficiently searches nodes based on tag information and reduces search time, thereby providing a smoother virtualization network landscape. Through experimental evaluation, the proposed methods were found to better minimize search time and communication delay compared to the basic network slicing method.

Future research will proceed with optimization measures according to network slicing service scenarios. In addition, a large-scale test bed will be considered by expanding the number of controllers to the SDN test bed, and machine learning should be applied accordingly as an improvement to plan for more efficient network slicing services.

Acknowledgements. This research was supported by Culture, Sports and Tourism R&D Program through the Korea Creative Content Agency grant funded by the Ministry of Culture, Sports and Tourism in 2022 (Project name: Customized tourism through AI-based tourist situation recognition and tourism information curation development of itinerary recommendation platform technology, Project Number: R2022020116).

References

1. Ghosh, A., Maeder, A., Baker, M., Chandramouli, D.: 5G evolution: a view on 5G cellular technology beyond 3GPP release 15. IEEE Access **7**, 127639–127651 (2019)
2. Sherwood, R.: A network virtualization layer. OpenFlow Switch Consortium, Tech. Rep, **1**, 132 (2009)
3. Braun, W., Michael M.: Software-defined networking using OpenFlow: Protocols, applications and architectural design choices. Future Internet **6**(2), 302–336 (2014)
4. Nunes, B.A.A., Mendonca, M., Nguyen, X.N., Obraczka, K., Turletti, T.: A survey of software-defined networking: Past, present, and future of programmable networks. IEEE Commun. Surv. Tutorials **16**(3), 1617–1634 (2014)
5. Al-Heety, O.S., Zakaria, Z., Ismail, M., Shakir, M.M., Alani, S., Alsariera, H.: A comprehensive survey: Benefits, services, recent works, challenges, security, and use cases for SDN-VANET. IEEE Access **8**, 91028–91047 (2020)
6. Feamster, N., Jennifer, R., Ellen, Z.: The road to SDN: an intellectual history of programmable networks. ACM SIGCOMM Comput. Commun. Rev. **44**(2), 87–98 (2014)

7. Jain, R., Subharthi P.: Network virtualization and software defined networking for cloud computing: a survey. IEEE Commun. Mag. **51**(11), 24–31 (2013)
8. Han, K., et al.: Application-driven end-to-end slicing: when wireless network virtualization orchestrates with NFV-based mobile edge computing. IEEE Access **6**, 26567–26577 (2018)
9. Mijumbi, R., Serrat, J., Gorricho, J.L., Bouten, N., De Turck, F., Boutaba, R.: Network function virtualization: state-of-the-art and research challenges. IEEE Commun. Surv. Tutorials **18**(1), 236–262 (2015)
10. Ullah, I., Ahmad, S., Mehmood, F., Kim, D.: Cloud based LoT network virtualization for supporting dynamic connectivity among connected devices. Electronics **8**(7), 742 (2019)
11. Cicirelli, V.G.: Effects of mother and older sibling on the problem-solving behavior of the younger child. Dev. Psychol. **11**(6), 749 (1975)
12. Kaloxylos, A.: A survey and an analysis of network slicing in 5G networks. IEEE Commun. Stand. Mag. **2**(1), 60–65 (2018)
13. Zhang, S.: An overview of network slicing for 5G. IEEE Wireless Commun. **26**(3), 111–117 (2019)
14. Afolabi, I., Taleb, T., Samdanis, K., Ksentini, A., Flinck, H.: Network slicing and softwarization: a survey on principles, enabling technologies, and solutions. IEEE Commun. Surv. Tutorials **20**(3), 2429–2453 (2018)
15. Hussain, M., Shah, N., Amin, R., Alshamrani, S.S., Alotaibi, A., Raza, S.M.: Software-defined networking: categories, analysis, and future directions. Sensors **22**(15), 5551 (2022)
16. Mininet. http://mininet.org
17. Opendaylight. https://www.opendaylight.org/

Gamification of the Graph Theory Course. Finding the Shortest Path by a Greedy Algorithm

Roman Tsarev[1,2], Shahzool Hazimin Azizam[3](✉), Aleksei Sablinskii[4], Elena Potekhina[5], Irina Gogoleva[6], Irina Nikolaeva[7], and Oleg Ikonnikov[2,8]

[1] Bauman Moscow State Technical University, Moscow, Russia
[2] Siberian Federal University, Krasnoyarsk, Russia
[3] Universiti Sultan Zainal Abidin, Kuala Terengganu, Terengganu, Malaysia
shahzool@unisza.edu.my
[4] Kemerovo State University, Kemerovo, Russia
[5] Russian State Social University, Moscow, Russia
[6] Arctic State Agrotechnological University, Yakutsk, Russia
[7] North-Eastern Federal University, Yakutsk, Russia
[8] Reshetnev Siberian State University of Science and Technology, Krasnoyarsk, Russia

Abstract. Information technology has had a significant impact on modern education. E-learning has become an integral part of the modern educational environment. One of the elements of modern education is gamification, which can essentially increase motivation and efficiency of the learner. The article describes the game of finding the shortest path in the graph developed by the authors. It can be used for gamification of courses related to discrete mathematics and information technology. A computer program has been developed to implement the proposed game, which makes it possible to use it in e-learning. #CSOC1120.

Keywords: E-learning · Gamification · Information Technology · Graph · Shortest Path

1 Introduction

The modern educational process, especially after experiencing the worldwide pandemic of COVID-19, is almost impossible to imagine without the use of different kinds of e-learning formats [1, 2]. E-learning arose due to the development of the Internet and multimedia, the key points being consulting, content, technology, services and support. It implies the transfer of knowledge mainly through the Internet and other information technology. Nowadays, e-learning continues to develop successfully in many different directions, providing possibilities which are not available in classical classroom education [3–5].

There are main advantages of e-learning in comparison to traditional formats: its modernity, relevance, wide prospects for further development and continuous improvement.

R. Silhavy and P. Silhavy (Eds.): CSOC 2023, LNNS 723, pp. 209–216, 2023.
https://doi.org/10.1007/978-3-031-35317-8_18

The modern world requires the use of the fastest and cheapest ways of generating and transmitting knowledge. E-learning is a technique that corresponds to these goals.

The beginnings of e-learning date back to the 19th century, when what experts call "the first generation of distance learning" was formed in the United States and Canada [6]. It was based primarily on sending books and coursework between a student and a teacher by regular mail. In the 1950s and 1960s, with the active development of information and telecommunication technologies, the "second generation" arose, when lessons were created on CD-ROMs, using software packages and other forms of storing information on various electronic media.

However, there is not only a technological difference between the two generations of e-learning development, but also a social and psychological one. In particular, they view learning itself from different perspectives. In these two generations there is no talk of social progress, no dynamic interaction and cooperation between teachers and learners [7, 8].

The most important distinguishing feature of modern e-learning, without which its effectiveness is greatly questioned, is the interactive format. Without this format, it is impossible to implement, for example, communicative, social functions of the educational process, which are the most important and significant. Later there was that necessary transformation of distance learning from a personal and individual one to a social one, where new forms of information exchange and interaction, interpersonal communication appeared. This, in fact, is what modern e-learning is all about, which we are observing and have become accustomed to.

In 2010 another innovative form of education began to gain popularity - these are massive open online courses, which have a truly worldwide scope and allow for simultaneous training for tens and hundreds of thousands of students from all over the world [9–12]. E-learning is groundbreaking, unique, state-of-the-art learning through a wide variety of techniques and formats [13]. Consequently, the methods used by teachers in the new format of work with students differ significantly from traditional ones [14–16]. For example, teaching through special web-resources, creation of virtual personal offices and entire classrooms, extensive use of various interactive formats and electronic textbooks, etc. [17–20]. It is difficult to find a modern educational organization, regardless of its status or the proposed level of education, where Internet forums, chats, mailing lists and LMSs are not used [3, 4, 13, 21].

One of the modern formats actively used by teachers in education nowadays is gamification.

2 Gamification in E-Learning

Gamification in its essence was used in the educational process earlier as well, but it had less complex, sophisticated internal structure - these were educational games, also used in practice. From pedagogical and psychological point of view gamification is a method positively affecting students and being perceived with interest and that is why it is one of the most effective methods for achieving the expected learning results [22, 23].

Gamification as a pedagogical technique implies realization of such educational process functions as teaching, communicative, compensatory, relaxing and even entertaining. In addition, gamification often helps students in self-expression, especially given the active development of modern cyberspace. Nevertheless, it is important to understand and remember that all this being said, the key goal of gamification methodology use is to obtain and apply in practice the knowledge obtained by students.

A distinctive feature of gamification is to preserve the content of the activity while changing the way of its organization [24]. Gamification differs from other game forms by the fact that in this way of organizing the learning process the connection with reality is preserved, and game elements are integrated into it.

The game attracts learners because it is clear, there are rules to achieve the goal, the goal itself and the path to it are clearly defined [25]. In a gamified course, the learner remains himself or herself without assuming any roles, and moves only on the basis of his or her motivation and internal goal, for example, to learn a foreign language.

Contrary to the similarity of the game shell, a gamified learning course is not just a computer game. As the student progresses through this course, he or she performs the assigned game and educational tasks, with the educational ones always taking precedence, while the game ones are an addition that helps to keep the learners motivated.

Not only scientists and teachers, but also students confirm the usefulness and effectiveness of gamification methods in the educational process [26, 27]. The latter note the qualitatively new nature of learning in the gamification format, ease of use and wide access to learning resources, as well as a clear psychological predisposition to the educational process implemented in this format. More than half of the students also state that they are motivated to succeed by the competitive nature of such educational courses [28].

Teachers, who also participated in the study, as noticeable results of the educational process implemented in the gamification format, name, in particular, an increase in students' grades, an improvement in actual knowledge, skills, as well as a significant growth in students' motivation and the degree of retention of their attention in the learning process [29, 30].

Thus, the effectiveness of gamification in the modern educational process is confirmed not only by the results of sociological research, empirical data, but also by specific numerical indicators, the comparison of which allows us to make conclusions about the quality of such a technique [28].

3 The Game "Finding the Shortest Path by a Greedy Algorithm"

The shortest paths problem implies that the result of an algorithm for finding the shortest path is a sequence of edges connecting given two vertices and having the shortest length among such sequences [31, 32]. Greedy algorithm is the algorithm that strives to make the best choice at each moment of time [33–35]. At each step the best choice is made without thinking about the future.

When gamifying a Graph Theory course, we designed a game called "Finding the Shortest Path by a Greedy Algorithm".

Assume that we have the original 10×10 graph (Fig. 1). The intersection of horizontal and vertical lines represents the vertices of the graph.

Initially the players are located in the vertices on the left side of the graph. The goal is to reach the right side of the graph. The players are evenly distributed on the left side

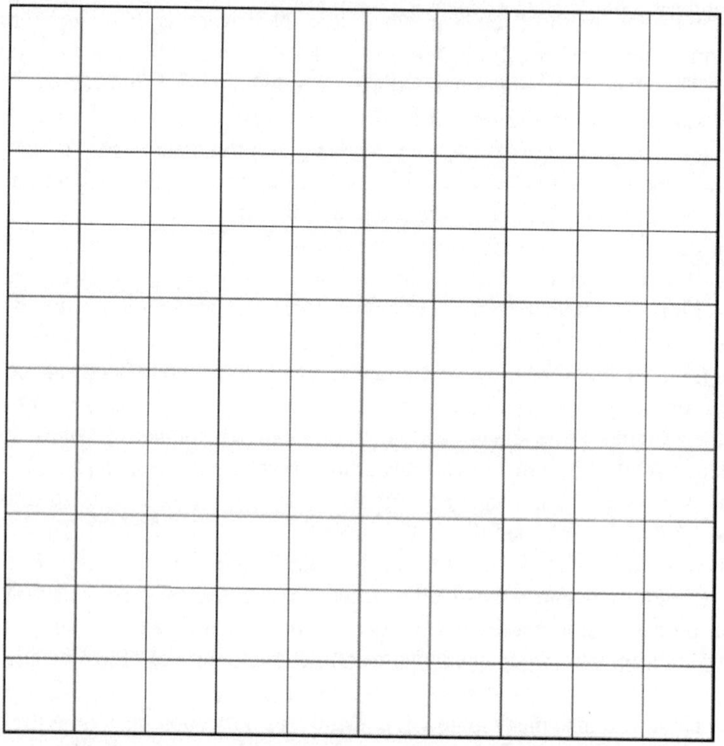

Fig. 1. Original graph.

(Fig. 2). If there is one player, he stands in the middle of the left side; if there are two players, the left side is divided into three roughly equal parts so that the players are equidistant from each other and from the edge nearest to them, and so on.

The maximum number of players is 11. Initially, each player has zero points. The weights on the edges are arranged randomly in the range from 1 to 100 (that is why it is reasonable to implement the game on a computer). Further, when the player moves to the next vertex, the weights of adjacent edges in three previously unknown directions appear (Fig. 3). The player moves along the edge with less weight (because the algorithm is greedy and chooses the edge with the shortest path). The player does not move along the edge with less weight, if he has already moved along it.

The weight of the edge that a player has passed through is added to the player's points. The game ends when all players reach the right side of the graph (in another variation it could be some particular vertex on the right side). Each player's points are then counted. The player with the fewest points at the end wins.

The game described above was implemented by us as a computer program. Playing it, students practically studied graphs and the problem of finding the shortest path.

Further, after studying the section related to the path in a graph, the students were tested on this topic. The test results show a good understanding of the subject. The test

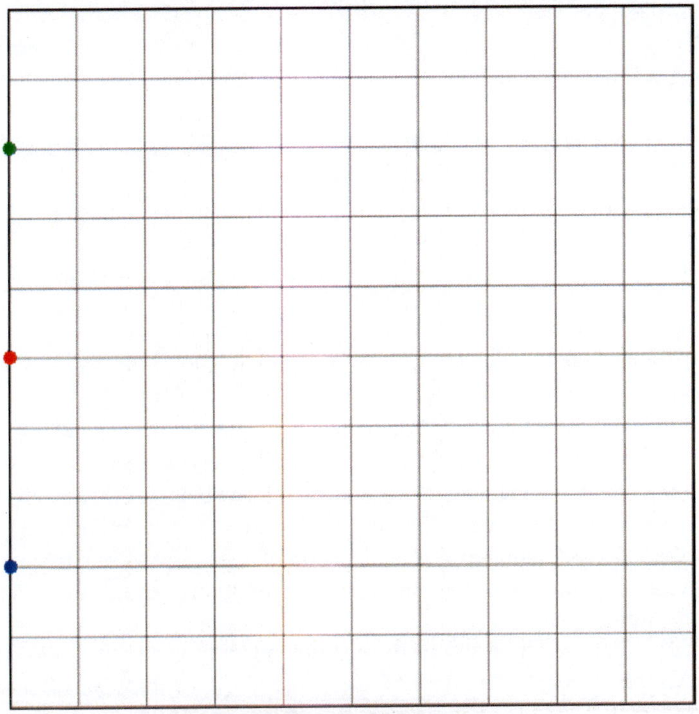

Fig. 2. Initial position of three players.

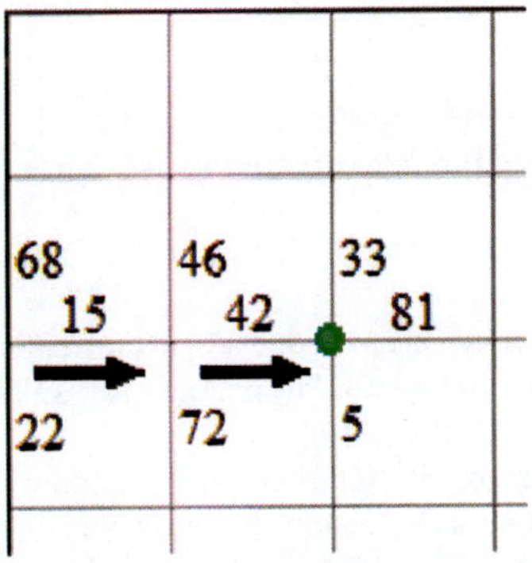

Fig. 3. Weighting coefficients on edges adjacent to vertices, on which the player passed.

results coincide with the normal distribution. This also indicates that the test is well balanced and relevant to the topic being studied.

Certainly, the results were also influenced by studying theoretical material and solving practical problems [13, 36]. However, the influence of gamification on the results of learning the problem of finding the shortest path in the graph cannot be avoided.

4 Conclusion

E-learning is a relevant and integral part of the modern educational process. Numerous teachers, psychologists, and other specialists speak and write about the urgency and importance of e-learning. At the present time e-learning is actively applied and it is assumed that in the future will develop and improve.

As one of the methods of e-learning, the object of our study was gamification. As noted in the paper, it is an effective pedagogical technique for use, recommended by experts for wide application in the educational process of a variety of educational organizations. A great number of different sociological and other research works are regularly carried out on the subject of gamification efficiency in the educational process which all confirm the positive effect of the method under consideration.

As an example of game-based methodology we gamified the course "Graph Theory" and developed the game "Finding the Shortest Path by a Greedy Algorithm". The results of practical application of the method in the educational process confirmed its relevance, increased interest in the method among students, as well as its efficiency.

Thus, we used a specific example to investigate how innovative methods of implementing the educational process in particular gamification allow us to achieve results efficiently and effectively. Theoretical assumptions were confirmed by practical implementation.

The game proposed in the article is recommended for use in the gamification of the courses "Graph Theory", "Discrete Mathematics", "Algorithms and Data Structures" and other disciplines related to mathematics and information technology.

References

1. Raza, S.A., Qazi, Z., Qazi, W., Ahmed, M.: E-learning in higher education during COVID-19: evidence from blackboard learning system. J. Appl Res. High. Educ. **14**(4), 1603–1622 (2022). https://doi.org/10.1108/JARHE-02-2021-0054
2. Li, S., Zhang, C., Liu, Q., Tong, K.: E-Learning during COVID-19: perspectives and experiences of the faculty and students. BMC Med. Educ. **22**(1), 328 (2022). https://doi.org/10.1186/s12909-022-03383-x
3. Hassouni, B.E., et al.: Realization of an educational tool dedicated to teaching the fundamental principles of photovoltaic systems. J. Phys: Conf. Ser. **1399**(2), 022044 (2019). https://doi.org/10.1088/1742-6596/1399/2/022044
4. Kononenko, A., Kravchenko, M., Nedospasova, L., Fedorovich, E.: E-Learning Online Platforms for Educational Approach. Lecture Notes in Networks and Systems **510**, 1089–1096 (2023). https://doi.org/10.1007/978-3-031-11051-1_111
5. Verma, S., Kaur, J.: A paradigm shift from classroom teaching to digital e-learning during Covid-19 pandemic: Study from teacher's perspective. In: Virtual learning: insights and perspectives, pp. 131–150. Taylor and Francis, London, UK (2022)
6. Nicholson, P.: A history of E-learning: echoes of the pioneers. In: Fernández-Manjón, B., Sánchez-Pérez, J.M., Gómez-Pulido, J.A., Vega-Rodríguez, M.A., Bravo-Rodríguez, J. (eds.) Computers and Education: E-learning, From Theory to Practice, pp. 1–11. Springer, Berlin (2008).https://doi.org/10.1007/978-1-4020-4914-9_1
7. Gupta, A., Motwani, S., Agarwal, A., Udandarao, V., Chakraborty, T.: Changing landscape of technical education pedagogy from traditional to practical e-learning. Computer **55**(11), 16–28 (2022). https://doi.org/10.1109/MC.2022.3164231

8. Souza, F.V., Motoki, F.Y.S., Mainardes, E.W., Azzari, V.L.: Public corporate e-Learning: antecedents and results. Public Organ. Rev. **22**(4), 1139-1156 (2022). https://doi.org/10.1007/s11115-021-00588-4
9. Ahmad, I., Sharma, S., Singh, R., Gehlot, A., Priyadarshi, N., Twala, B.: MOOC 5.0: a roadmap to the future of learning. Sustainability (Switzerland) **14**(18), 11199 (2022). https://doi.org/10.3390/su141811199
10. Perifanou, M., Economides, A.A.: The landscape of MOOC platforms worldwide. Int. Rev. Res. Open Dist. Learn. **23**(3), 104–133 (2022). https://doi.org/10.19173/irrodl.v23i3.6294
11. Shah, J., Khanna, M.: What determines MOOC success? Validation of MOOC satisfaction continuance model. Vision (2022). https://doi.org/10.1177/09722629221131386
12. Wu, B., Wang, Y.: Formation mechanism of popular courses on MOOC platforms: a configurational approach. Comput. Educ. **191**, 104629 (2022). https://doi.org/10.1016/j.compedu.2022.104629
13. Tsarev, R.Y., et al.: An approach to developing adaptive electronic educational course. In: Silhavy, R. (ed.) CSOC 2019. AISC, vol. 986, pp. 332–341. Springer, Cham (2019). https://doi.org/10.1007/978-3-030-19813-8_34
14. Aljarbouh, A., et al.: Application of the K-medians Clustering Algorithm for test analysis in E-Learning. Lecture Notes in Networks and Systems **596**, 249–256 (2023). https://doi.org/10.1007/978-3-031-21435-6_21
15. Baig, M.I., Shuib, L., Yadegaridehkordi, E.: E-learning adoption in higher education: a review. Inf. Dev. **38**(4), 570–588 (2022). https://doi.org/10.1177/02666669211008224
16. Huang, Y., Sun, Y.-Y.: A Meta-analysis of the Impact of E-learning on learner learning outcomes. In: Proceedings of the 2022 6th International Conference on E-Education, E-Business and E-Technology (ICEBT), 22–29. ACM, New York, NY, US (2022). https://doi.org/10.1145/3549843.3549847
17. Yu, X., Bo, L., Xin, C.: Low light combining multiscale deep learning networks and image enhancement algorithm. Mod. Innov. Syst. Technol. **2**(4), 0214–0232 (2022). https://doi.org/10.47813/2782-2818-2022-2-4-0215-0232
18. Zenyutkin, N.V., Kovalev, D.I., Tuev, E.V., Tueva, E.V.: On the ways of forming information structures for modeling objects, environments and processes. Mod. Innov. Syst. Technol. **1**(1), 10–22 (2021). https://doi.org/10.47813/2782-2818-2021-1-1-10-22
19. Akhmetjanov, M., Ruziev, P.: Fundamentals of modeling fire safety education informatics. Econ. Manage. **1**(2), 0301–0308 (2022). https://doi.org/10.47813/2782-5280-2022-1-2-0301-0308
20. Medvedeva, A., Veretekhina, S., Potekhina, E., Simonov, V.: The usage of the content management system technology wordpress for creating online stores. AIP Conf. Proc. **2442**, 040021 (2021). https://doi.org/10.1063/5.0076012
21. Poondej, C., Lerdpornkulrat, T.: Gamification in E-learning: a moodle implementation and its effect on student engagement and performance. Interact. Technol. Smart Educ. **17**(1), 56–66 (2019). https://doi.org/10.1108/ITSE-06-2019-0030
22. Alalgawi, D., Sadkhan, S.B.: Gamification trends in E-learning - a Survey. In: Proceedings of the 5th International Conference on Engineering Technology and its Applications (IICETA), pp. 193–198. IEEE, Al-Najaf, Iraq (2022). https://doi.org/10.1109/IICETA54559.2022.9888570
23. Kamunya, S.M., Oboko, R.O., Maina, E.M., Miriti, E.K.: A systematic review of gamification within e-learning. In: Handbook of Research on Equity in Computer Science in P-16 Education, pp. 201–218. IGI Global, Hershey, Pennsylvania, US (2020). https://doi.org/10.4018/978-1-7998-4739-7.ch012
24. Lunev, D.V., Poletykin, S.K., Kudryavtsev, D.O.: Brain-computer interfaces: technology overview and modern solutions. Mod. Innov. Syst. Technol. **2**(3), 0117–0126 (2022). https://doi.org/10.47813/2782-2818-2022-2-3-0117-0126

25. Bennani, S., Maalel, A., Ben, G.H.: Adaptive gamification in E-learning: a literature review and future challenges. Comput. Appl. Eng. Educ. **30**(2), 628–642 (2022). https://doi.org/10.1002/cae.22477
26. Pheng, H.S., Chin, T.A., Lai, L.Y., Choon, T.L.: E-Learning as a supplementary tool for enhanced students' satisfaction. AIP Conf. Proc. **2433**, 030005 (2022). https://doi.org/10.1063/5.0072901
27. Masa'deh, R., Almajali, D.A., Majali, T., Hanandeh, A., Al-Radaideh, A.: Evaluating e-learning systems success in the new normal. Int. J. Data Netw. Sci. **6**(4), 1033–1042 (2022). https://doi.org/10.5267/j.ijdns.2022.8.006
28. Maksimova, A.I.: Gamification in modern education on the example of the moodle platform. Young Sci. **23**(365), 322–324 (2021)
29. Leoste, J., et al.: Environment challenges of E-Learning in higher education—the teachers' perspective. Smart Innov. Syst. Technol. **908**, 143–156 (2023). https://doi.org/10.1007/978-981-19-5240-1_10
30. Nikolaeva, I., Sleptsov, Y., Gogoleva, I., Mirzagitova, A., Bystrova, N., Tsarev, R.: Statistical hypothesis testing as an instrument of pedagogical experiment. AIP Conf. Proc. **2647**, 020037 (2022). https://doi.org/10.1063/5.0104059
31. Aldoshkin, D.N., Tsarev, R.Y.: Evaluation of two-dimensional angular orientation of a mobile robot by a modified algorithm based on Hough Transform. Cybern. Inf. Technol. **18**(2), 1314–4081 (2018). https://doi.org/10.2478/cait-2018-0032
32. Rachmawati, D., Gustin, L.: Analysis of Dijkstra's algorithm and A* Algorithm in shortest path problem. J. Phys: Conf. Ser. **1566**(1), 012061 (2020)
33. Aldoshkin, D.N., Yamskikh, T.N., Tsarev, R.Y.: Mobile robot motion estimation using Hough transform. J. Phys: Conf. Ser. **1015**(3), 032161 (2018). https://doi.org/10.1088/1742-6596/1015/3/032161
34. Azis, H., Mallongi, R.D., Lantara, D., Salim, Y.: Comparison of floyd-warshall Algorithm and Greedy Algorithm in determining the shortest route. In: Proceedings of the 2nd East Indonesia Conference on Computer and Information Technology (EIConCIT), pp. 294–298. IEEE, Makassar, Indonesia (2018). https://doi.org/10.1109/EIConCIT.2018.8878582
35. Wayahdi, M.R., Ginting, S.H.N., Syahputra, D.: Greedy, A-Star, and Dijkstra's Algorithms in finding shortest path. Int. J. Adv. Data Inf. Syst. **2**(1), 45–52 (2021)
36. Rasheed, H.M.W., He, Y., Khalid, J., Khizar, H.M.U., Sharif, S.: The relationship between e-learning and academic performance of students. J. Public Aff. **22**(3), e2492 (2022). https://doi.org/10.1002/pa.2492

Check for updates

Integrated Privacy Preservation with Novel Encoding and Encryption for Securing Video in Internet-of-Things

Ramesh Shahabadkar[1](✉), Sangeetha Govinda[2], and Salma Firdose[2]

[1] Department of CSE, AMC Engineering College, Bengaluru, India
ramesh.shahabadkar@amceducation.in

[2] Department of CSE, Christ Academy for Advanced Studies, Bengaluru, India
{sangeetha,salmaf}@caias.in

Abstract. With the modernization of communication and security processes, the mechanism of video security system has also undergone a tremendous transformation leading to intelligence-incorporated video surveillance system. Review of existing literatures shows the presence of some of the concrete evolving security methodologies; however, they are never assessed on bigger scale of deployment scenario like that in Internet-of-Things (IoT). Therefore, the proposed scheme presents a novel computational analytical model where a unique encoding and simplified encryption scheme is introduced. The raw video feed from multiple location is stored in cloud server where they are indexed and subjected to sequences of security processing. The study outcome shows that proposed system to offer a potentially enhanced performance in comparison to existing scheme of video security.

Keywords: Video Security · Internet-of-Things · Video Surveillance System · Encryption · Cloud Server

1 Introduction

In majority of smart video security system, the system captures various private information apart from the video feed itself viz. Confidential details of user, location of individual, time, path of travel, etc. [1]. When this information is stored in one place than it is required to be encrypted for restoring maximum privacy preservation [2]. The benefit of storing encrypted video is that an attacker can only gain an access to encrypted feed in case of intrusion or security breach where they cannot have an access to original information [3]. However, this is quite challenging to be implemented especially in case of large and heterogeneous network of Internet-of-Things (IoT) based security surveillance [4]. This is because IoT integrates various applications and services along with video security system in smart city where it is computationally challenging task to find the breach origination point as well as to classify the type of breach [5]. At present, there are various approaches used by machine learning as well as deep learning

R. Silhavy and P. Silhavy (Eds.): CSOC 2023, LNNS 723, pp. 217–227, 2023.
https://doi.org/10.1007/978-3-031-35317-8_19

approaches towards magnifying the security of video surveillance system especially in wireless connected network [6–8]. Out of all this, the most important part of the threat is to secure the private information stored over distributed cloud servers. Developing a single algorithm to secure video surveillance system in IoT where there are various heterogeneous video security services running in parallel is quite a complex task [9, 10]. Therefore, the proposed study addresses the above-stated challenges by introducing a novel computational model which is lightweight, free from sophisticated and iterative approaches as seen in conventional studies. The paper contributes towards i) a novel encoding mechanism where strings of the video information are stored in higher levels of encoding process, and ii) a simple public key encryption process to secure the encoded video indexed private information. The organization of the paper is as follows: Sect. 2 discusses existing literatures, Sect. 3 outlined the identified research issues, adopted methodology is briefed in Sect. 4, system design is illustrated in Sect. 5, accomplished result is discussed in Sect. 6, while conclusive remarks is stated in Sect. 7.

2 Related Work

This section extends the discussion carried out in our prior work [11, 12] where recent schemes on video security is further updated. The discussion of recent techniques towards video security was carried out by Yu et al. [13] which stated the inclined trends towards usage of intelligence-based techniques towards video security. The work carried out by Zhang et al. [14] have developed an entropy-based model towards securing video surveillance system in public using sensor network. A secure cross-layer based approach is witnessed in work of Zhao et al. [15] where an iterative greedy algorithm has been used to improve the security strength. Study towards privacy preservation of streaming video is carried out by Tian et al. [16] where features are extracted via compressive sensing followed by adaptive clustering and segmentation. The study also deploys encryption followed by using Kalman filter for secured object tracking. Meulemeester et al. [17] have developed a scheme to protect the video information from leakage. Adoption of Convolution Neural Network is used for securing the video from anomalies. The study uses facial land mask to protect from outliers. Another unique form of study towards video security is formulated by Aloraini et al. [18] where a patch analysis is carried out towards securing the frames from video forgery attacks. Chen et al. [19] have developed an algorithm which can safeguard the discrete feature associated with facial videos. Adoption of Artificial Intelligence-based approach was reported in work of Dong et al. [20] where the multiple semantic features are extracted layer wise to be used for security monitoring. Further video steganography-based approach was seen in work of Fan et al. [21] where transform-based technique and modulation of quantized index is carried out. Current, there is also an increasing level of interest towards investigation of video security from the perspective of large-scale implementation scenario e.g., cloud, or Internet-of-Thing (IoT) system [22]. Various methodologies evolved in IoT-based video security are as follows: Campus surveillance using genetic algorithm and Bayesian network [23], Artificial Intelligence based security [24], encryption of image using probability [25–27], usage of access control protocol [28], usage of cryptosystem [29], ranking security using analytical hierarchy method [30], offloading security [31]. The next section highlights the issues.

3 Problem Description

The identified research problems from existing literatures are as follows: i) Existing schemes mainly uses Artificial Intelligence or sophisticated learning-based methods for introducing enhanced security features; however they are devoid of any cost effective computational features, ii) there is no studies reported for distributed security system over aggregated video from multiple location as they are mainly based on single location perspective, iii) adoption of encryption and key management based approach suffers from higher dependencies of key generation and complex key management in plain-text which can be easily compromised in large network, and iv) majority of the existing schemes has been implemented considering singular form of video attacks while network-based penetrating attacks are overlooked.

4 Proposed Methodology

The prime purpose of proposed study model is to develop a novel computational model which is capable of performing privacy preservation to optimized level along with an effective data integrity for securing multimedia files in IoT. A use case of video captured feed is considered in proposed study which is subjected for proposed security approach. The architecture of proposed scheme in Fig. 1 is as follows:

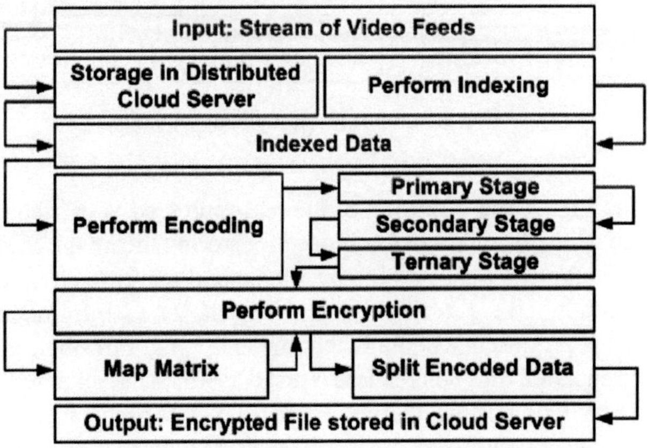

Fig. 1. Adopted Architecture of Proposed Video Security Model

The implementation of the proposed scheme is carried out considering analytical research methodology. Initially a topology of an IoT is constructed where the visual sensor is integrated with sensing layer while all the sensed data from multiple location are forwarded to cloud storage in network layer. The proposed algorithm is implemented over a data processing layer of IoT connected where the encryption is carried out. The query initiation and processing is carried out via application layer. A novel encoding and simplified encryption mechanism is presented in this paper to fulfill the security goal of privacy preservation. The next section discusses about system design adopted for implementation.

5 System Design

The prime purpose of the proposed research work is to design a secure multimedia surveillance system which is specifically designed to support smart video surveillance system in IoT. According to this model exhibited in Fig. 2, multiple feeds of video is captured from varied camera which are then stored in distributed cloud server where the encryption operation is carried out. The outcome of the encryption operation results in secure indexed data which are further protected in a separate database system where proposed algorithm is applied.

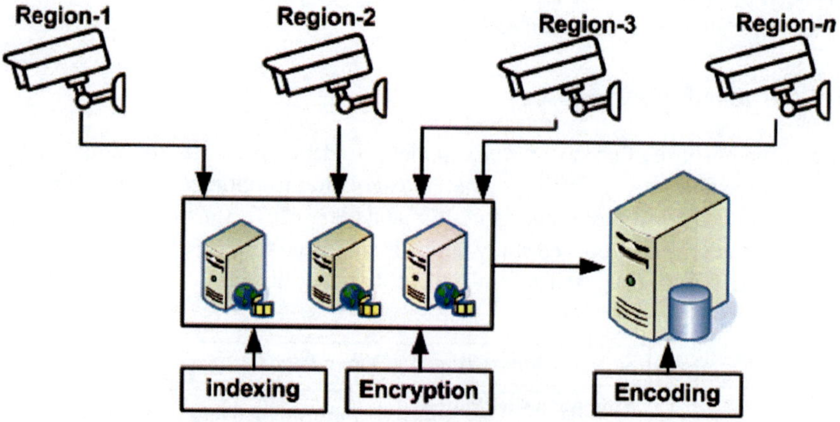

Fig. 2. Proposed deployment scenario

According to Fig. 2, the capturing of the real-time feed is carried out by the video capturing device while the cloud server is used for storing the image frames of the video. The process of encryption is carried out over the cloud server itself. A unique encoding process is further applied for storing the indexed data for each frame of the video. The contribution of the proposed scheme is that the encrypted information is stored in the form of indexed encoded files where there is no scope for an attacker to look for any form of information by giving normal queries. There is no trace associated with the plain-text indexed data as well as raw feed of data in plain-text formation which increases the privacy of the proposed scheme. It will eventually mean that it would be possible for an attacker to captured the raw feed or capture the encrypted feed; however, they won't be able to capture the plain-text files ever. Following are description of different phases of system design implementation:

5.1 Encoding Process

According to proposed encoding scheme, a unique function is designed to perform conversion of strings in original files with the unique set of single character. The proposed scheme looks for individual order of sorting of the converted strings as well as specific strings; however, it is not mandatory for the plain-text files to possess similar order of domain of encoded strings. The complete operation of encoding is carried out in three sub-stages of implementation viz. i) the primary stage is about converting all the single character to numerical values, ii) the secondary stage is about evaluating random estimates using pseudo-random numbers on the basis of domain of strings of character, and iii) evolving into novel estimates from arbitrary estimates followed by transforming scales into character.

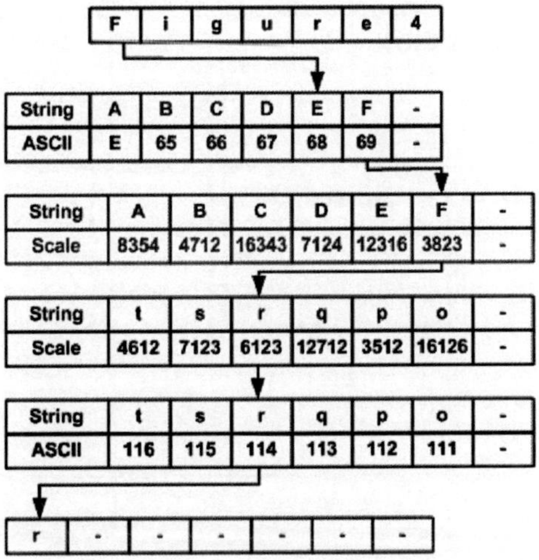

Fig. 3. Proposed Encoding Operation

According to Fig. 3, the study model considers that input of frames from where strings are maintained in the form of an array thereby implementing primary stage. Further, the secondary stage consists of linking the character from input array with the new array position specific to that character. In this stage, a sequential value is assigned for this specific character. In the ternary stage, the random values are obtained for that specific character in new array from the array in secondary stage. Finally, the transformed value of strong corresponding to this character is linked with an array of scale transformation of domain followed by obtained its respective ASCII value. The outcome of the encoding process is the transformed string itself. A closer look at this encoding process will show that it is nearly similar to encryption operation but much light weight operation with non-iterative nature in its process.

5.2 Encryption Stage

The proposed scheme uses Elliptical Curve Encryption (ECC) in order to carry out ciphering process of raw feeds in cloud storage unit. An identifier is allocated to the encrypted data of the video as well as the complete video file is further splitted into smaller segment where each segment also receives an allocated identity. Figure 5 highlights this process more elaborately where the module of transformed encoded index data module interacts with the cloud server for encryption stage. In this case, the identity of vide present in mapping matrix is checked for its similarity with that of encrypted identity of same video similarly, the identity of chunk in map matrix is checked for all the respective chunks of video. Apart from this, the proposed scheme also increases the encryption efficiency by securing the information of tag search that matches with identity of chunk of video as well as the complete video too. Further, encryption is carried out only for those segments of files where it matches with the identity of chunk (Fig. 4).

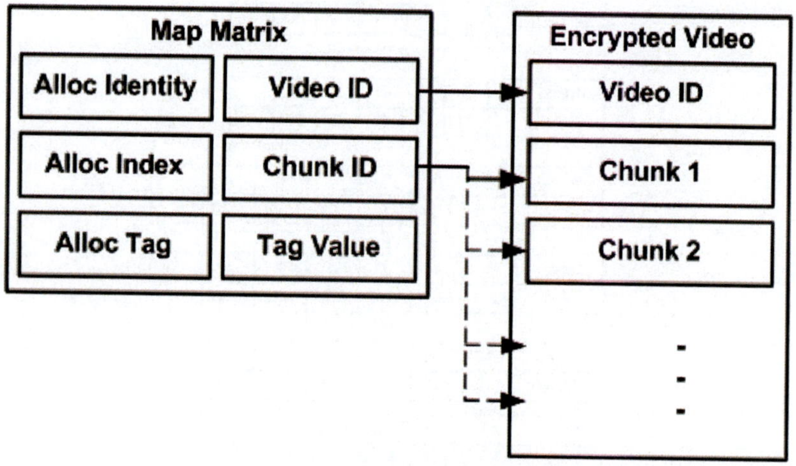

Fig. 4. Encryption Stage

The contribution of this stage of operation is that the system doesn't store any form of indexed data in the form of plain-text anywhere in the cloud server and hence all the private information of user or owner is secured. Another significant contribution of proposed video security scheme is that it is possible for making a direct query on the indexed data in-spite of the fact that they are not in the form of plain-text data in cloud server. It should be noted that in large scale IoT environment with distributed data, it is not feasible to carry out proper utilization of unique index associated with searching of range or right truncation. It can only facilitate for matching query which degrades the efficiency of querying data in cloud-based storage in IoT reported widely practiced in existing schemes. The novelty of proposed scheme is that there is no requirement of any form of information furnishing activity towards plain-text data. The next section discusses about the outcome.

6 Result Analysis

The development of the proposed logic is carried out in MATLAB considering a streamed video from 100 nodal points. The video feeds are converted to frames and subjected to the indexing and sequential security proceedings as discussed in prior section. In order to carry out benchmarking, the proposed system's outcome is compared with three existing schemes of security-based approaches noted from frequently adopted standards in literatures viz. Advanced Encryption Standard (AES), deep learning based Convolution Neural Network (CNN), and conventional SSL/TLS protocols in IoT ecosystem. The comparison is carried out over a standard performance metric of throughput, accuracy in identifying the threat, memory consumption, and overall execution time of the complete process. The idea of assessment is showcasing balanced outcomes between security and surveillance service performance in IoT.

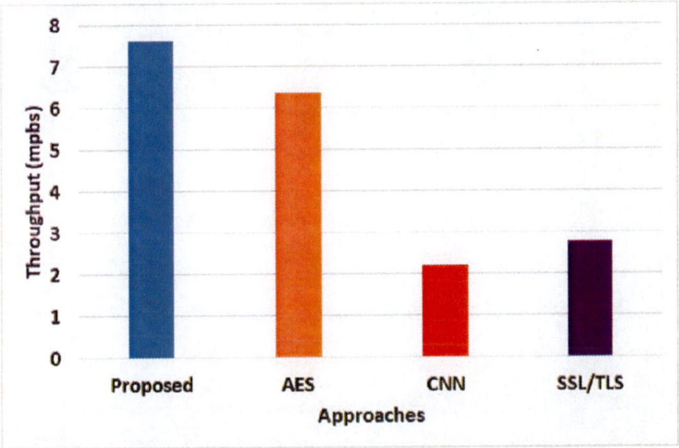

Fig. 5. Comparative Analysis of Throughput

Figure 5 highlights that proposed scheme exhibits enhanced throughput of approximately 55% in comparison to existing schemes. Although AES supports enhanced hardware acceleration, but still its key-dependent approach. CNN is characterized by higher degree of sophisticated processing while conventional SSL/TLS will need more resources for data delivery. On the contrary, proposed scheme presents a highly distributed mechanism where encoding and encryption works in parallel leading to no significant affect in data transmission.

Figure 6 highlights that CNN offers higher accuracy in threat detection. However, on the flipside, its accuracy depends upon availability of higher trained dataset, which could not be ensured round the clock. Hence, proposed scheme performs better threat identification as its modelling is carried out considering pattern of intruders while AES and SSL/TLS based approach could be used only towards resisting specific form of attacks. Hence, proposed scheme offers approximately 39% of enhanced accuracy from existing system.

Figure 7 showcase that proposed scheme accomplishes approximately 85% of reduction in memory consumption in contrast to existing approach. The higher memory consumption is by CNN approach owing to storage of information in its max pool layers, while SSL/TLS has higher implementation cost and network complexity. However, proposed scheme secured the complete indexed information of video feed in distributed manner followed by encryption and arbitrary assignment of string values. This makes it very lightweight and impossible for attacker to even guess the values.

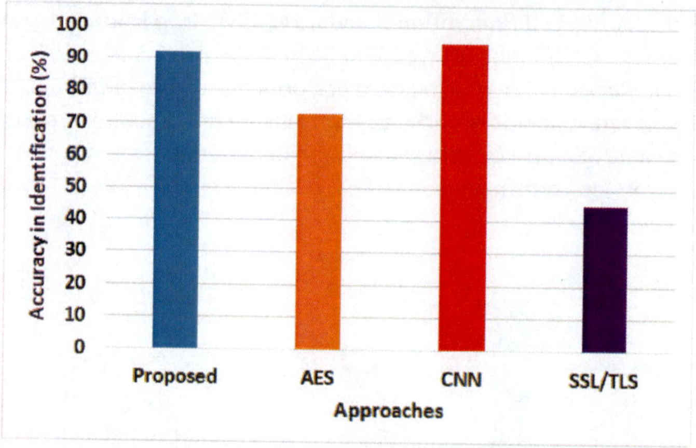

Fig. 6. Comparative Analysis of Accuracy

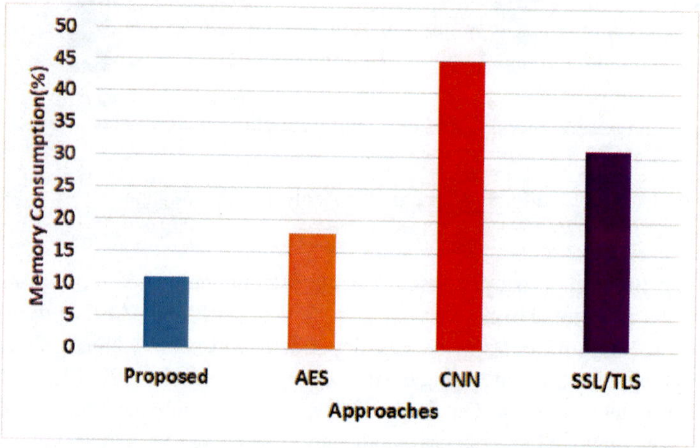

Fig. 7. Comparative Analysis of Memory Consumption

Figure 8 highlights that proposed scheme offers approximately 78% of faster execution time in contrast to existing approaches with similar justification stated in prior outcomes. Owing to sequential processing in encoding and non-iterative approaches, the execution time is highly improved. Apart from this, encryption is not carried out for all data which further makes the process faster unlike AES, CNN, SSL/TLS based approach.

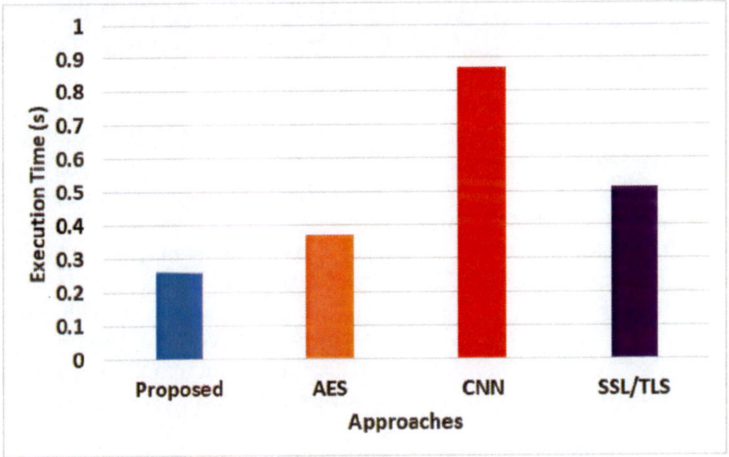

Fig. 8. Comparative Analysis of Execution Time

7 Conclusion

This paper has discussed about a solution towards securing the video feeds with respect to its private information while adopting smart video security system deployed in large area of IoT. The difference between conventional video security and smart video security system is that former stores only raw video feed while the latter also store raw video feed but along with various private information about the user and location. In scale of large deployment, such information grows and complexity which offers an imminent threat. The proposed study model offers a unique encoding scheme which increases the efficiency of indexed private information management in most cost-effective secured way. The scheme doesn't offer any form of dependency towards retaining plain-text data and thereby the private information are highly secured. Another novelty is that proposed scheme is designed to stop any form of intruder or dynamic attacker with malicious intention whereas the existing schemes are mainly designed towards stopping singular forms of threats. The future work of proposed study will be carried out further by considering the impact of network traffic towards the security performance to offer more sustainable security solution.

References

1. Zahra, A, Ghafoor, M., Munir, K., Ullah, A., Ul Abideen, Z.: Application of region-based video surveillance in smart cities using deep learning. Multimedia Tools Appl. (2021)https://doi.org/10.1007/s11042-021-11468-w
2. Gupta, A., Prabhat, P.: Towards a resource efficient and privacy-preserving framework for campus-wide video analytics-based applications. Comp. Intell. Syst. (2022). https://doi.org/10.1007/s40747-022-00783-w
3. Hosny, K.M., Zaki, M.A., Lashin, N.A., Hamza, H.M.: Fast colored video encryption using block scrambling and multi-key generation. Vis. Comp. (2022). https://doi.org/10.1007/s00371-022-02711-y

4. Barrera, D., Bellman, C., van Oorschot, P.C.: Security best practices: a critical analysis using IoT as a case study. ACM Trans. Priv. Secur. (2022)https://doi.org/10.1145/3563392
5. Alaghbari, K.A., Md Saad, M.H., Hussain, A., Alam, M.R.: Complex event processing for physical and cyber security in datacentres - recent progress, challenges and recommendations. J. Cloud Comput. **11**, 65 (2022). https://doi.org/10.1186/s13677-022-00338-x
6. Lohani, D., Crispim-Junior, C., Barthélemy, Q., Bertrand, S., Robinault, L., Rodet, L.T.: Perimeter intrusion detection by video surveillance: a survey. Sensors **22**(9), 3601 (2022). https://doi.org/10.3390/s22093601
7. Kalbo, N., Mirsky, Y., Shabtai, A., Elovici, Y.: The security of IP-based video surveillance systems. Sensors **20**(17), 4806 (2020). https://doi.org/10.3390/s20174806
8. Li, H., Xiezhang, T., Yang, C., Deng, L., Yi, P.: Secure video surveillance framework in smart city. Sensors **21**(13), 4419 (2021). https://doi.org/10.3390/s21134419
9. Khan, P., Byun, Y.-C., Park, N.: A data verification system for CCTV surveillance cameras using blockchain technology in smart cities. Electronics **9**(3), 484 (2020). https://doi.org/10.3390/electronics9030484
10. Socha, R., Kogut, B.: Urban video surveillance as a tool to improve security in public spaces. Sustainability **12**(15), 6210 (2020). https://doi.org/10.3390/su12156210
11. Shahabadkar, R., V.Pujeri, R.: Secure multimedia transmission in P2P using recurence relation and evolutionary algorithm. In: Thampi, S.M., Atrey, P.K., Fan, C.-I., Perez, G.M. (eds.) SSCC 2013. CCIS, vol. 377, pp. 281–292. Springer, Heidelberg (2013). https://doi.org/10.1007/978-3-642-40576-1_28
12. Shahabadkar, R., Pujeri, R.V.: Article: stratum based approach for securing multimedia content transmission over large scale P2P. Int. J. Comput. Appl. **89**(5), 43–48 (2014)
13. Yu, J.-Y., Kim, Y., Kim, Y.-G.: Intelligent video data security: a survey and open challenges. IEEE Access **9**, 26948–26967 (2021). https://doi.org/10.1109/ACCESS.2021.3057605
14. Zhang, H., Li, P., Du, Z., Dou, W.: Risk entropy modeling of surveillance camera for public security application. IEEE Access **8**, 45343–45355 (2020). https://doi.org/10.1109/ACCESS.2020.2978247
15. Zhao, T., He, L., Huang, X., Li, F.: QoE-driven secure video transmission in cloud-edge collaborative networks. IEEE Trans. Veh. Technol. **71**(1), 681–696 (2022). https://doi.org/10.1109/TVT.2021.3123787
16. Tian, X., Zheng, P., Huang, J.: Robust privacy-preserving motion detection and object tracking in encrypted streaming video. IEEE Trans. Inf. Forensics Secur. **16**, 5381–5396 (2021). https://doi.org/10.1109/TIFS.2021.3128817
17. Minari, G.H., et al.: Anomalies identification in images from security video cameras using mask R-CNN. IEEE Lat. Am. Trans. **18**(03), 530–536 (2020). https://doi.org/10.1109/TLA.2020.9082724
18. Aloraini, M., Sharifzadeh, M., Schonfeld, D.: Sequential and patch analyses for object removal video forgery detection and localization. IEEE Trans. Circuits Syst. Video Technol. **31**(3), 917–930 (2021). https://doi.org/10.1109/TCSVT.2020.2993004
19. Chen, M., Liao, X., Wu, M.: PulseEdit: editing physiological signals in facial videos for privacy protection. IEEE Trans. Inf. Forensics Secur. **17**, 457–471 (2022). https://doi.org/10.1109/TIFS.2022.3142993
20. Dong, Z., Wei, J., Chen, X., Zheng, P.: Face detection in security monitoring based on artificial intelligence video retrieval technology. IEEE Access **8**, 63421–63433 (2020). https://doi.org/10.1109/ACCESS.2020.2982779
21. Fan, P., Zhang, H., Zhao, X.: Adaptive QIM with minimum embedding cost for robust video steganography on social networks. IEEE Trans. Inf. Forensics Secur. **17**, 3801–3815 (2022). https://doi.org/10.1109/TIFS.2022.3215901

22. Lee, E., Seo, Y.-D., Oh, S.R. Kim, Y.-G.: A survey on standards for interoperability and security in the Internet of Things. IEEE Commun. Surv. Tutor. **23**(2), 1020–1047, Secondquarter (2021). https://doi.org/10.1109/COMST.2021.3067354
23. Zhou, Z., Yu, H., Shi, H.: Optimization of wireless video surveillance system for smart campus based on Internet of Things. IEEE Access **8**, 136434–136448 (2020). https://doi.org/10.1109/ACCESS.2020.3011951
24. Ahmed, A.A., Echi, M.: Hawk-Eye: an AI-powered threat detector for intelligent surveillance cameras. IEEE Access **9**, 63283–63293 (2021). https://doi.org/10.1109/ACCESS.2021.3074319
25. Khan, J., et al.: SMSH: secure surveillance mechanism on smart healthcare IoT system with probabilistic image encryption. IEEE Access **8**, 15747–15767 (2020). https://doi.org/10.1109/ACCESS.2020.2966656
26. Xu, C., Ren, W., Yu, L., Zhu, T., Choo, K.-K.R.: A hierarchical encryption and key management scheme for layered access control on H.264/SVC Bitstream in the internet of things. IEEE Internet Things J. **7**(9), 8932–8942 (2020). https://doi.org/10.1109/JIOT.2020.2997725
27. Ullah, F.U.M., et al.: AI-assisted edge vision for violence detection in IoT-based industrial surveillance networks. IEEE Trans. Industr. Inf. **18**(8), 5359–5370 (2022). https://doi.org/10.1109/TII.2021.3116377
28. Bera, B., Das, A.K., Garg, S., Jalil Piran, M., Hossain, M.S.: Access control protocol for battlefield surveillance in drone-assisted IoT environment. IEEE Internet Things J. **9**(4), 2708–2721 (2022). https://doi.org/10.1109/JIOT.2020.3049003
29. Alarifi, A., Sankar, S., Altameem, T., Jithin, K.C., Amoon, M., El-Shafai, W.: A novel hybrid cryptosystem for secure streaming of high efficiency H.265 compressed videos in IoT multimedia applications. IEEE Access **8**, 128548–128573 (2020). https://doi.org/10.1109/ACCESS.2020.3008644
30. Allifah, N.M., Zualkernan, I.A.: Ranking security of IoT-based smart home consumer devices. IEEE Access **10**, 18352–18369 (2022). https://doi.org/10.1109/ACCESS.2022.3148140
31. Zhao, T., He, L., Huang, X., Li, F.: DRL-based secure video offloading in MEC-enabled IoT networks. IEEE Internet Things J. **9**(19), 18710–18724 (2022). https://doi.org/10.1109/JIOT.2022.3161680

Luminance Histogram as a Tool for Examining the Image of a Hazardous Production Facility

O. S. Logunova(✉), M. Yu Narkevich, V. D. Kornienko, V. V. Kabanova, K. E. Shakhmaeva, and S. I. Chikota

Nosov Magnitogorsk State Technical University, Magnitogorsk 455000, Russia

Abstract. The purpose of the study is to determine the quality of video material obtained using unmanned aerial vehicles for their subsequent use in the system of expert assessment of the technical condition of hazardous production facilities. According to the results of earlier research, a generalized structure of images with the allocation of the main elements is proposed. The work justified the use of luminance histograms as the main tool for building an automatic image classifier. The functioning technology of the image classifier and the developed software product that performs the basic functions of the construction of histograms and its use in the work of the classifier are described. As a result, the suggested technology made it possible to process 106 images of hazardous industrial objects in an automated mode and to identify three main classes among them: 5 images with multiple objects and not suitable for further automated processing; 22 images which show only the object under study and do not require any preliminary preparation for further investigation; 70 images which require interactive identification of the study areas. It has been found that with the known structure of images it is possible to determine the image processing trajectory, to determine the quality of the obtained graphic material and the function of technological preparation of images. #CSOC1120.

Keywords: Image · Hazardous Production Facility · Image Histogram · Image Structure · Irregular Shape Object

1 Introduction

The modern level of computer technology and software development has made it possible to introduce computer vision and pattern recognition systems into industrial sectors. More and more often tasks in which decision-making is performed using visual information that is evaluated and classified by experts on the basis of a photo and video stream appear. The main devices for information gathering are photo and video cameras in the hands of researchers or installed on unmanned aerial vehicles (UAVs).

The most well-known in the industrial field are the tasks associated with the study of the quality of products or objects operating in the production area. In [1, 2] the problem of assessing the quality of continuously cast billets produced in the conditions of a large Russian enterprise, using the software "Q&S" to reduce the impact of the human factor.

R. Silhavy and P. Silhavy (Eds.): CSOC 2023, LNNS 723, pp. 228–236, 2023.
https://doi.org/10.1007/978-3-031-35317-8_20

In the researches conducted by the authors [3] a solution of a problem of obtaining and collecting information about technical condition of hazardous production facilities: buildings, structures and technical devices is considered.

As a result of inspection of objects a unique video material is collected which is subjected to subsequent processing: improvement, segmentation and recognition. One of the powerful tools for processing the obtained images is a luminance histogram [4–11]. By examining and transforming the histogram one can change the image contrast, change the sharpness of photos, hide and extract hidden information.

2 Methods

2.1 Images of Hazardous Production Facilities and Their Structure

When conducting research on the technical condition of hazardous production facilities, an unmanned aerial vehicle was used to collect visual information. With its help more than 1000 photo and video information units were collected [12, 13]. Examples of images are shown in Fig. 1. The obtained information is planned to be used for intellectual support in the system of making decisions on compliance of technical condition of hazardous production facilities with the requirements of normative documentation for their operation.

(a) (b)

Fig. 1. Examples of images of hazardous production facilities: a – image of the facade of an industrial building; b – image of a chimney.

For the collected video material there is a problem of assessing its quality for use in the system. An image is considered to be of high quality if the conditions are met:

- a large area of the image is occupied by the object to be studied – Cond1;
- the image of the object of examination differs from the luminance of the background and the surrounding objects – Cond2.

Studying the structure of obtained images (see Fig. 2) showed that:

1) in the image it is observed: 1 – general background; 2 – the main object of examination; 3 – violation of the continuity of the object of examination (see Fig. 2);

Table 1. Description of objects characterizing continuity violations.

№	Notation	Description
1	1	The main background of the image (photo)
2	2	Image of the examined object
3	3	Elements of continuity violation of the examined object
4	3.1	Recognition of regularly shaped objects
5	3.2	Recognizable objects of irregular shape
6	3.3	Recognizable objects of irregular shape with a complex structure
7	3.4	Thin objects of regular line form
8	3.5	Fine irregular objects
9	3.6	Regular and irregular shaped objects that do not have a recognition function assigned

2) violations of integrity of the examined object have a classification according to the form given in Table 1.

Figures 2b–e show typical structural elements of images obtained in the course of pilot inspection of technical condition of hazardous production facilities.

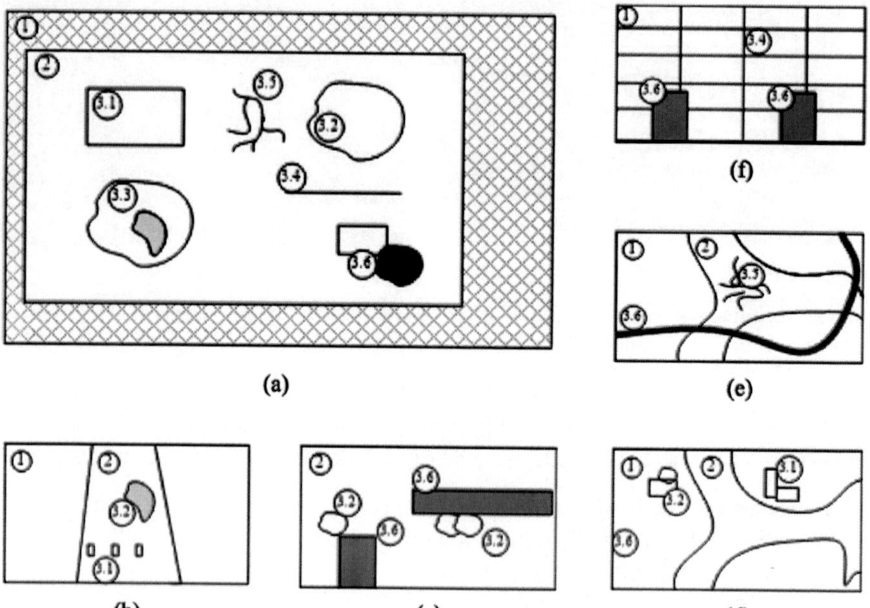

Fig. 2. Structure of the image of a hazardous production facility.

In Fig. 2 the following designations are introduced:

a – generalized image structure;
b – image of a facade fragment with rust or violation of paint coating;
c – image of a facade with icing;
d – image of a road with foreign objects on the roadside;
e – image of a road with cracks;
f – image of panels with crumbling connection joints.

2.2 Luminance Histogram as a Tool for Assessing the Structure of the Image of Hazardous Production Facilities

As one of the tools for assessing the quality of the obtained information material we have chosen a technique based on the image luminance histogram.

The image luminance histogram is a bar diagram, which shows the dependence of the frequency of occurrence of points with a given luminance. For the study we built luminance histograms for the channels Gray, Red, Green, Blue.

Table 2. Characteristic forms of histograms of the image of a panel building.

#	Source image	Histogram	Characteristics
1		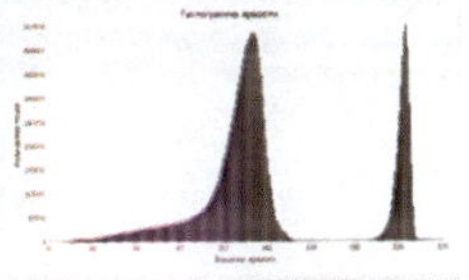	The image shows the main background of the object under study and the element(s) with high luminance that should be excluded from consideration
2		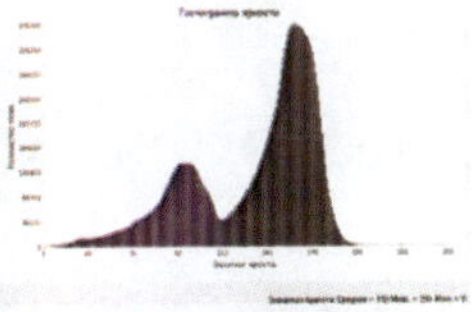	The image shows the main background of the object of interest and low element(s) that should be excluded
3		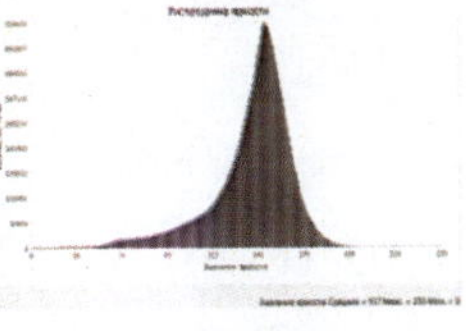	The entire image is almost completely filled with the background of the object of interest for which continuity discontinuity is being searched for
4		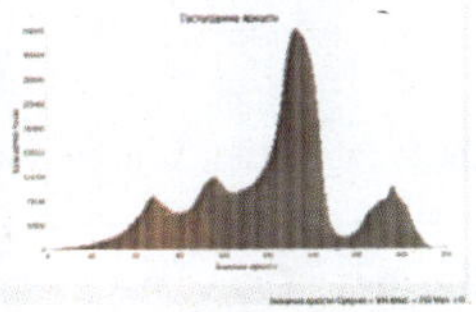	The image contains many objects with different colors and contrasts
5		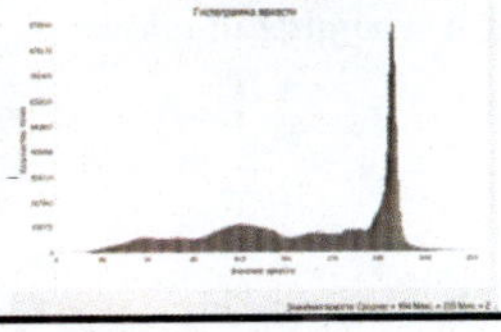	The image shows many objects with different colors and contrasts on an overall almost monotonous background

As a result of a visual study of the histograms of one of the tasks of estimating the proportion of joint failure between the building panels, the characteristic shapes of the histogram were revealed (see Table 2).

The ideal case, which excludes interactive user intervention, is the third histogram in Table 2. Cases 1 and 2 require exclusion of areas of increased or decreased luminance in the interactive mode. Cases 4 and 5 define images that are of "poor quality" for automatic processing and should be excluded from sampling.

2.3 Automated Image Sorter Based on Luminance Histogram Shape

In order to determine a typical image processing trajectory in production conditions, it is necessary to develop an automated sorter of initial data. The technology of the image sorter operation is presented in Fig. 3.

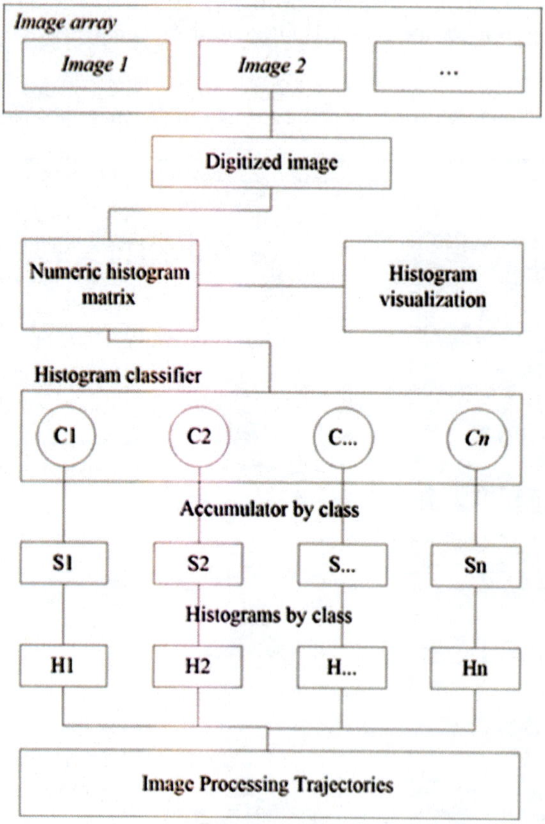

Fig. 3. Image sorter operation technology: *Ci* – image classes; *Si* – accumulated histogram frequencies; *Hi* – group histograms; *n* – number of classes.

According to the given technology, an *Image Array* of the studied images is formed. Each image is extracted from the Image Array one by one and digitized. On the basis of the obtained matrix the construction of a one-dimensional array with the frequency of occurrence of points with a given luminance (*Numeric histogram matrix*) is carried out and on its basis the image histogram is visualized (*Histogram visualization*). *Histogram classifier* makes a decision on assigning the image to predetermined classes based on the shape of the histogram. Histogram classifier leads to the task of one-dimensional array processing, in which it is required to determine the number and position of global extrema in one-dimensional array.

At the same time it is necessary to assign thresholds and concessions for the interval width and the difference between the extremum levels. For each class, frequency values of histograms are accumulated (*Accumulator by class*) and group histograms are constructed, the form of which determines the trajectory of image processing in the class (*Histograms by class*).

3 Results

The results of the research showed the need to develop software to build a histogram and study the image with its help. Figure 4 shows the view of the working window in the image processing mode.

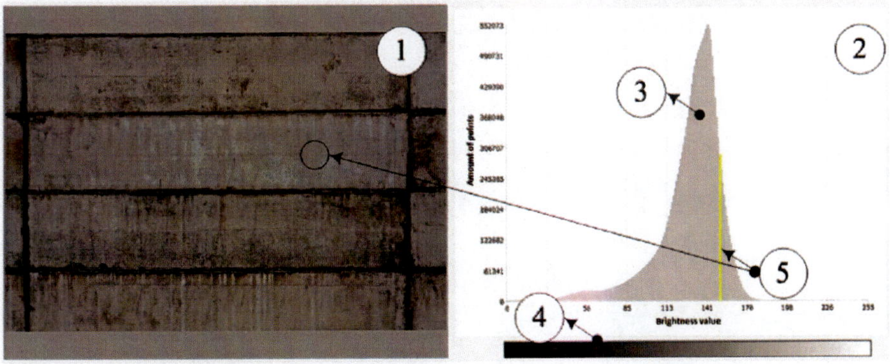

Fig. 4. View of the working window of the software product for building and examining the image: 1 – the initial image loaded; 2 – area for building the histogram; 3 – histogram; 4 – luminance scale; 5 – study point.

The software product includes features for examining the histogram:

- opening and loading an image in area (1);
- construction of the histogram (3) of the opened image in the area (2);
- displaying the luminance scale (4);
- interactive point indication on the image and its brightness search on the histogram (5);

- calculation of statistical characteristics of the image: average luminance of the image, minimum and maximum luminance values;
- unloading one-dimensional luminance array to a spreadsheet file;
- construction of channel histograms (See Fig. 5).

With the help of the developed program product we have made an automated classification of the images received during examination of interpanel seams damage of an industrial building. We have processed 106 images among which there were detected: 5 images with multiple objects and unsuitable for further automatic processing; 22 images which show only the object of research and do not require preliminary preparation for further research; 70 images which require interactive detection of the research fields.

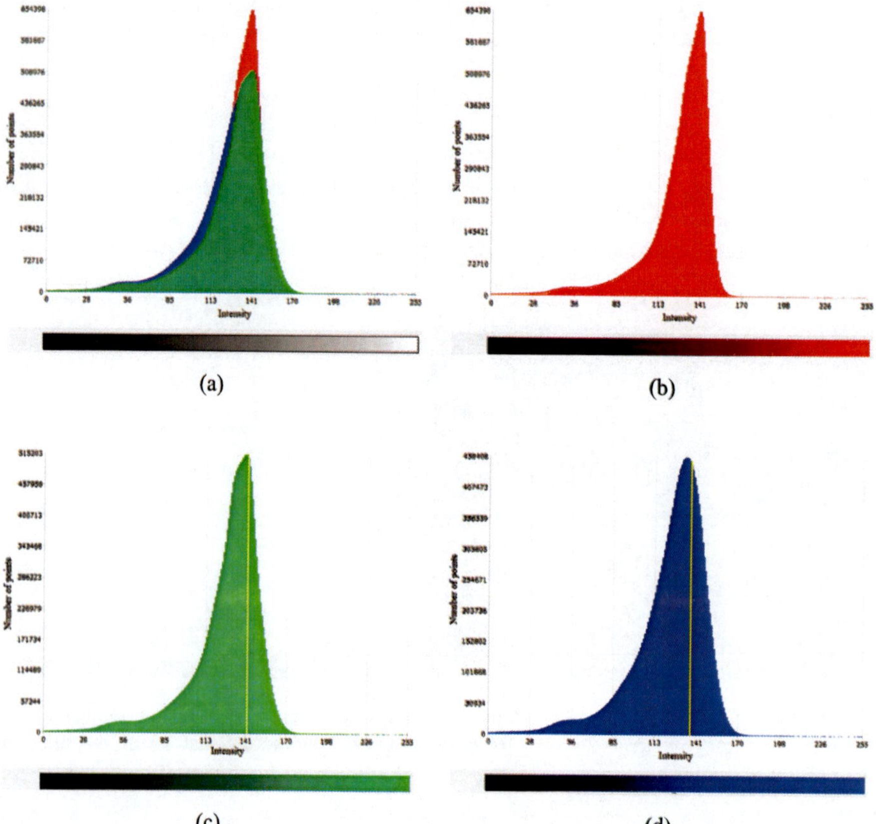

Fig. 5. Histograms by channel: a – RGB channels; b – Red channel; c – Green channel; d – Blue channel.

4 Conclusions

According to the results of the study, it was obtained that:

- the use of graphic information in the process of expert assessment of objects is a priority, which reduces the influence of human factor on the results and is currently promising for the development of technology of expert assessment;
- the use of luminance histograms requires the development of classifiers, which automatically determine the structure of images;
- with the known structure of the images can be determined the image processing trajectory, the quality of the resulting graphic material and the function of technological preparation of images.

A description of the mathematical basis for determining the shape-forming characteristics of an image is beyond the scope of this work and will be given in subsequent publications.

Acknowledgements. The study is performed under the contract №247715 dated 05.07.2021 between PJSC MMK and FGBOU VPO "Nosov Magnitogorsk State Technical University" "Development and application of methods of control of territories, buildings and facilities of PJSC MMK using unmanned aerial vehicles (UAV)".

References

1. Logunova, O.S., Devyatov, D.K., Nurov, K.K.: Computerized quality estimates of continuous-cast billet. Steel in Transl. **35**(9), 36–42 (2005)
2. Logunova, O.S., Matsko, I.I., Posohov, I.A., Luk'ynov, S.I.: Automatic system for intelligent support of continuous cast billet production control processes. Int. J. Adv. Manuf. Technol. **74**(9–12), 1407–1418 (2014)
3. Narkevich, M., et al.: Results of a pilot experiment on monitoring the condition of buildings and structures using unmanned aerial vehicles. IOP Conf. Ser.: Earth Env. Sci. **939**, 012030 (2021)
4. Vasilyev, V.N., Gurov, I.P., Volkov, M.V.: Image segmentation methods based on analysis and modification of local histograms. Sci. Tech. Bull. St. Petersburg State Univ. Inform. Technol., Mech. Optics **16**, 108–112 (2004)
5. Belozersky, L.A., Oreshkina, L.V.: Automation of histogram processing and analysis in space image recognition tasks. Earth Explor. Space **3**, 47–54 (2009)
6. Emelyanov, E.G., Komkov, V.S., Khanykov, A.E.: Research of efficiency of segmentation of complex-structured images by finding the mode of sliding window histogram. Russ. Sci. Educ. Today: Probl. Prospects **2**(5), 155–159 (2015)
7. Zeno, B., Yudin, D.A., Alkhatib, B.: Event recognition on images based on multi-level histograms of local patterns. In: International Conference Actual Problems of Robotics and Automation. October 08–09, 2015, pp. 67–74. Belgorod State Technological University named after V.G. Shukhov, Belgorod (2015)
8. Pyataeva, A.V., Favorskaya, M.N.: Application of evolutionary algorithm for smoothing histograms for early smoke detection in open spaces. DSPA Questions Dig. Signal Process. Appl. **6**(4), 813–816 (2016)

9. Klyuev, V.V.: Detection of objects in images using the histogram of directed gradients. Alley Sci. **2**(29), 913–917 (2019)
10. Raukhvarger, A., Poshekhonov, N.: Bringing luminance histograms of low-contrast digital images to two-level piecewise uniform distribution. Vestnik of Astrakhan State Tech. Univ. Series: Manage., Comput. Sci. Inform. **2020**(1), 57–63 (2020). https://doi.org/10.24143/2072-9502-2020-1-57-63
11. Raukhvarger, A.B., Poshekhonov, N.A.: Digital image processing by transformation of a luma gistogram to a two_level piecewise uniform distribution. Math. Methods Tech. Technol. **6**, 42–45 (2019)
12. Logunova, O.S., Kabanova, V.V., Kocherzhinskaya, Y.V., Ataullayev, S.N.: Results of a pilot experiment-study with continuous monitoring of the atmosphere over an industrial enterprise. IOP Conf. Ser.: Earth Env. Sci. **1043**, 012008 (2022). https://doi.org/10.1088/1755-1315/1043/1/012008
13. Yu Narkevich, M., et al.: Results of experimental tests of building samples. IOP Conf. Ser. Earth Env. Sci. **939**, 012031 (2021). https://doi.org/10.1088/1755-1315/939/1/012031

Optomechanical Manipulation of Nanoparticles in Hybrid Anapole State

Nikita Babich[1], Alexey Kuznetsov[1,2,3], Vjaceslavs Bobrovs[2(✉)], and Denis Kislov[2,3]

[1] ITMO University, St. Petersburg 197101, Russia
[2] Institute of Telecommunications, Riga Technical University, Riga 1048, Latvia
vjaceslavs.bobrovs@rtu.lv
[3] Center for Photonics and 2D Materials, Moscow Institute of Physics and Technology, Dolgoprudny 141700, Russia

Abstract. The report analyzes the optomechanical interaction of a Gaussian beam and a silicon nanoparticle in a hybrid anapole state. To study the particle scattering in the Gaussian beam waist, as well as to confirm the appearance of a hybrid anapole state, a multipole decomposition was carried out. We made calculations of the optical forces acting on the particle along different directions depending on the relative position of the light beam and the particle. The pressure and gradient components of force were analyzed separately. In addition, the torque acting on the anapole and scattering particles at the center of the beam waist was calculated. It is shown that the torque can change sign if the particle is at anapole. #CSOC1120.

Keywords: Anapole · Optical Force · Multipole Decomposition

1 Introduction

The phenomenon of the mechanical action of light on various objects [1] has been widely used in science and technology [2–5]. On its basis, an optical tweezer was created, which revolutionized biology [6–8]. With its help, cells and proteins are manipulated, which is impossible with mechanical tools. This technology allows us to study cells and bacteria at a qualitatively new level, as well as to study intracellular processes. In addition to biology, optical tweezers have been widely used in physics [9–11] and chemistry [12–16]. The adaptation of this technology made it possible to obtain a Bose condensate, which greatly advanced the study of quantum effects.

Turning to the nature of the phenomenon, we should note that when a wave interacts with a particle, two components of forces of different nature can be distinguished. The pressure component of the force is caused by the transmission of the wave momentum to the object, and the gradient component arising from the inhomogeneous distribution of the amplitude of the light intensity along the wave profile draws the particle into the region of maximum intensity [4]. In this paper, we wondered whether the hybrid anapole state [17–21] has advantages in terms of optical force [22–28]. The purpose of this report

R. Silhavy and P. Silhavy (Eds.): CSOC 2023, LNNS 723, pp. 237–243, 2023.
https://doi.org/10.1007/978-3-031-35317-8_21

was to investigate how the ratio of the components of the pressure and gradient forces acting on a particle in a hybrid anapole state [29–32] will change.

In the course of the work, a hypothesis was put forward about the existence of the hybrid anapole state for a cylindrical silicon nanoparticle in a linearly polarized Gaussian beam. The goal of the report was to study the optical forces and torque arising in a focused Gaussian beam for various configurations of the system.

2 Methods

We For the numerical experiment, simulations were carried out in the COMSOL Multiphysics package. To calculate the components of the forces $\boldsymbol{F}$, a time-averaged Maxwell stress-tensor $\overleftrightarrow{\mathbf{M}}$ is integrated over a surface around the nanoparticle. The similar way was to calculate the torque $\boldsymbol{T}$ arising in the system relative to the origin – integrating the cross product of the radius-vector and Maxwell stress-tensor. Integrating nanoparticles by volume, electric and magnetic dipole and quadrupole moments were obtained, using them, the scattering coefficients of each component were calculated, that are necessary to determine the existence of a hybrid anapole state. Here and further by forces and torques meant time-averaged forces and torques:

$$\langle \mathbf{F} \rangle = \int_S \left\langle \overleftrightarrow{\mathbf{M}} \right\rangle \cdot \mathbf{n}\, d\sigma \tag{1}$$

$$\langle \mathbf{T} \rangle = -\int_S \mathbf{n} \cdot \left[\left\langle \overleftrightarrow{\mathbf{M}} \right\rangle \times \mathbf{r} \right] d\sigma \tag{2}$$

where $\left\langle \overleftrightarrow{\mathbf{M}} \right\rangle = \frac{1}{2}\mathrm{Re}\left[\varepsilon_0 \mathbf{E} \otimes \mathbf{E}^* + \mu_0 \mathbf{H} \otimes \mathbf{H}^* - \frac{1}{2}\left(\varepsilon_0 |\mathbf{E}|^2 + \mu_0 |\mathbf{H}|^2 \right) \overleftrightarrow{\mathbf{I}} \right]$.

The system we are exploring is the cylindrical silicon resonator with special size parameters. The height of the cylinder is h = 369 nm, the radius is r = 127 nm. These parameters have been chosen to have a resonance wavelength of 750 nm where hybrid anapole state is observed with linearly polarized plane wave. The cylinder is placed in a Gaussian beam, and its center of mass is in the waist plane of the beam as it can be mentioned from Fig. 1. The default beam's waist radius was $w_0 = \lambda$.

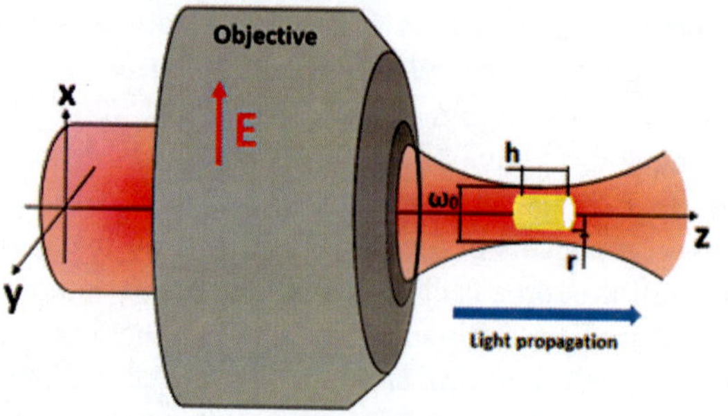

Fig. 1. The geometry of the problem.

3 Results

To confirm the existence of the hybrid anapole state (HAS) the linearly polarized wave was simulated. The scattering coefficient from the wavelength was calculated. Hybridity at λ ~ 750 nm was clearly confirmed, which allowed us to proceed to a further solution of the problem.

Then during the numerical experiment, the hypothesis of the existence of a HAS for a cylindrical nanoparticle in a Gaussian beam was tested, and the properties of a HAS were analyzed. The primary conclusion of the experiment is the fact of the destruction of the HAS as the nanoparticle moves away from the optical axis, which can be seen in Fig. 2.

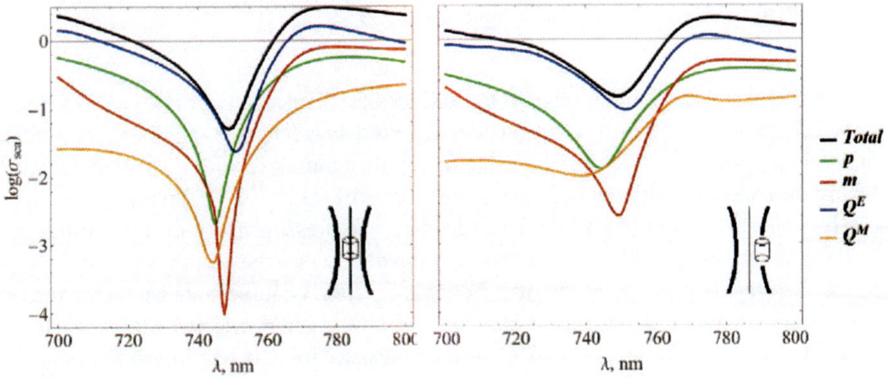

Fig. 2. Logarithm of scattering cross-section from the wavelength. HAS destruction demonstration with moving nanoparticle from the optical axis of the Gaussian beam.

Total is the sum of every multipole component of the coefficient (p – electric dipole, m – magnetic dipole, Q^E – electric quadrupole, Q^M – magnetic quadrupole). The minima of multipoles cease to be at the same wavelength. This is explained by the fact that at a distance from the optical axis, even in the focal plane of the beam, the wave front is curved and it cannot be approximated as a plane wave.

To obtain a gradient force comparable to the pressure force, a large curvature of the beam is required. This can be achieved by reducing the waist w_0 to values comparable to half the wavelength. During the simulation, an additional factor was revealed that strongly affects the optical forces in the system. The minimum waist radius is comparable to the radius of the cylinder itself. Therefore, the profiles of forces in our experiment for differently focused, will be different, which can be seen in Fig. 3.

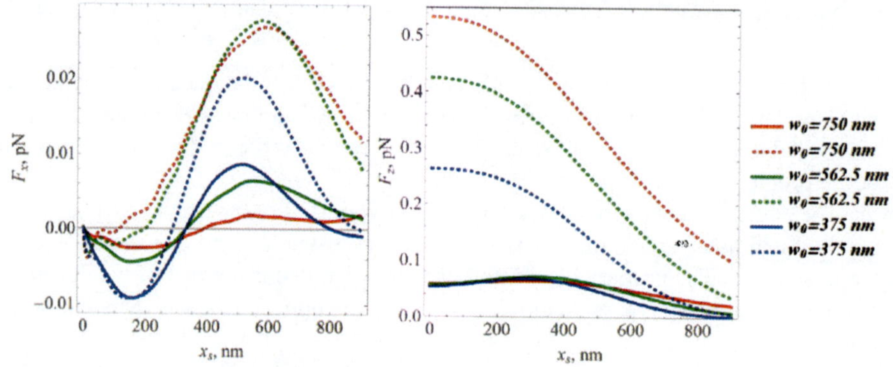

Fig. 3. The forces dependence on the position of nanoparticle. Dotted lines – non-anapole state ($\lambda_1 = 770$ nm). Solid graphs – anapole state ($\lambda_2 = 750$ nm).

During the experiment, three configurations of a Gaussian beam ($w_1 = 375$ nm, $w_2 = 562.5$ nm, $w_3 = 750$ nm) were considered at two wavelengths – anapole $\lambda_1 = 750$ nm and non-anapole $\lambda_2 = 770$ nm. Since the forces are symmetric with respect to the optical beam, they were considered only for a one-way shift – $x_s \in [0, 0.9]$ μm.

It is clearly proved that in the anapole state, the pressure force on the particle is an order of magnitude smaller. Let us consider separately the case of the waist radius equal to the wavelength ($w_0 = \lambda = 750$ nm). When the hybrid anapole state exists the pressure force on the cylinder placed on the optical axis of the beam $F_{z,anap.}(0) \sim 0.08$ pN. If HAS do not exists ($\lambda = 770$ nm, $w_0 = 750$ nm) the pressure force is maximum $F_{z,no\ anap.}(0) \sim 0.53$ pN. The relation of these forces exactly repeats relation of the scattering coefficients.

4 Discussions

In this report we also explore the rotation of the nanoparticle around the y-axis and torques arising in this system. It was important to understand how the hybrid anapole state behaves during the rotation of the cylinder in the Gaussian beam. It was clearly shown that HAS destructs and the scattering properties of the particle aims to non-anapole.

If the cylinder is coaxial to the optical beam, apparently there are no moments and there is only one component of the optical power (F_z is pressure). In our case, there is only one moment component (T_y), with T_x, $T_z = 0$. As expected, in the non-anapole system, there is one equilibrium position when the cylinder axis coincides with the beam axis (Fig. 4, dotted curve).

In the case when a hybrid anapole state is formed in the particle, there is one-point $\alpha = 40°$, when the torque changes sign. Thus, it was obtained that there are two equilibrium positions 0° and 90° in HAS. This is not a typical behavior compared to "ordinary" particles.

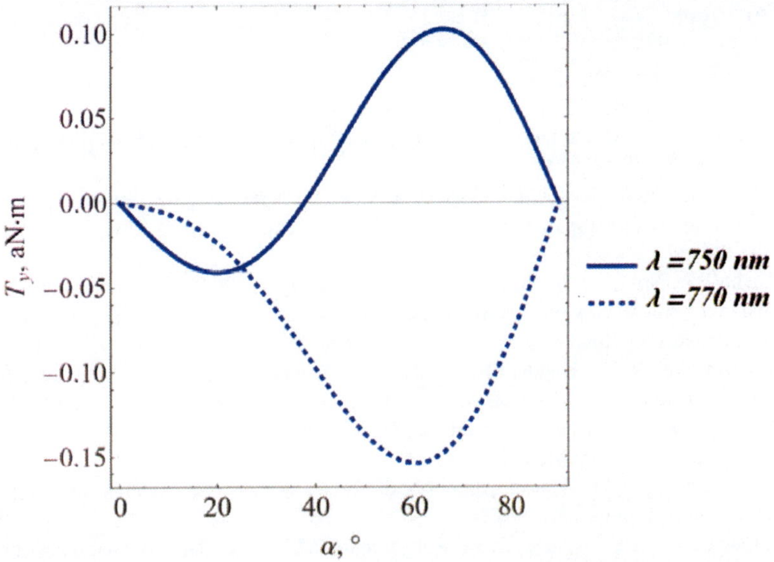

Fig. 4. Torque arising in the system out of rotating the cylinder in the Gaussian beam.

5 Conclusion

Summing up the results of the work, we can first of all highlight the fact of confirmation of the hypothesis of a decrease in the pressure force in the hybrid anapole state for a Gaussian beam. Some nontrivial features of the HAS were also discovered. The first of them is the fact of the destruction of the hybrid anapole state in the particle as it moves away from the optical axis. This is explained by the curvature of the beam front. Another fact highlights the dependence of the pressure force on the waist radius of the beam. This phenomenon is due to the fact that the waist radius of the beam is comparable with the radius of the cylinder. The beam intensity is distributed non-linearly at the end of the cylinder.

Moreover, torques arising when cylinder is tilted relative to the optical axis of the Gaussian beam were studied. There is a balance in the general case if the particle is not coaxial to the optical axis. However, for the anapole particles there is an angle ($\alpha = 40°$) when $\mathbf{T} = 0$ and torque changes sign.

Acknowledgements. The authors gratefully acknowledge the financial support from the Ministry of Science and Higher Education of the Russian Federation (Agreement No. № 075-15-2022-1150). D.K. acknowledges the support of the Latvian Council of Science, project: PHOTON, No. Lzp-2022/1-0579.

References

1. Ashkin, A., Dziedzic, J.M., Bjorkholm, J.E., Chu, S.: Observation of a single-beam gradient force optical trap for dielectric particles. Opt Lett. **11**(5), 288 (1986)
2. Tkachenko, G., Brasselet, E.: Optofluidic sorting of material chirality by chiral light. Nat. Commun. **5**, 3577 (2014)
3. Canós Valero, A., Kislov, D., Gurvitz, E.A., et al.: Nanovortex-driven all-dielectric optical diffusion boosting and sorting concept for lab-on-a-chip platforms. Adv. Sci. **7**(11), 1903049 (2020)
4. Kislov, D., Novitsky, D., Kadochkin, A., Redka, D., Shalin, A.S., Ginzburg, P.: Diffusion-inspired time-varying phosphorescent decay in a nanostructured environment. Phys. Rev. B **101**, 035420 (2020)
5. Kucherenko, M.G., Kislov, D.A.: Plasmon-activated intermolecular nonradiative energy transfer in spherical nanoreactors. The J. Photochem. Photobiol. A: Chem. **354**, 25 (2018)
6. Favre-Bulle, I.A., et al.: Optical trapping in vivo: theory, practice, and applications. Nanophotonics **8**, 1023 (2019)
7. Liu, X., et al.: All-optical targeted drug delivery and real-time detection of a single cancer cell. Nanophotonics **9**, 611 (2020)
8. Izmodenova, S.V., Kislov, D.A., Kucherenko, M.G.: Accelerated nonradiative electron-excitation energy transfer between molecules in aqueous pools of reverse micelles containing encapsulated silver nanoparticles. Colloid J. **76**(6), 683–693 (2014). https://doi.org/10.1134/S1061933X14060088
9. Kostina, N., et al.: Optical binding via surface plasmon polariton interference. Phys. Rev. B **99**, 125416 (2019)
10. Kislov, D.: Effect of plasmonic silver nanoparticles on the photovoltaic properties of Graetzel solar cells. Phys. Procedia **73**, 114 (2015)
11. Vestler, D., Shishkin, I., Gurvitz, E.A., et al.: Circular dichroism enhancement in plasmonic nanorod metamaterials. Opt Express **26**(14), 17841–17848 (2018)
12. Kostina, N.A., Kislov, D.A., Ivinskaya, A.N., et al.: Nanoscale tunable optical binding mediated by hyperbolic metamaterials. ACS Photonics **7**(2), 425–433 (2020)
13. Ivinskaya, A., et al.: Optomechanical manipulation with hyperbolic metasurfaces. ACS Photonics **5**, 4371 (2018)
14. Ivinskaya, A., Petrov, M.I., Bogdanov, A.A., Shishkin, I., Ginzburg, P., Shalin, A.S.: Plasmon-assisted optical trapping and anti-trapping. Light Sci. Appl. **6**, e16258 (2017)
15. Novitsky, D., Karabchevsky, A., Lavrinenko, A., Shalin, A., Novitsky, A.: PT symmetry breaking in multilayers with resonant loss and gain locks light propagation direction. Phys Rev B **98**, 125102 (2018)
16. Novitsky, D.V., Shalin, A.S., Redka, D., Bobrovs, V., Novitsky, A.V.: Quasibound states in the continuum induced by PT symmetry breaking. Phys Rev B **104**(8), 85126 (2021)
17. Baryshnikova, K., et al.: Giant magnetoelectric field separation via anapole-type states in high-index dielectric structures. Phys. Rev. B **98**, 165419 (2018)
18. Barhom, H., et al.: Biological kerker effect boosts light collection efficiency in plants. Nano Lett. **19**(10), 7062 (2019)
19. Terekhov, P.D., Baryshnikova, K.V., Greenberg, Y., et al.: Enhanced absorption in all-dielectric metasurfaces due to magnetic dipole excitation. Sci. Rep. **9**, 3438 (2019)
20. Kozlov, V., Filonov, D., Shalin, A.S., Steinberg, B.Z., Ginzburg, P.: Asymmetric backscattering from the hybrid magneto-electric meta particle. Appl. Phys. Lett. **109**, 203503 (2016)
21. Terekhov, P.D., et al.: Broadband forward scattering from dielectric cubic nanoantenna in lossless media. Opt. Express **27**(8), 10924 (2019)

22. Xu, X., et al.: Kerker-type intensity-gradient force of light. Laser Photonics Rev. **14**, 1 (2020)
23. Länk, N.O., Johansson, P., Käll, M.: Directional scattering and multipolar contributions to optical forces on silicon nanoparticles in focused laser beams. Opt. Express **26**(22), 29074 (2018)
24. Liu, H., Panmai, M., Peng, Y., Lan, S.: Optical pulling and pushing forces exerted on silicon nanospheres with strong coherent interaction between electric and magnetic resonances. Opt. Express **25**(11), 12357 (2017)
25. Salandrino, A., Fardad, S., Christodoulides, D.N.: Generalized Mie theory of optical forces. J. Opt. Soc. Am. B **29**(4), 855 (2012)
26. Duan, X.Y., Wang, Z.G.: Fano resonances in the optical scattering force upon a high index dielectric nanoparticle. Phys. Rev. A **96**(5), 053811 (2017)
27. Chen, J., Ng, J., Lin, Z., Chan, C.T.: Optical pulling force. Nat. Photonics **5**, 531 (2011)
28. Nieto-Vesperinas, M., Sáenz, J.J., Gómez-Medina, R., Chantada, L.: Optical forces on small magnetodielectric particles. Opt. Express **18**(11), 11428 (2010)
29. Terekhov, P.D., Evlyukhin, A.B., Redka, D., Volkov, V.S., Shalin, A.S., Karabchevsky, A.: Magnetic octupole response of dielectric quadrumers. Laser Photon Rev **14**(4), 1900331 (2020)
30. Shamkhi, H.K., Sayanskiy, A., Valero, A.C., et al.: Transparency and perfect absorption of all-dielectric resonant metasurfaces governed by the transverse Kerker effect. Phys. Rev. Mater. **3**(8), 85201 (2019)
31. Canós Valero, A., Gurvitz, E.A., Benimetskiy, F.A., et al.: Theory, observation, and ultrafast response of the hybrid anapole regime in light scattering. Laser Photon. Rev. **15**(10), 2100114 (2021)
32. Kuznetsov, A.V., Canós Valero, A., Tarkhov, M., Bobrovs, V., Redka, D., Shalin, A.S.: Transparent hybrid anapole metasurfaces with negligible electromagnetic coupling for phase engineering. Nanophotonics **10**(17), 4385–4398 (2021)

Analytical Condition for Toroidal Anapole

Dmitrii Borovkov[1,2] and Vjaceslavs Bobrovs[2(✉)]

[1] Center for Photonics and 2D Materials, Moscow Institute of Physics and Technology, Dolgoprudny 141700, Russia

[2] Institute of Telecommunications, Riga Technical University, Riga 1048, Latvia

vjaceslavs.bobrovs@rtu.lv

Abstract. Multipole decomposition of the localized current plays an essential role in a wide range of disciplines in modern physics, starting from the design of nanoantennas in photonics to the description of exotic states of matter. In the latter context, a third group of multipoles, toroidal multipoles, complementing the electric and magnetic ones, have been thoroughly investigated due to their lead to the formation of so-called anapole states, non-radiating sources. And further investigation shed light on the existence of higher-order toroidal dipoles, or equivalently, mean-square radii. Here we develop the classical anapole state obtained by destructive interference of electric and toroidal dipoles into the toroidal anapole which arises by destructive interference of electric toroidal dipole with first mean-square radius of electric toroidal dipole, presenting an analytical condition of such a state. #CSOC1120.

Keywords: Anapole · Toroidal Dipole · Multipole Decomposition

1 Introduction

Nowadays multipole expansion is an indispensable instrument in many areas of theoretical physics, from the description of condensed matter systems to antenna design [1–3]. In recent years, it has been actively used to understand and effectively tune the electromagnetic response of artificial metamaterials and metasurfaces for a wide range of optical applications, and now it plays an important role in the development of a whole emerging field of all-dielectric nanophotonics [4–7]. Multipole analysis of dielectric nanoantennas reveals amusing optical phenomena such as directional scattering [8, 9] or Fano resonances [10].

The multipole moments are commonly classified in terms of electric (oscillating charges) and magnetic (circulating currents) multipoles. They are usually obtained from an expansion of the current in a series of form factors, or equivalently from a small argument expansion of the vector potential. Using this method, a third class of multipoles with the same radiation patterns as the conventional ones can be obtained, known as toroidal multipoles [11]. Interference between the conventional electric dipole and the toroidal dipole leads to the formation of non-radiating states called ‘anapoles’ which has prompted intensive research in the photonics community. It has been shown that

R. Silhavy and P. Silhavy (Eds.): CSOC 2023, LNNS 723, pp. 244–248, 2023.
https://doi.org/10.1007/978-3-031-35317-8_22

complex distribution of currents within dielectric particles produces high-order [12] and hybrid [13, 14] anapole states with mixed electric and magnetic character. Understanding this unusual type of non-radiative current distribution requires a higher-order toroidal multipole [10] or mean-square radii [15, 16] (which are equivalent [17]). They appear as higher order terms in the expansion of the vector potential, thus increasing the retardation effects arising within the source.

Here we perform a theoretical analysis of anapole state obtained with destructive interference of electric toroidal dipole (TD) and first mean-square radius (MSRs) of electric toroidal dipole. Properties of such toroidal anapole (TA) should be enhanced compared to the classic anapole being the destructive interference of electric and toroidal dipole.

2 Methods

We begin with the expression of spherical (or exact) electric dipole [18]:

$$\mathbf{p} = -\frac{1}{i\omega}\left\{\int d^3\mathbf{r}\mathbf{J}_\omega j_0(kr) + \frac{k^2}{2}\int d^3\mathbf{r}\Big[3(\mathbf{r}\cdot\mathbf{J}_\omega)\mathbf{r} - r^2\mathbf{J}_\omega\Big]\frac{j_2(kr)}{(kr)^2}\right\} \tag{1}$$

This expression contains cartesian electric dipole (**d**), toroidal (**T**) and higher-order toroidal ($\mathbf{T}^n$) dipoles in a way that $\mathbf{p} = \mathbf{d} + \mathbf{T} + \mathbf{T}_2 + \ldots$. Now to obtain an analytical expression for the MSR we expand the exact electric dipole in a Taylor series for a small argument of kr. First step is to expand the Bessel functions $j_0(kr)$ and $j_2(kr)$ and then combine them into one sum:

$$\mathbf{p} = -\frac{6}{i\omega}\int d^3\mathbf{r}\sum_{m=0}^{\infty}\frac{(-1)^m(m+1)(kr)^{2m}}{(2m+3)!}\Big[(m+1)\mathbf{J}_\omega - m(\mathbf{r}\cdot\mathbf{J}_\omega)\frac{\mathbf{r}}{r^2}\Big]. \tag{2}$$

For the current as starting point we choose the divergence-less poloidal current describing TD and developed in [19], where the dependence on ρ is described by δ-function:

$$\mathbf{J}^{(TD)} = \frac{K\delta(\rho-\rho_0)}{R+\rho\cos\theta}\mathbf{e}_\theta. \tag{3}$$

We are using here the toroidal coordinates:

$$\mathbf{e}_\rho = \begin{pmatrix}\cos\theta\cos\phi\\ -\cos\theta\sin\phi\\ \sin\theta\end{pmatrix},\quad \mathbf{e}_\theta = \begin{pmatrix}-\sin\theta\cos\phi\\ \sin\theta\sin\phi\\ \cos\theta\end{pmatrix},\quad \mathbf{e}_\phi = \begin{pmatrix}-\sin\phi\\ -\cos\phi\\ 0\end{pmatrix} \tag{4}$$

Before we add to the current a second term corresponding to additional torus, for simplicity and clarity we will express current in Eq. (3) in cylindrical coordinates r_c, ψ, z:

$$\begin{aligned} r_c &= R+\rho\cos\theta,\\ \psi &= -\phi,\\ z &= \rho\sin\theta. \end{aligned} \tag{5}$$

The cartesian components of vector $\mathbf{e}_\theta$ [Eq. (4)] then are expressed in cylindrical coordinates:

$$\mathbf{e}_\theta = \frac{1}{\sqrt{(r_c - R)^2 + z^2}} \begin{pmatrix} -z\cos\psi \\ -z\sin\psi \\ r - R \end{pmatrix}. \tag{6}$$

Now to the current in Eq. (3) we can add a similar one pertaining to the second torus:

$$\mathbf{J}_\omega = \frac{K_0\delta(\rho(r_c, z) - \rho_0)}{r_c\sqrt{(r_c - R_0)^2 + z^2}} \begin{pmatrix} -z\cos\psi \\ -z\sin\psi \\ r_c - R_0 \end{pmatrix} - \frac{K_1\delta(\rho(r_c, z) - \rho_1)}{r_c\sqrt{(r_c - R_1)^2 + z^2}} \begin{pmatrix} -z\cos\psi \\ -z\sin\psi \\ r_c - R_1 \end{pmatrix}, \tag{7}$$

Calculating the expressions of $\mathbf{r} \cdot \mathbf{J}_\omega$ and $|\mathbf{r}|^2$ one gets:

$$\mathbf{r} \cdot \mathbf{J}_\omega = -z \sum_{j=0}^{1} (-1)^j \frac{K_j R_j \delta(\rho(r_c, z) - \rho_j)}{r_c\sqrt{(r_c - R_j)^2 + z^2}}, \tag{8}$$

$$|\mathbf{r}|^2 = r_c^2 + z^2. \tag{9}$$

Now we note that all x and y components of ED-like moments [Eq. (2)] for the current in Eq. (7) are zero because of sin or cos-like ψ-dependence of these components. For z components, on the other hand, we analyze TD and MSR separately.

TD:

$$(\mathbf{T})_\alpha = \frac{1}{10}\frac{k^2}{i\omega} \int d^3\mathbf{r}\Big(2|\mathbf{r}|^2 J_\alpha - (\mathbf{r} \cdot \mathbf{J}) r_\alpha\Big), \tag{10}$$

$$(\mathbf{T})_z = \frac{1}{10}\frac{k^2}{i\omega}(A_{0z} - A_{1z}), \tag{11}$$

where for $j = 0, 1$,

$$A_{jz} = \int d^3\mathbf{r}\Big(2(r_c^2 + z^2)(r_c - R_j) + z^2 R_j\Big) \frac{K_j\delta(\rho(r_c, z) - \rho_j)}{r_c\sqrt{(r_c - R_j)^2 + z^2}}. \tag{12}$$

Let's not forget that $r_c = r_c(x, y) = \sqrt{x^2 + y^2}$.

At this step we can for each torus separately go back to the toroidal coordinates and integrate δ-function. Integrals of δ-function leave us surface integral with local coordinates u and v so that:

$$\begin{aligned} x &= (R_j + \rho_j\cos u)\cos v, \\ y &= -(R_j + \rho_j\cos u)\sin v, \\ z &= \rho_j \sin u. \end{aligned} \tag{13}$$

With these replacements, the integral can be rewritten into the form:

$$A_{jz} = \iint_\Omega dudv\Big(2\Big((R_j + \rho_j\cos u)^2 + \rho_j^2\sin^2 u\Big)\rho_j\cos u + R_j\rho_j^2\sin^2 u\Big) \frac{K_j}{\rho_j(R_j + \rho_j\cos u)}\sqrt{EG - F^2}, \tag{14}$$

where

$$E = (x'_u)^2 + (y'_u)^2 + (z'_u)^2 = \rho_j^2,$$
$$F = x'_u x'_v + y'_u y'_v + z'_u z'_v = 0, \quad (15)$$
$$G = (x'_v)^2 + (y'_v)^2 + (z'_v)^2 = (R_j + \rho_j \cos u)^2.$$

And the resulting expression for integral in Eq. (14) is:

$$A_{jz} = 10\pi^2 K_j R_j \rho_j^2, \quad (16)$$

Now if we put this result into Eq. (11) we will get the final result for the z component of TD:

$$(\mathbf{T})_z = \frac{\pi^2 k^2}{i\omega}\Big(K_0 R_0 \rho_0^2 - K_1 R_1 \rho_1^2\Big). \quad (17)$$

In a similar way using the following formula, which received from Eq. (2):

$$(\mathbf{T}_2)_\alpha = -\frac{1}{280}\frac{k^4}{i\omega}\int d^3\mathbf{r}|\mathbf{r}|^2\Big(3|\mathbf{r}|^2 J_\alpha - 2(\mathbf{r}\cdot\mathbf{J})r_\alpha\Big), \quad (18)$$

one obtains z component of 1st MSR:

$$(\mathbf{T}_2)_z = -\frac{1}{10}\frac{\pi^2 k^4}{i\omega}\Big(K_0 R_0 \rho_0^2 (R_0^2 + \rho_0^2) - K_1 R_1 \rho_1^2 (R_1^2 + \rho_1^2)\Big). \quad (19)$$

The condition for toroidal anapole we are interested in is the following: $(\mathbf{T})_z + (\mathbf{T}_2)_z = 0$, which gives:

$$K_0 R_0 \rho_0^2\left(1 - \frac{k^2}{10}(R_0^2 + \rho_0^2)\right) - K_1 R_1 \rho_1^2\left(1 - \frac{k^2}{10}(R_1^2 + \rho_1^2)\right) = 0. \quad (20)$$

3 Conclusion

We analytically described the condition of the anapole state obtained with the destructive interference of electric toroidal dipole and first mean-square radii of electric toroidal dipole. We believe that further investigation of such toroidal anapoles will be proved useful, enhancing properties of anapoles.

Acknowledgements. The authors gratefully acknowledge the financial support from the Ministry of Science and Higher Education of the Russian Federation (Agreement No. № 075-15-2022-1150). V.B. acknowledges the support of the Latvian Council of Science, project: DNSSN, No. Lzp-2021/1-0048.

References

1. Terekhov, P.D., Evlyukhin, A.B., Redka, D., Volkov, V.S., Shalin, A.S., Karabchevsky, A.: Magnetic octupole response of dielectric quadrumers. Laser Photon. Rev. **14**(4), 1900331 (2020)
2. Novitsky, D.V., Shalin, A.S., Redka, D., Bobrovs, V., Novitsky, A.V.: Quasibound states in the continuum induced by PT symmetry breaking. Phys. Rev. B **104**(8), 85126 (2021)
3. Novitsky, D., Karabchevsky, A., Lavrinenko, A., Shalin, A., Novitsky, A.: PT symmetry breaking in multilayers with resonant loss and gain locks light propagation direction. Phys. Rev. B **98**, 125102 (2018)
4. Shalin, A.S.: Broadband blooming of a medium modified by an incorporated layer of nanocavities. JETP Lett. **91**(12), 636–642 (2010)
5. Vestler, D., Shishkin, I., Gurvitz, E.A., et al.: Circular dichroism enhancement in plasmonic nanorod metamaterials. Opt. Express **26**(14), 17841–17848 (2018)
6. Kostina, N.A., Kislov, D.A., Ivinskaya, A.N., et al.: Nanoscale tunable optical binding mediated by hyperbolic metamaterials. ACS Photonics **7**(2), 425–433 (2020)
7. Canós Valero, A., Kislov, D., Gurvitz, E.A., et al.: Nanovortex-driven all-dielectric optical diffusion boosting and sorting concept for lab-on-a-chip platforms. Adv. Sci. **7**(11), 1903049 (2020)
8. Barhom, H., Machnev, A.A., Noskov, R.E., et al.: Biological kerker effect boosts light collection efficiency in plants. Nano Lett. **19**(10), 7062–7071 (2019)
9. Shamkhi, H.K., Sayanskiy, A., Valero, A.C., et al.: Transparency and perfect absorption of all-dielectric resonant metasurfaces governed by the transverse Kerker effect. Phys. Rev. Mater. **3**(8), 85201 (2019)
10. Gurvitz, E.A., Ladutenko, K.S., Dergachev, P.A., Evlyukhin, A.B., Miroshnichenko, A.E., Shalin, A.S.: The high-order toroidal moments and anapole states in all-dielectric photonics. Laser Photon. Rev. **13**(5), 1800266 (2019)
11. Dubovik, V.M., Tugushev, V.V.: Toroid moments in electrodynamics and solid-state physics. Phys. Rep. **187**(4), 145–202 (1990)
12. Zenin, V.A., Evlyukhin, A.B., Novikov, S.M., et al.: Direct amplitude-phase near-field observation of higher-order anapole states. Nano Lett. **17**(11), 7152–7159 (2017)
13. Canós Valero, A., Gurvitz, E.A., Benimetskiy, F.A., et al.: Theory, observation, and ultrafast response of the hybrid anapole regime in light scattering. Laser Photon. Rev. **15**(10), 2100114 (2021)
14. Kuznetsov, A.V., Canós Valero, A., Tarkhov, M., Bobrovs, V., Redka, D., Shalin, A.S.: Transparent hybrid anapole metasurfaces with negligible electromagnetic coupling for phase engineering. Nanophotonics **10**(17), 4385–4398 (2021)
15. Radescu, E.E., Vaman, G.: Exact calculation of the angular momentum loss, recoil force, and radiation intensity for an arbitrary source in terms of electric, magnetic, and toroid multipoles. Phys. Rev. E Stat. Phys. Plasmas Fluids Relat. Interdiscip. Topics **65**(4), 47 (2002)
16. Nemkov, N.A., Basharin, A.A., Fedotov, V.A.: Electromagnetic sources beyond common multipoles. Phys. Rev. A **98**(2), 023858 (2018)
17. Borovkov, D., Valero, A.C.: On the link between mean square-radii and high-order toroidal moments. J. Phys. Conf. Ser. **2015**(1), 012021 (2021)
18. Alaee, R., Rockstuhl, C., Fernandez-Corbaton, I.: An electromagnetic multipole expansion beyond the long-wavelength approximation. Opt. Commun. **407**, 17–21 (2018)
19. Valero, A.C., Borovkov, D., Sidorenko, M., et al.: On the existence of pure, broadband toroidal sources in electrodynamics. Arxiv 2208.02908 (2022)

Point Magnetic Dipoles Forming Toroidal Anapole

Dmitrii Borovkov[1,2(✉)] and Vjaceslavs Bobrovs[2]

[1] Center for Photonics and 2D Materials, Moscow Institute of Physics and Technology, Dolgoprudny 141700, Russia
dmitriiborovkov94@gmail.com

[2] Institute of Telecommunications, Riga Technical University, Riga 1048, Latvia

Abstract. The role of multipolar expansion of the localized current is essential in a broad range of disciplines in the physics of nowadays. It is actively used in such problems as the design of nanoantennas in photonics or the description of exotic states of matter. The toroidal multipoles are the third group of multipoles which complements the electric and magnetic ones. The investigation of toroidal multipoles is important because of their role in the formation of non-radiating sources, so-called anapole states. The further study showed that there are also exist the higher-order toroidal dipoles or mean-square radii, which are equivalent. Here we suggest the structure which consists of point magnetic dipoles and supports the toroidal anapole state, which is obtained with the destructive interference of electric toroidal dipole with the first mean-square radius of the electric toroidal dipole. We also present an analytical condition for toroidal anapole with this system of point magnetic dipoles placed in two circles. #CSOC1120.

Keywords: Anapole · Toroidal Dipole · Multipole Decomposition

1 Introduction

Multipolar decomposition plays an important role in many areas of modern theoretical physics, e.g., the antenna design, the description of condensed matter systems and many others [1–3]. Especially significant role the multipolar decomposition plays in studying the electromagnetic properties of artificial metamaterials and metasurfaces for a broad range of optical applications, and nowadays it is actively used in the field of all-dielectric nanophotonics [4–7]. Multipolar analysis of dielectric nanoantennas allows to witness and study such optical phenomena as Fano resonances [8] or directional scattering [9, 10].

The classification of multipolar moments commonly distinguishes them into electric and magnetic ones. The regular way to obtain the multipolar decomposition is based on the Taylor series of the current or equivalently of the vector potential. This method reveals the existence of a third class of multipolar moments – the toroidal moments [11]. The toroidal multipoles give a rise to the so-called 'anapoles' – non-radiating states. The simplest example of those states is obtained due to the destructive interference between

R. Silhavy and P. Silhavy (Eds.): CSOC 2023, LNNS 723, pp. 249–253, 2023.
https://doi.org/10.1007/978-3-031-35317-8_23

electric and toroidal dipoles. The more complex examples are high-order [12] and hybrid [13, 14] anapole states. The latter is anapoles with mixed electric and magnetic character. To understand the former type the higher-order toroidal moments [8] or equivalently mean-square radii [15, 16] (their equivalence were shown in [17]) are needed.

2 Methods

Here we suggest a construction of anapole state obtained with destructive interference of electric toroidal dipole (TD) and first mean-square radius (MSR) using point magnetic dipoles (PMDs) and perform a theoretical analysis of this structure. Due to higher-order essence we expect the properties of the toroidal anapole (TA) to be enhanced compared to the classical anapole which is based on destructive interference between electric and toroidal dipoles.

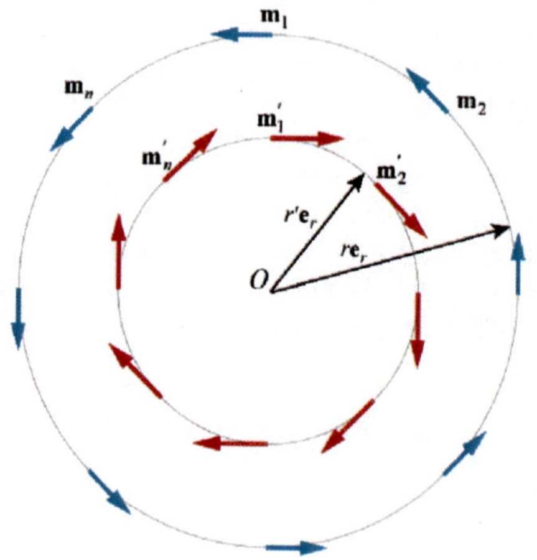

Fig. 1. The schematic of the geometry. Structure of 2 circles with 8 PMDs each. PMDs in each circle have equal magnitude and placed equidistantly.

The structure consists of 2 circles of 8 equidistantly placed PMDs each (Fig. 1). PMDs directed along the circles and in opposite directions compared to another circle. And the magnitude of PMDs are the same within each circle. The electric dipole of each PMD is zero and magnetic dipoles of opposite PMDs compensate each other, so overall electric and magnetic dipoles of structure are zero.

We derive Cartesian electric dipole (**d**), toroidal (**T**) and higher-order toroidal (**T**″) dipoles from the exact electric dipole [18]:

$$\mathbf{p} = -\frac{1}{i\omega}\left\{\int d^3\mathbf{r}\mathbf{J}_\omega j_0(kr) + \frac{k^2}{2}\int d^3\mathbf{r}\Big[3(\mathbf{r}\cdot\mathbf{J}_\omega)\mathbf{r} - r^2\mathbf{J}_\omega\Big]\frac{j_2(kr)}{(kr)^2}\right\} \quad (1)$$

To do this we will expand Eq. (1) into Taylor series for a small argument of kr. The mentioned ED, TD and MSR are the terms in this sum $\mathbf{p} = \mathbf{d} + \mathbf{T} + \mathbf{T}_2 + \ldots$. The expansion itself is based on combining the expansions of the Bessel functions $j_0(kr)$ and $j_2(kr)$ into one sum. The result is:

$$\mathbf{p} = -\frac{6}{i\omega}\int d^3\mathbf{r}\sum_{m=0}^{\infty}\frac{(-1)^m(m+1)(kr)^{2m}}{(2m+3)!}\Big[(m+1)\mathbf{J}_\omega - m(\mathbf{r}\cdot\mathbf{J}_\omega)\frac{\mathbf{r}}{r^2}\Big]. \quad (2)$$

The current of the system can be seen as the sum of currents of each PMD and those are defined as a curl of product of magnetization **m** and δ-function:

$$\mathbf{J} = \sum_j \mathbf{J}_j = \sum_j \nabla \times (\mathbf{m}_j\delta(\mathbf{r}-\mathbf{r}_j)). \quad (3)$$

Putting Eq. (3) into Eq. (2) one can find the arbitrary term in that sum with number j. Ignoring the multiplier for simplicity we get the following expression:

$$(\mathbf{T}_n)_\alpha \int d^3\mathbf{r} r^{2n-2}\Big((n+1)r_\mu r_\mu \varepsilon_{\alpha\beta\gamma}\partial_\beta m^j_\gamma \delta(\mathbf{r}-\mathbf{r}_j) - n r_\mu r_\alpha \varepsilon_{\mu\beta\gamma}\partial_\beta m^j_\gamma \delta(\mathbf{r}-\mathbf{r}_j)\Big) \quad (4)$$

For the TD ($n = 1$) the integral in Eq. (4) gives $-5\big[\mathbf{r}_j \times \mathbf{m}_j\big]_\alpha$, which after adding the multiplier gives:

$$\mathbf{T} = -\frac{1}{2}\frac{k^2}{i\omega}\sum_j [\mathbf{r}_j \times \mathbf{m}_j]. \quad (5)$$

The analogic result for the 1st MSR from Eq. (4) is $14r_j^2\big[\mathbf{r}_j \times \mathbf{m}_j\big]_\alpha$. And with the necessary multiplier:

$$\mathbf{T}_2 = \frac{1}{20}\frac{k^4}{i\omega}\sum_j r_j^2[\mathbf{r}_j \times \mathbf{m}_j]. \quad (6)$$

One can even calculate the arbitrary MSR for such system:

$$(\mathbf{T}_n)_\alpha = \frac{(-1)^n}{i\omega}\frac{6n(n+1)k^{2n}}{(2n+2)!}\sum_j r_j^{2n-2}\big[\mathbf{r}_j \times \mathbf{m}_j\big]_\alpha. \quad (7)$$

Since the electric and magnetic toroidal dipoles are already zero by construction for the structure showed on Fig. 1 the condition for toroidal anapole is: $(\mathbf{T} + \mathbf{T}_2) + \big(\mathbf{T}\prime + \mathbf{T}'_2\big) = 0$, where all values related to the smaller circle of PMDs are primed. The result is the following:

$$rm\Big(1 - \frac{1}{10}k^2r^2\Big) - r\prime m\prime\Big(1 - \frac{1}{10}k^2r\prime^2\Big) = 0, \quad (8)$$

r,$r\prime$ are radii and m, $m\prime$ are magnitudes of PMDs for bigger and smaller circles respectively.

We can find the solutions of Eq. (8) for $m/m\prime$ and for k:

$$\frac{m}{m\prime} = \frac{r\prime}{r}\frac{1 - \frac{1}{10}k^2 r\prime^2}{1 - \frac{1}{10}k^2 r^2};\ k = \left(\frac{10(rm - r\prime m\prime)}{r^3 m - r\prime^3 m\prime}\right)^{\frac{1}{2}}. \tag{9}$$

For practical applications the rare earth ions can serve as PMDs.

Acknowledgements. The authors gratefully acknowledge the financial support from the Ministry of Science and Higher Education of the Russian Federation (Agreement No. № 075-15-2022-1150). V.B. acknowledges the support of the Latvian Council of Science, project: DNSSN, No. Lzp-2021/1-0048.

References

1. Terekhov, P.D., Evlyukhin, A.B., Redka, D., Volkov, V.S., Shalin, A.S., Karabchevsky, A.: Magnetic octupole response of dielectric quadrumers. Laser Photon. Rev. **14**(4), 1900331 (2020)
2. Novitsky, D.V., Shalin, A.S., Redka, D., Bobrovs, V., Novitsky, A.V.: Quasibound states in the continuum induced by PT symmetry breaking. Phys. Rev. B **104**(8), 85126 (2021)
3. Novitsky, D., Karabchevsky, A., Lavrinenko, A., Shalin, A., Novitsky, A.: PT symmetry breaking in multilayers with resonant loss and gain locks light propagation direction. Phys. Rev. B **98**, 125102 (2018)
4. Shalin, A.S.: Broadband blooming of a medium modified by an incorporated layer of nanocavities. JETP Lett. **91**(12), 636–642 (2010)
5. Vestler, D., Shishkin, I., Gurvitz, E.A., et al.: Circular dichroism enhancement in plasmonic nanorod metamaterials. Opt. Express. **26**(14), 17841–17848 (2018)
6. Kostina, N.A., Kislov, D.A., Ivinskaya, A.N., et al.: Nanoscale tunable optical binding mediated by hyperbolic metamaterials. ACS Photon. **7**(2), 425–433 (2020)
7. Canós Valero, A., Kislov, D., Gurvitz, E.A., et al.: Nanovortex-Driven All-Dielectric Optical Diffusion Boosting and Sorting Concept for Lab-on-a-Chip Platforms. Advanced Science **7**(11), 1903049 (2020)
8. Gurvitz, E.A., Ladutenko, K.S., Dergachev, P.A., Evlyukhin, A.B., Miroshnichenko, A.E., Shalin, A.S.: The high-order toroidal moments and anapole states in all-dielectric photonics. Laser Photon. Rev. **13(5)**, 1800266 (2019)
9. Barhom, H., Machnev, A.A., Noskov, R.E., et al.: Biological Kerker effect boosts light collection efficiency in plants. Nano Lett. **19**(10), 7062–7071 (2019)
10. Shamkhi, H.K., Sayanskiy, A., Valero, A.C., et al.: Transparency and perfect absorption of all-dielectric resonant metasurfaces governed by the transverse Kerker effect. Phys. Rev. Mater. **3**(8), 85201 (2019)
11. Dubovik, V.M., Tugushev, V.V.: Toroid moments in electrodynamics and solid-state physics. Phys. Rep. **187**(4), 145–202 (1990)
12. Zenin, V.A., Evlyukhin, A.B., Novikov, S.M., et al.: Direct amplitude-phase near-field observation of higher-order anapole states. Nano Lett. **17**(11), 7152–7159 (2017)
13. Canós Valero, A., Gurvitz, E.A., Benimetskiy, F.A., et al.: Theory, observation, and ultrafast response of the hybrid anapole regime in light scattering. Laser Photon. Rev. **15**(10), 2100114 (2021)
14. Kuznetsov, A.V., Canós Valero, A., Tarkhov, M., Bobrovs, V., Redka, D., Shalin, A.S.: Transparent hybrid anapole metasurfaces with negligible electromagnetic coupling for phase engineering. Nanophotonics **10**(17), 4385–4398 (2021)

15. Radescu, E.E., Vaman, G.: Exact calculation of the angular momentum loss, recoil force, and radiation intensity for an arbitrary source in terms of electric, magnetic, and toroid multipoles. Phys. Rev. E Stat. Phys. Plasmas Fluids Relat. Interdiscip. Topics **65**(4), 47 (2002)
16. Nemkov, N.A., Basharin, A.A., Fedotov, V.A.: Electromagnetic sources beyond common multipoles. Phys. Rev. A **98**(2), 023858 (2018)
17. Borovkov, D., Valero, A.C.: On the link between mean square-radii and high-order toroidal moments. J. Phys. Conf. Ser. **2015**(1), 012021 (2021)
18. Alaee, R., Rockstuhl, C., Fernandez-Corbaton, I.: An electromagnetic multipole expansion beyond the long-wavelength approximation. Opt. Commun. **407**, 17–21 (2018)

Superscattering Regime for Si Conical Nanoparticles for the Different Directions of Excitation

Alexey V. Kuznetsov[1,2](✉) and Vjaceslavs Bobrovs[2]

[1] Moscow Institute of Physics and Technology, Dolgoprudny, Russia
alexey.kuznetsov98@gmail.com
[2] Riga Technical University, Riga, Latvia

Abstract. Through all-dielectric nanophotonics, it becomes possible to obtain a huge variety of new optical phenomena based on semiconductor dielectric nanoresonators. In this study, we investigate easy-to-make truncated-cone resonators and achieve superscattering regime due to the inherent property of cones – broken symmetry along the main axis without involving complex geometry or structured beams. Additional degrees of freedom due to the conicity allow for more fine-tuning of the interaction of light with matter at the nanoscale. #CSOC1120.

Keywords: Superscattering · All-Dielectric Nanophotonics · Truncated Cone

1 Introduction

In recent years, more and more relevant to the optical properties of subwavelength dielectric nanoparticles. With the help of such elements, it is possible to create platforms that exceed existing analogues in their properties, and sometimes even belong to a new class of devices. The undeniable advantages of such elements include high efficiency, low losses and the ability to control light at the nanolevel. Light control devices [1, 2], nanolasers [3, 5], nanoantennas [6, 8], optomechanical manipulation [9, 10], sensors [11, 14], metamaterials and metasurfaces [15–19] are driven by such resonators for different photonic application.

Interest in this direction grew significantly when it became possible to spectrally overlap resonances, which leads to various nanophotonic effects. Besides the completely non-scattering nanoobjects, researchers are also interested in the cases, when nanoparticles become anomalously strong scatterers. Such states are the opposite of an anapole [17, 18], because at a certain frequency scattering maxima of several multipoles appear; this state is called superscattering [20]. Nanoscatterers supporting this feature can find their application in a number of applications, such as sensing, optical communication, and other emerging areas [21, 22].

R. Silhavy and P. Silhavy (Eds.): CSOC 2023, LNNS 723, pp. 254–258, 2023.
https://doi.org/10.1007/978-3-031-35317-8_24

2 Superscattering Regime in Cone Silicon Particles

The study of various shapes of nanoscatterers has become increasingly popular in recent years due to the possibility of obtaining new effects and more precise tuning of multipole contributions to scattering pattern.

The most convenient in manufacturing geometry shape of scatterer with a large number of degrees of freedom is a truncated cone. Due to directed etching, a truncated cone is most likely to be obtained during sample production; therefore, the study of this shape of nanoresonators is important for optimizing the fabrication process and getting new effects [23, 24].

It is because of the symmetry breaking in the case of truncated nanocones along the z axis, it is important to study the multipole content of nanoscatterers depending on the direction of radiation (Fig. 1). By changing the direction, the multipole content of the nanoscatterer changes, while the total cross-section does not change.

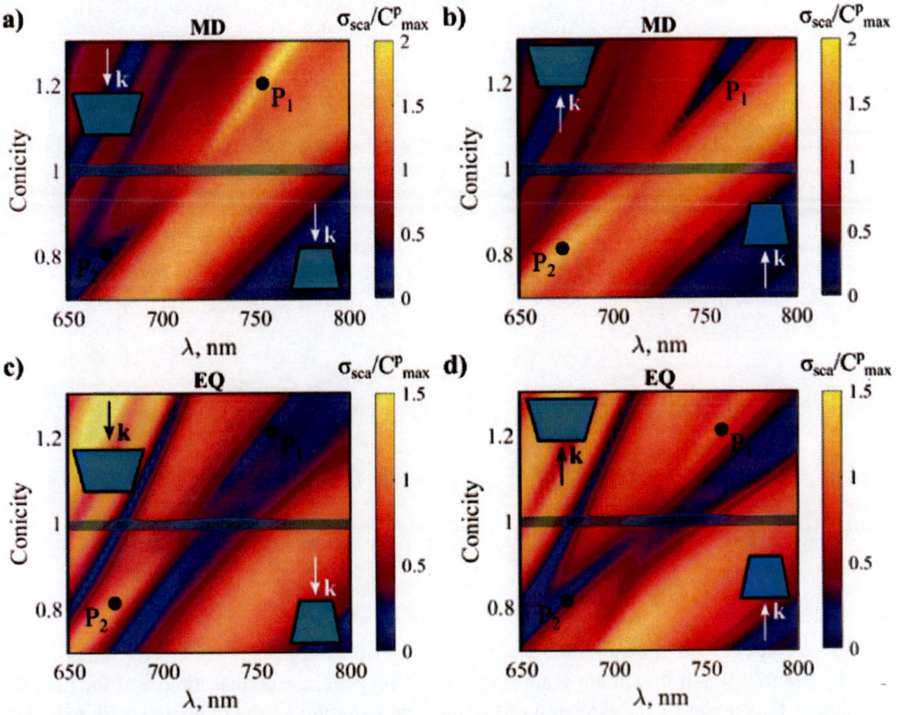

Fig. 1. Magnetic dipole and electric quadrupole contributions to the total scattering cross-section of the truncated cone. (a), (b) Multipole contributions for normally-incident plane wave for propagating along the z-direction (forward). (c), (d) Multipole contributions of the scattering for the opposite (backward) excitation. Geometric parameters: P_1: H = 415 nm, R_{top} = 216 nm, R_{bottom} = 180 nm, λ = 760 nm, P_2: H = 415 nm, R_{top} = 144 nm, R_{bottom} = 180 nm, λ = 673 nm

As depicted in Fig. 1, radiation occurs through the magnetic dipole (MD) and electric quadrupole (EQ) channels depending on the direction of the incident radiation. Interestingly, if the propagation of the light has a different direction the magnetic dipole channel displays strong superscattering in the lower branch of the interference (see Fig. 1(a), (b)).

A similar but opposite statement can be made for the electric quadrupole channel (see Fig. 1(c), (d)). Considering the structure is assumed to be lossless then the total scattering is conserved upon a reversed excitation direction. Notably, the two odd multipolar channels offset the difference in the scattering whereas their summed contribution to scattering and the overall bandwidth is nearly identical for both the forward and backward incident light propagation.

In Fig. 2 the near-field and far-field scattering signatures are shown for the two points P_1 and P_2 of the scattering maxima. The figure shows how a particle in the superscattering regime can significantly change the field around it. By carefully tuning the geometry of the nanoresonator, it is possible to obtain the necessary radiation patterns, which makes such particles a promising platform for devices capable of controlling light at the nanoscale.

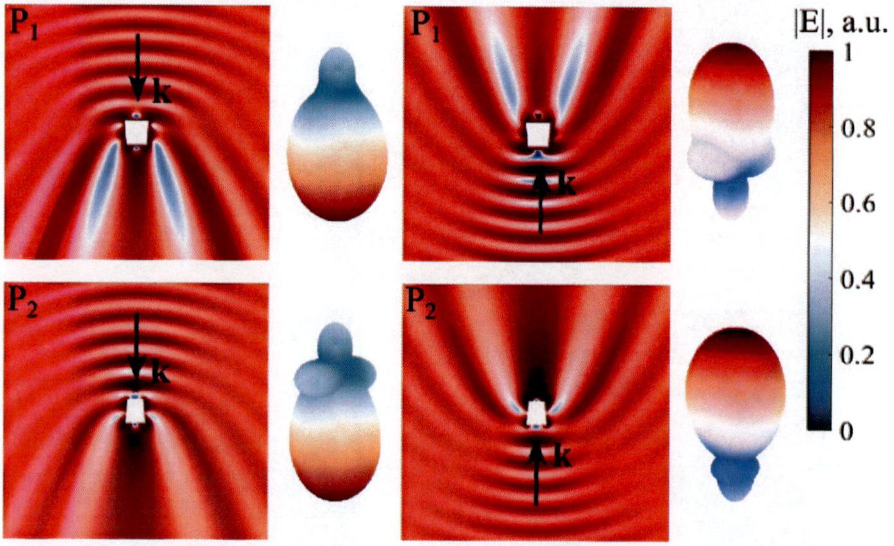

Fig. 2. Strong near-field distribution and the correspondent scattering pattern for the points (P_1 and P_2 (see Fig. 1)) of maximum scattering. Scattering patterns and distribution of the near field are shown depending on the direction of the incident radiation on the truncated cone particle. A plane wave was used as the incident radiation.

Although the total scattering remains identical upon reversed excitation direction there are noticeable differences in scattering patterns. While the presence of the scattering side lobes evidences a strong presence of electric quadrupole radiation the overall directivity at P_2 is higher compared to P_1. This can be noticed by the scaled backscattering-caused shadow at P_1 for the case of the backward excitation. However, when the magnetic dipole dominates the scattering, the situation is reversed where the scattering is more directive under the forward excitation.

In such way, by changing the direction of the incident light onto a cone particle, one can obtain different radiation patterns in the particle superscattering regime. This opens up new horizons for designing various structures with the ability to control optical properties depending on the direction of radiation.

3 Conclusion

Thus, we have demonstrated a remarkable control over scattering afforded by the conical geometry. We have demonstrated that conicity can be utilized to enhance scattering strongly and to control scattering linewidth in a broad range.

We showed that the behavior of different multipoles, which can be controlled by conicity, influences certain outcome in scattering for instance. It indeed leaves plethora of opportunities to optimize "the practical" cone geometry for intended applications.

Acknowledgments. The authors gratefully acknowledge the financial support from the Ministry of Science and Higher Education of the Russian Federation (Agreement No. № 075-15-2022-1150). V.B. acknowledges the support of the Latvian Council of Science, project: NEO-NATE, No. Lzp-2022/1-0553.

References

1. Barhom, H., et al.: Biological Kerker effect boosts light collection efficiency in plants. Nano Lett. **19**, 7062–7071 (2019)
2. Shamkhi, H.K., et al.: Transparency and perfect absorption of all-dielectric resonant metasurfaces governed by the transverse Kerker effect. Phys. Rev. Mater. **3**, 1–10 (2019)
3. Xu, K., Fang, M., Huang, Z.: Compact unidirectional laser based on all-dielectric metasurface with high quality factor. IEEE Photon. J. **13**, 1–9 (2021)
4. Shankhwar, N., Sinha, R.K., Kalra, Y., Makarov, S., Krasnok, A., Belov, P.: High-quality laser cavity based on all-dielectric metasurfaces. Photon. Nanostruct. **24**, 18–23 (2017)
5. Totero Gongora, J.S., Miroshnichenko, A.E., Kivshar, Y.S., Fratalocchi, A.: Anapole nanolasers for mode-locking and ultrafast pulse generation. Nat Commun. **8**, 1–9 (2017)
6. Sain, B., Meier, C., Zentgraf, T.: Nonlinear optics at all-dielectric nanoantennas and metasurfaces. ArXiv. **1**, 1–14 (2020)
7. Krasnok, A.E., Miroshnichenko, A.E., Belov, P.A., Kivshar, Y.S.: All-dielectric optical nanoantennas. Opt. Express. **20**, 20599 (2012)
8. Li, N., Lai, Y., Lam, S.H., Bai, H., Shao, L., Wang, J.: Directional control of light with nanoantennas. Adv. Opt. Mater. **9**, 1–32 (2021)
9. Canós Valero, A., et al.: Nanovortex-driven all-dielectric optical diffusion boosting and sorting concept for lab-on-a-chip platforms. Adv. Sci. **7**, 1903049 (2020)
10. Dienerowitz, M., Mazilu, M., Dholakia, K.: Optical manipulation of nanoparticles: a review. J. Nanophoton. **2**, 021875 (2008)
11. Novitsky, D.V., Karabchevsky, A., Lavrinenko, A.V., Shalin, A.S., Novitsky, A.V.: PT symmetry breaking in multilayers with resonant loss and gain locks light propagation direction. Phys. Rev. B. **98**, 125102 (2018)

12. Terekhov, P.D., Evlyukhin, A.B., Redka, D., Volkov, V.S., Shalin, A.S., Karabchevsky, A.: Magnetic octupole response of dielectric quadrumers. Laser Photon. Rev. **14**, 1900331 (2020)
13. Vestler, D., et al.: Circular dichroism enhancement in plasmonic nanorod metamaterials. Opt. Express. **26**, 17841 (2018)
14. Novitsky, D.V., Shalin, A.S., Redka, D., Bobrovs, V., Novitsky, A.V.: Quasibound states in the continuum induced by PT symmetry breaking. Phys. Rev. B. **104**, 085126 (2021)
15. Kozlov, V., Filonov, D., Shalin, A.S., Steinberg, B.Z., Ginzburg, P.: Asymmetric backscattering from the hybrid magneto-electric meta particle. Appl. Phys. Lett. **109**, 203503 (2016)
16. Terekhov, P.D., et al.: Enhanced absorption in all-dielectric metasurfaces due to magnetic dipole excitation. Sci. Rep. **9**, 1–9 (2019)
17. Canós Valero, A., et al.: Theory, observation, and ultrafast response of the hybrid anapole regime in light scattering. Laser Photon. Rev. **15**, 2100114 (2021)
18. Kuznetsov, A.V., Canós Valero, A., Tarkhov, M., Bobrovs, V., Redka, D., Shalin, A.S.: Transparent hybrid anapole metasurfaces with negligible electromagnetic coupling for phase engineering. Nanophotonics **10**, 4385–4398 (2021)
19. Kostina, N.A., et al.: Nanoscale tunable optical binding mediated by hyperbolic metamaterials. ACS Photon. **7**, 425–433 (2020)
20. Ruan, Z., Fan, S.: Superscattering of light from subwavelength nanostructures. Phys. Rev. Lett. **105**, 1–4 (2010)
21. Krasikov, S.D., et al.: Superscattering for non-spherical objects. J. Phys. Conf. Ser. **2015**, 012073 (2021)
22. Liu, W., Lei, B., Shi, J., Hu, H.: Unidirectional superscattering by multilayered cavities of effective radial anisotropy. Sci. Rep. **6**, 1–9 (2016)
23. Wang, J., Du, J.: Plasmonic and dielectric metasurfaces: design, fabrication and applications. Appl. Sci. **6**, 239 (2016)
24. Su, V.-C., Chu, C.H., Sun, G., Tsai, D.P.: Advances in optical metasurfaces: fabrication and applications [Invited]. Opt. Express. **26**, 13148 (2018)

A Delay and Energy-Aware Task Offloading and Resource Optimization in Mobile Edge Computing

Ducsun Lim and Inwhee Joe(✉)

The Department of Computer and Software, Hanyang University, Seoul 04763, Republic of Korea
{imcoms,iwjoe}@hanyang.ac.kr

Abstract. The rapid development and diffusion of the Internet of Things (IoT) have increased the growth of smart devices (SDs). SDs may run computation-intensive and delay-sensitive applications. However, SDs with limited computation power and a small battery cannot execute these applications. To solve this problem, task offloading can be used to transfer compute-intensive tasks to an external server or processor in mobile edge computing (MEC). There are many studies on task offloading or resource allocation in MEC systems, but few researches consider selecting an appropriate edge node (EN) as the optimal cost. In this paper, we propose an MEC system to minimize SDs' delay and energy consumption costs by offloading tasks to the adjacent ENs. We compared our proposed method with currently used techniques. The simulation results demonstrate the effectiveness of this proposed algorithm in terms of the delay cost and energy costs.

Keywords: smart device · Internet of Things · offloading · energy consumption · mobile edge computing

1 Introduction

The rapid development and diffusion of smart devices (SDs) have increased the growth of wearables, smartphones, tablet computers, and connected devices [1]. SDs with limited computation power and a small battery cannot execute these all application [2, 3]. These tasks could be addressed with mobile cloud computing (MCC) because most of them were compute-intensive tasks [4]. However, the unacceptable high latency communications with MCC is the dilemma for real-time and latency-sensitive applications. Thus, the use of mobile edge computing (MEC), also known as multi-access computing [5] was proposed for efficient task processing [6]. Unlike MCC, MEC provides an effective solution for computation offloading by placing a small server near users. In many studies, people only paid attention to the optimization of task offloading, because choosing a proper edge node (EN) for task offloading can determine the quality of experience (QoE) [7]. A number of methods have been proposed to overcome these challenges [8,

R. Silhavy and P. Silhavy (Eds.): CSOC 2023, LNNS 723, pp. 259–268, 2023.
https://doi.org/10.1007/978-3-031-35317-8_25

9]. However, these studies did not consider the profit of using nearby other edges to provide offload tasks when the nearest edge server cannot process these tasks. In addition, solving the two costs separately is not appropriate, because the decisions in the two procedures will affect each other. In other words, task offloading and resource allocation jointly determine the MEC. Actually, there are differences between task offloading and resource allocation. Firstly, the task offloading needs to select an appropriate EN to offload, and how many resources should be allocated to each task. Secondly, we need to make an exploration to search for a more suitable EN in the process of task offloading, while we must precisely determine the optimal cost value. Therefore, task offloading and resource allocation should be considered respectively, but previous studies did not solve this issue.

In addressing these situations, this paper proposes an offloading algorithm to consider delay and energy consumption together. At first, the problem of task offloading and resource allocation in MEC is formulated into a cost minimization. And the cost minimization considers both delay and energy consumption costs by offloading tasks to the adjacent ENs. The results of simulation-based tests show that compared with the used techniques, our proposed algorithm can make better than the existing task offloading method.

The rest of this article is organized as follows. Section 2 introduces related works and investigates previous research. Section 3 describes the mathematical models for components, local computing, and offloading. Section 4 problem modeling and algorithm and explains its operation. Section 5 presents the performance evaluation result. Finally, Sect. 6 summarizes the study and suggests future research directions.

2 Related Work

With the rapid advancement of network communication, MEC is emerging as a promising paradigm. As a supporting technology for IoT applications, MEC has received a lot of attention from industry and academia. Many methods related to task offloading and resource allocation have been studied.

Li et al. [10] focus on the delay of computing tasks in the MEC system and devise a particle swarm optimization-based energy-aware edge server placement algorithm to find the optimal solution. Samanta et al. [11] considered both delay-tolerant and delay-sensitive tasks to achieve optimized service delay and minimize the average response time by offloading the computation-intensive tasks of applications. Thai et al. [12] proposed solutions to solve the issue of optimizing the workload in the MEC but they only pursue a single optimization goal. However, these above studies consider optimization only one of energy and delay. In the real environment, objectives would be considered together. In [13], a joint communication and computation resource allocation algorithm was proposed to minimize energy consumption and delay.

Zhang et al. [14] considered optimization schemes for minimizing load balancing and energy consumption in scenarios. Eshraghi et al. [15] proposed an algorithm that optimized the decision-making on resource allocation and task offloading. Bi et al. [16] proposed the Simulated Annealing algorithm that joint optimization of the computing offload between the cloud data center and the MDs. These above studies have some shortcomings. For example, the task offloading and resource allocation are solved respectively. Also, these researches are not considered the multiple MEC in task offloading and resource allocating. There are many optimization methods adopted in MEC. Exactly, multiple optimization aims are always more difficult but practical to study compared with a single optimization aim.

3 System Model

In this section, first, we will be introduced the system model. Then, we explained the transmission model, delay, and energy consumption model. Finally, the minimization problem for the weighted sum of delay and energy is formulated. In addition, Table 1 lists the main notations.

3.1 Components

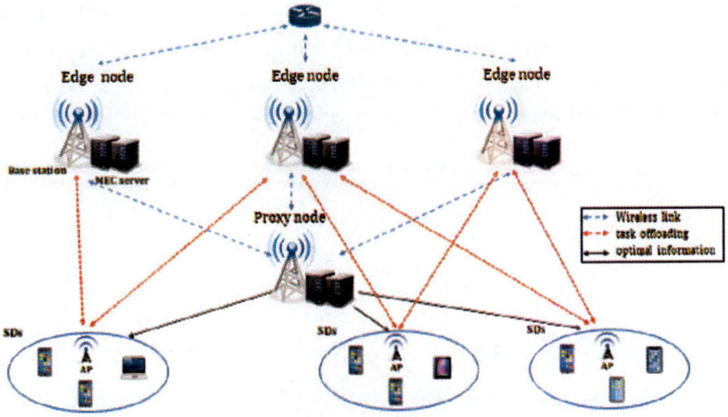

Fig. 1. Network architecture for MEC system

As shown in Fig. 1, we assumed a MEC system with multiple SDs, and multiple ENs [17]. SDs are responsible for generating the tasks and executing the tasks locally. One or more SDs were connected to the base station through wireless communication technology. One or more SDs were connected to the base station through wireless communication technology.

The workflow of the system is described as follows. First, SD receives a request for EN information from PN. The set of SDs M = {1, 2,..., m} processes tasks generated locally and offloads them to a remote MEC server through a base station (BS). Next, each SD sends the task to EN based on the receipt of offloading and server allocation strategy. Finally, each EN executes the offloaded task and returns the results to the corresponding SD. In this paper, time is separated into many time slots. We consider a set of ENs $M = \{1, 2, ..., M\}$ and a set of SDs $S = \{1, 2, ..., S\}$. And the task of each SD s is donated as $K = \{1\ 2, ..., K\}$, and each task is represented as a tuple $\{\rho_s, C_s\}$, where h and m denote the size of the task and the number of required CPU cycles.

3.2 Transmission Model

Each SD generates only one computation-intensive task in each time slot. Compared with locally, offloading to EN can reduce task delay and energy consumption. However, the process of offloading tasks will occur additional delay and energy. For transmission, the maximum achievable uplink and downlink data rate for an additive white Gaussian noise (AWGN) channel [18] can be derived as follows:

$$r_{ul}^{tr} = n\frac{B}{N}\log_2\left(1 + \frac{P_m|h_{ul}|^2}{\omega(\delta_{ul})d^{\iota}\eta}\right) \tag{1}$$

$$r_{dl}^{tr} = n\frac{B}{N}\log_2\left(1 + \frac{P_s|h_{dl}|^2}{\omega(\delta_{dl})d^{\iota}\eta}\right) \tag{2}$$

We assume the equal noise in transmission for uplink and downlink. Where B is the bandwidth, and N is the set of subcarriers. d denotes the distance between SD and EN, η is the noise power, P_m and P_s refer to the transmission power of SD and EN, respectively. ι is the path loss exponent, while δ_{ul} and δ_{dl} are the required bit error rate for uplink and downlink, respectively. And h_{ul} and h_{dl} are the channel fading coefficient while the $\omega(\delta_{ul}) = \frac{-2\log(5\delta_{ul})}{3}$ represents the SNR margin to meet the required bit error rate. We have considered Rayleigh-fading in the scenario. The energy consumption of SD, due to transmission and reception, depends upon the amount of data to be transferred in offloading.

3.3 Local Computing

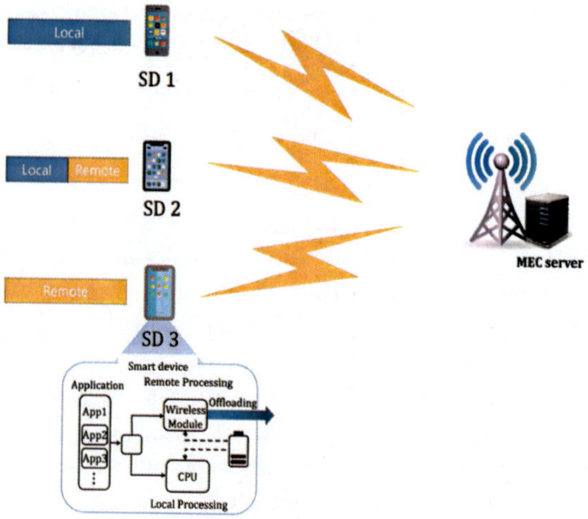

Fig. 2. MEC System considering partial offloading.

The SD has a computational capability (CPU cycles per second), which is denoted as f^{loc} (Fig. 2). The cycle denotes the required number of cycles per bit of task. These values may vary with the task type. Therefore, the local computing delay, T^{loc}, required to run K is expressed as follows:

$$T_s^{loc} = \frac{\lambda C_s}{f^{loc}} \tag{3}$$

where λ is the ratio for the non-offloaded task. Therefore, the number of task bits executed locally is $\lambda_s \rho_s$, and the task size executed by the EN is $(1 - \lambda_s)\rho_s$.

According to [19], the energy consumption of SD s performed locally is as follows:

$$E_s^{loc} = \mu (f^{loc})^2 \lambda_s C_s \tag{4}$$

where μ is the effective switching capacity set to $\mu = 10^{-9}$ [20]. Therefore, the cost of local is the combination of the local time and the local energy consumption and is defined as

$$C_s^{loc} = \sum_{s=1}^{S} (\alpha T_s^{loc} + \beta E_s^{loc}) \tag{5}$$

3.4 Edge Computing

In edge computing, the task transmission time to complete the task on the EN can be calculated as follows:

$$T_{ul}^{tra} = \frac{(1 - \lambda_s)\rho_s}{r_{ul}^{tr}} \tag{6}$$

$$T_{dl}^{tra} = \frac{(1 - \lambda_s)\rho_s}{r_{dl}^{tr}} \tag{7}$$

As for the EN, we define F as the computing frequency of the EN allocated to the task of SD. Thus the task data processing time on the EN is:

$$T^{pro} = \frac{(1 - \lambda_s)F_m}{f^{mec}} \tag{8}$$

Then, the task time consumed on EN is:

$$T_{en}^{tot} = T_{ul}^{tra} + T_{aen}^{pro} + T_{dl}^{tra} \tag{9}$$

The energy consumption in the offloading task can be as follows:

$$E^{mec} = P_m T_{en}^{tot} \tag{10}$$

Therefore, the cost of the edge node is the combination of edge computing time and energy consumption, as follows:

$$C_s^{en} = \sum_{s=1}^{S} (\alpha T_s^{en} + \beta \ E_s^{en}) \tag{11}$$

3.5 Adjacent Edge Computing

When the decision decides to offload a task to an adjacent EN can be calculated as follows:

$$T^{tot}_{aen} = T^{tra}_{ul} + T^{pro}_{aen} + T^{tra}_{dl} \tag{12}$$

In the adjacent edge node, the energy consumption in the offloading task can be as follows:

$$E^{mec} = P_m T^{tot}_{aen} \tag{13}$$

Therefore, the cost of the adjacent edge node is the combination of edge computing time and energy consumption, as follows:

$$C^{aen}_s = \sum_{s=1}^{S} (\alpha T^{aen}_s + \beta E^{aen}_s) \tag{14}$$

The total cost of the system can be expressed as follows:

$$C^{tot} = \sum_{s=1}^{S} \left(\frac{X_S(1 - X_S)(2 - x_S)C^{loc}_s}{2} + x_S(2 - x_S)C^{en}_s + x_S(x_S - 1)C^{aen}_s \right) \tag{15}$$

In total cost (15), that is, $x_s = 0$ chooses to execute tasks locally. Also, $x_s = 1$ chooses to offload tasks to the edge node, and, $x_s = 2$ chooses to offload tasks to an adjacent edge.

4 Problem Formulation and Algorithm

The binary variable $x_s \in \{0, 1, 2\}$ 0 denotes whether task is processed locally or offloaded to EN or adjacent EN. We set x = 0 indicating the task is processed locally, and x = 1 and x = 2 means the task is allocated to EN, and adjacent EN respectively. Then the cost minimization problem can be formulated using the following notation:

$$\textit{minimize } C^{tot} = \sum_{s=1}^{S} \left(\frac{X_S(1 - X_S)(2 - x_S)C^{loc}_s}{2} + x_S(2 - x_S)C^{en}_s + x_S(x_S - 1)C^{aen}_s \right) \tag{16}$$

subject to:

$$x_s \in \{0, 1, 2\}, \forall_s \in S \tag{17}$$

$$f^{mec} \leq F_m, \forall_m \in M \tag{18}$$

$$B_{m,s} \leq B_m, \forall_m \in M \tag{19}$$

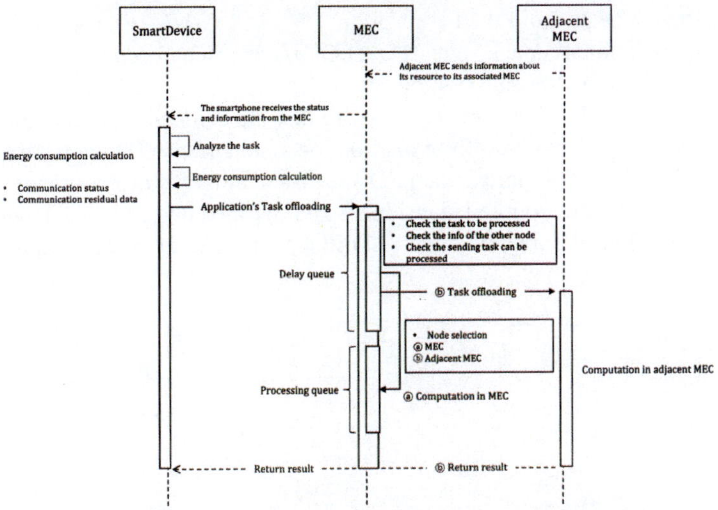

Fig. 3. Sequence diagram of the proposed offloading

where the objective functions defined by Formulas (16) purpose to search the task offloading and resource allocation strategy that minimizes the delay and energy consumption. Constraint (17) restricted the value of x_s and constraint (18) ensures that the computing capability of tasks executed on the EN does not exceed the maximum value F_m. In addition, Constraint (12) ensures that the bandwidth between EN and the task on the EN is less than the maximum B_m. When an SD has completely solved its optimization problem, the optimal MEC server can be executed the task.

Algorithm 1

Input: f^{loc}, f^{en}, f^{aen}, B_m, E, D, ω, λ, P
Output: C^{tot}
Initialize: f^{loc}, f^{en}, f^{aen}, P, λ
repeat
  Calculate the optimal delay f^{loc}, f^{en}, f^{aen} (3), (9), (12);
  Calculate the optimal energy consumption (4), (10), (13);
  Get the minimal cost from (15)
until C^{tot*} ← total cost
return C^{tot}

---

5 Experimental Results

The foundation simulation settings for performance evaluation are as follows. We performed simulation on a computer with a Intel i7 CPU and a 3.7-GHz, 32-GB RAM processor. For simulation, we assume a small cell with bandwidth B = 10 MHz. In the simulation, we assume that multiple MEC servers and randomly scattered SDs are within

200 m distance away from the base station. $f^{mec} = 5$ GHz/s is representing the MEC servers computational capacity and $f^{loc} = 1$ GHz/s represents each SD's CPU frequency. $P_m = 500$ mW is the transmission power of SDs while $P_m = 100$ mW is the set value as idle power [21]. Furthermore, considering that the computation offloading task size ρ_s (in kbits) complies with uniform distribution among (200, 400), and the CPU cycles of C_s comply 1000. For clarity, each SD's decision weight is taken as $\alpha, \beta = 0.5$. The next is to compare the proposed algorithm with full offloading and local where the tasks are executed by local computing and full offloading refers to the offloading of the task to the edge server.

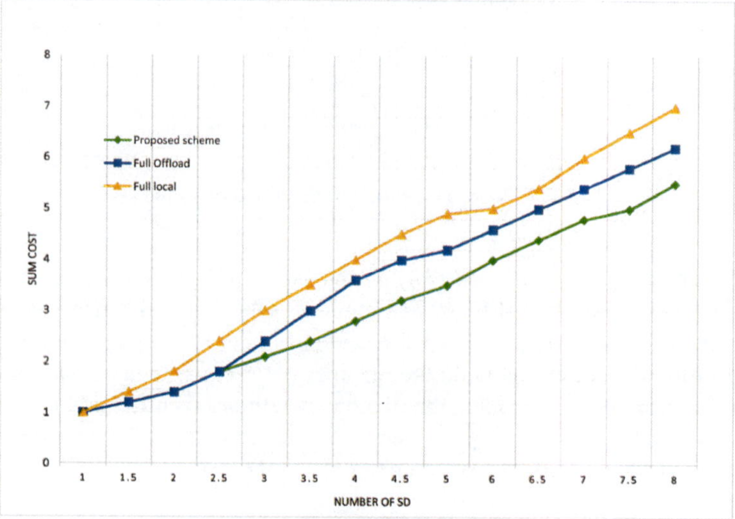

Fig. 4. Sum cost versus of SDs.

In Fig. 4 total cost versus is increased number of SD is presented. It can be shown from the graph that the total cost of every method increases while the number of SDs is increasing. The full offloading curve is comparatively higher than the proposed algorithm. With the increasing number of SDs it increases rapidly. The reason behind this is with a large number of SDs the MEC server capacity is not that enough to endeavor offload computation for all SDs. Herefore, it requests an adjacent server to process the task.

In Fig. 5, while MEC server computational capacity is increasing the total cost is also increasing. The proposed algorithm has the best simulation result in Fig. 3. In Fig. 3, the curve for Local only is unchanged even though the MEC server capacity is increasing because local does not use the computational resources of MEC servers. Other curves decrease with the increase of MEC servers computational because the processing time decreases as each SD is allocated to a more resource. In addition. It is noticed that the Offloading and proposed algorithm total costs decrease slowly and the performance of these offloading is nearly the equivalent. The result is a total cost of more locally than EN.

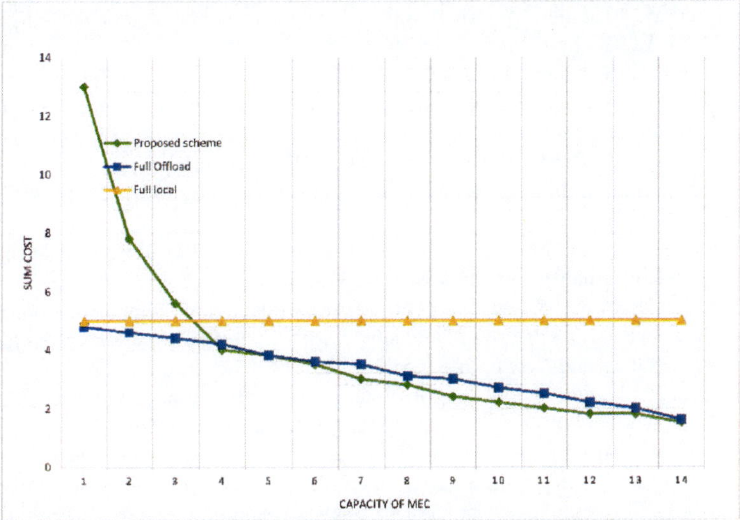

Fig. 5. Sum cost versus capacity of the edge node.

6 Conclusion

This study proposed algorithm, which uses optimal cost to select between methods based on local computing, and offloading. We investigated the computation task offloading problem for delay-sensitive tasks and computation-intensive in a MEC environment. We proposed the offloading decision-making to minimize costs in terms of both delay and energy of the SD. We considered formulating the optimization algorithm strategy. The simulation result confirmed that the proposed algorithm guaranteed a lower average delay than other methods. In future, we will consider the code offloading on edge server and cloud computing that are connected with more SD.

Acknowledgements. This research was supported by Culture, Sports and Tourism R&D Program through the Korea Creative Content Agency grant funded by the Ministry of Culture, Sports and Tourism in 2022 (Project name: Customized tourism through AI-based tourist situation recognition and tourism information curation development of itinerary recommendation platform technology, Project Number: R2022020116).

References

1. John Dian, F., Vahidnia, R., Rahmati, A.: Wearables and the Internet of Things (IoT), applications, opportunities, and challenges: a survey. IEEE Access 8, 69200–69211 (2020)
2. Zheng, J., Cai, Y., Wu, Y., Shen, X.: Dynamic computation offloading for mobile cloud computing: a stochastic game-theoretic approach. IEEE Trans. Mobile Comput. **18**(4), 771–786 (2019)
3. Nguyen, Q.-H., Dressler, F.: A smartphone perspective on computation offloading—a survey. Comput. Commun. **159**, 133–154 (2020)

4. Othman, M., Madani, S.A., Khan, S.U.: A survey of mobile cloud computing application models. IEEE Commun. Surv. Tutor. **16**(1), 393–413 (2013)
5. Porambage, P., Okwuibe, J., Liyanage, M., Ylianttila, M., Taleb, T.: Survey on multi-access edge computing for Internet of Things realization. IEEE Commun. Surv. Tuts. **20**(4), 2961–2991, 4th Quart. (2018)
6. Mao, Y., You, C., Zhang, J., Huang, K., Letaief, K.B.: A survey on mobile edge computing: the communication perspective. IEEE Commun. Surv. Tuts. **19**(4), 2322–2358, 4th Quart. (2017)
7. Hossain, M.D., et al.: Fuzzy decision-based efficient task offloading management scheme in multi-tier MEC-enabled networks. Sensors **21**(4), 1484 (2021)
8. Dolezal, J., Becvar, Z., Zeman, T.: Performance evaluation of computation offloading from mobile device to the edge of mobile network. In: 2016 IEEE Conference on Standards for Communications and Networking (CSCN). IEEE (2016)
9. Kamoun, M., Labidi, W., Sarkiss, M.: Joint resource allocation and offloading strategies in cloud enabled cellular networks. In: 2015 IEEE International Conference on Communications (ICC). IEEE (2015)
10. Li, Y., Wang, S.: An energy-aware edge server placement algorithm in mobile edge computing. In: 2018 IEEE International Conference on Edge Computing (EDGE). IEEE (2018)
11. Samanta, A., Chang, Z.: Adaptive service offloading for revenue maximization in mobile edge computing with delay-constraint. IEEE Internet Things J. **6**(2), 3864–3872 (2019)
12. Thai, M., Lin, Y., Lai, Y., Chien, H.: Workload and capacity optimization for cloud-edge computing systems with vertical and horizontal offloading. IEEE Trans. Netw. Service Manag. **17**(1), 227–238 (2020)
13. Hu, Y.C., Patel, M., Sabella, D., Sprecher, N., Young, V.: Mobile edge computing—a key technology towards 5G. ETSI White Paper **11**(11), 1–16 (2015)
14. Zhang, F., Wang, M.M.: Stochastic congestion game for load balancing in mobile-edge computing. IEEE Internet Things J. **8**(2), 778–790 (2021)
15. Eshraghi, N., Liang, B.: Joint offloading decision and resource allocation with uncertain task computing requirement. In: IEEE INFOCOM 2019-IEEE Conference on Computer Communications. IEEE (2019)
16. Bi, H. Yuan, S. Duanmu, M. Zhou and A. Abusorrah.: Energy-Optimized Partial Computation Offloading in Mobile-Edge Computing With Genetic Simulated-Annealing-Based Particle Swarm Optimization. IEEE Internet Things J. **8**(5), 3774–3785 (2021)
17. Lim, D., Lee, W., Kim, W.T., Joe, I.: DRL-OS: a deep reinforcement learning-based offloading scheduler in mobile edge computing. Sensors **22**(23), 9212 (2022)
18. Yu, S., Wang, X., Langar R.: Computation offloading for mobile edge computing: a deep learning approach. In: Proc. Int. Symp. Pers., Indoor, Mobile Radio Commun. (PIMRC), Oct., pp. 1–6 (2017)
19. Nath, S., et al.:.Delay and energy efficiency tradeoff for information pushing system. IEEE Trans. Green Commun. Netw. **2**(4), 1027–1040 (2018)
20. Zhao, P., Tian, H., Qin, C., Nie, G.: Energy-saving offloading by jointly allocating radio and computational resources for mobile edge computing. Proc. ICLR (2016)
21. Cao, Y., Jiang, T., Wang, C.: Optimal radio resource allocation for mobile task offloading in cellular networks. IEEE Netw. **28**(5), 68–73 (2014)

Scaling Networks and Capturing Keys Using Combined Combinatorial Block Designs

Alexander Frolov(✉), Natalya Kochetova, and Anton Klyagin

National Research University Moscow Power Engineering Institute, Moscow, Russian Federation
{frolovab,kochetovanp,kliaginao}@mpei.ru

Abstract. This paper discusses a united approach to detecting blocks containing an uncompromised key in wireless sensor networks that have a structure of combinatorial block designs and scaling those networks. This approach consists in using both individual blocks of combinatorial designs and collections of blocks containing a certain element. Such collections of blocks are called dual blocks. If the elements are interpreted as key identifiers, then knowledge of the composition of dual block, combined with information about the placement of blocks, allows planning operations to capture blocks with the key of interest to the invader. On the other hand, the paper proposes to construct a network with a combinatorial design structure, consisting of subnets corresponding to blocks, and possessing connections corresponding to dual blocks, thereby providing a scaling step. The resulting network has a structure of combined combinatorial design that along with blocks contains dual blocks, its nodes are pairs of blocks and dual blocks corresponding to elements of this block. The paper gives examples and estimates of scaling when converting a combinatorial design to a combined combinatorial design. It is shown that, on the basis of such systems, it is possible to significantly increase the scale of the network structure and control the search for an uncompromised key in the WSN. Examples of the application of the proposed scaling methods and key capturing in WNS of smart house are given. #CSOC1120.

Keywords: system of sets · combinatorial design · combined combinatorial design · combinatorial network structure · internal subnet · external subnet · selective node capture · network scaling · smart house

1 Introduction

In the era of Industry 4.0, the Internet of Things appeared and is developing based on the widespread use of secure wireless sensor network (WSN) structures. At the same time, resources are used that are limited both in terms of power consumption and in terms of the required memory [1, 2]. The security of such systems can be ensured by symmetric encryption systems with the preliminary distribution of keys in the network nodes. Sensor network key management includes [3]: preliminary key distribution, shared key detection, key delivery path determination. Currently, algorithms for these steps are created using some types of combinatorial block designs, which are systems of sets defined

R. Silhavy and P. Silhavy (Eds.): CSOC 2023, LNNS 723, pp. 269–289, 2023.
https://doi.org/10.1007/978-3-031-35317-8_26

by a set of elements and a set of its subsets (blocks) [4, 5]. In parallel, also using set systems, a number of papers study the structures of computer networks [6, 7]. A combinatorial configuration is used, which for a given combinatorial block design is built on the set of all pairs consisting of a block and a dual block corresponding to the element of this block. It is called a combined combinatorial block design. Its blocks are subsets with the selected first element of the pair, and dual blocks are subsets with the selected second element of the pair. Blocks and dual blocks of the combined combinatorial block design are mapped to interconnected internal and external subnets of the scaled WSN, as well as subgraphs of the key-sharing graph in it. The edges of this graph connect vertices with the same first or second elements. Taking into account a number of well-known approaches to scaling networks by modifying known combinatorial block designs, we propose to build and use combined analogs of such structures as interconnected internal and external subnets, any two nodes of which are connected along a route of no more than three, and evaluate the scale of the combined structure and the required amount of key memory of the node in comparison with the scale of the original structure and the memory capacity of its node. At the same time, the task naturally arises of constructing key distribution schemes and key spaces for networks with a combined combinatorial block design structure, assessing their security and analyzing the capabilities of the enemy to capture nodes for compromising keys. The aim of the work is to present this new way of scaling WSNs and schemes for the preliminary distribution of key spaces in them, as well as to analyze the security of such WSNs when capturing nodes and substantiate the applicability of this scaling method when designing the WSN of the Internet of Things.

2 Methods

2.1 Some Combinatorial Block Designs Used in WSN

Combinatorial block designs are systems of sets. The system of sets (**X, B**) is given by a finite set of **X** elements and a set **B** of its subsets [3]. Varieties of set systems currently used for the preliminary distribution of keys are balanced incomplete block designs (BIBD) with parameters (v,b,r,k), where v is the number of elements, b is the number of blocks, r is the number of blocks containing a given element, k is the number of elements in each block. In addition, the fifth parameter λ is used as the number of blocks, each of which contains a given pair of elements. If $v = b$, the BIBD is symmetric with parameters (v,k) and an additional parameter λ. It is abbreviated as SBIBD. SBIBD with parameters $(n^2 + n + 1, n + 1, 1)$ is called a projective plane (or geometry) of order n [4] and is denoted PG(2,n). Algorithms for constructing PG(2,n) for any prime degree n are known.

Example 1. Consider the cyclic projective plane $PG(2,2) = (\mathbf{X},\mathbf{B})$ [4], where

$$\mathbf{X} = \{0, 1, 2, 3, 4, 5, 6\},$$
$$\mathbf{B} = \{\{1, 0, 5\}, \{2, 1, 6\}, \{3, 2, 0\}, \{4, 3, 1\}, \{5, 4, 2\}, \{6, 5, 3\}, \{0, 6, 4\}\}.$$

It can be interpreted as a chordal computing network [7], assuming that the elements of the set **X** are nodes, and the elements in the blocks are ordered in such a way $[x_1,$

x_2, x_3] that define a pair of connections: x_1, x_2 are connected as neighboring nodes of a ring subnet, and x_2 and x_3 are connected by a chord. This simple rule defines the routing algorithm at each node. In another network interpretation, blocks are mapped to nodes, and two blocks having a common element are connected by a communication line. By the property of the projective plane, any two blocks have a common element, and in this interpretation, the projective plane of order n defines a fully connected network with the structure of the complete graph K_n. The network interpretation of this block design is not the only one. In the key interpretation [8], blocks correspond to nodes, there are as many of them as there are elements, and the elements of the blocks form a key ring space i.e. a set of $n + 1$ keys loaded into the key memory. By the property of the projective plane, any two nodes, in this case, have a common key, and the key-sharing graph is complete.

BIBD (n^2,n,1) is called an affine plane of order n, it is a residual BIBD obtained from PG(2,n) by removing any block and removing elements of this block from the remaining blocks [4]. Unlike the projective plane, not every pair of blocks has a common element, and a network in which nodes correspond to blocks and connections correspond to common elements of two blocks is not fully connected, but the length of the shortest path between two nodes does not exceed two.

2.2 Modified Combinatorial Block Designs

From each projective plane PG(2,n) it is possible to obtain $n^2 + n + 1$ different affine planes. Each block is present in n affine planes. The union of all affine planes corresponding to a given PG(2,n) forms a set of blocks of a modified block design [9].

Example 2. There are 7 affine planes. Their combination includes 21 blocks of the modified residual block design MRPG(2,2) corresponding to PG(2,2) on the same set **X.**

$$\mathbf{B} = \{\{2,6\}, \{3,2\}, \{4,3\}, \{4,2\}, \{6,3\}, \{6,4\}, \{0,5\}, \{3,0\}, \{5,4\}, \{5,3\}, \{0,4\}, \{1,5\}, \{1,6\}, \{4,1\}, \{6,5\}, \{2,0\}, \{5,2\}, \{0,6\}, \{1,0\}, \{3,1\}, \{2,1\}\}.$$

The number of MRPG blocks(2,n) is determined by the formula [9]

$$|\mathbf{B}| = \left(n^2 + n + 1\right)(n + 1)$$

(instead of ($n^2 + n + 1$) PG(2,n) blocks). This is the effect of scaling the network based on the affine plane. In a scaled network, nodes correspond to blocks, and connections correspond to common elements of blocks. The blocks either do not intersect or have from one to n–1 common elements. If all such connections are involved, then the length of the shortest path between the nodes will be no more than two. In this case, the network is formed by combining ($n^2 + n + 1$) fully connected subnets with the structure of the complete graph K_{n^2+n}. In the key interpretation, blocks correspond to nodes, and key rings correspond to block elements. Blocks do not either have common keys or have up to n–1 common keys [9]. The length of the shortest path in the key-sharing graph does not exceed two. The amount of key memory is k units.

Next, consider the transversal combinatorial block design. The linear transversal combinatorial block design TD(*k*,*n*) [3, 4] is constructed from elements of *k* pairwise disjoint sets w_i containing *n* elements in each, and consists of n^2 sets $\mathbf{Y}_j$, $j=\left(\overline{1, n^2}\right)$ containing *k* elements in each. In this case, each set $\mathbf{Y}_j$ has exactly one element from each set $\mathbf{W}_i$, $i = 0,\ldots,k-1$, and if $j \neq s$, then $\mathbf{Y}_j$ and $\mathbf{Y}_s$ have at most one element in common. Thus, the combinatorial block design TD(*k*,*n*) is defined on the set **X** of power *kn*. If *n* is a prime or a prime power, then TD(*k*,*n*) exists.

Then in the algebraic interpretation, the set **X** is the set of pairs (y,x), $x \in \mathbf{X}_1$, where $\mathbf{X}_1$ is the set of *k* elements of the field F_n, $y \in F_n$; blocks of linear transversal combinatorial block design TD(*k*,*n*) are defined by pairs of elements $(a,b) \in \mathrm{GF}(n^2)$ [3]:

$$\mathbf{B}_{a,b} = \{(ax + b, x) : x \in \mathbf{X}_1\}.$$

To each block from TD(*k*,*n*), we associate the set of non-empty subsets of lower cardinality obtained from other blocks by extracting this block. Combining the resulting sets, we obtain the residual transversal combinatorial block design RTD(*k*,*n*), corresponding to TD(*k*,*n*) [10].

Example 3. Consider a transversal combinatorial block design TD(3,3) = (**X**,**B**), where.

$$\begin{aligned}\mathbf{X} &= \{00, 01, 02, 10, 11, 12, 20, 21, 22\},\\ \mathbf{B} &= \{\{00, 10, 20\}, \{01, 11, 22\}, \{02, 12, 21\}, \{00, 11, 21\},\\ &\{01, 12, 20\}, \{02, 10, 22\}, \{00, 12, 22\}, \{01, 10, 21\}, \{02, 11, 20\}\}.\end{aligned}$$

The set of blocks of the corresponding residual transversal combinatorial block design RTD (3, 3) (only blocks of the same rank are used) is as follows

$$\begin{aligned}\mathbf{B} = \{&\{11, 21\}, \{01, 12\}, \{02, 22\}, \{12, 22\}, \{01, 21\}, \{02, 11\}, \{00, 21\},\\ &\{12, 20\}, \{02, 10\}, \{00, 12\}, \{10, 21\}, \{02, 20\}, \{00, 11\}, \{01, 20\},\\ &\{10, 22\}, \{00, 22\}, \{01, 10\}, \{11, 20\}, \{10, 20\}, \{01, 22\}, \{02, 12\},\\ &\{00, 10\}, \{11, 22\}, \{02, 21\}, \{00, 20\}, \{01, 11\}, \{12, 21\}\}.\end{aligned}$$

The number of blocks of the RTD(*k*,*n*) residual transversal combinatorial block design defined in this way corresponding to TD(*k*,*n*) is equal to n^3 instead of n^2 blocks of the original TD(*k*,*n*). The blocks either have no common elements or have up to $k - 2$ common elements.

The network interpretation consists in assigning blocks to nodes and connecting nodes whose blocks have a common element with communication lines. The length of the shortest path between nodes does not exceed two.

In the key interpretation, blocks also correspond to nodes, the key rings of blocks are determined by sets of their elements. The amount of key memory corresponds to $k - 1$ key units. The length of the shortest path in the key-sharing graph does not exceed two.

Note that the use of the projective plane for the preliminary distribution of keys is limited by the fact that the number of elements in the block is greater than the square root of the number of nodes. This is the reason for the preference for using transversal combinatorial block designs. To increase the resistance to key compromise in [11, 12], the network is divided into cells in which different key pools are used. At the same time, a reduction in the number of keys in the nodes is achieved. The projective plane is used to distribute keys in cells, and transversal combinatorial block designs are used to communicate between cells. In the key systems proposed below, the amount of key memory does not depend on the parameters of combinatorial designs.

3 Results

3.1 Combined Combinatorial Block Designs

To the system of sets (**X**,**B**), there is corresponding the combined system of sets (**X**,**B**,**D**,**M**), in which there are two more sets: the set **D** of dual blocks and the main set **M**. The set **D** is in one-to-one correspondence with the set **X**. Dual blocks are sets of blocks containing an element corresponding to a given dual block. The set **M** is the set of all different pairs of the form (block, block element) [13].

Example 4. The combined projective plane PG(2,2) = (**X**,**B**,**D**,**M**) corresponding to PG(2,2) is defined by the sets.

$$\mathbf{X} = \{0, 1, 2, 3, 4, 5, 6\},$$
$$\mathbf{B} = \{\{0, 1, 5\}, \{1, 2, 6\}, \{2, 3, 0\}, \{3, 4, 1\}, \{4, 5, 2\}, \{5, 6, 3\}, \{6, 0, 4\}\},$$
$$\mathbf{D} = \{\{\{0, 1, 5\}, \{2, 3, 0\}, \{6, 0, 4\}\}, \{\{0, 1, 5\}, \{1, 2, 6\}, \{3, 4, 1\}\},$$
$$\{\{1, 2, 6\}, \{2, 3, 0\}, \{4, 5, 2\}\}, \{\{2, 3, 0\}, \{3, 4, 1\}, \{5, 6, 3\}\},$$
$$\{\{3, 4, 1\}, \{4, 5, 2\}, \{6, 0, 4\}\}, \{\{0, 1, 5\}, \{4, 5, 2\}, \{5, 6, 3\}\},$$
$$\{\{1, 2, 6\}, \{5, 6, 3\}, \{6, 0, 4\}\}\}.$$
$$\mathbf{M} = \{(\{0, 1, 5\}, 0), (\{0, 1, 5\}, 1), (\{0, 1, 5\}, 5), (\{1, 2, 6\}, 1),$$
$$(\{1, 2, 6\}, 2), (\{1, 2, 6\}, 6), (\{2, 3, 0\}, 2),$$
$$(\{2, 3, 0\}, 3), (\{2, 3, 0\}, 0), (\{3, 4, 1\}, 3), (\{3, 4, 1\}, 4),$$
$$(\{3, 4, 1\}, 1), (\{4, 5, 2\}, 4), (\{4, 5, 2\}, 5),$$
$$(\{4, 5, 2\}, 2), (\{5, 6, 3\}, 5), (\{5, 6, 3\}, 6), (\{5, 6, 3\}, 3),$$
$$(\{6, 0, 4\}, 6), (\{6, 0, 4\}, 0), (\{6, 0, 4\}, 4)\}.$$

The set **D** of dual blocks of the combined modified residual block design CMRPG(2,2) corresponding to the modified residual block design MRPG(2,2) from Example 2 has the form.

$$\mathbf{D} = \{\{0, 1\}, \{0, 2\}, \{0, 3\}, \{0, 4\}, \{0, 5\}, \{0, 6\}\}, \{\{0, 1\}, \{1, 2\}, \{1, 3\}, \{1, 4\}, \{1, 5\}, \{1, 6\}\},$$
$$\{\{0, 2\}, \{1, 2\}, \{2, 3\}, \{2, 4\}, \{2, 5\}, \{2, 6\}\}, \{\{0, 3\}, \{1, 3\}, \{2, 3\}, \{3, 4\}, \{3, 5\}, \{3, 6\}\},$$
$$\{\{0, 4\}, \{1, 4\}, \{2, 4\}, \{3, 4\}, \{4, 5\}, \{4, 6\}\}, \{\{0, 5\}, \{1, 5\}, \{2, 5\}, \{3, 5\}, \{4, 5\}, \{5, 6\}\},$$
$$\{\{0, 6\}, \{1, 6\}, \{2, 6\}, \{3, 6\}, \{4, 6\}, \{5, 6\}\}.$$

The main set of this combined modified residual block design CMRPG(2,2) corresponding to MRPG(2,2) is as follows:

$\mathbf{M} = \{(\{0,1\},0), (\{0,1\},1), (\{0,2\},0), (\{0,2\},2), (\{0,3\},0), (\{0,3\},3), (\{0,4\},0), (\{0,4\},4), (\{0,5\},0), (\{0,5\},5), (\{0,6\},0), (\{0,6\},6), (\{1,2\},1), (\{1,2\},2), (\{1,3\},1), (\{1,3\},3), (\{1,4\},1), (\{1,4\},4), (\{1,5\},1), (\{1,5\},5), (\{1,6\},1), (\{1,6\},6), (\{2,3\},2), (\{2,3\},3), (\{2,4\},2), (\{2,4\},4), (\{2,5\},2), (\{2,5\},5), (\{2,6\},2), (\{2,6\},6), (\{3,4\},3), (\{3,4\},4), (\{3,5\},3), (\{3,5\},5), (\{3,6\},3), (\{3,6\},6), (\{4,5\},4), (\{4,5\},5), (\{4,6\},4), (\{4,6\},6), (\{5,6\},5), (\{5,6\},6)\}$.

In relation to the main set of the combined modified residual block design CMRPG(2,*n*), the formula is valid

$$|\mathbf{M}| = \left(n^2 + n + 1\right)\left(n^2 + n\right).$$

The power of the main set of the combined modified transversal combinatorial block design CRTD(*k*,*n*) corresponding to the transversal combinatorial block design TD(2,*k*,*n*) is determined by the formula.

$$|\mathbf{M}| = kn^3.$$

Note that the algorithms for constructing blocks and dual blocks of a series of combinatorial block designs by their numbers are presented in [13, 14].

3.2 Scaling Based on Combined Combinatorial Block Designs

In Sect. 2, a well-known scaling method based on the use of residual combinatorial block designs is applied. We show that a significant increase in scaling is possible on the basis of combined combinatorial block designs. We will build a network on a set of nodes corresponding to the elements of the main set of the combined combinatorial block design. Then the number of nodes will be k times more than the number of blocks in the original block design, where k is the number of block elements. Let's build a network structure as a composition of fully or partially connected networks of two types. Subnets of the first type, let's call them internal networks, are built on a set of nodes represented by pairs with the same first elements, and subnets of the second type (external) are built on sets of nodes with the same second elements. These internal networks correspond to the blocks of the original block design, and the external subnets correspond to its dual blocks.

Example 5. A network can be constructed on the main set (1) of the combined cyclic projective plane CPG(2,2) (see Fig. 1) with 7 internal subnets with the K_3 structure (numbers are in italics) (see Table. 1) and 7 external subnets with the same structure (in Fig. 1, their nodes are connected by a dotted line).

Table 1. Subnets of the combined cyclic projective plane CPG(2,2).

Internal subnets	External subnets
{({0,1,5},0),({0,1,5},1),({0,1,5},5)}	{({0,1,5},0),({0,1,5},1),({0,1,5},5)}
{({1,2,6},1),({1,2,6},2),({1,2,6},6)}	{({1,2,6},1),({1,2,6},2),({1,2,6},6)}
{({2,3,0},2),({2,3,0},3),({2,3,0},0)}	{({2,3,0},2),({2,3,0},3),({2,3,0},0)}
{({3,4,1},3),({3,4,1},4),({3,4,1},1)}	{({3,4,1},3),({3,4,1},4),({3,4,1},1)}
{({4,5,2},4),({4,5,2},5),({4,5,2},2)}	{({4,5,2},4),({4,5,2},5),({4,5,2},2)}
{({5,6,3},5),({5,6,3},6),({5,6,3},3)}	{({5,6,3},5),({5,6,3},6),({5,6,3},3)}
{({6,0,4},6),({6,0,4},0),({6,0,4},4)}	{({6,0,4},6),({6,0,4},0),({6,0,4},4)}

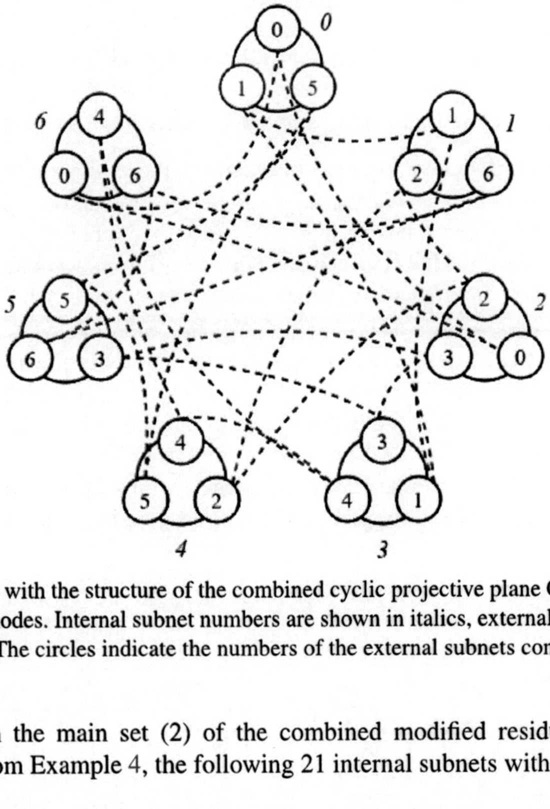

Fig. 1. A network with the structure of the combined cyclic projective plane CPG(2,2) (example 4). It contains 21 nodes. Internal subnet numbers are shown in italics, external subnets are shown with dotted lines. The circles indicate the numbers of the external subnets containing this node

Example 6. On the main set (2) of the combined modified residual block design CMRPG(2,2) from Example 4, the following 21 internal subnets with the K_2 structure.

$$\{(\{0,1\},0),(\{0,1\},1)\},\{(\{0,2\},0),(\{0,2\},2)\},\{(\{0,3\},0),(\{0,3\},3)\},$$
$$\{(\{0,4\},0),(\{0,4\},4)\},\{(\{0,5\},0),(\{0,5\},5)\},\{(\{0,6\},0),(\{0,6\},6)\},$$
$$\{(\{1,2\},1),(\{1,2\},2)\},\{(\{1,3\},1),(\{1,3\},3)\},\{(\{1,4\},1),(\{1,4\},4)\},$$
$$\{(\{1,5\},1),(\{1,5\},5)\},\{(\{1,6\},1),(\{1,6\},6)\},\{(\{2,3\},2),(\{2,3\},3)\},$$
$$\{(\{2,4\},2),(\{2,4\},4)\},\{(\{2,5\},2),(\{2,5\},5)\},\{(\{2,6\},2),(\{2,6\},6)\},$$
$$\{(\{3,4\},3),(\{3,4\},4)\},\{(\{3,5\},3),(\{3,5\},5)\},\{(\{3,6\},3),(\{3,6\},6)\},$$
$$\{(\{4,5\},4),(\{4,5\},5)\},\{(\{4,6\},4),(\{4,6\},6)\},\{(\{5,6\},5),(\{5,6\},6)\}$$

and following 7 external subnets of the structure K_6

$$\{(\{0,1\},0),(\{0,2\},0),(\{0,3\},0),(\{0,4\},0),(\{0,5\},0),(\{0,6\},0)\},$$
$$\{(\{0,1\},1),(\{1,2\},1),(\{1,3\},1),(\{1,4\},1),(\{1,5\},1),(\{1,6\},1)\},$$
$$\{(\{0,2\},2),(\{1,2\},2),(\{2,3\},2),(\{2,4\},2),(\{2,5\},2),(\{2,6\},2)\},$$
$$\{(\{0,3\},3),(\{1,3\},3),(\{2,3\},3),(\{3,4\},3),(\{3,5\},3),(\{3,6\},3)\},$$
$$\{(\{0,4\},4),(\{1,4\},4),(\{2,4\},4),(\{3,4\},4),(\{4,5\},4),(\{4,6\},4)\},$$
$$\{(\{0,5\},5),(\{1,5\},5),(\{2,5\},5),(\{3,5\},5),(\{4,5\},5),(\{5,6\},5)\},$$
$$\{(\{0,6\},6),(\{1,6\},6),(\{2,6\},6),(\{3,6\},6),(\{4,6\},6),(\{4,6\},6)\}.$$

The network structures described in this section have important properties:

1. sets of nodes of internal subnets form a partition of the set of all nodes into subsets of the same power;
2. similarly, the sets of nodes of external subnets also form a partition of the set of all nodes into subsets of the same power;
3. each network element is an element of some internal and some external subnets.

From the point of view of keys pre-distribution, these properties allow different key pools and different combinatorial models to distribute keys across subnets. For example, smaller subnets may use projective planes, while larger subnets use transversal combinatorial block designs (as in [11, 12]). Each node uses two sets of keys to communicate in the corresponding internal and external subnets. This reduces the amount of memory needed to store keys, as well as protection against compromising keys. At this stage of scaling, nodes use key pools of the same power and key pools are protected from mutual compromise.

Tables 2 and 3 show the number of nodes in the networks under consideration and their individual subnets. The scaling level of the CMRPG(2,*n*) network is the same as the scaling level of the residual RPG(2,*n*) subnet, which differs from the MRPG(2,*n*) network by an *n*-fold repetition of blocks, but blocks are not duplicated in it.

Table 2. Network PG(2,*n*), CPG(2,*n*), MRPG(2,*n*) and CMRPG(2,*n*) scales and their subnets sizes

	PG(2,*n*)	CPG(2,*n*)	MRPG(2,*n*)	CMRPG(2,*n*)
Net	n^2+n+1	$(n^2+n+1)\times(n+1)$	$(n^2+n+1)\times(n+1)$	$(n^2+n+1)\times(n^2+n)$
Internal subnet		$n+1$		n
External subnet		$n+1$		(n^2+n)

3.3 New Schemes for Pre-distribution of Keys

Let's consider the features of key distribution in networks with a combined combinatorial block design structure. For certainty, we will further keep in mind the combined

Table 3. Network D(2,k,n), CTD(k,n), RTD(k,n), and CRTD(k,n) scales and their subnets sizes

	TD(2,k,n)	CTD(k,n)	RTD(k,n)	CRTD(k,n)
Net	n^2	kn^2	n^3	kn^3
Internal subnet	–	k	–	$k-1$
External subnet	–	n	–	$(k-1)\,n$

projective plane CPG(2,n). Let's take two key pools $\mathbf{K}_{in}$ and $\mathbf{K}_{out}$ of power b and v and use them to assign keys. There are $v = n^2 + n + 1$ internal subnets of the K_n structure and the same number of external subnets of the same structure. We will distribute the same key k_{in},i $\in K_{in}$ to the elements (i,x) of the internal subnet $N_{in},_i$. To the elements (y,j) of the external subnet $N_{out,j}$ we will distribute the same key $k_{out},_j \in \mathbf{K}_{out}$. Each node (i,j) is an element of the external subnet $N_{out,j} = (y,j)$ and at the same time an element of the internal subnet $N_{in,i} = (i,x)$. It contains a key pair $(k_{in,i}, k_{out,j})$ and is the vertex of the internal subnet key-sharing graph and the vertex of the external subnet key-sharing graph. The subgraphs $G_{in,j}$ and $G_{out,i}$ of the key-sharing graph G corresponding to the internal and external subnets are thus complete K_n graphs. Therefore, there is one step between any two vertices (i_1,j_1) and (i_2,j_2) (if the vertices are in the same subnet, i.e. $i_1 = i_2$ or $j_1 = j_2$), a two-step chain (if $i_1 \in (y,j_2)$, $i_2 \in (y,j_1)$, $j_1 \in (i_2,x)$ or $j_2 \in (i_1,x)$) or a three-step chain (in other cases).

Example 7. Some chains in the key-sharing graph of network CPG(2,2) (Fig. 1):

$$(2,3)\overset{k_{in,2}}{\rightarrow}(2,0),$$

$$(2,2)\overset{k_{in,2}}{\rightarrow}(2,3)\overset{k_{out,3}}{\rightarrow}(5,3),$$

$$(2,2)\overset{k_{in,2}}{\rightarrow}(2,0)\overset{k_{out,0}}{\rightarrow}(6,0)\overset{k_{in,6}}{\rightarrow}(6,4).$$

The keys (one each for the internal subnet and the external subnet) in nodes $u_{(i,i)}$ are stored in the node's key memory and occupy two units of key information. Any node (j,i) belongs to the internal subnet (x,i) and external subnet (j,y).

The compromise of a network node is understood as the capture of both of its keys simultaneously. In this case, the direct links of the nodes of the corresponding internal and external subnets are compromised.

Example 8. Applied to the combined projective plane CPG (2,3) from Example 5, the compromise of the node (1,0) entails the compromise of the internal subnet (the nodes (0,0),(1,0),(5,0)) and the external subnets ((1,0),(1,1),(1,3)). Applied to the combined projective plane CPG(2,9), the network will have $v = 91 \times 10 = 910$ nodes. Compromise of one node causes compromise of direct links of the internal subnet of 10 nodes and compromise of direct links of the external subnet of 10 nodes and external subnets (one node is shared).

Keys $k_{in,i}$ and $k_{out,j}$ of the compromised node can not be replaced using the keys available in the network, and the keys of other nodes of these subnets can be restored using the uncompromised node of this subnet: if keys ($k_{in,i}$, $k_{out,j}$) are compromised by the capture of node (i,j), then the compromised key $k_{in,i}$ of another node (i,j') of the same internal subnet can in principle be replaced using key $k_{out,j'}$ of this node. Similarly compromised key $k_{out,j}$ of another node (i',j) of the same external subnet can in principle be replaced using the key $k_{in,i'}$ of this node. This kind of replacement of keys is also possible at the end of the period of use of the keys. Then all the internal subnet keys are replaced using the external subnet keys they contain, or all the external subnet keys are replaced using the internal subnet keys they contain, at different points in the usage period. These replacements may be provided for in the key management policy.

At the same time, the dual principle of constructing a key distribution scheme allows the adversary to purposefully choose nodes that have not yet been compromised.

When a node is selectively captured, at each stage, the attacker knows a set of K compromised keys and a set of T key numbers that have not yet been compromised [11, 12]. To compromise some key whose number belongs to T, the attacker must capture the node containing this key. Thus, he can capture any suitable node from the dual block, corresponding to the number of this key, as the nodes are compromised, the adversary expands the known subsets $\mathbf{K'}_{in}$ and $\mathbf{K'}_{out}$ of key pools $\mathbf{K}_{in}$ and $\mathbf{K}_{out}$. They correspond to the sets **I** and **J** first i and second j identifier elements (i,j) of compromised nodes. When a node (i',j'), $i' \in \mathbf{I}$, $j' \notin \mathbf{J}$, is compromised, the set $\mathbf{K'}_{in}$ (and respectively **I**) does not change, $\mathbf{K'}_{out}$ is completed with the key $k_{out,j'}$. If $i' \notin \mathbf{I}$, $j' \in \mathbf{J}$, the set $\mathbf{K'}_{out}$ (and respectively **J**) does not change, $\mathbf{K'}_{in}$ is completed with the key $k_{in,i'}$. If $i' \notin \mathbf{I}$, $j' \notin \mathbf{J}$, sets $\mathbf{K'}_{out}$ (and respectively **J**) and $\mathbf{K'}_{in}$ (and respectively **I**) are completed with compromised keys $k_{out,j'}$ and $k_{in,i'}$.

On the other hand, for a legal participant between two uncompromised nodes (i,j) and (i,j') of a compromised internal subnet $N_{in,i}$, secure commutation is possible through intermediary nodes $(i_1,j),(i_1,j_1),(i_2,j_1),(i_2,j')$ using uncompromised keys $k_{out,j}, k_{in,i_1}, k_{out,j_1}, k_{in,i_2}, k_{out,j}$. Similarly, between two uncompromised nodes (i,j) and (i',j') of the compromised external subnet $N_{out,j}$ secure commutation is possible through four intermediary nodes using five uncompromised keys.

Example 9. When capturing a node (0, 0) of the network in Fig. 1. The key $k_{in,0}$ is compromised and, consequently, all connections of the internal subnet (nodes (0,0),(0,1),(0,5)), but the possibility of commutation remains, for example, between its nodes (0,1) and (0,5):

$$(0,1)\overset{k_{out,1}}{\to}(1,1)\overset{k_{in,1}}{\to}(1,2)\overset{k_{out,2}}{\to}(4,2)\overset{k_{in,4}}{\to}(4,5)\overset{k_{out,5}}{\to}(0,5).$$

It is also clear that when node (i,j) is compromised, there is no commutation between nodes (i,j_1) and (i_1,j).

Direct connections between nodes of the internal or external subnet, if some of their nodes are compromised, can be preserved using the well-known technique of converting the key distribution scheme to the key space distribution scheme (multiple space pre-distribution schemes) [15, 16]. To do this, on the sets of vertices of the key-sharing graph, to which the same keys are assigned, a key space is created according to the (2,w)-Blom scheme [17]. In this case, each node receives a pre-shared key in the form of a set of w + 1 elements of the field on which this scheme is defined. In this case of key distribution schemes based on the combined block design, key spaces according to the Blom scheme

are created on the vertex sets of each internal and each external subnet. For each subnet, symmetric polynomials $f_{in,i}(x,y), f_{out,j}(x,y)$ of degree w are chosen in each variable over the large field. The i and j coordinates of the node (i,j) receive a second application as Blom identifiers of nodes (i,j) of the external subnet $N_{out,j}$ and the internal subnet $N_{in,i}$. Nodes (i,j) are assigned pre-shared keys
$k_{in(i,j)} = f_{in,i}(j,y)$ and $k_{out(i,j)} = f_{out,j}(i,y)$. In two nodes (i,j_1) and (i,j_2) of the internal subnet $N_{in,i}$ can be calculated the common key $k_{in,i(j_1,j_2)} = f_{in,i}(j_1,j_2)$ and in two nodes (i_1,j) and (i_2,j) of the external subnet $N_{out,j}$ can be calculated the common key $k_{out,j(i_1,i_2)} = f_{out,j}(i_1,i_2)$. In the scheme of distribution of key spaces constructed in this way, if no more than w *of* any nodes in each internal and each external subnet are compromised, the possibility of direct connections between any two other nodes of these subnets remains possible. Such schemes are consistently called w-secure: capture of any w nodes does not violate the security of communication between any remaining nodes [16]. In such a scheme, each node must store $2(w + 1)$ units of key information.

Example 10. When using the (2, 1)-Blom scheme in the scheme in Fig. 1. When a node (0, 0) is compromised, commutation is possible

$$(0, 1)\overset{k_{in,0}}{\rightarrow}(0, 5)$$

on the key $k_{in,0} = k_{in,0(1,5)} = k_{in,0(5,1)}$ of the Blom scheme, which is calculated at the nodes (0,1) and (0,5) using pre-shared keys $k_{in,0,(1,5)} = f_{in,0}(1,y)$ $k_{in,0,(5,1)} = f_{in,0}(5,y)$.

The method described above for changing compromised keys is also applicable for changing compromised pre-shared keys of key spaces according to the Blom scheme.

4 Discussions

4.1 WSN of "smart home" Based on the Combined Cyclic Projective Plane CPG(2,n)

Based on the established practice and technology for building a WSN, in particular for a "smart" home, based on the IEEE 802.15.4 standard [18], the routing protocol in ZigBee [19 – 21], we will assume that a "smart" home is an administrator-controlled system of instruments or devices. Smart home devices are data sources or data receivers and performers. The source can also be the receiver. There is a source-destination mapping table T that is created and modified by the administrator. The functions of the latter also include the division of many devices into zones of radio accessibility of routers, as well as key administration.

There is a coordinator program that creates, based on the table T, a combinatorial WSN model based on the combined cyclic projective plane of the – symmetric $(n^2 + n + 1, n + 1, 1)$-BIBD with blocks of $k = n + 1$ elements. The main set M of the combined projective plane CPG(2,n) has $(n^2 + n + 1)(n + 1)$ elements of the form (block, block element). Blocks and dual blocks are calculated by their numbers on the basis of the Singer difference set – a set of k numbers, which makes it possible to represent any other positive number not exceeding v-1 as the difference of a pair of numbers from this set. It allows to calculate any block or dual block of the combined cyclic projective plane by its number. In this case, the identifiers of all elements are calculated. Elements that make up block $\mathbf{B}_i$ of CPG(2,n) are assigned to devices of the i-th accessibility zone as identifiers (i,j), where i is the block number, j is the block element. Table T is encoded by replacing devices names with identifiers. Let it be stored in the coordinator, and routers have access to it, i.e. by the identifier of the source node (i,j) can determine $(s,t) = T(i,j)$ – the identifier of the destination node.

The construction of the WSN is completed at the direction of the administrator by placing devices with identifiers (i,j) as destination nodes, as well as routers R_i in the corresponding "visibility" zone and loading the lists of child node identifiers into the routers by the coordinator. In this case, the router can "serve" devices corresponding to several blocks of the combinatorial block design, but below, for simplicity, it is assumed that devices corresponding to one such block are in the same "visibility" zone of the router.

Sleeping nodes are located in the "visibility" zone of the parent router and exchange messages with it at low emitter power. The routers are in the "visibility" zone of the coordinator and mutually exchange messages. There are two types of router broadcast: "broadcast" and "broadcast for router". Accordingly, the router accepts messages broadcasted by its child nodes and any other router. These two types of broadcasts differ in frequency ranges and emitter power. It is assumed that the routers have sufficiently powerful power supplies and function constantly. Nodes "wake up" upon receiving a signal in the form of their identifier (i,j) from the parent router, which periodically broadcasts the identifiers of its child nodes. Each node (i,j) in the zone of router R_i stores two pre-shared keys $(k_{in,i(j,y)}, k_{out,j(i,y)})$ in key memory.

The router R_i for each child node (i,j) receives from the coordinator a pair ((i,j),T(i,j)). There are operations $C_B(i)$ of calculating the block $\mathbf{B}_i$ and $C_D(j)$ of calculating the dual block $\mathbf{D}_j$ by their numbers i and j, as well as operations of calculating the intersection of two blocks or two dual blocks. Their composition makes it possible to obtain a common element $b_{i_1,i_2} = C_B(i_1) \cap C_B(i_2) = \mathbf{B}_{i_1} \cap \mathbf{B}_{i_2}$ of two blocks $\mathbf{B}_{i_1}$ and $\mathbf{B}_{i_2}$ or common element $d_{j_1,j_2} = C_D(j_1) \cap C_D(j_2) = \mathbf{D}_{j_1} \cap \mathbf{D}_{j_2}$ of two dual blocks $\mathbf{D}_{j1}$ and $\mathbf{D}_{j2}$. Using these operations and the mentioned pairs ((i,j),T(i,j)) the router R_i calculates for each source node (i,j) chains of two, three, or four nodes starting with this source node and ending with the destination node (s,t) according to the following rule:

If $i = s$ or $j = t$, then this is a chain

a) $(i,j) \rightarrow (s,t)$;

Otherwise, two variants of chains of four nodes are possible:

b) $(i,j) \rightarrow (k,l) \rightarrow (m,n) \rightarrow (s,t)$, where

$l = n = b_{i,s}$ and $k = d_{j,l}$ and $m = d_{n,t}$;

c) $(i,j) \rightarrow (k,l) \rightarrow (m,n) \rightarrow (s,t)$

$k = m = d_{j,t}$ and $l = b_{i,k}$ and $n = b_{m,s}$;

In cases where $(k,l) = (m,n)$ they are converted into chains of three nodes:

d) $(i,j) \rightarrow (k,l) \rightarrow (s,t)$, where

$l = b_{i,s}$ and $k = d_{j,t}$; $j = l$ and $k = s$ or $i = k$ and $l = t$.

It is assumed that routers keep track of the execution of each chain, creating it when the protocol is initialized, transforming it as iterations of the protocol are executed, and broadcasting it in full for another router, and also, possibly, partially for a child node; the identifier of this child node is the first in the chain, and the index of this another router is equal to the first index of this identifier. Upon receiving a message from the router, nodes perform actions that depend on this message, which complete the execution of the protocol along this chain or generate a node response that is broadcast to the parent router for the next iteration.

Example 11. Protocol for a chain like $(i,j) \rightarrow (k,l) \rightarrow (s,t)$; $i = k$ and $l = t$.

$$\begin{gathered} \substack{i=k \\ l=t} : R_{i}{}^{((i,j)(k,l)(s,t))}_{((i,j)(k,l)(s,t))} \rightarrow (i,j)m \div (i,j)^{k_{in,i(j,l)}}_{(i,j)e(m)} \\ \rightarrow R_{i}{}^{((k,l)(s,t)(i,j)e(m))}_{((k,l)(s,t)(i,j)e(m))} \rightarrow (k,l)^{k_{in,k(l,j)}}_{m=d(e)} \div (i,j)^{k_{out,l(k,s)}}_{(k,l)e(m)} \\ \rightarrow R_i((s,t)(k,l)e(m)) \Rightarrow R_{s}{}^{((s,t)(k,l)e(m))}_{((s,t)(i,j)e(m))} \\ \rightarrow (s,t)^{k_{out,k(s,k)}}_{m=d(e)} \div (s,t). \end{gathered}$$

Router R_i by record $T(i,j) = [(i,j),(s,t)]$ in the table T creates a chain and translates its fragment for the child node (i,j). In node (i,j) a message is created, it is encrypted with the key $k_{in,i(j,l)}$ and the cryptogram $e(m)$ is broadcast. The router R_i(in this case i $= k$) changes the chain upon receiving the cryptogram $e(m)$ and broadcasts its fragment and the cryptogram for the child node (k,l). The node (k,l) decrypts the cryptogram e with the key $k_{in,i(l,j)}$, encrypting the message with the key $k_{out,l(k,s)}$, and broadcasting the cryptogram $e(m)$. The router R_k transforms the chain upon receiving the cryptogram $e(m)$ from the child node to broadcast to the router R_s. The router R_s changes the chain upon receiving the cryptogram $e(m)$ and broadcasts its fragment and the cryptogram for the child node (s,t). The node (s,t) decrypts the cryptogram e with the key $k_{out,l(s,k)}$. At this stage, the length of the chain fragment is 2, the destination node has received the message.

We concretize the execution of this protocol in relation to the WSN in Fig. 1. Let the message generated in the node be (0, 0) transmitted to the node (1, 1).

$$\begin{gathered} R_{o}{}^{((0,1)(0,1)(1,1))}_{((0,1)(0,1)(1,1))} \rightarrow (0,0)m \div (0,0)^{k_{in,0(0,1)}}_{(0,0)e(m)} \rightarrow R_{0}{}^{((0,1)(1,1)(0,0)e(m))}_{((0,1)(0,1)(0,0)e(m))} \\ \rightarrow (0,1)^{k_{in,0(1,0)}}_{m=d(e)} \div {}^{k_{out,1(0,1)}}_{(0,1)e(m)} \rightarrow R_0((1,1)(0,1)e(m)) \\ \Rightarrow R_{1}{}^{((1,1)(0,1)e(m))}_{((1,1)(0,1)e(m))} \rightarrow (1,1)^{k_{out,1(1,0)}}_{m=d(e)} \div (1,1). \end{gathered}$$

A compromise of any of the two intermediate nodes in a chain of length 3 (the vast majority of such chains) is compensated by a chain of the same length with two other intermediate nodes (see option c) above), but a compromise of a single intermediate node of a chain of length 2 is compensated by a chain of length 4 with three intermediate nodes. In this paper, the algorithms for their construction are not considered, but the authors know and implement them.

Example 12. We use the cyclic projective plane PG(2,9) – (91,10,1)-BIBD based on the Singer difference set [0,1,11,15,31,36,43,65,83,89]. This is a numeric set containing numbers whose differences represent any other number not exceeding 90. It allows routers to calculate blocks (« +» « −» are operations in the ring Z_{91}).

$$B_i = [0+i, 1+i, 11+i, 15+i, 31+i, 36+i, 43+i, 65+i, 83+i, 89+i], i = 0, \ldots, 90$$

and dual blocks

$$D_j = [0 - j, 1 - j, 11 - j, 15 - j, 31 - j, 36 - j, 43 - j, 65 - j, 83 - j, 89 - j], j = 0, \ldots, 90$$

of the combined projective plane CPG(2,9) as described above.

The order $n = 9$ of the combinatorial design is chosen so that the number of WSN nodes is equal to $(n^2 + n + 1)(n + 1) = 910$, so that the administrator can group them according to radio accessibility zones into clusters from 10 to 30 devices served from 31 to 91 routers. Table T source-destination correspondence must contain 910 records and be stored in the coordinator, while routers must have access to it, store and periodically broadcast up to 30 such records. Routers and coordinator must be in a single radio accessibility zone. The key memory of the nodes must store two keys each. If the key distribution scheme is used, then these are two units of key information, occupying 16, 32 or 64 bytes when using the Magda or Grasshopper encoders [22]. When implementing a key space distribution scheme based on the (2,w)-Blom scheme in a network, it is necessary to store $2 \times (w + 1)$ units of key information in the key memory, occupying (when using the same encoder) $16 \times (w + 1)$, $32 \times (w + 1)$ or $64 \times (w + 1)$ bytes. With $w = 3$, this is 64, 128, or 256 bytes, which will ensure that if three nodes are compromised, direct communications between any uncompromised nodes will be secure. When using the Blom scheme to calculate the keys of the specified encoders, the implementation of the modular arithmetic of the large prime field. Nodes should be able to compare the node identifier and the destination node identifier, represented as 8-bit numbers.

The formalized and numerical data given in paragraphs 4 and 5 were obtained using a modified author's electronic educational resource [23].

4.2 Using Duality Blocks for Capturing a Node with Uncompromised Key

Consider the key pre-distribution scheme possessing structure of projective plane PP(2,n). Let the set **B** of blocks corresponds to the set of nodes of the network and defines a set of key identifiers assigned to nodes. Duality blocks correspond to key identifiers and define the sets of blocks to which the key corresponding to this duality block is allocated. An element (x,**D**(x)), $x \in$ **B** of the main set defines the duality block **D**(x) to which block x belongs.

During selective node capture the attacker compromises those nodes whose keys have not already been compromised. Let's consider two ways to calculate the key for communication between two WSN nodes [11, 12].

Way 1. In key pre-distribution scheme [11] the nodes possess identifiers $(a,b,c) \in$ GF(p^3), and the keys possess identifiers (x,y,z). The sets of keys assigned to given node (the set of $B = (a,b,c)$ elements identifiers (x, y, z)) are computed by algorithm in Table 4 that followed from Algorithm 1. Key Pre-distribution Using PG(2,p)[11]. Any two nodes broadcast their node identifiers during the shared key discovery phase [11]. Authors note that "the key identifiers are not broadcasted. Thus at no stage can the attacker know what key identifier is present in which node. Thus there is no way of knowing which nodes are left to be compromised. Thus, unless the attacker compromises the node, she cannot choose a node for compromise to maximize the number of keys compromised." As a result authors concluded, that their "scheme is secure against selective node capture."

Table 4. Algorithm for block of GF(2,*p*) computing.

```
Input: block identifier (a,b,c)
Output: the set of block B=(a,b,c) elements identifiers (x, y, z)
------------------------------------------------------------------
B=[]
if (a,b,c)=(0,0,1):
  for x in GF(p):
    B.append (x,1,0)
  B.append (1,0,0)
if (a,b,c)=(0,1,c):
    for x in GF(p):
      B.append (x,–c,1)
    B.append (1,0,0)
if (a,b,c)=(1,b,c):
    for y in GF(p):
      x=–(c+by)
      B.append (x,y,1)
    B.append (–b,1,0)
```

However, it should be noted that using broadcasted identifiers of two nodes it is possible to compute by Algorithm 2. Shared Key Discovery Using Symmetric Design [11] common key identifier and then computer duality block, i.e. the set of all blocks containing that key. Corresponding algorithm is given in Table 5. Existence of this algorithm contradicts security scheme against selective node capture.

Way 2. In key pre-distribution scheme [12] the composition of blocks (i.e. sets of keys distributed to blocks) corresponds to the cyclic projective plane and is displayed by the f-strings of the Blom filter [24]. The only information needed for shared key discovery is broadcasted Bloom-filter f-bit string by all the nodes [12]. The authors concluded that "even if adversary is able to obtain all the f-bit strings broadcasted in the cell, he cannot obtain the shared key within any two sensor nodes". Using a Bloom.

Table 5. Algorithm for duality block of GF(2,p) computing.

Input: element identifier (x,y,z)

Output: the set of blocks containing keys with identifier (x,y,z)

```
D=[]
if (x,y,z)=(1,0,0):
  for c in GF(p):
    D.append (0,1,c)
  D.append (0,0,1)
if (x,y,z)=(x,1,0):
  for c in GF(p):
    D.append (1, –x,c)
  D.append (0,0,1)
if (x,y,z)=(x,y,1):
  for b in GF(p):
    c=–x–by
    D.append (1,b,c)
  D.append (0,1,–y)
```

filter hides the algebraic dependence of key identifiers, and calculating from the common bits of the two f-bit strings of a double block without knowledge of hash functions is almost impossible. However, it should be noted that the construction of an f -bit string establishes a correspondence between the set of key sets in the nodes and the set of f -bit string. If the hash functions used are well known, it is possible to represent block compositions in f -bit string notation. The intersection of two f-bit strings is a common key f -bit string. A simple brute force search can determine the set of blocks containing a given string, i.e., the corresponding dual block. Consequently, assuming that the information about the distribution of keys on the network nodes as well as the hash functions are public knowledge, in a selective node capture at each stage the attacker knows the set **K** of compromised keys and the set **T** numbers of keys that have not yet been compromised, to compromise some key whose number belongs to **T**, the attacker must capture the node containing that key. Thus, he can capture any suitable node from the duality block corresponding to that key number, i.e., its f-bit string. In particular, to compromise the key of the current commutation, it will calculate the duality block as a set of blocks containing a common t-bit string.

To complete arithmetic of combined PG(2,*p*) Algorithm for computing unique node containing two given keys, i.e. common element of two duality blocks is represented in Table 6, and Algorithm for computing common key of two blocks (nodes) is represented in Table 7. The algorithms in Tables 6 and 7 as well as the algorithms in Tables 4 and 5 are mutually dual.

The functionality of the algorithm in Table 7 differs from the functionality of the Algorithm 2. Shared Key Discovery Using Symmetric Design [11] with the exception of division by 0.

Blocks and elements (and hence dual blocks) can be numbered with non-negative integers:

$$n(0,0,1) = 0, n(0,1,c) = 1 + c, n(1,b,c) = 1 + p + pb + c,$$
$$m(1,0,0) = 0, m(x,1,0) = 1 + x, m(x,y,1) = 1 + p + py + x.$$

Then these algorithms can be applied in the protocols of Subsect. 4.1, instead of algorithms based on a cyclic projective plane with a decrease in computational complexity.

5 Conclusion

A method for scaling a network using fully connected subnets corresponding to all blocks of the original block design and fully connected subnets corresponding to its duality blocks is proposed. It is shown that the elements (block, duality block corresponding to this block element) of the main set can be interpreted as network nodes of increased scalability. This allowed, by scaling the sets of elements of two well-known combinatorial designs, to build novel larger-scale network structures that have properties characteristic of the practice of pre-distribution of keys. We also obtained the number of nodes in such network structures and in their subnets corresponding to blocks and dual blocks.

When scaling the network, the key-sharing graph is also scaled, thus generating a regular key-sharing graph, whose vertex degrees are defined as the sum of the order and rank of the original combinatorial block design. As for the structure of the combined projective plane, the length of the shortest route between pairs of vertices in the key-sharing graph does not exceed three, and such routes with two intermediaries are typical for the combined projective plane, while in the original graph pairs of vertices are connected directly.

We propose to use mutually dual node identifiers both for routing and for selecting the indexes of encryption and decryption keys in nodes. Each node belongs to the internal subnetwork $N_{in,i}$ and external subnetwork $N_{out,j}$ defined by pair of identifiers (*i,j*) and supplied with a pair of keys ($k_{in,i}$, $k_{out,j}$), i.e. the amount of key memory is limited to two key units and does not depend on the block design parameters, while the use of combinatorial block design instead of the combined block design is associated with the

Table 6. Algorithm for intersection of duality blocks of PG(2,p) computing.

Input: (x_i, y_i, z_i) and (x_j, y_j, z_j), the identifiers of duality blocks i and j respectively.

Output: identifier of common block of input duality blocks

---

if $z_i = z_j = 0$ *and* $y_i = 1$ *or* $y_j = 1$:

  return (0, 0, 1)

if $y_i = z_i = 0$ *and* $z_j = 1$:

  return $(0, 1, -y_j)$

if $y_j = z_j = 0$ *and* $z_i = 1$:

  return $(0, 1, -y_i)$

if $y_i = y_j = 1$ *and* $z_i = z_j = 0$:

  return (0, 0, 1)

if $z_i = y_j = 1$ and $z_j = 0$:

  return $(1, -x_j, -x_i+x_jy_i)$

if $z_j = y_i = 1$ and $z_i = 0$:

  return $(1, -x_i, -x_j+x_iy_j)$

if $z_i = z_j = 1$ *and* $y_i = y_j = y$:

  return $(0, 1, -y)$

$b = \frac{x_j - x_i}{y_i - y_j}$

return $(1, b, -x_i - by_i)$

---

use of memory size that increases with increasing order of block design. To ensure the w-secure property, it has been proposed to create key spaces on the sets of nodes of single-key subnets according to (2,w)-Blom scheme, which would require $(w + 1)$-fold increase in key memory size. This provides a counteraction to the capture of keys in the conditions of publication of their identifiers. The latter, as it was shown, allows you to define a set of nodes containing a key with this identifier.

The proposed protocols for secure communication between end-nodes using routers and algorithms for routing and defining indexes of keys in use, meet the IEEE 802.15.4 standard and ZigBee technology.

Table 7. Algorithm for intersection of blocks of PG(2,*p*) computing.

Input: (a_i, b_i, c_i) and (a_j, b_j, c_j), the identifiers of blocks (nodes) i and j respectively.

Output: identifier of common element of input blocks

if $a_i = a_j = 0$ *and* $b_i = 1$ *or* $b_j = 1$:

  return (1, 0, 0)

if $b_i = a_i = 0$ *and* $a_j = 1$:

  return $(-b_j, 1, 0)$

if $b_j = a_j = 0$ *and* $a_i = 1$:

  return $(-b_i, 1, 0)$

if $b_i = b_j = 1$ *and* $a_i = a_j = 0$:

  return (1, 0, 0)

if $a_i = b_j = 1$ and $a_j = 0$:

  return $(-c_i+c_jb_i, -c_j, 1)$

if $a_j = b_i = 1$ and $a_i = 0$:

  return $(-c_j+c_ib_j, -c_i, 1)$

if $a_i = a_j = 1$ *and* $b_i = b_j = b$:

  return $(-b, 1, 0)$

$$y = \frac{c_j - c_i}{b_i - b_j}$$

return $(-c_i - yb_i, y, 1)$

Acknowledgement. The presented paper is an extended version of [25].

References

1. Solari Esfehani, N., Haj Seyyed Javadi, H.: A survey of key pre-distribution schemes based on combinatorial designs for resource-constrained devices in the IoT network. Wireless Netw. **27**(4), 3025–3052 (2021). https://doi.org/10.1007/s11276-021-02629-8
2. Bahrami, P N., Javadi, H.H.S., Dargahi, T., Dehghantanha, A., Choo K.K.R.: A hierarchical key pre-distribution scheme for fog networks. Concurr. Comput.: Pract. Exper. **31**(22) (2019). https://doi.org/10.1002/cpe.4776
3. Lee, J., Stinson, D.R.: On the construction of practical key predistribution schemes for distributed sensor networks using combinatorial designs. ACM Trans. Inf. Syst. Secur. **11**(2), 1–35 (2008). https://doi.org/10.1145/1330332.1330333
4. Stinson, D.R.: Combinatorial Designs: Constructions and Analysis. Springer, Berlin (2007)
5. Tarannikov, Y.V.: Combinatorial properties of discrete structures and applications to cryptology. MCCME, Moscow (2014). ISBN 978–5–4439–2045–0. (In Russian)
6. Parkhomenko, P.: Algorithmizing design of a class of combinatorial block diagrams. Autom. Remote. Control. **77**(7), 1216–1224 (2016). https://doi.org/10.1134/S0005117916070080
7. Mozharov, G.P.: Fault-tolerant computer networks constructed on the basis of combinatorial block designs, Vestn. Mosk. Gos. Tekh. Univ. im. N.E. Baumana, Priborostr [Herald of the Bauman Moscow State Tech. Univ., Instrum. Eng.] **6**, 41–53 (2016). https://doi.org/10.18698/0236-3933-2016-6-41-53. (In Russian)
8. Çamtepe, S.A., Yener, B.: Combinatorial design of key distribution mechanisms for wireless sensor networks. In: Samarati, P., Ryan, P., Gollmann, D., Molva, R. (eds.) ESORICS 2004. LNCS, vol. 3193, pp. 293–308. Springer, Heidelberg (2004). https://doi.org/10.1007/978-3-540-30108-0_18
9. Modiri, V., Javadi, H.H.S., Anzani, M.: A novel scalable key pre-distribution scheme for wireless sensor networks based on residual design. Wireless Pers. Commun. **96**(2), 2821–2841 (2017). https://doi.org/10.1007/s11277-017-4326-9
10. Masaeli, N., Javadi, H.H., Erfani, H.: Key pre-distribution scheme based on transversal design in large mobile fog networks with multi-clouds. J. Inf. Secur. Appl. 54 (2020). https://doi.org/10.1016/j.jisa.2020.102519
11. Ruj, S., Roy, B.: Key predistribution using combinatorial designs for grid-group deployment scheme in wireless sensor networks, ACM Trans. Sens. Netw. **6**(1) (2009). https://doi.org/10.1145/1653760.1653764
12. Kumar, A., Pais, A.R.: A new combinatorial design based key pre-distribution scheme for wireless sensor networks. J. Ambient. Intell. Humaniz. Comput. **10**(6), 2401–2416 (2018). https://doi.org/10.1007/s12652-018-0902-4
13. Frolov, A., Kochetova, N.: Networks with the structure of a combined combinatorial block design. In: International Conference on Electrical, Computer, Communications and Mechatronics Engineering (ICECCME), pp. 1–6 (2021). https://doi.org/10.1109/ICECCME52200.2021.9591136
14. Frolov, A., Klyagin, A., Kochetova, N., Temnikov, D.: On the synthesis of systems possessing the structure of some combinatorial designs. J. Comput. Syst. Sci. Int. **60**(4), 585–594 (2021). https://doi.org/10.1134/S1064230721040067
15. Du, W., Deng, J., Han, Y., Varshney, P., Katz, J., Khalili, A.: A pairwise key pre distribution scheme for wireless sensor networks. ACM Trans. Inform. Syst. Secur. **8**, 228–258 (2005). https://doi.org/10.1145/948109.948118
16. Urivskiy, A.: Key Predistribution scheme using affine planes and Blom's scheme. In: Proceedings International M*athematical Conference* "50 Years OF IITP" (2011). ISBN 978–5–901158–15–9

17. Blom, R.: An optimal class of symmetric key generation systems. In: Beth, T., Cot, N., Ingemarsson, I. (eds.) EUROCRYPT 1984. LNCS, vol. 209, pp. 335–338. Springer, Heidelberg (1985). https://doi.org/10.1007/3-540-39757-4_22
18. Gholami, K.El., Yassine, M.Y., Fatani, I.: The IEEE 802.15.4 standard in industrial applications: a survey. J. Theor. Appl. Inf. Technol. **99**(15), 1–17 (2021)
19. Kim, S.H., Chong, P.K., Kim, T.: Performance study of routing protocols in ZigBee wireless mesh networks. Wireless Pers. Commun. **95**(2), 1829–1853 (2017). https://doi.org/10.1007/s11277-017-3996-7
20. Kim, T., Kim, S.H., Yang, J., Se, Y., Kim, D.: Neighbor table-based shortcut tree routing in ZigBee wireless networks. IEEE Trans. Parallel Distrib. Syst. **25**(3), 706–716 (2014)
21. Gomez, C., Paradells, J.: Wireless home automation networks: a survey of architectures and technologies. IEEE Commun. Mag. **48**(6), 92–101 (2010)
22. GOST R (Russian: ГОСТР) 34.12–2015. Information technology. Cryptographic data security. Block ciphers. Moscow, Standartinform (2015)
23. Frolov, A., Klyagin, A, Kochetova, N., Vinnikov, A.: Educational desktop for remote modeling of wireless computer networks subject to constraints. In: 2022 VI International Conference on Information Technologies in Engineering Education (Inforino), pp. 1–5. Russian Federation, Moscow (2022). https://doi.org/10.1109/Inforino53888.2022.9782913
24. Bloom, B.H.: Space/time trade-offs in hash coding with allowable errors. Commun ACM **13**(7), 422–426 (1970)
25. Frolov, A., Kochetova, N.: Scaling networks and capturing keys using combined systems of sets. In: Kravets, A., Bolshakov, A., Shcherbakov, M. (eds.) Society 5.0: Cyber-Solutions for Human-centric technologies. Studies in Systems, Decision and Control. Springer, Cham. (Accepted)

Measurement of Transmission Characteristics of LiFiMAX

David Hecl, Martin Koppl, Viktor Szitkey, Andrej Grolmus, Matus Hozlar, Rastislav Roka, Ivan Baronak, Stefan Pocarovsky, Milos Orgon, and Petr Blazek(✉)

FEI STU Bratislava, Ilkovicova 2961, 84104 Karlova Ves, Slovakia
{xhecl,xszitkey,rastislav.roka}@stuba.sk,
{xkoppl,xhozlar,xpocarovsky}@is.stuba.sk, ivan.baronak@ut.fei.stuba.sk,
orgon@ktl.elf.stuba.sk, blazekpetr@vut.cz

Abstract. Optical Wireless Communication (OWC) is one of the possible technologies that will come into wider use in the coming years. Its more significant influence is described by the authors in connection with the sixth-generation (6G) networks. Today, we can come across various Light Fidelity (LiFi) or Visible Light Communication (VLC) applications. The subject of this article is primarily the LiFi technology. The article investigates the transmission characteristics of OLEDCOMM's LiFiMAX devices, which are publicly available and aim to replace WiFi radio communication. The article evaluates the achieved results and outlines the further possible direction of studies of these devices.

Keywords: LiFi · OWC · IR · wireless technology

1 Introduction

In recent years, there has been a significant increase in devices connected to the Internet using various forms of wireless technology for different types of applications such as smart grid, smart city, smart factory, smart health and many others [1–3]. This increase is primarily related to people's need to be online all the time and to the arrival of new IoT (Internet of Things) devices [4]. This particular need put very high expectations on reliability, performance and security [5–7]. Moreover, demand for wireless connection growth is closely related to the availability of increasingly higher frequency bands reserved for operators' needs. The radio signal band below 10 GHz is becoming increasingly crowded and often needs to reach sufficient transmission parameters to receive or send comprehensive documents or multimedia content quickly [8].

The arrival of the 5th generation of mobile networks made the above mentioned frequency band available to ordinary users. Millimeter waves, i.e. high-band frequencies of the 5G spectrum above 24 GHz [9] offer sufficient transmission speeds and bandwidth, meeting people's needs and society's ever-increasing demands. According to [10], a frequency spectrum is highly affected by losses

R. Silhavy and P. Silhavy (Eds.): CSOC 2023, LNNS 723, pp. 290–299, 2023.
https://doi.org/10.1007/978-3-031-35317-8_27

during signal propagation, has low to none obstacle penetration, and requires a Line-f-Sight (LoS) communication. However, it is not just the radio spectrum of the electromagnetic radiation. It is also possible to use Optical Wireless Communication (OWC), including Light-Fidelity (LiFi) technology. Reconfigurable Intelligent Surfaces (RIS) is a promising trend in the future [11] and one of the applications is RIS-aided communication in VLC [12].

LiFi is a direct response to the current trend of using high frequencies for sending and receiving data. As stated in [10], LiFi is a type of communication that uses the so-called nano waves, as the frequency spectrum is shifted to the order of hundreds of THz. LiFi is a communication technology that uses the light spectrum of electromagnetic radiation for high-speed data transmission. The LiFi communication system consists of sending and receiving parts. The sending part is mainly made up of LEDs or Laser Diodes (LD), enabling the high-speed transmission of information itself. Due to suitable modulation, the LED works on the principle of transmitting light pulses, which are invisible to the human eye due to their speed. Thus, possible unwanted health effects that could have an impact on users of LiFi networks are prevented. On the other hand, the receiving part is built on a suitably chosen photodetector, most often in the form of a photodiode, which allows for capturing light pulses from the transmitter and interpreting them into the necessary data.

The optical spectrum of electromagnetic radiation represents excellent potential for the future in eliminating the need to use the radio spectrum. Precisely, the optical spectrum consists of three parts - Infrared Radiation (IR, 0.3–394.7 THz), Visible Light (VL, 394.7–833.3 THz), and Ultraviolet Radiation (UV, 750 THz – 30 PHz) [13]. In many studies, LiFi uses visible light in the downlink (DL) direction and IR, VL, or UV radiation in the uplink (UL) direction to transmit information, depending on the specific application of the technology. However, it is essential to add that UV radiation, also due to its possible negative impact on human health, is only used in marginal situations [14]. A schematic sketch of a LiFi network with connected devices is outlined in Fig. 1.

Compared to existing technologies based on the radio spectrum of electromagnetic radiation, LiFi enables more efficient use of energy resources, higher transmission speeds, higher security and lower costs for creating new infrastructures. When using energy-efficient LEDs, LiFi can reduce the total energy consumption to cover a specific area compared to radio communication [15]. Due to the possibility of using high frequencies, it can achieve data transmission many times faster than radio technologies. According to studies [14,16], the theoretical maximum transmission rate is 100 Gbit/s or 224 Gbit/s using appropriate technology at the transmitter and receiver sides. Networks built on LiFi are safe due to LoS communication's necessity. Prevention against interference by potential attackers is also ensured, as the light used to transmit information with LiFi does not have sufficient penetration to pass through obstacles [17]. In addition, the costs of creating a complex infrastructure are minimal since the room lighting can be used as a LiFi signal transmitter with a suitable control unit.

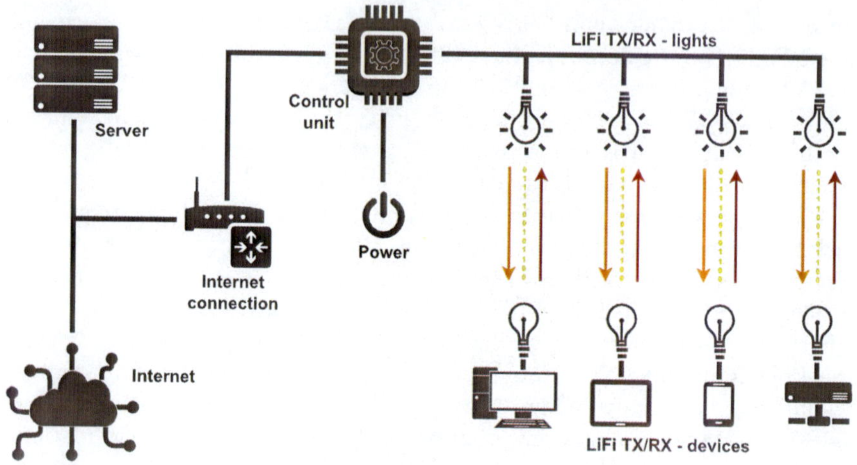

Fig. 1. Illustration of a LiFi network [14]

2 LiFiMAX AP and Dongle Overview

Measurements were conducted to determine the transmission characteristics of OLEDCOMM's LiFiMAX devices using a pair of network LiFi components. Specifically, it was the LiFiMAX access point on the server side and the LiFiMAX dongle on the client side. Figure 2 shows a more accurate sketch of the network and individual devices.

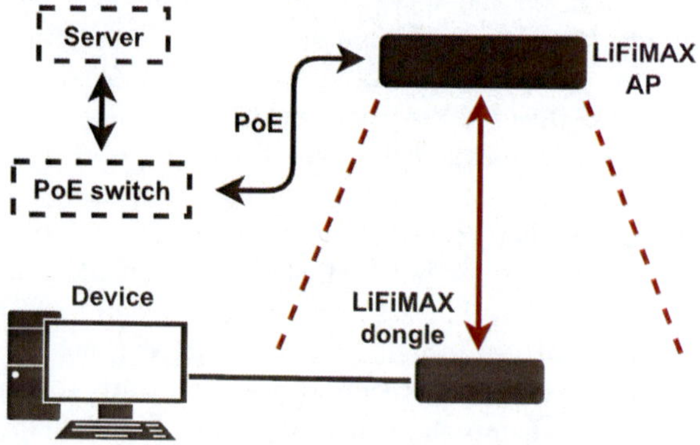

Fig. 2. Schematic diagram of the used LiFiMAX network

The LifiMAX devices used in the test network enable communication in the downlink direction with a speed of 100 Mbit/s. In the uplink direction, information exchange also occurs with a transmission speed of 100 Mbit/s. For data transmission, LiFiMAX devices use a light wavelength of 940 nm. This is infrared radiation, which does not fundamentally affect human health and is invisible to the human eye. In comparison, security cameras often use infrared radiation with a wavelength of 850 nm, which can be perceived by the human eye directly on illuminated LEDs under certain circumstances. According to [18], infrared light with a wavelength of 940 nm has up to 50% less range compared to the same radiation with a wavelength of 850 nm. The LiFiMAX technology can therefore work at frequencies that are not dangerous for the user's health.

2.1 LiFiMAX Access Point

It is a device that allows a user's dongle to access the relevant network. LiFiMAX Access Point (AP) covers an area of 10 m^2 with IR radiation at a mounting height of 2 m, according to [19]. The radiation angle of the access point is 90° [20]. In addition, the device allows the simultaneous connection of up to 16 separate users' LiFiMAX dongles. Therefore, the device meets the IEC/EN 62471 standard dealing with the photobiological safety of light sources and light source systems. According to the assessment of the relevant standard, the LiFiMAX AP has no photobiological risk and is additionally equipped with an automatic shutdown in case of an interruption of communication with the LiFiMAX dongle. Also, this device meets the requirements of IEC/EN 61000 standardization for electromagnetic compatibility (EMC). More detailed characteristics are described in Table 1.

Table 1. LiFiMAX AP specifications [19]

Characteristics	Values
Resistance class	IP30
Power	PoE (IEEE 802.3af) - plug-and-play
Connector	RJ45
Optical technology	IR (940 nm) - LED
Standardization	ITU-T G.vlc
Security	Advanced Encryption Standard 128 (AES-128)
Maximal transmission speeds	100 Mbit/s uplink, 100 Mbit/s downlink
Power consumption	<7 W
Maximal number of users	16
Certifications	CE marking, IEC/EN 61000, IEC/EN 62471

2.2 LiFiMAX Dongle

The LiFiMAX dongle connects to the user's end device with a plug-and-play system. It is supported by Linux, Windows, macOS and Android operating systems, allowing the device to power itself. However, when using iOS and the

Lightning connector, it is necessary to use an external power source. Like the LiFiMAX AP, the LiFiMAX dongle complies with the requirements of the CE mark, IEC/EN 62471, and IEC/EN 61000 specifications. More detailed characteristics are described in Table 2.

Table 2. LiFiMAX dongle specifications [19]

Characteristics	Values
Resistance class	IP30
Connector	RJ45
Interface	USB 2.0
Supported OS	Linux, Windows (from 7), macOS, Unix, Android
Optical technology	IR (940 nm) - LED
Standardization	ITU-T G.vlc
Security	Advanced Encryption Standard 128 (AES-128)
Maximal transmission speeds	100 Mbit/s uplink, 100 Mbit/s downlink
Power consumption	<2 W
Installation	plug-and-play
Certifications	CE marking, IEC/EN 61000, IEC/EN 62471

3 Measurement of Transmission Properties

Measurements of the LiFiMAX transmission characteristics were performed at three different transmitter placement heights and several different positions between the LiFiMAX access point and the LiFiMAX dongle. The experiment consisted of setting the access point successively at three heights, namely 100, 150, and then 200 cm above the LiFiMAX dongle. During testing, the LiFiMAX dongle gradually moved relative to the access point. Figure 3 shows the individual measurements of the LiFiMAX transmission properties. The receiver gradually moved along pre-defined paths corresponding to 30° segments of a circle. At a mounting height of the access point of 100 cm, there was a gradual shift in individual steps with a length of 10 cm until the connection was lost. At the height of 150 cm, the shift between individual measurements was 15 cm, and at the height of 200 cm, the individual measurement points were 20 cm apart. The measurements continued until the LiFiMAX dongle was disconnected from the LiFiMAX AP. The LiFiMAX dongle was connected to the laptop during testing, used to power the device and measure the transmission speeds. The iperf 3.0.7 software was used to measure transmission speeds, which measured the maximum transmission speed between two clients connected via the LiFi environment. There was no other source of infrared radiation in the room during the testing of the transmission characteristics, which could negatively affect the transmission characteristics.

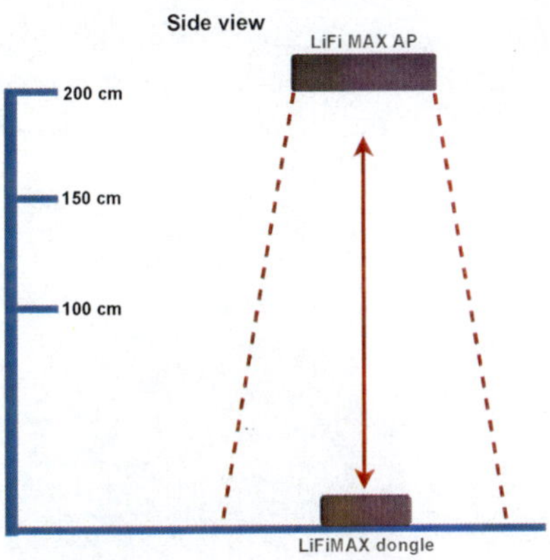

Fig. 3. Illustration of individual measurements of LiFiMAX transmission properties

4 Results

To evaluate the measured data, we used MATLAB and its extension [21] to create area plots from the data. According to the LiFiMAX AP specifications, the infrared radiation angle of the access point is 90°. It follows that the assumed irradiation area should have a circular shape with a radius of 100 cm at a mounting height of the access point of 100 cm, 150 cm at the height of 150 cm, and finally, 200 cm at a mounting height of the access point of 200 cm. However, the measured data show that the shape of the irradiated surface does not correspond to a circle of the corresponding radius. At the same time, it is clear from the results that the limit values of individual measurements at specific heights at some points exceed the maximum assumed radiation surface. In some points, the distance from the center of the radiation surface to its edge is smaller.

4.1 Height 100 cm

A summary of 107 measurements was taken when the AP was mounted 100 cm above the ground. The average transfer speed was 80.84 Mbit/s. At an angle of 120° at an initial distance of 20 cm from the centre of the irradiated surface, the transmission speed reached the highest value (96.54 Mbit/s). On the contrary, we recorded the lowest transmission speed at an angle of 210° at a distance of 100 cm. Its value was 4.12 Mbit/s. Figure 4 shows the measured LiFiMAX transmission speed data for access point mounting height of 100 cm.

Based on the obtained values of transmission speeds, we can conclude that the radiating surface has an irregular shape with the maximum edge coverage at a distance of 110 cm from the center, which was recorded on the tracks lying at 300° and 330°. Conversely, the shortest range of coverage was on tracks lying at 120°, 150°, and 180°, when the edge of coverage was 90 cm. In our evaluations, the quadrant number 2 appeared to have the smallest covered area, even from other measured data.

If the expected edge of the surface coverage is located 100 cm from the center, we can conclude from the results that most of the tracks at the respective angles reached at least the expected distance. Therefore, the average transmission speed at the assumed edge of the irradiated surface is 33.06 Mbit/s.

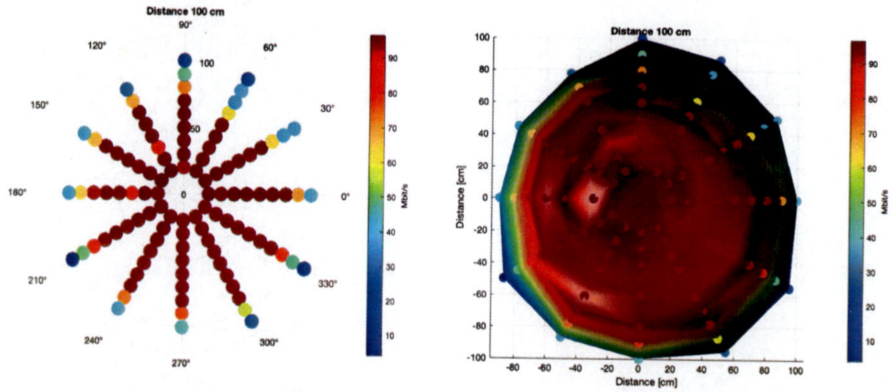

Fig. 4. Display of measured LiFiMAX transmission speed data for access point mounting height of 100 cm

4.2 Height 150 cm

For the second installation of the LiFiMAX AP at 150 cm height, 87 measurements were recorded. Their average transmission speed was 80.30 Mbit/s. The path of 210° at a distance of 65 cm corresponded to the maximum achieved. On the other hand, the minimum transmission speed falls on a path of 270° at an edge distance of 105 cm. Figure 5 shows the measured LiFiMAX transmission speed data for access point mounting height of 150 cm.

Similar to the previous measurements, we can conclude from the measured values that the radiating surface is not a regular circle. The most significant distance from the center when it was still possible to establish a connection between the LiFiMAX AP and the LiFiMAX dongle was 120 cm on the tracks 0°, 60°, 270°, 300° and 330°. On the other hand, on the 150° and 180° tracks, communication was already disabled at a distance of more than 90 cm. Therefore, it can be concluded that, as in previous measurements, the second quadrant was the weakest in terms of coverage area.

None of the measurements reached the expected edge of coverage, which should have been at a distance of 150 cm from the center. The average transmission speed at the edge distance of 120 cm was 35.59 Mbit/s, while the average transmission speed at all achieved edge points was 31.53 Mbit/s.

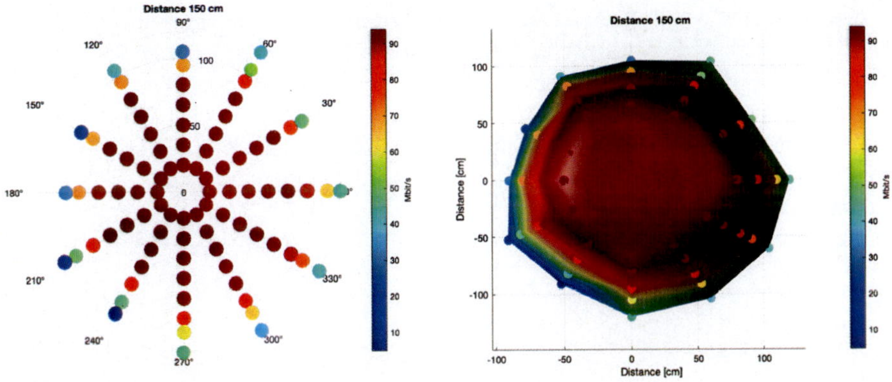

Fig. 5. Display of measured LiFiMAX transmission speed data for access point mounting height of 150 cm

4.3 Height 200 cm

The last series of measurements was at a mounting height of LiFiMAX AP above the ground surface of 200 cm. In total, 82 measurements were made in this series. The average transmission speed was 66.89 Mbit/s. The maximum value of the transmission rate was on the 240° path at a distance of 20 cm. On the same path at a distance of 120 cm from the center was the lowest value of the transmission speed (1.34 Mbit/s). Figure 6 shows the measured LiFiMAX transmission speed data for access point mounting height of 200 cm.

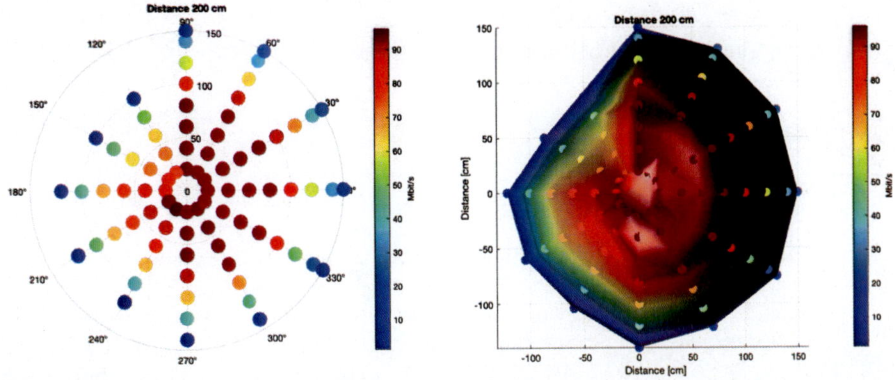

Fig. 6. Display of measured LiFiMAX transmission speed data for access point mounting height of 200 cm

The maximum distance at which it was possible to establish a connection between the LiFiMAX AP and the LiFiMAX dongle and subsequently perform measurements was 150 cm. It was possible to achieve it on tracks of 0°, 30°, 60°, 90° and 330°. The average transmission speed at these points was 16.95 Mbit/s. The overall average transmission speed at the edge of coverage was 10.97 Mbit/s at the height of 200 cm.

The last series of measurements also confirmed that the irradiated surface is not in the shape of a circle. Furthermore, it confirmed that the smallest transmitter range was in the second quadrant, where the LiFiMAX access point could communicate with the LiFiMAX dongle within 100 cm for 120° and 150° and within 120 cm for 180°.

5 Conclusion

This article aimed to determine the transmission properties of LiFiMAX devices from OLEDCOMM. We can conclude from the measurements that the surface was not irradiated by infrared radiation with a wavelength of 940 nm in a continuous circle. Three measurements revealed that the weakest coverage values are located in the second quadrant, specifically on the paths we set at 120°, 150°, and 180°. The shortest distances were covered on these tracks. From this, we can conclude that the radiation of the LiFiMAX AP device is the weakest in these directions. A more detailed analysis of the shape of the irradiated surface during data transmission through the mentioned devices will be the subject of further investigation.

Acknowledgments. This article is created with the support of the KEGA agency project - 034STU-4/2021 "Utilization of Web-based Training and Learning Systems at the Development of New Educational Programs in the Area of Optical Wireless Technologies" and COST Action CA19111 (European Network on Future Generation Optical Wireless Communication Technologies, NEWFOCUS), supported by COST (European Cooperation in Science and Technology).

References

1. Fujdiak, R., Masek, P., Mlynek, P., Misurec, J., Olshannikova, E.: Using genetic algorithm for advanced municipal waste collection in smart city. In: 10th International Symposium on Communication Systems. Networks and Digital Signal Processing (CSNDSP) 2016, pp. 1–6 (2016)
2. Mlynek, P., Fujdiak, R., Misurec, J., Sláčik, J.: Experimental measurements of noise influence on narrowband power line communication, vol. **10**, pp. 94–100 (2016)
3. Fujdiak, R., Misurec, J., Mlynek, P., Janer, L.: Cryptograph key distribution with elliptic curve Diffie-Hellman algorithm in low-power devices for power grids. Rev. Roum. Sci. Tech. - Serie Électrotechnique et Énergétique **61**, 84–88 (2017)
4. Masek, P., Fujdiak, R., Zeman, K., Hosek, J., Muthanna, A.: Remote networking technology for IoT: cloud-based access for alljoyn-enabled devices (2016)
5. Mlynek, P., et al.: Performance testing and methodology for evaluation of power line communication. Elektron. ir Elektrotechnika **24**(3), 88–95 (2018)

6. Krivtsova, I., et al.: Implementing a broadcast storm attack on a mission-critical wireless sensor network. In: Mamatas, L., Matta, I., Papadimitriou, P., Koucheryavy, Y. (eds.) WWIC 2016. LNCS, vol. 9674, pp. 297–308. Springer, Cham (2016). https://doi.org/10.1007/978-3-319-33936-8_23
7. Khan, S.Z., Jamal, S.S., Sajid, A., Rashid, M.: FPGA implementation of elliptic-curve Diffie Hellman protocol. Comput. Mater. Continua **73**(1), 1879–1894 (2022)
8. Stava, M.: A synthesis of reversible digital circuits to solve the boolean satisfiability. In: Nath, V., Mandal, J.K. (eds.) Proceedings of the Fourth International Conference on Microelectronics, Computing and Communication Systems. LNEE, vol. 673, pp. 233–246. Springer, Singapore (2021). https://doi.org/10.1007/978-981-15-5546-6_19
9. Nokia: 5G spectrum bands explained-low, mid and high band, Nokia. https://www.nokia.com/networks/insights/spectrum-bands-5g-world/. Accessed 28 Nov 2022
10. Haas, H., Yin, L., Wang, Y., Chen, C.: What is LiFi? J. Lightwave Technol. **34**(6), 1533–1544 (2016)
11. Le, C.B., Do, D.T., Li, X., Huang, Y.F., Chen, H.C., Voznak, M.: Enabling NOMA in backscatter reconfigurable intelligent surfaces-aided systems. IEEE Access **9**, 33782–33795 (2021)
12. Munochiveyi, M., Pogaku, A.C., Do, D.T., Le, A.T., Voznak, M., Nguyen, N.D.: Reconfigurable intelligent surface aided multi-user communications: state-of-the-art techniques and open issues. IEEE Access **9**, 118584–118605 (2021)
13. RF Wireless World: Difference between Infrared, Visible light and Ultraviolet. https://www.rfwireless-world.com/Terminology/Difference-between-Infrared-Visible-Light-and-Ultraviolet.html. Accessed 12 Dec 2022
14. Popadić, M., Kočan, E.: LiFi Networks: concept, standardization activities and perspectives. In: 2021 25th International Conference on Information Technology (IT), pp. 1–4 (2021)
15. Stava, M.: On precise fault localization and identification in NoC architectures. In: 2019 22nd Euromicro Conference on Digital System Design (DSD), pp. 478–484 (2019)
16. LiFi.co: What is LiFi? https://lifi.co/what-is-lifi/. Accessed 28 Nov 2022
17. Stava, M.: Efficient error recovery scheme in fault-tolerant NoC architectures. In: 2019 IEEE 22nd International Symposium on Design and Diagnostics of Electronic Circuits & Systems (DDECS), pp. 1–4 (2019)
18. AXTON: IR Wavelengths, infrared - 850 nm vs 94 nm. https://axtontech.com/infrared-850nm-vs-940nm-wavelength/. Accessed 9 Dec 2022
19. Oledcomm: LiFiMAX cybersecurity Solutions Technical DATASHEET - Ajour. Media. http://www.ajour.ca/LiFiMAXcybersecuritySolutionsTechnicalDATASHEET.pdf. Accessed 9 Dec 2022
20. LiFi.co: Brochure LiFiMax. https://lifi.co/wp-content/uploads/2021/08/Brochure-LiFiMax-pdf-HD-24062018.pdf. Accessed 12 Dec 2022
21. Mecklai, Z.: Making surface plots from scatter data. https://www.mathworks.com/matlabcentral/fileexchange/5105-making-surface-plots-from-scatter-data. Accessed 12 Dec 2022

Daily Routine Monitoring Using Deep Learning Models

Al Jizani Mohammed Kadhim Salman[1] and Humam K. Majeed Al-Chalabi[2](✉)

[1] University Politehnica of Bucharest, 060042 Bucharest, Romania
[2] University of Craiova, 200585 Craiova, Romania
hemoomajeed@gmail.com

Abstract. Today's healthcare systems are up against a huge challenge in terms of patient control. Especially since the COVID-19 epidemic when hospital beds have been confined to emergency cases. In many circumstances, however, hospitalization is essential when there is insufficient information about the patient. Frequently, physicians leave patients in hospitals for a few days to check their condition, which can be costly. Furthermore, acquiring reliable data from patients when symptoms are no longer present is difficult. This paper looks at how different sensors, such as the gyroscope and accelerometer present in smartwatches and smartphones, can be used to monitor patient activity. These sensors can measure properties like the angular velocity of human motions and use them to train algorithms that can recognize human movement. Furthermore, identifying a patient's diet and activities might help get reliable data on them throughout their time away from the hospitals, a process known as remote patient health monitoring. Using the WISDM dataset, which is freely available online for study, this work proposes a CNN model to identify and forecast patient diet and activity.

Keywords: Human Activity Recognition · Deep Learning · Sensor · Convolutional neural network · LSTM (ConvLSTM2D)

1 Introduction

Human Activity Recognition (HAR) is a branch of study concerned with the automatic recognition of people's everyday routine activities based on time series recordings made with sensors. The interconnected sensing technology such as sensors, IoT, the Internet of Things, cloud, and edge computing have all seen significant developments in the previous decade [1]. Sensors are low-cost devices that can be easily incorporated or embedded in both portable and non-portable devices, hence sensor technology has become the focus of most of the HAR research [2]. Wearable devices with sensors, which are a common IoT application, may be utilized to quickly capture different body movements for human activity detection. In recent years, the rapid proliferation of affordable smartphones and smartwatches embedded with wearable Inertial Measurement Units (IMU) sensors, comprising accelerometers and gyroscopes that can easily identify and locate human movements, has enabled the development of a plethora of applications

R. Silhavy and P. Silhavy (Eds.): CSOC 2023, LNNS 723, pp. 300–315, 2023.
https://doi.org/10.1007/978-3-031-35317-8_28

in branches such as healthcare, biometrics, personal fitness trackers, sports analytics, elderly care, security and surveillance applications [3]. The significance of HAR based on wearable sensors becomes evident from the fact that it can be used for classifying and documenting numerous daily activities such as eating, drinking, brushing teeth, detecting sleep anomalies, and is not confined to exercise-related activities. There are two types of actions in HAR: simple and complex. Complex human activities require executing a simple human activity followed by a specific transition action [4, 5]. The primary goal of this work is to develop CNN/LSTM deep learning models capable of reliably classifying 18 everyday routine human activities from the WISDM dataset into three groups: a) Ambulation Oriented Activities such as walking, jogging, and so on. b) Hand Oriented Activities (In General): writing, typing, etc. c) Hand Oriented Activities (Eating) such as eating chips, eating pasta, and so on. Derived from low-level time-series sensor data collected using the tri-axial accelerometer and gyroscope of smartwatches and smartphones, which can capture the acceleration or vibration and angular velocity of human body actions [6]. The following are the research questions developed in response to the study's goal: • Question No. 1: How do you handle the collected data and perform the pre-processing procedure so it can improve the model performance as better as possible? • Question No. 2: How the CNN (Conv2D) and LSTM (ConvLSTM2D) models train and get to learn from the data? and how the Evaluation of these parts perform? • Question No. 3: Despite having similar gyroscope and accelerometer sensors, which of the two wearable devices, a smartphone or a wristwatch, delivers more accuracy when performing HAR?

2 Dataset and Methods

2.1 Data Description

The activity "WISDM" Smartphone and Smartwatch and Biometric Data Set" represents the data that we obtained from 51 persons who completed 18 tasks for a total of 3 min [7, 8]. Each person wears a smartwatch on his or her dominant wrist and carries a smartphone in his or her pocket. An app that measures and runs on the smartphone and connected watch was used to control the data. The sensor data from the smartphone and smartwatch's accelerometer and gyroscopic sensors. With a 20 Hz frequency (50 MS). Table 1 shows the general characteristics of the data and data collection process [9].

At a rate of 20 Hz, the raw accelerometer and gyroscope sensor data are gathered from the wristwatch and smartphone. It is gathered from 51 test participants as they complete 18 tasks for three minutes each. Each device's (phone, watch) and sensor type's (accelerometer, gyroscope) sensor data is kept in a separate directory (so there are 4 data directories). There are 51 files in each directory, one for each of the 51 test subjects. Each entry follows the same format:. Along with the attribute information, these attributes' descriptions are supplied. We produce examples using a 10-s frame that describe the sensor data in addition to the raw time-series sensor data.

High-level features are created based on the 200 (10 s $\times$ 20 readings/s) readings contained in each segment of the raw time series data for each sensor (per person and per activity), which are separated into 10-s nonoverlapping segments. We assumed that a 10-s frame would provide for enough time to record numerous repetitions of actions involving repetitive movements while being brief enough to allow for fast responses (i.e., a prediction every 10 s).

Table 1. Dataset Information Summary

Information Type	Value
Number of subjects	51
Number of activities	18
Minutes collected per activity	3
Sensor polling rate	20 Hz
Smartphone used	Google Nexus 5/5x or Samsung Galaxy S5
Smartwatch used	LG G Watch
Number raw measurements	15, 630, 426

2.2 The Activities

There are 18 activities that have been completed. Actual data files use the table below code to specify activities (Table 2):

Table 2. Dataset Information Summary

Activity	Code
Walking	A
Jogging	B
Stairs	C
Sitting	D
Standing	E
Typing	F
Brushing Teeth	G
Eating Soup	H
Eating Chips	I
Eating Pasta	J
Drinking from Cup	K
Eating Sandwich	L
Kicking (Soccer Ball)	M
Playing Catch w/Tennis Ball	O
Dribblinlg (Basketball)	P
Writing	Q
Clapping	R

Human Activity Recognition systems and can be divided into two categories:

- vision-based.
- Sensor-based.

3 Data Pre-processing

3.1 Data Visualization

The gyroscope and the accelerometer sensor data will be merged in each of the data frames. Participants' data was first aggregated into 51 data frames based on the timestamp column for each of the sensors (gyroscope and accelerometer).

Table 3. Dataset Attributes

Attribute	Description
Subject-id	Value from 1600 to 1650 which represents the activity code of 51 participants. Activity The character between 'A' and 'S' (no 'N') identifies
Activity	The different activities code performed by each participant
Timestamp	Unix style timestamp
x	x dimension sensor reading (accelerometer or gyroscope
y	y dimension sensor reading (accelerometer or gyroscope)
z	dimension sensor reading (accelerometer or gyroscope)

And then merged individually for the smartwatch and the smartphone. A visualization in Table 3 shows the final shape of the panda's data frames after the merger for two 2 activities by the 22 sensors employed. The dataset contains noise that can be traced to human activity.

The table below illustrates the fluctuation of x, y, and z for a subject who was walking, this data has been collected by the phone (Fig. 1).

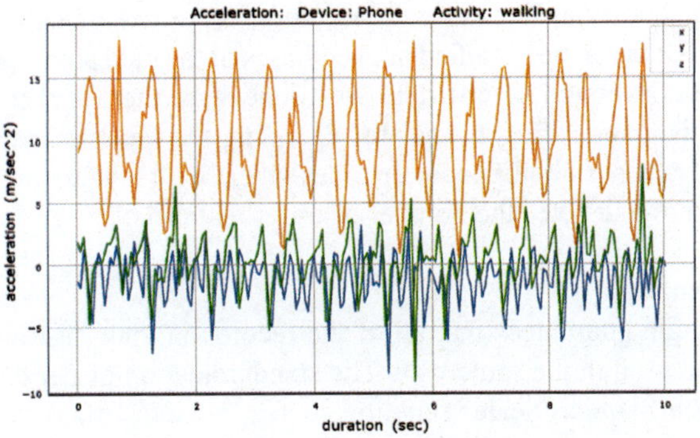

Fig. 1. The fluctuation of x, y and z for the Walking activity by Phone

The following table shows the variation of the variables x, y and z, for a subject who was brushing teeth, the dataset here is collected by the watch (Fig. 2).

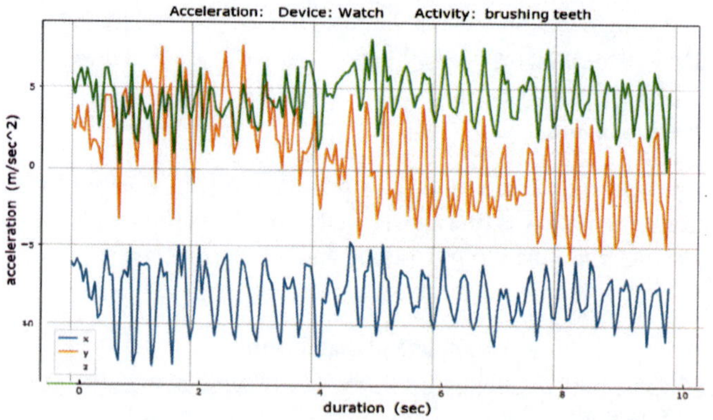

Fig. 2. The fluctuation of x, y and z for the brushing teeth activity by Watch

3.2 Data Cleaning

The technique of "cleaning" data before evaluating it is known as data cleaning. This is accomplished by altering or removing data from the dataset that is erroneous, incomplete, irrelevant, corrupt, duplicate, or incorrectly formatted. The environment for performing this task remains python using panda's library.

Dropping the User ID Column
The approach during this analysis are going to be to visualize all the participants as a closed united sample population. The aim of this analysis isn't to inquire the repose participant variability. Also, the participants variety is just too tiny to considerably represent the variability of the overall population. Therefore, The participant column will be dropped.

Data Transformation
The process of modifying the format, structure, or values of data is known as data transformation. Data can be modified at two phases of the data pipeline for data analytics initiatives. The conversion of our date was done with the LabelEncoder function in Python, which is a very efficient tool for encoding the levels of categorical variables into numerical values. LabelEncoder encodes labels with a value ranging from zero to n-classes-1, where n denotes the number of different labels.

Data Normalization
Data normalization guarantees that all of the records in your customer database look, read, and can be used in the same way. The standardization of our data was performed using the Python StandardScaler function, which removes the mean and scales each

feature/variable to unit variance. This operation is carried out in a feature-by-feature manner. Because StandardScaler involves the calculation of the empirical mean and standard deviation of each feature, it can be influenced by outliers (assuming they exist in the dataset). The normalization in this project was done for 3 columns: x, y, z, as we can see in this table below.

Table 4. Dataset Information Summary

Label	X	Y	z
0	−0.132022	−1.810972	−0.076826
0	−0.091579	−1.794148	−0.543316
0	0.2.59619	−1.059228	−1.176500
0	0.534876	−1.194265	−0.471217
0	−0.048671	−1.766297	0.567825

The Time Series dataset exemplifies the measurement of a quantity through time in the form of data points. The problem of classifying data points over time based on how they behave is addressed by Time Series Classification (TSC) [10]. TSC is a method for predicting the class labels of a new time series based on the learning from an existing time series with labeled data in supervised learning. Deep learning-based Feature extraction approaches were utilized to classify diverse human activities in this study. The two deep learning models employed in this study, Conv2D (CNN) and ConvLSTM2D, are explored in detail in the following sections.

Because this is a multiclass classification problem, categorical cross-entropy is used as the loss function:

$$-\sum\nolimits_{i=1}^{N} ylog(\hat{y}) \tag{1}$$

In all of the models when N is the output size or the number of scaler values in the output.

3.3 CNN Model

A convolutional neural network (CNN) is a sort of artificial neural network model for supervised learning that uses perceptrons, a machine learning unit method, to analyze data. Image processing, natural language processing, and other cognitive tasks can all benefit from CNNs. The equation below depicts how CNN works layer by layer. The x 1 symbol represent the input for first layer whereas the inbox W1 represent the first layer processing tensor. The data travels from x 1 and pass-through w 1 reaching to x L − 1 and w L − 1 which is before the last layer where data preparation is performed at this layer. The final layer x L represent the output of the last layer.

$$x1 \rightarrow w1 \rightarrow x2 \rightarrow \ldots xL - 1wL - 1 \rightarrow xL \rightarrow wL \rightarrow z \tag{2}$$

Keras Conv2D
is a two-dimensional convolution layer that generates a tensor of outputs by winding a convolution kernel with the input layers (Fig. 3).

```
Model: "sequential"
_________________________________________________________________
 Layer (type)                Output Shape              Param #
=================================================================
 conv2d (Conv2D)             (None, 79, 2, 256)        1280

 dropout (Dropout)           (None, 79, 2, 256)        0

 conv2d_1 (Conv2D)           (None, 78, 1, 512)        524800

 dropout_1 (Dropout)         (None, 78, 1, 512)        0

 flatten (Flatten)           (None, 39936)             0

 dense (Dense)               (None, 1024)              40895488

 dropout_2 (Dropout)         (None, 1024)              0

 dense_1 (Dense)             (None, 18)                18450

=================================================================
Total params: 41,440,018
Trainable params: 41,440,018
Non-trainable params: 0
_________________________________________________________________
None
```

Fig. 3. CNN Model Summary

3.4 LSTM Models

Long short-term memory (LSTM) is a deep learning (DL) architecture based on artificial recurrent neural networks (RNNs). Unlike typical feedforward neural networks, LSTM has feedback connections. It can handle not only individual data points (such as photos), but also entire sequences of data (such as speech or video).

ConvLSTM2D
It's a recurrent layer like the LSTM, but instead of internal matrix multiplications, convolution operations are used. As a result, instead of being a one dimension vector containing features, the data that flows through the ConvLSTM cells retains the input dimension (2D in our case) (Fig. 4).

```
Model: "sequential"
_________________________________________________________________
Layer (type)                 Output Shape              Param #
=================================================================
conv_lstm2d (ConvLSTM2D)     (None, 1, 79, 2, 128)     264704

dropout (Dropout)            (None, 1, 79, 2, 128)     0

conv_lstm2d_1 (ConvLSTM2D)   (None, 78, 1, 256)        1573888

dropout_1 (Dropout)          (None, 78, 1, 256)        0

flatten (Flatten)            (None, 19968)             0

dense (Dense)                (None, 512)               10224128

dropout_2 (Dropout)          (None, 512)               0

dense_1 (Dense)              (None, 18)                9234

=================================================================
Total params: 12,071,954
Trainable params: 12,071,954
Non-trainable params: 0
_________________________________________________________________
None
```

Fig. 4. CNN-LSTM Model Summary

4 Evaluation of the Model

To evaluate the proposed ConvLSTM2D model, a comparison between our model and other popular model is made. in this study, a comparison between the Conv2D model and ConvLSTM2D, Random Forest, and KNN models is made. The criteria of comparison are train and testing accuracy and model train and test losses. However, for the Machine learning models, only the testing accuracy is considered as they differ from the neural network in the way of training. The specific parameters are not known like the neural network therefore, its not possible to track the other parameters. Table 4 summarizes the results. Figure 4 presents a visual summary of the model's comparison accuracy experiment. The figure clearly shows the proposed CNN-LSTM model provide superior results compared to other models (Fig. 5).

Table 5. Dataset Information Summary

Model	Train Accuracy		Test Accuracy	
	Phone	Watch	Phone	Watch
Conv2D	91.47%	95.64%	81.92%	13.69%
ConvLSTM2D	91.69%	86.50%	84.86%	84.09%
Random Forest	99.99%	99.99%	75.38%	61.09%
KNN	72.07%	12.63%	65.26%	12.40%

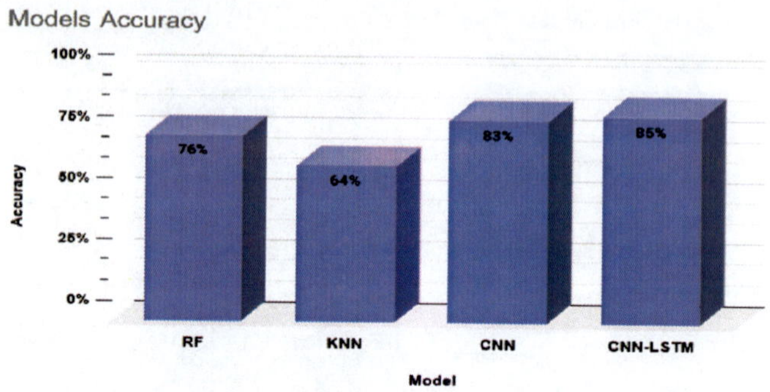

Fig. 5. Summary of Models Accuracies

To generate the results, experimental enactment is conducted using Python 3.9, Tensorflow, Keras, Pandas, Numpy and scipy was used to run the experiments for evaluation of the performance of Deep Learning models used in this work. Technique of PAF signal, the small 2–5 min episodes of AF have no relevance for our study and the obtained results.

Conv2D Model Evaluation

This CNN model was developed using Keras in TensorFlow, the parameters used in this model are:

- Adam algorithm for optimization: Adam is a replacement optimization algorithm for stochastic gradient descent

  for training deep learning models:

- Loss function: sparse categorical cross-entropy function.
- There are 35 Epochs in total. This table below show the Evaluation of the accuracy during the training for 35 epochs (Fig. 6):

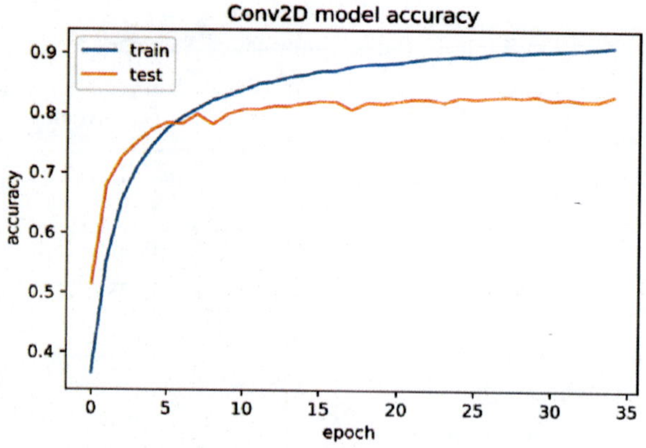

Fig. 6. Conv2D model accuracy Evaluation

This table below show the Evaluation of the loss function during the training for 35 epochs (Fig. 7):

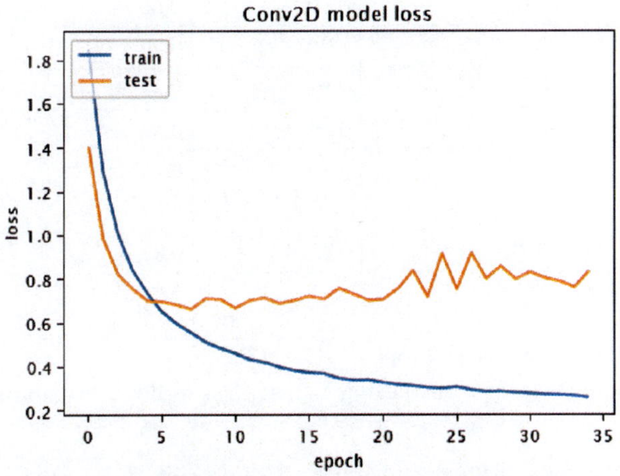

Fig. 7. Conv2D model Loss function Evaluation

Table 6. CNN Classification Report

Label	Precision	Recall	F1-score	Support
0	0.95	0.84	0.89	1174
1	0.97	0.95	0.96	1105
2	0.82	0.76	0.79	1028
3	0.85	0.89	0.87	958
4	0.86	0.85	0.86	1019
5	0.89	0.90	0.89	850
6	0.87	0.85	0.86	1077
7	0.85	0.75	0.80	1109
8	0.76	0.79	0.77	939
9	0.81	0.84	0.82	881
10	0.76	0.84	0.80	901
11	0.83	0.83	0.83	933
12	0.61	0.73	0.67	831

(continued)

Table 6. (*continued*)

Label	Precision	Recall	F1-score	Support
13	0.62	0.70	0.66	899
14	0.79	0.70	0.74	1169
15	0.89	0.87	0.88	977
16	0.86	0.89	0.88	940
17	0.74	0.76	0.75	932
Accuracy			0.82	17722
Macro avg	0.82	0.82	0.82	17722

ConvLSTM2D Model Evaluation

This LSTM model was developed using Keras in Tensorflow, the parameters used in this model are:

- Changing the input data dimension to 5 dimension due to the LSTM algorithm structure, so it can work without errors.
- Adam algorithm for optimization: Adam is a replacement optimization algorithm for stochastic gradient descent for training deep learning models
- Loss function: sparse categorical crossentropy function.
- Epochs: 25 epochs
- This table below show the Evaluation of the accuracy during the training for 25 epochs (Fig. 8):

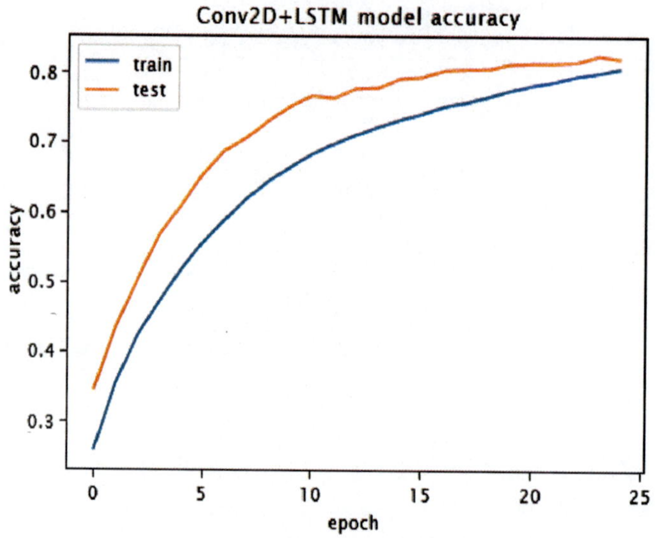

Fig. 8. ConvLSTM2D model accuracy Evaluation

Note: We can notice that while increasing the number of epochs in our training part can affect the accuracy but it will also increase the model complexity in a significant way.

This table below show the evaluation of the loss function during the training for 25 epochs (Fig. 9 and Table 7):

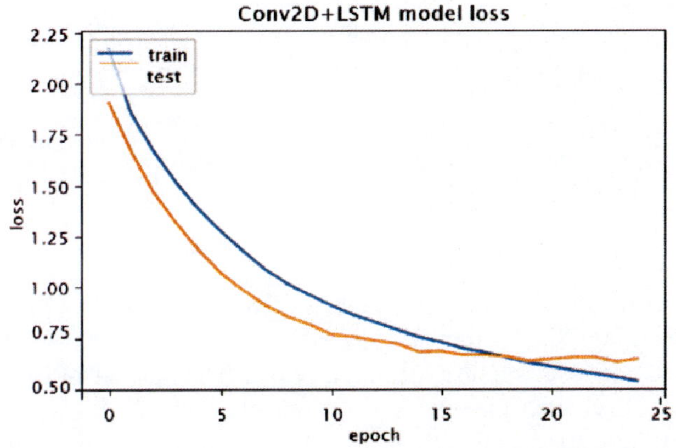

Fig. 9. ConvLSTM2D model loss function Evaluation

Table 7. CNN Classification Report

Label	Precision	Recall	F1-score	support
0	0.95	0.86	0.90	1193
1	0.98	0.96	0.97	1128
2	0.83	0.81	0.82	988
3	0.85	0.92	0.88	928
4	0.84	0.86	0.85	963
5	0.91	0.93	0.92	907
6	0.88	0.88	0.88	1052
7	0.87	0.78	0.82	1129
8	0.80	0.81	0.80	918
9	0.81	0.85	0.83	855
10	0.80	0.81	0.80	895
11	0.87	0.86	0.86	1004
12	0.75	0.71	0.73	1051

(continued)

Table 7. (*continued*)

Label	Precision	Recall	F1-score	support
13	0.67	0.73	0.70	946
14	0.77	0.80	0.79	977
15	0.91	0.91	0.91	900
16	0.87	0.89	0.88	947
17	0.76	0.79	0.77	941
Accuracy			0.84	17722
Macro avg	0.84	0. 84	0.84	17722

RF Model Evaluation

Random Forest is a more advanced method than the popular decision tree algorithm in machine learning. Simply its many decision trees aggregated in a final decision tree which will provide results to the specific problem. The number of trees applied in this experiment is 30, whereas the randomness applied is 100. Table 5 shows the classification report of this experiment. The label column represent each activity label and the precision, recall, f1-score, and support is presented for each activity (Table 8).

Table 8. Random Forest Classification Report

Label	Precision	Recall	F1-score	support
0	0.64	0.90	0.75	550
1	0.79	0.97	0.87	489
2	0.49	0.52	0.51	455
3	0.86	0.89	0.87	492
4	0.83	0.83	0.83	544
5	0.87	0.86	0.86	459
6	0.83	0.83	0.83	528
7	0.84	0.82	0.83	504
8	0.81	0.83	0.82	522
9	0.82	0.77	0.80	458
10	0.86	0.81	0.83	558
11	0.88	0.82	0.85	510
12	0.50	0.42	0.86	528

(*continued*)

Table 8. (*continued*)

Label	Precision	Recall	F1-score	support
13	0.58	0.59	0.59	501
14	0.58	0.45	0.51	543
15	0.87	0.84	0.85	457
16	0.85	0.84	0.84	441
17	0.70	0.62	0.66	483
Accuracy			0.76	9022
Macro avg	0.76	0.76	0.75	9022

KNN Model Evaluation

K-nearest neighbors algorithm is another possible approach to classify patient activities. However, the accuracy of this approach is the lowest compared to CNN, CNN-LSTM and RF methods. The parameters configurations set in this experiment is 5 neighbors, uniform weights, and best fit algorithm is used whether its ballTree or KdTree to automatically calculate the best suited one and use it which can give the best performance and finally, Leaf size parameter used is 30 and metric is minkowski. Table 6 shows the report of this experiment, where label column is the index of each activity in the dataset and the report for precision, recall, f1-score, and support (Table 9).

Table 9. Knn Classification Report

Label	Precision	Recall	F1-score	support
0	0.95	0.83	0.88	550
1	1.00	0.81	0.90	489
2	0.68	0.15	0.25	455
3	0.68	0.85	0.75	492
4	0.45	0.88	0.59	544
5	0.69	0.81	0.74	459
6	0.64	0.76	0.69	528
7	0.65	0.78	0.74	522
8	0.71	0.78	0.74	522
9	0.64	0.70	0.67	458
10	0.67	0.83	0.74	558
11	0.67	0.83	0.74	510
12	0.57	0.08	0.14	528

(*continued*)

Table 9. (*continued*)

Label	Precision	Recall	F1-score	support
13	0.45	0.32	0.37	501
14	0.57	0.23	0.23	543
15	0.72	0.77	0.74	457
16	0.57	0.67	0.62	441
17	0.43	0.48	0.46	483
Accuracy			0.64	9022
Macro avg	0.65	0.64	0.62	9022

5 Conclusion

This study paved the way for the creation of a CNN deep learning model that classifies human activities. The WISDM data that we used in this investigation comes from the data collected from the specified sensors. The sliding window technique was used to alter data in a pre-processing step. Basic models like CNN (Conv2D) and LSTM (ConvLSTM2D) simplified the process of creating CNN models. It's also worth noting that the results have been confirmed through tests and validation, based on the findings of this study. In terms of precision, smart watches outperform smartphones when it comes to recognizing complicated human behaviors. Overall, the findings showed that LSTM deep learning models can extract spatio-temporal characteristics from raw sensor data to identify human behaviors more effectively and accurately than other models. In addition to CNN and LSTM, we will investigate more complicated deep neural network models in the future. Deep Transformer models might be used in the future to classify time series of human activities from the WISDM dataset. Transformers are self-attention neural networks that may be used to investigate and learn relationships in raw sensor data input sequences. When an enlarged WISDM dataset with additional activities and participants becomes available, it may be utilized to categorize new activities.

References

1. Dumoulin, V., Visin, F.: A guide to convolution arithmetic for deep learning, pp. 1–31, March 2016. arXiv:1603.07285. http://arxiv.org/abs/1603.07285
2. Agarwal, P., Alam, M.: A lightweight deep learning model for human activity recognition on edge devices. Procedia Comput. Sci. **167**(2019), 2364–2373 (2020). ISSN: 18770509. arXiv:1909.12917. https://doi.org/10.1016/j.procs.2020.03.289
3. Lockhart, J.W., Weiss, G.M., Xue, J.C., Gallagher, S.T., Grosner, A.B., Pulickal, T.T.: Design considerations for the WISDM smart phone-based sensor mining architecture. In: Proceedings of the Fifth International Workshop on Knowledge Discovery from Sensor Data - SensorKDD 2011, pp. 25–33. ACM Press, New York (2011). ISBN: 9781450308328. https://doi.org/10.1145/2003653.2003656. http://portal.acm.org/citation.cfm?doid=2003653.2003656
4. Khan, I.U., Afzal, S., Lee, J.W.: Human activity recognition via hybrid deep learning based model (2022). https://doi.org/10.3390/s22010323

5. Zhang, H. , Xiao, Z., Wang, J., Li, F., Szczerbicki, E.: A novel IoT-perceptive human activity recognition (HAR) approach using multihead convolutional attention. IEEE Internet Things J. **7**(2), 1072–1080 (2020). ISSN: 2327-4662. https://doi.org/10.1109/JIOT.2019.2949715. https://ieeexplore.ieee.org/document/8883222/
6. Sherstinsky, A.: Fundamentals of recurrent neural network (RNN) and long short-term memory (LSTM) network. Physica D Nonlinear Phenomena **404**, 132306 (2020). ISSN: 01672789. arXiv:1808.03314. https://doi.org/10.1016/j.physd.2019.132306
7. Weiss, G.M., Yoneda, K., Hayajneh, T.: Smartphone and smartwatch-based biometrics using activities of daily living. IEEE Access **7**, 133190–133202 (2019)
8. Walse, K.H., Dharaskar, R.V., Thakare, V.M.: Performance evaluation of classifiers on WISDM dataset for human activity recognition. In: Proceedings of the Second International Conference on Information and Communication Technology for Competitive Strategies - ICTCS 2016, pp. 1–7. ACM Press, New York (2016). ISBN: 9781450339629. https://doi.org/10.1145/2905055.2905232. http://dl.acm.org/citation.cfm?doid=2905055.2905232
9. Esfahani, P., Malazi, H.T.: PAMS: a new positionaware multi-sensor dataset for human activity recognition using smartphones. In: 2017 19th International Symposium on Computer Architecture and Digital Systems (CADS), pp. 1–7. IEEE, December 2017. ISBN: 978-1-5386-4379-2. https://doi.org/10.1109/CADS.2017.8310680. http://ieeexplore.ieee.org/document/8310680/
10. Shocher, A., Feinstein, B., Haim, N., Irani, M.: From discrete to continuous convolution layers, Technical report. arXiv:2006.11120v1

A Consolidated Strategy for Organizational Information Privacy Protection

Hanifa Abdullah(✉)

School of Computing, College of Science, Engineering and Technology (CSET), University of South Africa (UNISA), Florida, Roodepoort, Johannesburg, South Africa
abdulh@unisa.ac.za

Abstract. Data breaches have increased substantially over the years, having an immense effect on the reputation and finances of impacted organizations. A major concern for any organization is their reputation and finances and a poorly devised information privacy protection strategy will adversely impact both these elements. It is therefore imperative that information privacy protection should be implemented as a fundamental aspect of any organization's strategy. Developing a solid information privacy protection strategy is however a major challenge for most organizations. This study seeks to propose a consolidated organizational information protection strategy by combining the components of data, people, process, technology, and rules with the parameters of framework, vision, scope, objectives, actions, and governance. This is vital as there exists a critical requirement for clarity and structure in managing an organizational information privacy program in a coordinated manner.

The ever-changing state of information privacy protection has amplified pressure on those tasked with information privacy protection in their organizations. The strategy posed in this study will serve to alleviate this concern and provide any organization with a comprehensive organizational information privacy protection strategy.

Keywords: Data Breaches · Privacy · Information Privacy · Information Privacy Protection Strategy

1 Introduction

The data privacy landscape is growing at an escalating pace commensurate with advances in technology which alters the manner whereby personal information is gathered, stored, shared, examined, and circulated [1]. The extensive adoption of Information and Communication Technologies (ICTs) have escalated the volume of personal information gathered by government and commercial bodies [2]. In the big data period, the data of users are collected both deliberately and inadvertently by both public and private organizations alike [3]. Some of the collected data is used effectively to provide a superior level of service to users but can also place a user's sensitive information at risk [3]. Users can interrelate with one another via ICTs such as social media and use ICTs such as fitness trackers to assist them in their fitness journey but these ICTs present privacy risks [2].

R. Silhavy and P. Silhavy (Eds.): CSOC 2023, LNNS 723, pp. 316–327, 2023.
https://doi.org/10.1007/978-3-031-35317-8_29

Data analysis is profoundly entrenched in our systems ranging from e-commerce (electronic commerce) to medical and research [1]. The data analysis provides comprehensive information such as data about one's traits, one's interests and preferences which alone or when merged with other analyses can disclose subtle information of individuals such as one's health or financial status [1]. This can have a detrimental impact on a person's reputation, employability and may even cause criminal penalties and bodily harm [1]. Privacy anxieties in respect to the gathering, storage and usage of personal information are a persistent subject of public debate [1].

Data breach statistics illustrate that cybercriminals are sophisticated and driven by the rewards that emanate from personal and financial data [4]. A data breach occurs when personal, exclusive, inaccessible, or sensitive information is disclosed to a person who does not have the authority to access this information [5]. This can manifest from an unintentional event or a deliberate attempt to steal information from either an individual or an organization [5]. The average cost of a data breach in 2021 has risen to R46 million according to an IBM Security research [6].

Malware, phishing, ransomware, man-in-the-middle, and denial of service attacks are the most prevalent cyberthreats [7]. South Africa was most affected by ransomware since the first quarter of 2021 [7]. Transnet and Transnet Port terminals were the victim of a ransomware attack that took place in July 2021, resulting in the malfunctioning of critical systems [7]. This affected the operation of the ports of South Africa and impacted food security, the economy and ultimately societal instability [7].

Sophos, a software security company, surveyed South Africans and in the "State of Ransomware 2021" report, revealed that 24% of the participants encountered a ransomware attack in the preceding twelve months [8]. The report further highlighted that South Africa and Zimbabwe are the countries that have experienced the most ransomware attacks [8]. The average cost to recover stolen data in industrial ransomware attacks in South Africa amounted to $447,097 [8].

The above statistics clearly articulate the need for an organizational information privacy protection strategy, the objective of this study. To attain this objective, this paper is structured as follows: Sect. 2 provides the theoretical underpinning to the study. The research context and research methodology are discussed in Sects. 3 and 4 respectively. Section 5 provides an overview of the consolidated organizational information privacy protection strategy. Section 6 lists the limiting factors of the study. Section 7 summarizes the study highlighting potential areas for prospective research.

2 Background

According to [2], privacy is a dominant subject of the information age. Privacy is an intricate concept having numerous interconnected descriptions and understandings [9]. From the perspective of law, [10] defines privacy as the "right to be alone". From a philosophical viewpoint, [11] defines privacy as a "state of limited access". In the realm of social sciences, [12], defines privacy as "control over information about one's self". Varying descriptions of privacy have made it evident that there lacks a consensus on what privacy connotes [9].

According to [13], information privacy is a subset of privacy and is concerned with a person's right to exercise control over the use of one's personal data. Personal data encompasses both physical and digital data that can uniquely identify an individual [13].

The focus of this study is on information privacy, more specifically on organizational information privacy protection. It is important to define the concepts data privacy and information privacy. This is essentially the difference between the terms data and information. Data is "raw unorganized facts that require processing" [14]. Upon data processing, data is organized and structured to make it useful, and this is called information [14]. The emphasis of this study is on information privacy but the concept data privacy (to denote unprocessed but important data) and information privacy (referring to data that has been processed) will both be used.

The concept of information privacy originated in the 1960's and 1970's at the time the word "data protection" (derived from the German, datenschutz) formed part of the terminology of European experts [15]. Information privacy issues are intricate and interdependent and can be studied from several angles, including law, economics, psychology, management, marketing and information technology [16].

Information privacy infers the prohibition of any activity entailing the collection, storage, distribution, or use of a person's private life data without the permission of the individual concerned [17]. In its most basic form, information privacy implies controlling the manner whereby one's personal information is obtained and utilized. According to [16], information privacy refers to the longing of a person to manage or have some authority over their data usage which entails the protection against illicit or random access by authorities and police.

As discussed in Sect. 1, organizations can be subjected to data breaches which can severely impact their financial and reputational standing. Considering the risks to individuals, organizations tasked with personal information management, deploy numerous measures to protect individuals' personal information [1]. However, [4] states that a large number of companies are ill prepared for the manifestation of a data breach, in spite of the fact that data breaches are a common occurrence.

This study seeks to propose a proactive strategy that any organization that processes large volumes of personal information can use to assist them in personal information privacy protection. The proposed strategy will assist in the mitigation of privacy risks that culminate in breach occurrences.

3 Research Context

The approach in this study is not directed to a particular type of industry but has been created to be sufficiently generic and applicable to any organization that processes large volumes of personal information. The study is approached from a South African perspective and therefore all statistics and legislation are discussed in a South African context.

4 Research Methodology

A research literature is a methodical, clear, and reproducible means of finding, analyzing, and synthesizing the prevailing body of accomplished work presented by scholars and researchers [18]. The research methodology for this study is a literature review to study what has been researched within the field of information privacy protection strategies with a view of developing a consolidated information privacy protection strategy.

5 A Consolidated Strategy for Organizational Information Privacy Protection

A strong information privacy protection strategy is fundamental for any organization. Data protection strategies encompass several steps that describe how security measures should be enforced and maintained [19]. The objective of a data protection strategy is to lessen the trail of sensitive information and prevent threat actors from illicitly accessing data [19].

A data protection strategy goes beyond mere compliance with data protection legislation [20]. The data protection strategy is a framework that highlights how the business operates and manages both the organizations and customers data [20].

There exist several data protection strategies [19–25] with varying guidelines/steps/components and an organization is at liberty to use any of the proposed strategies. The five-lens strategic perspective combined with overall privacy strategy considerations [26] is considered for this study as this strategy has well-defined components and parameters that can be combined to create a consolidated information privacy protection strategy.

Ref. [26] lists several parameters of a privacy strategy. These parameters are listed in Table 1 below.

Table 1. Parameters of a Data Strategy [26]

1.	Establish the privacy framework
2.	Create the vision
3.	Agree on the scope
4.	Develop the strategic objectives
5.	Agree on the actions
6.	Provide good governance

Ref. [26] lists the five-lens *components* of the privacy strategy. These components are discussed in Table 2 below.

Sections 5.1, 5.2, 5.3, 5.4, 5.5 and 5.6 examine the various parameters and components listed in Tables 1 and 2 respectively with a view of consolidating the parameters and components to create an organizational information privacy protection strategy. The parameters and components will be explained where applicable in terms of examples and additional literature. The strategy will be streamlined to eliminate parameters or components that are not required.

Table 2. Components of a Data Strategy [26]

1.	Data
2.	People
3.	Process
4.	Technology
5.	Rules

5.1 Privacy Framework

Standards, the privacy framework, and best practices are dictated by the terms of reference of the organization [26]. Key privacy frameworks the organization can consider include the NIST Privacy Framework [27], ISO/IEC 29100:2011 [28] and ISO/IEC 27701:2019 [29].

5.2 Vision

The vision of the organization leads to the derivation of the objectives for privacy and data protection [26]. The goals are commensurate on the size of the organization, the privacy maturity level, compliance requirements, organizational priorities, and the level of competition [26].

An example of vision in a cloud computing environment could be: "Our organization endeavors to be the partner of choice for cloud computing solutions to the public sector in South Africa. The vision of the organization, underscored by its values of passion and professionalism, is to protect the privacy of data subjects and foster transparency for the data subjects and stakeholders concerned. The business brings together IT, security, marketing, procurement, legal, compliance and HR to break the privacy silos within the organization. Our organization aims to generate a huge profit in an ethical and responsible manner by delivering far-reaching solutions to all clients, thereby maximizing value, building a strong brand reputation, fostering greater data subject trust and confidence, and earning a competitive edge in the marketplace."

5.3 Scope

The distinct vision can set the boundaries for the activities that the organization should embark on [26]. For example, if the organization is a cloud computing organization, all the activities in terms of information privacy protection should revolve around this environment.

5.4 Strategic Objectives

This phase entails identifying the elements required to attain the defined vision within the designated scope [26]. When constructing a data protection strategy, the aim is to safeguard data from destruction by inside or outside threats [21]. A well-constructed data protection strategy is a critical part of the general data management program [21].

These objectives should be relevant and achievable based on the components listed in Table 2. An example of defining the strategic objectives could be as follows:

To have a robust integrated organizational information privacy protection, the organization should follow a rigid and methodological approach to organizational information privacy protection to obtain the following objectives:

- Attain, maintain, and evaluate compliance with privacy laws and regulations.
- Develop a record-keeping system for privacy issues for the use of the organisation, public requests for information and for audit purposes.
- Ensure organizational information privacy risks are identified, assessed, evaluated, and treated accordingly.
- Ensure privacy breach incidents are recorded and dealt with efficiently.
- Promote awareness, education, and training in terms of privacy by means of privacy training materials.
- Build relationships with stakeholders regarding information privacy protection within the organisation.
- Invest in human capital for a Chief Privacy Officer (CPO).

5.5 Actions

Part of the process of developing a strategy is to outline the required result of the strategy and then identify the required actions to attain the objective [21].

In this phase, the organization specifies actions that must be executed to attain the strategic objectives identified in Phase D. This entails identifying who is accountable (people), what activity must be carried out (program) and the due date for activity completion (deadline) [26]. The performance indicators for each outcome are decisive for the accomplishment of this process and the strategy as a whole [26]. An excellent point for an organization to commence is by using the list of comprehensive organizational information privacy protection elements identified by [30] as depicted in Fig. 1 together with the components of a data privacy strategy outlined in Table 2.

The Action phase is discussed in terms of the three action requirements of *people, program,* and *deadline.* These requirements are discussed by combining the organizational information privacy protection elements (Fig. 1) and the data privacy components (Table 2).

- People – This refers to the stakeholders involved in the privacy process of the organization [26]. Since this study is from an organizational perspective, the stakeholders will be the privacy personnel tasked with the protection of personal information within the organization. Referring to Fig. 2, the organizational issues element (E9) must be addressed in its entirety. This means that the aspects of privacy awareness, training and education, roles and responsibilities, organizational culture and change management must be considered for the people requirement of the Action phase.
- Program – In the context of this study, the activities that must be carried out pertain to the process of attaining information privacy protection as this is the strategic objective as stated in Sect. 5.4. Figure 1 illustrates the multiple elements that are required for organizational information privacy protection. Some of these elements are mentioned as parameters and components in Tables 1 and 2 respectively. The activity list will

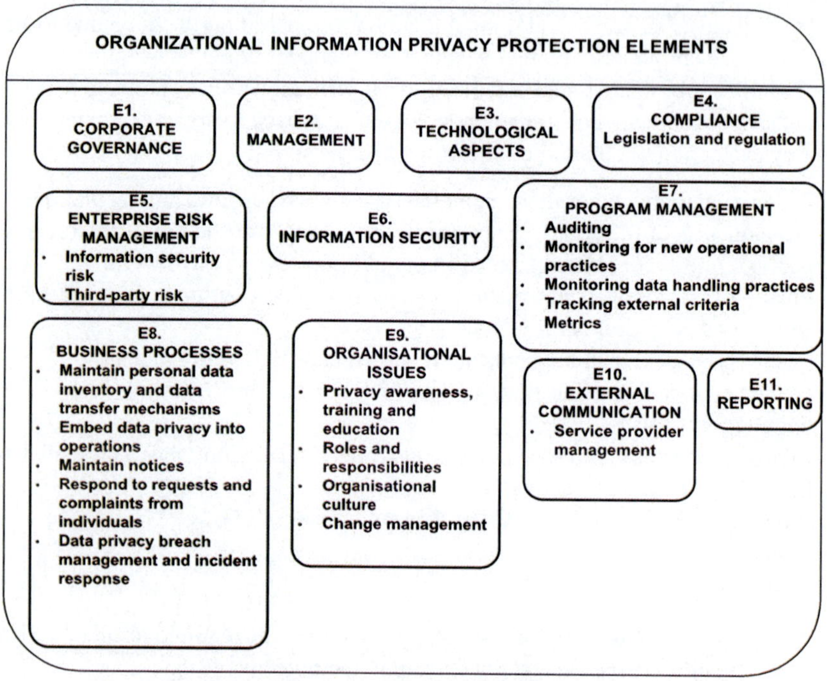

Fig. 1. Organizational information privacy protection elements [30]

merge the components listed in Table 2 with the elements identified in Fig. 1. To attain organizational information privacy protection, an organization must address:

1. Corporate Governance and Management (E1 and E2: Fig. 1) → Good Governance (Table 1). In terms of governance and management, the organization should refer to the King IV Report on Corporate Governance for South Africa [31]. A governance system generates valuable accountability levels for oversight and coordination [26]. The governance system has review stages to affirm that privacy activities are functioning as planned and where there is a deviation in the planed actives, remedial plans can be enforced [26].
2. Technological Aspects (E3) → Technology (Table 2). This refers to the required technical safety measures for the protection of personal information [26].
3. Compliance (E4) → Rules (Table 2) – This refers to the privacy laws and standards whereby an organization must abide to [26]. In South Africa, the data legislation is in terms of the Protection of Personal Information Act (POPIA) [32].
4. Enterprise Risk Management (E5) –ERM embraces the idea that risk management and analysis is performed across the enterprise and the objective is to determine a business's ability to identify key risks and to manage the exposure of risk within acceptable levels of management [33]. For the effective management of risks, Privacy Risk Management (PRM) endeavours to eradicate risks prior to a breach implying that PRM is a preventative approach [34].

5. Information Security (E6) - Information security implies the protection of data in any form (physical or electronic) or the protection of any form of data (personal or non-personal), whereas data protection protects the processing of personal data only [35]. The focal point of discussion of this paper concerns organizational information privacy protection. To achieve privacy, there is fundamental need to first have good security. While you can have security without privacy, it is impossible to maintain privacy if you have poor security. Even running encrypted systems cannot guarantee privacy without security. Without security, attackers will ultimately obtain the encryption and decryption keys leading to a complete loss of privacy. Information security and information privacy are two separate concerns and implementing one does not automatically ensure the other will follow. Instead, they must both be used in tandem to allow both to work better.
6. Program Management (E7) – For program management, the organization should several processes as depicted in Fig. 1. These include auditing, the monitoring of both operational and data handling processes, the tracking of external criteria and using privacy metrics. The data privacy metrics must be aligned to the organizational objectives and goals.
7. Business Processes (E8) – The business processes include a host of activities. These include the implementation of privacy policies and procedures, creation of a privacy statement and an inventory for personal data maintenance. Data privacy must be incorporated within the operations of the organization and there should be a means to maintain notices to individuals consistent with the data privacy policy, legal requirements, and operational risk tolerance. There should be a means to react to complaints from individuals regarding data breaches and a strong breach management and incidence response plan.
8. External Communication (E9) – This refers to a clear plan of communication with external stakeholders.
9. Reporting (E10) – Reporting can be done on many levels. There should be a reporting structure for employees reporting a data breach incident. There should also be reporting to the board.

- Deadline – The organization should have a clearly defined timeline to indicate the implementation dates of the various actions to attain organizational information privacy protection.

The consolidated strategy should include an additional parameter, namely *Monitoring and Reviewing* as this has been identified as an important component in a number of data protection strategies [19, 21, 22]. This additional parameter is discussed below.

5.6 Monitoring and Review

Monitoring and reviewing assists organizations in scrutinizing their data activities, risks, and controls, thereby enhancing protection and response to threats and anomalies [19]. Monitoring and reviews may assist in satisfying compliance requirements [19].

All the parameters and components in Tables 1 and 2 have been discussed and integrated to formulate a consolidated strategy for organizational information privacy protection apart from data and process in Table 2. This is because data refers to multiple means whereby the organization processes personal data and process refers to the different methods of processing personal data, technology oriented and not [26]. The focus of this study is exclusively on how information should be protected and not how or who should process the information. Therefore, defining data and process is not discussed.

Figure 2 depicts a consolidated strategy for organizational information privacy protection. This strategy is presented as a conceptual framework that has been build up by combing the components of data, people, process, technology, and rules with the parameters of framework, vision, scope, objectives, actions, and governance.

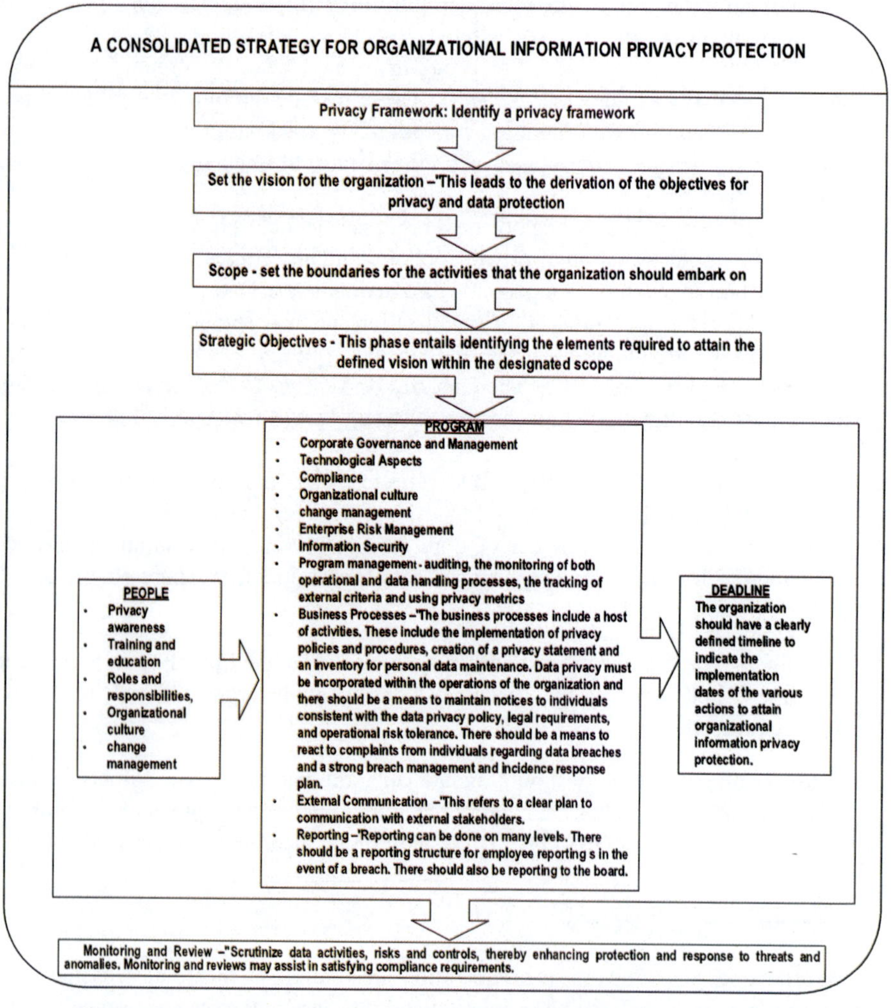

Fig. 2. Consolidated strategy for organizational information privacy protection

6 Limitations

This study has provided an overview of the essential parameters and components required for organizational information privacy protection by merging these parameters and components to create a consolidated strategy for organizational information privacy protection. This strategy does not prescribe the actual manner whereby each action should be executed as this is dependent on each organization according to their vision, scope, and objectives. This strategy however comprehensively lists the actions that every organization should execute to attain a high-level of information privacy protection. Organizations can refer to alternate data protection strategies to build up on each activity listed in the action parameter (Sect. 5.5). A few data protection strategies have been provided in Sect. 5. These data protection strategies have varying elements that can be embedded within the core action activities to keep the strategy streamlined.

The focus of this study is on the organizational practices of handling personal information in response to concerns regarding information mishandling. This study does not focus on how users can protect their individual personal information but instead how organizations can protect their information.

7 Conclusion and Future Research

Developing an organizational information privacy protection strategy is a complex endeavor. Every organization should understand the essential aspects of the strategy and follow this for their environment which may be a financial organization, a healthcare organization, or an educational institution. There can never be a guarantee that an organization will never be a victim of a data breach but by following the comprehensive consolidated strategy posed in this study will give an organization a high level of information privacy protection attainment.

Future research necessitates developing the consolidated strategy into a prototype which organizations can populate for their environment. The artefact will be field tested. The artefact will web-based so it would be relatively easy to validate making use of an expert review or Delphi technique. Eternal validation will also be used to refine the artefact. The prototype should also have a link to Project Management software to detail the tasks and timelines of achieving information privacy protection.

References

1. Nissim, K., Wood, A.: Is privacy privacy? Philos. Trans. R. Soc. A Math. Phys. Eng. Sci. **376**, 20170358 (2018)
2. Wu, P.F., Vitak, J., Zimmer, M.T.: A contextual approach to information privacy research. J. Am. Soc. Inf. Sci. **71**, 485–490 (2020)
3. Jiang, B., Li, M., Tandon, R.: Local information privacy and its application to privacy-preserving data aggregation. IEEE Trans. Dependable Secure Comput. (2020)
4. Must-Know Data Breach Statistics for 2022. https://www.upguard.com/blog/data-breach-statistics. Accessed 2 Aug 2022
5. What is a Data Breach? https://www.fortinet.com/resources/cyberglossary/data-breach. Accessed 6 June 2022

6. Data Breaches in South Africa: What You Need To Know. https://isite.co.za/data-breaches-south-africa/. Accessed 5 June 2022
7. Cyberattacks: personal data and money are vulnerable. https://www.iol.co.za/business-report/opinion/cyberattacks-personal-data-and-money-are-vulnerable-8901e6ae-a604-46f8-a2ed-03e1903b6e41. Accessed 25 May 2022
8. Look out for these cybersecurity trends in 2022. https://www.defenceweb.co.za/cyber-defence/look-out-for-these-cybersecurity-trends-in-2022/. Accessed 15 May 2022
9. Chantal, M., Twizeyimana, J., Viberg, O.: Students' information privacy concerns in learning analytics: towards model development. In: 2021 Nordic Learning Analytics (Summer) Institute, NLASI 2021, 23 August 2021, Stockholm, Sweden. CEUR-WS (2021)
10. Warren, S.D., Brandeis, L.D.: The right to privacy. Harv. Law Rev. **4**, 193–220 (1890)
11. Schoeman, F.D.: Philosophical Dimensions of Privacy: An Anthology. Cambridge University Press, Cambridge (1984)
12. Kokolakis, S.: Privacy attitudes and privacy behaviour: a review of current research on the privacy paradox phenomenon. Comput. Secur. **64**, 122–134 (2017)
13. Presthus, W., Vatne, D.M.: A survey on facebook users and information privacy. Procedia Comput. Sci. **164**, 39–47 (2019)
14. Difference Between Data and Information. https://byjus.com/biology/difference-between-data-and-information/. Accessed 15 May 2022
15. Bennett, C.J.: The European general data protection regulation: an instrument for the globalization of privacy standards? Inf. Polity **23**, 239–246 (2018)
16. Lehner, F., Dzepina, A.: Information privacy in a digitalized world: private issue or public matter?. In: Harteis, C. (ed.) The Impact of Digitalization in the Workplace. Professional and Practice-Based Learning, vol. 21, pp. 183–195. Springer, Cham (2018). https://doi.org/10.1007/978-3-319-63257-5_12
17. Serohin, V.: Information privacy: a conceptual approach (2020)
18. Fink, A.: Conducting Research Literature Reviews: From the Internet to Paper. Sage publications (2019)
19. Data Protection Strategy: 10 goals of an Effective Strategy. https://cloudian.com/guides/data-protection/data-protection-strategy-10-components-of-an-effective-strategy/. Accessed 11 Apr 2022
20. What is Data Protection & Why Should You Care About It? https://www.virtru.com/blog/data-protection-benefits-strategy. Accessed 3 Apr 2022
21. 20 keys to a successful enterprise data protection strategy. https://www.techtarget.com/searchdatabackup/tip/20-keys-to-a-successful-enterprise-data-protection-strategy. Accessed 15 June 2022
22. Essentials Every Data Protection Strategy Requires. https://www.polar.security/post/7-essentials-every-data-protection-strategy-requires. Accessed 29 Apr 2022
23. Components of an Effective Data Protection Strategy. https://wickr.com/7-components-of-an-effective-data-protection-strategy/. Accessed 4 May 2022
24. How to setup a data protection strategy. https://fraudwatch.com/how-to-set-up-a-data-protection-strategy-our-guide/. Accessed 18 Apr 2022
25. What is Data Protection: Everything You Need to Know. https://rewind.com/blog/data-protection-everything-you-need-to-know/. Accessed 21 June 2022
26. Barclay, C.: What is your privacy and data protection strategy? ISACA J. **2** (2021)
27. Boeckl, K.R., Lefkovitz, N.B.: NIST privacy framework: a tool for improving privacy through enterprise risk management, version 1.0 (2020)
28. ISO/IEC: ISO/IEC:29100:2011, Information Technology – Security Techniques – Privacy Framework. International Organization for Standardization and the International Electrotechnical Commission (2011)

29. ISO/IEC: ISO/IEC 27701:2013, Security techniques — Extension to ISO/IEC 27001 and ISO/IEC 27002 for privacy information management — Requirements and guidelines. International Organization for Standardization and the International Electrotechnical Commission (2019)
30. Abdullah, H.: A structural and navigational method for integrated organizational information privacy protection. In: 2018 International Conference on Advances in Big Data, Computing and Data Communication Systems (icABCD), pp. 551–559. IEEE (2018)
31. King IV Report on Corporate Governance for South Africa. https://www.adams.africa/wp-content/uploads/2016/11/King-IV-Report.pdf. Accessed
32. South Africa: Act No. 4 of 2013: Protection of Personal Information Act, 2013. Government Gazette, vol. 581, pp. 1–149 (2013)
33. Serfontein, H.: A Critical Evaluation of the Importance of a Governance, Risk and Compliance Software in the GRC Process. Stellenbosch University, Stellenbosch (2012)
34. Privacy Risk Management. https://www.ipc.on.ca/wp-content/uploads/2010/04/Privacy-Risk-Management-Building-privacy-protection-into-a-Risk-Management-Framework-to-ensure-that-privacy-risks-are-managed.pdf. Accessed 1 Aug 2022
35. Guamán, D.: Privacy vs. data protection vs. information security. Softw. Serv. Eng. **2022** (2016)

Three «Great Challenges» of Medical Informatics

V. A. Galkin[1], T. V. Gavrilenko[1(✉)], V. M. Eskov[1], and A. Yu Kukhareva[2]

[1] Surgut Branch of Federal State Institute "Scientific Research Institute for System Analysis of the Russian Academy of Sciences", 4, Energetikov Street, Surgut 628426, Russia
taras.gavrilenko@gmail.com

[2] Surgut State University, 1, Lenin Avenue, Surgut 628403, Russia

Abstract. The first quarter of the 21st century is coming to an end, but the problem of reduction has not become the central problem of the 21st century, as Nobel laureate V.L. Ginzburg. At the end of the 20th century, he identified three "great" problems of physics and all of science, and they are discussed today in the framework of the new theory of chaos-self-organization, the foundations of which we now present. These great problems led to the uniqueness of the samples, the loss of homogeneity of any groups, and the appearance of type 1 uncertainty (statistics do not work). We have proved the reality of these three "great" problems experimentally. #CSOC1120.

Keywords: Stochastics · Chaos · Eskov-Zinchenko Effect

1 Introduction

In 1999, Nobel laureate V.L. Ginzburg, closing the 20th century, presented three "great" problems for all physicists [1]. These problems were associated not only with physics and modern mathematics, but also with biology, with all the sciences of living systems. These are indeed fundamental problems of all science [2–6].

We are talking about the irreversibility of processes, the role of consciousness in a quantum experiment, and the possibility of reduction (is it possible to describe living systems by the laws of physics?). In fact, now we can talk about the cognizability of biosystems from the standpoint of modern deterministic and stochastic science.

Recall that one of the founders of information theory W. Weaver in 1948 proposed to bring all biosystems (systems of the third type, STT) beyond the limits of modern deterministic and stochastic science [7]. As a result, we are now raising the question of the boundaries of deterministic and stochastic science in the description and prediction of biosystems. Why cannot the theory of dynamical systems and all mathematical statistics describe STT? We now present the answer to this question in this report. In fact, this means the completion of deterministic and stochastic science methods in the study of STT, which is due to the Eskov-Zinchenko effect, i.e. the uniqueness of any sample of parameters of the human body [2–6].

R. Silhavy and P. Silhavy (Eds.): CSOC 2023, LNNS 723, pp. 328–337, 2023.
https://doi.org/10.1007/978-3-031-35317-8_30

2 Deterministic and Stochastic Science in the Description of Systems of the Third Type

In the entire history of the development of science, only three people really tried to raise the issue objective study of biosystems. Is it possible to describe STT within the framework of methods and models of deterministic and stochastic science? And if not, then why and how do restrictions arise?

Recall that in 1947 the outstanding specialist in biomechanics of the 20th century N.A. Bernstein put forward the hypothesis of "repetition without repetition" in the organization of movements. In his work "on the construction of movements", he proved the existence of at least five different motion control systems (these are systems A, B, C, D, E). It follows from the logic of its work that the inclusion (strength, duration) of any of these 5 systems in the organization of any movement occurs chaotically. The strength and duration of such inclusion are not regulated by anything, and therefore any movement must occur without repetition.

A year later, one of the founders of information theory, W. Weaver, contrasted all science with a special complexity that included all biosystems. He indicated this in the very title of his article (Science and complexity) [7], and then proposed to remove all STT outside the deterministic and stochastic science. Weaver pointed directly to the need to create a new (third, after deterministic and stochastic science) science. This science is needed to describe biosystems. At the same time, Weaver did not provide evidence for the special properties of STT (why STT are not an object of modern deterministic and stochastic science), but he talks about their features of self-organization [7].

For 70 years no one has put these two hypotheses together (Bernstein and Weaver). Everyone considered the Bernstein hypothesis (about repetitions without repetitions) and the Weaver hypothesis (STT is not an object of deterministic and stochastic science, a third science is needed) as some kind of eccentricity of two geniuses. This is a very strange neglect on the part of the scientists of the world. Obviously, this was the loss of sensitivity of science to new phenomena [8–14].

However, in 1999 Nobel laureate V.L. Ginzburg returns to this problem. He raises three "great" problems of physics (and of all science) in his review article [1]. Ginzburg directly says: "... The question of reductionism is both a great physical and biological problem, and, as I am convinced, it will be one of the central ones in the science of the 21st century!". He convinced of this, and the main thing is the special properties of biosystems [8–14].

In the first "great" problem, Ginzburg speaks of the irreversibility (and hence non-equilibrium) of all biosystems. I.R. Prigogine was firmly convinced of this, but he did not go beyond deterministic and stochastic science in his thermodynamics of nonequilibrium systems [12]. He really hoped (together with other Nobel laureates R. Penrose, V.L. Ginzburg, M. Gell-Mann) on the possibility of dynamic chaos in the description of STT. However, the hopes of these Nobel laureates did not come true [9–13].

In the second "great" problem, Ginzburg talks about the relationship of consciousness during a quantum experiment with the processes of quantum entanglement. In this field we can cite works of Michael B. Mensky on the quantum theory of consciousness. In the new theory of chaos-self-organization, these ideas received experimental evidence based on instability of electroencephalogram samples.

Finally, Ginzburg's third "great" problem has to do with reductionism. In this problem, the question remains open: can physics describe biosystems? V. Ginzburg does not answer this fundamental question [1]. However, he, along with Weaver, questions the use of deterministic and stochastic science in the description of specific STT. As a result, Ginzburg writes: "...is it possible to consider that the possibility of reducing biology to modern physics is beyond doubt? It seems to us that it would be wrong to give a positive answer" [1].

Ginzburg is not just expressing doubt. He is convinced that "until the job is done, can't we rule out the possibility that we don't yet know something necessary for reduction, even at a fundamental level?" [1]. Let's present our answer to these questions.

3 Fundamental Level for Biosystems Within Theory of Chaos-Self-organization

For about 200 years, all life sciences have been using statistics to describe and predict any biosystems. For example we measure some parameter x_1 during time interval Δt_1 and obtain the sample, and then repeat this over the interval Δt_2. In all 100 years of using mathematical statistics, no one in biomedicine, psychology, ecology and other life sciences has asked elementary questions: what happens to the biosystem before Δt_1, between Δt_1 and Δt_2 and after Δt_2. No one has answered these questions in these 100 years. Moreover, and this is very surprising, no one posed the elementary question: can two adjacent samples coincide statistically, can they have one common population?

In our researches we tested these hypotheses 20 years ago, first on tremorograms (TMG) and tappinggrams (TPG). It turned out that after 100 repetitions of paired TMG measurements (Δt_1 and Δt_2 side by side), we obtained the frequency of statistical coincidence of these two samples of TMG $P^*_{i,i+1} < 0.05$. For TPG (voluntary vertical movements of the subject's finger), this $P^*_{i,i+1} \leq 0.15$. But this is all very low values from the standpoint of statistics, i.e. statistics does not work, the samples are unique [9–15].

Recall that in statistics one usually requires $P_{i,i+1} \geq 0.95$. Moreover, evidence-based medicine requires $P_{i,i+1} \geq 0.99$ or more. For biomechanics, these are already fantastic values. Two neighboring samples coincide very rarely. Statistics in biomechanics does not work [2–6].

Moreover, if we measure 15 samples of cardiointervals (heart beat-to-beat RR interval values, RR intervals) in a row in the same person (at rest) and then compare all these 15 samples in pairs with each other, we will get Table 1 (and hundreds of similar ones).

It turned out that if we register a sample of RR intervals (one subject) for 5 min and make a matrix of paired comparisons of all these samples, then in such a matrix the number k of pairs of RR intervals samples (which have a common population) with the Wilcoxon signed-rank test $P_{ij} \geq 0.05$ (Table 1) will be a very small number, usually $k \leq 15\%$ [14–17].

For example, we present a typical matrix (one of many hundreds similar to it) in Table 1. We recorded all (for pairs of i-th j-th samples) P_{ij} and got the number $k_i < 15\%$. This proves that the probability of matching any i-th and j-th RR intervals sample (for one person at rest) is very low (less than 0.15). For statistics, this is a very low number.

Table 1. Matrix of pairwise comparison of samples of RR intervals of the same person (without load, the number of repetitions of registration of RR intervals $n = 15$), the Wilcoxon signed-rank test was used (significance level $p < 0.05$, number of matches $k = 12$).

N	1	2	3	4	5	6	7	8	9	10	11	12	13	14	15
1	–	0.00	0.16	0.00	0.00	0.00	0.00	0.00	0.00	0.02	0.00	0.05	0.19	0.00	0.00
2	0.00	–	0.00	0.00	0.00	0.66	0.00	0.00	0.00	0.00	0.00	0.00	0.00	0.00	0.00
3	**0.16**	0.00	–	0.00	0.00	0.00	0.00	0.00	0.00	0.00	0.00	0.42	0.00	0.00	0.00
4	0.00	0.00	0.00	–	0.00	0.00	0.08	0.00	1.00	0.00	0.00	0.00	0.00	0.00	0.00
5	0.00	0.00	0.00	0.00	–	0.00	0.00	0.00	0.00	0.00	0.00	0.00	0.00	0.03	0.13
6	0.00	**0.66**	0.00	0.00	0.00	–	0.01	0.00	0.00	0.00	0.00	0.00	0.00	0.00	0.00
7	0.00	0.00	0.00	0.08	0.00	0.01	–	0.00	0.16	0.00	0.00	0.00	0.00	0.00	0.00
8	0.00	0.00	0.00	0.00	0.00	0.00	0.00	–	0.00	0.00	0.00	0.00	0.00	0.75	0.96
9	0.00	0.00	0.00	**1.00**	0.00	0.00	**0.16**	0.00	–	0.00	0.00	0.00	0.00	0.00	0.00
10	0.02	0.00	0.00	0.00	0.00	0.00	0.00	0.00	0.00	–	0.00	0.00	0.12	0.00	0.00
11	0.00	0.00	0.00	0.00	0.00	0.00	0.00	0.00	0.00	0.00	–	0.00	0.00	0.00	0.00
12	**0.05**	0.00	**0.42**	0.00	0.00	0.00	0.00	0.00	0.00	0.00	0.00	–	0.01	0.00	0.00
13	**0.19**	0.00	0.00	0.00	0.00	0.00	0.00	0.00	0.00	**0.12**	0.00	0.01	–	0.00	0.00
14	0.00	0.00	0.00	0.00	0.03	0.00	0.00	**0.75**	0.00	0.00	0.00	0.00	0.00	–	0.58
15	0.00	0.00	0.00	0.00	**0.13**	0.00	0.00	**0.96**	0.00	0.00	0.00	0.00	0.00	**0.58**	–

As a result, any two RR intervals samples (at any intervals Δt_1 and Δt_2) do not coincide statistically (doesn't have common population) with probability $P_{ij} \geq 0.85$. This implies the uniqueness of the samples of parameters of the human body. Any person on our planet generates a sample of RR intervals (TMG, TPG, EEG, EMG, etc.), which is difficult to repeat. This is called the Eskov-Zinchenko effect for any subject. This effect shows uniqueness for TMG, TPG, EEG, EMG and other parameters [2–6, 9–17].

When analyzing a sample of RR intervals for a group of different (for example, 15) subjects, the situation becomes even more complicated. If we register samples for 15 subjects (supposedly the same physiologically, mentally, physically, etc.) and then compare them, we will get a heterogeneous group.

This means that out of 105 different comparison pairs of these 15 different RR intervals samples, we can get no more than 15–20% of pairs for which the Newman-Keuls test (and all other tests for unrelated samples) $P_{ij} \geq 0.05$. The remaining 80–85% of RR intervals pairs of comparison cannot have a common population (they have $P_{ij} < 0.05$). For example, we will present a matrix for a group of 15 different people (we compared their samples in pairs) [2–8, 14–18].

If almost all samples of 15 different subjects do not statistically match it, it will not be a homogeneous group. We do not know why almost all samples of RR intervals of

Table 2. Pairwise comparison matrix of 15 samples of RR intervals of a group of women. Newman-Keuls test was used (significance level $p < 0.05$, number of matches $k = 15$).

–	1	2	3	4	5	6	7	8	9	10	11	12	13	14	15
1	–	0.00	0.00	1.00	0.00	0.00	0.00	0.00	0.00	0.00	0.00	0.00	1.00	1.00	0.00
2	0.00	–	0.00	0.00	0.00	0.00	0.00	0.00	0.00	0.00	0.00	0.00	0.00	0.00	0.00
3	0.00	0.00	–	0.00	0.00	1.00	0.00	0.00	0.00	0.00	0.00	0.00	0.00	0.00	0.00
4	**1.00**	0.00	0.00	–	0.00	0.00	0.00	0.00	0.00	0.00	0.00	0.00	1.00	1.00	0.00
5	0.00	0.00	0.00	0.00	–	0.00	0.00	0.28	0.00	0.00	0.00	0.00	0.00	0.00	1.00
6	0.00	0.00	**1.00**	0.00	0.00	–	0.00	0.00	0.00	0.00	0.86	0.00	0.00	0.00	0.00
7	0.00	0.00	0.00	0.00	0.00	0.00	–	0.00	1.00	0.00	0.00	1.00	0.00	0.00	0.00
8	0.00	0.00	0.00	0.00	**0.28**	0.00	0.00	–	0.00	1.00	0.00	0.00	0.00	0.00	0.00
9	0.00	0.00	0.00	0.00	0.00	0.00	**1.00**	0.00	–	0.19	0.00	1.00	0.00	0.00	0.00
10	0.00	0.00	0.00	0.00	0.00	0.00	0.00	**1.00**	**0.19**	–	0.00	0.00	0.00	0.00	0.00
11	0.00	0.00	0.00	0.00	0.00	**0.86**	0.00	0.00	0.00	0.00	–	0.00	0.00	0.00	0.00
12	0.00	0.00	0.00	0.00	0.00	0.00	**1.00**	0.00	**1.00**	0.00	0.00	–	0.00	0.00	0.00
13	**1.00**	0.00	0.00	**1.00**	0.00	0.00	0.00	0.00	0.00	0.00	0.00	0.00	–	1.00	0.00
14	**1.00**	0.00	0.00	**1.00**	0.00	0.00	0.00	0.00	0.00	0.00	0.00	0.00	**1.00**	–	0.00
15	0.00	0.00	0.00	0.00	**1.00**	0.00	0.00	0.00	0.00	0.00	0.00	0.00	0.00	0.00	–

the group belong to different populations. However, these subjects cannot be combined into a homogeneous group (see Table 2, more than 85% of discrepancies).

Mathematical statistics categorically prohibits working with heterogeneous groups. And this is the second (after Eskov-Zinchenko effect) problem for all biomedicine. Any group according to the parameters of body functions cannot be homogeneous. All these 150–200 years in biomedicine have worked with heterogeneous groups. As a result, any sample will be unique, and any group of subjects will be heterogeneous (see Table 2). This is a consequence of Eskov-Zinchenko effect for a group of different subjects.

4 Uncertainty of the 1st Type and Deterministic and Stochastic Science

There is third problem in biomedicine, for which the further use of statistics is very problematic. Very often, samples show statistical coincidences, but the organism is in different conditions. For example, statistics show no difference between a group of ill and healthy people, or before and after some exposure.

However, biology, physiology and all medicine prove their difference on the basis of biological criteria. For example, between a group of ill and healthy patients, statistics may show the coincidence of samples, but biologically the groups differ significantly [3–8, 18].

In theory of chaos-self-organization, this situation was designated uncertainty of the 1st type, when statistics cannot show differences between samples, and the clinic (biology, physiology) clearly demonstrates the differences between the studied groups [3–7].

In this case, we need to use artificial neural networks in two special modes. These regimes follow from the Eskov-Zinchenko effect and from the entire theory of chaos-self-organization [3–7]. In particular, the study of thousands of electroencephalograms (EEG) showed that the EEG of one subject cannot coincide when EEG is recorded in a row at rest. Statistical chaos arises, which is actually the Eskov-Zinchenko effect for the EEG [6–11, 15, 16, 18].

EEG samples are not statistically repeated, and we designate this as chaos in the work of brain neural networks. This is a basic property of the brain of any person. Moreover, brain biopotentials (EEG) - $x_1(t)$ can never be at rest.

The second basic property of the brain, after EEG chaos, is the absence of rest modes. EEG always shows $dx_1/dt \neq 0$. The brain continuously generates biopotentials (oscillations $x_1(t)$). They can be referred to as reverberation. This is the second basic property of human brain neural networks.

We have introduced these two fundamental properties of brain neuron network into the operation of artificial neural networks. First, for each new setting of the artificial neural networks, we randomly set the initial weights w_{i0} (diagnostic signs x_{ij}, i.e. $w_{i0} \in (0.1)$).

Next, we make the artificial neural networks programmatically reverberate, i.e. repeat your setup many times. As a result, the artificial neural networks perform a number of iterations (more than 100 and even 1000). As a result, we obtain samples of finite values w_i features x_i. These samples w_i are statistically processed.

For example, we will present the uncertainty of the 1st type, from the field of rehabilitative medicine. Here, 300 schoolchildren were transported from the north of Russia (point 1) to the south (measurement point 2), then for almost 3 weeks they were healed (before the start of recovery - point 2 and after their treatment - point 3) and returned back to the North of Russia (return to the North-point 4).

At all these 4 points, we measured six parameters of the heart (x_1- sympathetic activity index (a.u.), x_2- parasympathetic activity index (a.u.), x_3- heart rate (beat/m), x_4- standard deviation of all RR intervals in sample (s), x_5- Baevsky's stress index (a.u.) and x_6- blood oxygen saturation level (%)). The samples of these six parameters were statistically compared for all six different pairs (points 1–2, 2–3, 3–4, 1–3, 1–4, 2–4) [2, 5, 13, 14, 17].

If the Wilcoxon signed-rank test P_{ij} for the i-th and j-th pair (each diagnostic feature x_1) was $P_{ij} \geq 0.05$, then such a pair was considered statistically identical. This pair could have a common population. In Table 3 we present the results of such calculations.

At the same time, schoolchildren (boys and girls, same age) were divided into separate groups. In Table 3 we present results of calculation. Obviously, out of 36 different pairs of comparison of these six parameters, only 10 pairs showed statistical differences (the remaining 26 pairs are the same) in Table 3 (boys). For groups of girls only 6 pairs are different.

Table 3. Significance levels *(p)* for pairwise comparisons parameters derived from time-domain heart rate variability analysis of cardiovascular system in schoolchildren with latitudinal movements in 4 related samples using the Wilcoxon signed-rank test ($p < 0.05$).

Comparison group	Significance levels					
Boys ($n = 25$)	Sympathetic activity index	Parasympathetic activity index	Heart rate	Standard deviation of all RR intervals in sample	Baevsky's stress index	Blood oxygen saturation level
1 and 2	0.50	0.37	0.19	0.09	0.07	**0.00**
1 and 3	0.40	0.97	0.85	0.68	0.92	**0.00**
1 and 4	0.08	**0.01**	**0.00**	**0.01**	**0.04**	0.66
2 and 3	1.00	0.79	0.79	0.77	0.65	0.57
2 and 4	0.16	*0.06*	***0.02***	0.15	**0.04**	0.07
3 and 4	0.24	**0.03**	**0.04**	0.14	0.13	**0.03**
Girls ($n = 30$)	–	–	–	–	–	–
1 and 2	0.47	0.24	0.28	0.07	0.16	0.84
1 and 3	0.10	**0.02**	**0.01**	**0.03**	0.06	0.47
1 and 4	0.87	0.13	0.36	0.29	0.63	0.66
2 and 3	0.26	0.11	0.09	0.20	0.19	**0.03**
2 and 4	0.85	0.79	0.46	0.65	0.69	0.68
3 and 4	**0.02**	0.12	**0.02**	0.11	0.27	0.78

Note: p - the achieved level of significance in pairwise comparison using the Wilcoxon signed-rank test, n is the number of subjects; 1st point of the study - before the departure of the children to the health camp; 2nd point - upon arrival to the health camp; 3rd point – 3 weeks past (the end of rest); 4th point directly upon arrival to the North of Russia.

This result proves that statistics cannot give us a difference in the change in the cardiovascular system of schoolchildren during the flight, as well as after treatment. It turns out that their cardiovascular system remains almost unchanged, girls have even fewer such pairs (6 of them), and this is already a strongly pronounced uncertainty of the 1st type [2–5, 9, 12].

However, the use of artificial neural networks immediately makes differences between the samples in all comparison pairs. Moreover, we can find the main diagnostic parameters, i.e. order parameters. And this is already a solution to the problems of system synthesis. This cannot be solved by statistical methods (statistics does not work) [9–17].

The Nobel laureate V.L. Ginzburg presented three "great problems of physics" and of all science. These problems concern the study of biosystems. STT are unique, there are no statistical repetitions of any samples. Then physics does not describe biosystems.

We have now transformed these three great problems into the real problems of all modern science, since science must study biosystems. First, every sample is unique and cannot be repeated. Secondly, there are no homogeneous groups in biomedicine. Thirdly, type 1 uncertainty arises, when statistics do not show differences, but artificial neural networks distinguishes them. A special work of the artificial neural networks is needed; it follows from the Eskov-Zinchenko effect and the special properties of biosystems.

As a result, we come to the denial of the possibility of further use of deterministic and stochastic science in the study of the biosystem. These STT cannot be objects of the theory of dynamical systems and statistics. We need the 3rd science, which was mentioned by W. Weaver [7].

5 Discussion

Unfortunately, in 1999 the Nobel laureate V.L. Ginzburg presented the "three great problems" of physics, which did not receive a worthy assessment among all scientists of the world. The arrow of time, the quantum theory of consciousness and the problem of the reduction of biosystems (is it possible to describe biological systems within the framework of physics) remain without attention even now.

However, after the discovery of the Eskov-Zinchenko effect, it became clear why any biosystems are unique (it is statistically very difficult to repeat them), why there are no homogeneous groups in biomedicine, and why statistics cannot describe STT (type 1 uncertainty). All these problems are being examined in the new theory of chaos-self-organization [9–17].

We emphasize that Eskov-Zinchenko effect and the loss of homogeneity of groups are global properties of any biosystem. All this is now in the new theory of chaos - self-organization is called uncertainty of the second type. For this uncertainty, an analogue of the Heisenberg principle is introduced into theory of chaos-self-organization [2–6, 14–17]. An analog of quantum uncertainty for STT appears [9–17].

However, there is also uncertainty of the 1st type, when statistics do not show differences, and theory of chaos-self-organization clearly distinguishes between samples and the state of biosystems. As a result, we come to uncertainties of the 1st and 2nd types in theory of chaos-self-organization, which limits the statistics.

For their disclosure, special theory of chaos-self-organization methods are used (calculation of the parameters of pseudo-attractors, the use of artificial neural networks in two special modes). All these are already new concepts and methods; they are not in the deterministic and stochastic science [9–17].

It is obvious that further use of statistics may be limited for the study of biosystems. We need a new (third) science; we need new methods and models [2–6, 9–17]. All this now provided by theory of chaos-self-organization [2–6, 9–17].

6 Conclusion

Three geniuses of the 20th century (N.A. Bernstein, W. Weaver, V.L. Ginzburg) tried to raise the problem of special properties of all biosystems. However, neither they nor other scientists (until the beginning of the 21st century) proved the special properties of biosystems.

Over the past 20 years, we have shown the uniqueness of any sample of any parameter of all functions of the human body (in the form of Eskov-Zinchenko effect). This was the 1st first basic problem of all science and living systems. This is the property of irreversibility (the arrow of time).

Next, we proved the loss of homogeneity of any group of subjects (their samples do not have common populations). Finally, we proved the reality of type 1 uncertainty (statistics do not show differences between samples). As a result, three "great" problems of biomedicine arise and they really complete the further use of statistics in the study of biosystems. It is obvious that the great problems of Ginzburg are turning into new great problems of medical informatics and all modern science.

References

1. Ginzburg, V.L.: What problems of physics and astrophysics seem now to be especially important and interesting (thirty years later, already on the verge of XXI century)? Physics-Uspekhi **42**, 353–73 (1999). https://doi.org/10.1070/PU1999v042n04ABEH000562
2. Eskov, V.V., Gazya, G.V., Bashkatova, Yu.V., Filatova, O.E.: Systems synthesis: environmental factors impact assessment in non indigenous women living in the North. IOP Conf. Ser. Earth Environ. **981**, 032089 (2022). https://doi.org/10.1088/1755-1315/981/3/032089
3. Gazya, G.V., Eskov, V.V., Gavrilenko, T.V., Stratan, N.F.: Artificial intelligence systems based on artificial neural networks in ecology. In: Silhavy, R. (ed.) CSOC 2022. LNNS, vol. 503, pp. 149–158. Springer, Cham (2022). https://doi.org/10.1007/978-3-031-09073-8_14
4. Eskov, V.V., Orlov, E.V., Gavrilenko, T.V., Manina, E.A.: Capabilities of artificial neuron networks for system synthesis in medicine. In: Silhavy, R. (ed.) CSOC 2022. LNNS, vol. 503, pp. 171–179. Springer, Cham (2022). https://doi.org/10.1007/978-3-031-09073-8_16
5. Gazya, G., Eskov, V., Bashkatova, Yu., Stratan, N.: Research of the industrial electromagnetic field influence on heart state in oil and gas workers of the Russian federation. Ecol. Ind. Rus. **26**(5), 55–59 (2022). https://doi.org/10.18412/1816-0395-2022-5-55-59
6. Eskov, V.M., Filatova, O.E., Eskov, V.V., Gavrilenko, T.V.: The evolution of the idea of homeostasis: determinism, stochastics and chaos–self-organization. Biophysics **62**(5), 809–820 (2017)
7. Weaver, W.: Science and complexity. Am. Sci. **36**(4), 536–544 (1948)
8. Eskov, V.M., Filatova, O.E.: Problem of identity of functional states in neuronal networks. Biophysics **48**(3), 497–505 (2003)
9. Eskov, V.V., Filatov, M.A., Galkin, V.A., Filatova, O.E.: New computational methods for investigation of the third type of systems. AIP Conf. Proc. **2402**, 050017 (2021). https://doi.org/10.1063/5.0073431
10. Eskov, V.M.: Hierarchical respiratory neuron networks. Model. Meas. Control **48**(2), 47–63 (1995)
11. Khadartsev, A.A., Eskov, V.V., Pyatin, V.F., Filatov, M.A.: The use of tremorography for the assessment of motor functions. Biomed. Eng. **54**(6), 388–392 (2021)
12. Filatova, O.E., Galkin, V.A., Eskov, V.V., Filatov, M.A., Gavrilenko, T.V.: Warren weaver's complexity and fuzziness of Lotfi A. Zadeh leading to uncertainty in biosystem study. AIP Conf. Proc. **2467**, 060046 (2022). https://doi.org/10.1063/5.0092442
13. Gazya, G.V., Eskov, V.V., Filatov, M.A.: The State of the cardiovascular system under the action of industrial electromagnetic fields. Int. J. Biol. Biomed. Eng. **15**, 249–253 (2021). https://doi.org/10.46300/91011.2021.15.30

14. Filatova, D., Bashkatova, Y., Filatov, M.A., Ilyashenko, L.K.: Parameter evaluation of cardiovascular system in schoolchildren under the conditions of latitudinal displacement. Hum. Ecol. **4**, 30–35 (2018)
15. Filatov, M.A., Ilyashenko, L.K., Kolosova, A.I., Makeeva, S.V.: Stochastic and chaotic analysis of students' attention parameters of different ecological zones. Hum. Ecol. **7**, 11–16 (2019)
16. Kolosova, A.I., Filatov, M.A., Maistrenko, E.V., Ilyashenko, L.K.: An analysis of the attention indices in students from Surgut and Samara oblast from the standpoint of stochastics and chaos. Biophysics **64**(4), 827–832 (2019)
17. Khadartseva, K.A., Filatov, M.A., Melnikova, E.G.: The problem of homogenous sampling of cardiovascular system parameters among migrants in the Russian North. Hum. Ecol. **7**, 27–31 (2020)
18. Eskov, V.M.: Cyclic respiratory neuron network with sub cycles. Neural Netw. World **4**(4), 403–416 (1994)

Analysis of Multi-layer Information Flow in a Four-Layer Blockchain Model

Fernando Rebollar, Marco A. Ramos, J. R. Marcial-Romero, and J. A. Hernández-Servín(✉)

Facultad de ingeniería, Universidad Autónoma del Estado de México, Cerro de Coatepec S/N, Ciudad Universitaria, 50110 Toluca, Mexico
joseph_servin@uaemex.mx
https://www.uaemex.mx/

Abstract. The blockchain technology allows nowadays to decentralise information and the use of cryptographic techniques make this technology secure and reliable thus adding functionality for sharing information through smart contracts among other functionalities. One of the main disadvantages of this process is scalability in the speed of transactions making the systems slow for practical applications. In this paper, a multilayered approach is proposed in order to address the problem of improving speed in transactions; the method is presented by means of a case study analysis, by dividing the information in four-layers based on blockchain which allow users to participate and contribute to decentralise information thus making the service more reliable. The main contribution in this paper is the analysis that shows interaction for sharing information between different layers thus monitoring the behaviour when transactions are being made.

Keywords: Blockchain · decentralization · smart contracts · government services · digital services

1 Introduction

The offer and need of digital services, in the public and private sector has steadily grown [1]; being an example the governments of developed countries that have increased the number of digital services offered to their citizens. These services provide some advantages such as reduced costs, time and bureaucracy reduction [2]. These types of services operate in a centralised manner and the administrators depend on the same entity, in which trust plays a crucial role. To keep an optimal functionality of the system, is that honesty of a centralised entity must be present at all costs. Since, in real life this is normally not the case, it is necessary to decentralise the information and mechanisms that prevents a dishonest entity to be able to corrupt the system and make malicious changes.

Blockchain can provide decentralisation of technological services [3], while keeping information verified at each step of the process by providing full transparency, integrity and immutability of that information [4]. As an example, smart

R. Silhavy and P. Silhavy (Eds.): CSOC 2023, LNNS 723, pp. 338–348, 2023.
https://doi.org/10.1007/978-3-031-35317-8_31

contracts using block-chain [5], which one can loosely define as being a digital assets containing secure digital agreements that are important to the actors involved in the transactions. These one of the key features where cryptography techniques play an important role in the development of secure and reliable digital transactions. On the other hand, decentralisation becomes important since multiple computers (or nodes) store information at different locations thus using a consensus algorithm to keep information uncorrupted; making almost impossible to modify.

The advantages of blockchain had already been recognised by the government of different countries around the world. To some extent, some services are already in use such as money remittance in the United Arab Emirates to Nepal [6]. Another successful example is property registration in Rwanda which has been reported to avoid legal disputes among their citizens by keeping a clean track of the titles. This saves money, which is normally spend in legal disputes for rigged transactions in purchasing land [7]. Nonetheless, difficulties arise in block-chain technologies when the number of transactions increases per second in various digital services [8], thus the need for alternative improvements of the systems. The proposal to address these problems in this paper, is a multi-layer block-chain system that divides the information into four-layers [9]. This way, each layer specialises in a certain type of information. The model has the capacity to support several digital services, in order to decentralise and monitor the information handled by the system, to allow voluntary nodes to participate in order to validate blocks and help to keep secure the information. This article presents an analysis of a case study using the four-layers of the proposal. In Sect. 2, we describe a multi-layer blockchain for government services, using four-layers and how in this model is possible to decentralise information by levels, defining some properties and its operation by identifying the general architecture. In Sect. 3, a case study of a four-layer blockchain model is analysed, reviewing the information flow between the layers that make up the model and finally in Sect. 4 the conclusions are presented.

2 Four-layer Blockchain Model

The blocks that make up a blockchain contain different types of information. Examples of this information are hash values table, index tables, stored data (usually encrypted), transaction information, smart contracts as well as files, although files are not usually stored in blockchain, because the size of the blocks will increase the speed of transactions [10] by saturating the network with large files. The model separates the information typically contained in a block [9]. As a result, an analysis is performed on each piece of information contained within the block. The properties that have the highest priority are obtained. Based on these properties, the information placed in a layer is intended to guarantee the chosen properties. In similar models, it has been shown that separating the information into layers allows the optimisation of the blockchain-based service. The model, after analysing the key requirements of blockchain-based applications, typically uses four types of data: (1) index-keys to identify hashes, (2) transactions that are performed, (3) smart contracts to enforce autonomous execution and (4) log files that are evidence of application-related processes (when required). The four-layer model is described in detail as follows:

1-Index-Keys. This layer is dedicated to storing index tables containing the hash of all pieces of information within the layers. Therefore, it will be implemented using a public blockchain using the PoW (Proof of Work) consensus protocol [11] which corresponds to a public blockchain type. By only storing indexes, this layer will be the one with the smallest block size and the highest level of decentralisation. It is also possible for citizens to participate in the validation of these blocks, generating greater trust in the information stored. The blocks in this layer are interconnected in a staggered order with the other three layers.

2-Transactions. Dedicated to store transactions, i.e., it stores the information necessary for all transactions that take place. To improve the speed of transactions, this layer is less decentralised than layer 1 by implementing a private blockchain using the PBFT (Practical Byzantine Fault Tolerance) consensus protocol [12], which corresponds to a type of private blockchain. Layer 2 queries information from layer 1 to validate that transactions are only from previously registered and valid users.

3-Smart Contracts. Stores all information related to smart contracts. This layer also uses the PBFT consensus protocol, which corresponds to a type of private blockchain. Layer 3 interacts with layer 1 to validate users who can create smart contracts. It also interacts with layer 2, as smart contracts execute one or more transactions.

4-Files. Stores files in the formats required by the application. This layer uses the Interplanetary File System (IPFS), a P2P protocol and network that is used to store and share data in a distributed manner [13].

Depending on the application, a node can give access to a file by adding the content identifier in a transaction or in a smart contract. Layer 4 does not use a consensus protocol as it is not necessary.

2.1 Properties of the Four-Layer Model

The following properties describe the generation of the blocks and the connection between layers. In order to specify and explain the behaviour of the model, its attribute on each layer is structured in a unique way so that the desired results can be obtained later on. The transaction rate is the capacity to process different amounts of information in a given period of time (t); the greater the amount of information, the more difficult it is to maintain and reduce the time required for processing. That is why the transaction rate is affected by the size of the blocks, the level of decentralisation and speed of generation of the blocks. The following properties refer to each one of these.

To describe the transactions some notation is first introduced. Let L_j be the jth layer, where j is the number of the layers, and L_j consist of a set blocks indexed by x_{ij} where $i = 1, ..., n_j$ and n_j is the number of blocks within the layer j, since every layer has a different number of blocks that is changing from layer to layer in a descending way. That is, if n_j is the total number of blocks within the layer j which at the same time takes t_{ij} time to generate each block in any given layer j. Here, $t_j = \sum_i t_{ij}$ is the total time that it takes to generate all blocks within the j layer.

Remark 1 (Block generation speed). Depending on the digital service an appropriate block generation must be configured, however, the following must taken into account: In the layer $L_j = \{x_{1j}, x_{2_j}, ..., x_{n_j j}\}$ such that j is the layer number, so in the case of a four-layer system we have that $j = 1, 2, 3, 4$. The x_{1j} block is the first one of each layer and t_{ij} is the unit of time in which each x_{ij} is generated and it is satisfied that $t_{ij} \leq t_{i(j+1)}$. The above means that every block on each layer, with the same index takes less time to be generated that the block in the next layer with the same index. Therefore, a layer generates a larger amount of blocks in a given period of time compared to the next layer. If we denote by f a process applied to an specific block which outputs always a new block, thus we can represent this formally as $\forall x_{ij} : f(x_{ij}) = x_{(i+1)j}$ for $j = 1, 2, 3, 4$; That is, we must always generate a new block in each layer, so all four blockchain layers will keep generating new blocks whenever they are required. However, it is also true that there is a block at different layer that comes from previous layers with already generated blocks. If we denote this process by g we can formally expressed this as $\exists x_{ij} : g(x_{ij}) = x_{(i+1)j+1}$ for $j = 1, 2, 3$. Except for layer 4 which cannot unchained any new block since it is the end of process. The level of decentralisation refers to two aspects: the first one is the number of nodes that hold a complete copy of the information (architectural decentralisation), and the second one corresponds to the organisation that control the nodes (political decentralisation). Political decentralisation increases the integrity and security of information. This is reason why layer 1 is public and allows any anonymous node to contribute to the process; whereas layers 2, 3 and 4 are private with defined nodes.

Remark 2 (Level of decentralisation). Let K_{ij} be the number of nodes that hold a copy of the blockchain for each block i and layer $j = 1, 2, 3, 4$. The higher the number of nodes that hold a copy of a layer, the higher the level of decentralisation that the system complies with. This requirement is satisfied with $K_{ij} > K_{i(j+1)}$. The above guarantees that level of decentralisation keeps going down, where the first layer is the one with highest decentralisation compared to any other layer.

The block size has a significant influence on the fast distribution of the block to each node, therefore, depending on the digital service an appropriate block size must be configured. The following must be fulfilled in order to achieve that.

Remark 3 (Block Size). The block size of each layer must satisfy $|x_{ij}| < |x_{i(j+1)}|$ for $j = 1, 2, 3, 4$. Here $|\cdot|$ measures the size of the block in whatever units the size of a single character is measured which can be bytes. Thus, blocks in layer 1 are smaller compared to blocks in lower layers, in order to be inversely proportional to remark (2).

With above remarks in mind, the behaviour of block generation can accurately be monitored for step of the process or layer. The maximum block size can also be controlled to ensure an optimal decentralisation level. These properties all together have an impact on the transaction rate for each layer in the whole system.

By making all properties hold in the system, one can be sure that the 4-layer model is able to distribute blocks to the nodes efficiently. This also will keep a good transfer speed, as each layer has an adequate block size in accordance with their level of decentralisation. Also, a consensus algorithm, can be used to obtain the best possible performance when implemented in the system.

3 Analysis of a Case of Study

The four-layer model presented can be seen as a reference framework that can be used on different scenarios. It is possible to use as many layers as necessary, as an example four possible scenarios are described in [9]. It could be the case that service does not need to store files so a four-layer system is unnecessary (scenario 2). Another case is when a service lacks or does not provide smart contracts so a three layer system (scenario 3) may not be necessary. The other scenario is when a service does not need neither smart contracts nor to store files, so it does not require layer 3 and 4 (scenario 1).

Scenario number 4 takes into consideration all of its functionalities, that means all four-layers are required. Layer labelled 1 generates their respective cryptographic keys to subsequently generate transactions in layer 2. These, in turn, create smart contracts in layer 3, where files are generated and stored in layer 4. This creates one or more transactions at layer 2, and these transactions are store their respective indexes at layer labelled 1 in our case. We have chosen this scenario as it is the most complete scenario to perform the analysis, see the system sequence diagram, Fig. 1.

The following paragraph is a detailed description of diagram of Fig. 1 on each step of the process.

1. The administrator registers and configures the set of rules, policies, users and processes of the service at layer 1.
2. At layer 1 the necessary cryptographic keys are generated, such as hashes, public and private keys needed to ensure the integrity and security of the information.
3. The users authorised to perform signed transactions at layer 2, therefore one or more transactions are generated with the parameters of digital signatures, transaction identifiers and transaction data.
4. Transactions can generate smart contracts at layer 3 to automate future transactions in the services, which as parameters carry the smart contract identifier, the transaction information that generates the smart contract and the smart contract data.

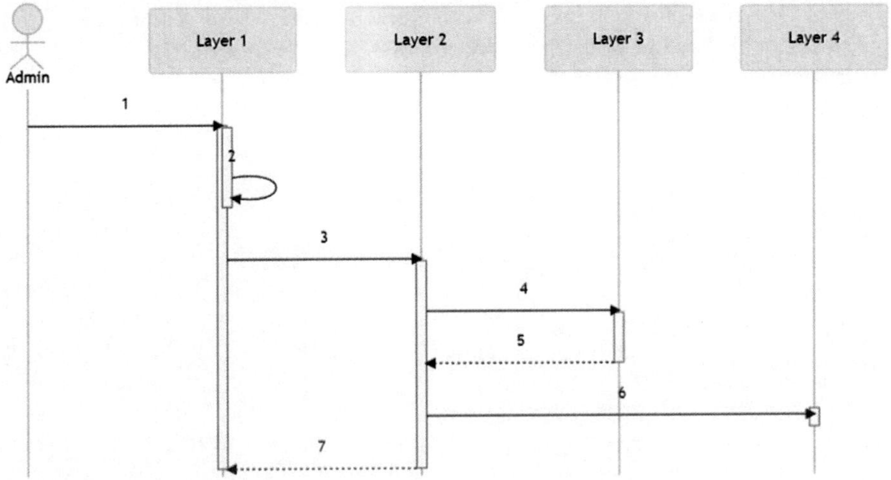

Fig. 1. When all 4 layers are used: keys and indexes, transactions, smartcontracts and files

5. Smart contracts can generate transactions in layer 2 when executed, for each transaction that is generated, the transaction identifier, the smart contract information that generates the transaction and the transaction data are stored.
6. Transactions can generate files in layer 4, which as parameters carry the file identifier, the information of the transaction that generates it and the file data.
7. Once the transactions are completed, the transaction identifiers, smart contract identifiers and file identifiers are returned to layer 1 to be saved.

3.1 Fines System

This section describes a possible case in actual production and what information flow might look like. The case of study for the four-layer model might be well applied to a system for charging fines when an administrative rule is violated such as traffic light rule violation. In governments, the fine system is administered by a specific police department. This department has control of the system which in theory a fine cannot be overruled unless is authorised by a law enforcement entity. It is also possible to keep track of revenues from collection of fines, and the information may also be readily available to other government entities.

The multi-layer blockchain control system for tracking fines is decentralised to one department. The patterns stored in the blockchain blocks cannot longer be modified or deleted. Theoretically, this is true but the system may have flaws and that depends on the level of security implemented using cryptography. Particularly, smart contracts have the property of automate the transaction of paying the fine, its status and can also be automatically linked to the collection department. One possible configuration is as shown in Fig. 2 for a four-layer model. The regional division of a country into states and municipalities is highlighted. Different countries could have more regional divisions where the model can be adapted.

Fig. 2. Possible nodes that should participate in each layer, the size of the figures represents the storage capacity required by the nodes and the number of figures in each layer is the recommended number of nodes to have.

In Fig. 2, layers 4, 3 and 2 as described in the model, they work as a private blockchain so the nodes are pre-established and identified; whereas layer 1, the nodes work as a public blockchain so they can be anonymous and not necessarily identified. This allows anonymous nodes to participate to further decentralise the information. The four-layer model describes, that it can work with few nodes in each layer, although the more nodes the more decentralised and secure the information is. Layer 4 is responsible for storing the documents evidence of the fine payment in the form vouchers and invoices. Layer 3 contains smart contracts for the payment of fines or cancellation of fines in case the fine is invalid due to legal issues. It could also contain smart contracts to link automatic collections as desired. Resources to storage requires less capacity and can be more decentralised than layer 4. Layer 2 is responsible for issuing the system's fines, which in this case can only be issued by a registered, authenticated and authorised agents. It is also responsible for fine payments confirmation. Layer 2 blocks require authorised nodes when is necessary to have a high transfer rate per second to handle the fines and payments. Layer 1 contains transaction, smart contract and document identifiers. It also contains information of the authorised agents to create the fines (names, electronic signatures, etc.). Layer 1 consists of blocks that any authorised or anonymous node can consult and verify the information, which in our case of study could be any citizen. They are the blocks that operate with the highest transparency to the public and are therefore easy to audit. Also voluntary nodes are allowed to contribute computing power to validate the blocks of this layer and contribute to make the information more reliable.

This case of study should be adapted taking into account differences between locations and policies where is to be implemented. But it could also unify fine systems at different locations, e.g. municipal, state and federal systems. The flow of information as shown in Fig. 3, where rules, policies, users, etc. of digital services are configured in layer one. The authorised officers involved to issue fines can be registered. Once they are registered, their cryptographic keys are generated and they are authorised to issue fines. The fines are registered in layer 2. In this case, smart contracts can also be generated in layer 3 every time a fine is issued. This allows automatisation to enforce a fine. In this case a fine can be paid or not in case the corresponding authority decides not to do so. The advantage of the smart contract is that automatically detects payments or revoked obligation to do, issued by the corresponding authority. After the occurrence of either event, a proof file is saved in layer 4 and the compliance transaction is also generated, which in turn saves the indexes in layer 1 where the corresponding evidence is stored.

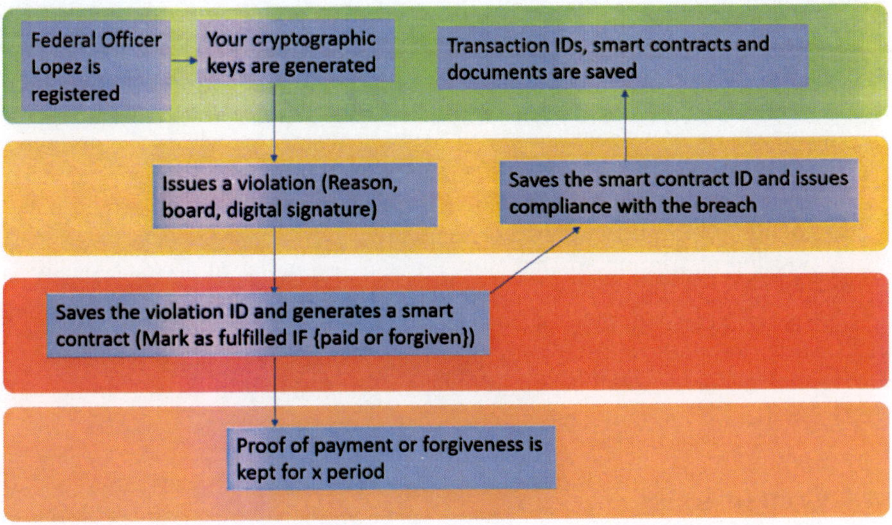

Fig. 3. Information flow using the 4 layers of the case study in a fine system.

In case someone wants to look up a particular violation in the past, it is very simple to track it down since the properties of a blockchain keeps the violation history. This is stored and protected against any future changes.

3.2 Some Considerations When Implementing the Case Study

The blockchain-based multi-layer proposal model could be used in some of the digital services of governments, to make the process of sharing information transparent. There are some considerations to take into account in order to increase effectiveness because there are risks that can affect the objective. Some of the considerations are as follows:

1. The hardware and software requirements are unrestricted for team participation. Teams with heterogeneous characteristics are eligible to participate. Any team that wants to participate can join the network. Computers with a higher capacity to calculate hashes per second will contribute more to the network.
2. External participants will not be able to make changes to the data contained in the blocks.
3. Volunteers participating in the network may or may not be rewarded. By contributing computational power, governments may set positions to reward participants or not, but it is desirable that there is no reward at all so as not to generate interests unrelated to the original purpose. However, it is not excluded that rewards for participation could be distributed.
4. Multiple individuals and non-governmental organisations may be interested in contributing even if there is no reward. Public universities are candidates for decentralised testing. They may even benefit from teaching blockchain courses, where a viable practice is to ask students to voluntarily participate in such a network to supplement their learning.
5. Participating teams could be restricted to being located within the same country. The use of VPNs would allow teams from other countries to participate in the system, but if they wanted to restrict this, a mechanism would have to be found and implemented to prevent this.
6. A possible restriction so that only citizens belonging to a country can participate is to administer an access control to the blockchain by means of an official ID of the country that implements it.

Depending on regional laws and different types of digital services to be added to the system, there may be extra considerations to analyse.

4 Conclusions

The case of study shown is a practical example of the four-layer blockchain-based model that could be implemented, the design of the model is specifically created to be used by governments for greater control, transparency and functionality in sharing information.

The case study shows how it divides and organises the different types of information needed in the four-layers with a focus on delivering an efficient digital service with a high transaction rate. This separation supports file storage without affecting transaction speed or smart contract execution times, as opposed to using a single layer.

For the case of study to work as described, it is assumed that the authorities in charge of avoiding and paying an infringement use digital signatures. This automate and make it possible to use layer 3 as smart contracts, but it is also possible to adapt it to scenarios 3, 2 or 1 described in [9], because of the flexibility of the four-layer model. This can be adapted to different requirements, unlike similar works that are associated to a single use case. Multiple services could coexist in a four-layer model implementation, so the infringement system could work and coexist with other services. The case study does not contemplate yet, non-human autonomous agents (IoT and smart city environment) to issue fines but it is compatible with that environment. It is possible to upgrade the systems and add these entities to issue fines. There are several challenges to improve inherited from blockchain [14–16], some important challenges are regarding the efficiency of transactions as they increase in number. The multi-layer model aims to decrease some of the disadvantages. Currently, the proposal uses elliptic curve cryptography and we are exploring to switch to hyperelliptic curve cryptography (or genus 2 elliptic curves) which could have a shorter key size compared to elliptic curve cryptography and could have better performance [17,18].

Finally, to increase trust and transparency, it is proposed to let people know the precise implementation of the system via a sharing model such as open source, available to the community to avoid any suspicious of hidden backdoor to corrupt the system. In this way, everyone can review and validate their implementation and operation, and be able to download and compare versions using hashing to validate that the code that was deployed is the same code that is running.

References

1. Lipschultz, J.: Free Expression in the Age of the Internet: Social and Legal Boundaries. Routledge (2020)
2. Gabison, G.: Policy considerations for the blockchain technology public and private applications. SMU Sci. Tech. L. Rev. **19**, 327–336 (2016)
3. Crosby, M., Pattanayak, P., Verma, S., Kalyanaraman, V., et al.: Blockchain technology: beyond bitcoin. Appl. Innov. **2**(6–10), 71 (2016)
4. Wüst, K., Gervais, A.: Do you need a blockchain?. In: 2018 Crypto Valley Conference on Blockchain Technology (CVCBT), pp. 45–54. IEEE (2018)
5. Buterin, V.: A next-generation smart contract and decentralized application platform (2014). https://github.com/ethereum/wiki/wiki/White-Paper
6. OECD: The development potential of remittances using blockchain technology in Nepal (2021). https://oecd.org/finance/blockchain/The-development-potential-of-remittances-using-blockchain-technology-in-Nepal.pdf
7. T. new times: How RWANDA uses blockchain technology to ease land transactions (2021). https://newtimes.co.rw/news/how-rwanda-uses-blockchain-technology-ease-land-transactions
8. Jun, M.: Blockchain government-a next form of infrastructure for the twenty-first century. J. Open Innov. Technol. Mark. Compl. **4**(1), 7 (2018)

9. Rebollar, F., Aldeco-Perez, R., Ramos, M.A.: Modeling a multi-layered blockchain framework for digital services that governments can implement. J. Intell. Fuzzy Syst. **42**(5), 4551–4562 (2022)
10. García-Morales, E.: Luces y sombras sobre el impacto del blockchain en la gestión de documentos. Anuario ThinkEPI **12**, 345–351 (2018)
11. Kaur, S., Chaturvedi, S., Sharma, A., Kar, J.: A research survey on applications of consensus protocols in blockchain. In: Security and Communication Networks, vol. 2021 (2021)
12. Cachin, C.: Architecture of the hyperledger blockchain fabric. Workshop Distrib. Cryptocurr. Consens. Led. **310**, 2–15 (2016)
13. Steichen, M., Fiz, B., Norvill, R., Shbair, W., State, R.: Blockchain-based, decentralized access control for IPFS. In: 2018 IEEE iThings and IEEE GreenCom and IEEE CPSCom and IEEE SmartData, pp. 1499–1506. IEEE (2018)
14. Peck, M.E.: Blockchain world-do you need a blockchain? This chart will tell you if the technology can solve your problem. IEEE Spectr. **54**(10), 38–60 (2017)
15. Yang, R., Yu, F.R., Si, P., Yang, Z., Zhang, Y.: Integrated blockchain and edge computing systems: a survey, some research issues and challenges. IEEE Commun. Surv. Tutor. **21**(2), 1508–1532 (2019)
16. Moin, S., Karim, A., Safdar, Z., Safdar, K., Ahmed, E., Imran, M.: Securing IoTs in distributed blockchain: analysis, requirements and open issues. Futur. Gener. Comput. Syst. **100**, 325–343 (2019)
17. Martínez, V.G., Hernández-Álvarez, L., Encinas, L.H.: Analysis of the cryptographic tools for blockchain and bitcoin. Mathematics. **8**(1) (2020)
18. Frey, G., Shaska, T.: Curves, Jacobians, and cryptography. In: Algebraic Curves and Their Applications, pp. 279–344. American Mathematical Society Providence (2019)

FIDO2 Passwordless Authentication for Remote Devices

Sumedh Ashish Dixit, Arnav Gupta, Ratnesh Jain, Rahul Joshi, Sudhanshu Gonge(✉), and Ketan Kotecha

Department of Computer Science and Engineering, Symbiosis Institute of Technology, Pune, India
{ashish.dixit.btech2019,arnav.gupta.btech2019, ratnesh.jain.btech2019,rahulj,sudhanshu.gonge}@sitpune.edu.in, head@scaai.siu.edu.in

Abstract. Today, passwords are a very popular mode of authentication in the modern web and applications. But as we have known for a long time, passwords are not secure. To improve its security, OTP and other mechanisms are used. One of the newer methods is the WebAuthn protocol, developed by the FIDO alliance. However, this is not widely applicable for all devices due to hardware limitations. This paper provides a methodology for remote WebAuthn authentication using QR and smartphone based authentication. With this method of passwordless authentication, we can ensure an additional layer of security that is easily accessible even to remote devices like TVs.

Keywords: Authentication · Security · WebAuthn · Password-less · Proxy

1 Introduction

1.1 Purpose

Authentication is the process of identifying the user who is requesting our services. A good authentication system aims to have features like accurately identifying the user using our service, verifying the user's authenticity and whether or not it is the user himself who is authorizing him/herself.

Passwords were not secure for a long time. They are hard to remember and quickly misplaced. In addition, they are the most common target for cybercriminals. So much so that 81% of all breaches are caused by weak or stolen passwords.

To validate a user's identity, passwordless authentication can be utilized. It employs more secure equivalents, such as registered cellphones, one-time passwords [OTPs], or biometrics - fingerprint and retina scans. The current industry is moving away from text-based passwords, using alternatives such as 2FA (2-factor authentication) with OTPs and MFA (Multi-Factor Authentication).

Passwordless authentication completely eliminates the need for a password, giving immediate security against brute force and phishing efforts, two of the most dangerous

R. Silhavy and P. Silhavy (Eds.): CSOC 2023, LNNS 723, pp. 349–362, 2023.
https://doi.org/10.1007/978-3-031-35317-8_32

and prevalent assaults. Employees are not allowed to provide their passwords, even if they get phishing emails or SMS. Because there are no more pairs of users and passwords to steal, brute force assaults are also no longer feasible. Furthermore, it is very challenging to forge fingerprints, push notifications, unique links, OTPs, or other forms of authentication.

Passwordless authentication mechanisms like OTPs and Magic Links are far superior to traditional password structures, but not a perfect solution. One-time passcodes (OTPs) are particularly vulnerable to hacker installation of malware, and Magic Links typically need the installation of a pre-authenticated application (for example, discords authentication) (Fig. 1).

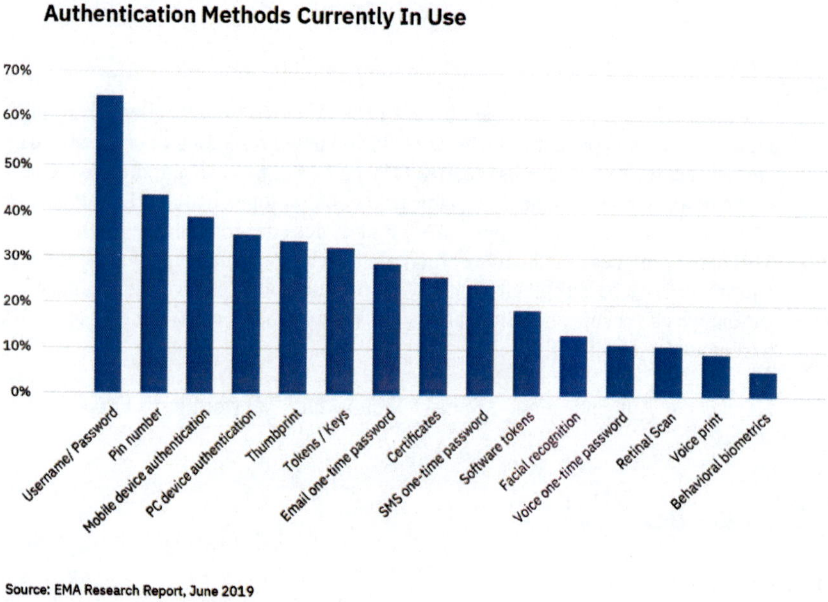

Fig. 1. Authentication Methods Currently in use.

2 Literature Survey

Michitomo Morii et al. (2017), in [3] explains that they have used packet capture to examine the data flow along the communication path and determine that there is no chance of an unauthorized login.

Sanam Ghorbani Lyastani et al. (2020), in [2] explains in order to learn more about how end users perceive, accept, and have concerns about passwordless authentication, this article presents the first extensive lab investigation of FIDO2 single-factor authentication.

Matthew Casey et al. (2020), in [4] explains how they have put out a password-less authentication architecture that is intended to be more deployable and usable. Their architecture, which facilitates registration and login to web services, is built on the WebAuthn standards.

Paul Wagner et al. (2021), in [1] explains how they have implemented passwordless authentication using Remote WebAuthN by separating the FIDO2 client across multiple systems.

Johannes Kunke et al. (2021), in [5] explains how user accounts can be retrieved if the security token is rendered permanently inaccessible will determine if FIDO2 is widely accepted. Because of this, the authors offer a heuristic assessment of 12 account recovery methods in terms of their suitability for FIDO2 passwordless authentication.

2.1 Existing Systems

The Existing methods of authentication include

Passwords (1FA). Passwords are the traditional way of authentication where a password is validated against the one stored in the database, to provide access.

OAuth (1FA). OAuth is a standard which allows users to login in into the given service without sharing their account passwords.

OTPs (2FA). OTPs are sent to the user through his/her registered email and/or mobile numbers. The user is expected to approve the login access from them. This adds a second layer of security.

Passwords (1FA). Passwords are the traditional way of authentication where a password is validated against the one stored in the database, to provide access.

In the limitations of these systems, Passwords, as we have already discussed, are prone to being breached. With OAuth, the task of authentication and identity management is delegated to some other party, but this does not solve the problem and can create dependency on other systems.

OTPs are a safe solution, they can prove that the login attempt using the password, but they are slow to use and depend on the SMS service, they too have some vulnerabilities that can be used by hackers (OTP reusability. OTP Brute Force. OTP Bypass).

3 Proposed Work

The project idea is to develop a platform that authenticates the users using the WebAuthn standard with biometrics. The solution also works on platforms that lack native biometric authentication hardware. This is a platform independent solution that increases the accessibility of biometric authentication with the WebAuthn standard. Seeing the existing solutions for secure authentication, there is a new technology that is being introduced. WebAuthn is a new web standard that enables platform dependent user authentication by biometrics (or security USB keys) in web applications. The WebAuthn standard is developed and maintained by the FIDO2 alliance, and currently implemented in major web browsers including chrome (and by extension all the chromium based browsers), Firefox and safari. WebAuthn provides a simplified API and gives a detailed specification on how to manage the authentication workflow.

Currently, the WebAuthn standard is limited to devices with in-built hardware authentication like fingerprint or other biometrics or using of security keys. Our solution is a proxy based remote WebAuthn authentication, that can be used on devices that lack the above functionality. The normal WebAuthn workflow can be seen in the Fig. 2.

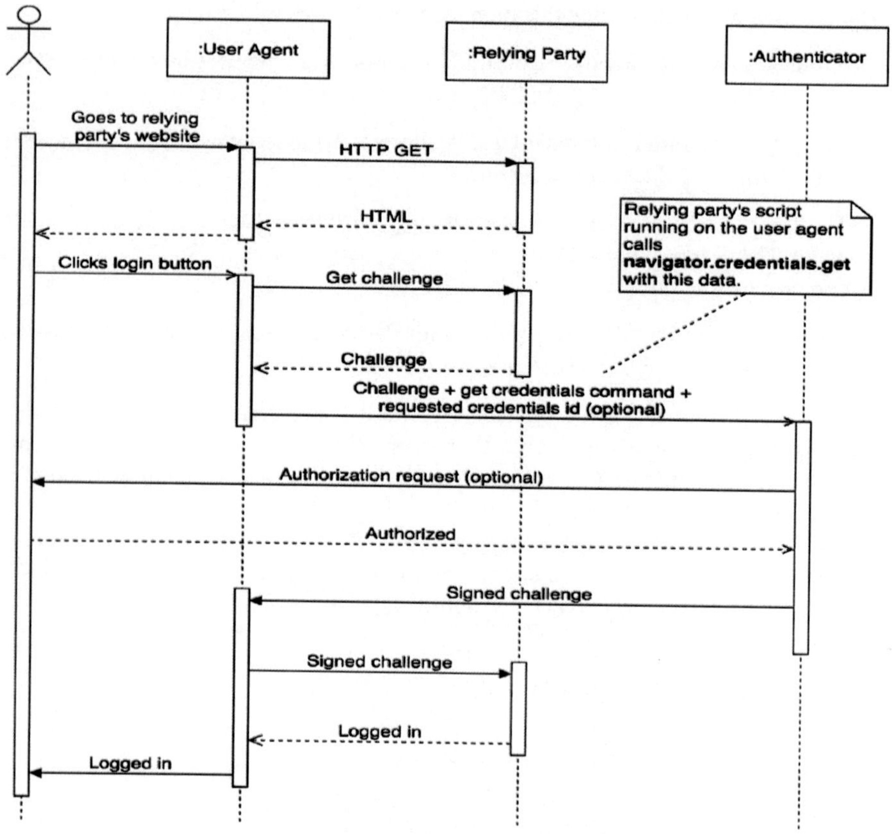

Fig. 2. WebAuthn Specification Workflow

4 System Design and Workflow

As seen in Fig. 3 and Fig. 4 authentication flow of our system is separated into two workflows.

1. The Registration Process
2. The Login Process

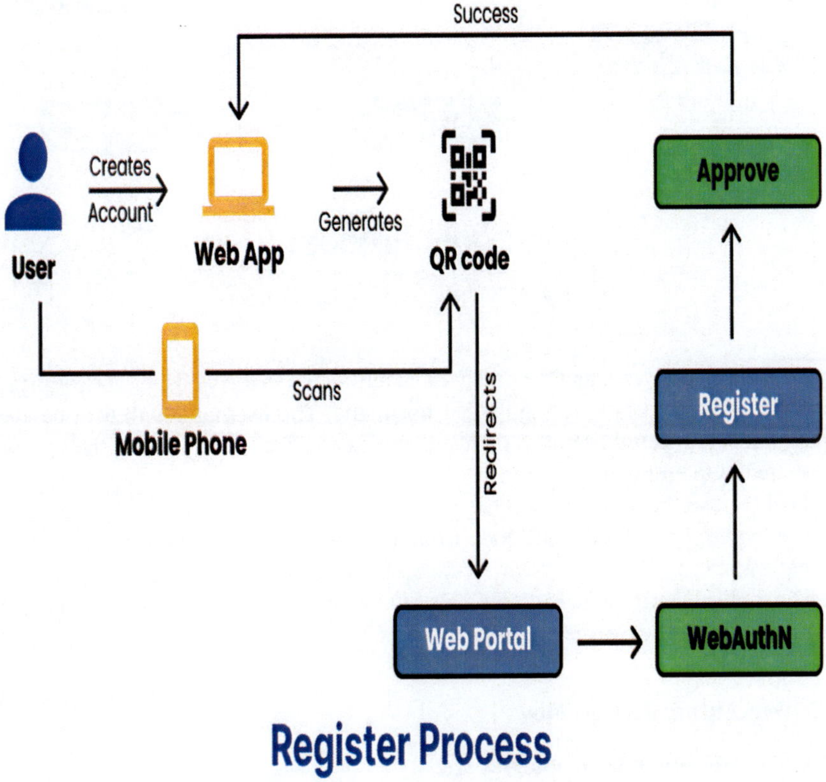

Fig. 3. Overview of Register Process

The workflow from Fig. 4. Explains the registration process is initiated by the user, when the user creates a new account. The user is asked to create an account with his username and password. Whenever the user chooses to log on to a particular site or service, the website will generate a unique QR code. To register a device user scans this QR with their phone and a biometric confirmation page will be popped up, authentication is complete this mobile is linked with the user account.

The login must be scanned by the user's registered mobile phone. Upon scanning the QR, the phone will redirect the user to a webpage where a biometric scan will be triggered. The phone will use its pre-registered biometrics to verify the user. For this process, the pre-registered user biometrics will be made use of. Upon approval, the user

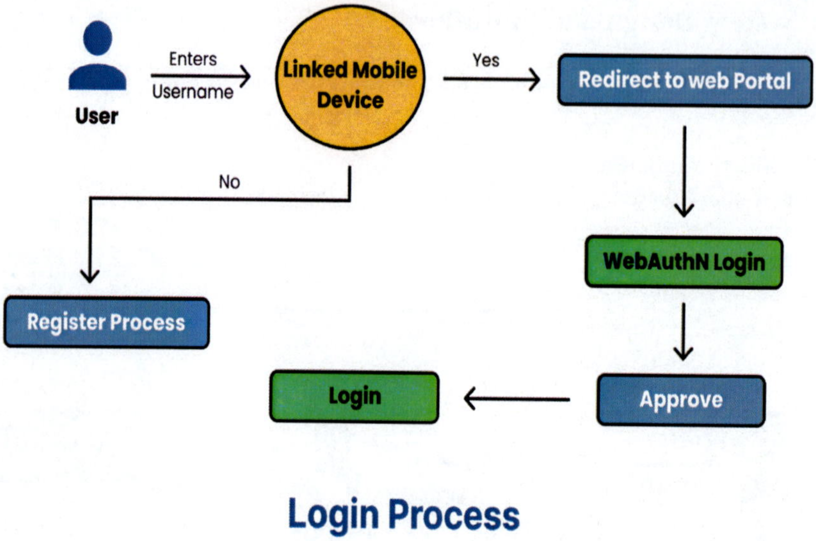

Fig. 4. Overview of Register Process

phone will be approved, and the user will be redirected to the website. The login process will start with the user entering his/her username. The username will then be checked whether it is registered or not. If the username is not already registered, the user will be redirected to the registration page.

Had the user been already registered, the website would show a QR code. The user will be required to scan the QR code from his mobile, which would redirect the user to the authentication web portal, where the biometric scan will be triggered. The user can then prove his identity with his biometrics. On approval by the mobile, the user will then be authenticated and verified.

4.1 Web Authentication Flow

Important WebAuthn terminologies are -

Relying Party. An programmed that performs user authentication. The application that will request that the user connect with the authenticator is known as the relying party (or rp).

Client/User-Agent. Software that uses the Web Authentication API is known as a Client (or WebAuthn client). The majority of the time, this will be a JavaScript application that is exposed to the API through a compliant web browser.

Authenticator. A device called an authenticator is used to generate and save user credentials. In a password-based authentication, the user's brain stores the credentials (the passwords). The credentials in a WebAuthn scenario are kept on a piece of hardware. It may also be integrated into a user agent or the operating system, such as Windows

Hello. An authenticator can verify user credentials using interfaces to fingerprint readers or facial recognition sensors.

OAuth (1FA). OAuth is a standard which allows users to login in into the given service without sharing their account passwords.

4.2 Web Authentication Flow

As shown in Fig. 5. The WebAuthn client receives a request from the Relying Party (RP), which includes numerous parameters such as the Relying Party's details, the authenticating user's ID, and a challenge that will be used to check the authenticator's response. Before allowing the user to register new WebAuthn credentials, the RP must first authenticate the user using a traditional method if they already have an account registered with the Relying Party.

The client forwards the request to an authenticator. A list will be shown to the user by the client if more than one authenticator is available. A crucial part of the client's job is to add details about the request's creation history to it.

The user gives permits the creation of new credentials in the selected authenticator for the specified Relying Party.

A new set of credentials, consisting of a pair of private and public cryptographic keys, is now created by the authenticator. Along with other information (most notably the origin of the request), the public key is included in the response, and the entire response is signed. The response is returned to the relying party by the authenticator. The client is once more employed as a proxy.

The authenticator's response is verified by the relying party. It can determine whether the response has been tampered with and whether the authenticator is genuine. The authenticator's origin is also confirmed by the relying party. It determines if the origin's value matches what it anticipates. This crucial component of the ceremony serves to guard against phishing attempts. A request sent from a fake website will have a different origin, and the relying party will thus reject it. When the relying party accepts the response as legitimate, it stores the user data along with the public key that the authenticator returned. The registration process is now finished.

4.3 For Verifying Stored Credentials

Once more, the relying party starts the ceremony by sending a GET request to the Web Authentication API. An option object that contains data identifying the relying party as well as additional fields is supplied by the relying party. The client contributes details about the request's origin, similar to the registration ceremony, so that they may be later confirmed to thwart phishing. The user is prompted by the authenticator to confirm their desire to authenticate with the requesting Relying Party. The authenticator replies with a response which is signed using the private key generated during the registration ceremony after receiving confirmation. The fundamental distinction between authentication and registration is that this response doesn't include any details about the previously generated public key.

After receiving the response, the relying party verifies it. In order to validate the response's signature and ultimately the authenticating user, it employs the public key it has stored. Once more, the Relying Party's crucial responsibility is to confirm the provenance of the response. To foil any phishing attempts, it should be consistent with an expected source.

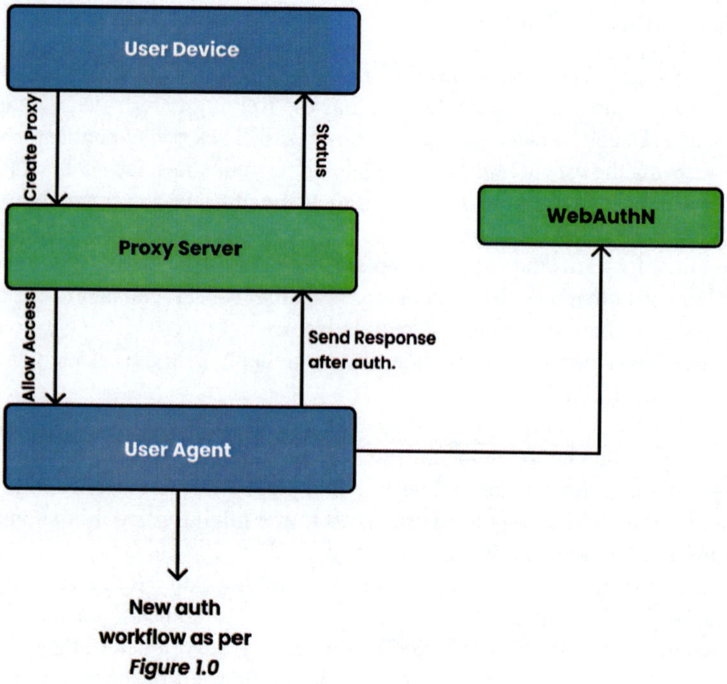

Fig. 5. WebAuthn Mechanism using Proxy

4.4 For Verifying Stored Credentials

Our webauthn workflow for registering can be rescued here, but here the relying party and the client are on different devices and possibly on separate networks. We introduce a proxy endpoint through which these two can communicate the necessary information securely, privately and reliably.

The user-agent initiates the communication by creating a proxy session, the server creates a proxy session and adds the client to it, this session is monitored and has assigned permission for every action taken through it. The user permission are capped to either registering or verifying in a single session. The relying party uses this proxy endpoint and starts the webauthn registering or verifying process in the device the endpoint is accessed on. Once the process is completed the state (success or failure) is relayed to the client.

5 Methodology

5.1 For Verifying Stored Credentials

We used Express JS for building our API (Application Programming Interface). Express is an easy-to-use and flexible Node.js web application framework that provides all the features needed to build online and mobile applications.

It makes it possible to quickly create Web applications using the Node framework. Some of the key components of the Express framework include the capacity to configure middleware's to respond to HTTP requests, the definition of a routing table that is used to carry out various HTTP based method operations and URL, and the capacity to dynamically render HTML pages by passing arguments to Templates.

The API consists of 3 major endpoints -

/auth endpoint. It consists of controllers that handle user registration, user login, session creation, and user logout. This endpoint deals with user sessions and the authorization permissions associated with each particular session. Security is a major concern here, all the inputs must be sanitized, and edge cases must be accounted for. SHA512 with salt algorithm is used to form hashing and verifying passwords.

Average Request Time: 9 ms.
(SHA-512 - (Secure Hash Algorithm 512) The Secure Hash Algorithm (SHA-512) may hash text of any length to produce a fixed-size string. Each output creates an SHA-512 length of 512 bits (64 bytes). This algorithm is often used for password hashing, email address hashing, and digital record verification.)

/webauthn endpoint. It is a collection of controllers that handle the WebAuthn integration of the platform. This endpoint is according to the specification of WebAuthn protocol and deals with the registration and verification of authenticators like PIN, USB key and biometrics (fingerprint etc.). Average Request Time: 12 ms.

/proxy endpoint. This is the most important part of the project. This endpoint deals with creating and verifying proxy sessions and handling data between them, for example between a laptop and a mobile phone for accessing and registering the devices as authenticators. As discussed in the workflow, all the logic of WebAuthn authentication using a remote device like a smartphone, is implemented here. Security and speed are of major importance in this endpoint. Average Request Time: 15 ms.

5.2 Frontend

For the frontend, we have made use of React.js along with Tailwind CSS. React.js is used to build modular user interface elements. The creation of modular user interfaces makes use of the React library. It encourages the creation of reusable user interface elements that show dynamic data. The V in MVC is typically represented with React. React separates the DOM from you, allowing for simpler development and faster performance.

In essence, Tailwind CSS is a utility-first CSS framework for quickly creating unique user experiences. Tailwind CSS is essentially a utility-first CSS framework for fast designing distinctive user experiences. It is a well-liked and effective CSS framework that provides you with all the tools necessary to create original designs without requiring you to put up with obnoxious, opinionated styles. The main advantage of Tailwind is that it doesn't impose design standards or dictate how your website should look; rather, it gives you the freedom to quickly combine small components to create a distinctive user experience. Simply put, Tailwind offers the results after processing a raw CSS file through a configuration file.

The whole frontend is divided into two main pages. The default route leads us to a login and registration page where the user is expected to enter his username and password (Fig. 6).

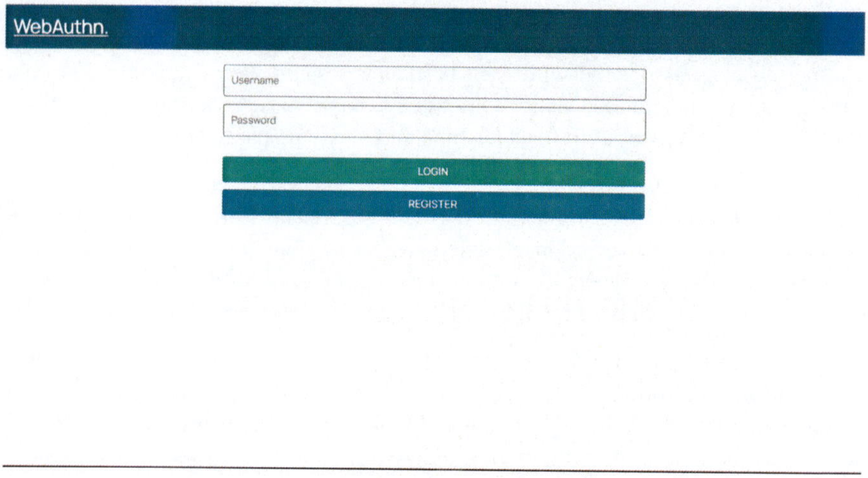

Fig. 6. Login Page

Figure 4 and Fig. 5 will be the landing page of the application where a new user has to register first by entering the username & password and after registering the user can login and add their passwordless credentials.

The '/webauthn' route leads to the WebAuthn authentication page. Here the user will be provided with two buttons namely, 'Add a credential' to add more credentials under the current user account and a 'Try ReAuth'. This button allows the user to trigger authentication. Upon login, the user details and credentials will be visible on the top.

The entire frontend is mobile view compatible as the users will be using it on mobile as well. The pages are completely mobile friendly and have responsive views for different screen sizes (Fig. 7).

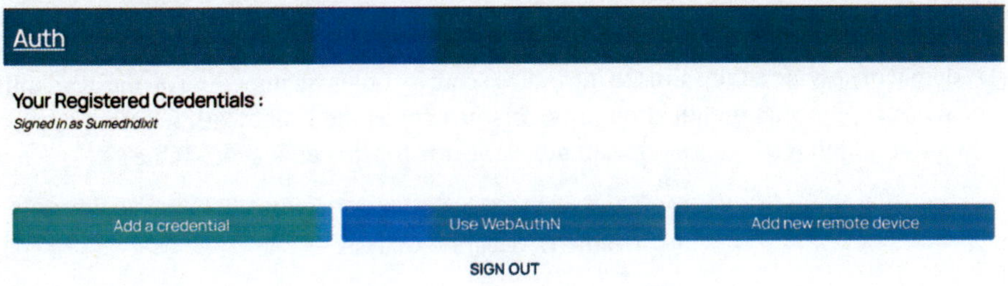

Fig. 7. Authentication Page

6 Results

In the current scenario there are many attacks, of which the following are quite famous-

6.1 Spoofing

Spoofing is a type of fraud in which the sender's identity is forged by someone or something in order to collect personal information, money, transmit malware, or steal data while posing as a reliable source, company, colleague, or other trusted contact.

6.2 Tampering

Data tampering is the intentional change of data via illegal means (e.g., deletion, manipulation, or editing). Data exists in two states: at rest and in transit. In any setting, data can be intercepted and manipulated. The core of digital communications is data transport.

6.3 Repudiation

The victim, in this particular scenario is also the attacker, denies being accountable for an actual authentication that was carried out.

6.4 Information Disclosure

Information Disclosure attacks are aimed at the confidentiality protection purpose. Possible weak spots in this scenario include authentication and request in transit. In general, it may be classified into two categories: whether the stolen data can be utilized directly for a subsequent assault or merely for identification inference.

6.5 Denial of Service

A denial of service attack aims to introduce security holes by impeding on the availability of services. Since authentication providers just act as the framework for other apps, the threat of an outage is not a relevant attack vector for this endeavor (Table 1).

Table 1. Security Analysis.

0: Very Low, 1: Low, 2: Normal, 3: High, 4: Very High

Types of Authentication	Spoofing	Tampering	Repudiation	Information Disclosure	Denial of Service
Passwords (Plain Text)	2	4	2	4	2
Password (Hashed)	2	1	2	0	2
OTPs	1	0	1	2	2
Oauth	1	1	1	3	1
WebAuthn	1	0	1	0	2

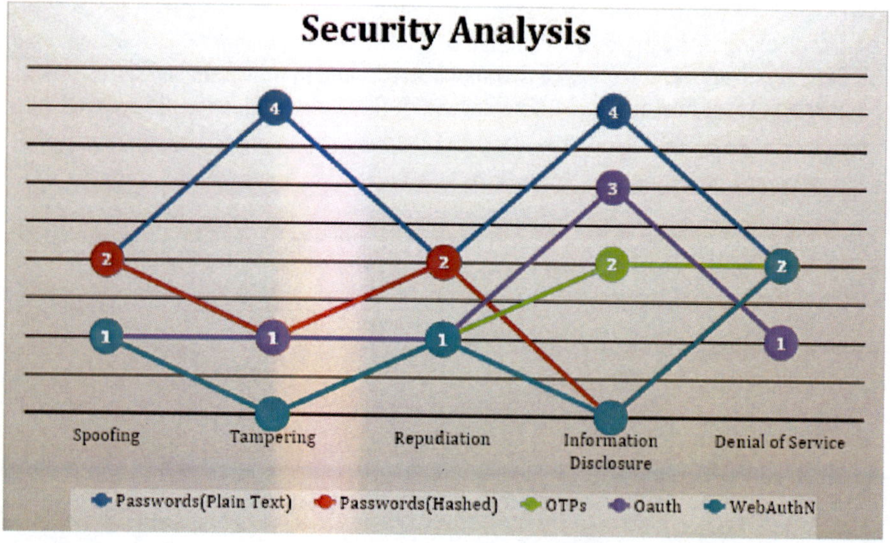

Fig. 8. Security Analysis Graph

7 Conclusion

As shown in Fig. 8. Various authentication mechanisms are compared with respect to various security attacks, we can conclude that password based mechanisms are less secure particularly in fields like tampering and information disclosure, when compared to passwordless mechanisms. We have studied and implemented Remote WebAuthn, a brand-new FIDO2-related authentication technique. On gadgets like Smart TVs that have constrained capabilities or interfaces and little software support, Remote WebAuthn focuses on delivering security similar to FIDO2. Given that most of the web has adopted better authentication methods and best practices like 2FA and OTPs, these have some limitations. The focus of our proposed solution was implementing a secure authentication method with focus on ease of use for the users and easy adaptability in various web applications. WebAuthn seems to be the future of this trend of secure passwordless web. The current implementation already proves to be practical, the future seems promising in terms of large scale adoption.

References

1. Wagner, P., Heid, K., Heider, J.: Remote WebAuthn: FIDO2 authentication for less accessible devices (2021). https://www.scitepress.org/Papers/2021/101927/101927.pdf. Acessed 24 Nov 2022
2. Lyastani, S.G., Schilling, M., Neumayr, M., Backes, M., Bugiel, S.: Is FIDO2 the kingslayer of user authentication? A comparative usability study of FIDO2 passwordless authentication (2020). https://ieeexplore.ieee.org/document/9152694
3. Morii, M., et al.: Research on integrated authentication using passwordless authentication method (2017). https://ieeexplore.ieee.org/document/8029677/
4. Casey, M., Manulis, M., Newton, C.J.P., Savage, R., Treharne, H.: An interoperable architecture for usable password-less authentication. In: Saracino, A., Mori, P. (eds.) ETAA 2020. LNCS, vol. 12515, pp. 16–32. Springer, Cham (2020). https://doi.org/10.1007/978-3-030-64455-0_2
5. Kunke, J., Wiefling, S., Ullmann, M., Iacono, L.L.: Evaluation of account recovery strategies with FIDO2-based passwordless authentication. https://arxiv.org/abs/2105.12477
6. WebAuthn explanation by auth0. https://auth0.com/blog/introduction-to-web-authentication/
7. Playground for webauthn testing. https://webauthn.io/
8. Bonneau, J., Herley, C., Van Oorschot, P.C., Stajano, F.: The quest to replace passwords: a framework for comparative evaluation of web authentication schemes. In: IEEE S&P (2012). https://ieeexplore.ieee.org/document/6234436
9. Bonneau, J., Preibusch, S.: The password thicket: Technical and market failures in human authentication on the web. In: WEIS (2010). https://dl.acm.org/doi/abs/10.1007/978-3-642-25867-1_8
10. FIDO-specs (2020). Fido specifications overview. https://www.fidoalliance.org/specifications
11. Firefox Browser side API docs. https://developer.mozilla.org/en-US/docs/Web/API/Web_Authentication_API
12. Lindemann, R.: Fido security reference (2018). https://fidoalliance.org/specs/fido-v2.0-rd-20180702/fido-security-ref-v2.0-rd-20180702.pdf
13. Reese, K., Smith, T., Dutson, J., Armknecht, J., Cameron, J., Seamons, K.: A usability study of five two-factor authentication methods. In: SOUPS (2019)

14. Analysis of web auth technology. https://sec.okta.com/articles/2020/04/webauthn-great-and-it-sucks
15. Microsoft: Passwords-less protection
16. Yubico: Yubico — YubiKey, strong two-factor authentication (2020). https://www.yubico.com/
17. World Economic Forum: Passwordless authentication: The next breakthrough in secure digital transformation (2020). http://www3.weforum.org/docs/WEF_Passwordless_Authentication.pdf
18. Microsoft Announces FIDO Support Coming to Windows 10
19. De Cristofaro, E., Du, H., Freudiger, J., Norcie, G.: A comparative usability study of two-factor authentication. In: Workshop on Usable Security and Privacy (USEC 2014) (2014)
20. The Path to Passwordless Authentication is Shorter Than We Thought (2019). https://securityintelligence.com/posts/the-path-to-passwordless-authentication-is-shorter-than-we-thought/
21. Webauthn workflow algorithm (2019). https://curity.io/resources/learn/webauthn-overview/

Using a Digital Educational Environment to Monitor Student Learning Results

Rukiya Deetjen-Ruiz[1], Irina Yarygina[2], Ikhfan Haris[3], Michael Sabugaa[4], Aleksey Losev[5](✉), and Valentina Everstova[6]

[1] Zayed University, Abu Dhabi, United Arab Emirates
officialrukiyadeetjen@gmail.com
[2] MGIMO University, Moscow, Russia
[3] Universitas Negeri Gorontalo, Gorontalo, Indonesia
[4] Agusan del Sur State College of Agriculture and Technology, Bunawan, Philippines
[5] Russian State Agrarian University - Moscow Timiryazev Agricultural Academy, Moscow, Russia
al.losev@rambler.ru
[6] North-Eastern Federal University, Yakutsk, Russia

Abstract. The use of digital educational environments contributes to improving the quality of education, forms information and communication competence of students and teachers, optimizes the process of control and evaluation of the results obtained, makes it automated and objective. Since testing is one of the most effective ways to check the level of competence of students, the problem of creating a high-quality digital test is still a burning issue of e-learning. One of the main methodological tasks of the teacher is to create a valid, objective, representative and reliable digital test that reflects the real knowledge of students. This article deals with digital testing in the digital educational environment. The quality of the test, mainly determined by the quality of the test questions. Test questions have to meet the requirements for their level of difficulty. In this paper, an analytical expression is used to determine the degree of difficulty of test questions. Using this expression allows the teacher to analyze the questions, identify and eliminate their imperfections in order to create an optimal digital test. #CSOC1120.

Keywords: E-Learning · Digital Learning Environment · Digital Testing · Test Question · Validity · Objectivity · Distractor

1 Introduction

Currently, the trend towards informatization is one of the leading directions of modern education [1–5]. Under the conditions of informatization, digital learning (e-learning) is actively introduced [6–11]. E-learning implies the use of information and communication technologies, digital educational resources, educational and methodical multimedia materials, digital libraries, and digital educational environments [12, 13]. Digital learning environments, in turn, constitute an integral part of the information and educational space [14].

R. Silhavy and P. Silhavy (Eds.): CSOC 2023, LNNS 723, pp. 363–370, 2023.
https://doi.org/10.1007/978-3-031-35317-8_33

Digital learning environment is a set of information and communication technologies that provide effective training and interaction between students and the teacher [15, 16].

Digital learning environments provide:

- independence from the place and time of learning (organization of distance, individual learning);
- access to curricula, working programs of disciplines, lectures;
- storage of materials for educational activities;
- implementation of modern technological and methodological approaches to the provision of information using interactive teaching methods;
- monitoring the results of student learning with the help of digital tests;
- project and research work of students using digital educational resources.

The use of digital learning environment has a number of benefits for all participants in the educational process.

Students. A flexible learning program allows building an individual educational trajectory [17]. Students always have access to modern educational resources, develop their ICT-competence and research skills.

Teachers. First of all, of course, this includes the facilitation of the learning process itself, teachers can save a significant part of the classroom time or time devoted to controlling learning results. Thus, there is an opportunity for more creative types of pedagogical activities. Formation of new opportunities of the educational process creates new conditions for the students' motivation in performing tasks. Use of digital educational environment provides visualization of complex learning processes.

University. The digital educational environment and the use of modern interactive technologies enhance the quality of education, and hence the status of the educational organization.

The most significant advantage of the digital educational environment is the recording of the educational process. Thanks to the digital educational environment, a teacher can monitor the knowledge, skills and abilities of students, and assess the level of their competences.

Digital testing within the digital educational environment optimizes the process of training and control of the obtained knowledge of students [18–20]. Automatic checking and evaluation of assignments saves time of the teacher and eliminates subjective factor. In addition, digital tests allow to create questions of different types, including those with visual support. Digital tests require no additional software and can be conducted as in-person classes in the classroom, and remotely. Digital testing on the basis of an digital educational environment makes it possible to prevent test takers from having complexes and fears associated with possible poor results. Each individual student has the opportunity to remotely see his score, the number of points, mistakes made.

The main methodological task in creating test items is to meet all the necessary criteria to ensure its quality: objectivity, validity, reliability, representativeness and scientificity.

Test validity is a characteristic that determines its suitability as a whole. The purpose of an digital test should correspond to its content, that is, it should reflect what it is supposed to assess. All quality criteria of digital tests are inextricably related to validity. High validity of a test determines a high level of objectivity, reliability and usefulness.

The criterion of objectivity is that test results should not depend on who is administering the test. Reliability is a measure of the degree of accuracy and consistency of an digital test.

The representativeness of the digital test reflects compliance with testing norms. Thanks to this criterion, students' results can be classified into levels of competence (high, average and low). The test should perform this function also outside the sample, i.e. it should be applied to other students of the educational institution.

In addition, testing norms and standards should be observed. A digital test should be accompanied by a specification that includes the name of the discipline, specialty, the protocol of the methodological commission, the author, the type of test (input, current or final control of learning outcomes), the principle of creating test questions (types of questions) and their allotted time, the scale and rules for evaluation.

Digital tests have a high objectivity, because the automatic verification of results and assessment provides intersubjectivity of digital testing. In addition, the teacher should be distinguished by a high level of professional competence.

Thus, it is necessary to meet a number of requirements for the creation of digital test questions:

- The content of tests should be aimed at checking the studied material of a particular course, it is unacceptable to use terminology that goes beyond the discipline.
- Exclude or minimize the probability of "blind guessing". Development of correct distractors and different types of questions, including input of own variant of the answer.

Distractor is a false version of the test question. The correct distractor should be a non-obvious, distracting alternative, a very similar option to the correct answer. A strong distractor is considered to be an answer option that, statistically, is chosen by many test takers. An inadequate distractor is one that is chosen by students with very high levels of preparation. The lowest selection threshold for a distractor is 5% of the total number of test takers, and up to 80% is the upper limit of distractor selection.

- Ensuring the stability of the digital educational environment and digital testing without technical failures, high speed of information processing and performing all actions on different devices.
- Design of digital testing. For example, the test question should be placed completely on the screen without scrolling.
- Test questions should be clearly stated in the form of short judgments or clear questions without cues, slang, quotes, repetitions, or double negations.
- It is necessary to use different types of questions to reduce student fatigue and increase the objectivity of testing (but the teacher should be guided by set goals, implemented with the help of an digital test).
- The use of graphic and multimedia components to create a test question or an answer to the question.
- Providing diversification of students into those who have mastered the material and those who have difficulties.

- Providing for the degree of fatigue and emotional load of students.

The following rules should be followed to increase the objectivity of the test results:

1. Random set and order of test questions.
2. Random order of answers in the question.
3. Time limit for one test question.
4. Time limit for testing as a whole on all questions.

After preparing the content of the digital test, as well as the design and technical support of digital testing comes the stage of experimental verification - validation. Only through empirical testing, the teacher can calculate the optimal average time of the student's response to the test question and the total time spent by the student on all the questions of the digital test. In addition, at this stage an analysis of test questions in terms of their level of difficulty is carried out.

Testing within the digital educational environment provides an opportunity to analyze the results of student work. Thus, the teacher needs to understand how difficult or easy this or that test question was.

2 Materials and Methods

During the research we used digital educational environment, including components that allow digital testing. Tests included questions:

- with the choice of one correct answer from the proposed list;
- with an expanded answer;
- pairwise matching;
- determining a true or false statement;
- with a numerical answer.

Test questions were characterized by a high degree of interactivity, visual support was used, some questions required a creative approach and activation of critical thinking.

In the experimental quality control of digital testing involved four groups of students with a total of 97 people. The digital testing was conducted in a computer lab in an digital learning environment for 60 min.

The teacher receives information about the results of digital testing in the section "Testing results" of the digital educational environment. According to the results of the first test too many students got good and excellent marks. Thus, it was necessary to analyze the test questions in terms of their ease or difficulty.

The following equation was used to determine the level of difficulty of a test question:

$$\frac{\sum_{i=1}^{n} x_i}{n}, \tag{1}$$

where n - the number of students who answered the question; x_i - the evaluation of the result of the answer to the question by the i-th student. The result of the calculation of formula (1) is a value in the range from 0 to 1. The closer the value is to one -

the easier the question is and the less information it carries about the differentiation of students. Conversely, the closer to zero - the more difficult the question and the more differentiating features the test question has.

Figure 1 presents a graphical representation of test difficulty. The meaning of the curves is as follows:

(1) too difficult test - high scores are few, low scores are many;
(2) normal test - optimal for statistical processing;
(3) too easy test - high results are few, high results are many;
(4) poor test - all students perform almost identically.

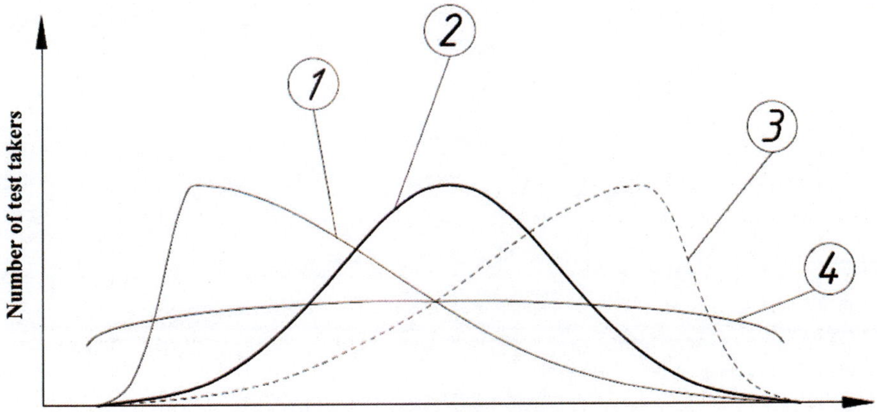

Fig. 1. Levels of difficulty of test questions.

Thus, to create a normal test that meets all necessary quality criteria, it is essential to exclude too easy questions (those that were solved by an absolute number of students) and too difficult ones (test questions that were not solved by any of the test takers). In case all students show almost the same results a digital test can be considered poor.

3 Results and Discussion

We conducted an digital test in which four groups, totaling 97 people, participated. The results of the test are presented in Fig. 2.

In Fig. 2 one can see that the curve line goes to the right, which indicates too easy a level of test, a large number of easy test questions. All easy questions (60% of the total number) for which the value of formula (1) is greater than 0.9 were removed or corrected. The content and types of some questions were adjusted, questions with only one correct answer choice were changed to questions with several answer choices to exclude “blind guessing”, and some questions with an extended answer were removed in order to increase objectivity.

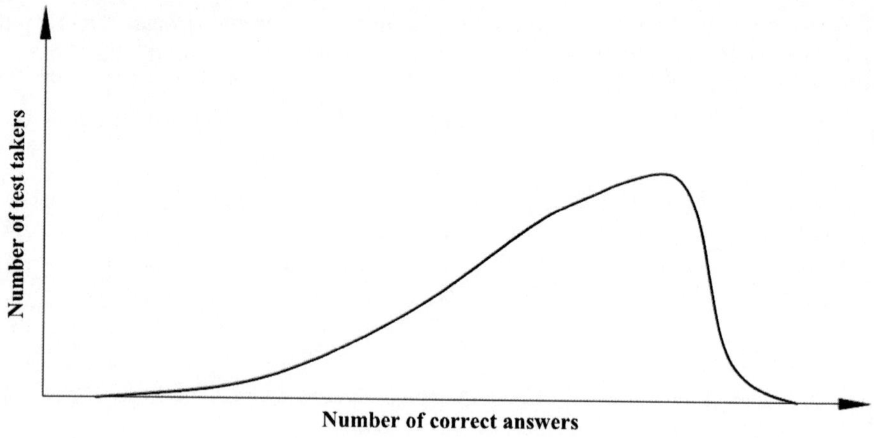

Fig. 2. Results of the first testing.

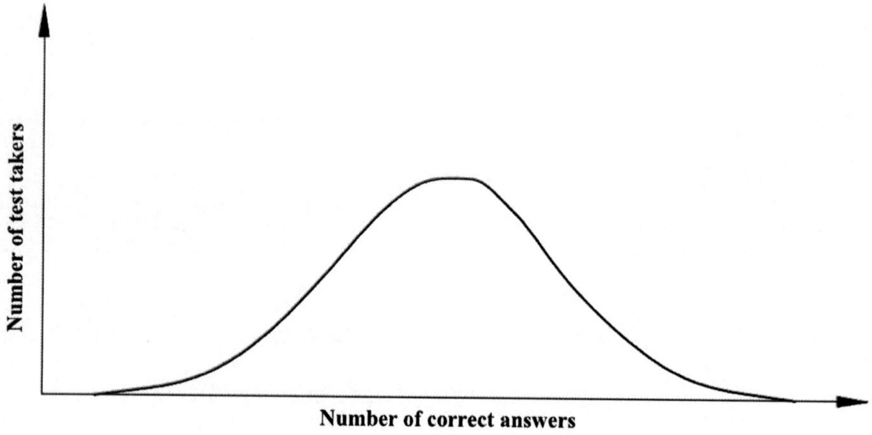

Fig. 3. Results of the second testing.

In order to confirm the quality of the digital test, taking into account the calculations and changes made, the other four groups, whose number exceeded 97 people, took part in the test. However, in order to compare the results of the previous and this test, 97 people were randomly selected for the second test. The testing conditions did not differ. The results of the second test are presented in Fig. 3.

So, the second digital test is optimal. The test results show that it has the greatest distinguishing features, and it is also optimal for statistical processing. The results also demonstrate that calculation of test question difficulty according to formula (1), on the basis of which too easy or too difficult questions can be excluded or corrected, allows objective, valid and reliable digital testing.

4 Conclusion

Digital learning environment is one of the most advanced methods of organizing modern education. Digital educational environment in the context of e-learning has many advantages, which distinguish it from traditional methods and forms of learning. Digital educational environments allow to organize and control the learning process remotely from anywhere in the world through access to the Internet. It is thanks to digital learning environments that a teacher can make a record of students' mastering of educational programs and save class time and expendable resources. Automatic assessment of test results contributes to the intersubjectivity of the test task, which increases its objectivity as a whole.

However, in order to create a valid, objective, reliable, and useful digital test, a number of requirements must be met, including those concerning the test questions themselves. Test questions should be clearly formulated and focused on checking the studied material. Different forms of questions, the use of visual support and well-chosen distractors increase the objectivity of the digital test. In addition, the test question itself should not have too easy or too hard a level of difficulty. That is why it is necessary to test questions, the results of which can determine the difficulty of the question and assess their suitability. Also, at this stage there is an analysis of the norms of test questions and questions that have significant weaknesses, the time parameters of the test questions are determined, random faults, "blind guesses" and inaccuracies are identified.

In order to determine how easy or difficult a particular test question is, a formula was used to determine the quality of the test and the adequacy of the results obtained by students. The present formula allows to determine whether a digital test is too easy, normal, too difficult. Through this formula, we were able to identify that the first digital test that was conducted contained questions that were too easy. Corrected test questions were offered to other groups of students. The analysis of their results, using the formula, showed that the second test was objective and adequate.

Thus, the use of digital educational environment is an effective means of controlling the results of student learning. Compliance with all the necessary rules and regulations for creating test questions guarantees its adequacy. And the application of the formula proposed in the paper allows to calculate the difficulty of test questions and ensure the high quality of digital testing, and hence the quality of students' acquired knowledge.

References

1. Malysheva, O., Tokareva, E., Orchakova, I., Smirnova, Y.: The effect of online learning in modern history education. Heliyon **8**(7), e09965 (2022). https://doi.org/10.1016/j.heliyon.2022.e09965
2. Meirbekov, A., Maslova, I., Gallyamova, Z.: Digital education tools for critical thinking development. Think. Ski. Creat. **44**, 101023 (2022). https://doi.org/10.1016/j.tsc.2022.101023
3. Pikhart, M.: Intelligent information processing for language education: the use of artificial intelligence in language learning apps. Procedia Comput. Sci. **176**, 1412–1419 (2020). https://doi.org/10.1016/j.procs.2020.09.151
4. Zhao, Y., Llorente, A.M.P., Gómez, M.C.S.: Digital competence in higher education research: a systematic literature review. Comput. Educ. **168**, 104212 (2021). https://doi.org/10.1016/j.compedu.2021.104212

5. Kovalev, I., Kuznetsov, A., Voroshilova, A.: Overview of the V International Scientific Workshop on Modeling, Information Processing and Computing - MIP: Computing-V 2022 (2022). Modern Innovations, Systems and Technologies **2**(2), 0215–0230. https://doi.org/10.47813/2782-2818-2022-2-2-0215-0230
6. Bossman, A., Agyei, S.K.: Technology and instructor dimensions, e-learning satisfaction, and academic performance of distance students in Ghana. Heliyon **8**(4), e09200 (2022). https://doi.org/10.1016/j.heliyon.2022.e09200
7. Hu, X., Zhang, J., He, S., Zhu, R., Shen, S., Liu, B.: E-learning intention of students with anxiety: evidence from the first wave of COVID-19 pandemic in China. J. Affect. Disord. **309**, 115–122 (2022). https://doi.org/10.1016/j.jad.2022.04.121
8. Maqbool, S., et al.: Student's perception of E-learning during COVID-19 pandemic and its positive and negative learning outcomes among medical students: a country-wise study conducted in Pakistan and Iran. Ann. Med. Surg. **82**, 104713 (2022). https://doi.org/10.1016/j.amsu.2022.104713
9. Nikolaeva, I., Sleptsov, Y., Gogoleva, I., Mirzagitova, A., Bystrova, N., Tsarev, R.: Statistical hypothesis testing as an instrument of pedagogical experiment. AIP Conf. Proc. **2647**, 020037 (2022). https://doi.org/10.1063/5.0104059
10. Stecuła, K., Wolniak, R.: Advantages and disadvantages of E-learning innovations during COVID-19 pandemic in higher education in Poland. J. Open Innov. Technol. Mark. Complex. **8**(3), 159 (2022). https://doi.org/10.3390/joitmc8030159
11. Lunev, D., Poletykin, S., Kudryavtsev, D.: Brain-computer interfaces: technology overview and modern solutions. Mod. Innov. Syst. Technol. **2**(3), 0117–0126 (2022). https://doi.org/10.47813/2782-2818-2022-2-3-0117-0126
12. Hassouni, B.E., et al.: Realization of an educational tool dedicated to teaching the fundamental principles of photovoltaic systems. J. Phys. Conf. Ser. **1399**(2), 022044 (2019). https://doi.org/10.1088/1742-6596/1399/2/022044
13. Maulana, F.I., Febriantono, M.A., Raharja, D.R.B., Khaeruddin, Herasmara, R.: Twenty years of e-learning in health science: a bibliometric. Procedia Comput. Sci. 216, 604–612 (2023). https://doi.org/10.1016/j.procs.2022.12.17
14. Zenyutkin, N., Kovalev, D., Tuev, E., Tueva, E.: On the ways of forming information structures for modeling objects, environments and processes. Mod. Innov. Syst. Technol. **1**(1), 10–22 (2021). https://doi.org/10.47813/2782-2818-2021-1-1-10-22
15. Dong, Y., Shao, B., Lou, B., Ni, C., Wu, X.: Status and development of online education platforms in the post-epidemic era. Procedia Comput. Sci. **202**, 55–60 (2022). https://doi.org/10.1016/j.procs.2022.04.008
16. Veeramanickam, M.R.M., Ramesh, P.: Analysis on quality of learning in e-Learning platforms. Adv. Eng. Softw. **172**, 103168 (2022). https://doi.org/10.1016/j.advengsoft.2022.103168
17. Tsarev, R.Y., et al.: An approach to developing adaptive electronic educational course. In: Silhavy, R. (ed.) CSOC 2019. AISC, vol. 986, pp. 332–341. Springer, Cham (2019). https://doi.org/10.1007/978-3-030-19813-8_34
18. Aljarbouh, A. et al.: Application of the K-medians clustering algorithm for test analysis in E-learning. In: Silhavy, R., Silhavy, P., Prokopova, Z. (eds.) CoMeSySo 2022. LNNS, vol. 596, pp. 249–256. Springer, Cham (2023). https://doi.org/10.1007/978-3-031-21435-6_21
19. Moravec, T., Štěpánek, P., Valenta, P.: The influence of using E-learning tools on the results of students at the tests. Procedia. Soc. Behav. Sci. **176**, 81–86 (2015). https://doi.org/10.1016/j.sbspro.2015.01.446
20. Akhmetjanov, M., Ruziev, P.: Fundamentals of modeling fire safety education. Informat. Econ. Manag. **1**(2), 0301–0308 (2022). https://doi.org/10.47813/2782-5280-2022-1-2-0301-0308

Social Cybernetics in Human Resource Management

Alla Subocheva[(✉)]

Financial University Under the Government of the Russian Federation, Moscow, Russian Federation
subotscheva2003@mail.ru

Abstract. This paper explores the application of social cybernetics to human resource management. It is shown that social cybernetics is a social direction and its application refers to the field of social management. The content of social cybernetics and its stratification are disclosed. The distinction between cybernetics of the first, second, third and fourth order is shown. The factors of development of social cybernetics are described. It is shown that models of social systems differ from models of living systems. The main systems studied in social cybernetics are distinguished: observable systems, observational systems, cognitive systems, self-developing systems. It is shown that fourth-order cybernetics studies the mechanisms of information interaction and tacit knowledge. Fourth order cybernetics studies rational choice and rational behavior. The concept of rationality in social cybernetics is disclosed and its difference from the concept of optimality is shown. The connection between heuristics and social cybernetics is shown. Reflexivity as a factor of fourth-order cybernetics is described. Stratification of human resource management from the standpoint of social cybernetics is described. The specificity of human resource management for cybernetics of different levels is shown. A number of cybernetic models related to human resources is described. A cybernetic analysis of the Big Five model is given. #CSOC1120.

Keywords: Social Management · Technical Cybernetics · Social Cybernetics · Social Cybernetics Stratification · Optimal Solution · Rational Solution · Cybernetic Models

1 Introduction

Initially, cybernetics was associated with the concept of N. Wiener [1] that it was the science of control in living systems and machines. It was regarded as a technical science. Later it was interpreted as a science about the regularities of obtaining, storing, transforming and transmitting information. The first problem arose from the fact that many kinds of information for which there are no general laws of control were identified. The Encyclopedia Britannica distinguishes 17 types of information. For example, information in mass media does not correspond to information in computer programs, information transmitted by insects is not identical to information in the theory of communication, information used by flocks of birds is not identical to information used by

R. Silhavy and P. Silhavy (Eds.): CSOC 2023, LNNS 723, pp. 371–377, 2023.
https://doi.org/10.1007/978-3-031-35317-8_34

schools of fish, information transmitted by bees is different from the information of ants in an ant colony. Human genetic information differs from the information transmitted by flowers to insects, information contained in musical works differs from the information in works of painting, and so on. Man perceives information through six channels of communication. The information he receives through his sense of smell differs from visual information. The diversity of information determined the diversity of its perception and excluded the general laws of its processing and management.

The second problem arose from the diversity of control objects and the corresponding diversity of control technologies. At one time there was a tendency to transfer the laws of living organisms to human society and social groups. But even this analogy proved to be inaccurate. This has led to the fact that cybernetics has come to be perceived as a principle that has different implementations in different conditions. Transferring the principles of cybernetics to the field of sociology led to the emergence of social cybernetics. Social cybernetics is neither technical cybernetics nor artificial intelligence theory. Social cybernetics is an attempt to describe the different variety of management technologies and management patterns in the social sphere.

2 Stratification of Social Cybernetics

Initially, social cybernetics developed from psychology with gradual use of systems theory in the social sphere. Social cybernetics [2–4] (SC) refers to the direction of sociology. SC in its simplest form is defined as "systems science in sociology and other social sciences". SC tried to use for research and management in society. As a generalization of experience, SC uses the methods of psychology, for example, behaviorism. Unlike pure psychology, SC uses logic, principles of systems theory [5] and social systems theory [6, 7]. The International Sociological Association, emphasizing the importance of SC, has created a special committee in this field, designated RC51. There is a journal Sociocybernetics and other journals on this topic. The concept of "Sociocybernetics" appeared, which is synonymous with the concept of "social cybernetics. Niklas Luhmann [8] applies the term "social" to any complex social system. As such, SC has for quite a long time considered the model of society as a model of a living system [9]. The idea of functional differentiation (Auguste Comte) led to the division of SC into four levels according to the subjects of research.

3 Factors in the Development of Social Cybernetics

The development of social cybernetics (SC) is based on different models. These are models of behavior of living systems, informative, descriptive and prescriptive models [10], and models in sociological research [11]. One of the peculiarities of SC is the formation of systems of decision-making rules, without bringing this system to the construction of an algorithm of the first kind. The object of SC research is the regularities of human society development. The development of computer technology does not directly affect the development of SC, but expands the possibility of modeling, which is used in SC.

For a long time the hypothesis was accepted that social systems behave similarly to biological systems [12]. Work by the Santa Fe Institute [13] revealed that some social systems, such as cities, do not behave completely like organisms. In the development of many social systems there is a theory of social cycles [14] that is not inherent in living systems. This circumstance has led to the rejection of the viewpoint that many SC models are full analogues of living systems.

In-depth studies of SC have revealed its stratification. This led to the emergence of "cybernetics of different orders". Modern SC includes cybernetics of the first, second, third and fourth orders. It became clear that there are different types of control objects, which are characterized by different complexity and require the use of different technologies to control level objects.

First-order cybernetics is associated with the study of "observable systems". Second-order cybernetics is associated with the study of "observing systems", that is, such systems that themselves investigate other systems. First-order SC deals with allopoietic machines [15]. Hereinafter a machine is a cybernetic or cognitive system. The notion of machine was introduced by Wiener [1]. Allopoietic systems cannot produce their components. These systems are used for their intended purpose and are therefore regarded as controlled machines or systems. This direction is close to technical cybernetics.

Research on autopoietic machines is performed in second-order cybernetics. Autopoietic systems can create their own components. They possess knowledge and can control other systems. Therefore autopoietic systems are regarded as controlling systems. Second-order cybernetics autopoietic or self-reproducing machines, second-order SC investigates the dependence of the knowledge obtained on the observer and on scientific theory.

Developments in epistemology and systems science [16] have identified third- and fourth-order cybernetics. Third-order cybernetics [17] studies "mutually observable systems," which are characterized by factors of observable and observing systems. Third order cybernetics deals with cognitive systems. It examines cognitive language as part of a communication system. This communication system creates a common field of interaction between living systems and organizational-technical systems [18]. An example of a cybernetic model of the third order is society. It uses some laws of biological development of living systems, but is not a living system itself [19]. A cognitive machine, according to Wiener's terminology, is an analogue of human decision-making. It serves as an object of research in cybernetics of the third order.

Fourth-order cybernetics [20] shifts research from the dynamics of large social systems to the dynamics of individual human systems or social groups. It studies the "systems of self-observation and self-awareness" that underlie human development. Fourth-order SC studies the mechanisms of information interaction that are contained in human intelligence. From the position of epistemology it is the study of implicit knowledge [21] and methods of transformation of implicit knowledge into explicit knowledge. Fourth-order SC studies rational choice and rational behavior. It also studies cognitive language, but not as the main object of research, but as a cognitive link between human systems, computer systems and intelligent systems. Fourth-order SC develops cognitive logic.

Rational choice and rationality of the cognitive system is the main result of fourth-order cybernetics research. Rationality is related to local discrete optimization. It eliminates global optimization on an infinite sample and replaces it with a rational solution on a bounded choice. Man is limited in time and in choice, so he cannot search infinitely long for the best solution. Hence, the object of research in fourth-order social cybernetics (SC-4) is the methodology of rationality selection. It is not based on analytics, but on the mechanism of heuristics and metaheuristics [22]. Heuristics are referred to empirical methods of problem solving and learning. As part of SC-4, heuristics are interpreted as a method of human interaction to achieve order in the environment.

The main product of fourth-order cybernetics is the rationality of the intelligent or cognitive machine. For the executive, it is rational decision-making. For human resources, it is the rational choice of personnel and their rational use, including training. Fourth-order cybernetics leads to the notion of continuing education.

For the performer, reflexivity in the framework of fourth-order cybernetics can be seen as a process of self-observation and the choice of a supervisor or place of work. Reflexivity leads to conditionally rational behavior. But the evaluation of rational behavior depends on the criterion of rationality. Because of this, rational behavior from the position of the supervisor is almost always different from the viewpoint of the executor.

Rationality as a coherence of thought is not completely related to the individual human mind. It depends on awareness and even on emotionality of perception. Example, the slogan "let me be worse off, but I will do it my way". The rationality of the individual differs from the rationality of the group and from the rationality of the leader. The task of fourth-order cybernetics is to reconcile the different rationalities. The different kinds of rationalities: personal, group, leader, and the reconciliation between them depend on the language of communication, reasoning, and intelligence.

The term "systems science" in social cybernetics stems from the fact that social cybernetics is not limited to theory, but includes applications, empirical research, methodology, axiology and epistemology. In general, "systems theory" and "cybernetics" are often interchangeable or found in combination. In some cases, but not always, they can be regarded as synonyms, although the two terms come from different traditions and are not used uniformly in different languages and national traditions.

From an interdisciplinary perspective, sociology is clearly at the center of interest in social cybernetics; at the same time, other social sciences such as psychology, anthropology, political science, economics influence social cybernetics.

4 Stratification of Human Resource Management from the Perspective of Social Cybernetics

Stratification of social cybernetics provides a basis for stratifying human resources. SC-1 is associated with the study of "observable systems". In terms of resources, these are the doers and low-skilled workers. In the military, they are soldiers who unconditionally obey orders. These human resources can be considered first-level resources. Management at the SC-1 level is directive management.

SC-2 is associated with the study of "observing systems". In the aspect of human resources, they are performers who have the ability to observe, remember and accumulate experience to improve the quality of work. They are also highly skilled workers with the ability to improve the quality of work independently. In the military, they are junior and mid-level officers who take initiative in the task at hand. These human resources can be considered second-tier resources. Directive management at the SC-1 level is not effective. Subsidiary management should be applied at this level [23].

SC-3 is associated with the study of "communication in social systems". In the aspect of human resources it is the study of the organization of groups or teams, in which due to the organization there is a synergistic effect in work. These organizational resources or human resources of the manager can be considered as resources of the second level. At this level, the organization of employee compatibility or complementarity plays an important role [24]. In essence, this is "soft" management.

SC-4 is associated with the study of "self-development in social systems". In the aspect of human resources it is the study of self-development of human resources. A study conducted at Cornell University [25] found that a significant number of hotel companies use innovative methods of human resource development at the SC-4 level. Several firms have developed a set of practices that include elements such as employee empowerment, interactive CD-ROM training, performance evaluations, and guest surveys.

Increased self-development and human resource levels are facilitated by machine learning of staff [26].

Analysis of human resource competence in small and medium-sized enterprises [27] has provided a new formulation of human resource competence. It is that competence is the ability to perform a job or task based on skills and knowledge supported by a work attitude. Competence shows certain knowledge, skills, and attitudes about a profession in certain skill characteristics. The level of competence depends on the level of self-development.

One of the joint tasks of SC-3 and SC-4 is to find a cybernetic model of work-life balance over time [28]. This work-life balance (WLB) model relies on four components: person-environment fit; psychological time as a contextual factor in perceived work-life balance; work-life balance, including the personal domain in addition to the work-family domain; and current, anticipated, and retrospective assessments of subject suitability.

The cybernetic model assumes that people will seek to restore balance when they notice a negative imbalance. Any cybernetic model is based on feedback. The ability to attract and retain talent and human resources is an important factor in the competitiveness of any organization. Going from the opposite direction, it is important to find out the causes of talent and resource leakage. These processes are interrelated and form another cybernetic model. This model relies on four components: market factors - a shortage of workers with the required qualifications; internal problems of job attractiveness (material - compensation, intangible - career prospects, flexible working hours, legal employment); internal factors of inefficient personnel management, including recruitment; psychophysiological factors of job satisfaction.

The Big Five model is reflected in cybernetic theory [29]. The model of the Big Five assumes that human personality includes five general and relatively independent traits (dispositions): extraversion, benevolence, conscientiousness, neuroticism, openness to experience. The connection between the components of the Big Five creates a cybernetic model.

Cybernetic Big Five theory (CB5T) attempts to explain personality in cybernetic terms, conceptualizing personality traits as manifestations of variations in the parameters of neural mechanisms that have evolved to facilitate cybernetic control. Cybernetic Big Five theory attempts to provide a comprehensive, synthetic and mechanistic explanatory model. It is therefore auxiliary to the aspect of human resources. The constructs describing psychological individual differences are divided into personality features, reflecting variability in the parameters of evolved cybernetic mechanisms, and characteristic adaptations, representing goals, interpretations, and strategies defined in relation to the specific life circumstances of the individual.

The theory identifies mechanisms in which variation is responsible for features at the top three levels of a hierarchical taxonomy of features, based on the Big Five, and describes the causal dynamics between features and characteristic adaptations. Finally, the theory links function and dysfunction in features and character adaptations to psychopathology and well-being. This theory is more psychological than cybernetic. But this once again emphasizes social cybernetics' focus on sociology rather than technology.

5 Conclusion

Social cybernetics is a social direction. Its main tasks are to "understand," "model," and "use". Social cybernetics, as the study of purposive systems, offers a basis for developing an integrative theory of personality. Personality traits describe individual differences in the frequency and intensity of various emotional, motivational, cognitive and behavioral states. All of these are part of social cybernetics. The development of computing technology and the application of complex computational models indirectly influences social cybernetics as an explanatory or supportive factor. Social cybernetics as social management sciences investigates decision support models in cognitive machines and mechanisms of rationality formation. A decision-making model in social cybernetics can be viewed from the perspective of a combination of information units [29]. Social cybernetics as an object metamodel absorbs and generates cybernetics of different levels, solving special problems. For human resource management, social cybernetics provides the tools to analyze and develop managerial decisions.

References

1. Wiener, N.: Cybernetics: or Control and Communication in the Animal and the Machine. Science, p. 344 (1983)
2. Scott, B.: The sociocybernetics of observation and reflexivity. Curr. Sociol. **67**(4), 495–510 (2019)
3. Hornung, B.R.: The challenges for sociocybernetics. Curr. Sociol. **67**(4), 511–526 (2019)
4. Kudzh, S.A., Tsvetkov, V.Y.: Development of social cybernetics. Slavic Forum **3**(29), 96–106 (2020)
5. Tsvetkov, V.Y.: Fundamentals of Complex Systems Theory: Textbook, p. 152. Lan' Publisher, St. Petersburg (2019)
6. Hoche, M.: Social Theory. Technical Report Stanford (2020)
7. Condorelli, R., et al.: Complex systems theory: some considerations for sociology. Open J. Appl. Sci. **6**(7), 422 (2016)

8. Luhmann, N.: Ystemtheorie, Evolutionstheorie und Kommunikations theorie. Soziologische Gids **22**(3), 154–168 (1975)
9. Tsvetkov, V.Y., Kozlov, A.V.: Use of models of living organisms for analysis of evolution of complex organizational and technical systems. Educ. Resour. Technol. **4**(29), 68–76 (2019)
10. Kozlov, A.V.: Descriptive and prescriptive logical models. Inf. Technol. Sci. Educ. Manag. **6**(10), 3–8 (2018)
11. Tsvetkov, V.Y.: Information modeling by social research. Int. J. Appl. Fundam. Res. **4**, 87–90 (2013)
12. Verwey, S., Davis, C.: Sociocybernetics and autopoiesis-new laws of organisational form? Communicare J. Commun. Sci. Southern Africa **30**, 1–26 (2011)
13. Bettencourt, L.M., Lobo, J., Helbing, D., Kühnert, C., West, G.B.: Growth, innovation, scaling and the pace of life in cities. http://www.pnas.org/cgi/content/abstract/0610172104v1
14. Korotaev, A.V., Malkov, A.S.: Introduction to Social Macrodynamics: Secular Cycles and Millennial Trends. Editorial URSS (2006)
15. Shabalina, O.A., et al.: Categorizations of cybernetics. In: VI International Conference Information Technologies in Science, Management, Social Sphere and Medicine, pp. 436–440. TPU, Tomsk (2018)
16. Almaguer-Kalixto, P.E., Giglietto, F.: Steering the world from where we are: an introduction to the sociocybernetics perspective. Curr. Sociol. **67**(4), 479–494 (2019)
17. Mancilla, R.G.: Introduction to sociocybernetics (part 1): third order cybernetics and a basic framework for society. J. Sociocybern. **9**(1/2) (2011)
18. Rosenberg, I.N., Solov'ev, I.V., Tsvetkov, V.Y., Yakunin, V.I.: Complex innovations in the management of complex organizational and technical systems. Feoria, Moscow, p. 248 (2010)
19. Mancilla, R.G.: Introduction to sociocybernetics (part 2): power, culture and institutions. J. Sociocybern. **10**(1/2) (2012)
20. Mancilla, R.G.: Introduction to sociocybernetics (part 3): fourth order cybernetics. J. Sociocybern. **11**(1/2) (2013)
21. Sigov, A.S., Tsvetkov, V.Y.: Tacit knowledge: oppositional logical analysis and typologization. Herald of the Russ. Acad. Sci. **85**(5), 429–433 (2015). https://doi.org/10.1134/S1019331615040073
22. Talbi, E.G.: Machine learning into metaheuristics: a survey and taxonomy. ACM Comput. Surv. (CSUR) **54**(6), 1–32 (2021)
23. Tsvetkov, V.Y., Kozlov, A.V.: Principles of subsidiary management. The State Counselor **4**(24), 20–28 (2018)
24. Tsvetkov, V.Y.: Complementary relationships. Sci. Bull. NSTU **2**(75), 101–114 (2019)
25. Enz, C.A., Siguaw, J.A.: Best practices in human resources. Cornell Hotel Restaurant Adm. Q. **41**(1), 48–61 (2000)
26. Garg, S., et al.: A review of machine learning applications in human resource management. Int. J. Product. Perform. Manag. **71**(5), 1590–1610 (2022)
27. Sedyastuti, K., et al.: Human resources competency at micro, small and medium enterprises in Palembang Songket industry. In: 2nd Annual Conference on Social Science and Humanities. ANCOSH 2020, pp. 248–251. Atlantis Press (2021)
28. Adkins, C.L., Premeaux, S.F.: A cybernetic model of work-life balance through time. Hum. Resour. Manag. Rev. **29**(4), 100680 (2019)
29. DeYoung, C.G.: Cybernetic big five theory. J. Res. Pers. **56**, 33–58 (2015)
30. Ozhereleva, T.A.: Systematics for information units. Eur. Res. **11/1**(86), 1894–1900 (2014)

A Critical Review of Success Models for Measuring Information System

Mkhonto Mkhonto(✉) and Tranos Zuva

Department of Information and Communication Technology, Vaal University of Technology, Gauteng, South Africa
acemkhonto@gmail.com, tranosz@vut.ac.za

Abstract. Information System (IS) success may be the most arguable and important dependent variable in the IS field. The purpose of the present study is to find the evolution of success models from 1992 when it was proposed by DeLone and McLean. Information quality and system quality are across all the models. This are the core variables for success models. Most of the variables that are added are environment dependent. In all the variables that was proposed by DeLone and McLean in 1992 has been modified depending on the environment that it has been used.

Keywords: Information System Success · Success Models

1 Introduction

Numerous models relating to information system (IS) success have been proposed along with those elements that influence information system success. Several IS studies used these models without any extension, whereas others combined various previous models or extended them to include new variables to develop these models and perform their studies. These models help to provide a comprehensive understanding of IS success by identifying and explaining the relationships among the most critical dimensions of success along which IS are commonly evaluated.

To measure the success of these various IS, organisations are moving beyond traditional financial measures, such as return on investment [1]. According to [2] did a research and pointed out that measuring success in general is a very difficult process as there are no structured frameworks that guide such process. This is supported by [3] who emphasized that the IS researchers have addressed different aspects when evaluating information systems success models.

2 Literature Review

Defining Information Systems Success
Information systems (IS) success is defined as the measure of the degree to which the

R. Silhavy and P. Silhavy (Eds.): CSOC 2023, LNNS 723, pp. 378–388, 2023.
https://doi.org/10.1007/978-3-031-35317-8_35

person evaluating the system believes that the stakeholder is better off [4]. According to [5], Information systems success can be defined as the new system that is accepted in the work place as the way of doing the tasks that the system has been designed to carry out or is invoked as a support tool on a regular basis. [6] further highlighted that in most cases, determination of IS success has been done through evaluation of user satisfaction by way of using surveys.

3 The Problem of Measuring Success of Information System

Research in measuring information system success and investment has been on-going since the DeLone and McLean developed an IS success model in 1992. However, there is no standard model that is used regularly to measure IS success in different business domains since researchers addressed various aspects of measurement. The problem of measuring success of information systems is recognized by many researchers as a difficult concept to define. [7] discusses the concept of IS success and define a legitimate measure that can clearly differentiate the relative levels of success in the firms studied. They define the effectiveness of an information system as a common description of the success of an information system, which can be described as the degree of contribution of the information system to the achievement of organisational goals, that is, its impact on organisational performance. However, there are many controversies about the impact of information systems on organisational performance. Stakeholder expectations can vary widely and are not easy to assess accurately. An information system is an organisational resource that serves certain stakeholders but no other stakeholders. Understand IS problems and find solutions to them [8].

4 Information System Success Models

Information system (IS) success models have been used over the years by various researchers to measure the complex dependent variable in IS research. Information system success is one of the most researched topics in IS literature [9]. Measuring IS success is believed to be a critical issue in the IS field. Several studies have been conducted and considerable attention was paid to this issue; due to the amounts of money, time and efforts spent on IS projects. These models have been developed over the years and resulted from the extension of each other. The next section provides an overview of the information system success models that have been proposed along with those factors that influence IS success. The purpose of this section is to review IS success models that contributed to our understanding of IS.

4.1 DeLone and McLean IS Success Model

The initial model (DeLone and McLean's information systems success model) was published in 1992 and was developed because of a collection of previous studies conducted on information systems success. It is one of the most popular models for measuring information systems success within organisations. It was developed in the field of information systems. This model is deemed to be a comprehensive model because it does not depend on a single variable; also it is an attempt to create an integrated framework to measure information systems success.

Delone and McLean (D and M model) developed a model, that measured six dimensions of IS success as system quality, information quality, use, user satisfaction, individual impact, and organizational impact. These six variables are not independent success measures but are interdependent variables. This model's independent variables are system quality, information quality, and service quality. Information quality is defined as "the desirable characteristics of the system outputs' [10] while system quality is defined in terms of the system's reliability, ease of use, and response time [11]. Information quality measures the semantic dimension of information and system quality measures technical success. The independent variable affects the intention to use and user satisfaction. Use and user satisfaction assess the overall system effectiveness. The diagrammatic representation of the model is given in Fig. 1.

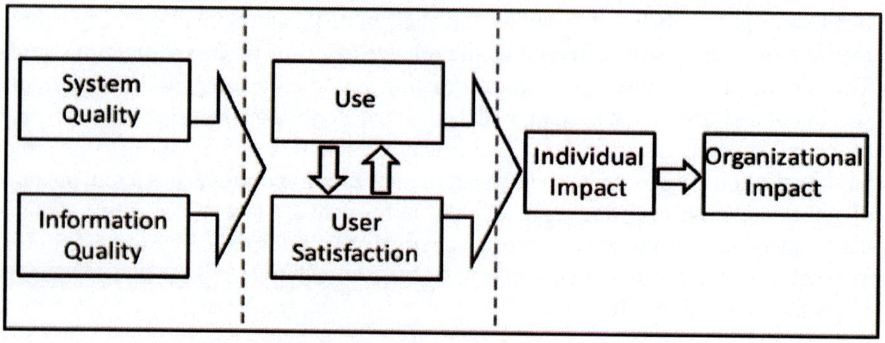

Fig. 1. DeLone and McLean's IS Success Model [3].

4.2 Seddon's is Success Model

Seddon's IS success model [12] is one of the first researchers to critique Delone and McLean IS success model and propose an alternative model. The model was developed by [12]. It was developed in the field of information systems. It consists of the following categories of measures of IS success: system quality, information quality, perceived usefulness, user satisfaction and IS use. [12] views IS use to be a behaviour and not a measure of IS success. This definition of IS use means it is seen as an outcome of IS success rather than part of success. The IS conceptual model is presented in Fig. 2 below.

4.3 System Success Model

Another model is the system success model (depicted in Fig. 3 below) proposed by [13]. This model has been built on established IS success models including the well-known D&M IS success (1992) model. It was developed by [13]. It was developed in the field of information systems. The model consists of environmental variables related to the organisation, users, IS operation, IS development, IS and external environment.

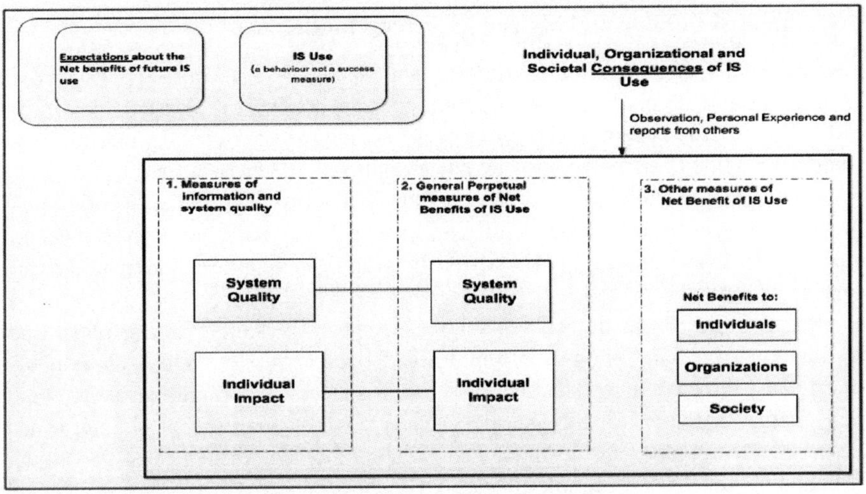

Fig. 2. Seddon's IS Success Model [12].

These variables have a direct influence on the use, satisfaction of the individual and the organisational impact of IS. The 3-D model of IS success consists of five environmental variables that interact with IS. Environmental variables and the D&M success model variables are combined to form System Success Model.

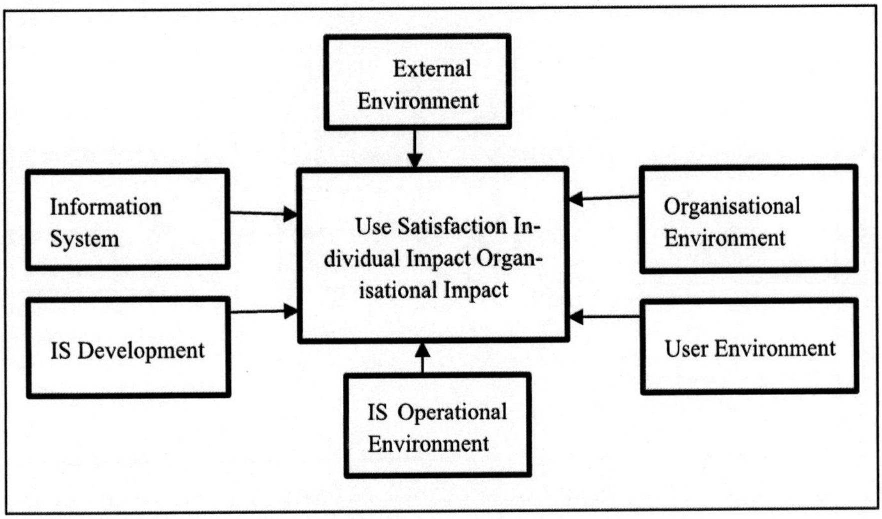

Fig. 3. System success model [13].

4.4 Updated DeLone and Mclean IS Success Model

A decade later after developing the original model of IS success, [14] proposed an updated model to extend, validate and strengthen the original one. Based on prior studies, [14] propose an updated model of IS success by adding a service quality measure as a new dimension of the IS success model and by grouping all the impact measures into a single impact or benefit category called net benefit [15]. This model provides a comprehensive understanding of IS success by identifying and explaining the relationships of six critical variables for IS success [16]. [14] updated their model to include an independent variable service quality.

Based on the suggestions from some researchers, the Author of this paper updated IS Success Model and name it reformulated IS success model. Which consists of nine interrelated dimensions of information systems success: Information quality, Technology quality, Technical quality, service quality, Output quality, Use, Intention to use, User satisfaction, and Net benefits. The "System quality" variable was substituted by "Technology quality". Technology quality is defined in terms of the system's reliability, ease of use, functionality and response time. The new "Technical quality", and "output quality" variables were added to the reformulated IS success model in Fig. 4. The Technical quality is used to check if the technology is performing according to specifications whereas the output quality is the considerations of what tasks the technology or a system is capable of performing.

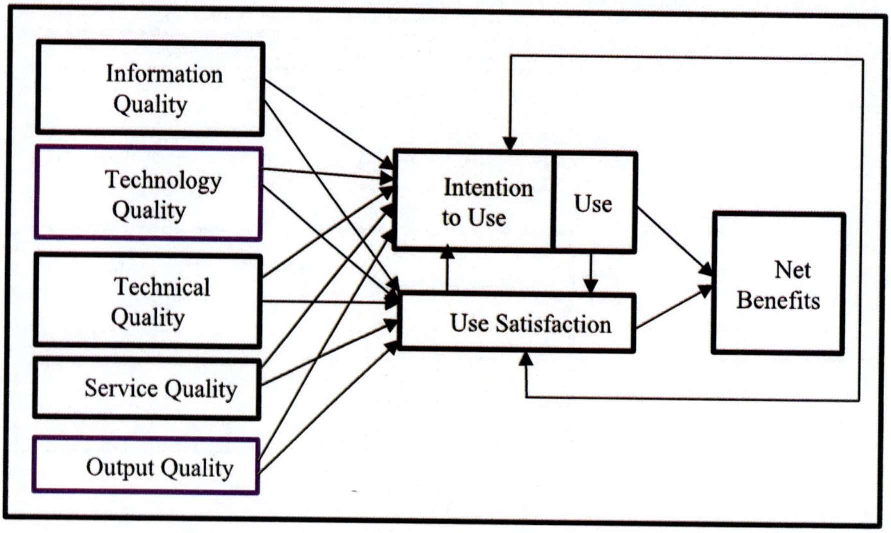

Fig. 4. The Reformulated IS success model (Adapted from Delone and Mclean). [Author]

4.5 Elpez and Fink's IS Success Model

[17] proposed another model for evaluation of IS success in the public sector. It was developed by [17]. It was developed in the field of information systems. According to

[17], evaluation of IS success has mainly been performed in the private sector while IS success in the public sector has been neglected. When designing an IS success model for the public sector, researchers need to take into account public sector characteristics like expenditure control, accountability and long term perspective [17].

[17]) identified the following variables of IS success in order of importance: meeting user requirements, system usability and performance, information quality, use, user acceptance and IS ownership and interaction with IT infrastructure. Public sector characteristics include expenditure control, accountability and long term perspective. The model as presented in Fig. 5 below, aligns IS success variables with public sector characteristics.

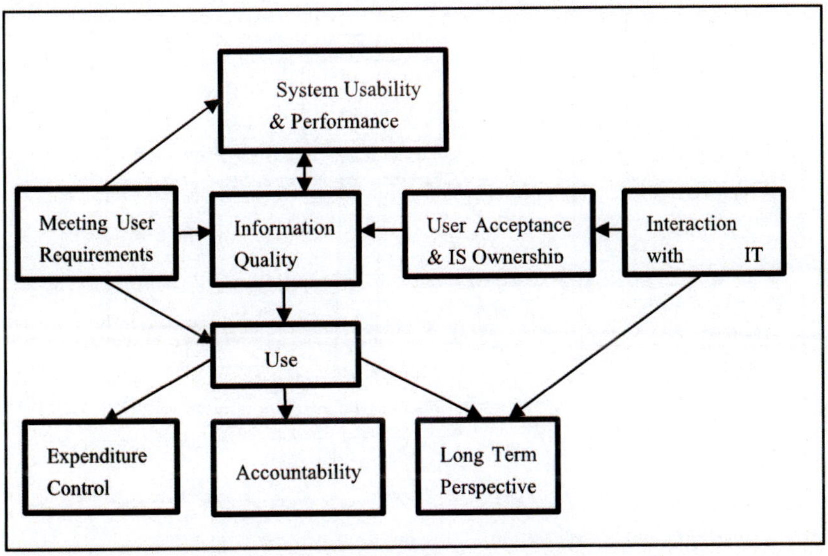

Fig. 5. Elpez and Fink's IS Success Model [17].

4.6 Information Systems Acceptance Model

Another model, Information System Acceptance (ISA) model, was proposed by [18]. The ISA model was developed based on well- established IS frameworks and IS success models namely: diffusion of innovation, information system success and information system implementation. According to [18], by combining these frameworks it enables ISA to provide richer issues to consider when evaluating IS. By evaluation of the system in terms of technology characteristics, management actions and service quality, the ISA model can be used to identify potential challenges to IS success.

The ISA model, as presented in Fig. 6 below, can be interpreted as follows; a system possesses certain technology characteristics, namely, relative advantage, compatibility, simplicity and trialability. Users will have expectations of using the system that will have an influence on their intention to use the system. If the expectations are positive, use of the system will increase. Once software is used over a period of time, users will develop opinions of using the system based on their experiences [18].

Management actions have an impact on "intention to use", for example not sharing the motivation and objectives of the system may potentially reduce the intention to use the system. Rewards, managing expectations and leading by example may positively impact "Intention to Use." Resources in the form of funds and personnel will help in the implementation process. All components of management actions i.e. communication, rewards, resourcing and implementation of management will also influence "User Satisfaction." Service quality whether provided internally or externally will influence "Intention to Use" and "User Satisfaction."

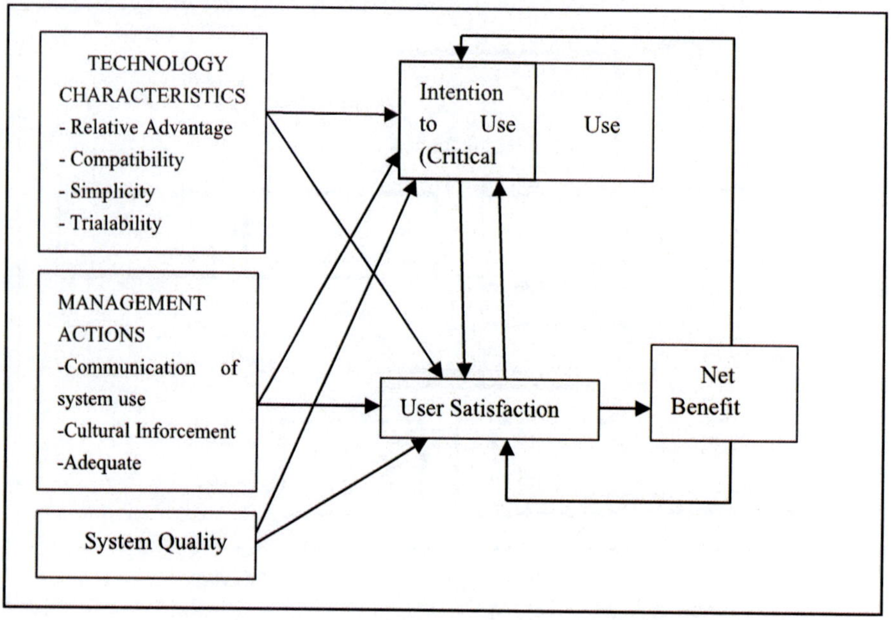

Fig. 6. Information systems acceptance model [18].

4.7 Gable et al.'s (IS-Impact) Measurement Model

Another model to measure IS success or impact is the IS-Impact Measurement Model introduced by [19]. It was developed in the field of information systems. [19] define the IS-impact of an Information System (IS) as *"a measure at a point in time, of the stream of net benefits from the IS, to date and anticipated, as perceived by all key-user groups"*.

The IS-Impact model, which is based in DeLone and McLean's work, overcomes many concerns with past IS Success models. [20] pointed out that the IS-Impact Model deviates from the traditional DeLone and McLean model in the following ways: (a) it depicts a measurement model and does not purport a causal/process model of success, (b) it omits the use construct, (c) satisfaction is treated as an overall measure of success, rather than as a construct of success, (d) new measures were added to reflect the contemporary IS context and organisational characteristics, and (e) it includes additional measures to probe a more holistic organisational impacts construct.

The IS-Impact model has been extensively validated statistically and uses mainly perceptual measures. According to [19], the model was developed in two phases: the exploratory phase and the confirmatory phase. Two surveys were conducted in the exploratory phase where the purpose of the first one is identify success measures and the purpose of the second is to test what is called a priory model. In the confirmatory phase, the model was tested for reliability and validity using different data set. The IS-Impact Measurement Model is presented in Fig. 7 below (Table 1).

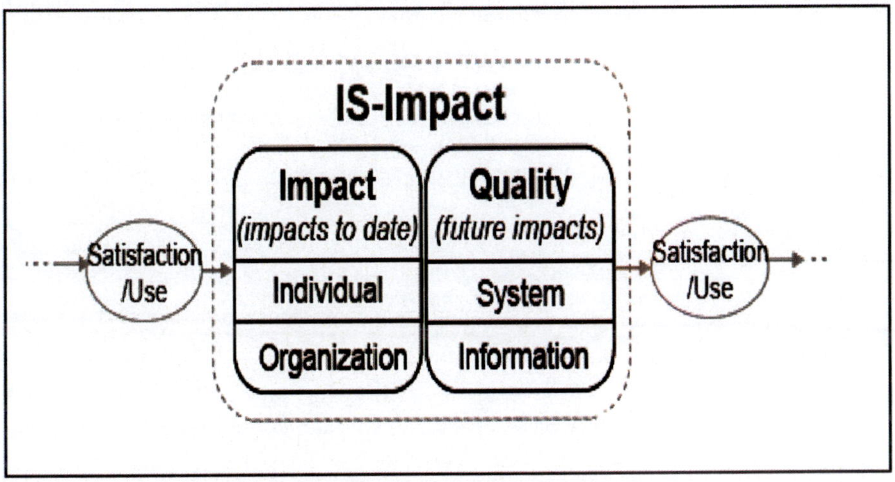

Fig. 7. The IS-Impact measurement model [19].

Table 1. Summary of Information System Success Models.

No	Model	Variables included in Model	Model developed by	Year Model	Proposed area of Application
1	DeLone & McLean (D and M) Model	1.System quality, 2.Information quality, 3.Use, 4.User satisfaction, 5.Individual impact 6.Organisational impact	DeLone-McLean	1992	Information System
2	Seddon Model of IS Success	1. System quality, 2.Information Quality, 3. Perceived usefulness 4.User satisfaction, 5. Benefits to Individuals, 6. Benefits to organisation, 7. Benefits to society	Peter B. Seddon	1997	Information System

(*continued*)

Table 1. (*continued*)

No	Model	Variables included in Model	Model developed by	Year Model	Proposed area of Application
3	System Success Model	1.Information System, 2.IS development Environment, 3.External Environment, 4.Use satisfaction individual impact organisational impact, 5.IS operations Environment, 6.Organisational Environment 7.User Environment	Hwang, Windsor and Pryor	2000	Information System
4	Updated DeLone & McLean Model	1.System quality, 2.Information quality, 3.Service quality, 4.User satisfaction, 5.Intention to use and use, 6.Net benefits	DeLone-McLean	2003	Information System
5	Elpez and Fink's IS success model	1. Meeting user requirements, 2. System usability and performance, 3. Information quality, 4.Use, user acceptance, 5. IS ownership, 6.Interaction with infrastructure	Elpez & Fink	2006	Information System
6	Information Systems Acceptance Model	1,Technology characteristics, 2.Management actions, 3.System Quality, 4,Intention to use, 5.User satisfaction, 6.Use, 7.Net benefits	Seen, Rouse and Beaumont	2007	Information System
7	Gable et al.'s (IS-Impact) model	1. System quality, 2.information quality, 3.satisfaction,4.individualimpact 5.organisational impact	Gable *et al*	2008	Information System

Source: Adopted by the researcher from the literature

5 Results Discussion

There are factors that are core to success, namely information quality and system quality. Information quality and system quality are the most important variables for information system success. Net benefit is the end product of information system success. In order for the system to be success, the quality of the information should be good and the technology or system must be of higher quality. Service is used to get benefits. Benefits is the outmost goal of success therefore the variable that influence success should contribute to the benefit of the organisation.

6 Conclusion

This paper presents review of various information system success models that are classified based on 1992 to 2008. Seven models in the field of information systems were critically reviewed. These models are: DeLone and McLean (D and M) Model, The Seddon model, System Success model, Updated DeLone and McLean model, Wix, Elpez and Fink's IS success model, Information systems acceptance model and Gable et al.'s (IS-Impact) model. The researcher however recommends the use of DeLone and McLean model because it is one of the most popular models for measuring information systems success within organisations. It can be noted that the two models, DeLone and McLean and Seddon Model of IS Success have some commonalities and several important distinctions. Both models integrate and interrelate multiple constructs of IS success. Most of the models has been modified depending on the environment that it has been used. The study will provide future researchers an opportunity to adopt a convenient success model for conducting their empirical studies on different technologies. Furthermore, Researchers can benefit by applying the IS success model in the conduct of similar research in other organisational settings.

Acknowledgment. I wish to express my sincere gratitude and appreciation to my Co-author, Prof Tranos Zuva whose input helped us to produce a paper of good standard.

References

1. Rubin, H.: Into the light. In CIO Magazine (2004). http://www.cio.com.au/index.php/id;1718970659
2. Gable, G.: Measuring enterprise systems success: a preliminary model. In: Proceedings of the of the 9th AMCIS conference, Tempa, Florida, pp. 576–591 (1991)
3. DeLone, W.H., McLean, E.R.: Information systems success: the quest for the dependent variable. Inf. Syst. Res. **3**(1), 60–95 (1992)
4. Garrity, E.J., Sanders, G.L.: Information Systems Success Measurement. Idea Group Publishing, Hershey (1998)
5. Land, F.: The management of change: guidelines for the successful implementation of information systems. In: The Multimedia and Networking Paradigm on Computer-Supported Work, pp. 273–286. Unicom, Aldershot (1994)
6. Grover, V., Jeong, S.R., Segars, A.H.: Information systems effectiveness: the construct space and patters of application. Inf. Manag. **31**(4), 177–191 (1996)
7. Caldeira, M.M., Ward, J.M.: Understanding the successful adoption and use of IS/IT in SMEs: an explanation from Portuguese manufacturing industries. Inf. Syst. J. **12**, 121–152 (2002)
8. Zwikael, O., Meredith, J.R.: Evaluating the success of a project and the performance of its leaders. IEEE Trans. Eng. Manage. **1**, 1–13 (2019)
9. Roldan, J., Leal, A.: Executive information systems in Spain: a study of current practices and comparative analysis. In: Mora, M., Forgionne, G., Gupta, J., (eds.) Decision Making Support Systems: Achievements, Trends, and Challenges for the New Decade, pp. 272–286. Idea Group Publishing, London (2003). https://doi.org/10.4018/978-1-59140-045-5.ch018 [Citation Time(s):1]
10. Petter, S., DeLone, W., McLean, E.R.: Information systems success: the quest for the independent variables. J. Manag. Inf. Syst. **29**(4), 7–62 (2013). https://doi.org/10.2753/MIS0742-1222290401

11. Rai, A., Lang, S.S., Welker, R.B.: Assessing the validity of IS success models: an empirical test and theoretical analysis. Inform. Systems Res. **13**(1), 50–69 (2002)
12. Seddon, P.: A respecification and extension of the DeLone and McLean model of IS success. Inf. Syst. Res. **8**(3), 240–253 (1997)
13. Hwang, M.I., Windsor, J.C., Pryor, A.: Building a knowledge base for MIS research: a meta analysis of a systems success model. Inf. Resour. Manag. J. **13**(2), 26–32 (2000)
14. DeLone, W.H., McLean, E.R.: The DeLone and McLean model of information systems success: a ten-year update. J. Manag. Inf. Syst. **19**(4), 9–30 (2003)
15. Wang, A.P., Liao, H.Y., Chang, T.H.: Adaptive neuro-fuzzy inference system on downstream water level forecasting. In: 2008 Fifth International Conference on Fuzzy Systems and Knowledge Discovery, vol. 3, pp. 503–507 (2008)
16. Yu, P., Qian, S.: Developing a theoretical model and questionnaire survey instrument to measure the success of electronic health records in residential aged care. PLoS One Measuring Success of EHR in Residential Aged Care **13**, 1–18 (2018)
17. Elpez, I., Fink, D.: Information systems success in the public sector: stakeholders' perspectives and emerging alignment model. Issues Inf. Sci. Inf. Technol. **3**(2), 219–231 (2006)
18. Seen, M., Rouse, A., Baeumont, N.: Explaining and predicting information systems acceptance and success: an integrative model. In: Proceedings of the 15th European Conference on Information Systems (ECIS 07). University of St. Gallen (2007)
19. Gable, G., Sedera, D., Chan, T.: Re-conceptualizing information system success: the IS-impact measurement model. J. Assoc. Inf. Syst. **9**(7), 377–408 (2008)
20. Gable, G., Sedera, D., Chan, T.: Enterprise systems success: a measurement model. In: Proceedings of the 24th International Conference on Information Systems (ICIS 03), 14–17 December, Seattle, Washington (2003)

Gamification of E-Learning Based on Information Technology

Shokhida Irgasheva[1(✉)], Maksim Mastepanenko[2], Ivetta Varyan[3,4], Ivan Otcheskiy[5], Edwin Daniel Félix Benites[6], and Juan Carlos Orosco Gavilán[7]

[1] Tashkent Institute of Finance, Tashkent, Uzbekistan
irgasheva_shoxida@tfi.uz
[2] Stavropol State Agrarian University, Stavropol, Russian Federation
[3] N.M. Emanuel Institute of Biochemical Physics, Russian Academy of Sciences, Moscow, Russia
[4] Plekhanov Russian University of Economics, Moscow, Russia
[5] Tyumen State University, Tyumen, Russian Federation
[6] Universidad Nacional José María Arguedas, Andahuaylas, Perú
[7] Universidad Privada del Norte, Lima, Perú

Abstract. E-learning has become an integral part of the educational environment. It makes it possible to significantly expand the methods and techniques used in traditional education. One of the elements that are actively used in the modern educational process is gamification. Gamification allows to increase students' motivation and involvement. It allows to improve the learning of theoretical material, gives an opportunity to improve practical skills in the study of a discipline. The paper presents the process of creating a project-game when reconstructing mountainous terrain, using 3D computer graphics game engine Unreal Engine. Creation of the mountain terrain simulator by students allows students to get acquainted with computer modeling in a game form, as well as to master practical skills and abilities in the field of information technology.

Keywords: E-Learning · Gamification · Educational Environment · Students' Motivation · Project-Game · Unreal Engine 4

1 Introduction

An integral part of improving the modern educational process is the introduction of information technology. Information technologies increase students' interest in the study of academic disciplines, enable teachers and students to communicate outside the classroom and help develop students' skills of independent work, help develop cognitive activity and apply their knowledge in practice [1–6]. Information technologies give a chance to all interested persons, including students, to prove themselves in the direction of information technologies and increase the level of practice-oriented training [7, 8].

Information technologies allow the teacher to combine fundamental professional knowledge with new skills of the digital world, such as digital innovative thinking,

R. Silhavy and P. Silhavy (Eds.): CSOC 2023, LNNS 723, pp. 389–399, 2023.
https://doi.org/10.1007/978-3-031-35317-8_36

practice-oriented and research approaches, mastery of new technologies and methods such as design thinking, pedagogical design, flexible skills to solve specific educational goals and objectives [9–12].

Application of information technologies in general is focused on the development of such qualities in a teacher and student as high self-organization, continuous education, ability to work in different educational spaces.

The following e-learning trends have become popular in the last few years and are becoming more and more widespread every year:

1. Adaptive learning [11, 13]. Personalization, as a necessary condition of the modern educational process, artificial intelligence, which organizes the interaction with students, adjustment to the abilities, knowledge, skills and even his emotional state.
2. Microlearning. In a general form it is learning a small amount of educational material in a short time interval in a digital format. It is based on the following principles: the length of each part of the content for 1–15 min, the focus on the result of the training, the ability to transfer content to any platform and device. At such training brain of the student is not overloaded, and the training material is absorbed better [14, 15].
3. Development of soft skills [16]. These skills which can be useful for everyone: the management of personal development, the ability to manage their time, negotiation skills, the ability to work in a team.
4. Gamification is the process of incorporating game elements into a training course. This technology involves the adaptation of game methods to non-game processes for deeper student involvement in the learning process [17–19]
5. Virtual reality [20–24]. Here the world is simulated by computer technology, in which the user can immerse themselves with special sensory devices.
6. Augmented reality is a technology that allows a computer or other device to augment the world with a digital object [25, 26]. Augmented reality in education is used to enhance learning. According to researches, students using augmented reality remember new information better.
7. Artificial Intelligence - the ability of artificially created systems and programs to perform creative functions that are traditionally considered the prerogative of humans. The most common use of artificial intelligence are chatbots - programs that figure out users' needs and then help meet them [27, 28].

This paper proposes a gamification approach, which allows to significantly increase educational opportunities for students, to implement the choice and implementation of an individual approach in an open educational environment. The implementation of gamification elements in the educational training process contributes to the motivation of students to master the material independently. This approach is the most relevant direction of educational technology development.

2 Methods and Software Implementation

In order to develop practical skills, increase motivation and involvement in the educational process, in the classes of the discipline of information technology students developed a game on Unreal Engine 4, a simulation of mountainous terrain, which allows students to learn a computer model to master the practical skills and abilities in professional activities. Thanks to this approach students all over the world can practice their theoretical knowledge, making the educational process and every lesson more exciting.

The game is developed on Unreal Engine 4, a game engine created back in 1998 and still being updated today. The language used was an internal one - Blueprint. It is a full analogue of C++ with the same syntax, only the code is written in the form of "blueprint" blocks [29].

For the game the terrain elevation maps were used. The terrain map is a view of the area from above in black and white format; by it one can know the "height" of each pixel based on the gradient from black to white (Fig. 1).

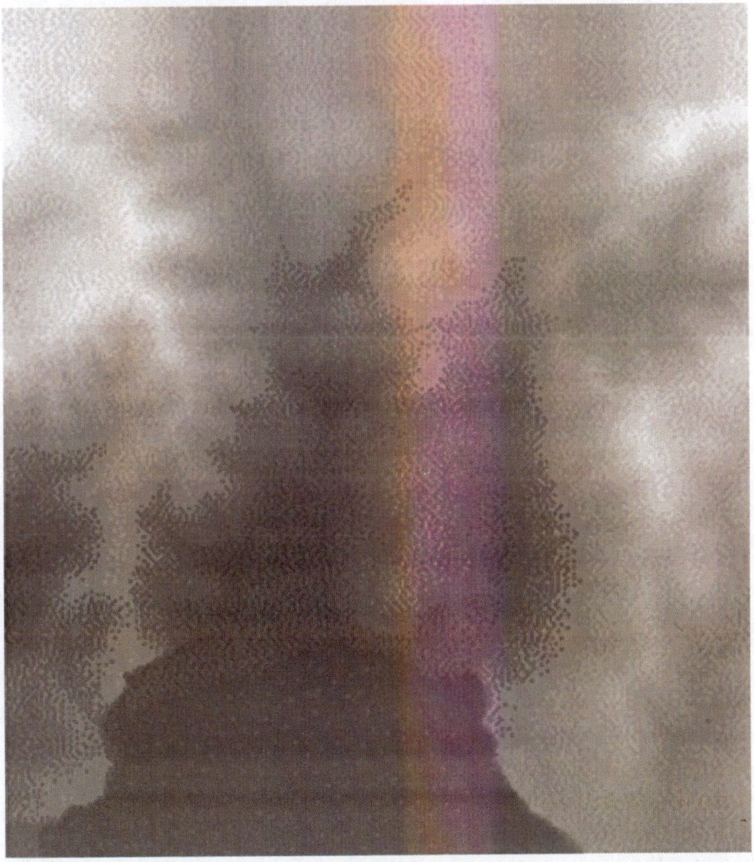

Fig. 1. Elevation map.

After creating a landscape plane and applying a elevation map to it, the terrain appears. Next we need to add a number of details [30]. The first is the surface texture. There can be several terrain textures, so the built-in Landscape element has an array of these details, including grass, grounds, trees, etc. Separate elevation maps need to be created for individual elements, such as grass (Fig. 2).

After the placing of vegetation and adjusting the lighting (Fig. 3) is required to simulate the landscape for the current purposes and the placement of appropriate models of the environment, such as tables, cars, etc. (Fig. 4, 5).

Fig. 2. Grass elevation map.

Next, we begin programming the logic of the game. First, we need to create the characters that will move and interact with the world.

To do this, create a class Character. This class already contains some methods to speed up the process of creating the character, such as: walking, jumping, camera view with the mouse. After a little tuning of walking speed variables, jump height, etc., the interaction logic is created.

The player needs to pick up the camera to take pictures of objects in the game, so the PlayerStart component is placed, which is responsible for the appearance of the character exactly where the component itself is now. Next, the camera model is added.

The code for recognizing an object in the player's direct field of view is given in Fig. 6.

The principle of object recognition is that with the help of the line, which is drawn from the camera and ends at 250 cm from the character, the impact of this line on the object is registered. If this line meets an object, it will be processed and we will get a link to it through the break hit result. In this case the beam itself will be considered as a hit, i.e. collided. Then we put this object into the Active item variable and have a direct

Fig. 3. Terrain relief.

Fig. 4. Alignment of terrain with the help of Sculpt tool.

reference to the interactive item class, which so far only includes the camera. Now you can fully control the camera model, i.e. change its location, assign a location to another object, etc. To make a simulation of holding the camera, it is assigned the location of the sphere added to the character (see Fig. 7).

Further, for a more immersive effect in the created virtual world, "aiming" has been added. The character looks through the viewfinder of the camera to take a photo. To do

Fig. 5. Placing objects in the environment.

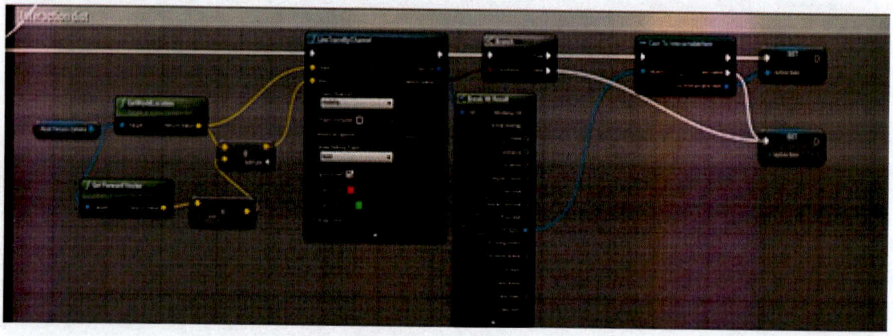

Fig. 6. Object recognition.

this, the camera moves from the sphere to the character's camera, the "character's eyes". The Timeline time curve is used for this, with which you can get a smooth curve from 0 to 1 in the float type, which makes it possible to move the camera smoothly from the starting point to the end point (Fig. 8). However, this is not enough, so we added an effect on the player interface itself (Fig. 9). Namely, we have made an "animation" of the screen darkening (from the beginning to the 1st second of the animation), and after its appearance (from the 1st second to the end).

Application of the developed game as part of the implementation of digital technologies in the educational process allows to increase interest in learning, students better master the theoretical material and consolidate practical skills.

Gamification is applicable at any stage of the lesson, the essence of this development is to use the natural human tendency to play as a key to involvement in the processes of exchange and consumption of information in training.

We can conclude that e-learning is a modern approach to the organization of the educational process. In the future, interest in this kind of learning will only increase,

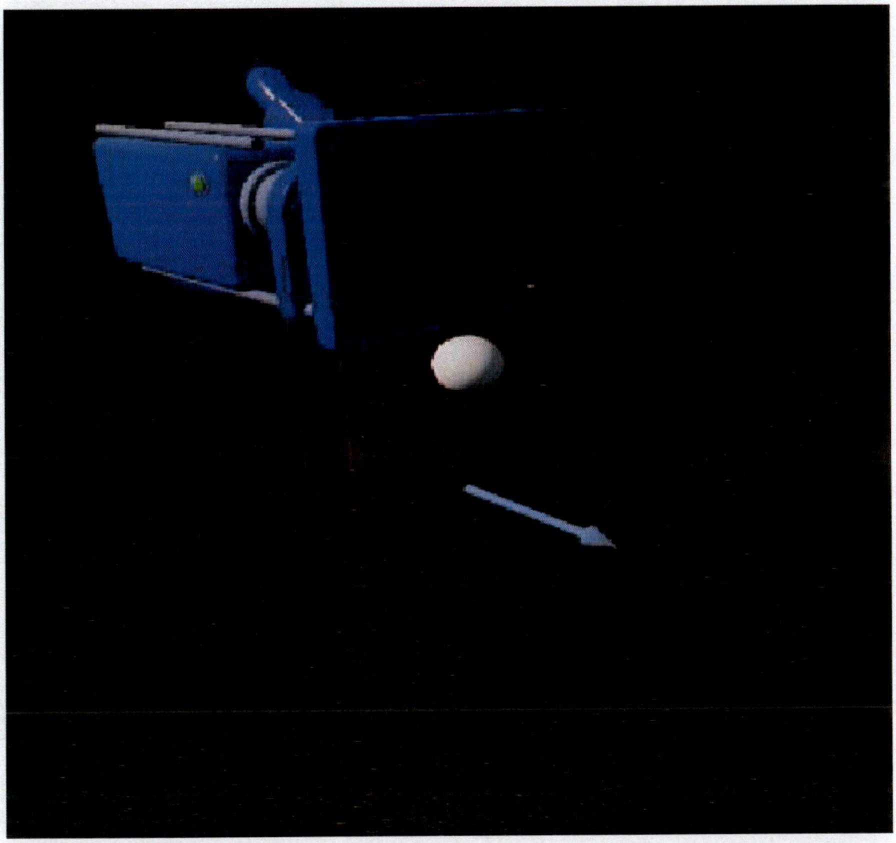

Fig. 7. Representation of the character holding the camera as a sphere.

and the development of digital technology will give impetus to the emergence of new learning tools. Now educational organizations are making many efforts to improve the quality and accessibility of education. E-Learning is a powerful and progressive method for achieving these goals with ease.

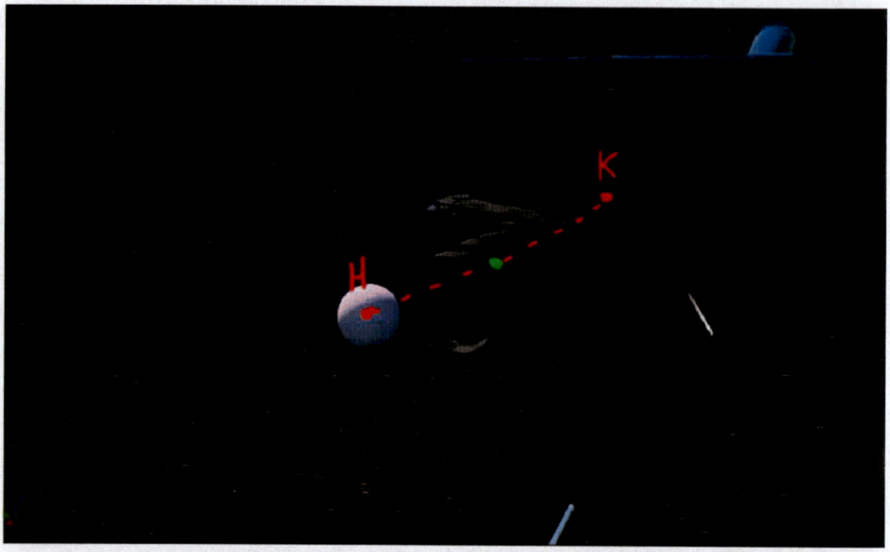

Fig. 8. Moving the camera from the character's "hand" to his "eyes".

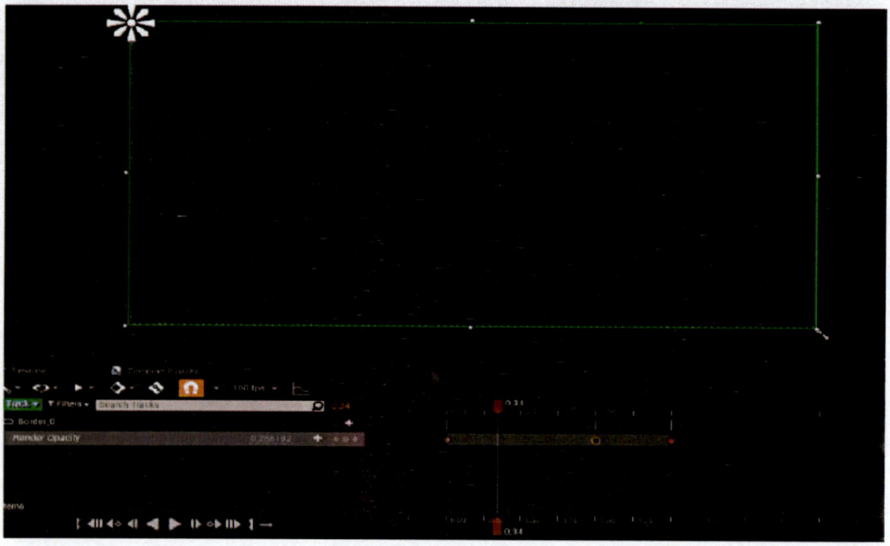

Fig. 9. Fade effect.

3 Conclusion

We can conclude that digital educational resources have great potential applications by increasing motivation to learn different disciplines. Gamification technology is one of the trends in education today. While e-learning in education is fully used for many objective reasons, elements of gamification in the form of didactic techniques and elements are present almost in the work of many professionals. This once again confirms that the new generations have a game mentality from an early age, computer applications are becoming an everyday reality, and in the future they can expect the same gamified interaction in all fields, even those where play is not originally planned.

The idea of creating a digital environment as a comfortable and safe educational environment underlies the organization of project activities. This approach allows solving one of the main tasks - formation of students' general and professional competences, sustainable motivation, which induces them to productive and systematic activity.

Thus, the introduction of game solutions is a trend in education due to the need to meet the needs of modern generations. Modern digital opportunities show that gamification as a method and tool of learning becomes more interesting, productive in terms of educational effect. The main reason for the increased attention to these technologies is associated with new opportunities to change classical forms of learning, make them easy, get useful information, while having fun with the learning process.

References

1. Aljarbouh, A., et al.: Application of the K-medians clustering algorithm for test analysis in E-learning. Lect. Notes Netw. Syst. **596**, 249–256 (2023). https://doi.org/10.1007/978-3-031-21435-6_21
2. Dehghan, H., Esmaeili, S.V., Paridokht, F., Javadzade, N., Jalali, M.: Assessing the students' readiness for E-Learning during the Covid-19 pandemic: a case study. Heliyon **8**(8), e10219 (2022). https://doi.org/10.1016/j.heliyon.2022.e10219
3. Nikolaeva, I., Sleptsov, Y., Gogoleva, I., Mirzagitova, A., Bystrova, N., Tsarev, R.: Statistical hypothesis testing as an instrument of pedagogical experiment. AIP Conf. Proc. **2647**, 020037 (2022). https://doi.org/10.1063/5.0104059
4. Tan, L.S., Kubota, K., Tan, J., Kiew, P.L., Okano, T.: Learning first principles theories under digital divide: effects of virtual cooperative approach on the motivation of learning. Educ. Chem. Eng. **40**, 29–36 (2022). https://doi.org/10.1016/j.ece.2022.04.003
5. Akhmetjanov, M., Ruziev, R.: Fundamentals of modeling fire safety education. Inform. Econ. Manag. **1**(2), 0301–0308 (2022). https://doi.org/10.47813/2782-5280-2022-1-2-0301-0308
6. Lunev, D., Poletykin, S., Kudryavtsev, D.O.: Brain-computer interfaces: technology overview and modern solutions. Mod. Innov. Syst. Technol. **2**(3), 0117–0126 (2022). https://doi.org/10.47813/2782-2818-2022-2-3-0117-0126
7. Hassouni, B.E., et al.: Realization of an educational tool dedicated to teaching the fundamental principles of photovoltaic systems. J. Phys: Conf. Ser. **1399**(2), 022044 (2019). https://doi.org/10.1088/1742-6596/1399/2/022044
8. Mitra, N.K., et al.: Improving the learning process in anatomy practical sessions of chiropractic program using e-learning tool. Transl. Res. Anat. **23**, 100100 (2021). https://doi.org/10.1016/j.tria.2020.100100
9. Çınar, M., Ekici, M., Demir, Ö.: A snapshot of the readiness for e-learning among in-service teachers prior to the pandemic-related transition to e-learning in Turkey. Teach. Teach. Educ. **107**, 103478 (2021). https://doi.org/10.1016/j.tate.2021.103478
10. Nakajima, T.M., Goode, J.: Transformative learning for computer science teachers: examining how educators learn e-textiles in professional development. Teach. Teach. Educ. **85**, 148–159 (2019). https://doi.org/10.1016/j.tate.2019.05.004

11. Tsarev, R.Y., et al.: An approach to developing adaptive electronic educational course. In: Silhavy, R. (ed.) Cybernetics and Automation Control Theory Methods in Intelligent Algorithms. Advances in Intelligent Systems and Computing, vol. 986, pp. 332–341. Springer, Cham (2019). https://doi.org/10.1007/978-3-030-19813-8_34
12. Ung, L.-L., Labadin, J., Mohamad, F.S.: Computational thinking for teachers: development of a localised E-learning system. Comput. Educ. **177**, 104379 (2022). https://doi.org/10.1016/j.compedu.2021.104379
13. Boussakssou, M., Hssina, B., Erittali, M.: Towards an adaptive E-learning system based on Q-learning algorithm. Procedia Comput. Sci. **170**, 1198–1203 (2020). https://doi.org/10.1016/j.procs.2020.03.028
14. Wakam, G.K., Palmon, I., Kulick, A.A., Lark, M., Sonnenday, C.J., Waits, S.A.: Adapting to the times: combining microlearning videos and twitter to teach surgical technique. J. Surg. Educ. **79**(4), 850–854 (2022). https://doi.org/10.1016/j.jsurg.2022.02.001
15. Wang, C., Bakhet, M., Roberts, D., Gnani, S., El-Osta, A.: The efficacy of microlearning in improving self-care capability: a systematic review of the literature. Public Health **186**, 286–296 (2020). https://doi.org/10.1016/j.puhe.2020.07.007
16. Rebele, J.E., St. Pierre, E.K.: A commentary on learning objectives for accounting education programs: the importance of soft skills and technical knowledge. J. Account. Educ. **48**, 71–79 (2019). https://doi.org/10.1016/j.jaccedu.2019.07.002
17. Alsadoon, E., Alkhawajah, A., Suhaim, A.B.: Effects of a gamified learning environment on students' achievement, motivations, and satisfaction. Heliyon **8**(8), e10249 (2022). https://doi.org/10.1016/j.heliyon.2022.e10249
18. Jääskä, E., Lehtinen, J., Kujala, J., Kauppila, O.: Game-based learning and students' motivation in project management education. Proj. Leadersh. Soc. **3**, 100055 (2022). https://doi.org/10.1016/j.plas.2022.100055
19. Tsarev, R.: Application of Kahoot! in education gamification. Int. J. Adv. Stud. **7**(1), 9–17 (2017). https://doi.org/10.12731/2227-930X-2017-1-9-17
20. Ifanov, Jessica, P., Salim, S., Syahputra, M.E., Suri, P.A.: A systematic literature review on implementation of virtual reality for learning. Procedia Comput. Sci. **216**, 260–265 (2023). https://doi.org/10.1016/j.procs.2022.12.135
21. Suri, P.A., Syahputra, M.E., Amany, A.S.H., Djafar, A.: Systematic literature review: the use of virtual reality as a learning media. Procedia Comput. Sci. **216**, 245–251 (2023). https://doi.org/10.1016/j.procs.2022.12.133
22. Won, M., et al.: Diverse approaches to learning with immersive virtual reality identified from a systematic review. Comput. Educ. **195**, 104701 (2023). https://doi.org/10.1016/j.compedu.2022.104701
23. Jahidul, I.M., Faruq, O.: Further exploration of deep aggregation for shadow detection. Mod. Innov. Syst. Technol. **2**(3), 0312–0330 (2022). https://doi.org/10.47813/2782-2818-2022-2-3-0312-0330
24. Yu, X., Bo, L., Xin, C.: Low light combining multiscale deep learning networks and image enhancement algorithm. Mod. Innov. Syst. Technol. **2**(4), 0214–0232 (2022). https://doi.org/10.47813/2782-2818-2022-2-4-0215-0232
25. Baabdullah, A.M., Alsulaimani, A.A., Allamnakhrah, A., Alalwan, A.A., Dwivedi, Y.K., Rana, N.P.: Usage of augmented reality (AR) and development of e-learning outcomes: an empirical evaluation of students' e-learning experience. Comput. Educ. **177**, 104383 (2022). https://doi.org/10.1016/j.compedu.2021.104383
26. Neffati, O.S., et al.: An educational tool for enhanced mobile e-Learning for technical higher education using mobile devices for augmented reality. Microprocess. Microsyst. **83**, 104030 (2021). https://doi.org/10.1016/j.micpro.2021.104030

27. Huang, A.Y.Q., Lu, O.H.T., Yang, S.J.H.: Effects of artificial intelligence-enabled personalized recommendations on learners' learning engagement, motivation, and outcomes in a flipped classroom. Comput. Educ. **194**, 104684 (2023). https://doi.org/10.1016/j.compedu.2022.104684
28. Song, Y., Chen, B., Liu, X., Weijun, H., Xiangyu, X., Yuqi, Y.: Audio and video editing system design based on OpenCV. Inform. Econ. Manag. **1**(2), 0101–0120 (2022). https://doi.org/10.47813/2782-5280-2022-1-2-0101-0120
29. Unreal Engine 4.27. https://docs.unrealengine.com/4.27/en-US/InteractiveExperiences/Tracing/HowTo/SingleLineTraceByChannel/. Accessed 23 Jan 2023
30. Landscape Technical Manual. https://docs.unrealengine.com/4.27/en-US/BuildingWorlds/Landscape/TechnicalGuide/. Accessed 23 Jan 2023

Check for updates

Sorting the Quality Criteria of an ServiceDesk in a State Public Court of Brazil Using Macbeth

Cristiano Henrique Lima de Carvalho, Plácido Rogério Pinheiro(✉), Flávia Montenegro de Albuquerque, and Paulo Victor Xavier Pinheiro

University of Fortaleza, Fortaleza, Brazil
placido@unifor.br

Abstract. Understanding the dimensions and criteria of quality in Information Technology (IT) care is one of the key factors in achieving competitive advantage. Therefore, the expectations associated with providing IT services are very challenging. In this context, there was a need to identify the criteria that permeate a successful service and its association with the SERVQUAL model, in order to monitor the quality of this service. The objective of this article is to organize the criteria and dimensions of the IT service, using a multicriteria decision model. We contextualize the IT service process, the available frameworks, the use of the brainstorm tool, the cognitive mapping and the application of the multicriteria model to carry out the ordering, making it possible to identify the most relevant criteria that affect the quality of service.

Keywords: Information technology service · SERVQUAL model · Multiple criteria decision aid (MCDA) · Multiple attribute utility theory methods (MAUT) · Macbeth · ServiceDesk · M-Macbeth

1 Introduction

With the growing demand for Information Technology (IT) service, which includes support for equipment (hardware) and systems (software) and IT infrastructure, there is a continuous need to provide a service desk solution to ensure the quality and satisfaction of users and the proper functioning of the various services of the organization.

The technical support of IT is a service offered to meet the demands of internal and external users, providing solutions ranging from simple technical consultations, operation of technological infrastructure and management of processes and procedures. Among the main features, we highlight the unique point of IT service and uninterrupted operation ("follow the sun") for users of IT.

Due to the complexity that involves serving the public, it is essential that this service be performed effectively and efficiently. For the correct execution of institutional activities, it is essential to identify, monitor and categorize service indicators, fundamental foundations for the correct provision of the service.

According to [1], the day-to-day in the area of ICT technical support services is to work without respite to be successful in the solution of each request, of each doubt,

R. Silhavy and P. Silhavy (Eds.): CSOC 2023, LNNS 723, pp. 400–414, 2023.
https://doi.org/10.1007/978-3-031-35317-8_37

guidance or recovery of an incident or failure of operation of a service asset delivered to the user.

Thus, IT management is an increasingly important aspect for organizations seeking a strategic alignment between business and IT. This maturing allows companies to integrate the IT department with other areas, resulting in benefits such as cost reduction, improved quality of products and services, increased competitiveness and risk reduction.

Second [2], highlight as positive results from the management of the IT area, increased efficiency of procedures, increased user satisfaction and improved quality.

Second [3], when analyzing the literature and publications, identified that the contributions of IT governance in order to reduce human dependence in the stages of business processes, in order to make IT proactive, present and appropriate to the objectives intended for the organization.

The use of IT in the provision of public services is not recent. In Brazil we have success cases, such as the capture, transmission and totalization of elections through electronic ballot box technology; the sending, calculation of amounts and the payment of annual income tax; in addition to one of the safest, fastest and most innovative bank clearing system.

[4], highlights as a consequence of the transformation that technology has been providing, we are being able to reap great benefits that contribute to facilitate the construction of organized digital environments capable of enabling cooperation and collaboration between the most varied sectors of society.

In 2006, Law No. 11,419 – Law of Computerization of the Digital Process was promulgated, which enabled the implementation of the Digital Judicial Process (PROJUDI) system, which initiated the effective digitization of Brazilian justice.

With the advancement of Information and Communication Technology (ICT) as an indispensable tool for the operation of judicial and administrative units, it gave the IT sector of this State Court of Justice a great responsibility for the ICT products and services provided to society.

2 Maturity Models in Information Technology (IT)

IT maturity is an increasingly important aspect for organizations looking to achieve strategic alignment with IT. This maturing allows companies to integrate IT with other areas, resulting in benefits such as cost reduction, improved quality of products and services, increased competitiveness and reduced risk.

According to [2], highlight as positive results from it management, increased efficiency of procedures, increased user satisfaction and improved quality.

According to [3], found that in general, the distance observed between IT and business decisions impacts on services aimed at meeting current problems, instead of planning solutions appropriate to the desired future environment and that IT sometimes delivers deficient services because it contains risks associated with human dependence in the stages of business processes.

In this way, the maturity of IT is related to the company's ability to manage it effectively and efficiently, maximizing its resources to meet the needs of the company for better alignment to the business, greater and better productivity (efficiency and effectiveness), knowledge management, elimination (or reduction) of waste and increased quality of products and services delivered by the IT area.

According to [5], the Information Technology Infrastructure Library (ITIL) was developed in the late 1980s by the Office of Government Commerce (OGC) with the aim of developing it management best practices and standardizing IT documentation for the British government.

According to [6], the fourth edition of the library was made available in early 2019. Unlike previous versions, where the focus was on the concepts of "process maturity" and "service lifecycle," ITIL 4 addresses a new Service Value System model. Still according to [6], ITIL 4 has incorporated 4 (four) new dimensions in service management and 5 (five) key elements, seeking to implement and strengthen quality in the provision of services and not just follow a set of rules.

The Control Objectives for Information and Related Technology (COBIT), according to Junior (2021) points out that COBIT first launched in 1996 and was designed as a set of IT control objectives to help the financial auditing community better navigate the growth of IT environments.

According to [7], this methodology also allows organizations to map their processes, assess risks and improve their efficiency in terms of IT management. COBIT is strongly based on good international practices and helps organizations be more competitive and accountable.

According to [8] and [9], note that COBIT 2019 was introduced to build governance strategies that are more flexible, collaborative and address new and ever-changing technologies and that new theories and concepts were introduced in the latest COBIT update, which now includes 40 goals for building a governance and management program.

According to [10], Capability Maturity Model (CMM) emerged during the 1980s as a model for risk assessment in the hiring of software companies by the U.S. Department of Defense (DoD) that wanted to be able to evaluate the development processes used by companies that competed in bids as an indication of predictability of quality, costs and deadlines in the projects contracted.

Currently, CMMI, the evolution of CMM, is managed by the Information Systems Audit and Control Association (ISACA). According to [11], there are few studies that present proposals related to the implementation of quality models. In addition, CMMI-DEV V2.0 (presented in March 2018), because it is a new model, there are few bibliographic references and less than one hundred companies worldwide have implemented this new version. CMMI V2.0 is an important step in the maturity of process improvement for organizations striving to improve their quality processes and achieve internationally recognized performance.

The International Organization for Standardization (ISO) was created in 1947 to facilitate the coordination and unification of industrial standardization processes. According to [12], the leading international standard technical standard for it management services administration is ISO 20000 (ISO/IEC 20000–1:2005). It was developed in 2005, defining a set of best practices for it service administration.

According to [12], the standard provides a framework for providing quality IT services and helps organizations meet customer requirements and expectations. It is based on the ISO 9,000 quality management standard, the British Standards Institution's BS 15,000, published at the beginning of the 21st century and compatible with ITIL.

According to [13], the latest ISO 20000 update (ISO/IEC 20000–1:2018) was released in September 2018, highlighting improvements in the quality and management of IT services, including clarification of terminology, guidance on assessment, risk mitigation, technology changes and alignment with the new versions of ISO 9001 (technical standards that establish a quality management model) and ISO 27001 (standard for security management system of information).

3 Quality Models

Quality is a characteristic or attribute that is perceived. It is often used to refer to the degree of excellence, that is, as perfection, improvement or satisfaction of some product or service. Some attributes can contribute to quality include: design, materials, labor, service, durability, among others.

According to [5], defines the quality of a service can be determined by the level of customer satisfaction in relation to it, that is, how the customer perceives the service provided/delivered.

However, according to [14], defines that quality is often determined by the degree in and something meets or exceeds customer expectations. Still according to [14], the customer's perception of a company and its service is as fundamental as the reality of what the company offers, since it can establish a differentiating differentiation to competing organizations.

According to [15], highlights the importance of customer perceptions and how they influence purchasing choices. According to [13], quality is the degree to which a set of inherent characteristics meets the requirements.

According to [16], have described that quality is one of the main factors influencing customer satisfaction, along with price, promotion, distribution and design. According to [17] and [18], described customer satisfaction can be seen as a function of perceived quality. They claim that there are three types of quality: basic quality, attractive quality and emotional quality. These three levels of quality (functional, cosmetic and emotional) represent different aspects of the perception of quality by consumers and each level that should be considered in the search for excellence in quality.

The delivery of quality products and services is essential for customer satisfaction ensuring the repetition of business. It is necessary to constantly seek ways to improve quality and find new ways to meet and exceed customer expectations.

The SERVQUAL scale was initially described by A Parsu Parasuraman, Leonard Berry and Valerie Zeithaml, as "A Conceptual Model of Service Quality and Its Implications for Future Research" in a 1988 article. Later, in 1991, the authors of the scale wrote and published the article: "A scale for measuring customer perceptions of service quality", a refinement and consolidation of the initial proposal. The SERVQUAL model is a widely used tool for measuring and evaluating service quality. The model is based on the premise that service quality is a construct made up of five dimensions: Reliability, Responsiveness, Assurance, Empathy and Tangibility, according to Fig. 1.

Fig. 1. Five dimensions of SERVQUAL model

The SERVQUAL model has been widely used in a variety of service providers such as banking, healthcare, retail and tourism. Over the years, several studies were made to validate the model and its dimensions, thus confirming that the quality of the service was indeed multidimensional and that the five dimensions of quality of service identified in the SERVQUAL model were relevant in the context of service delivery.

According to [19], demonstrated a literature review of the SERVQUAL model, identifying and compiling the number of published articles and the countries with more publications in the studied period in Fig. 2.

It is noteworthy that the SERVQUAL model has a broad theoretical and practical basis, making the union between the dimensions identified, expectations, gaps and the perception of how customers organize in their mind the quality of the service provided. By considering the five dimensions of the model, companies can offer high-quality IT service and ensure customer satisfaction.

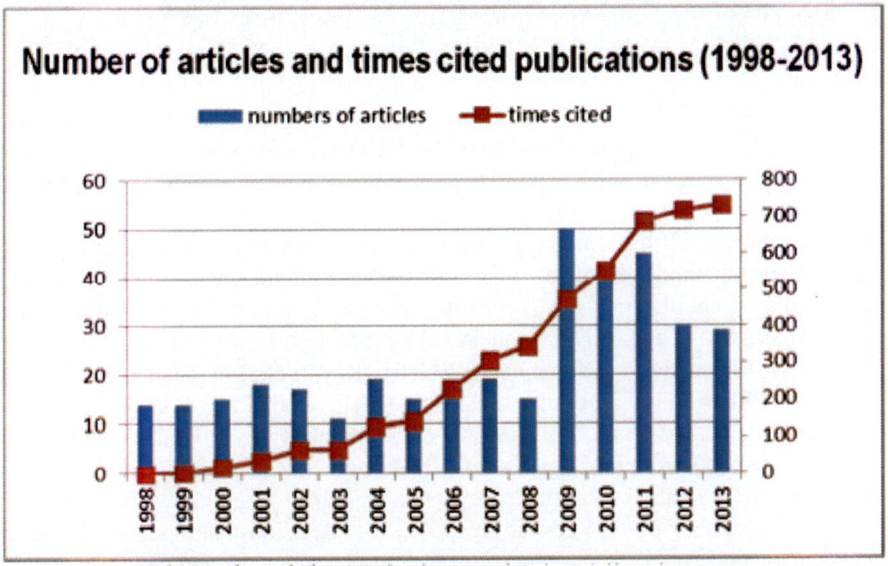

Fig. 2. Graphic of number of articles of SERVQUAL and publications cited per year

4 Multi-criteria Decision Analysis Method

The multicriteria methodology is an operational research field that deals with decision-making problems in the presence of multiple conflicting criteria. It is also known as Multiple Criteria Decision Analysis (MCDA). This methodology was developed to deal with the increasing complexity of real problem analysis.

The process of choice is natural and constant since the beginning of human history. It is the way to identify the best option within the available alternatives. According to [20], states that not all decisions are easy to resolve, as some entail great risks, negotiations, costs and insatiss. Common sense tells us that most of the time decision-making is made in the relationship between emotion and reason.

According to [21], describes that real-life decision problems generally involve several conflicting points of view, which must be taken into account jointly in order to reach a reasonable decision. According to [22], reinforces that MCDA is a systematic approach to help decision in complex situations where various criteria must be balanced to achieve optimal outcome.

In the increasingly complex world, with its ever-changing markets and technologies, using a decision-making process that helps individuals and organizations make better choices, taking into account multiple factors can be the best way to ensure the best possible decision.

The multicriteria methodology is a tool that can be used to make decisions in situations where there are multiple options or criteria to consider, taking into account multiple criteria and alternative options.

According to [23], the MCDA methodology allows a rigorous and fair evaluation of options, facilitating decision making. Therefore, the application of the multi-criteria methodology can help organizations make more informed and well-founded decisions.

MCDA is a useful tool for decisions involving trade-offs between different objectives, constraints and preferences. In addition, the MCDA methodology is widely used in a wide range of industries, including IT, project management, environment, health and transportation.

According to [24], even though there are many different ways of classifying the multi-criteria method, one of the main is the one that divides it between an American and a French school. From the American school we can note the AHP multi-criteria method (Saaty, 1980), the UTA (Jaquet-Legreze e Siskos, 1982) and MAUT (Keeney e Raiffa,1993). From the French school: ELECTRE (Roy, 1968, 1978; and Roy and Shalka, 1984) and PROMÉTHÉE (Brans et al., 1985), (Mello et al., 2003). Additionally, some multi-criteria methods can't be placed into only one school because they contain mixed technical elements from both, such as: TODIM (Gomes, Araya and Cariggnano, 2004) and MACBETH method (Bana e Costa, Corte and Vansnick, 2005).

The decision support process is an approach to decision-making that ensures that all stakeholders are considered, that data is collected and analyzed correctly, and that the best alternative is chosen, in line with objectives and available data. As a way to structure this model, According to [23] and [25], highlight the 3 (three) main phases of the decision support process. They are: Structuring, Evaluation, Prescription.

However, it is necessary to describe the activities of each profile in the performance of the multicriteria analysis process to help choose a particular alternative or to evaluate an existing situation, developing and deepening the importance of each criterion in the decision-making process. According to [27, 28] and [29], the performance profiles are highlighted below: Decision-Makers, Specialist, Model and Alternatives.

According to [30], explains that when we model a multicriteria decision problem, several problems (or formulation of problems) can be considered. According to [31], distinguished four basic problems:

- Description Issue (P.δ) – the alternatives are described;
- Classification Issue (P.β) – The alternatives are ranked;
- Order Issue (P.γ) – The alternatives are put in order;
- Choice Issue (P.α) – The best alternative is chosen.

According to [32], it describes that the identification of the Primary Elements of Evaluation (EPAs) is necessary for the subsequent construction of the cognitive map, making it possible to identify the Fundamental Points of View (FPVs), and from these, the criteria for evaluation are derived.

According to [33], informs that FPVs explain the values that decision-makers consider important in that context, while defining the characteristics of actions that are of interest to decision makers. According to [34], complements the description highlighting that the tree structure uses the decomposition logic, in which a more complex criterion to be measured is decomposed into subcriteria of easier measurement, elementary viewpoints (EPVs). Therefore, the value tree provides a (graphical) view of the organization of FPVs and EPVs at their various levels to the authors of the process.

5 Methodology

The methodology refers to the set of methods and procedures used in conduction, knowledge foundation and application of research methods, including sampling techniques and data collection methods. According to [35], there are three main approaches to studying them: Observe the behavior that occurs naturally in the real scope; Ask people about their behaviour, what they do and do and about their subjective states, what, for example, they think; Create artificial situations and observe behavior in the face of tasks defined for these situations.

Initially, it will be necessary to identify the criteria that should be analyzed, through interviews and group activities, coating, at this moment, the work of a qualitative character. According to [24, 26], there are many techniques that help structuring the problems such as: brainstorm, priority matrix, fish-diagram spine, decision tree, cognitive mapping.

Cognitive maps have a hierarchical structure in the form of means-ends and help to express the objectives that are important by the actors making it easier to understand the problem. Cognitive maps may generate relevant interpretation of the problem through the identification of element connections related to the subject.

Among the tools or techniques used in this phase, brainstorming techniques will be used, a discussion oriented to produce the free formation of ideas with the technical team of the IT area and with the main managers of the State Court of Justice, to identify, consolidate and refine the criteria that will be used and their connection with the dimensions described in the SERVQUAL model.

The identification and ordering of the criteria that should be monitored as a priority will provide, through the multicriteria methodology, the reduction of subjectivity in decision-making regarding the subject. In this way, the following question is asked: "How to sorting the quality criteria of a ServiceDesk, using the multicriteria methodology?".

6 Multicriteria Paradigm for Sorting Criteria Applied to a ServiceDesk

The quality of customer service is a critical success factor in organizations and can be the decisive factor in the perception of the institution's image. Companies that offer quality customer service are typically more successful than those that don't, and happy customers are more likely to continue doing business with the company, and they are more likely to recommend it to others. A quality ServiceDesk can help increase sales, build customer loyalty, and improve your results. It demands meet, within a ServiceDesk, takes place, in general, according to Fig. 3.

The structuring of the research was planned with group interviews. It was informing the initial context of the research, that is, ordering the quality criteria of the ServiceDesk and supporting a future questionnaire to improve the quality of care. The duration was from 60 (sixty) to 90 (ninety) minutes and, during the interviews, the technique used was brainstorming. It was using the Microsoft Teams web conference tool. The interview script consisted of the participants discussing four (4) main questions:

1. What are the strengths of Servicedesk?
2. What are the weaknesses of Servicedesk?

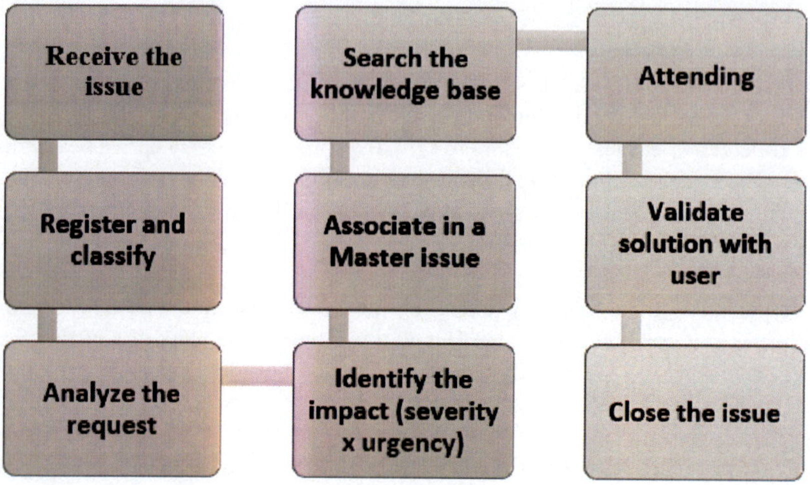

Fig. 3. ServiceDesk service flow

3. What are servicedesk threats?
4. What are servicedesk opportunities?

During this activity and as a way to better develop the boiling of opinions, this researcher only guided the filling based on the indications of strengths, weaknesses, weaknesses and opportunities, thus translating an outline of the Analysis of SWOT. This interview methodology was used in the following meetings, as highlighted by the Table 1, with the following quantitative participants and their duration:

Table 1. Interviews

Order	interview month	Number of participants	Occupation area	Duration
1	December 2021	15	ServiceDesk	90 min
2	December 2021	15	ServiceDesk	60 min
3	January 2022	8	Human Resources and Financial Management	60 min
4	January 2022	8	Judicial Management and Strategy Management	40 min
5	February 2022	5	Administration Area (Maintenance - Engineering - Assets - Logistics - Warehouse)	40 min

After the first stage of information collection, that is, the interviews, where it was possible to identify the strengths, weaknesses, threats and opportunities, the functional hierarchy of the criteria occurred, through a value tree, according to Fig. 4.

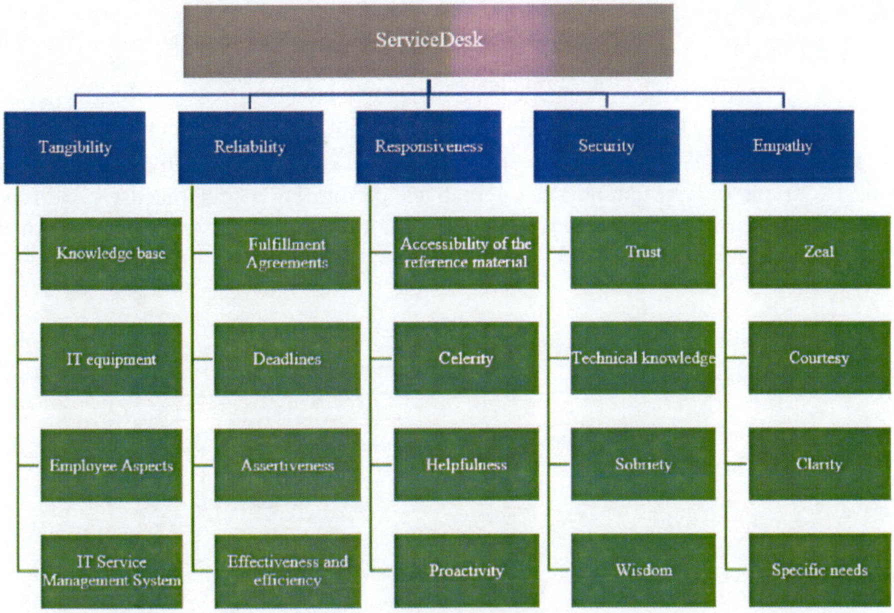

Fig. 4. Value tree

After the construction of value tree, it was possible to identify the Fundamental Points of View (FPV) – the dimensions of the SERVQUAL model and the Elementary Points of View (EPV) – the criteria identified.

7 Measuring Attractiveness by a Categorical Based Evaluation Technique (MACBETH)

The MACBETH method is an approach derived from the Multiattribute Utility Theory (MAUT) that requires interactive judgments of qualitative bias to help a decision-making or decision support group quantify the relative attractiveness of options. It was developed in 1994, after the publication of the article "MACBETH – An interactive path towards the construction of cardinal value functions" by professors Carlos Bana e Costa and Prof. Dr. Jean Claude Vansnick.

According to [36], MACBETH transforms ordinal scales into cardinals, based on absolute judgments on the difference in attractiveness between ordinal scale levels, initially applying an interactive questioning protocol that comems two elements at a time, only requiring a qualitative preference judgment, according to [37], with the following options: null, very weak, moderate, strong, too strong or extreme. After applying the predicted mathematical model, it is possible to perform the following analyses:

- detection of inconsistencies;
- value functions;
- value tree;

- matrix of judgments;
- performance descriptors;
- sorting of the options (thermometer).

The MACBETH decision-making methodology is a mature, well-known and widely used approach to dealing with complex decisions. It can be adapted to different types of decisions and different number of alternatives, performing a comprehensive and transparent comparison of options, either with quantitative criteria or qualitative criteria according to Fig. 5.

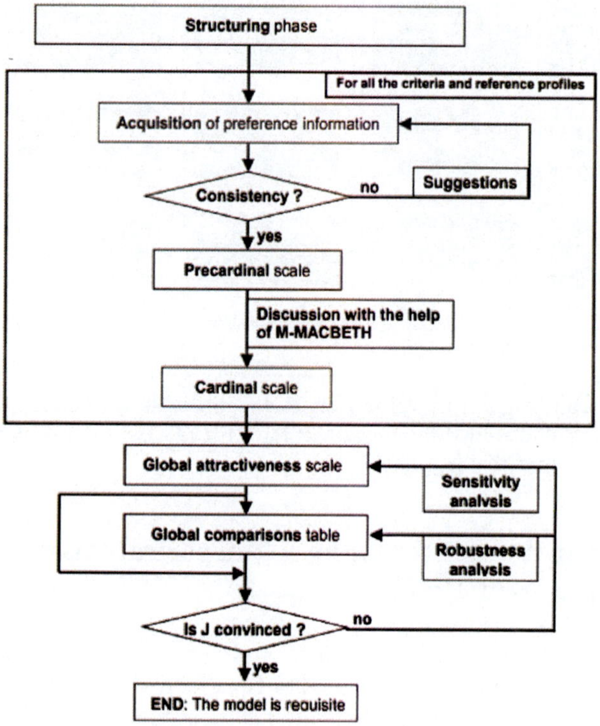

Fig. 5. Summary of the MACBETH decision aid process

After identifying the dimensions (FPV) and criteria (EPV), it is necessary to make their semantic comparison, the confirmation of the judgment, the generation of current scale and anchorage functions. Once the activity of constructing and entering the value functions for each criterion was performed and filling them in the M-Macbeth application, the overall weighting and substitution rates (weights) of the EPVs and PVs were performed.

According to [32], substitution rates, also known as weights, are parameters that decision makers deem appropriate to add, in a compensatory way, local performances (in the criteria) in a global performance. After the research data were tabulated and treated in the M-Macbeth application it was possible to calculate the weighted weights for each of the PVs and EPVs, thus being structured according to Fig. 6 and 7.

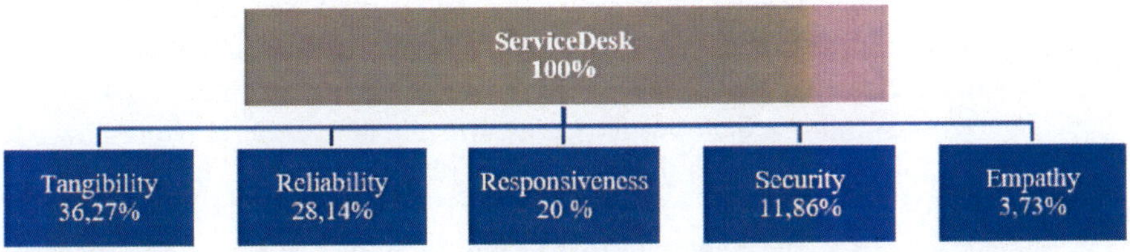

Fig. 6. Value tree

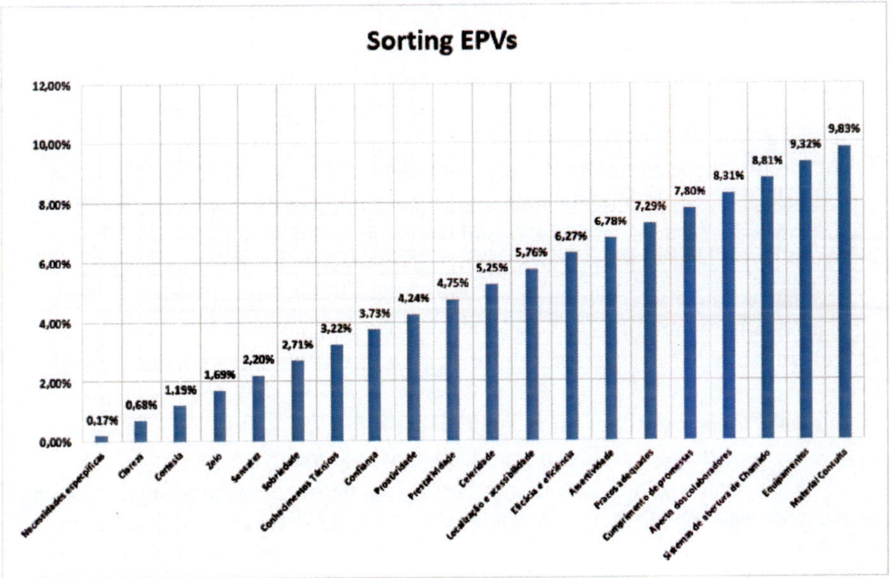

Fig. 7. Sorting of EPVs

Based on the approaches followed in this study, which incorporate the semantic judgments of the EPVs descriptors, the global semantic judgments among all EPVs and the intercession of the FPVs and EPVs in the performance Table, the overall score was obtained as a result (duly calculated weights) and the ordering of each criterion suggested by the M-Macbeth application.

8 Final Considerations

With the exponential growth of demands for IT services, ensuring the quality of the service provided, the satisfaction of users and the proper functioning of the IT sector corroborates in the achievement of the strategic vision of this State Justice Court which is to be a Court of Justice of national reference to speed and efficiency, recognized for being accessible and contributing to the reduction of inequalities.

In this scenario, for the correct execution of institutional activities, it is essential to identify, follow-up, or to order the criteria perceived by users, fundamental foundations for the correct provision of the service, reducing decisions involving conflicts of choice (trade-offs).

It is important to highlight the participation and contribution of the technical team of 30 (thirty) technical analysts and the 21 (twenty-one) senior managers of various areas of activity within the State Justice Court, which with the plurality of knowledge, which allowed the identification of criteria, their relationships and the union with the dimensions of the SERVQUAL model.

Another important and important role in this study was that of the "Decider" or "stakeholder" who acted, in the structuring of the model, in the identification of performance levels, in semantic judgment, performing the comparison along side the impact levels of the same descriptor. This task is dependent on human activity and sensitivity, technical knowledge and maturity of the "Decider" or "stakeholder", where value judgment plays a fundamental role in the validation of the weights obtained, to ensure the reflection of opinion and, consequently, on a list of criteria and dimensions, ordered and weighted as to the degree of impact.

In addition to the use of the Macbeth methodology, the availability of the M-Macbeth tool, which allowed the attractiveness of the descriptors of the same criterion, the attractiveness between criteria, the analysis of semantic judgments, the proportionality relationships and the analysis of robustness for the levels of local and global information for the evaluated model are performed.

The results obtained allowed a review of the service strategy, applying improvements in the current IT Service Center, reinforcing the profiles of the attendants, enabling the review and prioritization of current and future acquisitions, as well as the optimization of human and technological resources applied in the service, significant gains, confirming the relevance and applicability of the Macbeth methodology.

As a suggestion of future work, it is considered pertinent to carry out the application of the proposed model in a more comprehensive and diverse way, such as the states justice court and private companies, regardless of size and area of activity. With the advent of "personal assistants" such as Alexa, Siri, Google Assistant and Cortana that use machine learning, artificial intelligence and virtual/augmented reality, it will also be possible, in an integrated and disruptive way, to provide technical service anywhere, anytime, in any application and on any computing device.

Acknowledgements. The authors thank the Edson Queiroz Foundation/University of Fortaleza for all the support provided.

References

1. Melendez, R.: Service desk corporativo: solução com base na ITIL V3. Novatec Editora, São Paulo (2011). ISBN 978-85-7522-269-0
2. da Silva, E.D., Gaspar, M.A., Costa, I., de Magalhães, F.L.F.: Gestão do conhecimento de service desk: estudo de caso da integração em empresas adquiridas por uma corporação multi negócios. Revista Gestão & Tecnologia, Pedro Leopoldo, **18**(1), 140–170 (2018)

3. Coser, T.: Governança de tecnologia da informação para reduzir problemas de agência no nível operacional. Revista Gestão & Tecnologia, Pedro Leopoldo, **17**(3), 85–118 (2017)
4. Vinhas, C.T., Sosa, R.C., Dilli, C., Silva, G.L., Fiori, A.S., Silva, M.O.S.: A dimensão da tecnologia da informação e comunicação na administração pública. Revista Ibero-Americana de Humanidades, Ciências e Educação **8**(1), 1083–1095 (2022). https://doi.org/10.51891/rease.v8i1.3939
5. Magalhães, I.L.: Gerenciamento dos serviços de TI na prática: uma abordagem com base na ITIL. Novatec Editora, São Paulo (2007)
6. Axelos: ITIL® Foundation. ITIL, 4. ed. (2019). (TSO - The Stationery Office)
7. ISACA: A business framework for the governance and management of enterprise IT (2012). https://static1.squarespace.com/static/56b3cadb59827ecd82b02b43/t/56d8c0d84d088e673055c308/1457045725120/COBIT-5_res_eng_1012.pdf. Accessed 25 Abr 2022
8. Chari, R.: COBIT 2019 e ITIL4: uma comparação. Revista Online: ITSM na Prática (2019)
9. White, S.K.: What is COBIT? A framework for alignment and governance. CIO United States (2019)
10. CMMI Institute: CMMI model V2.0. CMMI Institute an ISACA Enterprise (2022). Disponível em https://cmmiinstitute.com/cmmi. Accessed 10 Abr 2022
11. Machado, E.M.M., Mexas, M.P., de Oliveira, S.B.: Proposta para a implantação do CMMI-DEV V2.0 ML3 em empresas de pequeno e médio porte de desenvolvimento de software. Revista Ibérica de Sistemas e Tecnologias de Informação, **E41**, 83–97 (2021)
12. ISO: ISO/IEC 20000-1:2005. ISO (2005). (Especificação e Guia de Práticas)
13. ISO: ISO 9000:2015. ISO (2022). https://www.iso.org/obp/ui/#iso:std:iso:9000:ed-4:v1:en. Accessed 14 Mar 2022
14. Porter, M.E.: Vantagem competitiva: criando e sustentando um desempenho superior. 20. ed. Rio de Janeiro: Campus (1989)
15. Kotler, P.: Marketing management. 14. ed. Prentice Hall (2011). ISBN 978-0132102926
16. Kotler, P., Armstrong, G.: Princípios de marketing. 15. ed. Prentice Hall (2014). ISBN 978-8543004471
17. Crosby, P.B.: Quality is Free: The Art of Making Quality Certain. McGraw-Hill, New York (1979)
18. Kano, N.: Attractive quality and must-be quality. J. Japanese Soc. Qual. Control **31**(4), 147–156 (1984)
19. Wang, Y.L., Luor, T., Luarn, P., Lu, H.P.: Contribution and trend to quality research – a literature review of SERVQUAL model from 1998 to 2013. Informatica Economica, **19**(1), (2015). https://doi.org/10.12948/issn14531305/19.1.2015.03
20. da Costa, R.F.P.: Utilização de metodologias multicritério de apoio à decisão como ferramenta de suporte numa empresa de serviços energéticos. Tese (Doutorado em Engenharia Eletrotécnica e de Computadores) – Instituto Superior de Engenharia do Porto, Porto (2012)
21. Wang, J.J., Jing, Y.Y., Zhang, C.F., Zhao, J.H.: Review on multi-criteria decision analysis aid in sustainable energy decision-making. Renew. Sustain. Energy Rev. **13**(9), 2263–2278 (2009)
22. Keeney, R.L.: Value-Focused Thinking: A Path to Creative Decision Making. Harvard University Press, Cambridge (1992)
23. Brans, J.P., Mareschal, B., Vincke, P.: PROMETHEE: a new family of outranking methods in multicriteria analysis. Oper. Res. Int. Journal **3**, 477–490 (1984)
24. Barreto, T.B.O., Pinheiro, P.R., Silva, C.F.G.: The multicriteria model support to decision in the evaluation of service quality in customer service. In: Silhavy, R. (ed.) CSOC2018 2018. AISC, vol. 763, pp. 158–167. Springer, Cham (2018). https://doi.org/10.1007/978-3-319-91186-1_17
25. Bana e Costa, C.A., Meza, L.A., Oliveira, M.D.: O método MACBETH e aplicação no Brasil. Engevista, 15(1), 3–27 (2013)

26. Ensslin, L., Ensslin, S.R., Rocha, S., Marafon, A.D., Medaglia, T.A.: Modelo multicritério de apoio à decisão construtivista no processo de avaliação de fornecedores. Produção **23**(2), 402–421 (2013). https://doi.org/10.1590/S0103-65132012005000065
27. Campos, V.R.: Modelo de apoio à decisão multicritério para priorização de projetos em saneamento. Tese (Doutorado em Economia, Organizações e Gestão do Conhecimento) – Escola de Engenharia de São Carlos, Universidade de São Paulo, São Carlos (2011)
28. Treinta, F.T., Farias Filho, J.R., Sant'anna, A.P., Rabelo, L.M.: Metodologia de pesquisa bibliográfica com a utilização de método multicritério de apoio à decisão. Production, **24**(3), 508–520 (2014). https://doi.org/10.1590/S0103-65132013005000078
29. Mello, J.C.C.B.S., Gomes, E.G., Leta, F.R., Pessolani, R.B.V.: Conceitos básicos do apoio multicritério à decisão e sua aplicação no projeto aerodesign. Engevista, **5**(8), 22–35 (2003)
30. Miranda, C.M.G., Almeida, A.T.: Visão multicritério da avaliação de programas de pós-graduação pela capes: o caso da área engenharia III baseado nos métodos ELECTRE II e MAUT. Gestão e Produção **11**(1), 51–64 (2004)
31. Roy, B.: Multicriteria Methodology for Decision Aiding. Kluwer Academic Publishers, Boston (1996)
32. Ensslin, L., Montibeller Neto, G., Noronha, S.M.: Apoio a decisão – metodologia para estruturação de problemas e avaliação multicritério de alternativas. Florianópolis: Insular (2001)
33. Bana E Costa, C.A.: Structuration, construction et exploitation dun modele multicritered'aide a la decision. Thesis (Doctorate in Ingenierie de Systemes) – Instituto Técnico Superior, Universidade Tecnica de Lisboa, Lisboa (1992)
34. Tezza, R., Zamcopé, F.C., Ensslin, L.: A metodologia multicritério de apoio à decisão construtivista para a identificação e avaliação de habilidades para o setor de estamparia têxtil. Universidade Federal de Santa Catarina, Florianópolis (2009)
35. Günther, H.: Pesquisa qualitativa versus pesquisa quantitativa: esta é a questão? Psicologia: Teoria e Pesquisa, **22**(2), 201–209 (2006). https://doi.org/10.1590/S0102-37722006000200010. Accessed 19 Mar 2021
36. Ensslin, L., et al.: Identification of costumers needs in the products development process: an innovative proposal illustrated for the automotive industry. Produção **21**(4), 555–569 (2011)
37. Bana E Costa, C.A., Sanchez-Lopez, R., Vansnick, J.C., De Corte, J.M.: Analisis multicriterio para la toma de decisiones: metodos y aplicaciones. In: Leyva Lopez, J.C. (ed.) Introduccion a MACBETH, pp. 233–241. Plaza y Valdes, Mexico (2011)

Performance of DNS Over HTTPS Implementation on Low-Power Devices

Ladislav Huraj(✉), Dominik Hrinkino, and Marek Simon

Department of Applied Informatics, University of Ss. Cyril and Methodius in Trnava, 91701 Trnava, Slovak Republic

{ladislav.huraj,hrinkino1,marek.simon}@ucm.sk

Abstract. Privacy and anonymization of network communications are very important in modern computer networks. This is mainly because communication takes place using computer networks on a daily basis. The DNS system is one of the oldest protocols in computer networking, making it easy to monitor, read, or manipulate DNS queries from the local network. An example of this could be blocking a website at the DNS level from a provider or system administrator, or redirecting DNS communication to an attacker's or system administrator's DNS server.

In this paper, we discuss the possibilities of enhancing the security of DNS communication using DNS over HTTPS and its implementation on a low-power device. We also investigate by measuring the speed of obtaining responses to uncached requests against other open-source or commercial implementations of DNS over HTTPS and compare them against traditional unencrypted DNS communication. The goal is to highlight the use of low-power devices in network security and compare their performance against similar open-source or commercial implementations.

Keywords: DNS · DNS over HTTPS · DNS Proxy · DNScrypt-proxy · Raspberry Pi

1 Introduction

Due to technological advances in the world of computing, it is now possible to communicate seamlessly and securely between companies, government offices, or even just with family and friends. Data encryption and security have never been so inflated as they are today in an era full of attacks and data leaks. Many protocols have already responded to these events and have implemented encryption for their communications. DNS, the Domain Name System, is one of the keystones of modern network communications, without which many users cannot imagine functioning and navigating computer networks. This protocol has its roots in the ARPANET, and it can be said that the DNS protocol is one of the few protocols that has not changed during its existence [1]. However, today's requirements exceed the capabilities of DNS, especially when network security and privacy are taken into account. Traditional DNS requests are fully transparent and easy to read, either for the network administrator, the ISP, or even for an attacker

R. Silhavy and P. Silhavy (Eds.): CSOC 2023, LNNS 723, pp. 415–425, 2023.
https://doi.org/10.1007/978-3-031-35317-8_38

[2]. Using DNS, it is possible to build a profile of a user. The profile shows which sites the user visits, at what times, and how often. Such information is very precious and especially sensitive as it represents personified activity on computer networks, especially on the Internet.

During the existence of DNS, there have been several projects that have aimed to increase anonymity when communicating with DNS servers. Perhaps the most popular is DNS over TLS, which uses a TLS tunnel to create an encrypted channel to communicate with a DNS over TLS server that can respond to requests. The main drawback of DNS over TLS is that it has an explicitly defined TCP port of 853. It is very easy for a network administrator to filter this communication on the network and completely disable all DNS over TLS communication [3]. The DNS over HTTPS standard can solve this problem. The present study implements the DNS over HTTPS standard on a low-power Raspberry Pi 4 Model B device. The goal is to show that even such devices have sufficient performance and can increase security and anonymity in computer networks without major speed differences compared to commercial products or robust servers found in data centers [4].

Furthermore, the tested Raspberry Pi-based solution can be deployed alongside IoT devices, with the implementation cost not exceeding the value of low-cost IoT devices. It should be noted that the DNS server operator and any entity that monitors DNS traffic can misuse the information obtained to track users within the websites they visit. Additionally, previous research shows that an attacker can infer what IoT devices are present and also some of the activities that are carried out at home [5].

2 DNS Over HTTPS

The DNS over HTTPS is one of the newest forms of communication with a DNS server. Unlike traditional DNS, which uses unencrypted communication using UDP, DNS over HTTPS uses the existing HTTPS standard, using TCP on port 443. HTTPS or also Hyper Text Transport Protocol Secure is primarily used for web browsing. DNS over HTTPS benefits from this use case. DNS over HTTPS uses standard HTTPS communication to retrieve responses from the DNS server. The primary advantage over implementations that have previously addressed DNS communication encryption, such as DNS over TLS, is that DNS over HTTPS uses HTTPS for communication. Therefore, blocking port 443 on a computer network is counterproductive and would cause many websites to be blocked [6, 7].

The DNS over HTTPS proxy concept is based on the DNS proxy concept. The DNS proxy is built in such a way that it presents itself to the devices as a fully functional DNS server; however, such a proxy does not take care of any translation of requests, but these requests are forwarded to the selected recursive DNS server, which already performs the translation, and such a response is returned to the DNS proxy. Some DNS proxies may also store translated records in their cache memory, but RFC5625 does not explicitly require this [8]. The DNS over HTTPS proxy adopts the above features of a standard DNS proxy, but compared to a DNS proxy, the requests sent are not directed to a traditional recursive DNS server but are sent to the DNS over HTTPS server as valid DoH requests. Likewise, the DoH server response for the DNS over HTTPS proxy is formulated as a valid DoH response. However, DNS over HTTPS proxy still communicates with end stations using traditional DNS. DNS proxy, or even DNS over HTTPS proxy, is located directly inside the local network or before the last router that borders the provider network and the local or organization network [8], Fig. 1.

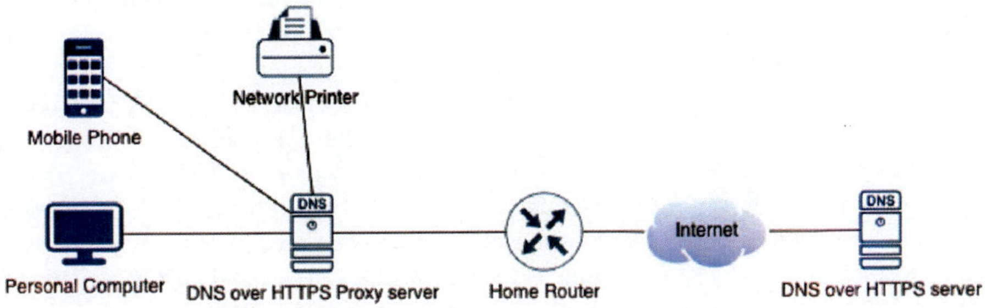

Fig. 1. Topology of possible DNS over HTTPS Proxy implementation in a local home network.

Many modern web browsers, such as Mozilla Firefox or Google Chome, can natively use DNS over HTTPS; the same cannot be said for operating systems. Microsoft with their Windows 11 operating system added support for using DNS over HTTPS for system communication as well, but older operating systems or many GNU/Linux distributions do not support such native functionality [9]. To implement DNS over HTTPS on systems that do not support its functionality, the implementation is provided by a service that handles the translation. However, the installation of the service can be tedious for multiple devices and in the final analysis decentralizes the management and creates multiple points of failure. Other devices that are more critical are mobile phones, game consoles, smart TVs, and a wide variety of IoT devices. These devices usually do not provide a comprehensive network configuration, and their entire network configuration depends a lot on DHCP. A possible solution is provided by DNS over HTTPS proxies, where no major changes need to be made to the local network. Simply change the DNS server at the end stations either manually or by using a protocol such as DHCP to a device that implements DNS over HTTPS proxy.

3 Experiment Test Environment

The main objective of the experiment was to show the benefits of using low-power devices in network security, in particular by deploying DNS over HTTPS resolver implementations, and to make a comparison with a similar price range of commercial products. For a better comparison, measurements of the same model implemented on a standard server were also performed; in addition, DoH testing was also done using a local DoH resolver implemented directly on the device. In addition to testing DNS over HTTPS, testing was also performed using traditional unencrypted DNS communication. Measurements were made for all methods that were also used for DNS over HTTPS and additionally by direct communication with the DNS resolver in the Internet. A public DNS server from Cloudflare was used for testing. The choice of Cloudflare was mainly due to the fact that the company offers a standard DNS resolver and also a DNS over HTTPS resolver. The resulting data are shown in the graphs below.

For the implementation of the experiments, a special network connected to the Internet was set up; the selected device that performed the test measurements and the model under test were on the same L2 network, connected by a switch. The topology of the experiment was chosen in order to avoid possible deviations in the measurements produced by the router used, Fig. 2. Mikrotik HaP Lite devices, Raspberry Pi 4 4GB version and Dell PowerEdge R210 II server were used for the implementation of the models and Dell PowerEdge R210 II server was used for the virtual machines. The Mikrotik HaP Lite device was chosen due to its great popularity in homes but also due to its scalability in SOHO (Small Office, Home Office) environments. Mikrotik devices are frequently used in home and SOHO environments because of their simplicity, flexibility, and affordability. Mikrotik devices also offer a wide range of features for network management, including firewall policies, VPNs, hotspots, DNS and DHCP servers, and more. Mikrotik devices in their RouterOS operating system support native implementation of DNS over HTTPS along with standard DNS for upstream DNS. The offered Mikrotik device configuration matches the experiment model created on the Raspberry Pi device.

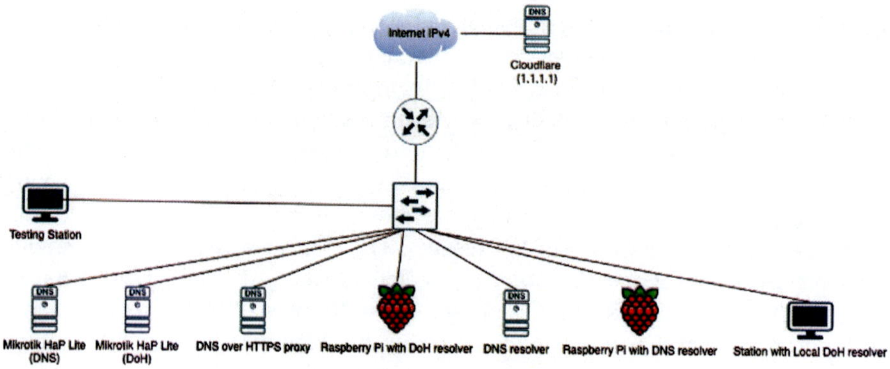

Fig. 2. Logical network scheme.

Each device was assigned its own private static IP address, and each device had its own 1GBps Full-Duplex link from the switch; however, the Mikrotik hap lite only used a 100Mbit FullDuplex connection due to its NIC limitations. During all tests, the cache memory function was intentionally disabled in the test station and in all test models; the cache memory function could affect the measurement results. If the cache memory function was used, either at the DNS resolver level or at the system level, the cache memory would subsequently no longer request domain name resolution if it had the requested record. Consequently, when the DNS resolver is queried, the answer is returned from the cache. In real operation, the cache memory functionality is commonly used, which speeds up the overall network communication process.

The implementation of the DoH proxy on the Raspberry Pi was done using the DNScrypt-proxy program. DNScrypt-proxy allows primary communication with an external DNS resolver using their DNScrypt protocol. DNScrypt-proxy works on the principle that it pretends to be a standard DNS resolver for the device, which receives unencrypted DNS requests using UDP port 53. It processes the requests internally and communicates encrypted with the external resolver defined in the configuration file using the chosen protocol, in the experiment implemented using DNS over HTTPS [10]. The advantage of the above implementation is that there is no need to make major changes to the end stations inside the network or to change the end stations for newer ones with DNS over HTTPS support. All that is needed is to change the DNS server on the end-station directly to a DNScrypt-proxy device. Implementation is very easy on Raspberry Pi devices due to their support of GNU/Linux distributions. The official Raspberry Pi OS Lite distribution was used in the model, which shares many similarities with the GNU/Linux Debian distribution.

The DNS over HTTPS proxy model on the x86_64 device represents the identical DNScrypt-proxy configuration as implemented on the Raspberry Pi. GNU/Linux Debian was chosen as the operating system. This choice was made because the operating system used for the "DNS over HTTPS proxy on Raspberry Pi" is the Raspberry Pi OS, which is based on the GNU/Linux Debian distribution. Due to this choice, the configuration is identical. The deployment on an x86_64 device was chosen as a direct comparison to the Raspberry Pi. The identical implementations allowed one to create measurements that directly describe the difference in performance between a small Raspberry Pi device and a standard server that can be found in a data center.

The Mikrotik HaP Lite TC, somewhere referred to as RB941-2nD-TC, is a Mikrotik SOHO router that offers multiple configuration options. Due to the variety of configurations, this router can be used not only as a traditional SOHO router, but also as a DNS resolver, WAP, firewall, and also as a switch. The operating system of the device, RouterOS offers the use of a DNS over HTTPS upstream server. Due to this, it was possible to create a model similar to the one implemented on the Raspberry Pi device. In the model on the Mikrotik device, all router options were disabled and a static address and gateway were set up for network access. The appropriate certificate needed to authenticate and encrypt communication with Cloudflare's DoH server was uploaded to the device. The DoH server was then defined using the DoH URL. Due to the fact that switching between a traditional upstream DNS server and a DoH upstream server is relatively easy, there was only one Mikrotik HaP Lite TC device in the network, and during the measurement, the configuration was only adjusted according to the type of measurement. The Mikrotik itself does not allow to completely disable the cache memory for storing the results. However, for the purpose of the measurements, the cache memory was required to be disabled due to the possible influence of the measurements. Mikrotik supports manual definition of the maximum TTL value for the cache. To simulate the cache memory off, a TTL setting of 0 s was chosen and the difference between requests sent was defined as 2 s. This setting allowed bypassing the internal cache memory of the device [11].

Since the DNScrypt-proxy does not support the ability to send DNS requests using the traditional method, the model needs to be redesigned and a new implementation created. For the DNS resolver model, the Dnsmasq service was used. Dnsmasq implements a DNS cache and a DHCP server within itself. Only the DNS cache implementation offered by Dnsmasq was used for the experiment, as the main goal was to implement a DNS forwarder. Dnsmasq, unlike Mikrotik routers, supports the disabled local cache [12]. As a result, the model behaves identically to the Mikrotik model with DNS upstream.

The Dnsmasq implementations for Raspberry Pi and for GNU/Linux Debian are identical. Therefore, it was sufficient to copy the same configuration from the Raspberry Pi model to the new model. The functionality remained identical without any changes as on the Raspberry Pi model.

As the next model, a local DNS Over HTTPS resolver was implemented directly on the test device. The implementation of a local DoH resolver is already present in, for example, the Microsoft Windows 11 operating system; however, the testbed in the experiment was based on GNU/Linux Debian, which allowed greater access to network settings. The implementation of the local DoH translator consisted of deploying a DNScrypt proxy with the same configuration as the model deployed on the Raspberry Pi. However, with this model, DNS requests did not leave the device's network stack, and all traditional DNS communication was done internally on the device, and the device communicated over the Internet using DNS Over HTTPS.

As a final model for comparison, sending requests directly to the IP address of a DNS resolver on the Internet network was added. Again, the Cloudflare DNS resolver was used, with the local DNS cache turned off for the measurement. This ensured that each request was sent to the DNS server and that no responses were stored.

The test device that sent all DNS requests to perform the translation was an x86_64 virtual machine running the GNU/Linux Debian 11 operating system. The choice of operating system was due to the greater ability to configure the network stack and the greater control over communication on the device. The GNU/Linux Debian operating system is based on the systemd init system, and this init system also provides DNS caching at the system level. For this reason, system modifications were needed and the systemd-resolved service was disabled, which takes care of DNS translation. After disabling this service, the system directly used the GNU/Linux Debian network manager, while no local cache was used [13].

4 Test Scenarios and Results

Four test scenarios were created for the purpose of the experiment. The first scenario dealt with translating the domain "kai.fpv.ucm.sk" and retrieving its A record. For testing, a special Python script was created that was able to invoke a DNS request to translate the domain name using the Python library socket. The time from the start of the request to its successful translation was calculated by detecting the time before the translation was invoked and detecting the time after the response was obtained. Data were recorded in milliseconds, Table 1. Testing took place in each of the models mentioned above, by performing one hundred measurements on each model. A delay of 2 s was added between measurements, primarily to simulate standard communication and to avoid potential cache usage in the model with a Mikrotik device that does not explicitly allow cache disabling. For models that required external devices, DNS requests were sent from the test device. For the local DoH translator model, testing was performed directly on the device where the model was implemented. The results of each model were averaged by arithmetic mean. All values are reported in milliseconds rounded to two decimal places for better representation. The results are recorded in the bar graph in Fig. 3, where the X-axis shows the time in milliseconds of each model tested.

Table 1. Results of the translation measurement of the A record of the domain kai.fpv.ucm.sk

Device with service provided	milliseconds
Mikrotik DoH	111.32
Raspberry Pi DoH	15.75
Mikrotik standard DNS	14.25
x86_64 DoH resolver	12.91
Local DoH resolver	12.42
Raspberry Pi DNS resolver	11.53
x86_64 DNS resolver	11.48
Standard DNS	10.75

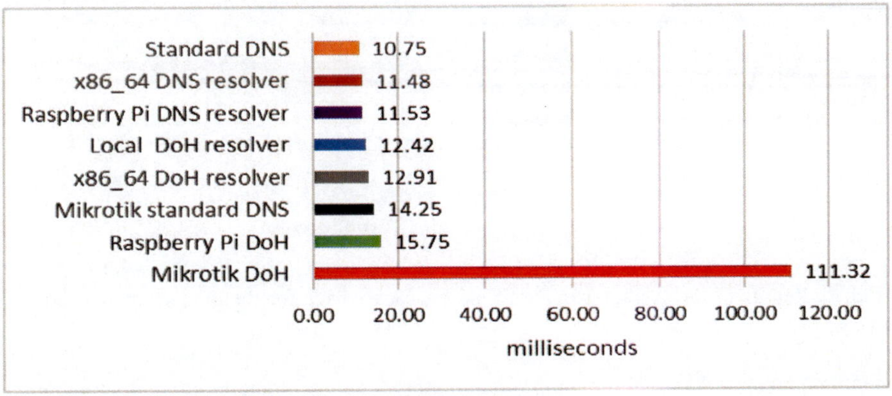

Fig. 3. Results of the average response time for each device for the domain kai.fpv.ucm.sk (A records).

The second scenario dealt with measuring the speed of retrieving all TXT records from the fpvucm.sk domain. The measurement was similar to the first scenario, but here the "dnspython" library was used, since the "socket" library did not support explicit retrieval of TXT records only. For this scenario, measurements were also taken one hundred times for each model with a delay of 2 s, Table 2. The values were then averaged and recorded in the table in milliseconds and displayed on a graph, Fig. 4.

In the third scenario, the acquisition rate of MX records was measured. MX records are primarily used for mail servers to retrieve the SMTP server that handles mail services for the domain. The testing was exactly the same as in the second scenario. Again, the retrieval of all MX records for the domain fpvucm.sk was measured, one hundred

Table 2. Results of TXT records retrieval speed measurement from fpvucm.sk domain in milliseconds

Device with service provided	milliseconds
Mikrotik DoH	113.55
Mikrotik standard DNS	13.59
Raspberry Pi DoH	13.13
x86_64 DoH resolver	12.98
Local DoH resolver	12.02
x86_64 DNS resolver	11.51
Raspberry Pi DNS resolver	11.41
Standard DNS	10.80

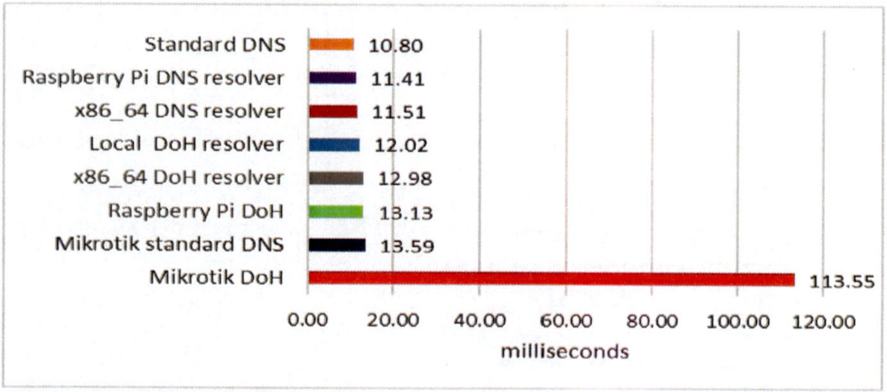

Fig. 4. Results of the average response time for each device for the domain fpvucm.sk (TXT records).

measurements were taken for each model with a delay of two seconds, Table 3. These values were averaged using an arithmetic mean and the data were recorded in a table and graph in milliseconds, Fig. 5.

The last fourth scenario dealt with measuring the translation and retrieving the A records of one hundred selected domain names. These were mainly academic websites, various government offices, and root name servers. The measurement was performed using a Python script with the Python library socket. In the measurement, for each domain name model, the response to ten requests for that domain name was measured with a two-second interval between each request. From these measured translation times, an arithmetic mean is created and stored in an array with the other values. Unlike the previous scenarios, in this scenario the data is represented using a line graph as a time increment where the time of the average responses is gradually added up and at the end represents the total execution time of all responses, Fig. 6.

Table 3. Results from measuring the speed of retrieving MX records from the domain fpvucm.sk in milliseconds

Device with service provided	milliseconds
Mikrotik DoH	111.88
Mikrotik standard DNS	15.00
Raspberry Pi DoH	14.93
x86_64 DoH resolver	13.54
Local DoH resolver	13.52
x86_64 DNS resolver	12.71
Raspberry Pi DNS resolver	11.50
Standard DNS	11.22

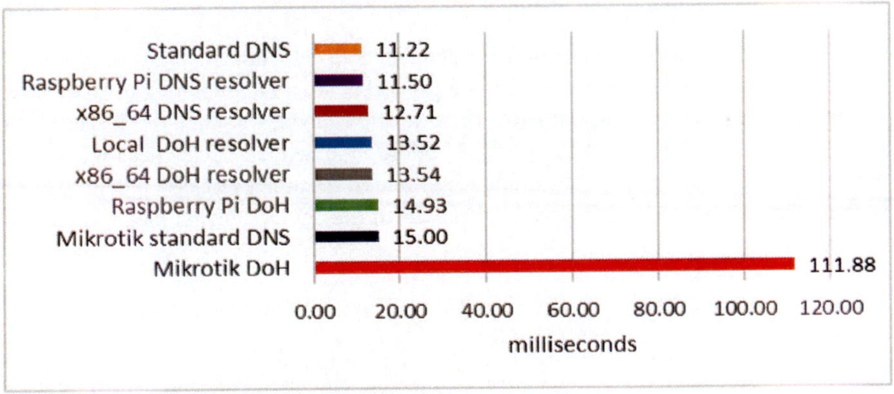

Fig. 5. Results of the average response time for each device for the domain fpvucm.sk (MX records).

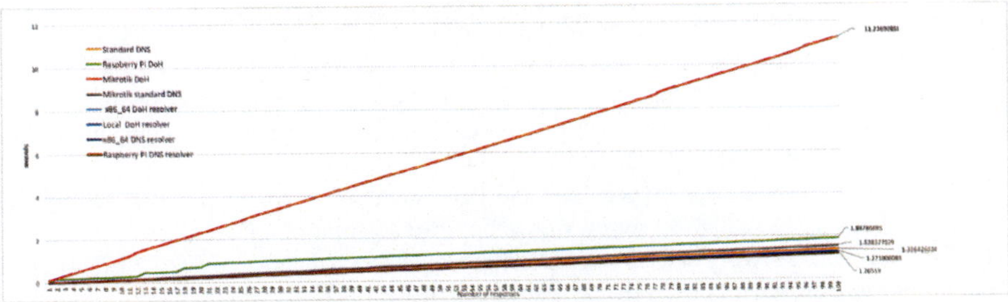

Fig. 6. Results of measuring the average response rates for 100 domain names in a time sequence.

5 Conclusions

Experimental results show that the Raspberry Pi device, which belongs to the category of low-power devices, can match or even outperform commercial and other open-source implementations. A simple model can increase network anonymity by encrypting DNS requests using DNS over HTTPS. Testing has shown that the use of DNS over HTTPS does not significantly affect the speed of communication on a computer network and is very similar in speed to traditional DNS proxies, or traditional DNS servers, which can be found implemented on many home routers. In the DNS over HTTPS proxy results measured, the proposed implementation on Raspberry Pi was the best, along with the implementation on the x86_64 virtual machine. On the other hand, the worst results were measured on the Mikrotik HaP Lite TC device with the LTS version of the RouterOS 6.48.6 operating system; for the Mikrotik device, DNS proxies without DNS over HTTPS were also measured where the results were more favorable. It is important to work with the information that the tests completely bypassed the caching process; in typical communications, caching can significantly reduce the time to retrieve requests that are repeated multiple times.

The testing performed may be complemented in the future by testing Oblivious DNS over HTTPS [14], which serves as a proxy between DoH client and DoH server communication. As a result, the network administrator does not see the IP address of the DoH server, but only sees the proxy. In addition, extended device tests will also be performed in the future under the influence of one of the attacks on computer networks, such as the DDoS attack [15], and the behavior of the devices during the attack will be investigated.

Acknowledgements. The work was funded by the grant KEGA 012UCM-4/2021 *Modern technologies and innovations in network security education.*

References

1. Rader, R.: One history of DNS. Byte. org (2006)
2. Mockapetris, P.V.: Rfc1035: Domain names-implementation and specification (1987)
3. Hu, Z., et al.: Specification for DNS over transport layer security (TLS). No. rfc7858 (2016)
4. Skrinarova, J., Dudáš, A.: Optimization of the functional decomposition of parallel and distributed computations in graph coloring with the use of high-performance computing. IEEE Access **10**, 34996–35011 (2022)
5. Nisenoff, A., et al.: User expectations and understanding of encrypted DNS settings. In: Proceedings of NDSS DNS Privacy Workshop. Virtual Event (2021)
6. Csikor, L., et al.: Privacy of DNS-over-HTTPS: requiem for a dream? In: 2021 IEEE European Symposium on Security and Privacy (EuroS&P). IEEE (2021)
7. Hynek, K., et al.: Summary of DNS over https abuse. IEEE Access **10**, 54668–54680 (2022)
8. Bellis, R.: DNS Proxy implementation guidelines. No. rfc5625 (2009)
9. Mitsuhashi, R., et al.: Malicious DNS tunnel tool recognition using persistent DoH traffic analysis. IEEE Trans. Netw. Serv. Manag. (2022)
10. Singanamalla, S., et al.: Oblivious DNS over https (ODOH): a practical privacy enhancement to DNS. arXiv preprint arXiv:2011.10121 (2020)

11. Wiki.mikrotik.com (2018). Manual:IP/DNS. https://wiki.mikrotik.com/wiki/Manual:IP/DNS. Accessed 16 Feb 2023
12. Rankin, K.: Hack and: localhost DNS cache. Linux J. **2014**(247), 7 (2014)
13. Klein, A.: Cross layer attacks and how to use them (for DNS cache poisoning, device tracking and more). In: 2021 IEEE Symposium on Security and Privacy (SP). IEEE (2021)
14. Kinnear, E., et al.: RFC 9230 Oblivious DNS over HTTPS (2022)
15. Šimon, M., Huraj, L., Čerňanský, M.: Performance evaluations of IPTables firewall solutions under DDoS attacks. J. Appl. Math. Stat. Inform. **11**(2), 35–45 (2015)

The Future of Next Generation Web: Juxtaposing Machine Learning and Deep Learning-Based Web Cache Replacement Models in Web Caching Systems

Elliot Mbunge[1](✉), John Batani[2], Stephen Gbenga Fashoto[1], Boluwaji Akinnuwesi[1], Caroline Gurajena[3], Ogunleye Gabriel Opeyemi[4], Andile Metfula[1], and Zenzo Polite Ncube[1]

[1] Department of Computer Science, Faculty of Science and Engineering, University of Eswatini, P Bag 4 Kwaluseni, Eswatini, Swaziland
mbungeelliot@gmail.com

[2] Faculty of Engineering and Technology, Botho University, Maseru 100, Lesotho

[3] Department of Computer Science, University of Fort Hare, Alice, South Africa

[4] Department of Computer Science, Federal University Oye-Ekiti, Oye, Nigeria

Abstract. Massive data generated by connected smart devices, particularly in distributed computer networks, contributed to the large network traffic burden caused by the ever-increasing use of the Internet infrastructure and web caching systems. Such connected smart devices generate massive data, shared and synchronized together in real-time over the connected nodes using various emerging technologies through the Web. However, the performance of classical web caching methods usually degrades when caching web objects due to various factors. Therefore, this study presents a comprehensive review of machine learning and deep learning-based web cache replacement models that could be effectively used to improve the performance of web caching systems. The study revealed that random forest, artificial neural networks, support vector machine, LSTM and fuzzy approaches are among web cache replacement models that have been used to improve the performance of web caching systems. However, due to the sparse use of web object features, some models are typically unable to handle today's unpredictable web caching demands. Therefore, to deal with rising web usage, high latency, increased user-perceived delays, web cache overload, network traffic congestion, increased web object size, real-time data sharing, and limited bandwidth size, various web object attributes should be included in the next-generation adaptive and robust machine and deep learning-based web cache replacement models. In addition to web object features such as recency, frequency, cost, hit rate, modification and expiration time, more features are required in developing adaptive web caching algorithms to improve the performance of web caching systems.

Keywords: Web caching · Machine learning · Web cache replacement · Deep learning · Industry 4.0

R. Silhavy and P. Silhavy (Eds.): CSOC 2023, LNNS 723, pp. 426–450, 2023.
https://doi.org/10.1007/978-3-031-35317-8_39

1 Introduction

The World Wide Web (WWW), also known as the Web, has significantly become the most crucial source of information on the Internet in various fields including education, health, entertainment, gaming, business, and telecommunications [1]. Internet usage is increasing exponentially due to the ever-growing demand for data sharing, transformation and transmission [2], especially in the era of artificial intelligence, big data, social networks, cloud computing, Internet of People (IoP), Internet of Things (IoT), edge computing systems [3] and the fourth industrial revolution (4IR). Such demand is transforming the Web environment into a multifaceted system with tight synchronization and coordination between machine learning and deep learning models, physical elements, communication protocols, computing devices and their virtual representation. This results in the integration of online technologies and communications solutions, ultimately leading to the Internet of Things [4] and industrial cyber-physical systems [5]. Industrial cyber-physical systems involve the global stringent connectedness between digital devices and the physical world using the Internet and the Web. These devices are interconnected through communication protocols to share, disseminate and transmit data using internet connections. Given the pervasiveness of IoT technologies [6], it is inexorably shifting in form and nature by integrating emerging advanced technologies, especially in the fourth industrial revolution (4IR) era.

The fourth industrial revolution, which is also known as Industry 4.0, involves the integration and use of environmentally friendly high-tech computing technologies including artificial intelligence, Internet of People, 3D printing, nanotechnology, quantum computing, big data, cloud computing, 5G technology and the Internet of Things which allows ordinary objects to communicate with each other online in real-time [7, 8]. The 4IR transforms the world by creating a close interaction and ubiquitous connection between humans, pervasive computing devices and smart household items, data and smart cars. Such interactions lead to the increased usage of the Web and subsequently overloaded Web servers, network traffic, access latency and congestion [9], high bandwidth usage, user-perceived delays and consequently poor response time [10]. This eventually degrades web servers' performance and causes an unreasonable amount of time when sending, uploading and retrieving documents from the web as well as storing documents in web servers [11]. This is also attributed to the use of conventional web cache replacement algorithms to deal with an online and real-time large collection, processing, and analysis of structured and unstructured data [12]. A web cache is a hardware device or software application for temporarily storing frequently accessed static content. Conventional web cache replacement algorithms are not a panacea and are unlikely to meet increasing users' demands, especially in the emotive Web and beyond Industry 4.0 because of variations of web objects size, bandwidth demand, networks traffic and web server overload as well as the problem of determining ideal web objects that are likely to be revisited in the future [13].

Contribution of the Study

Several traditional approaches, including web caching, have been implemented to overcome challenges encountered in the degradation of web servers' performance [14]. Web caching is one of the first major techniques implemented to reduce network traffic [15] and also to reduce the load from the original remote servers; proxy servers are used. A web cache is defined as "*web cache is a buffered repository of the Web objects that are most likely to be requested frequently*" [10]. There are two ways of improving web cache performance, namely, "increasing the cache size or decreasing latency time by prefetching, but its cost is very large" [15], and designing effective and efficient web cache replacement algorithms [16]. However, the performance of web cache replacement algorithms is affected by several factors including feature statistics or frequencies, previous access time, cost of fetching the object, cache block size [17] and cache object size [18]. Most of the web cache replacement algorithms focus more on the web object's age, frequency, and recentness, and ignore other factors that have an impact on the Web caching such as the objects' varying sizes [19].

There are three types of web caching: proxy caching [20], server-side caching [21], and client-side caching [22]. Client-side caching refers to the cache that is performed by the web browsers to store users' web page addresses, passwords, usernames and other information. It is done on the user side; however, client-side caching is for a single user at a time. Server-side caching is a process of establishing the cache on the web server side to reduce server load [23]. Proxy caching acts as an intermediator between the web user and the web server. It decides whether to store a copy in its cache or not when responding to and processing a user's request. Proxy caching is performed by the Proxy servers. The main functions of proxy servers include web caching, keeping clients anonymous and blocking unnecessary content from the network and logging and auditing client usage [24].

Several researchers have published systematic reviews on classes of web cache replacement algorithms from several different perspectives [6, 10–12, 20, 21, 25–28]. However, there is limited literature on the application of machine learning and deep learning models in web cache replacement, especially in the era of the 4IR. To bridge the gap, this study provides a comprehensive review of web caching, focusing on the performance of machine learning and deep learning-based web cache replacement models developed and applied across different web caching systems. Therefore, this study sought to achieve the following objectives:

- Identify and explain deep learning and machine learning-based web cache replacement models deployed in various web caching systems.
- Assess the performance of deep learning and machine learning-based web cache replacement models deployed in various web caching systems
- Identify research gaps that could be addressed when developing the next-generation data-driven web cache replacement models.

Therefore, the remaining sections of this paper are organized as follows: Sect. 2 provides an overview of the web evolution. Section 3 presents the methodology adopted to carry out this review. Section 4 provides a comprehensive analysis of deep learning and machine learning-based web cache replacement models deployed in various web caching systems. Finally, the conclusion and future work are presented in Sect. 5.

2 Related Work

2.1 Web Evolution

The advancement of the Internet has revolutionized how humans interact and how data is accessed, shared, transmitted and transformed. The Web uses the Internet as a medium to share information by using protocols such as Hypertext Transfer Protocol (HTTP), File Transfer Protocol (FTP), and Simple Mail Transfer Protocol (SMTP) to transmit data over the Internet [29]. The Web was introduced first by Tim Burners-Lee in 1989 [30] with the ascension to Web 1.0 as a web of human cognition [31]. Web 1.0 provides static web pages as "*read-only*" and is known as the first generation of the WWW which is an information space for content delivery only to publish information to anyone anytime [30] and to establish an online presence. According to [32], Web 1.0 uses Hypertext Markup Language (HTML), HTTP and Uniform Resources Identifier (URI) as core web protocols to display static web pages. The lack of dynamic and responsive representation of web pages is the tremendous challenge of Web 1.0. The abrupt change of web technology led to the official introduction of Web 2.0 by Dale Dougherty in 2004 as a business revolution that supports collaborative "*read-write web*" for user-centric, business-centric and technology-centric [29] moving from static era to dynamic on different platforms including Facebook and Twitter, and user-generated content such as blogs, wikis and YouTube [33]. Web 2.0 supports flexible web design, dynamic creative reuse, updates, and collaborative content creation that ultimately lead to increased bandwidth usage and web object size. In 2006, Tim Berners-Lee suggested the introduction of the third generation of the web as the "*Semantic Web*" [34], to enhance data sharing, transformation and transmission as well as increase efficiency in providing services and data processing.

Web 3.0 uses data mining, artificial intelligence web crawlers, spiders and scripts that provide automation, integration, recommendations, and personalized suggestions. [32] states that Web 3.0 supports Open Source Software Platforms, World Wide Databases, Resource Pooling, Intelligent Web and Web Personalization. However, Semantic Web faces several challenges including:

- the vastness of web pages that might lead to data redundancy which may not be able to eliminate all semantically duplicated terms [35],
- the vagueness of user queries which may be difficult to combine in different knowledge bases with overlapping but subtly different concepts,
- logical contradictions arising from using large different ontologies from unique and various sources [36].

Web 4.0 is the fourth generation of the Web, known as an ultra-intelligent electronic agent, webOS, symbiotic web and ubiquitous web. Web 4.0 supports a highly interactive ubiquitous web with the full pervasiveness of autonomous intelligent agents [37] and computing devices to read, navigate the contents of the web and decides what to execute and load to the web. According to [29], Web 5.0, also known as sensory-emotion Web, is the latest version of the Web that incorporates emotional components between human and machine interaction through the use of ultra-sensors. Such drastic advancement in Web technology evolves towards 4IR requires more advanced web cache replacement algorithms that can adapt to the increasing data and performance demand [38], and web objects size without degrading system performance.

2.2 History of Web Proxy Servers

The history of Web proxy servers dates back to the very beginning of networking and the Internet itself. Web proxies began to play a very important role around the years of 1994 [39]. A web proxy server is a neutral computer that serves as a gateway between local computer networks and external networks like the Internet [40]. Web proxy servers provide access to the Web for client users on closed subnets who can only access the Internet through a firewall machine [41]. The web proxy server acts as a middleman (see Fig. 1), for both client users and web servers with the main function of forwarding requests to the original servers on behalf of the clients and also fetching the results from the original servers back to the clients that requested that particular document or web object. Additional functions of the proxy servers include logging and auditing client usage, masking an Internet Protocol (IP) address to keep clients anonymous, blocking unwanted content on a network and also caching.

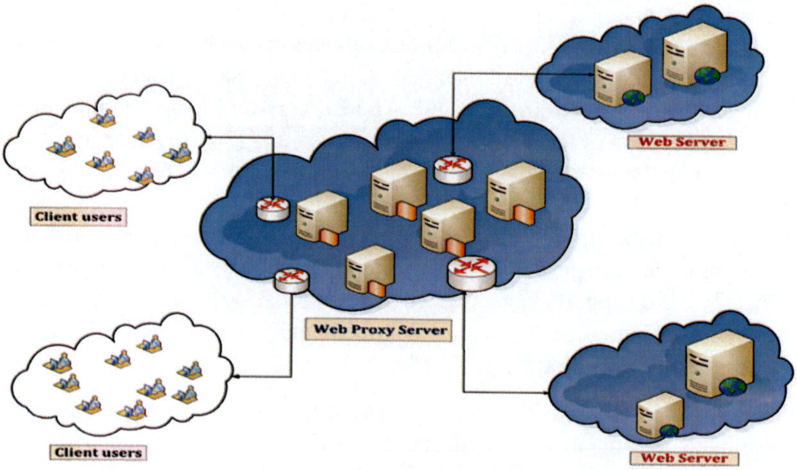

Fig. 1. : Web proxy server

Therefore, whenever a client user's request is satisfied by the web proxy server, it minimizes the response time and reduces the overload of the web origin server [10].

2.3 Why Web Caching

According to [17, 42], web caching reduces bandwidth consumption, which in turn decreases the network traffic and lessens network congestion. Web caching also reduces the access latency in the sense that the frequently accessed files or documents are fetched from the nearby web proxy caches [15] instead of the remote original servers which minimizes the transmission delay. The reduction in the network traffic causes documents that are not present in the cache to be retrieved relatively faster due to less congestion and minimized workload at the remote servers [43]. Web caching also serves as a backup in the case where the remote server has temporarily crashed, the client can obtain a cached copy from the proxy server.

2.4 Current Trends in Proxy Caching

There are different types of proxy caching but only two are considered in this study and they are as follows:

Reverse Proxy Caching: Reverse proxy caches are set up closer to the origin of the content being requested as opposed to the clients, as shown in Fig. 2. They work best for servers expecting a large number of requests and still need to ensure that they provide a high-quality level of service. This type of proxy caching is also important in aiding web hosting farms (virtual domains mapped to a single physical site) which is a very common service for many Internet Service Providers [44].

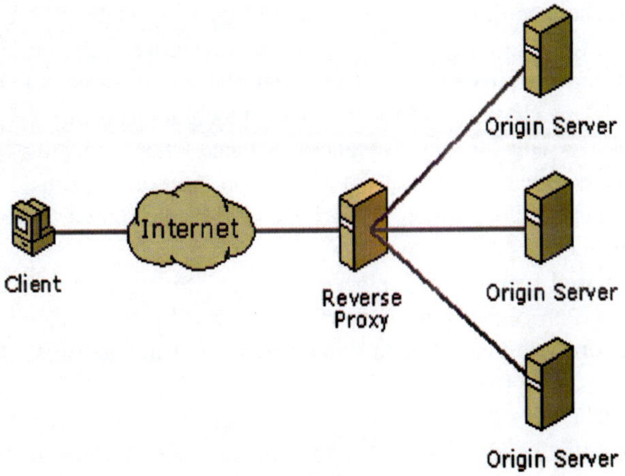

Fig. 2. Reverse proxy server

Transparent Caching: This is the type of proxy caching that eliminates the need for web browser configuration. The transparent proxy works by interfering with HTTP requests and averting them to web cache servers. This type of caching creates a point at which distinct kinds of administrative control are accessible. There are two ways of implementing this kind of proxy caching, router-based transparent proxy caching and switch-based transparent proxy cache [16]. With router-based transparent proxy caching, they use policy-based routing to direct requests to the desired cache (e.g. requests from certain clients can be associated with a particular cache). In the latter, the switch acts as a committed load balancer. This advent is favourable since it minimizes the overhead that is usually suffered by policy-based routing.

2.5 Current Trends in Cache Replacement Algorithms

In the case where the proxy server cache memory is full, a web cache replacement algorithm is invoked to determine which page to replace in the memory buffer. There are several classic cache replacement algorithms such as LRU (the least recently used algorithm) [10], LFU (the least frequently used algorithm) [45], the random number-based algorithm, the cache object size-based algorithm, and WRP (the cache weight-based algorithm) [46]. According to [27], the First in First out (FIFO) cache replacement algorithm replaces the oldest object in the cache memory. The Least Recently Used (LRU) algorithm replaces the object that has been used the least recently [47]. Now, the combined FIFO–LRU cache replacement algorithm is an algorithm that combines the FIFO cache replacement algorithm and LRU cache replacement algorithm to come up with an improved hybrid algorithm that has a better judgment on which objects to evict from the cache and makes space for new ones in the case where the cache memory is full. From previous research work done in this area, there are many web cache replacement algorithms classified based on recency-based techniques, frequency-based techniques, and a combination of recency-frequency-based techniques [17, 48].

Web cache replacement algorithms based on recency-based techniques include LRU [49], LRU-Threshold [50], Pitkow/Recker's strategy [20], SIZE [51], LOG2-SIZE [12], LRU-Min [52], Value-aging, HLRU, Pyramidal Selection Scheme [17]. These algorithms use temporal locality property to predict how likely an object is to appear again shortly. However, the challenge of recency-based techniques is that they do not combine the web object size with recency to create a balance when replacing web objects [10]. In addition to web object size, recency-based techniques do not consider the web object frequency information [17] which could be a vital factor to consider when improving web cache replacement performance.

Frequency-based web cache replacement algorithms use the web object popularity index to cache an object that appears the most [20]. Frequency-based web cache replacement algorithms that fall into this classification include Least Frequently Used (LFU) [37], LFU-aging [53], Least Frequently Used -Dynamic-Aging (LFU-DA) [28], ∞-Aging, and LFU* [54]. The outstanding advantage of this technique is that it keeps track of feature statistics of web objects and replaces the least frequently used object, thus reducing user-perceived delays and access time. However, frequency-based web cache replacement algorithms require additional priority queues to keep track of frequency which might cause cache pollution and complexity to the replacement process.

Hybrid frequency/recency-based cache replacement algorithms maintain the properties of recency-based and frequency-based techniques combined with temporal spatiality. Segmented LRU, Cubic Selection Scheme, Generational replacement, HYPER-G, and LRU with Swapping and Placeholders (LRU-SP) are some examples of Hybrid frequency/recency-based cache replacement algorithms [15, 54, 55]. Also, [26] and [56] posit that some web cache replacement algorithms are function-based, thus, they apply a general characteristic function to define a request value for an object to be replaced.

Function-based web cache replacement algorithms include Greedy-Dual (GD) Size [57], GD* [10], GreedyDual∗ [58], Tyler Series prediction, Least Normalized Cost Replacement for the Web-Proxy Caching (LNC-R-W3-PC) [59] and Least Unified-Value (LUV) [20]. The distinct advantage of this classification is on creating a distinct line between the recency and frequency of web objects, which is a paramount factor in web cache replacement algorithms. However, it is difficult to apply these classification techniques when dealing with randomized web cache replacement algorithms [49]. Function-based web cache replacement algorithms encounter challenges such as choosing the appropriate weights and the use of latency value which creates more problems that can lead to inferior object replacement decisions [17, 20]. A good web cache replacement algorithm adapts itself to changes in Web workload characteristics [12].

The Least Recently Used Algorithm Implemented Using the Counter Logical Clock
The LRU cache replacement algorithm uses a counter-logical clock that was developed to improve the number of cache hits and minimize the number of cache misses [25]. This algorithm works by removing the page that has been used the least recently and it can be implemented using a counter logical clock or stack logical clock. The counter-logical clock implementation works by adding a register to every object frame in the memory buffer. This register contains the last time each page was accessed known as the logical clock. The logical clock advances by one tick every time a memory reference is made [60], meaning if an object gets requested the logical clock gets incremented by one. Therefore, when replacement is done, the object that gets replaced is the one with the smallest logical clock value in its register. The major disadvantage of the LRU counter's logical clock implementation is the identification of the page to be replaced. The algorithm iterates through the whole memory buffer to find the minimum timestamp of a logical clock or object with the smallest logical clock value and replace it, which is a lot of work in cases of larger cache memories which may result in poor performance of the algorithm.

FIFO-LRU Web Cache Replacement Algorithm
Authors such as [61] developed a hybrid algorithm that applied FIFO and LRU to perform better in larger cache sizes, leveraging the advantages of both FIFO and LRU cache replacement algorithms. From their findings, they realized that the LRU algorithm performs better in deciding which objects to keep in the cache and which ones to replace but its major shortcoming is the time it takes in trying to find the object with the smallest time stamp so that it can be replaced [15, 62]. There is a need to speed up the search for the object to be replaced in the cache. This algorithm uses the LRU mechanism in deciding which object to replace in the cache but at the same time maintains the order in which the objects arrived in the cache using FIFO. Maintaining the order in which the objects arrived in the cache allows us to predict where the page with the smallest timestamp might be on the cache as opposed to iterating through the whole memory buffer to find the page to be replaced which may make the algorithm slower in case of larger cache sizes. When searching for the object to be replaced, it searches for the page with the smallest logical clock value starting from the oldest object up to the object in the middle of the memory buffer [63] instead of the whole memory buffer as LRU counter

logical clock implementation does. This is done to reduce the high overhead faced by the LRU algorithm.

3 Methodology

To provide a comprehensive systematic review of web cache replacement algorithms in the era of the 4IR, the study adopted the Preferred Reporting Items for Systematic Reviews and Meta-Analyses (PRISMA) model [64] to critically analyze the existing literature.

3.1 Search Strategy

Relevant journal articles from Google Scholar, Scopus, Science Direct, IEEE Xplore, Web of Science and other databases were searched for articles published between January 2005 and February 2023, using the following search keywords: "*Web Cache Replacement Algorithm*" OR "*machine learning-based web cache algorithms* "OR "*deep learning web cache*" AND "*fourth industrial revolution*" OR" *web Proxy caching algorithms*" OR" *enhanced web proxy*." The choice of these databases was made tenaciously and purposefully because they are the largest and most reliable online databases for scientific research that support multidisciplinary research [29].

3.2 Selection Criteria

Authors screened journal articles based on the following inclusion criteria: the previously published studies should be peer-reviewed; written in English; empirical; should be full papers; related to web cache algorithms. The screening process was done in two phases: screening using papers' abstracts; and screening by analyzing the full paper to determine its relevance to this study. A total of 47 journal articles were included after excluding duplicates and papers that did not match the study's inclusion criteria. All these journal articles were analyzed using the content analysis method whereby each paper was critically read and analyzed using the following thematic areas; performance of the web cache replacement algorithm, performance/ accuracy and limitations.

4 Results and Discussion

The findings of the systematic review reveal the categorization of web cache replacement algorithms based on the computational technique applied, performance and limitations. Table 1 shows deep learning and machine learning web cache replacement models that have been applied to replace web objects in the web cache.

Table 1. Deep learning and machine learning-based web cache replacement models

Model and reference	Model features	Performance	Limitations
Least Recently Used Algorithm and Neuro-Fuzzy System [65]	Proposes an intelligent scheme that is based on a neuro-fuzzy system. The proposed solution splits Web caches on the client computer into short-term and long-term caches. The solution uses LRU to move web objects that are more frequently visited than the pre-specified value and uses a neuro-fuzzy system to classify each web object in the long-term cache into either a cacheable or uncacheable object when the long-term cache becomes saturated	The solution improved the performance of client-side caching substantially compared to other common caching policies	The proposed solution has extra computational overhead in the training process. The classification technique applied encountered some issues when identifying ranking attributes and improving the accuracy of classification
Learning Cache Replacement (LeCaR) [66]	LeCaR manages two data structures of metadata for each cache entry	Results showed that LeCaR is competitive with Adaptive Replacement Cache (ARC) for relatively large cache sizes but is markedly superior when cache sizes become smaller	LeCaR doesn't consider semantic and contextual information in recently accessed pages and future web objects' requests
Semantics Aware Caching System (SACS) [67]	Use semantic information on cached Web objects to select objects to remove from the cache	Results for SACS showed that the proposed algorithm either matched or offered better performance compared to other cache replacement strategies	The size of Web objects was not considered in the model

(*continued*)

Table 1. *(continued)*

Model and reference	Model features	Performance	Limitations
Stimulable Neural Network (SNN)-cache [68]	SNN-cache is inspired by the biological neural network; it analyzes requests in real time to reduce the load on the content server	The algorithm reduces the load on content sever by 30.7%	The SNN-cache algorithm was compared with conventional approaches which are not adaptive such as LFU, LRU, SLRU, FIFO, and RANDOM
BackPropagation neural network [48]	The Web proxy cache replacement scheme used in this model is based on a back-propagation neural network	The model achieves hit rates of about 86.60% of the optimal in worst cases and 100% of the optimal in best cases	The algorithm was compared with LRU and LFU which have their weaknesses highlighted in Sect. 2
Greedy Web Proxy Caching [10]	Uses Weighted Random Indexing	Greedy Web Proxy caching improved the caching system performance by adapting the Data Mining Classifier model through Web User clustering and Weighted Random Indexing Methods	The proposed model suffers from caching pollution
SVM-cache [69]	Uses SVM regression for prediction in web prefetching and caching	The algorithm's performance is efficient with a hit rate between 67% and 89%. The byte hit rate varies between 67% and 98%	Download latency and Web object size were not considered
Adaptive probabilistic cache algorithm (*AProb)* [70]	AProb "is based on three main techniques: (1) dynamic probabilistic caching; (2) ghost list; and (3) adaptive probing and protection" [70]	*Aprob* improved web server hit ratio, footprint distance, and caching time	The model did not consider other web object features

(continued)

Table 1. (*continued*)

Model and reference	Model features	Performance	Limitations
Data Envelopment Analysis (DEA) [71]	Evaluates the efficiency of web caching object replacement strategies using DEA [71]	The proposed solution improved web object replacement efficiency	Large cache sizes may subvert DEA
SG-LRU [72]	Combined LRU and Score-Based Object Selection	The study showed that the combined solution has a better hit rate compared to LRU	The study does not consider the size of Web objects and prediction ability
Very Fast Decision Tree Classifier (VFDT) [73]	Improves the performance of replacement policies of Web proxy cache through the adaptation of VFDT	Improves LRU and Greedy Dual Size Frequency (GDSF) hit and byte hit ratio performance	Hard to implement on different web platforms such as on mobile devices
Multinomial Logistic Regression (MLR) [74]	Uses lightweight machine learning techniques to build semi-intelligent web cache systems	The results reflect that MLR performed better in terms of hit ratio and disk space as compared to LRU and LFU	The model considers recency and frequency to determine the web object to replace
Expectation Multinomial Naïve Bayes classifier [73]	Uses Expectation-Maximization and Naïve Bayes classifier to enhance proxy cache performance	The Expectation Multinomial Naïve Bayes classifier performed better than Expectation Maximization Naïve Bayes-LRU (EMNB-LRU) and GDSF	It has a low byte-hit ratio and doesn't take into consideration the predicted accesses [75]
Intelligent client-side Web caching scheme (ICWCS) [76]	Improves performance by spitting client user cache into the short-term cache and long-term cache	The proposed solution improves LRU and LFU Web caching performance for the hit ratio, byte hit ratio and latency saving ratio	The training process has extra computational overhead and the algorithm was implemented for the client only

(*continued*)

Table 1. (*continued*)

Model and reference	Model features	Performance	Limitations
Neural Network Proxy Cache Replacement-2 (NNPCR-2) [77]	Models NNPCR-2 algorithm in a real proxy server environment	Compared to LRU and GDSF, NNPCR-2 made better and more balanced decisions	The model did not consider size and object modification time
Integrated Proxy Prefetch-Cache System [59]	Integrated Proxy Prefetch-Cache System was developed for poor network communication and limited bandwidth in rural areas	The proposed framework can be incorporated into scalable hierarchical proxy server clusters and distributed geographical region-based proxy server clusters with enhanced content cooperation, efficient metadata management, and low access delay	Needs to be paired with Web caching technology to produce faster performance
Double Deep Q-Network (Double DQN) [78]	A double DQN learning framework was implemented to solve the edge caching optimization problem by maximizing the hit rate of user requests	The Double Deep Q-network was tested against FIFO and LRU and proved to be effective	The model was compared with convection web cache replacement models
Logistic Regression-LRU: A PACS-Oriented Intelligent Cache Replacement Policy [63]	LR-LRU algorithm uses medical data mining to identify variables that affect the probabilities of file access	LR-LFU policy significantly improves cache performance compared to the LRU, LFU, SIZE, GDS, and GDSF policies	LR-LRU cannot predict probabilities of future access [63]
C3C: A New Static Content-Based Three-Level Web Cache [49]	The alggrithm considers the dependencies among the cached content in different parts	C3C reduced duplicate hits while increasing the hit ratio [49]	It does not consider semantic and contextual information in recently accessed pages and future web objects' requests

(*continued*)

Table 1. (*continued*)

Model and reference	Model features	Performance	Limitations
An adaptive multi-level caching strategy for Distributed Database Systems [79]	Instead of the traditional LRU algorithm, the approach uses a more efficient cache replacement algorithm that considers data distribution and time-varying access rate	The approach improved data access performance by 20% compared to traditional caching strategies [79]	Suffers from caching pollution
Q-Learning based Collaborative Cache Allocation (Q-LCCA) [80]	Combines Baseband Units (BBU) pools and Mobile Edge Computing (MECs) into Intelligent BBU Pool (IBPs) to improve cache performance	Q-LCCA improves cache performance and prevents duplicate caches through collaborative cache	For collaborative cache, each Intelligent Baseband Units (BBU) Pool (IBP) must be connected to one or more other IBP
Generalization of LRU (gLRU) [81]	Proposes a solution for asymmetric file sizes through the adaptation of the LRU strategy in which each file is subdivided into chunks of equal sizes	The gLRU strategy showed a 32% improvement compared to LRU	gLRU requires an additional priority queue to keep track of frequency which might cause cache pollution
Decision tree and Naïve Bayes (NB) supervised machine learning algorithms [82]	The proposed approach incorporates the J48 decision tree and Naïve Bayes into an intelligent cooperative web caching approach	The proposed approach outperforms traditional web caching policies which include LRU, LFU, and GDS in terms of hit ratio and byte hit ratio	It does not consider semantic and contextual information in recently accessed pages and future web objects' requests
Fuzzy page replacement algorithm [83]	The model uses the reference rate of each page, frequency and recency cluster cache pages	The proposed algorithm outperforms Distributed Weight Replacement Policy (DWRP), Weight Replacement Policy (WRP), Most Frequently Used (MFU), LFU, LRU and FIFO	Cost and object size was not included in the model

(*continued*)

Table 1. (*continued*)

Model and reference	Model features	Performance	Limitations
FIFO-AR: Cache replacement algorithm based on association rules [84]	Determines the segment to be cached by combining the association rule with the user's play record and the popularity of the segment	Compared to traditional cache replacement algorithms; FIFO-AR offers a higher hit rate and a lower response time	It does not consider contextual information in recently accessed pages and future web objects' requests
Combined Greedy Dual Size-Frequency (GDSF) and Support Vector Machine (SVM) web cache intelligence approach [85]	Combines web replacement approach to use GDSF and SVM	Offers better performance for both the hit ratio and byte hit ratio when the cache size is 16%	The combined web replacement strategy only applied timestamp, elapsed time, size, and type in the URL object. More advanced web object features are required to further test the performance of the combined GDSF and SVM web cache intelligent approach
LR-LRU [63]	Combines logistic regression and the least recently used	The model provides better performance in picture archiving and communication systems	The model does not consider other important web object features

4.1 Deep Learning (DL) and Machine Learning (ML) Models for Web Cache Replacement

Machine learning and deep learning models are subsets of artificial intelligence that have been applied to solve web caching problems [86]. Machine learning models learn and extract meaningful insights from either structured or unstructured datasets. The performance of machine learning models is influenced by the training function and representative features in solving either classification or clustering problems [87]. However, deep learning models solve complex problems by exploiting three or more layers of representation to find a complex and inherent relationship in data to recognize special patterns with high accuracy [88, 89]. Both ML and DL techniques have been applied to solve web caching problems. Therefore, the following subsection discusses several DL and ML models that have been applied to solve web caching problems.

a. **Artificial Neural Network (ANN) and Support Vector Machine (SVM) based web cache methods**

ANN consists of interconnected neurons (processing units) that can learn from the datasets [90, 91]. ANN emulates the functioning of the biological neural network to make reasonable decisions after rigorous training. Each neuron is connected at various proceeding layers namely; hidden layer(s), and an output layer and then performs mathematical computations using activation function(s). A study by [68] applied Stimulable Neural Network (SNN)-cache replacement algorithm inspired by the human brain to analyze the inter-relationships among sequenced requests in real time. The algorithm reduces the load on the content server by 30.7%, however, its performance was only compared to conventional web cache algorithms which are not adaptive. For instance, [48] and [77] applied a web proxy cache replacement scheme based on a back-propagation neural network and achieved hit rates that are 86.60% of the optimal in the worst case. Also, [69] used the SVM model to yield useful information from the client user's request sequence and predict the web objects that the client user is likely to request shortly. The SVM model achieved a varied hit rate and byte hit rate, from 67% to 89% and from 67% to 98%, respectively [69], however, the model focuses only on hit ratio and byte ratio while excluding other vital performance metrics. Also [70] developed an adaptive probabilistic cache algorithm (AProb) using three techniques: dynamic probabilistic caching, ghost list; and adaptive probing and protection to improve web server hit ratio, footprint distance, and caching time without the involvement of the network administrators and synchronization. A study by [92] applied ANN to develop an algorithm for web cache replacement called Khalid Obaidat Replacement Algorithm (KORA) which can identify transient and shadow cache lines. Transient cache lines specify the new web object that arrives at the cache [10] and shadow lines are specified arrivals of the web object that were replaced recently from the cache to cater for transient cache lines. Moreover, [93] and [48]developed artificial neural network web cache replacement algorithms that use the Backpropagation learning rule, web log files and web access patterns to train the network and then predict the web object that is likely to be requested shortly. Backpropagation rule achieved 82% correct prediction using one-day training [93] while [48] achieved hit rates that are 86.60% of the optimal in the worst case. To improve the performance of neural network web proxy cache replacement algorithm, [77] proposed a Neural Network Proxy Cache Replacement (NNPCR) model which was trained with up to twelve times more data and shows a gain of about 5–10% increase in Correct Classification Ratio (CCR) and also obtained balanced output in terms of hit rate and byte hit rate. However, their model does not consider network latency and the size of web objects.

b. **Logistic Regression (LR) based web cache methods**

Authors such as [94] applied logistic regression (LR) in an adaptive Web cache to predict web objects that might be requested shortly. The LR-based web cache method replaces the Web object that has the lowest re-visited value [10]. However, LR can replace important web objects with low re-visited values without considering the priority level of web objects. To counter this problem, [95] suggested using Multinomial Logistic Regression (MLR) model which is trained to classify web objects using information extracted from weblogs. The MLR is only used for web cache object classification which ultimately resulted in low classification accuracy and reduced cache performance [10].

c. **Naïve Bayes based web cache methods**

Naïve Bayes is a machine learning model that applies the Bayes theorem in making classifications. According to [75], the Naïve Bayes classifier improves the performance of conventional Web proxy caching approaches such as LRU, LFU and function-based (Greedy-Dual-Size) by classifying web objects. Also, [96] combined the Naïve Bayes classifier and Expectation Maximization to develop an Expectation-Maximization Naive Bayes classifier (EMNB) that uses some aspects of EMNB-LRU and EMNB- Greedy-Dual-Size-Frequency to train from web proxy log files and predict the class of web objects to be revisited or not.

d. **Neuro-Fuzzy based web cache methods**

Fuzzy rules have been utilized to optimize the performance of web caching through a fuzzy inference engine or system. Fuzzy Inference Systems (FIS) consist of inputs, an inference engine and an output phase [97]. The input phase consists of input variables including web object frequency value, web object recency value converted using appropriate membership functions and truth values. These are fed into the inference engine which invokes fuzzy rules in the form of *IF-THEN* statements and later output results. The input variables in the input space describing the features of each web object are fuzzified using fuzzification membership functions such as Gaussian, triangular, or trapezoidal at the input phase, and the output is defuzzified using defuzzification methods at the output phase to replace the web page. For instance, [26] developed an adaptive neuro-fuzzy inference system (ANFIS) to classify web objects using LRU and Naïve Bayes. The web objects were classified into either cacheable or uncacheable objects, thus, Naïve Bayes is used to determining whether the web object will be re-visited shortly or not [13] and LRU was used to replace the web object based on the output of Naïve Bayes and its recency. In addition, [98] proposed a page replacement algorithm that uses fuzzy approaches to cluster pages. However, the ANFIS faces some challenges of LRU such as failing to consider the web object size and reducing access latency.

e. **Genetic Algorithm (GA) based web cache methods**

GA is an evolutionary approach to solving adaption and optimization computational problems in a dynamic environment. From a web cache perspective, GA generates an initial size of the web object population of each web object represented by a finite string of symbols encoding a possible solution in a given problem space [86]. Thus, the initial size of the web objects population of each web object is constantly updated at every evolutionary step which ultimately leads to a new generation of web objects, which is further processed by the fitness function [10]. Each web object's fitness function is of paramount importance in determining the web object to replace after the mutation and crossover process. For instance, [99] developed a GA-based cache replacement policy that takes into consideration of the web object size, and download time along with web object reference value but does not consider the recency and frequency access of web objects.

4.2 Identified Research Gaps in Computational Intelligence-Based Web Cache Replacement Models

The identified computational intelligent-based web cache replacement algorithms have numerous weaknesses the authors identified, which are presented in this section. These include non-consideration of semantic and contextual information in recently accessed web pages and future objects' requests [66], placing an extra computational overhead for the training process and issues in identifying attributes ranking and improving classification accuracy [65]. Moreover, some of the algorithms do not consider the size of web objects [67, 100], download latency [69] and predication ability [72], and their performance may deteriorate with huge cache sizes [71]. Additionally, some of the researchers, such as [68], compared the performance of their computationally intelligent models with conventional, non-adaptive approaches like LFU, LRU, FIFO, RANDOM and SLRU [48].

Additionally, some of the models are hard to implement on different web platforms, especially on mobile devices [75] because of the extra computational overhead required for the training process [101]. The LR-LRU, for instance, lacks the predicting capacity to predict the future access probabilities of the models of the cached items [63]. The gLRU model requires additional priority queues to keep track of frequency, which might cause cache pollution [62]. Moreso, the combined web replacement strategy by [85] only applied timestamp, elapsed time, size, and type in the URL object. More advanced web object features are required to further test the performance of the combined GDSF and SVM web cache intelligent approach and also add the security aspect [102]. While intelligent models are generally better than the traditional ones, the results of this review indicate that there are still gaps in the current models that need to be filled. There is a need to explore the performance of ensemble methods [103, 104] in web cache replacement instead of only focusing on classical machine learning models and then comparing their performance.

5 Conclusion

The exponential growth of web usage, web objects size, web objects sharing, transformation, transmission and limited bandwidth size have led to the continuous increase in latency, user-perceived delays, web cache overload and network traffic congestion degraded system performance and challenges in designing the next-generation robust web architectures. To alleviate these challenges, computational intelligent-based web caches replacement techniques such as DL and ML were proposed by several authors. This study revealed deep learning-based web cache replacement models perform better than machine learning models. Cost, size, frequency, recency, FIFO, the hit rate for each object, modification and expiration time are the most frequently implemented web object characteristics in the modern web object replacement models. The study revealed that there is significant progress in developing DL and ML web cache replacement models but there is also a need for developing adaptive next-generation web architectures to meet the web caching demands of emerging technologies beyond Industry 4.0.

References

1. Hou, R.: Performance analysis of cache replacement algorithm in information center network and construction of electronic music composition system. Alex. Eng. J. **61**, 863–872 (2022). https://doi.org/10.1016/J.AEJ.2021.04.082
2. Li, C., Zhang, Y., Luo, Y.: Intermediate data placement and cache replacement strategy under Spark platform. J. Parallel Distrib. Comput. **163**, 114–135 (2022). https://doi.org/10.1016/J.JPDC.2022.01.020
3. Wu, H., Luo, Y., Li, C.: Optimization of heat-based cache replacement in edge computing system. J. Supercomput. **77**, 2268–2301 (2021). https://doi.org/10.1007/S11227-020-03356-1/FIGURES/18
4. Atzori, L., Iera, A., Morabito, G.: The internet of things: a survey. Comput. Netw. **54**, 2787–2805 (2010). https://doi.org/10.1016/j.comnet.2010.05.010
5. Colombo, A.W., Karnouskos, S., Kaynak, O., Shi, Y., Yin, S.: Industrial cyberphysical systems: a backbone of the fourth industrial revolution. IEEE Ind. Electron. Mag. **11**, 6–16 (2017). https://doi.org/10.1109/MIE.2017.2648857
6. Hou, J., Qu, L., Shi, W.: A survey on internet of things security from data perspectives. Comput. Netw. **148**, 295–306 (2019). https://doi.org/10.1016/j.comnet.2018.11.026
7. Onik, M.M.H., Kim, C.-S., Yang, J.: Personal data privacy challenges of the fourth industrial revolution. In: 2019 21st International Conference on Advanced Communication Technology, pp. 635–638. IEEE (2019). https://doi.org/10.23919/ICACT.2019.8701932
8. Gubbi, J., Buyya, R., Marusic, S., Palaniswami, M.: Internet of Things (IoT): a vision, architectural elements, and future directions. Future Gener. Comput. Syst. **29**, 1645–1660 (2013). https://doi.org/10.1016/j.future.2013.01.010
9. Welsh, M., Culler, D.: Adaptive Overload Control for Busy Internet Servers (2016)
10. Pernabas, J.B., Fidele, S.F., Vaithinathan, K.K.: Enhancing greedy web proxy caching using weighted random indexing based data mining classifier. Egypt Inform. J. **20**, 117–130 (2019). https://doi.org/10.1016/J.EIJ.2019.01.001
11. Dehghan, M., Massoulie, L., Towsley, D., Menasche, D.S., Tay, Y.C.: A utility optimization approach to network cache design. IEEE/ACM Trans. Netw. **27**, 1013–1027 (2019). https://doi.org/10.1109/TNET.2019.2913677
12. Balamash, A., Krunz, M.: An overview of web caching replacement algorithms. IEEE Commun. Surv. Tutor. **6**, 44–56 (2009). https://doi.org/10.1109/COMST.2004.5342239
13. Waleed, A.A., Shamsuddin, S.M., Ismail, A.S.: Intelligent Naïve Bayes-based approaches for Web proxy caching. Knowl.-Based Syst. **31**, 162–175 (2012). https://doi.org/10.1016/j.knosys.2012.02.015
14. Ponnusamy, S.P.: Cache optimization on hot-point proxy caching using weighted-rank cache replacement policy. ETRI J. **35**, 687–696 (2013). https://doi.org/10.4218/etrij.13.0112.0606
15. Ma, T., Hao, Y., Shen, W., Tian, Y., Al-Rodhaan, M.: An improved web cache replacement algorithm based on weighting and cost. IEEE Access **6**, 27010–27017 (2018). https://doi.org/10.1109/ACCESS.2018.2829142
16. Barish, G., Obraczka, K.: World Wide Web caching: trends and techniques. IEEE Commun. Mag. **38**, 178–184 (2000). https://doi.org/10.1109/35.841844
17. Podlipnig, S., Böszörmenyi, L.: A survey of Web cache replacement strategies. ACM Comput. Surv. **35**, 374–398 (2003). https://doi.org/10.1145/954339.954341

18. Sheu, J.-P., Wang, P.-Y., Jagadeesha, R.: Wildcard-rule caching and cache replacement algorithms in software-defined networking. In: 2017 European Conference on Networks and Communications, pp. 1–6. IEEE (2017). https://doi.org/10.1109/EuCNC.2017.7980654
19. Sathiyamoorthi, V.: Web caching system: improving the performance of web-based information retrieval system. Des. Solut. User-Centric Inf. Syst. 184–204 (2017). https://doi.org/10.4018/978-1-5225-1944-7.ch011
20. ElAarag, H.: Web Proxy Cache Replacement Strategies: Simulation, Implementation, and Performance Evaluation. Springer, London (2012). https://doi.org/10.1007/978-1-4471-4893-7
21. Wang, B., Chen, L., Li, W., Sheng, J.: A caching strategy for transparent computing server side based on data block relevance. Information **9**, 42 (2018). https://doi.org/10.3390/info9020042
22. Tasnim, F., et al.: Towards personalized medicine: the evolution of imperceptible health-care technologies. Foresight **20**, 589–601 (2018). https://doi.org/10.1108/FS-08-2018-0075
23. Lu, H., Wang, M., Sangaiah, A.K.: Human emotion recognition using an EEG cloud computing platform. Mob. Netw. Appl. **25**(3), 1023–1032 (2018). https://doi.org/10.1007/s11036-018-1120-1
24. Fashoto, S.G., Adekoya, A., Owolabi, O., Ogunleye, O., Adediran, S., Tomori, R.: Development of an identity management system for a web proxy server in a tertiary institution using anonymity technology. Int. J. Phys. Sci. **11**, 157–167 (2016). https://doi.org/10.5897/IJPS2016.4482
25. Janapsatya, A., Ignjatovic, A., Peddersen, J., Parameswaran, S.: Dueling CLOCK: adaptive cache replacement policy based on the CLOCK algorithm. In: 2010 Design, Automation & Test in Europe Conference & Exhibition (DATE 2010), pp. 920–925. IEEE (2010). https://doi.org/10.1109/DATE.2010.5456920
26. Karami, A., Guerrero-Zapata, M.: An ANFIS-based cache replacement method for mitigating cache pollution attacks in Named Data Networking. Comput. Netw. **80**, 51–65 (2015). https://doi.org/10.1016/j.comnet.2015.01.020
27. Swain, D., Dash, B.N., Shamkuwar, D.O., Swain, D.: Analysis and predictability of page replacement techniques towards optimized performance. In: International Conference on Recent Trends in Information Technology and Computer Science, pp. 12–16 (2011)
28. Baugh, J.P., Guo, J.: A per-face popularity scheme to increase cache robustness in information-centric networks. Proc. Comput. Sci. **134**, 267–274 (2018). https://doi.org/10.1016/j.procs.2018.07.170
29. Tavakoli, R., Wijesinghe, S.N.R.: The evolution of the web and netnography in tourism: a systematic review. Tour Manag. Perspect. **29**, 48–55 (2019). https://doi.org/10.1016/j.tmp.2018.10.008
30. Aghaei, S., Nematbakhsh, M.A., Farsani, H.K.: Evolution of the World Wide Web: from Web 1.0 to Web 4.0. Int. J. Web Semant. Technol. (2012)
31. Nath, K., Dhar, S., Basishtha, S.: Web 1.0 to Web 3.0 - evolution of the Web and its various challenges. In: 2014 International Conference on Reliability Optimization and Information Technology, pp. 86–89. IEEE (2014). https://doi.org/10.1109/ICROIT.2014.6798297
32. Choudhury, N.: World Wide Web and Its Journey from Web 1.0 to Web 4.0 (n.d.)
33. Lee, C., Chen, Y.: A survey of the World Wide Web evolution with respect to security issues. PeerJ Prepr. (2017). https://doi.org/10.7287/peerj.preprints.2793v1
34. Patel, K.: Incremental journey for World Wide Web: introduced with Web 1.0 to recent Web 5.0 – a survey paper. Int. J. Adv. Res. Comput. Sci. Softw. Eng. **3** (2013)
35. Asim, M.N., Wasim, M., Ghani Khan, M.U., Mahmood, N., Mahmood, W.: The use of ontology in retrieval: a study on textual, multilingual, and multimedia retrieval. IEEE Access **7**, 21662–21686 (2019). https://doi.org/10.1109/ACCESS.2019.2897849

36. Shchekotykhin, K., Friedrich, G., Fleiss, P., Rodler, P.: Interactive ontology debugging: two query strategies for efficient fault localization. J. Web Semant. **12–13**, 88–103 (2012). https://doi.org/10.1016/j.websem.2011.12.006
37. Khaleel, M.S.A., Osman, S.E.F., Sirour, H.A.N.: Proposed ALFUR using intelligent agent comparing with LFU, LRU, SIZE and PCCIA cache replacement techniques. In: 2017 International Conference on Communication, Control, Computing and Electronics Engineering, pp. 1–6. IEEE (2017). https://doi.org/10.1109/ICCCCEE.2017.7867672
38. You, F., Liu, T., Peng, X., Liang, J., Zhang, B., Zhou, Y.: An efficient web caching replacement algorithm. In: Sun, X., Wang, J., Bertino, E. (eds.) ICAIS 2020. LNCS, vol. 12239, pp. 479–488. Springer, Cham (2020). https://doi.org/10.1007/978-3-030-57884-8_42
39. Wang, Y., Zhang, Z.-L., Du, D.H.C., Su, D.: A network-conscious approach to end-to-end video delivery over wide area networks using proxy servers. In: Proceedings of the IEEE INFOCOM 1998, Conference on Computer Communications. Seventeenth Annual Joint Conference of the IEEE Computer and Communications Societies. Gateway to the 21st Century (Cat. No. 98CH36169), vol. 2, pp. 660–667. IEEE (1998). https://doi.org/10.1109/INFCOM.1998.665087
40. Xuan, T.N., Thi, V.T., Khanh, L.H.: A design model network for intelligent web cache replacement in web proxy caching. In: Anh, N.L., Koh, SJ., Nguyen, T.D.L., Lloret, J., Nguyen, T.T. (eds.) Intelligent Systems and Networks. LNNS, vol. 471, pp. 591–600. Springer, Singapore (2022). https://doi.org/10.1007/978-981-19-3394-3_68
41. Luotonen, A., Altis, K.: World-Wide Web proxies. Comput. Netw. ISDN Syst. **27**, 147–154 (1994). https://doi.org/10.1016/0169-7552(94)90128-7
42. Wang, J.A.: A survey of web caching schemes for the Internet. Assoc. Comput. Mach. 36–46 (1999)
43. Mbunge, E., Jiyane, S., Muchemwa, B.: Towards emotive sensory Web in virtual health care: trends, technologies, challenges and ethical issues. Sensors Int. **3**, 100134 (2022). https://doi.org/10.1016/J.SINTL.2021.100134
44. Luo, Q., Naughton, J.F., Krishnamurthy, R., Cao, P., Li, Y.: Active query caching for database web servers. In: Goos, G., Hartmanis, J., van Leeuwen, J., Suciu, D., Vossen, G. (eds.) WebDB 2000. LNCS, vol. 1997, pp. 92–104. Springer, Heidelberg (2001). https://doi.org/10.1007/3-540-45271-0_6
45. Zhou, Y., Wang, F., Shi, Z., Feng, D.: An end-to-end automatic cache replacement policy using deep reinforcement learning. In: Proceedings of the International Conference on Automated Planning and Scheduling, vol. 32, pp. 537–545 (2022). https://doi.org/10.1609/ICAPS.V32I1.19840
46. Yang, C., Zhou, T., Qu, C.: Research on optimization strategy of web cache replacement algorithm, vol. 12500, pp. 1547–1552 (2022). https://doi.org/10.1117/12.2660948
47. Waleed, A.A., Siti, M.S.: Integration of least recently used algorithm and neuro-fuzzy system into client-side web caching. Int. J. Comput. Sci. Secur. **3**, 1–14 (2013)
48. Cobb, J., ElAarag, H.: Web proxy cache replacement scheme based on back-propagation neural network. J. Syst. Softw. **81**, 1539–1558 (2008). https://doi.org/10.1016/j.jss.2007.10.024
49. Trinh, T., Wu, D., Huang, J.Z.: C3C: a new static content-based three-level web cache. IEEE Access **7**, 11796–11808 (2019). https://doi.org/10.1109/ACCESS.2019.2892761
50. Fan, Y., Wang, Y., Ye, M., Lu, X., Huan, Y.: An improved method of cache prefetching for small files in Ceph system. In: Krömer, P., Zhang, H., Liang, Y., Pan, J.-S. (eds.) ECC 2018. AISC, vol. 891, pp. 381–389. Springer, Cham (2019). https://doi.org/10.1007/978-3-030-03766-6_43
51. Abrams, M., Standridge, C.R., Abdulla, G., Fox, E.A., Williams, S.: Removal policies in network caches for World-Wide Web documents. In: Conference Proceedings on Applications,

Technologies, Architectures, and Protocols for Computer Communications - SIGCOMM 1996, pp. 293–305. ACM Press, New York (1996). https://doi.org/10.1145/248156.248182
52. Aggarwal, C., Wolf, J.L., Yu, P.S.: Caching on the World Wide Web. IEEE Trans. Knowl. Data Eng. **11**, 94–107 (1999). https://doi.org/10.1109/69.755618
53. Koller, R., Mashtizadeh, A.J., Rangaswami, R.: Centaur: host-side SSD caching for storage performance control. In: 2015 IEEE International Conference on Autonomic Computing, pp. 51–60. IEEE (2015). https://doi.org/10.1109/ICAC.2015.44
54. Benhamida, N., Bouallouche-Medjkoune, L., Aïssani, D.: Simulation evaluation of a relative frequency metric for web cache replacement policies. Evol. Syst. **9**(3), 245–254 (2017). https://doi.org/10.1007/s12530-017-9194-8
55. Zhang, T., Luo, P., Cheng, Z., Li, J.: Efficient flash-aware page-mapping cache management for on-board remote sensing image processing. J. Syst. Archit. **88**, 1–12 (2018). https://doi.org/10.1016/j.sysarc.2018.04.003
56. Hoang, X.T., Bui, N.D.: An enhanced semantic-based cache replacement algorithm for web systems. In: 2019 IEEE-RIVF International Conference on Computing and Communication Technologies, pp. 1–6. IEEE (2019). https://doi.org/10.1109/RIVF.2019.8713680
57. Zhao, Z., Ma, Y., Cong, Q.: GDSF-based low access latency web proxy caching replacement algorithm. In: Proceedings of the 2018 2nd International Conference on Computer Science and Artificial Intelligence - CSAI 2018, pp. 232–236. ACM Press, New York (2018). https://doi.org/10.1145/3297156.3297237
58. Jin, S., Bestavros, A.: GreedyDual∗ Web caching algorithm: exploiting the two sources of temporal locality in Web request streams. Comput. Commun. **24**, 174–183 (2001). https://doi.org/10.1016/S0140-3664(00)00312-1
59. Ghosh, S.S., Patra, M., Jain, A.: Reduction of web latency: an integrated proxy prefetch-cache system framework. In: Abraham, A., Dutta, P., Mandal, J., Bhattacharya, A., Dutta, S. (eds.) Emerging Technologies in Data Mining and Information Security. AISC, vol. 814, pp. 621–633. Springer, Singapore (2019). https://doi.org/10.1007/978-981-13-1501-5_55
60. Aguado, J., Mendler, M., Pouzet, M., Roop, P., von Hanxleden, R.: Deterministic concurrency: a clock-synchronised shared memory approach. In: Ahmed, A. (ed.) ESOP 2018. LNCS, vol. 10801, pp. 86–113. Springer, Cham (2018). https://doi.org/10.1007/978-3-319-89884-1_4
61. Tanwir, G.H., Achmad, A.: Combination of FIFO-LRU cache replacement algorithms on proxy server to improve speed of response to object requests from clients. ARPN J. Eng. Appl. Sci. **12**, 710–715 (2017)
62. Friedlander, E., Aggarwal, V.: Generalization of LRU Cache Replacement Policy with Applications to Video Streaming (2018)
63. Wang, Y., Yang, Y., Han, C., Ye, L., Ke, Y., Wang, Q.: LR-LRU: a PACS-oriented intelligent cache replacement policy. IEEE Access **7**, 58073–58084 (2019). https://doi.org/10.1109/ACCESS.2019.2913961
64. Moher, D., Liberati, A., Tetzlaff, J., Altman, D.G.: Preferred reporting items for systematic reviews and meta-analyses: the PRISMA statement. PLoS Med. **6**, e1000097 (2009). https://doi.org/10.1371/journal.pmed.1000097
65. Waleed, A.A., Shamsuddin, S.M.: Integration of least recently used algorithm and neuro-fuzzy system into client-side web caching. Int. J. Comput. Sci. Secur. **3** (2009)
66. Vietri, G., et al.: Driving cache replacement with ML-based LeCaR. In: 10th USENIX Workshop on Hot Topics in Storage and File Systems HotStorage 2018, Co-Located with USENIX ATC 2018, Boston (2018)
67. Negrão, A.P., Roque, C., Ferreira, P., Veiga, L.: An adaptive semantics-aware replacement algorithm for web caching. J. Internet Serv. Appl. **6**(1), 1–14 (2015). https://doi.org/10.1186/s13174-015-0018-4

68. Im, Y., Prahladan, P., Kim, T.H., Hong, Y.G., Ha, S.: SNN-Cache: a practical machine learning-based caching system utilizing the inter-relationships of requests. In: 2018 52nd Annual Conference on Information Sciences and Systems. IEEE (2018)
69. Ujjainiya, B., Shrivastava, S.K.: SVM regression for web prefetching and caching. Int. J. Comput. Appl. **19**, 47–51 (2011)
70. Tarnoi, S., Kumwilaisak, W., Suppakitpaisarn, V., Fukuda, K., Ji, Y.: Adaptive probabilistic caching technique for caching networks with dynamic content popularity. Comput. Commun. **139**, 1–15 (2019). https://doi.org/10.1016/j.comcom.2019.03.001
71. Kastaniotis, G., Maragos, E., Douligeris, C., Despotis, D.K.: Using data envelopment analysis to evaluate the efficiency of web caching object replacement strategies. J. Netw. Comput. Appl. **35**, 803–817 (2012). https://doi.org/10.1016/j.jnca.2011.11.013
72. Hasslinger, G., Ntougias, K., Hasslinger, F., Hohlfeld, O.: Performance evaluation for new web caching strategies combining LRU with score based object selection. Comput. Netw. **125**, 172–186 (2017). https://doi.org/10.1016/j.comnet.2017.04.044
73. Benadit, P.J., Francis, F.S., Muruganantham, U.: Improving the performance of a proxy cache using tree augmented Naive Bayes classifier. Proc. Comput. Sci. **46**, 184–193 (2015). https://doi.org/10.1016/j.procs.2015.02.010
74. Sajeev, G.P., Sebastian, M.P.: Building semi-intelligent web cache systems with lightweight machine learning techniques. Comput. Electr. Eng. **39**, 1174–1191 (2013). https://doi.org/10.1016/j.compeleceng.2013.02.005
75. Waleed, A.A., Sulaiman, S., Ahmad, N.: Performance improvement of least-recently used policy in web proxy cache replacement using supervised machine learning. Int. J. Adv. Soft Comput. Appl. **6** (2014)
76. Waleed, A.A., Shamsuddin, S.M.: Neuro-fuzzy system in partitioned client-side Web cache. Expert Syst. Appl. **38**, 14715–14725 (2011). https://doi.org/10.1016/j.eswa.2011.05.009
77. Romano, S., ElAarag, H.: A neural network proxy cache replacement strategy and its implementation in the Squid proxy server. Neural Comput. Appl. **20**, 59–78 (2011). https://doi.org/10.1007/s00521-010-0442-0
78. Li, W., Wang, C., Li, D., Hu, B., Wang, X., Ren, J.: Edge caching for D2D enabled hierarchical wireless networks with deep reinforcement learning. Wirel. Commun. Mob. Comput. **2019**, 1–12 (2019). https://doi.org/10.1155/2019/2561069
79. Lu, F., Shi, Z., Gu, L., Jin, H., Yang, L.T.: An adaptive multi-level caching strategy for distributed database system. Future Gener. Comput. Syst. **97**, 61–68 (2019). https://doi.org/10.1016/j.future.2018.11.050
80. Chien, W.-C., Weng, H.-Y., Lai, C.-F.: Q-learning based collaborative cache allocation in mobile edge computing. Future Gener. Comput. Syst. **102**, 603–610 (2020). https://doi.org/10.1016/j.future.2019.08.032
81. Friedlander, E., Aggarwal, V.: Generalization of LRU cache replacement policy with applications to video streaming. ACM Trans. Model. Perform. Eval. Comput. Syst. **4**, 1–22 (2019). https://doi.org/10.1145/3345022
82. Ibrahim, H., Yasin, W., Udzir, N.I., Hamid, N.A.W.A.: Intelligent cooperative web caching policies for media objects based on J48 decision tree and Naïve Bayes supervised machine learning algorithms in structured peer-to-peer systems. J. Inf. Commun. Technol. **15**, 85–116 (2020)
83. Akbari Bengar, D., Ebrahimnejad, A., Motameni, H., Golsorkhtabaramiri, M.: A page replacement algorithm based on a fuzzy approach to improve cache memory performance. Soft Comput. **24**(2), 955–963 (2019). https://doi.org/10.1007/s00500-019-04624-w
84. Gui, Y.-Q., Ju, S.-S.: Cache replacement strategy based on association rules in P2P streaming media system. In: Proceedings of the 2019 2nd International Conference on Data Science and Information, pp. 49–53. ACM, New York (2019). https://doi.org/10.1145/3352411.3352420

85. Chao, W.: Web cache intelligent replacement strategy combined with GDSF and SVM network re-accessed probability prediction. J. Ambient Intell. Humaniz. Comput. **11**, 581–587 (2020). https://doi.org/10.1007/S12652-018-1109-4/TABLES/3
86. Mourad, A.A., Mesbah, S., Mabrouk, T.F.: A novel approach to cache replacement policy model based on genetic algorithms. In: Proceedings of the World Conference on Smart Trends in Systems, Security and Sustainability, WS4 2020, pp. 405–411 (2020). https://doi.org/10.1109/WORLDS450073.2020.9210347
87. Agbehadji, I.E., Awuzie, B.O., Ngowi, A.B., Millham, R.C.: Review of big data analytics, artificial intelligence and nature-inspired computing models towards accurate detection of COVID-19 pandemic cases and contact tracing. Int. J. Environ. Res. Public Health (2020). https://doi.org/10.3390/ijerph17155330
88. Yang, J., Zhang, J., Ma, C., Wang, H., Zhang, J., Zheng, G.: Deep learning-based edge caching for multi-cluster heterogeneous networks. Neural Comput. Appl. **32**, 15317–15328 (2020). https://doi.org/10.1007/S00521-019-04040-Z/FIGURES/7
89. Sanyanga, T.M., Chinzvende, M.S., Kavu, T.D., Batani, J.: Searching objects in a video footage. Int. J. ICT Res. Afr. Middle East **8**, 18–31 (2019). https://doi.org/10.4018/ijictrame.2019070102
90. Mbunge, E., Makuyana, R., Chirara, N., Chingosho, A.: Fraud detection in e-transactions using deep neural networks-a case of financial institutions in Zimbabwe cite this paper fraud detection in e-transactions using deep neural networks-a case of financial institutions in Zimbabwe. Int. J. Sci. Res. **17** (n.d.). https://doi.org/10.21275/ART20176804
91. Batani, J.: An adaptive and real-time fraud detection algorithm in online transactions. Int. J. Comput. Sci. Bus. Inform. **17**, 1–12 (2017)
92. Khalid, H., Obaidat, M.S.: KORA: a new cache replacement scheme. Comput. Electr. Eng. **26**, 187–206 (2000). https://doi.org/10.1016/S0045-7906(99)00041-5
93. Tian, W., Choi, B., Phoha, V.V.: An adaptive web cache access predictor using neural network. In: Hendtlass, T., Ali, M. (eds.) IEA/AIE 2002. LNCS (LNAI), vol. 2358, pp. 450–459. Springer, Heidelberg (2002). https://doi.org/10.1007/3-540-48035-8_44
94. Foong, A.P., Hu, Y.-H., Heisey, D.M.: Logistic regression in an adaptive Web cache. IEEE Internet Comput. **3**, 27–36 (1999). https://doi.org/10.1109/4236.793455
95. Sajeev, G.P., Sebastian, M.P.: A novel content classification scheme for web caches. Evol. Syst. **2**, 101–118 (2011). https://doi.org/10.1007/s12530-010-9026-6
96. Pernabas, J.B., Fidele, S.F.: Enhancements to greedy web proxy caching algorithms using data mining method and weight assignment policy. Int. J. Innov. Comput. Inf. Control **14**, 1311–1326 (2018). https://doi.org/10.24507/ijicic.14.04.1311
97. Sabeghi, M., Yaghmaee, M.H.: Using fuzzy logic to improve cache replacement decisions. IJCSNS Int. J. Comput. Sci. Netw. Secur. **6** (2006)
98. Akbari Bengar, D., Ebrahimnejad, A., Motameni, H., Golsorkhtabaramiri, M.: A page replacement algorithm based on a fuzzy approach to improve cache memory performance. Soft Comput. **24**, 955–963 (2020). https://doi.org/10.1007/S00500-019-04624-W/FIGURES/3
99. Chen, Y., Li, Z.Z., Wang, Z.W.: A GA-based cache replacement policy. In: Proceedings of 2004 International Conference on Machine Learning and Cybernetics (IEEE Cat. No. 04EX826), vol. 1, pp. 263–266. IEEE (2004). https://doi.org/10.1109/ICMLC.2004.1380674
100. Hasslinger, G., Ntougias, K., Hasslinger, F., Hohlfeld, O.: Performance evaluation for new web caching strategies combining LRU with score based object selection. In: 2016 28th International Teletraffic Congress (ITC 28), pp. 322–330. IEEE (2016). https://doi.org/10.1109/ITC-28.2016.150

101. Waleed, A.A., Shamsuddin, S.M.: Intelligent dynamic aging approaches in web proxy cache replacement. J. Intell. Learn. Syst. Appl. **07**, 117–127 (2015). https://doi.org/10.4236/jilsa.2015.74011
102. Mbunge, E., Muchemwa, B., Batani, J., Mbuyisa, N.: A review of deep learning models to detect malware in Android applications. Cyber Secur. Appl. **1**, 100014 (2023). https://doi.org/10.1016/J.CSA.2023.100014
103. Mbunge, E., et al.: Implementation of ensemble machine learning classifiers to predict diarrhoea with SMOTEENN, SMOTE, and SMOTETomek class imbalance approaches. In: 2023 Conference on Information Communications Technology and Society (ICTAS), pp. 1–6 (2023). https://doi.org/10.1109/ICTAS56421.2023.10082744
104. Mbunge, E., et al.: Application of machine learning techniques for predicting child mortality and identifying associated risk factors. In: 2023 Conference on Information Communications Technology and Society (ICTAS), pp. 1–5 (2023). https://doi.org/10.1109/ICTAS56421.2023.10082734

Solving the Global Optimization Problem with Swarm Intelligence

Ayman Aljarbouh[1](✉), Michael Sabugaa[2], Mohammed Ayad Alkhafaji[3], Ismail Keshta[4], Edwin Daniel Félix Benites[5], and Ashot Gevorgyan[6]

[1] University of Central Asia, Naryn, Kyrgyz Republic
ayman.aljarbouh@ucentralasia.org
[2] Agusan del Sur State College of Agriculture and Technology, Bunawan, Philippines
[3] National University of Science and Technology, Dhi Qar, Iraq
[4] AlMaarefa University, Riyadh, Saudi Arabia
[5] Universidad Nacional José María Arguedas, Andahuaylas, Perú
[6] Yerevan State University, Yerevan, Republic of Armenia

Abstract. Swam intelligence is based on the principles of behavior of living creatures. The formalized algorithmic behavior of these creatures allows solving various mathematical problems. This paper concerns the application of swarm intelligence in solving a global optimization problem. The paper gives an example of solving a global optimization problem and presents an analysis of obtained results. This approach can be used in a variety of applications requiring the solution of global optimization problems. #CSOC1120.

Keywords: Swarming Intelligence · Agent · Global Optimization · Global Extremum · Ackley Function

1 Introduction

Swarm intelligence describes the behavior of a decentralized self-organizing system consisting of elements called agents [1, 2]. Swarm intelligence is based on the behavior of a community of homogeneous living creatures, e.g. honey bees [3, 4], bird flocks [4–6], ant colony [7–9], fish schools [5, 10]. The actions of agents are described by a set of simple rules that each individual agent follows [11]. There is no single control center. Due to the effect of emergent nature, the system of agents can solve problems that individual agents cannot solve [12].

Swarm intelligence has a number of advantages, the main ones being the following:

Multitasking. Swarm self-organization and the independence of individual agents' actions make it possible to perform several tasks at once.

Scalability. A swarm can comprise a few agents or hundreds of thousands of agents. The number of swarmed agents can easily be changed. The problem will be solved in any case. A swarm of any scale is decentralized and solves all of its tasks as rationally as possible without centralized control.

R. Silhavy and P. Silhavy (Eds.): CSOC 2023, LNNS 723, pp. 451–457, 2023.
https://doi.org/10.1007/978-3-031-35317-8_40

Self-organization. This property is one of the main properties of swarm intelligence. Decentralization, the absence of a single center of control and independent decision-making by agents allows the swarm to make fast, correct and flexible decisions and to be fully independent from the quality of control.

Flexibility. Swarm agents are extremely responsive to changes in both the external environment and changes within the swarm.

Reliability. The problem the swarm is working on does not depend on the performance of the individual agent. Even if individual agents do something contrary to the overall goal or die, it does not affect the end result.

Mathematical and algorithmic description of agents' actions made it possible to apply the principles that underlie the behavior of living beings to various tasks of science and technology.

Nowadays, swarm intelligence is used in computer network security [13], in Data Science [14–16], web services [17], traffic optimization [18], Internet of Things [19], new materials discovery and design [20], cancer diagnostics [21], pharmacy [22] and healthcare [23], power optimization [24] and other practical applications.

In this paper, we aim to explore the potential of swarm intelligence in solving global optimization problems. Swarm intelligence can be used to solve global optimization problems, that is, problems in which there are several extrema and the minimum (or maximum) value among them must be found. Cooperating together, swarm agents can explore and use the search space to determine the global extremum. This approach is especially useful for complex and high-dimensional optimization problems, where finding an optimal solution with existing global optimization methods can be too time consuming. In general, the global optimization problem is characterized by high dimensionality, exponential growth of the number of solutions with increasing number of controlled variables, high computational complexity of calculations of criteria and constraints, multiple local extrema, and, in some cases, very stringent requirements on the time rate of solution [25–27].

Due to the fact that swarm intelligence is characterized by nondeterministic behavior and is based on heuristic principles of operation, it shows high efficiency, is able to escape from local extremes and find a global optimum or a solution close enough to it in an acceptable time.

2 Method

The problem of selecting indicators of agent performance involves multiple realizations, in which it is reasonable to consider both the success of each agent and the swarm as a whole [28]. In the determination of the variability indicator of the function, the heuristic control consisted in the selection of the basic swarm with the maximum Lipschitz constant [29]. Strengthening of the agents' function is possible through heuristic tuning of the algorithm parameters based on the changing properties of the environment, the parameters of which are the current coordinates of the agents, values of the target function, averaged coordinates of the agents, and estimated values of Lipschitz constants for each agent.

When solving global optimization problem, transition to new coordinates of the ith agent on the next iteration is defined as a function [28, 29]:

$$x^{(i)[t+1]} = x^{(i)[t]} + \alpha \cdot Y^{(i)[t]} + U[0,\ \beta_1] \otimes \left(x_b^{(i)[t]} - x^{(i)[t]}\right) + \ U[0,\ \beta_2] \otimes \left(x_g^{[t]} - x^{(i)[t]}\right)$$
$$+U[0,\ \beta_3] \otimes \left(x_{g*}^{[t]} - x^{(i)[t]}\right) + U[0,\ \gamma] \otimes \left(x^{(q(i))[t]} - x^{(i)[t]}\right) + U[0,\ \delta] \otimes \left(x_*^{[t]} - x^{(i)[t]}\right),$$

where $\alpha,\ \beta_1,\ \beta_2,\ \beta_3,\ \gamma,\ \delta$ are the tunable coefficients. The summand $\alpha \cdot Y^{(i)[t]}$ defines "inertial" component, $Y^{(i)[t]}$ is an agent's velocity. $U[0,\ \beta_1]\otimes\left(x_b^{(i)[t]} - x^{(i)[t]}\right)$ is called "cognitive" component, it expresses agent's tendency to return to position $x_b^{(i)[t]}$ with minimal value of objective function for given agent. $U[0,\ \beta_2]\otimes\left(x_g^{[t]} - x^{(i)[t]}\right)$ is called "social" component, it expresses tendency of ith agent to move to position $x_g^{[t]}$, which corresponds to coordinates of best agent in swarm. $U[0,\ \beta_3] \otimes \left(x_{g*}^{[t]} - x^{(i)[t]}\right)$ corresponds to the tendency of moving in the direction of the best agent over all considered swarms $x_{g*}^{[t]}$. $U[0,\ \gamma]\otimes\left(x^{(q(i))[t]} - x^{(i)[t]}\right)$ characterizes the movement of the agent associated with the direction of the maximum change of the target function in the direction of the agent $q(i)$ with coordinates $x^{(q(i))[t]}$. $U[0,\ \delta] \otimes \left(x_*^{[t]} - x^{(i)[t]}\right)$ corresponds to the movement of the agent in the direction of the found averaged coordinates $x_*^{[t]}$. The stochasticity of the method consists in the use of uniformly distributed random variables $U[0,\ \beta_1],\ U[0,\ \beta_2],\ U[0,\ \beta_3],\ U[0,\ \gamma],\ U[0,\ \delta]$ in the specified ranges of values; $\otimes$ is the symbol of the piecewise multiplication of vectors.

It may be noted that the strengthening of the coefficient γ corresponds to a decrease in the variability of the target function along the parameter $\frac{L_{pi}-L_{\min}}{L_{\max}-L_{\min}}$, where L_{pi}, $L_{\max}$, $L_{\min}$ represent the values of the estimates of the local Lipschitz constants in the ith agent in p^{th} swarm, the maximum and the minimum values of the Lipschitz constant across all agents, accordingly; coefficient α is included only when the variability of the target function is small; the tendency to aspire to the found averaged coordinates increases with parameter δ; coefficients β_2 and β_3 are computed according to the distance of the ith agent to the best agent in the swarm and the best agents in all swarms.

3 Results

Figure 1 shows the results of the search for the global minimum of the Ackley function:

$$f(x, y) = -20e^{-0.2\sqrt{0.5(x^2+y^2)}} - e^{0.5(\cos 2\pi x + \cos 2\pi y)} + e + 20.$$

Figure 2 shows changes in maximum values of agents' target function (top line), changes in values of target function in averaged coordinates (middle line), and changes in minimum values of agents' target function (bottom line). Global minimum was found in 4 iterations out of the provided 10 iterations.

Figure 3 shows the results of the process of calculating the global minimum of the Ackley function. Five swarms of 100 agents in each swarm were used. Figure 3 shows the distribution of target function values across the agents of one swarm. It can be seen that agent #21 reached the global minimum.

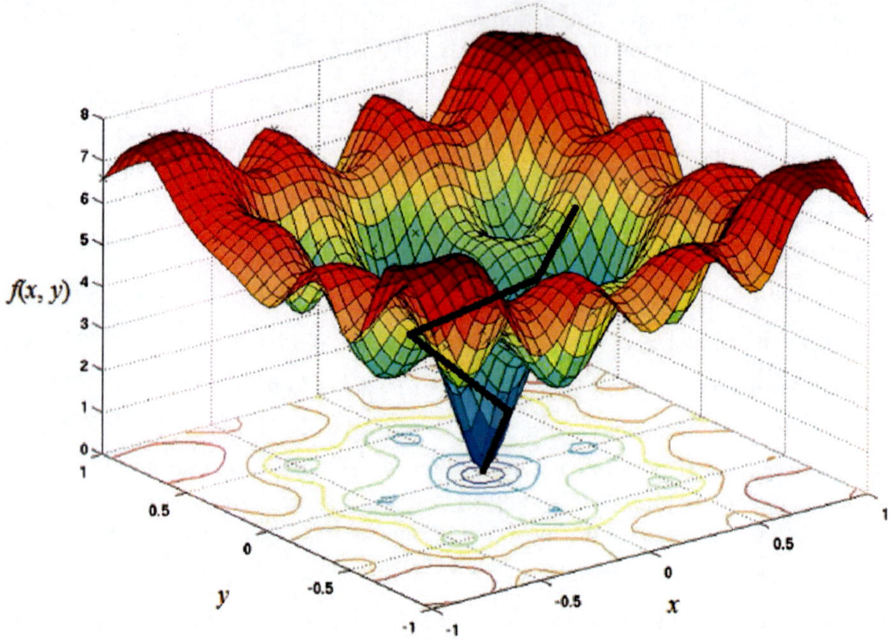

Fig. 1. Finding the global minimum of the Ackley function for 2 variables.

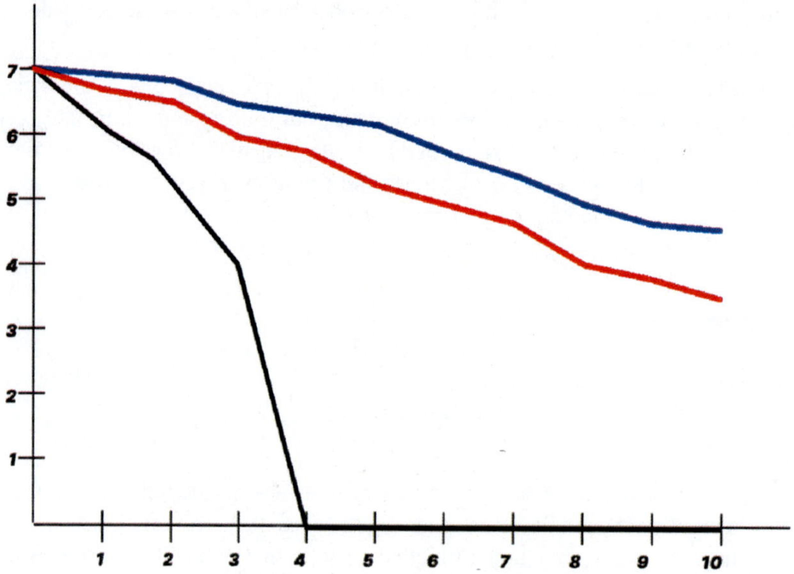

Fig. 2. Finding the global minimum of Ackley's function in 10 iterations.

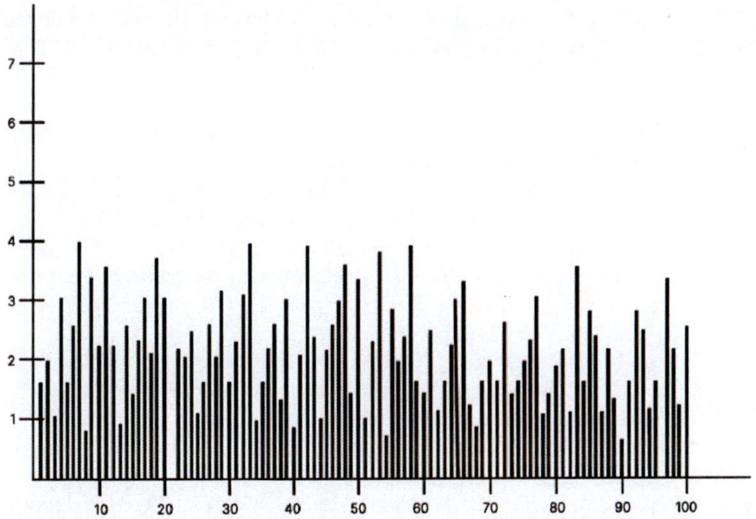

Fig. 3. Distribution of the values of the target function across the agents of one swarm.

4 Conclusion

This paper has examined the solution to a global optimization problem using swarm intelligence. Global optimization is a complex problem, the solution of which is quite resource consuming. The application of swarm intelligence can give better results compared to classical optimization methods.

The research results presented in the paper, in particular, comparison of different methods show the effectiveness and feasibility of this approach. The approach can compete with other global optimization methods, and the algorithmic complexity contributes to the simplicity of its implementation.

Swarm intelligence as a whole has sufficient potential to develop new techniques, methods and algorithms to solve a wide class of scientific and practical problems, including those related to global optimization. Further research can be devoted to these problems.

References

1. Schranz, M., et al.: Swarm intelligence and cyber-physical systems: concepts, challenges and future trends. Swarm Evol. Comput. **60**, 100762 (2021). https://doi.org/10.1016/j.swevo.2020.100762
2. Thrun, M.C., Ultsch, A.: Swarm intelligence for self-organized clustering. Artif. Intell. **290**, 103237 (2021). https://doi.org/10.1016/j.artint.2020.103237
3. Chen, X., Tianfield, H., Du, W.: Bee-foraging learning particle swarm optimization. Appl. Soft Comput. **102**, 107134 (2021). https://doi.org/10.1016/j.asoc.2021.107134
4. Lien, L.-C., Cheng, M.-Y.: A hybrid swarm intelligence based particle-bee algorithm for construction site layout optimization. Expert Syst. Appl. **39**(10), 9642–9650 (2012). https://doi.org/10.1016/j.eswa.2012.02.134

5. Kashani, A.R., Camp, C.V., Akhani, M., Ebrahimi, S.: Optimum design of combined footings using swarm intelligence-based algorithms. Adv. Eng. Softw. **169**, 103140 (2022). https://doi.org/10.1016/j.advengsoft.2022.103140
6. Yu, Y., et al.: Torsional capacity evaluation of RC beams using an improved bird swarm algorithm optimised 2D convolutional neural network. Eng. Struct. **273**, 115066 (2022). https://doi.org/10.1016/j.engstruct.2022.115066
7. González, P., Prado-Rodriguez, R., Gábor, A., Saez-Rodriguez, J., Banga, J.R., Doallo, R.: Parallel ant colony optimization for the training of cell signaling networks. Expert Syst. Appl. **208**, 118199 (2022). https://doi.org/10.1016/j.eswa.2022.118199
8. Gupta, A., Srivastava, S.: Comparative analysis of ant colony and particle swarm optimization algorithms for distance optimization. Proc. Comput. Sci. **173**, 245–253 (2020). https://doi.org/10.1016/j.procs.2020.06.029
9. Wu, L., Huang, X., Cui, J., Liu, C., Xiao, W.: Modified adaptive ant colony optimization algorithm and its application for solving path planning of mobile robot. Expert Syst. Appl. **215**, 119410 (2023). https://doi.org/10.1016/j.eswa.2022.119410
10. Jun, M.: Research on the fish behavior simulation based on swarm intelligence. Proc. Eng. **43**, 547–551 (2012). https://doi.org/10.1016/j.proeng.2012.08.095
11. Wei, D., Wang, Z., Si, L., Tan, C.: Preaching-inspired swarm intelligence algorithm and its applications. Knowl.-Based Syst. **211**, 106552 (2021). https://doi.org/10.1016/j.knosys.2020.106552
12. Pham, Q.-V., et al.: Swarm intelligence for next-generation networks: recent advances and applications. J. Netw. Comput. Appl. **191**, 103141 (2021). https://doi.org/10.1016/j.jnca.2021.103141
13. Nasir, M.H., Khan, S.A., Khan, M.M., Fatima, M.: Swarm intelligence inspired intrusion detection systems—A systematic literature review. Comput. Netw. **205**, 108708 (2022). https://doi.org/10.1016/j.comnet.2021.108708
14. Chopra, D., Arora, P.: Swarm intelligence in data science: challenges, opportunities and applications. Proc. Comput. Sci. **215**, 104–111 (2022). https://doi.org/10.1016/j.procs.2022.12.012
15. Figueiredo, E., Macedo, M., Siqueira, H.V., Santana, C.J., Gokhale, A., Bastos-Filho, C.J.A.: Swarm intelligence for clustering—A systematic review with new perspectives on data mining. Eng. Appl. Artif. Intell. **82**, 313–329 (2019). https://doi.org/10.1016/j.engappai.2019.04.007
16. Rostami, M., Berahmand, K., Nasiri, E., Forouzandeh, S.: Review of swarm intelligence-based feature selection methods. Eng. Appl. Artif. Intell. **100**, 104210 (2021). https://doi.org/10.1016/j.engappai.2021.104210
17. Chen, J., Mao, C., Song, W.W.: QoS prediction for web services in cloud environments based on swarm intelligence search. Knowl.-Based Syst. **259**, 110081 (2023). https://doi.org/10.1016/j.knosys.2022.110081
18. Rajput, N.S., Banerjee, R., Sanghi, D., Santhanam, G., Singhal, K.: Swarm intelligence inspired meta-heuristics for solving multi-constraint QoS path problem in vehicular ad hoc networks. Ad Hoc Netw. **123**, 102633 (2021). https://doi.org/10.1016/j.adhoc.2021.102633
19. Forestiero, A.: Heuristic recommendation technique in Internet of Things featuring swarm intelligence approach. Expert Syst. Appl. **187**, 115904 (2022). https://doi.org/10.1016/j.eswa.2021.115904
20. Liu, Z., et al.: Swarm intelligence for new materials. Comput. Mater. Sci. **214**, 111699 (2022). https://doi.org/10.1016/j.commatsci.2022.111699
21. Macedo, M., Santana, M., dos Santos, W.P., Menezes, R., Bastos-Filho, C.: Breast cancer diagnosis using thermal image analysis: a data-driven approach based on swarm intelligence and supervised learning for optimized feature selection. Appl. Soft Comput. **109**, 107533 (2021). https://doi.org/10.1016/j.asoc.2021.107533

22. Kılıç, F., Uncu, N.: Modified swarm intelligence algorithms for the pharmacy duty scheduling problem. Expert Syst. Appl. **202**, 117246 (2022). https://doi.org/10.1016/j.eswa.2022.117246
23. Ribino, P., Ciampi, M., Islam, S., Papastergiou, S.: Swarm intelligence model for securing healthcare ecosystem. Proc. Comput. Sci. **210**, 149–156 (2022). https://doi.org/10.1016/j.procs.2022.10.131
24. Yelisetti, S., Saini, V.K., Kumar, R., Lamba, R., Saxena, A.: Optimal energy management system for residential buildings considering the time of use price with swarm intelligence algorithms. J. Build. Eng. **59**, 105062 (2022). https://doi.org/10.1016/j.jobe.2022.105062
25. Chen, A., Ren, Z., Wang, M., Liang, Y., Liu, H., Du, W.: A surrogate-assisted variable grouping algorithm for general large-scale global optimization problems. Inf. Sci. **622**, 437–455 (2023). https://doi.org/10.1016/j.ins.2022.11.117
26. El Moumen, S., Ellaia, R., Aboulaich, R.: A new hybrid method for solving global optimization problem. Appl. Math. Comput. **218**(7), 3265–3276 (2011). https://doi.org/10.1016/j.amc.2011.08.066
27. Seyyedabbasi, A.: A reinforcement learning-based metaheuristic algorithm for solving global optimization problems. Adv. Eng. Softw. **178**, 103411 (2023). https://doi.org/10.1016/j.advengsoft.2023.103411
28. Koshur, V.D., Ryzhikova, E.O.: Strengthening swarm intelligence in the method of global optimization based on virtual particles. In: Proceedings of the XIII International Youth Scientific Conference "Intellect and Science", pp. 89–90. SibFU, Zheleznogorsk, Russia (2013)
29. Ryzhikova, E.O.: Multiagent approach and swarm intelligence enhancement in global optimization method. In: Proceedings of the X All-Russian Conference "Youth and Science", pp. 1–4. SibFU, Krasnoyarsk, Russia (2014)

Multicriteria Evaluation of Reinforced Concrete Slabs Using Analytical Hierarchy Process

Ítalo Linhares Salomão(✉) and Plácido Rogério Pinheiro

Graduate Program in Applied Informatics, University of Fortaleza, Fortaleza, Brazil
{italosalomao,placido}@unifor.br

Abstract. The typology regarding slabs to be adopted in a building is usually chosen by the design engineer based on the structural performance of each type of system depending on the limitations imposed by the architectural design, as well as the cost related to the amount of material, concrete, and steel used in each system. Due to the technical attributions of this professional, this decision process does not consider the other aspects that impact the construction, such as the productivity of each system, the aesthetics and appearance of the slab, the residue generation of each system, and the thermal/acoustic comfort. Seeking to meet all the demands cited, this paper pursues the development of a multicriteria method to assist the stakeholder in deciding the type of structural system to be adopted for the slabs that best fit the needs and objectives of the project in question. To this end, the Analytic Hierarchy Process (AHP) was adopted with information gathered from bibliographic research of articles that developed works related to the subject, the Delphi Method of research with experts, and the usage of a machine learning applied on a dataset from more than 2000 slabs already constructed.

Keywords: Reinforced Concrete Slab · Principal component analysis (PCA) · Analytic Hierarchy Process · Multicriteria decision analysis

1 Introduction

In in-built areas, the slab corresponds to 50% of the building's entire volume of reinforced concrete [1]. This element has a wide variation of typologies, such as slab on beams, flat slab, waffle slab, hardy slab, and, more recently, typologies such as bubble dec and hole deck. Usually, the choice of the slab type occurs based on the experience of the structure designer, who determines the adopted element from his field of work perspective, validating your choice based on the criteria related only to structural performance and cost of materials directly applied on the construction, such as concrete, steel, and formwork.

With the advancement of construction processes and the strength of materials, architectural projects have become increasingly audacious. Consequently, it is necessary to create design systems that seek solutions with attention to criteria beyond those mentioned above, such as thermal and acoustic comfort, productivity, appearance, and, more recently, the concern with environmental, economic, and social sustainability factors

R. Silhavy and P. Silhavy (Eds.): CSOC 2023, LNNS 723, pp. 458–465, 2023.
https://doi.org/10.1007/978-3-031-35317-8_41

[2]. There is an effort to systematize the process of choosing the appropriate type of structure, from which material to use to which system to be adopted and its construction process. Usually, this process is based on subjective preferences and the experience of the responsible engineer. Among the papers found in the bibliographical survey, we can highlight the concern with the selection of materials using multicriteria methods in the construction industry through works such as [3], which examines the selection of materials based on their performance when used in the construction of buildings.

Moreover, [4] present a model for selecting the foundation for a ground-floor dwelling. Also, [5] creates a model to assist in choosing the best pile technology to be used in a building. In addition, [6] create a model for selecting the design parameters to be adopted in one-way slabs.

The presented context generates a space for conducting research and creating a validation system through multicriteria decision processes, specifically, the Analytic Hierarchy Process (AHP). In this perspective, this article seeks to fill this gap by creating a model that helps professionals involved in a building project, regardless of their area of expertise, to have a more holistic view of all criteria and their respective impacts on the final choice of a slab system. This paper is configured as follows: Sect. 2 exposes the theoretical referential of reinforced concrete slabs; Sect. 3 presents the theoretical referential of the AHP method, the Delphi Method of research, the application of machine learning and Principal Component Analysis in decision problems; Sect. 4 explains the methodology adopted in this work; Sect. 5 presents the obtained result and the hierarchization of the alternatives of the slabs; finally, Sect. 6 concludes the study.

2 Slab Selection is Based on Their Types and Functionality

Slabs are linear plane elements responsible for transmitting loads from external actions to the beams or the columns. These elements are mostly subjected to bending stress and must be designed, detailed, and built to attend to this internal stress.

Although already exist several types of slabs, these elements have constant changes. They frequently evolve their construction process to generate the most outstanding possible economy of materials and archive the architectural demand. Thus, among the various possibilities, we can highlight the most common ones: the slab on beams, the waffle slab, and the flat slab, as shown in Fig. 1. Slabs on beams are concrete blocks supported on the beams and are usually executed in one single step. The waffle slabs are composed of the cover and the ribbed, formed by fillers that can be bricks, styrofoam blocks, or any forms that allow the execution of these elements with voids. This slab can be divided into one-way or two-way, depending on the direction that it will be supported on the beams, i.e., if the load is being transmitted to the shaft in only one direction, it is classified as one-way, if it is sent in two directions, it is called two-way. Finally, there are flat slabs, which are elements directly supported on the columns, and may or may not have a widening in the slab-column transition. These slabs can be formed by the two previous types, i.e., solid or waffle.

The building techniques for the three kinds of slabs and their subdivisions vary, which affects the productivity of each. Additional differences between the models, such as the quantity of material utilized and whether a model has better stiffness or less deformation,

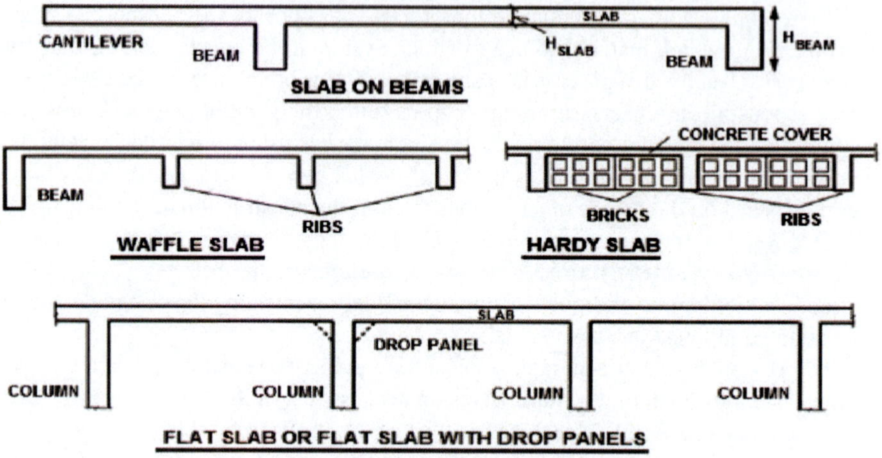

Fig. 1. Most common types of slabs [7].

may also be quantified statistically. There are also subjective differences, such as the aesthetic suitability of the slab compared to the architecture's ideal. These differences between the types impact the choice of which slab best suits a particular building.

3 Analytic Hierarchy Process (AHP)

Decision-making problems in which a single alternative must be chosen from a finite number of possibilities are part of every professional's daily routine. The final choice is usually based on previous knowledge or the experience of each manager or designer working with state-of-the-art. Among the multicriteria methods, there is the Analytic Hierarchy Process (AHP), in which a systematic approach is developed to help the decision maker choose, based on several criteria, the best alternative for the problem presented. As presented by [8], the AHP method can be divided into three stages, the first being the structuring of the problem to be solved, the second the evaluation of criteria and decision alternatives, composing a hierarchical structure or decision tree, finally, the third stage, which is the categorization, ranking, ordering or prioritization of other options.

For the initial stages, it is possible to adopt the Delphi Method, which "can be characterized as a method for structuring a group communication process in such a way that the process is effective in allowing a group of individuals as a whole to deal with complex problems" [9]. This method can be applied in several areas, such as planning and design, cost and schedule, and construction methods and materials [10].

To reduce the subjectivity of the choice of criteria, whether by the Delphi method or any other similar methods, when having a large volume of data, machine learning can be used to determine weights based on historical values. This area, called Large Scale Decision Making or LSDM, has become quite popular in decision problems and can follow the taxonomy presented by [11]. Datasets sometimes have many features that can produce several issues in the analysis. One way to minimize this is using dimensionally reduction techniques such as Principal Component Analysis (PCA). One of the PCA objectives is the "dimensionality reduction using the first principal components which reflect the structure in the data" [12].

4 Methodology

The development of this research began with a literature review aimed at analyzing and defining the criteria used in the decision-making process of the problem studied. This survey was carried out through the databases: Compendex, ScienceDirect, Google Scholar, Wiley, and SpringerLink, searched in English and Portuguese languages, with emphasis on scientific articles published between the years 2001 and 2022. As a research question of this stage, one can highlight: Which criteria professionals in the AEC Industry (Architecture, Engineering, and Construction) believe impact the decision of which slab to adopt in each project. This survey highlighted the criteria as follows:

- Structural performance: Capacity of slabs to support applied loads, considering the serviceability, resistant capacity, and durability.
- Productivity/Constructability: Related to the simplicity of the construction process, being directly proportional to the execution time and the labor resources allocated.
- Cost: Monetary value associated with the cost of the materials used in the slab's construction (concrete, steel, and formwork), as well as the value of the execution.
- Appearance: Related to the aesthetic when integrating the structural element and the architectural concept of the improvement.
- Residue Generation: Environmental impact generated by construction waste.
- Thermal/Acoustic Comfort: Related to the performance of the slab concerning the thermal comfort provided by the heat transmission of the slab and the acoustic comfort offered by the insulation of noise between floors.

With the above information and based on the five slab alternatives: slab on beams, one-way waffle slab, two-way waffle slab, solid flat, and waffle flat, it was possible to develop a criteria tree for the problem presented, as shown in Fig. 2.

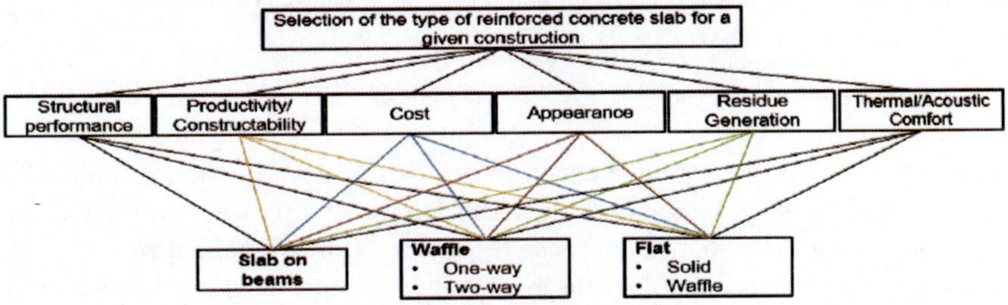

Fig. 2. Criteria tree of the problem.

For the judgment matrix construction, because the criteria are divided into two categories, quantitative (structural performance, productivity and cost, and residue generation) and qualitative (appearance and thermal/acoustic comfort), the analysis process was divided between the survey through machine learning and the Delphi methodology respectively. For the subjective criteria, to apply the Delphi research methodology, it is necessary to follow a sequence of scientifically validated steps that involves the elaboration of the system to be adopted and the flow that will be used, as presented in Fig. 3.

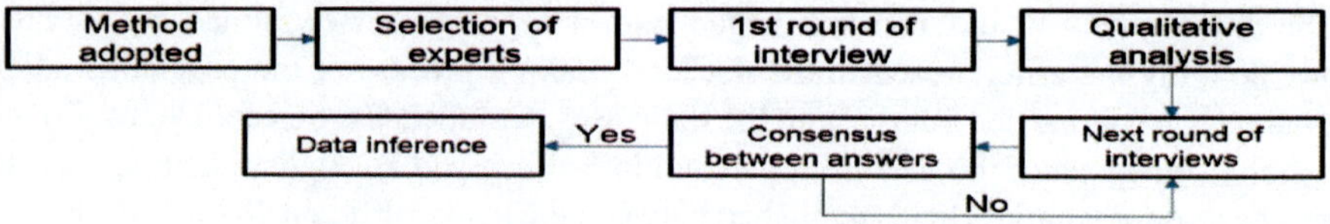

Fig. 3. Flowchart applied to the Delphi Method.

Once the method was chosen, it was selected 12 specialists divided as follows: 4 builders, four structural engineers, and four architects. This choice was because each of these professionals has different points of view and priorities regarding the decision process. For example, the engineer will prioritize the structural performance, the builder's cost, and the architect's appearance. The heterogeneity of this sample aims to give a more holistic view of the process.

After selecting the specialists, the first round of interviews consists of open questions aiming at the exactification of criteria and later compared with those raised from a previously executed bibliographical survey. With this data, the second-round initiates by performing a quiz assembled through Google Forms on a scale comparison from 0 to 9. Two different forms were carried out for the objective criteria, the first associated with the cost and structural performance. The calculation and determination of the values associated with each slab alternative were developed based on the man. On the other hand, the productivity criteria were made from a machine learning algorithm to determine an expected value based on the 2147 accurate execution data of the slab alternatives studied throughout the Brazilian territory. With all the data collected in the previous steps, the hierarchy calculation was developed using the AHP method. Finally, the alternatives of the slabs were developed and hierarchized based on what was obtained.

5 Results and Analysis

The questionnaire responses were analyzed in a pairwise comparison with the 12 specialists to achieve the proposed objective. The decision matrix was constructed for this stage through a geometric mean of these responses. Next, we normalized the decision matrix, i.e., dividing each component by magnitude. From the normalized matrix, the average value of each row was calculated to obtain the vector of relative priorities, also known as relative weight, as shown in Table 1.

To verify the priorities vector's consistency, [13] was used for the consistency ratio. This methodology considers the division of the consistency index and the random index for n, where n is the dimension of the vector studied. For the proposed case, this ratio

Table 1. Relative priority vector of the criteria.

Criteria	Relative Priority
Structural performance	0,40037
Productivity/Constructability	0,12018
Cost	0,16137
Appearance	0,03236
Residue Generation	0,06251
Thermal/Acoustic Comfort	0,22320

presented a value of 8%, implying that the priority vector is adequate (value lower than 10%).

Then comparison matrices of the five types of slabs were built for each of the six criteria. The data was also from the questionnaire applied with the experts on qualitative issues (appearance and thermal/acoustic comfort). In addition, these values were raised by calculating the geometric mean of their answers, producing the normalized matrix and its vector of relative priorities. In stratifying the responses by the segment of activity of the interviewees, it is possible to observe a more excellent uniformity of solutions in the criteria to which they have greater mastery. For example, architects answered with a lower standard deviation in the appearance item and a higher standard deviation in the structural performance item.

For the quantitative criteria, the structural performance and cost were calculated and classified according to the average data of 2147 slabs developed by a company with an excellent reputation in the market. For the productivity criteria, the database used had eight different features, being necessary to apply the Principal Component Factor Analysis (PCA) method to identify the most significant correlation with the slab types (targets). From this method, it is possible to classify the one that presents the best productivity. The same process was applied to the lowest waste generation based on the number of materials used for each slab type, having analyzed four features for applying PCA. It should be noted that because they are quantitative measures, consumption rates were used to determine the relative priority. Finally, the composite priorities were calculated by multiplying the preferences by the criteria average. The result is presented in Table 2.

Table 2. Composite priorities vector.

Alternatives	Priorities
Slab on beams	0,25392
One-way Waffle Slab	0,18287
Two-way Waffle Slab	0,21602
Flat Slab	0,17991
Flat Waffle Slab	0,16728

6 Conclusion

Due to the distinct vision of the professionals involved in the decision-making process of the structural typology, decision support methodologies help to choose the best type of slab to meet the priority points for each market segment, whether architect, structural designer, or builder. Thus, from a survey with 12 professionals and a database that contains information about 2147 slabs executed in their various types, it was possible to observe that among the five presented, the one that presents the priority choice is the conventional solid slab type, followed by two-way waffle slab, one-way waffle slab, solid flat slab, and waffle slab respectively. The application of the AHP method proposed in this paper successfully created a model to help companies and the several layers of the construction sector in this decision-making process, as well as in structuring its implementation based on each project's work type. Further criteria ordering research is found in [14–16].

Although applied in a generalist study, i.e., not restricting the model to a building class or occupation (commercial or residential, for example), the study used the multicriteria methodology based on academic works and publications on the subject. The limitations found are due to the non-holistic vision of the professionals, who are commonly restricted to their field of work, and consequent favoritism in the choices of criteria that most favor their area, making it challenging to build a measurement scale with subjective variables and complicated standardization. As a proposal for the continuity of this work, it is suggested to apply this method for each building class, which can be high-standard residential, low-standard, or commercial, among others. It is also recommended that the model presented in this work be applied to The AEC Industry to validate the model.

Acknowledgments. The second author thanks the National Council for Technological and Scientific Development (CNPq) through grant # 04272/2020-5. The authors thank the Edson Queiroz Foundation/University of Fortaleza for all the support provided.

References

1. Giongo, J.S.: Concreto Armado: Projeto Estrutural de Edifícios. USP, São Paulo (2007)
2. Cuadrado, J., Zubizarreta, M., Rojí, E., García, H., Larrauri, M.: Sustainability-related decision making in industrial buildings: an AHP analysis. Math. Probl. Eng. **2015**, 1–13 (2015). https://doi.org/10.1155/2015/157129
3. Lee, D., et al.: Analytic hierarchy process-based construction material selection for performance improvement of building construction: the case of a concrete system form. Materials **13**(7), 1738 (2020). https://doi.org/10.3390/ma13071738
4. Urskis, Z., Daniunas, A., Zavadskas, E.K., Medzvieckas, J.: Multicriteria evaluation of building foundation alternatives. Comput.-Aided Civ. Infrastruct. Eng. **31**(9), 717–729 (2016). https://doi.org/10.1111/mice.12202
5. Zavadskas, E.K., Sulinskas, S., Daniunas, A., Turskis, Z., Silvilevicius, H.: Multiple criteria selection of pile-column construction technology. J. Civ. Eng. Manag. **18**(6), 834–842 (2012). https://doi.org/10.3846/13923730.2012.744537
6. Fernandez-Ceniceros, J., Fernandez-Martinez, R., Fraile-Garcia, E., Martinez-de-Pison, F.J.: Decision support model for one-way floor slab design: a sustainable approach. Autom. Constr. **35**, 460–470 (2013)

7. Clímaco, J.C.T.S.: Estruturas de concreto armado: fundamentos de projeto, dimensionamento e verificação, 3rd edn. Editora Universidade de Brasília, Brasilia (2016)
8. Hummel, J.M., Bridges, J.F.P., Ijzerman, M.J.: Group decision making with the analytic hierarchy process in benefit-risk assessment: a tutorial. Patient - Patient-Cent. Outcomes Res. **7**(2), 129–140 (2014)
9. Chan, A.P.C., Yung, E.H.K., Lam, P.T.I., Tam, C.M., Cheung, S.O.: Application of Delphi method in selection of procurement systems for construction projects. Constr. Manag. Econ. **19**(7), 699–718 (2001)
10. Ameyaw, E.E., Hu, Y., Shan, M., Chan, A.P., Le, Y.: Application of Delphi method in construction engineering and management research: a quantitative perspective. J. Civ. Eng. Manag. **22**(8), 991–1000, 25 (2016)
11. Ding, R.-X., et al.: Large-scale decision-making: characterization, taxonomy, challenges and future directions from an artificial intelligence and applications perspective. Inf. Fusion **59**, 84–102 (2020)
12. Esbensen, K.H.: Multivariate Data Analysis: In Practice: An Introduction to Multivariate Data Analysis and Experimental Design, 5th edn. Camo Software (2010). 597 p.
13. Saaty, T.L.: How to make a decision: the analytic hierarchy process. Eur. J. Oper. Res. **48**, 9–26 (1990)
14. Carvalho, D., Pinheiro, P.R., Pinheiro, M.C.D.: A hybrid model to support the early diagnosis of breast cancer. Proc. Comput. Sci. **91**, 927–934 (2016)
15. Tamanini, I., Carvalho, A.L., Castro, A.K., Pinheiro, P.R.: A novel multicriteria model applied to cashew chestnut industrialization process. In: Mehnen, J., Köppen, M., Saad, A., Tiwari, A. (eds.) Applications of Soft Computing. AINSC, vol. 58, pp. 243–252. Springer, Heidelberg (2009). https://doi.org/10.1007/978-3-540-89619-7_24
16. Salazar, M.C., Pinheiro, P.R., de Castro, I.C.C.: Multicriteria model for evaluation of outsourcing services by logistics operators. In: Silhavy, R. (ed.) CSOC 2020. AISC, vol. 1224, pp. 554–566. Springer, Cham (2020). https://doi.org/10.1007/978-3-030-51965-0_49

Prediction of the Dependence of the Physico-chemical Properties of Water Using Artificial Intelligence

Burumbaev Adil Ilmirovich[1(✉)], Agapitov Denis Vadimovich[1], Burumbaev Danil Ilmirovich[1], Kuanishev Valery Taukenovich[1], Barbin Nikolai Mihailovich[2,3], and Minina Elena Aleksandrovna[1]

[1] The Ural Technical Institute of Communications and Informatics (Branch) of the Siberian State University of Telecommunications and Informatics (UrTISI SibGUTI), Yekaterinburg, Russia
adilburumbaev@mail.ru

[2] Department of VM&F, Ural Technical Institute of Communications and Informatics (Branch) of the Siberian State University of Telecommunications and Informatics (UrTISI SibGUTI), Yekaterinburg, Russia

[3] Urals State Institute of Fire Service, Yekaterinburg (UrI GPS MES of Russia), Yekaterinburg, Russia

Abstract. In the near future, it is expected that many areas of modern technology will work on the basis of artificial intelligence, including manufacturing. With the help of artificial intelligence, computers can learn to perform a task without even being explicitly programmed to do so. Artificial intelligence is, in fact, one of the computer-simulated human intelligence that can be programmed to make decisions, perform specific tasks, and learn from the results obtained.

In this paper, the existing approaches of using artificial intelligence and mathematical algorithms to find the dependencies of the physico-chemical properties of water in order to predict them based on each other are analyzed, and a conclusion is made about the feasibility of developing this direction.

The task of predicting the dependence of electrical conductivity on other physico-chemical properties of water became particularly relevant after a series of experiments and further processing of measurement results in the Statistica program, in order to establish the relationship between different parameters. To solve the forecasting problem, various models were used: a regressor based on gradient boosting, linear regression, an algorithm for averaging features, an algorithm for averaging the target feature. #CSOC1120.

Keywords: Artificial intelligence · linear regression · gradient boosting · forecasting · water properties

1 Introduction

Water is the most important component that determines the structure and properties of countless elements of living and inanimate nature. The ability to respond to even the most insignificant external influences makes it a unique object for study. In [1–5], the

R. Silhavy and P. Silhavy (Eds.): CSOC 2023, LNNS 723, pp. 466–477, 2023.
https://doi.org/10.1007/978-3-031-35317-8_42

influence of a low-intensity electromagnetic field of the radio frequency range (f = 100–200 MHz) on the physico-chemical properties of water, depending on the intermolecular interaction: heat and evaporation rate, cohesion, adhesion, electrical conductivity, etc. was studied.

Studies of the properties of water and aqueous solutions that change as a result of external influences of various physical nature [6–9], as well as electromagnetic radiation in the ultrahigh frequency range, electric and magnetic fields are presented in a number of papers [9–12].

Exposure to electromagnetic fields of both natural origin and man-made nature in public places and at work due to the growing trend in the use of electronic devices can have an adverse effect on human health. In [13–17], a possible mechanism for the effect of low-power electromagnetic radiation on a living organism through a liquid medium is proposed.

The study of the variability of the characteristics of the composition and properties of fresh natural waters is a poorly formalized task due to the random nature of the desired indicators. The patterns of their dynamics necessary to achieve water management goals are usually unknown [18], as a result of which water treatment systems operate under conditions of uncertainty. The increasing requirements for water and environmental safety and accessibility to clean water increasingly raise the question of taking into account the above patterns, the search for which turns from a cognitive task into one of the most urgent national economic goals. However, it is not easy to find the required patterns, as can be seen from the example of the task under study, based on data from the Ural Department of Roshydromet for the water quality of the Iset River in the Sverdlovsk region in the vicinity of Yekaterinburg, one of the largest industrial regions. At the same time, taking into account the difficulties of obtaining correct water-ecological information, it is proposed to carry out the so—called exploratory Data Analysis (RAD) - Exploratory Data Analysis (EDA) [19] before solving practical problems by checking the main hypotheses and preliminary description of water-ecological characteristics using an artificial neural network (INS).

To solve tasks like this, taking into account the complexity of traditional computing or their physical inadequacy, it is advisable to use RAD based on artificial intelligence methods. The basis of these methods are artificial neutron networks (INS), capable of learning from examples (learning with a teacher) and capturing hidden dependencies in experimental data [20].

2 Research Methods and Results Obtained

In this work, the results of measurements of water properties were processed using a professional water quality meter WMM 97 at a temperature of 21 °C (the pH of the medium corresponded to GOST 58144 5.0–6.6). The electrical conductivity of distilled water was measured for a week (electrical conductivity corresponded to GOST 58144, no more than 0.43 mSm/m), pH, redox potential, total oxygen and dissolved solids content, and readings were taken every second.

The hypothesis was to find a relationship between electrical conductivity and some other parameters. If this hypothesis is fulfilled, it would be possible to predict the value of electrical conductivity based on these parameters at a given time, without measuring the electrical conductivity itself with the device. The study was conducted in Jupiter Notebook in Python. As the study progressed, the data was preprocessed, then divided into a training and test sample. The test sample was the last fifty thousand measurements taken from the instruments. Pretreatment included the removal of anomalies and the selection of signs. In the preprocessing process, data normalization and standardization were not carried out, since this is not a general requirement for the selected machine learning algorithms. On the created samples, models were trained using cross validation and the final accuracy was evaluated. Two machine learning models, Gradient Boosting Classifier and Linear Regression, were tested with the selection of hyperparameters, and the models were checked for adequacy using mathematical algorithms. The entire research process is visualized on graphs. First of all, graphs of the kinetics of the process of each parameter for a given temperature were formed (see Figs. 1, 2, 3, 4, 5 and 6).

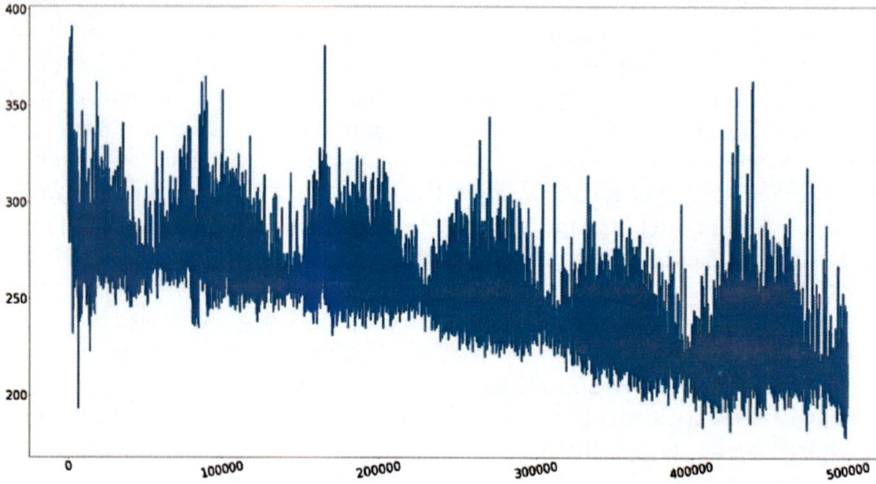

Fig. 1. Electrical conductivity

The graphs show that dissolved oxygen, total oxygen and pH remain constant after release, so these parameters are not used for analysis.

Next, the sec_from_start columns were introduced to indicate the start time of the study and a correlation matrix was constructed, which is shown in Fig. 7.

It can be seen from the correlation matrix that each parameter affects the target parameter cond (electrical conductivity), therefore, it makes sense to train the model on all signs.

The first model that has been tested is the Gradient Boosting model from the sklearn package, which is based on a random forest. This model was chosen because it is suitable for most classification and regression tasks, and the results of its work can be easily interpreted.

To find the best model of artificial intelligence, it is required to select hyperparameters. To do this, the GridSearchCV aglorithm is selected, which is built into the sklearn

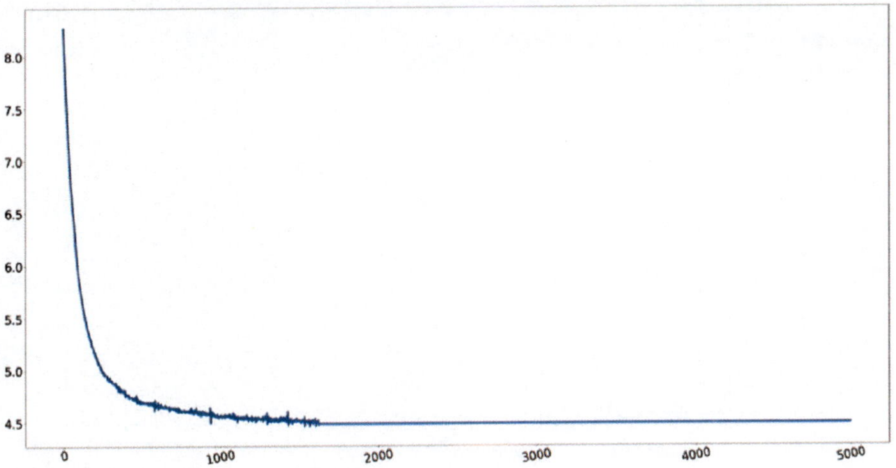

Fig. 2. Dissolved oxygen

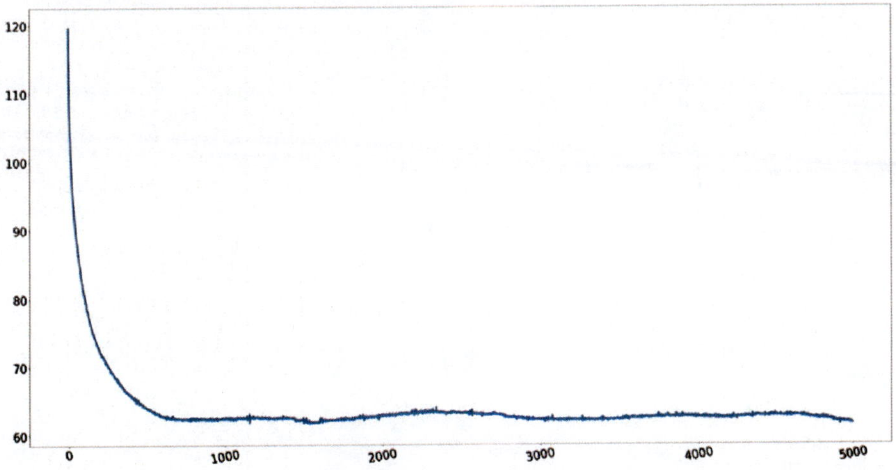

Fig. 3. Total oxygen

package. The algorithm is good because it implements cross-validation, which allows you to choose the most generalizing model.

First of all, we select the most resource-intensive hyperparameter for GradientBoostingRegressor - n_estimators in order to save time on selecting other hyperparameters. This hyperparameter shows how many decision trees are involved in the algorithm. The more of them, the higher the chance of retraining and a longer training time. The best accuracy on validation samples was recorded at n_estimators = 32 after testing the GridSearchCV algorithm. This means that the hyperparameter n_estimators = 32, corresponding to this coefficient, is the best for this model, and will be used as a fixed

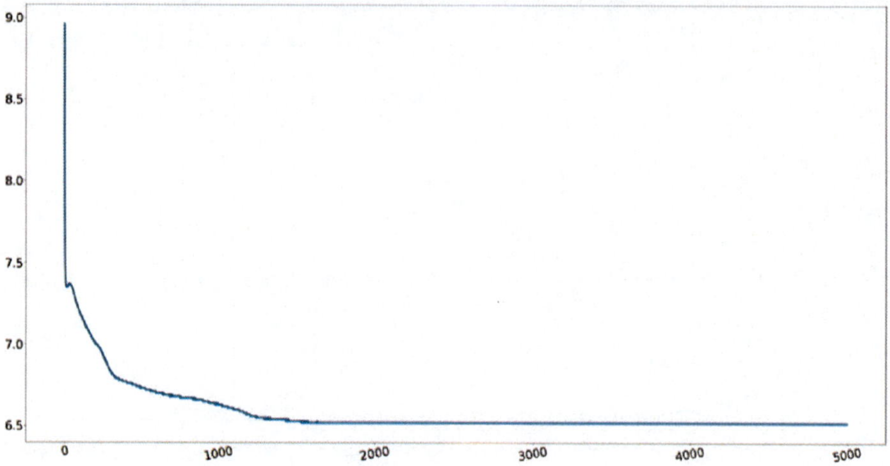

Fig. 4. pH

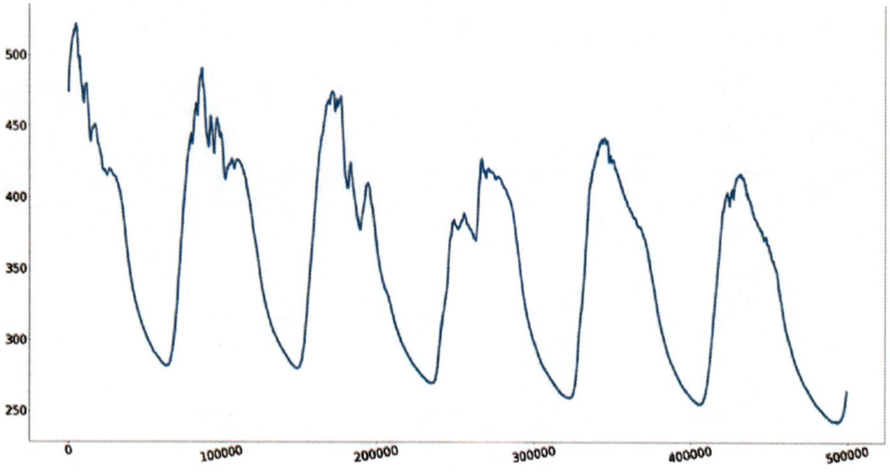

Fig. 5. Redox potential

one for further selection of hyperparameters. The coefficient of determination r2 on the training sample at n_estimators = 32 is 0.88.

Next, we select the remaining hyperparameters: max_depth, min_samples_leaf, min_samples_split. After running the GridSearchCV algorithm and interpreting its results, the following hyperparameter values were obtained: max_depth = 6, min_samples_leaf = 1, min_samples_split = 6, n_estimators = 32. The coefficient of determination r_2 turned out to be 0.912 in the training sample, which is the best indicator of all possible combinations of hyperparameters. After the selection of hyperparameters, the gradient boosting model was trained and tested. The prediction on the test sample is shown in Fig. 8, and the prediction on the training sample is shown in Fig. 9.

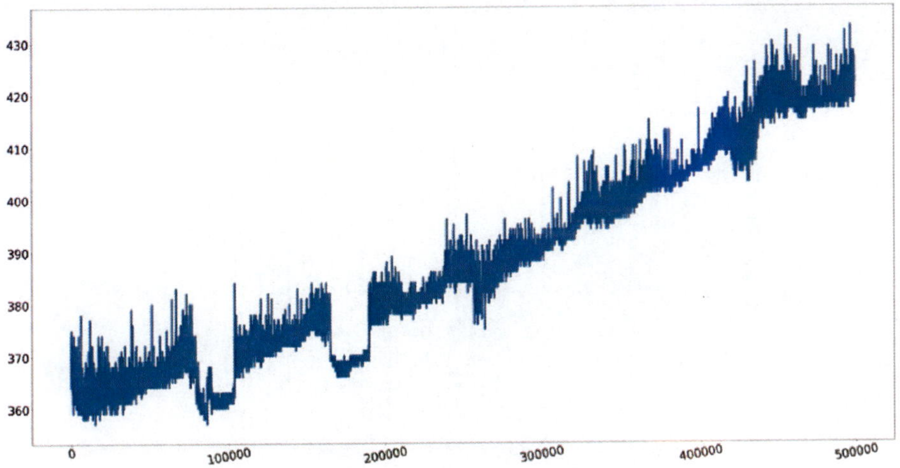

Fig. 6. Solid particle content

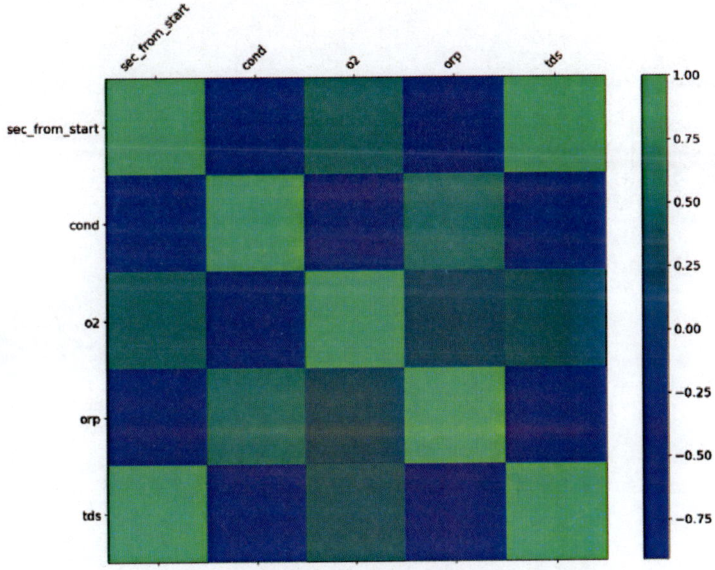

Fig. 7. Correlation matrix

According to the figures the gradient boosting model performed worse with the test sample, which indicates poor generalizing ability on new data. The coefficient of determination is negative and is equal to −0.43, which indicates that the model coped with the task worse than a trivial model could have done.

After evaluating the results of gradient boosting, a hypothesis was put forward that the model is too heavy for this task, therefore it is prone to retraining. Based on this, it

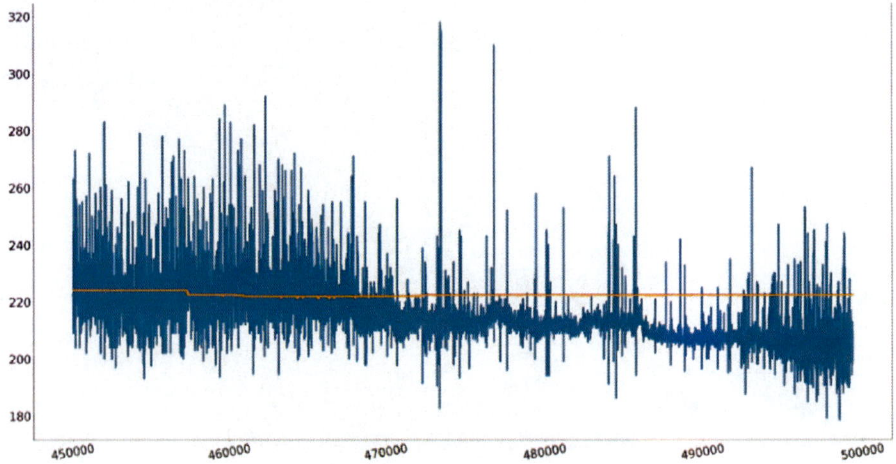

Fig. 8. Prediction on a test sample

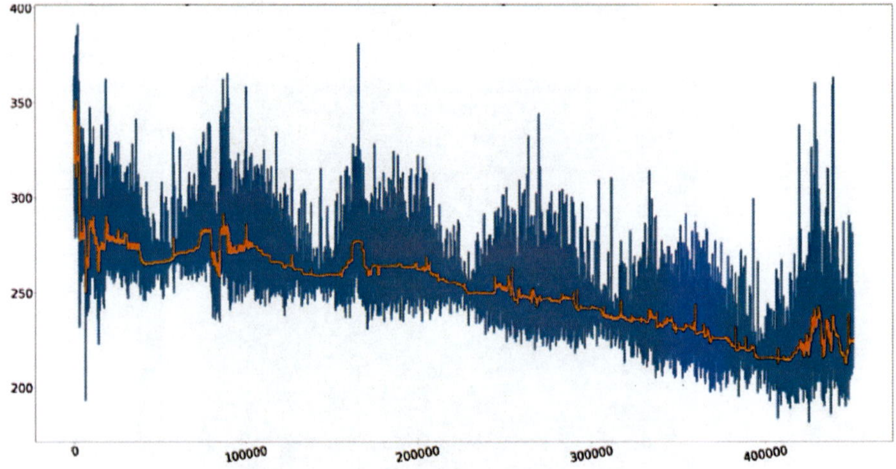

Fig. 9. Prediction on a training sample

was decided to train a linear regression model in order to compare it with the results of gradient boosting.

Linear regression is a model that is often used in practice due to the speed of its operation and the lack of effort to fine–tune. This model was chosen because it can only find linear dependencies in features, therefore it is less prone to retraining. Linear regression also does not require standardization or normalization of data, since an analytical solution method was chosen in the process of its implementation in the sklearn package. The prediction on the test sample is shown in Fig. 10, and the prediction on the training sample is shown in Fig. 11.

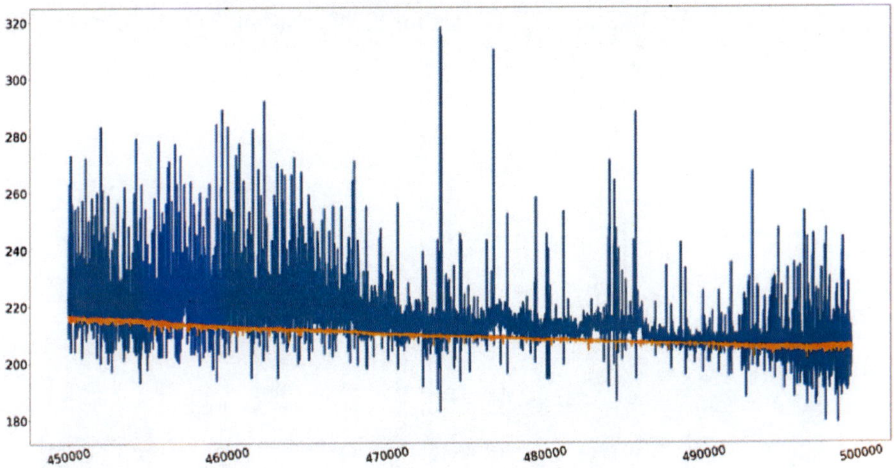

Fig. 10. Prediction on a test sample

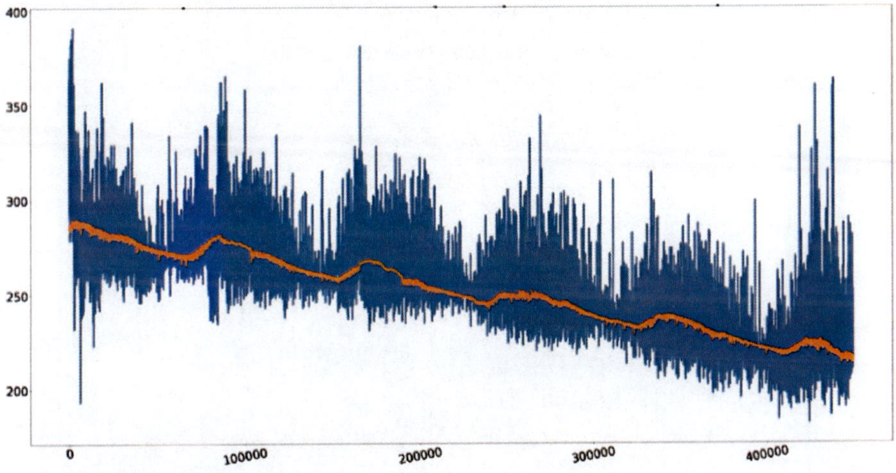

Fig. 11. Prediction on a training sample

The linear regression model performed worse than gradient boosting on the training sample, $r_2 = 0.818$, but better on the test sample, $r_2 = -0.08$. From this it can be concluded that linear regression has been retrained to a lesser extent on training data and is a more generalizing model than gradient boosting. After analyzing the results, it was hypothesized that the models are not suitable for this task, since they do not cope well with the test sample compared to the training one. Based on this, it was decided to test the models for adequacy using mathematical algorithms.

The third algorithm, which was modeled in order to test the adequacy of models, is the prediction of the target parameter based on the averaging of features. The predictions on the test sample are shown in Fig. 12, and the predictions on the training sample are shown in Fig. 13.

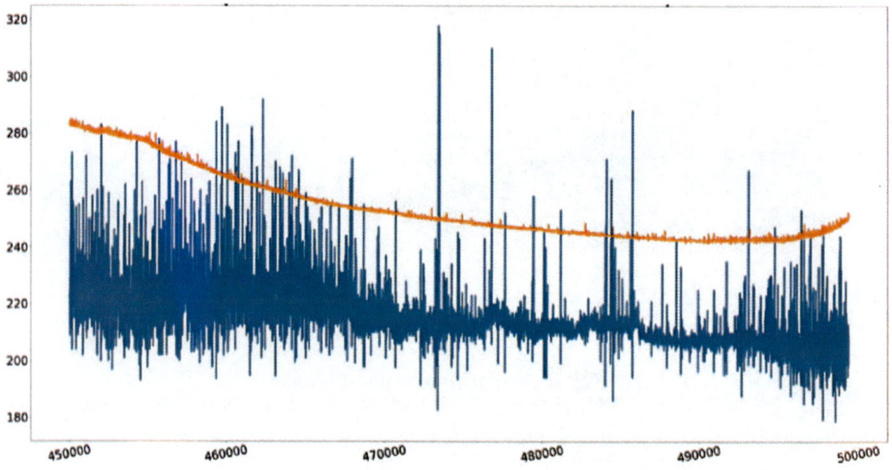

Fig. 12. Prediction on a test sample

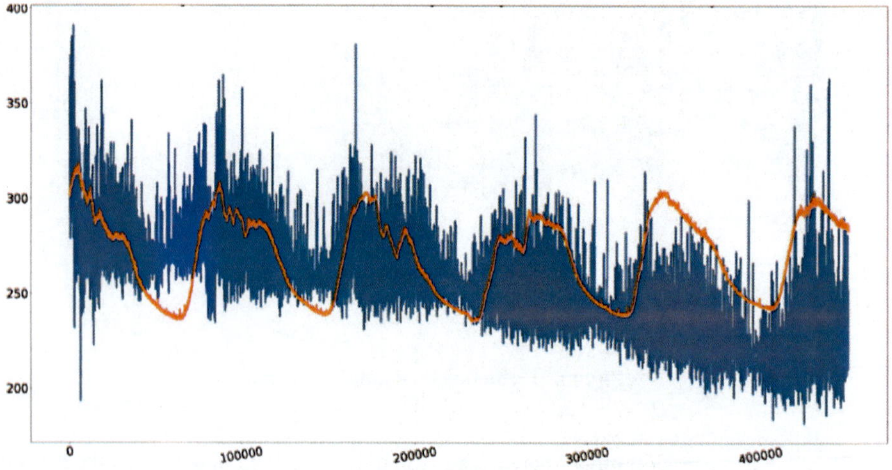

Fig. 13. Prediction on a training sample

After averaging the signs, the graph became similar to ph, since it is ph that has a larger scale, respectively, it affects the final result more strongly than other parameters. This algorithm performed worse than both of the above models, which confirms the coefficient of determination. On the training sample, $r_2 = -1.3$, and on the test sample, $r_2 = -14.7$

The fourth algorithm, which was also modeled in order to test the models for adequacy, is the prediction of a target feature based on its average value, that is, prediction by a constant. This algorithm is considered trivial for the coefficient of determination, so on the training sample $r_2 = 0$. The results on the test sample are shown in Fig. 14, and the results on the training sample are shown in Fig. 15.

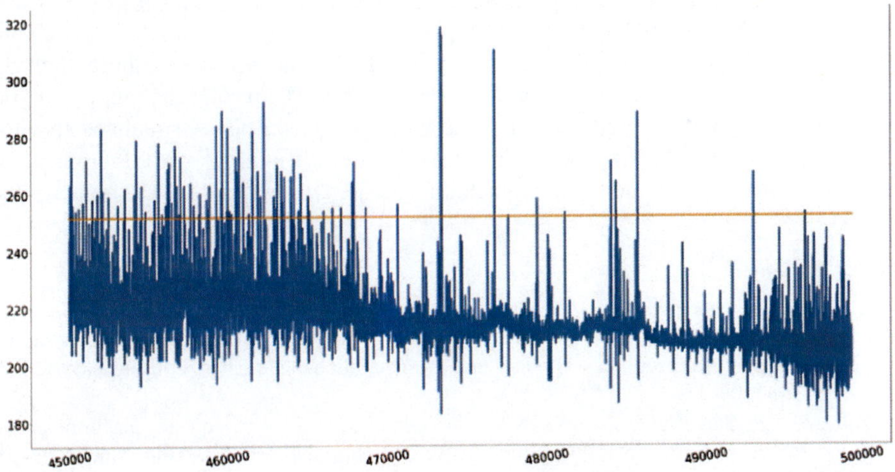

Fig. 14. Prediction on a test sample

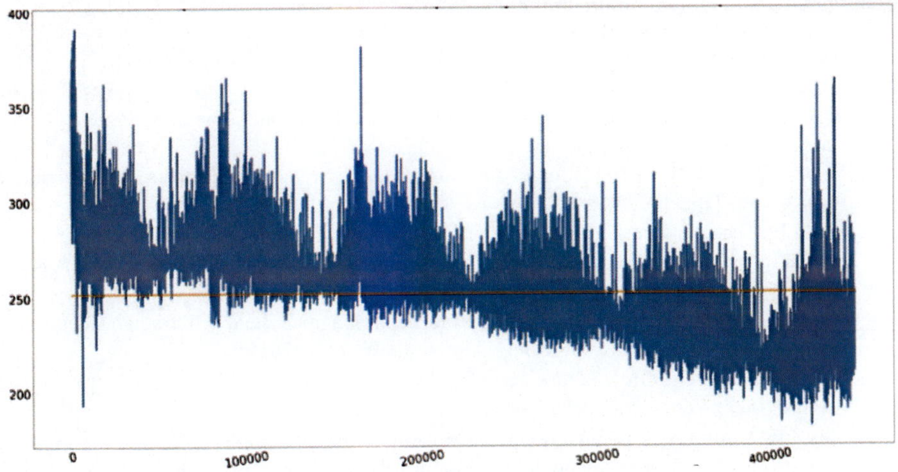

Fig. 15. Prediction on a training sample

The algorithm has a zero coefficient of determination on the training sample, which follows from the specifics of the calculations of this coefficient. But on the test sample, the algorithm coped worse than artificial intelligence models. On the training sample, $r_2 = 0$, and on the test sample, $r_2 = -12.8$.

3 Conclusion

As a result of the research, two artificial intelligence models were built and tested, the adequacy of which was verified using two mathematical algorithms. The study was conducted in the Python programming language in order to find the dependence of electrical conductivity on the physico-chemical properties of water. The linear regression model turned out to be the best algorithm for this task, which indicates the possibility of using artificial intelligence for research in the described area. The results can improve if you use more parameters for prediction and take readings from the device simultaneously for all parameters. Artificial intelligence in general has the potential to solve this problem, so it makes sense to test models further, including modern models and automl frameworks.

References

1. Bessonova, A.P., Stas, I.E.: The influence of a high-frequency electromagnetic field on the physico-chemical properties of water and its spectral characteristics. Polzunovsky Bull. **3**, 305–309 (2008)
2. Gerd, A.P., Stas, I.E., Aksenova, N.V.: The influence of a high-frequency electromagnetic field on the properties of solutions of alkali metal chlorides. Izv. Altai State Univ. **67**(3–2), 141–145 (2010)
3. Gulyaev, Yu.V., Meshchanov, V.P., Elkin, V.A.: Development of a complex of low-intensity microwave irradiation of water-containing biological materials and its application. Success. Mod. Radio Electron. **76**(6), 5–12 (2022)
4. Mikhailova, O.P., Stas, I.E., Shipunov, B.P.: The effect of a high-frequency electromagnetic field on the adsorption capacity of ionogenic surfactants. J. Phys. Chem. **84**(12), 2325–2329 (2010)
5. Yashkichev, V.I.: On the question of the influence of the magnetic field on the reactivity of water. J. Inorg. Chem. **25**(2) (1980)
6. Yemets, B.G.: Delayed relaxation of aqueous solutions exposed to electromagnetic action. J. Phys. Chem. **71**(6) (1997)
7. Stas, I.E.: The influence of a high–frequency electromagnetic field on the critical concentration of micelle formation of an aqueous solution of sodium dodecyl sulfate. J. Phys. Chem. **83**(2), 324–325 (2009)
8. Stas, I.E.: Physico-chemical processes in the electromagnetic field of ultrahigh frequencies. Monograph, p. 101. Publishing House Alt. Un-ta, Barnaul (2015)
9. Krasikov, N.N.: The effect of an electric field on the ionic composition of aqueous solutions. J. Phys. Chem. **76**(3), 567–568 (2002)
10. Stas, I.E., Chirkova, V.Y.: Integral molar heats of mixing propanol-1 and water exposed to a high-frequency electromagnetic field. VSU Bull. J. Ser.: Chem. Biol. Pharm. **2** (2017)
11. Pavlenko, V.I., Lapteva, S.N.: Study of the hardening process of gypsum binders sealed with water activated by an ultrahigh frequency electromagnetic field. Izv. Universities. Chem. Chem. Technol. **60**(8) (2017)
12. Chirkova, V.Y., Shalaeva, E.A., Stas, I.E.: Change of cohesive and adhesive characteristics of water as a result of electromagnetic influence. Izvestiya vuzov. Appl. Chem. Biotechnol. **9**(2) (2019)
13. Lifanova, R.Z., Orlova, V.S., Tsetlin, V.V.: The effect of electromagnetic radiation on the energy state of water molecules. Bull. Peoples' Friendsh. Univ. Russ.. Ser.: Ecol. Life Saf. **27**(2), 138–142 (2019)

14. Suslyaev, V.I., Pavlova, A.A.: Assessment of changes in the structure of water from external influences. Reports of TUSUR, No. 2 (22), Part 2, December 2010
15. Tseitlin, V.V., Stepanova, G.P.: Investigation of the impact of electromagnetic environmental factors on water and the internal environment of living organisms. J. Aerosp. Environ. Med. **53**(6), 2019 (2019)
16. Lunkov, A.E., Kulikova, L.N., Kovalev, D.G.: Intrinsic electrical conductivity of water induced by the influence of a high-frequency electric field. J. Butlerov Commun. **67**(8) (2021)
17. Myshkin, V.F., Vlasov, V.A., Khan, V.A.: Structure and properties of water irradiated by microwave radiation. KubGAU Sci. J. **81**(07) (2012)
18. Gossweiler, B., Joel, A.: Spatial and temporal variations in water quality and land use in a semi-arid catchment in Bolivia. Water **11**(11), 2227 (2019)
19. Tukey, J.: Analysis of the Results of Observations: Exploratory Analysis. Mir, Moscow (1981). 691 p.
20. Anthony, M., Bartlett, P.: Neural Network Learning: Theoretical Foundations. Cambridge University Press, Cambridge, New York (2002). 504 p.

The Use of Digital Educational Resources in the Educational Process

Darío Salguero García[1](✉), Ivan Otcheskiy[2], Mohammed Ayad Alkhafaji[3], Ismail Keshta[4], Shaliko Gabriyelyan[5], and Ashot Gevorgyan[6]

[1] Almería University, Almería, Spain
dariosalguero@ual.es
[2] Tyumen State University, Tyumen, Russian Federation
[3] National University of Science and Technology, Dhi Qar, Iraq
[4] AlMaarefa University, Riyadh, Saudi Arabia
[5] Stavropol State Agrarian University, Stavropol, Russian Federation
[6] National University of Architecture and Construction of Armenia, Yerevan, Republic of Armenia

Abstract. To date, the use of digital educational resources in various fields of modern education is large-scale and complex. The relevance of the topic is that today's changing world requires new approaches to education, improving its quality and accessibility. The paper describes the experience of digital resources getLocus, Yandex.Maps, Yandex.Constructor. An example of effective use of digital resources to improve educational outcomes of students is given. #CSOC1120.

Keywords: Digital Educational Resource · Adaptive Learning · GetLocus · Yandex.Constructor · Digital Educational Resource

1 Introduction

The use of modern digital learning environments in the educational process is an important factor in the transition of the educational system to a new level and improving the quality of education [1, 2]. Therefore, the main goal of teachers of educational institutions is to create a learning process using digital learning tools, as the main drivers for improving the quality of educational activities [3]. Non-standard forms of classes using the latest computer technology enhance the cognitive activity of students and help maintain a sustained interest in the learning process, as well as better mastering the material [4]. To achieve high results in learning, it is necessary to use new means of education information, which include digital educational resources [5].

Creating an e-learning environment using the most modern technologies allows to organize a quality education in an online environment [6]. The application of modern digital resources in the learning process helps to increase the cognitive potential of students, increase their independence, considering the motivation, providing a comprehensive differentiation of the training processes [7–9]. Not less important is the fact that

R. Silhavy and P. Silhavy (Eds.): CSOC 2023, LNNS 723, pp. 478–485, 2023.
https://doi.org/10.1007/978-3-031-35317-8_43

these programs take into account personal characteristics of students and the pace of their learning.

There is a number of promising areas of the use of digital education in the educational process [10, 11]. One of these directions is adaptive learning [12–14]. Adaptive learning is a technology of education on the basis of construction of an individual trajectory of the student in accordance with the current knowledge, abilities, motivation and other characteristics of students [15].

Active development of information and communication technologies has a great impact on the education system. Traditional educational technologies are being replaced, actively developing, by e-learning. The main idea of e-learning is to introduce adaptive learning in the modern educational process, and to be able to adapt to the knowledge of learners.

2 Designing the Learning Process Based on the Use of E-Learning Tools

Digital learning resources are becoming increasingly popular. They facilitate the work of teachers, save time, provide an opportunity to learn something new, go beyond the proposed material, thereby facilitating the formation of skills of independent and creative work of students.

The organization of education in the digital environment is a difficult instrumental-technical and pedagogical problem. At realization of electronic learning, unlike traditional, the direct participation in the educational process of the teacher considering individual features of students is not provided. Adaptive e-learning course can fulfill this function to a large extent.

To date, there are a significant number of developed programs and platforms for learning disciplines, which are available and actively used in e-learning. As a digital educational resource, these platforms must satisfy didactic, ergonomic and specific requirements. These requirements are based on the psychological and physiological patterns of human perception of information and must necessarily be taken into account by teachers when choosing one or another information tool for use in the educational process [16].

Designing the learning process on the basis of getLocus tools and Yandex services implies a change in the role of a teacher and a student.

First of all, the teacher's position is being reconsidered. He becomes more of a "coordinator" or "mentor" than a direct source of information and knowledge.

Also the student takes the position of an active participant of the learning process, independently studying the possibilities of getLocus and Yandex tools, using them in solving different types of tasks, including experimental with elements of research.

Based on the foregoing, it seems optimal to design and organize the learning process using e-learning environments, in which:

- the main accent is to organize an active type of cognitive activity of students, formation of "open" cognitive positions;
- the teacher is a pedagogue-manager and director of education, who is ready to offer students a necessary set of learning tools, and not just convey information;

- educational information is used as a means of organizing the cognitive activity and activity of the student, and not as a goal of learning;
- the student acts as a subject of activity together with a teacher and his or her personality-oriented development is one of the main educational tasks.

As a result, the student takes an active part in designing his or her individual route and mastering the learning material.

In the process of learning with the help of getLocus and Yandex tools, conditions are created for students to perform tasks both independently and in a group. Depending on the subject matter, students can use both traditional lab equipment and digital (computerized) measuring instruments, digital boards, maps and satellite images, etc.. Services like getLocus, Yandex.Forms, Yandex.Maps are available for everyone, which enables students and teachers to go beyond the classroom and take the educational process to a new level.

The getLocus service has some distinctive features that distinguish it from other services that help create a virtual educational environment.

An educational environment is created for each group of students, which students can join through a link or via social media. Both the teacher and students can see all assignments on the getLocus home screen. With a combination of teacher-created announcements and integrated assignment commenting capabilities, it is always possible to keep in touch and be aware of the status of each assignment. The system can be accessed through a browser. There are applications for iOS and Android operating systems to simplify the use of the system from mobile devices.

Practical assignments in getLocus, Yandex.Forms, and Yandex.Maps can be either assignments completed by a certain deadline or materials with no exact time limit.

We should separately mention the board tool used in getLocus. This is one of the simple and convenient tools of the service built into the educational environment [17]. Using the getLocus virtual board makes it possible to organize the collaborative work of a group. Students can create content together, print, write on the board, edit texts, draw, upload images and work on those images.

According to the planned goals and results, pedagogical technologies, methods and content of educational material are selected for the lessons. When conducting lessons, integration of elements of problem-based learning, level differentiation, development of critical thinking, developmental learning, information and communication technologies can be used.

Conducting classes online makes it possible to use the flipped class method of teaching. Studying new material can be carried out with direct contact between a teacher and students by developing a checklist for students where their activities are outlined step by step. The checklist can be made using a Yandex service called Yandex.Forms.

An important step is the practical application of the acquired knowledge when doing homework. Access to certain tasks that do not require specialized equipment allows students who are home-schooled to perform them, as well as during the suspension of educational activities of the educational organization for sanitary and epidemiological reasons, for example, during a pandemic.

The main didactic goals of using digital educational resources are to communicate information, form, improve and consolidate knowledge, skills and abilities, as well as to control the learning of material [18, 19]. Information technology makes it possible at any time and in any location to provide students with a guide to the task in text, audio or video format, as well as a table for entering the results of the task. In turn, the teacher also regardless of time and place can check the students' completion of assignments.

3 Using Yandex Map Constructor for Demonstration Examination

The new model of assessment in the form of a demonstration examination was developed on the basis of an in-depth analysis of international practice to assess the quality of implementation of educational programs. The use of information technology allows to simulate the processes of the real world, to present information in a detailed form, as well as to automate the process of evaluating the results of the exam. Demonstration examination using simulation fully meets the needs of finding more effective methods of assessing the quality of students' education.

The exam simulates a real-life situation in which the task of the student in practice to demonstrate their abilities and skills acquired during the acquisition of the educational program using the digital educational course. Therefore, the students' education is built on the acquisition of real practical skills, which helps graduates in further employment and professional development.

In computer science classes to form a student's personal competence profile as part of the interim attestation in the form of a demonstration exam, students developed a practical task to form a competence of information technology and practicing skills of working with information resources to make a social map of the area for the recipient applying for social assistance, taking into account those organizations that are offered for a particular citizen (Fig. 1) [20, 21].

The task is as follows, students need to find their district, create a new map. Using the tools offered in the constructor, draw an optimal route to the social security center, and specify its length in meters and the average time it will take to pass it [21]. The task is composed in such a way that it is as close as possible to the real working environment of a social worker.

When performing the task, system analysis, information technology, data processing algorithms, data analysis methods, and elements of topology were used. The practical achievements of the authors' team were used to create a real map of the area with individual nodes of interest for the solution of the task. When moving between these nodes, the algorithm for finding the shortest path was used. The results obtained in the performance of this task are of practical importance and can be used by the district administration.

Within the framework of implementation of professional module "Ensuring the implementation of the rights of citizens in the sphere of pension provision and social protection" [22], when developing this practical assignment, attention was paid to the high importance of the following didactic units for formation of professional competencies of type of activity: "To identify persons in need of social protection, and to conduct their registration, using information and computer technologies":

- creating a social map of the district using the interactive form Yandex.Constructor;

Fig. 1. Performing a practical assignment using Yandex.Constructor.

- usage of information graphic data, including the use of information and communication network Internet.

Experience in the use of digital resources in the educational process shows that students increase interest in the discipline studied, increasing attendance, in the course of the lesson everyone actively works. This contributes to the intensification of the learning process and effective learning material.

These digital educational resources are used mainly as supplementary material or as a system of students' independent work. Implementation of digital educational resources in the educational process allows for a significant expansion of the range of intellectual approach and forms of organization used in the educational process. This combination increases the efficiency of the educational process and contributes to improving the results of students' learning activities.

It can be assumed that the use of digital educational resources has a good potential for the implementation of blended learning, as these resources allow to organize independent work of students, the study of theoretical material, as well as the acquisition of some procedural skills and abilities, which makes it possible to significantly reduce the number of classroom hours.

The implementation of the educational process with the use of digital educational resources, will significantly expand the areas of educational activities without reducing the quality of students' education.

4 Discussion

Having considered the benefits of using digital platforms, it should be noted that the use of digital educational resources in the learning process has a positive impact on student motivation, as the desire to learn to use information technology tools is a powerful factor in enhancing learning motivation.

Moreover, these platforms assume the individual nature of the activity of the learner - each student works at a pace that is convenient for him, which corresponds to his general training, abilities, skills, and emotional state [15].

The use of digital educational resources in the education system makes the learning process much more effective due to the fact that it opens up opportunities:

- organizing a variety of forms of student activities for independent extraction and presentation of knowledge;
- application of the full range of possibilities of modern information and telecommunication technologies in the process of performing various activities, such as registration, collection, storage, processing of information, interactive dialogue, modeling of objects, phenomena, processes, etc.;
- creating conditions for the implementation of individual independent learning activities of students, forming skills of self-development, self-education and self-realization;
- promptly provide teachers and students with up-to-date information relevant to the goals and content of education.

Application of information technology in solving practical tasks allows u to increase student engagement in both traditional and distance form classes. E-learning involves the development and improvement of skills in handling computers and specialized software. Practical solution of tasks on the computer allows to develop systematic thinking of students, allows them to form the competencies necessary to automate their problem solving in the future. As a result of using specialized software students show significantly better results in solving real-world problems than those students who did not use it. Moreover, the first group has notably less forgetfulness of the studied material than the second group.

The current trend of active implementation of digital educational resources in the educational process, changing methods and forms of learning, allows to use the traditional model of learning along with innovative, distinguished by the active use of information technology.

5 Conclusion

Today, education is inextricably linked with digital educational resources. The modern labor market requires highly qualified professional specialists who can actively enter the work process, solve non-standard issues using critical thinking, which causes the need to use various digital resources in the educational process.

In connection with the formation of the information society digital educational resources are an integral part of the educational process, influencing the development of information competencies, contributing to the formation of a competitive personality in the labor market. Having additional qualities in comparison with traditional means, digital educational resources increase the importance of independent educational process of students, expand accessibility of educational material, assessment of its assimilation and correction of results, and also stimulate creative activity of both teachers and students.

The use of digital educational resources in education is extremely extensive. However, we can state that we are at the very beginning of the way to introduce new information technologies in the field of education, this promising direction is only being formed. Undoubtedly, many of the educational processes, as mentioned above, can be automated very successfully and positively influence the development of education and the educational process itself.

The profession of a teacher belongs to the category of those professions in which digital resources or software products, even the most high-tech ones, cannot fully replace humans. However, information technologies help to create a unique and diverse environment in which students and teachers can work together and develop. The introduction of digital educational resources in the modern educational system is not so much contrary to the traditional form of education, but expands it.

In general, speaking about the application of digital technologies in the field of education, we should expect that in the near future the roles of "teacher-student" will change, the latest e-learning platforms will appear, in which the educational process will remain a co-creativity of people, a platform of communication and interaction.

References

1. Hassouni, B.E., et al.: Realization of an educational tool dedicated to teaching the fundamental principles of photovoltaic systems. J. Phys.: Conf. Ser. **1399**(2), 022044 (2019). https://doi.org/10.1088/1742-6596/1399/2/022044
2. Masa'deh, R., Almajali, D.A., Majali, T., Hanandeh, A., Al-Radaideh, A.: Evaluating e-learning systems success in the new normal. Int. J. Data Netw. Sci. **6**(4), 1033–1042 (2022). https://doi.org/10.5267/j.ijdns.2022.8.006
3. Minamatov, Y.E.O.G.L., Nasirdinova, M.H.Q.: Application of ICT in education and teaching technologies. Sci. Prog. **3**(4), 738–740 (2022)
4. Tsarev, R.Yu.: Application of Kahoot! in education gamification. Int. J. Adv. Stud. **7**(1), 9–17 (2017). https://doi.org/10.12731/2227-930X-2017-1-9-17
5. Takamatsu, K., et al.: A new concept of ICT on eduinformatics in higher education. In: Yang, X.-S., Sherratt, S., Dey, N., Joshi, A. (eds.) Proceedings of Sixth International Congress on Information and Communication Technology. LNNS, vol. 235, pp. 693–700. Springer, Singapore (2022). https://doi.org/10.1007/978-981-16-2377-6_64
6. Baig, M.I., Shuib, L., Yadegaridehkordi, E.: E-learning adoption in higher education: a review. Inf. Dev. **38**(4), 570–588 (2022). https://doi.org/10.1177/02666669211008224
7. Pheng, H.S., Chin, T.A., Lai, L.Y., Choon, T.L.: E-learning as a supplementary tool for enhanced students' satisfaction. AIP Conf. Proc. **2433**, 030005 (2022). https://doi.org/10.1063/5.0072901
8. Rasheed, H.M.W., He, Y., Khalid, J., Khizar, H.M.U., Sharif, S.: The relationship between e-learning and academic performance of students. J. Public Aff. **22**(3), e2492 (2022). https://doi.org/10.1002/pa.2492
9. Sobolev, V.Y., Kiseleva, O.V.: Interactive methods of learning as a basis for competence formation. High. Educ. Today **9**, 70–74 (2014)
10. Medvedeva, A., Veretekhina, S., Potekhina, E., Simonov, V.: The usage of the content management system technology wordpress for creating online stores. AIP Conf. Proc. **2442**, 040021 (2021). https://doi.org/10.1063/5.0076012
11. Poondej, C., Lerdpornkulrat, T.: Gamification in E-learning: a Moodle implementation and its effect on student engagement and performance. Interact. Technol. Smart Educ. **17**(1), 56–66 (2019). https://doi.org/10.1108/ITSE-06-2019-0030

12. Jayasiriwardene, S., Meedeniya, D.: An adaptive and interactive learning toolkit (iLearn). Softw. Impacts **15**, 100471 (2023). https://doi.org/10.1016/j.simpa.2023.100471
13. Nabizadeh, A.H., Gonçalves, D., Gama, S., Jorge, J., Rafsanjani, H.N.: Adaptive learning path recommender approach using auxiliary learning objects. Comput. Educ. **147**, 103777 (2020). https://doi.org/10.1016/j.compedu.2019.103777
14. Normadhi, N.B.A., Shuib, L., Nasir, H.N.M., Bimba, A., Idris, N., Balakrishnan, V.: Identification of personal traits in adaptive learning environment: systematic literature review. Comput. Educ. **130**, 168–190 (2019). https://doi.org/10.1016/j.compedu.2018.11.005
15. Tsarev, R.Y., et al.: An approach to developing adaptive electronic educational course. In: Silhavy, R. (ed.) CSOC 2019. AISC, vol. 986, pp. 332–341. Springer, Cham (2019). https://doi.org/10.1007/978-3-030-19813-8_34
16. Vexler, V.A.: Ergonomic requirements for electronic educational resources. Psychol. Sociol. Pedagogy **5**, 1–8 (2015). https://psychology.snauka.ru/2015/05/4949. Accessed 14 Feb 2023
17. Virtual Board getLocus. https://getlocus.io/ru/. Accessed 14 Feb 2023
18. Aljarbouh, A., et al.: Application of the K-medians clustering algorithm for test analysis in e-learning. In: Silhavy, R., Silhavy, P., Prokopova, Z. (eds.) CoMeSySo 2022. LNNS, vol. 596, pp. 249–256. Springer, Cham (2023). https://doi.org/10.1007/978-3-031-21435-6_21
19. Nikolaeva, I., Sleptsov, Y., Gogoleva, I., Mirzagitova, A., Bystrova, N., Tsarev, R.: Statistical hypothesis testing as an instrument of pedagogical experiment. AIP Conf. Proc. **2647**, 020037 (2022). https://doi.org/10.1063/5.0104059
20. Service Yandex.Constructor. https://yandex.ru/map-constructor/. Accessed 14 Feb 2023
21. Practical assignment on making a social map on the service Yandex.Constructor. https://disk.yandex.ru/i/5rrhESDwlCjldw. Accessed 14 Feb 2023
22. Rutskiy, V., et al.: The relationship between gender inequality and the performance of large corporations. In: Silhavy, R., Silhavy, P., Prokopova, Z. (eds.) CoMeSySo 2022. LNNS, vol. 597, pp. 576–586. Springer, Cham (2023). https://doi.org/10.1007/978-3-031-21438-7_46

Check for updates

VirtualBox and Proxmox VE in Network Management: A User-Centered Comparison for University Environments

Marek Simon(✉) and Ladislav Huraj

Department of Applied Informatics, University of Ss. Cyril and Methodius in Trnava, 91701 Trnava, Slovak Republic
{marek.simon,ladislav.huraj}@ucm.sk

Abstract. During previous partial or complete lockdowns and isolations, there was an acute need for distance and mixed education. In the field of computer network management, a logical step was the intensive use of virtualization technologies. There are many studies that compare virtualization technologies according to different criteria. The view of students who gain new skills from these technologies is lacking. This paper focuses on a comparison of the two most commonly used technologies, Proxmox Virtual Environment and VirtualBox. The study incorporates the real perspectives of students who use these virtualization tools, which ensures a high level of relevance and reliability of the results. The results of the study can be useful for educators seeking to incorporate virtualization technologies into their teaching practices, as well as for IT professionals looking to implement virtualization solutions in their organizations.

Keywords: virtualization technology · Proxmox · VirtualBox · network management

1 Introduction

Virtualization technologies are compared according to different criteria. Whether it is performance parameters [1, 4, 5] or specific features and suitability for certain environments [2, 3, 6, 7]. This study compares two of the best virtualization technologies currently available in a university environment, namely VirtualBox and Proxmox Virtual Environment. These two technologies are compared from the perspective of the students who use these tools, which ensures a high level of relevance and reliability of the results.

The Proxmox Virtual Environment (PVE) is an open-source enterprise-class virtualization environment for servers [8]. It is based on the Debian operating system and uses the KVM hypervisor for Linux for virtualization. Proxmox VE also includes a well-designed web-based management interface that makes it easy to create and manage virtual machines, containers, storage, virtualized networks, and highly available clustering from any device with a Web browser or through a command-line interface. PVE has an integrated custom stateful firewall. Its use does not have a negative impact

R. Silhavy and P. Silhavy (Eds.): CSOC 2023, LNNS 723, pp. 486–495, 2023.
https://doi.org/10.1007/978-3-031-35317-8_44

on the execution of virtual machines running, and the VM load-balancing also does not affect the functionality of the PVE firewall [9].

VirtualBox is a free, open-source virtualization software that allows users to run multiple operating systems on a single host computer [11, 12]. VirtualBox was originally developed by Innotek, which was later acquired by Sun Microsystems and still later by Oracle, and is available for Windows, MacOS, and Linux [10]. VirtualBox uses a type-2 hypervisor, which means that it runs as an application on top of the host operating system, rather than being integrated into the host kernel like Proxmox VE.

Both Proxmox VE and VirtualBox have strengths and weaknesses. Proxmox VE uses KVM, which allows for higher performance and more advanced features such as live migrations and storage management. VirtualBox, on the other hand, is lighter and has a simpler user interface, making it easier for less experienced users to get started with virtualization.

Proxmox also has an HA (high availability) cluster feature, which means that if one node goes down, another node will automatically take over the VMs; this feature is not available in VirtualBox.

Another important functionality of Proxmox VE is support for live virtual machine migration. Moreover, Proxmox VE supports remote access to virtual machines using a Web interface, while Virtual Box does not have this feature; thus, Proxmox VE is more suitable for classroom use because it allows students to work with virtual machines from any location using a Web interface.

In addition, Proxmox VE supports the use of tools to automatically install and configure virtual machines, whereas Virtual Box does not have this feature; thus, Proxmox allows students to work with tools to automate processes.

VirtualBox is a cross-platform tool, while Proxmox VE is for Linux only. This means that VirtualBox can be used on different operating systems, while Proxmox VE is only available for Linux. Virtual Box has a simpler interface and is easier for beginners to work with virtualization. Proxmox VE has a more complex interface and requires more knowledge to manage virtual machines.

Both tools are currently available free of charge, allowing schools and universities to use them without much financial cost. In terms of teaching network management, both tools provide different options for managing and monitoring virtual network labs, as well as creating different types of network topologies and scenarios, allowing the teacher to tailor the classroom to the needs of the students. Both VirtualBox and Proxmox VE provide support for different operating systems, provide different options for using automation and scripting, and different options for configuring network protocols and security settings, allowing the teacher to teach different types of network security scenario. Additionally, both tools provide extensive documentation and community support, allowing students to easily get the information and assistance they need for teaching.

The aim was to obtain detailed information and student evaluations of VirtualBox and Proxmox VE virtual machines and their use in the network management classroom through an in-depth interview. An in-depth interview can help to gain a deeper understanding and more accurate information about students' experiences with these virtual machines. Additionally, such a comparison between VirtualBox and Proxmox VE will allow teachers to determine which of these tools is more appropriate for their teaching network management needs in different contexts [13].

2 Background

During their studies at university, students of computer science courses commonly work with virtualization, among other technologies. The advent of the global pandemic and the subsequent period of partial or complete lockdowns and isolation have extended the field of virtualization from a subject of study to a tool used in teaching. More precisely, as a tool for distance and blended education. In the subject of computer network management, a more intensive use of virtualization technologies was a logical step.

During the semester, students worked in VirtualBox with prepared VMs with pre-installed CentOS Linux, Fig. 1.

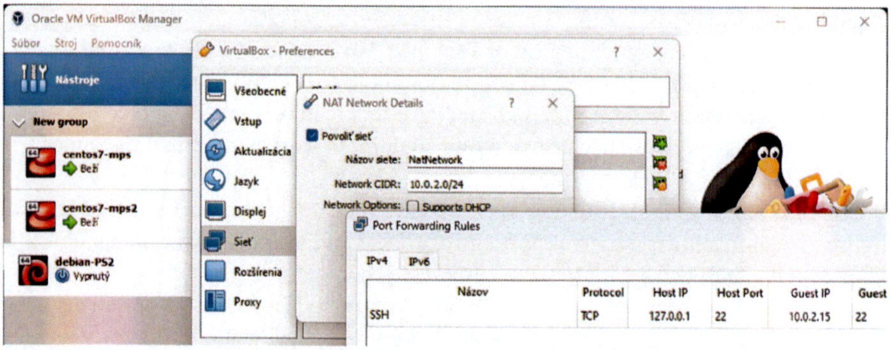

Fig. 1. Configuring port forward for ssh access to the VM command line.

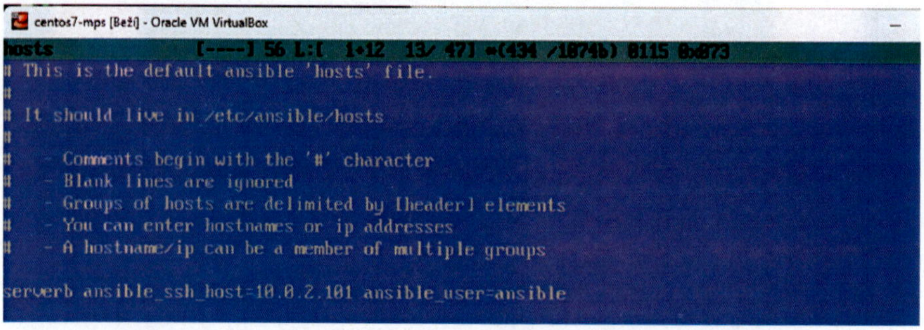

Fig. 2. Configuring Ansible in the VirtualBox application window.

Students learn to configure and manage the basic system. Then install, configure, and use tools to facilitate the management of larger networks of servers. Such as Nagios and Zabbix monitoring tools, which are capable of real-time monitoring of IT infrastructure and allow data collection and analysis; or IT automation tools such as Ansible and

Rundeck, which allow remote configuration and management of servers or management and scheduling of complex processes in IT infrastructure, Fig. 2.

The students had two virtual machines at their disposal, one always acting as the monitoring/management node, and the other being monitored or managed. Providing students with two virtual machines with different roles has several advantages in teaching network management. The first is to get closer to a real-world corporate environment, where students choose a suitable machine during their project, implement it, and then present and defend their project to their classmates. Another advantage is the testing of different scenarios, where students can try different configurations in different environments and analyze their effects. For example, how the network would behave if the monitoring/management VM was attacked or if incorrect parameters were set. Additionally, by having two virtual machines with different tasks, students can decide which task to perform and which to monitor, which allows for different types of task and enhances students' creativity. As mentioned above, this type of setup allows students to work with different monitoring and management tools and gain experience with more of them.

The final independent work used the Proxmox Virtualization Environment and the OPNsense firewall. The physical connection diagram is shown in Fig. 3.

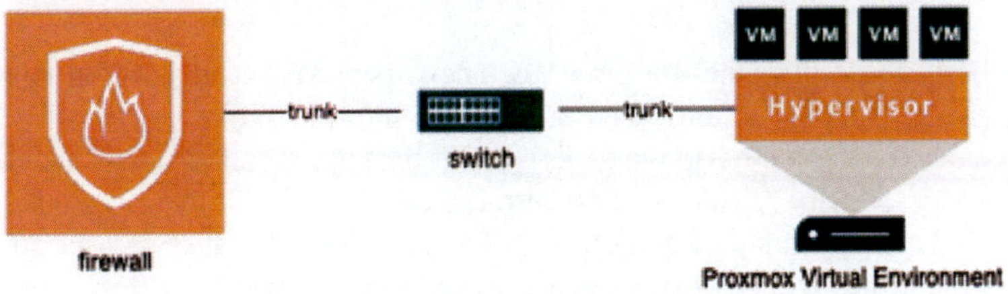

Fig. 3. PVE physical connection diagram.

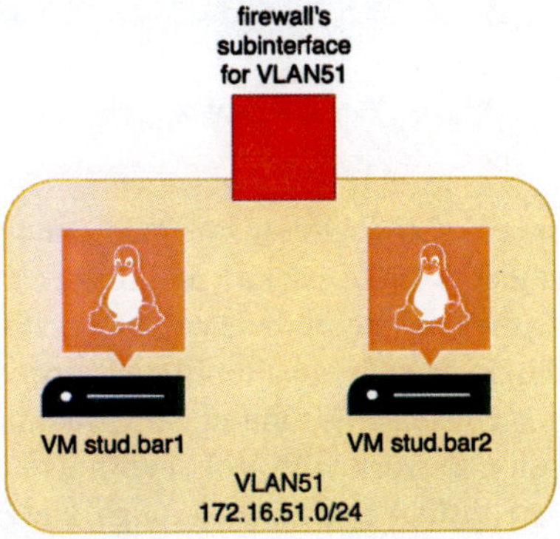

Fig. 4. VLAN level logic connection diagram.

Each student had two virtual machines available after logging into PVE. Both of them had to be configured and prepared for further work.

For each student, an associated VLAN was created in which the student had both VMs. The default gateway from the student's VLAN was the associated firewall subinterface connected in the associated VLAN, Fig. 4. The firewall also defined from which IP addresses (besides the university's) a particular student could connect.

All students had an account created on the Proxmox Virtualization Learning Environment. Each account was associated with a PVE Pool in which VMs were assigned to the corresponding student, Fig. 5. PVE Pools can be used to group sets of VMs and datastores; permissions can then be easily set for entire pools that all pool members inherit. Handling a Pool in this way allows access control to be simplified.

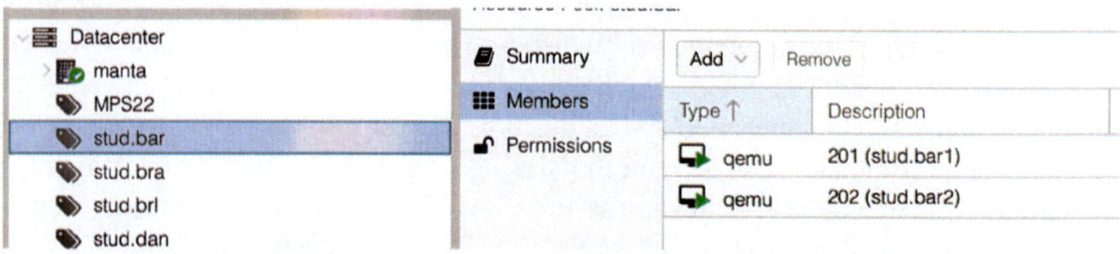

Fig. 5. Student PVE pools and an example of their content.

Similarly, to simplify the work in PVE, a new special role called MPS (abbreviation for the course used in the university's academic information system) was created for students, which enabled basic work, i.e., audit, backup, monitor, power management, snapshot, etc., Fig. 6. A role in PVE terminology is a simple list of privileges. The Promox VE contains a set of predefined roles that generally satisfy a need in a casual environment. Since specific privileges are needed when using PVE for education, it was not possible to use the predefined roles, and the MPS role mentioned above was created.

Fig. 6. Logical network scheme.

For students, the display of the virtual machine configuration has been enabled through the MPS role. Students could backup and restore their own virtual machines. They were given access to the console of their virtual machines. Students could turn on and off their virtual machines. In order to simplify and speed up students' experiments with different configurations of systems running on virtual machines, the MPS role was extended with the permission to work with VM snapshots.

Students accessed their virtual machines through the PVE console. The console terminal usually opens in another browser window. By default, the VNC console is set up for PVE. However, the VNC console does not support the commonly used copying of text to and from the console, as does the VirtualBox environment. The solution for PVE was to add a serial port to the VM and modify the guest's kernel configuration by simply adding line GRUB_CMDLINE_LINUX="quiet console=tty0 console=ttyS0,115200" in/etc/default/grub and update system loader Grub, Fig. 7.

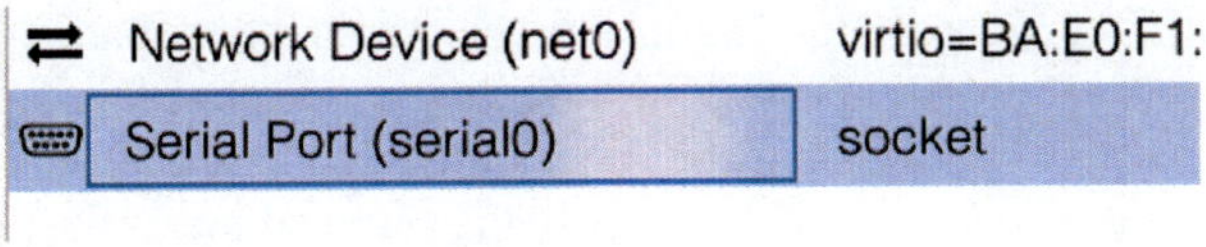

Fig. 7. Adding a serial port to a VM by simply using the hot keys ctrl+c and ctrl+v for text.

Then it was possible to select xterm.js as the console, which supports text copying; and it was already possible to use keyboard shortcuts to copy text in the console just as it is possible in VirtualBox, Fig. 8.

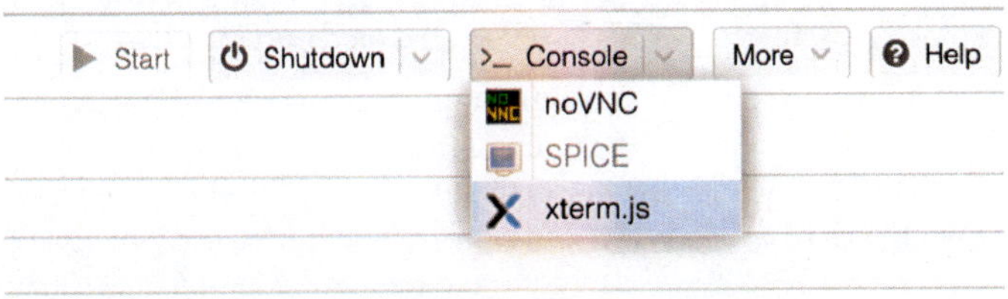

Fig. 8. Selecting the xterm.js console.

3 Experiment Setup

The intent of teaching the Network Management course is to provide students with theoretical and practical knowledge in the management and administration of computer networks and their elements, including the basics of network architecture, technologies and protocols, network security, network monitoring, backup and recovery, automation, virtualization, and cloud computing. In this course, students learn to use the tools and technologies used in today's enterprise environments to manage and administer networks. The goal is for students to be able to plan, design, implement, and manage computer networks and their components with an eye toward security, efficiency, and reliability. Students also learn to use a variety of tools and technologies to optimize networks and improve their performance and availability. In addition, the course also aims to develop the ability of students to analyze and solve practical network problems.

The aim of the experiment was to allow students to try out and compare two different virtualization tools, VirtualBox and Proxmox VE, in the context of network management, with the aim of gaining students' perspectives on the advantages and disadvantages of both virtualization environments. This allowed students to gain experience with both tools and decide which one they found most suitable for teaching the subject of network management. The choice of these two tools ensured that the students could see the differences between the desktop and server environments and learn how to operate them in different contexts. This allowed them to gain hands-on experience in real-world scenarios and to see what problems can occur when using these tools in real-world network environments.

At the end of the semester, the 12 students who had agreed to participate in the experiment were asked to implement any of the tools they had become familiar with during the semester in a network management course in a different virtualization environment. The idea was that students would be able to verify the practical use of the tools they learned during the semester in the different environment they worked in during the course, and thus be able to draw comparisons between the two environments by testing the performance and functionality of both solutions. Additionally, the above approach allowed students to better understand how different tools and technologies are used in different contexts and situations and to better understand the advantages and disadvantages of not only VB and PVE but also the individual tools used.

Since the students worked remotely in the virtual environment from a remote location, they could ask the teacher for consultation if they had a problem they did not know how to solve. This is where the advantage of PVE compared to VirtualBox in terms of teaching became apparent. The teacher had administrator access to PVE and could therefore work directly with the student's VMs. Then, using a simple video conference and zooming out the application window, the teacher could show and explain to the student where the problem was and how to solve it.

4 Discussion

The final stage of the teaching was the presentation of the students' own implementation of the chosen tool from network management. During the presentation, the students also commented on the usefulness of PVE for teaching and compared PVE and VB from this point of view.

An in-depth interview with the experiment participants was chosen as another method that contributed to gather information on the subjective experiences, opinions, and attitudes of the students toward both instruments. When comparing two different network management tools, as was the case here, it is difficult to obtain objectively measurable data on the performance of the tools. The in-depth interview with the students provided important information about their experiences of using each tool, including what they liked and disliked. This information is useful for future improvements in teaching and research on network management.

The following statements were summarized from the presentations and the in-depth interview that followed, Table 1.

Students appreciate PVE is opensource and licensing policy. During their studies, students are exposed to many different software programs. In the case of open source software, students are informed of this, as well as of the various licenses under which open source is released. It should be noted that VB is also open source. In addition, during the experiment to learn the Nagios monitoring system, they were alerted to the Icinga project, which is a fork of Nagios. If Nagios was not open source but only freeware, Icinga would not exist. Therefore, students are aware of the advantages of open source software. If the project were to stop further development (the lead author loses interest, the umbrella company goes out of business, etc.), the community around the development would simply continue on its own.

The students categorized PVE as a tool used in enterprises for professional enterprise solutions; PVE is an advanced solution for them, while VB is seen by the students as a tool for playing on the PC.

One of the tasks the students had to perform during the experiment was to study the features of both virtualization tools and compare them with each other based on the features provided. The students highlighted the robustness of PVE, as they also learned from the documentation about a number of other features of PVE that were not presented during the class. They also learned about other Proxmox projects, Proxmox Backup Server and Proxmox Mail Gateway, which extend the virtualization tool portfolio.

Students who excel in programming courses have appreciated the REST API at PVE, which as an application programming interface allows interaction with RESTful web services. They also appreciated the dynamic scaling of storage and execution. Within the curriculum, students can profile themselves professionally and choose different electives based on their personal preferences. A cast of students profile themselves as SW developers and choose subjects in this area. For them, Ansible IT automation technology was interesting precisely because of the programming of playbooks as blueprints for automation tasks. Therefore, they were interested in the possibility of REST APIs in Proxmox, even though the course did not directly address this.

Table 1. Advantages and disadvantages of the virtualization environments used mentioned by the students.

	VirtualBox	Proxmox VE
Operating system	cross-platform tool	for Linux only
Necessity of software installation by the user	yes	no
Uploading	no	yes
Open source	yes	yes
Interface	simpler interface	more complex interface
Purpose	personal use, testing	enterprise use, professional virtualization
Valued feature, function	fast testing; VirtualBox installs as an application in a regular operating system; virtual machines only; virtualization also on PCs with CPU without virtualization support	containers and virtual machines; dynamic storage and performance scaling; web GUI; clustering; high VM availability using CEPH distributed subcore system

The students stated that PVE was better for teaching because it was worked with via a web interface and there was no need to download additional software. In addition, the students felt that starting with PVE was easy. There is no need to install any additional menu tool. Everything can be done in a web browser (Google Chrome and Firefox are recommended) running on any device. It should be noted that VB itself is installed as an application and then handled as a regular application. The PVE web GUI can also be adapted to mobile devices. If a CLI (Command Line Interface) is needed on a PVE node, it is possible to open the HTML5 console directly in the browser.

Although all students appreciated the benefits of PVE in the classroom, some students worked better with VB than with PVE. A prerequisite to working with PVE was a fast and reliable Internet connection. Understandably, students in areas with a limited choice of ISPs preferred to work with VB running on their local computer. Thus, it can be concluded that for independent work on a student's computer with no or limited access to the Internet, VirtualBox is preferable to PVE.

5 Conclusions

Based on interviews with students, we can conclude that Proxmox VE is generally more suitable for teaching network management than VirtualBox. Students repeatedly appreciated the simplicity and speed of starting work on a virtual machine in the Proxmox Virtual Environment.

From the teacher's point of view, the burden of localizing the problem during the student's work in the virtual environment is removed. The teacher does not have to search whether the problem is in the operating system on the virtual machine, at the level of virtualization, or at the level of the operating system that runs on the student's computer. Simply because the teacher had administrator access to the PVE, the teacher could work directly with the student's VMs and debug any problems. Subsequently, using a simple video conference and zooming out of the application window, the teacher could show and explain to the student where the problem was and how to solve it.

The present study has some limitations. They can be seen in the application of the experiment to undergraduate computer science students at only one university in Slovakia and to a small number of students in one year; so the results may not be generalizable to other target groups. The results obtained represent the subjective opinions of the students about the two virtualization tools used. Future studies should include universities from more countries to achieve greater diversity in the groups addressed, as well as on other virtualization tools.

Acknowledgements. The work was funded by the grant KEGA 012UCM-4/2021 *Modern technologies and innovations in network security education.*

References

1. Đorđević, B., Timčenko, V., Kraljević, N., Jovičić, N.: Performance comparison of KVM and Proxmox Type-1 Hypervisors. In: 2022 30th Telecommunications Forum (TELFOR), pp. 1–4. IEEE (2022)
2. Gabriška, D.: Software requirements for the control systems according to the level of functional safety. J. Appl. Math. Stat. Inform. **12**(1), 25–32 (2016)

3. Li, J., Jin, J., Yuan, D., Zhang, H.: Virtual fog: a virtualization enabled fog computing framework for Internet of Things. IEEE Internet Things J. **5**(1), 121–131 (2017)
4. Padala, P., Zhu, X., Wang, Z., Singhal, S., Shin, K.G.: Performance evaluation of virtualization technologies for server consolidation. HP Labs Tec. Report 137 (2007)
5. Algarni, S.A., Ikbal, M.R., Alroobaea, R., Ghiduk, A.S., Nadeem, F.: Performance evaluation of Xen, KVM, and proxmox hypervisors. Int. J. Open Source Softw. Process. (IJOSSP) **9**(2), 39–54 (2018)
6. Arango, C., Dernat, R., Sanabria, J.: Performance evaluation of container-based virtualization for high performance computing environments. arXiv preprint arXiv:1709.10140 (2017)
7. Robles-Gómez, A., Tobarra, L., Pastor-Vargas, R., Hernández, R., Cano, J.: Emulating and evaluating virtual remote laboratories for cybersecurity. Sensors **20**(11), 3011 (2020)
8. Goldman, R.: Learning Proxmox VE. Packt Publishing Ltd., UK (2016)
9. Ariyanto, Y., Harijanto, B., Firdaus, V.A.H., Arief, S.N.: Performance analysis of Proxmox VE firewall for network security in cloud computing server implementation. In: IOP Conference Series: Materials Science and Engineering. IOP Publishing, p. 012081 (2020)
10. Li, P.: Selecting and using virtualization solutions: our experiences with VMware and VirtualBox. J. Comput. Sci. Coll. **25**(3), 11–17 (2010)
11. Watson, J.: Virtualbox: bits and bytes masquerading as machines. Linux J. **2008**(166), 1 (2008)
12. Younge, A.J., Henschel, R., Brown, J.T., Von Laszewski, G., Qiu, J., Fox, G.C.: Analysis of virtualization technologies for high performance computing environments. In: 2011 IEEE 4th International Conference on Cloud Computing, pp. 9–16. IEEE (July 2011)
13. Škrinárová, J., Vesel, E.: Model of education and training strategy for the high performance computing. In: 2016 International Conference on Emerging eLearning Technologies and Applications (ICETA), pp. 315–320. IEEE (2016)

Information and Analytical Support for Decision-Making on Resource Support of Fire Fighting Units

Olga Shikulskaya[1](✉), Timur Yesmagambetov[2], Tatyana Ten[2], Bakyt Spanova[3], and Mikhail Shikulskiy[4]

[1] Astrakhan State University of Architecture and Civil Engineering, Astrakhan, Russian Federation
shikul@mail.ru

[2] Karaganda University of Kazpotrebsouz, Karaganda, Republic of Kazakhstan

[3] Karaganda Buketov University, Karaganda, Republic of Kazakhstan

[4] Astrakhan State Technical University, Astrakhan, Russian Federation

Abstract. The study object is the Fire Fighting units' resource supply. The study subject is the apparatus for information and analytical support for decision-making on the resource support of operations. It was concluded that the reason for the low indicators of the public protection values from fires in Russia and Kazakhstan is determined by the irrational resources provision for fire departments. The studies relevance aimed at the efficiency increasing of the Fire Fighting units' resource supply is justified. An overview of research in the subject area has been made. The basic approach to solving the problem has been determined, which is based on the integrated use of the FTA method and the system approach with MSS system locations. The concept of "Area of acceptable values" has been offered. A computational procedure for calculating nodes and the root of the failure tree of the emergency response operation has been developed. The criteria system of the efficiency, reliability and quality of the operation plan has been developed, allowing to compare the solutions to the identified problems of non-compliance of the state of resources with the required one. The significant impact of the Arrival Time at Call Location factor on the Fire Fighting operation success is shown. The process choice for the practical implementation of theoretical studies is justified. The nodes and the root of the tree have been calculated. Solutions with the same reliability test values and different quality test values have been demonstrated. The studies will increase the decision-making efficiency on the Fire Fighting units' resource support.

Keywords: fire · emergency response · fault tree analysis · multi-state system · computational procedure · criteria system

1 Introduction

Based on the results of the official fire statistics generalization, it can be concluded that the reason for the low indicators of the public values protection from fires in Russia and Kazakhstan is determined by the irrational resources provision of fire departments.

R. Silhavy and P. Silhavy (Eds.): CSOC 2023, LNNS 723, pp. 496–507, 2023.
https://doi.org/10.1007/978-3-031-35317-8_45

Ensuring fire safety requires enormous costs. Financial resources, as a rule, are not enough. It is often impossible to provide fire departments with resources in accordance with regulatory requirements. And it is not always advisable. According to the modern approach based on object-oriented rationing, the cost of ensuring fire safety should not exceed the damage from a likely fire.

Works of European, Russian, Kazakh, Caucasian, Vietnamese and other researchers: A.V. Matveeva, Maksimova A.V., Menshikh V.V., Samorokovsky A.F., Matyushina A.A. Poroshina, E.V. Bobrineva, S.A. Oleinika, V.L. Semikova, Nguyen Ba Tuan, A.T. Berdibeko-va, A.V. Dolya, V.V. Gruzina, S.N. Nurakova, A.T. Berdibekova, D.A. Sabekiya, V.V. Terebneva, V.A. Gracheva, A.O. Semenova, V.T. Averyanova, E.S. Topilkina, V.Ya. Vilisova, and others are devoted to resource management. However, their work does not imply the possibility of using resources of inadequate condition and in insufficient quantity [4–6, 8–11, 15, 16].

Such a task is very difficult, and it requires information and analytical support.

Computer support is often used to solve fire safety problems [1, 2, 12–14, 17].

A team of authors led by N.N. Brushlinsky (E.M. Alekhine, P. Wagner, Yu.I. Kolomiets, S.V. Sokolov) proposed analytical dependencies describing the flows of emergency rescue services calls in time, and simulation modeling to take into account spatial characteristics, the processes of functioning of the city's emergency rescue services for design, organizational and management decisions on their operational activities [1]. However, their work also does not take into account the state of resources. In addition, the implementation of the proposed solutions requires significant costs.

French scientists Girard, C., Piatyszek, E., David, P., Flaus, J.-M. [7] proposed to take into account the different influence of the level of deviation from the requirements of the state of resources on the degree of implementation of the emergency response plan to carry out a priori analysis of the plan based on a combination of structural-functional interaction methods, failure tree analysis and a system with many states. It is advisable to develop and use their research in solving the problem of resource management in conditions of their quantitative and qualitative non-compliance with the requirements.

To solve the research problems, as a basic one, a mathematical formulation of the resource supply management problem was formulated, the information algorithmic scheme of which is shown in Fig. 1.

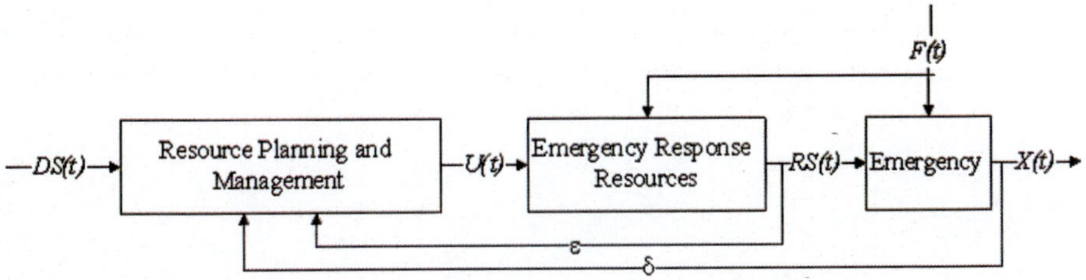

Fig. 1. Generalized diagram of the set-theoretic model for managing emergency response processes, taking into account the resources state

The set-theoretic model of resource support for emergency response operations (fire) can be represented by a tuple $RP(t) = \langle FS(t), F(t), U(t), RS(t), X(t)\rangle$, where $RP(t)$ - resource support for emergency response operations, $FS(t)$ – the need to ensure a safe environment (fire safety state of the municipal facility), $U(t)$ - vector of control actions aimed at stabilizing the resource supply of emergency response operations, $RS(t)$- variable state of resources, $F(t)$ - vector of external factors that degrade the state of resources, $X(t)$ - vector of state variables of the facility on which the emergency occurred, Ω – the area of permissible values of state variables of the facility, θ- the area of permissible values of resource variable states. In general, the task of managing an emergency comes down to determining such *U(t),* values in which the object state variables take only permissible values $X(t) \in \Omega$, while the resource state variables initially have permissible values $RS(t) \in \theta$.

Suppose that among the resources attracted there are those that do not meet the quantitative and/or qualitative indicators of the requirements for resources, i.e. $\exists RS(t) \notin \theta$, $X(t) \notin \Omega$. Due to the limited funding, there is no way to get the desired result: $\nexists U(t) : RS(t) \in \theta \wedge X(t) \in \Omega$. In this case, we are forced to reduce the requirements Ω to the range of acceptable values. $\tilde{\Omega}$. Let's define the range of acceptable values $\tilde{\Omega} \supseteq \Omega$. Need to find: $U(t) : X(t) \in \tilde{\Omega}$.

The main semantic above stated in the adopted mathematical statement of the problem of resource supply managing was the conclusion that in order to bring the indicators of the state of the object of protection into the area of acceptable values, it is necessary to make an appropriate financial resource distribution, which will be necessary, for this case, management impact (decision).

2 Methods

2.1 Method for Predicting the Effectiveness of the Fire Rescue Service in Fire Conditions Based on a Priori Assessment of the Degree of Implementation of Plans Based on the State of the Resources Involved

The method developed by the authors is the development of research by French scientists Girard, C., Piatyszek, E., David, P., Flaus, J.-M. It combines the fault tree (FTA) method and system analysis with multi-state systems (MSS).

The fault tree is a logical combination of events that can cause the function to fail. The fault tree must be created for each function. Function resource failures are the main events of the function failure tree.

In the fault tree, two process states are traditionally used: operation and failure. However, to comprehensively analyze the weaknesses of emergency response models and assess a wide range of occupied resources, it is advisable to use a system approach with many system states [7]. This approach allows us to identify a discrete relationship between the degree of achievement of the purpose of the operation and the state of resources.

2.2 Mathematical Description of the Emergency Response Process Model as a Multi-state System

The Multi-State System (MSS) is defined as a system with a finite number of performance levels with a finite number of resource performance levels.

For any *j* element in the system, there are k_j states corresponding to the performance level g_{ji},, where *i* is the state number $i \in \{1, 2, ..., k_j\}$. *Gj(t)* is the performance level of the element. The probability of being in state g_{ji} at time *t* may be represented by the set $p_j(t) \in \{p_{j,1}(t), p_{j,1}(t), ..., p_{j,k_j}(t)\}$, where $p_{j,1} = \Pr\{G_j(t) = g_{j,k_j}\}$.

If the system consists of n elements, the performance of the system is determined by the function of the performance levels of its elements: $G(t) = \varphi\{G_1(t), G_2(t), ..., G_n(t)\}$. The function output *G(t)* characterizes the system state.

We select 4 system states and 4 resource degradation levels. For each failure of resources and processes of the model, the failure degrees (degradation levels) are determined:

- Lvl 1: Not degraded ($\{g_{j,1} = 1, p_{j,1}\}$).
- Lvl 2: Rather not degraded ($\{g_{j,2} = 2, p_{j,2}\}$).
- Lvl 3: Rather degraded ($\{g_{j,3} = 3, p_{j,3}\}$).
- Lvl 4: Completely degraded ($\{g_{j,4} = 4, p_{j,4}\}$).

Specific values of the "degradation level" shall be adopted in accordance with the requirements of the analyzed scenarios. For example, in the evacuation function: full evacuation - level 1, partial evacuation (more than 3/4 of the evacuated population - level 2, partial evacuation (less than 3/4 of the evacuated population) - level 3, and total evacuation failure - level 4. This means that the levels are determined by the experts in relation to the acceptable performance of the function. This is comparable to the classical construction of the risk matrix. The failure method is characterized by a list of pairs {level, probability}. Thus, for each resource and function, 4 failure levels are defined - 4 pairs.

To ensure the automation of calculations of nodes and the root of the tree tree, the authors developed a computational procedure.

To analyze and compare various solutions, the authors developed effectiveness, reliability, quality criteria of the emergency response operation model and effectiveness of the solution for resourcing the activities of the fire rescue service, a priori assessment of the reliability of operational activity plans, a computational procedure for counting the elements of the failure tree [3].

2.3 Calculation Procedure for Tree Nodes Calculating

The fault tree describes the logical relationship between the state of the system and the state of its resources and it is represented as a binary tree.

The root of the tree is the process: $v_0 \in V$, the leaves of the tree are resources. Non-boundary vertices represent intermediate system failures. Each node in the tree, except the leaves, includes a boolean, such as <AND>, <OR>, and <PRIOR>. Each failure (tree element) is characterized by a list of pairs {level, probability} defined by the set $C_i = (G_i, P_i)$, where i is the number of the node of the tree, G_i is the level of deterioration of the resource state, P_i is the probability of this level. For leaves, the values of the elements of the set C_i are determined by experts or by a statistical method. For other nodes and roots, they are calculated through logical elements.

Let's define the functions of logical elements.

At the output of each logic element, there are two sets $\{C_1, C_2\}$, hat determine the probabilities of the degradation levels of the tree elements. At the input, there is one set $\{C\}$, that also determines the probabilities of the degradation levels of the preceding tree element. It is necessary to determine $\{C\}$ through $\{C_1, C_2\}$. The function of the logical element of the tree $\{C\} = f(\{C_1, C_2\})$.

The relation $R = C_1 \times C_2 = R(G_1, P_1, G_2, P_2)$ is the Cartesian product of the sets C_1 and C_2. Let's define the relationship R' through the relationship R, complementing it with the calculated attributes: $R'(G_1, P_1, G_2, P_2, G_3, P_3)$. $G_1 : \{g_{11}, ..., g_{1i}, ..., g_{1k}\}$, $G_2 : \{g_{21}, ..., g_{2i}, ..., g_{2k}\}$, $P_1 : \{p_{11}, ..., p_{1i}, ..., p_{1k}\}$, $P_2 : \{p_{21}, ..., P_{2i}, ..., P_{2k}\}$.

For i-th relationship tuple R': $g_{3m} = g_{2m} \# g_{2m}, p_{3m} = p_{2m} \cdot p_{2m}$, m is a - tuple number.

The # operator is defined by a logical element:

- for logic element <AND>: $g_{2m} \# g_{2m} = \min(g_{1,m}, g_{2,m})$,
- for logic element <OR>: $g_{2m} \# g_{2m} = \max(g_{1,m}, g_{2,m})$,
- for logic element <PRIOR>: $g_{2m} \# g_{2m} = g_{1,m}$.

Relation $C(G, P)$ determines the probabilities of the degradation levels of tree elements at the logical element output. It is derived from the relation R' using the SQL aggregate function:

$$G : \{g_1, \ldots, g_i, \ldots, g_k\},\ P: \{p_1, \ldots, p_i, \ldots, p_k\}, \text{где } p_i = \sum_{g_i} p_j .$$

TNow, knowing the functions of logical elements, you can calculate the entire tree to the root. We start traversing the tree from the root.

The arc along which we pass away from the root is denoted as 0, after returning as 1. On a node whose two outgoing arcs are denoted by 1, we start the function of the logical element and determine the set of probabilities of the deterioration levels of the states of the node element $C(G, P)$. Bypassing the entire graph we get a set of probabilities of the levels of deterioration of the states of the process (tree root) $C_0(G_0, P_0)$, by which we can estimate the process.

To calculate multilevel fault trees, it is necessary to determine the lower level probabilities. This can be done with the help of experts or based on statistical analysis of available data.

2.4 The Reliability and Quality Criteria of the Plan of the Fire Rescue Service Operation and the Solutions Effectiveness for Resource Support of Its Activities

The concept of effectiveness will be considered in two aspects: the goal effectiveness and the solution effectiveness taken to eliminate the problem.

We will understand the goal effectiveness as the degree of its compliance with real needs and problems. The decision effectiveness to eliminate the problem is determined by the ratio of damage from a likely fire and the cost of measures to reduce or eliminate this damage.

If, on the basis of the analysis, the partial non-compliance with the requirements of the indicators of the operation goal achievement due to inadequate quality and/or insufficient quantity of resources involved in its achievement is identified, the SCL must make a management decision, which can be of two types: either the restoration of resources that do not meet the requirements - bringing their parameters to the required ones (the effectiveness of the decision is assessed), or expanding the range of permissible values Ω (the effectiveness of the goal is assessed). Even if there are enough funds, the question arises about the validity of the requirements (the effectiveness of the goal). If the reduction in the results of the operation does not affect the number of victims, and the estimated economic damage is less than the required costs for eliminating resource mismatch, then such investments are not economically profitable - not effective. An appropriate control solution in this case would be to extend the range of acceptable values to the range of acceptable solutions $\tilde{\Omega}$: $\Omega \subseteq \tilde{\Omega}$. A management decision is considered effective if financial investments in elimination of non-compliance of values of resource parameters with requirements do not exceed economic damage from a probable fire in case of their non-elimination, provided that the risk of death in a fire does not change - human life is invaluable. Thus, the range of acceptable values $\tilde{\Omega}$ is determined based on the positions of the effectiveness of management decisions.

In case of insufficient funds to eliminate inconsistencies, the expansion of the permissible values range to the acceptable values range will be forced. But nevertheless, the SCL should strive to make the most effective decision. I.e. when choosing one of several solutions, the condition must be met: at the same financial costs, it should give the best result. If you do not want to extend the range of valid values, the range of acceptable values will be the same as the range of valid values: $\tilde{\Omega} \equiv \Omega$. The question arises of what to understand by the areas of permissible and acceptable values and how to evaluate the result.

The range of allowable values is a set of such system states characterized by a set of pairs {level, probability}, for which the state level is 1 (since by default the first level meets the requirements), and the probability of the first level is not lower than the one specified for the range of allowable values: $\Omega = \{\omega_i = (g_i, p_i) : g_i = 1, p_i \geq p_\Omega\}$, where p_Ω is the lowest allowable probability of the first state of the system.

The acceptable value area is a set of such system states characterized by a set of pairs {level, probability}, for which the state level is equal to the level received by the SCL (i.e. the parameters of this state level of the SCL system are considered acceptable), and the probability of this level is not lower than the value set by the SCL for the acceptable value area: $\tilde{\Omega} = \left\{\tilde{\omega}_i = (g_i, p_i) : g_i \leq g_{\tilde{\Omega}}, p_i \geq p_{\tilde{\Omega}}\right\}$, where $g_{\tilde{\Omega}}$ is the highest acceptable value of the system state level, $p_{\tilde{\Omega}}$ is the lowest acceptable probability of the first system state.

Criteria for the reliability of the emergency response plan. We define two reliability criteria: by the range of permissible values $K_R(\Omega)$ and by the range of acceptable values $K_R\left(\tilde{\Omega}\right)$. The system reliability criterion for the range of permissible values $K_R(\Omega)$ will mean the probability of reaching the first level of system state $K_R(\Omega) = \{g = 1, p_1\}$, The system reliability criterion for the range of acceptable value $K_R\left(\tilde{\Omega}\right)$ will mean the probability of reaching the level determined by the SCL for the range of acceptable values $K_R\left(\tilde{\Omega}\right) = \left\{g_{\tilde{\Omega}}, \sum_{i=1}^{g_{\tilde{\Omega}}} p_i\right\}$ the probability of reaching the level of state determined by the range of acceptable values: $K_R\left(\tilde{\Omega}\right) = \left\{g_{\tilde{\Omega}}, \sum_{i=1}^{g_{\tilde{\Omega}}} p_i\right\}$. The description of the indicators of the areas of permissible and acceptable values is summarized in Table 1.

Table 1. The indicators description of the permissible and acceptable values areas

Indicator	Range of valid values Ω	Range of acceptable values $\tilde{\Omega}$
Boundaries of the area	$\Omega = \langle g_1, p_\Omega \rangle$	$\tilde{\Omega} = \left\langle g_i, p_{\tilde{\Omega}} \right\rangle$
Reliability indicator	$K_R(\Omega) = \{g = 1, p_1\}$	$K_R\left(\tilde{\Omega}\right) = \left\{g_{\tilde{\Omega}}, \sum_{i=1}^{g_{\tilde{\Omega}}} p_i\right\}$
Area Membership Condition	$K_R(\Omega) \subseteq \Omega$, $p_1 \geq p_\Omega$	$K_R\left(\tilde{\Omega}\right) \subseteq \tilde{\Omega}$, если $\sum_{i=1}^{g_{\tilde{\Omega}}} p_i \geq p_{\tilde{\Omega}}$

Moreover, the probability of reaching an acceptable level is determined as the sum of the probabilities of this level with the probabilities of all previous levels. Systems with the same reliability criteria may differ in quality. Therefore, we will additionally introduce a quality criterion.

Quality Criterion of the Emergency Response Operation Plan. At the root of the logical tree is a set of degrees of achievement of the goal of the operation with their probabilities. Such a set can be considered as a discrete law of probability distribution of the degree of achievement of the goal. The same can be said of the knots and leaves of the tree. Sets with many system states in them are discrete laws of state level probability distribution. In this case, statistical methods can be applied. Then the quality criterion K_R can easily be calculated as a mathematical expectation. The mathematical expectation of a discrete random variable G: $K_Q = M(G) = \sum_{i=1}^{k} g_i \cdot p_i$. The result is a fractional value $K_Q = M(G) \subset [1, k]$. In our case, k = 4, since we consider four levels of resource state and 4 degrees of goal achievement. The level number to which this value is closest is the most likely state. Using this value, you can compare the result with the specified (required) indicator of the degree of achievement of the operation goal and compare the results of various decisions with each other.

3 Results

Quality criterion of the emergency response operation plan. At the root of the logical tree is a set of degrees of achievement of the goal of the operation with their probabilities. Such a set can be considered as a discrete law of probability distribution of the degree of achievement of the goal. The same can be said of the knots and leaves of the tree. Many with many states of the system in them "Time Factor" significantly affects the development of fire, the material and social damage caused to it. The time of arrival of units at the place of call is essential. Therefore, it is this process that is chosen to implement theoretical developments. The fault tree developed for this process is shown in Fig. 1.

Table 2 is filled in to determine tree leaf values based on statistical data and/or expert survey.

The calculation results are shown in Fig. 3.

4 Discussions

When a problem is identified and solutions are formed that bring the parameters of the protected object into the area of acceptable or permissible values, the SCL must select one of the solutions that most meets the requirements. The effectiveness of the solution is ensured by its belonging to the range of acceptable values. SCL shall define priorities for reliability and quality criteria. First, the results of decisions are ranked according to a higher priority criterion, and the second one. It is preferable to rank the solutions first according to the reliability criterion to reduce the risk of failure of the operation. Then solutions with the same reliability criterion values are ranked according to the quality criterion.

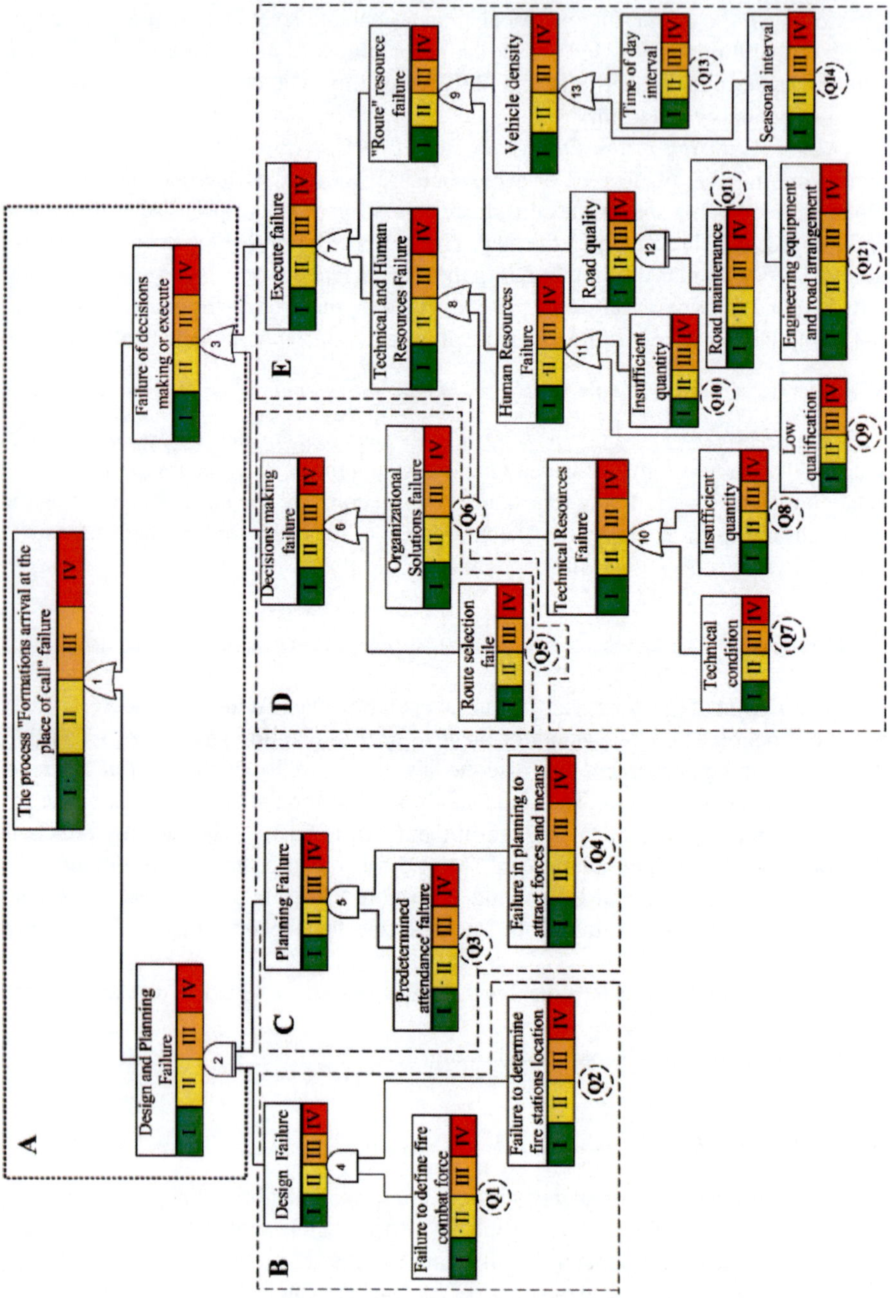

Fig. 2. Fire Brigade Arrival Failure Tree

The difference of solution with the same reliability criteria but different quality criteria is illustrated in Fig. 4. The most preferred solution is a) and the worst solution is c) (Fig. 2).

Table 2. Вероятности состояний ресурсов на нижнем уровне дерева (объект 1)

Question	Lvl 1	Lvl 2	Lvl 3	Lvl 4
Q1. Failure to define fire combat force	1			
Q2. Failure to determine fire stations location	1			
Q3. Predetermined attendance' falture	1			
Q4. Failure in planning to attract forces and means	1			
Q5. Route selection faile	0.7	0.3		
Q6. Organizational Solutions failure	0.9	0.1		
Q7. Technical Resources Failure. Technical condition	0.9		0.1	
Q8. Technical Resources Failure. Insufficient quantity	0.9	0.1		
Q9. Human Resources Failure. Low qualification	0.9	0.1		
Q10. Human Resources Failure. Insufficient quantity	0.9			0.1
Q11. Road quality. Road maintenance	0.75		0.25	
Q12. Road quality. Engineering equipment and road arrangement	0.5	0.2	0.3	
Q13. Vehicle density. Time of day interval	0.8		0.1	0.1
Q14. Vehicle density. Seasonal interval	0.9	0.1		

Failure levels g_i	Failure probability p_i
1	22%
2	32%
3	36%
4	10%

$$K_Q = \sum_{i=1}^{k} g_i \cdot p_i = 2{,}3$$

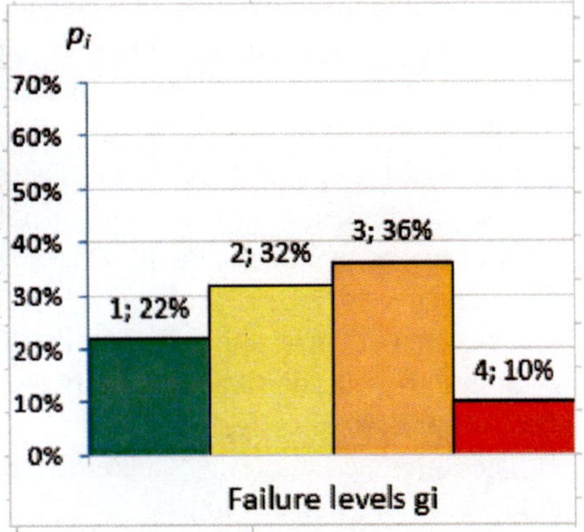

Fig. 3. Calculation Result of failure tree root of emergency fire response operation

Failure levels g_j	Failure probability p_j
1	50%
2	30%
3	12%
4	8%

Failure levels g_j	Failure probability p_j
1	36%
2	44%
3	11%
4	9%

Failure levels g_j	Failure probability p_j
1	21%
2	59%
3	9%
4	11%

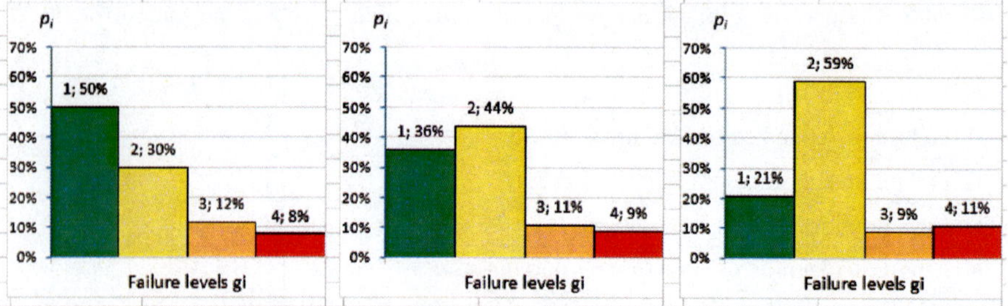

K_R=0.8	K_Q=1.8	K_R=0.8	K_Q=1.9	K_R=0.8	K_Q=2.1

a) Decision 1 b) Decision 2 c) Decision 3

Fig. 4. Illustration of systems with the same reliability criterion values and different quality criterion values

5 Conclusion

In the work, the analysis of the subject area was carried out. The need to take into account the impact of the level of non-compliance with the requirements of resources on the degree of achievement of the purpose of the operation is justified. A toolkit has been developed to solve the identified problem on the basis of the known approach with the integrated use of the (FTA) fault tree method and the system approach with multi-state system (MSS), which is supplemented by the computational procedure for calculating nodes and the root of the fault tree developed by the authors and a system of criteria.

The system of criteria includes criteria of efficiency, reliability and quality. The developed criteria allow you to compare the resulting solutions between failures.

The significant impact of the Arrival Time at Call Location factor on the success of the Fire Fighting operation is shown. These are justified by the choice of this process for the practical implementation of the given theoretical studies. On the basis of statistical data and survey of experts, the leaves of the failure tree are filled, and the nodes and the root of the tree are calculated.

The calculation result is graphically illustrated.

Solutions with the same reliability test values and different quality test values have been demonstrated.

The completed studies will increase the efficiency of decision-making on the resource provision of fire-fighting units.

References

1. Brushlinsky, N.N., Sokolov, S.V., Alekhine, M., et al.: City security. Imitation modeling of urban processes and systems. FASIS (2004)
2. Butuzov, S.Y., Korobko, V.B., Pranov, B.M., Stankevich, T.S., Chlenov, A.N.: A neuro-fuzzy model of support for fire extinguishing control in seaports. Contr. Syst. Inf. Technol. **4**(66) 91 (2016)
3. Esmagambetov, T.U., Shikulskaya, O.M.: Analysis of the reliability of the emergency population evacuation plan as a multi-state system based on the structure of the fault tree. Adv. Mod. Sci. **8**(4), 68–72 (2016)
4. Flaus, J.M.: A modelling framework for model based risk analysis, pp. 1533–1540. ESREL, Troyes France (2011)
5. Georgiadou, P.S., Papazoglou, I.A., Kiranoudis, C.T., Markatos, N.C.: Modeling emergency evacuation for major hazard industrial sites. Reliab. Eng. Syst. Safety **92**, 1388–1402 (2007)
6. Georgiadou, P.S., Papazoglou, I.A., Kiranoudis, C.T., Markatos, N.C.: Multi-objective evolutionary emergency response optimization for major accidents. J. Hazard. Mater. **178**, 792–803 (2010)
7. Girard, C., David, P., Piatyszek, E., Flaus, J.-M.: Emergency response plan: model-based assessment with multi-state degradation. Saf. Sci. **85**, 230–240 (2016)
8. Jain, S., McLean, C.: A framework for modeling and simulation for emergency response. In: Proceedings of the 2003 International Conference on Machine Learning and Cybernetics, pp. 1068–1076 (2003)
9. Jonkman, S.N., Lentz, A., Vrijling, J.K.: A general approach for the estimation of loss of life due to natural and technological disasters. Reliab. Eng. Syst. Safety **95**, 1123–1133 (2010)
10. Karagiannis, G.M., Piatyszek, E., Flaus, J.M.: Model-driven and risk-based performance analysis of industrial emergency plans. J. Conting. Crisis Manage. **21**, 96–114 (2013)
11. Kolen, B., Kok, M., Helsloot, I., Maaskant, B. EvacuAid: a probabilistic model to determine the expected loss of life for different mass evacuation strategies during flood threats. Risk Anal.: Off. Publ. Soc. Risk Anal. (2013)
12. Semikov, V.L., Nguyen, B.T.: Resource management model for ensuring the activities of the fire rescue service. Fire Saf.: Challenges Prospects T. 2. **1**(6), 110–113 (2015)
13. Stankevich, T.S., Kiper, A.V.: Intellectual decision support system based on fuzzy neural networks for fire extinguishing manager in the territory of "Kaliningrad Commercial Seaport". OJSC Certif. Stat. Registr. Comp. Progr. **12**(86), 2013661903 (2018)
14. Stankevich, T.S., Butuzov, S.Y., Ryzhenko, A.A.: Information and analytical support for management of fire extinguishing at highly dangerous and technically complex facilities. In: KnE Engineering/Russian Forum of Young Scientists (RFYS), pp. 229–237 (2018)
15. Terebnev, V.V., Grachev, V.A., Semenov, A.O.: Management of fire and rescue units resources and technological operations for extinguishing fires. Fire Explosion Saf. **15**(5), 60–65 (2006)
16. Vaez, N., Nourai, F.: RANDAP: an integrated framework for reliability analysis of detailed action plans of combined automatic-operator emergency response taking into account control room operator errors. J. Loss Prevent. Process Indust. **26**, 1366–1379 (2013)
17. Vilisov, V.Y.: Algorithm for optimal allocation of limited resources based on the method of game iterations. Inf. Technol. Bull. **2**(20), 89–99 (2019)

Informatization of Education in India

Darío Salguero García[1](✉), Indrajit Patra[2], Alexander Yanovskii[3], Vitaly Grinchenko[3], Natalia Bystrova[4], Samrat Ray[5], Belkadi Lamiaa[6], Khatori Youssef[6], and Nodira Safarova[7]

[1] Almería University, Almería, Spain
dariosalguero@ual.es
[2] NIT Durgapur, Durgapur, West Bengal, India
[3] Stavropol State Agrarian University, Stavropol, Russia
[4] Minin State Pedagogical University of Nizhny Novgorod, Nizhny Novgorod, Russia
nat.bystrova@rambler.ru
[5] Peter The Great Saint Petersburg Polytechnic University, Saint Petersburg, Russia
[6] Ibn Tofail University, Kenitra, Morocco
[7] Tashkent Institute of Finance, Tashkent, Uzbekistan

Abstract. Over the last 5 years, the educational environment in India has undergone significant changes. Information technology in education has begun to play a role that had never played before. This paper is devoted to the organization and development of educational environment, first of all, from the point of view of the application of information technology. The problems of organizing the information environment of the educational process are considered. The questions related to the creating and developing the distributed educational environment are described. The problems of interaction between the individual elements of the unified electronic educational environment were solved. Possible prospects for the development of the distributed electronic educational environment in India are outlined. #CSOC1120.

Keywords: E-learning · Online Education · Informatization · Electronic Educational Environment

1 Introduction

Despite intense economic development and scientific and technological progress, there is a huge illiteracy in India. At the same time India is one of the fastest developing countries in the world [1]. Notable development is taking place in areas related to technology and computerization. India is entering into the field of modern educational solutions, i.e. such as open education and distance learning using modern satellite solutions [2].

Information technology is being widely adopted throughout the country [3]. The most advanced state in this regard is Kerala, which has the highest literacy rate in the country.

Kerala was also the first state in the country to introduce information technology as a subject at the secondary school level back in the early 2000s [4, 5]. Less than a year

R. Silhavy and P. Silhavy (Eds.): CSOC 2023, LNNS 723, pp. 508–517, 2023.
https://doi.org/10.1007/978-3-031-35317-8_46

later, the government was forced to include free software in the curriculum because of enthusiastic protests and the stance taken by the teachers' association. Finally, starting in 2007, schools began to teach GNU/Linux, and all computers had only GNU/Linux installed. At that time, or even today, it was the most massive installation of GNU/Linux in educational institutions, and it made headlines even in other countries.

The country is now approaching an education in which information technology is integrated [3, 6]. Information technology will eventually not be taught as a separate course. Instead, all subjects will be taught using information technology, so that students, on the one hand, will learn how to use it and, on the other hand, can use educational applications (such as those mentioned below) and online resources (such as text materials, content from sources like Wikipedia, images, animations, and videos) to master the core curriculum.

Teachers and students have already begun using applications such as Dr. Geo, GeoGebra, and KtechLab to study geometry and electronics. Applications such as Sunclock, Kalzium and Ghemical are also popular with teachers and students.

The Indian government's new education policy (2020) talks about constructivism, IT-enabled education, free software, and educational resource sharing. Once a few large states successfully move to free software, it is believed that the whole country will follow suit in a relatively short time [7, 8]. When this happens, India may have the largest GNU/Linux and free software base in general.

2 Electronic Education in Universities

Modern information and education technology is widely used in higher education in India, such as virtual universities [9, 10]. The University of Madras has opened a Virtual University in partnership with universities in Mumbai, Calcutta, and beyond. Many Indian universities and professors share free lectures through The National Program on Technology Enhanced Learning (NPTEL), a government-supported cooperative program in engineering education.

Distance learning and open education are also very advanced [11, 12]. The most prominent example is Indira Gandhi National Open University (IGNOU), which has about 4 million students, and they live in different countries [13, 14].

The university was founded in 1985 by a parliamentary resolution to introduce and promote open university systems and distance learning in India. The main goal of the university is to make education more accessible. Student facilities include 48 regional centers, 6 sub-regional centers and 1271 science centers. The university's programs are broadcast on a special educational channel, Gyan Darshan. The university has launched 17 radio stations under the initiative of Gyan Vani FM.

At present IGNOU has established cooperation with educational institutions of many countries [15, 16]. The Distance Education Council, under the leadership of IGNOU, is responsible for coordinating and maintaining the standards of open and distance learning in India. This university has developed and implemented high-quality curricula in the humanities, exact and social sciences, as well as professional studies in information technology, education, engineering, management, nursing, and tourism [17, 18]. Most university programs offer certificates, diplomas and degrees.

IGNOU has proven that modern communications technology can be used effectively to expand access to education and that new technology does not have to be very expensive. The university has built its own audio-visual production facilities, with support from the British and Japanese governments.

The current Electronic Media Production Center is an extremely modern technological facility that enhances the University's educational opportunities [15]. The university's educational materials are broadcast nationwide through a television network as well as selected radio stations. A specially designed satellite-based teleconferencing network is being developed on a pilot basis in collaboration with the Indian Space Research Organization. The aim of this collaboration is to develop interactive solutions for distance education programmes offered by open universities in India.

IGNOU is also the national supreme authority for distance learning, and the Distance Education Council has statutory authority in this regard. It is responsible for promoting, coordinating, and constantly operating the standards of the open and distance education system throughout the country. Its role is also to create and develop a network of open universities, and to ensure the exchange of intellectual and physical resources in the distance education system between different institutions [18]. The Council is also responsible for improving the system by expanding its reach and ensuring compliance with quality standards of education and curriculum. As part of its responsibilities, it also financially supports the development of open universities and e-learning institutions from funds allocated by the central government for its own activities. It also supports public open universities. Currently there is one national open university, 11 state open universities, and a number of e-learning institutions in various parts of the country.

The Distance Learning Council has launched a number of initiatives to develop a network of open universities. Programs designed and developed by IGNOU are widely used by state open universities across the country. Efforts have also been made to develop programs, degree awards, exams and grading systems so that students can more freely choose programs. Steps have been taken to launch a diploma transfer system. Efforts have also been made to develop norms and standards for creating, designing, and implementing programs in specific fields and ensuring their high quality. Open university programs have proven to be effective at low cost. The costs of these institutions are much lower than standard comparable programs require.

The programs and courses offered by open universities in India are designed to meet the needs of the economy and labor market. Most programs differ significantly from those offered by traditional universities. These programs are designed and implemented in several modules, allowing students to choose a module according to their specific needs. Universities try to diversify their courses and develop quality open multimedia curricula that meet the scientific, technical, and professional needs of different groups of students. The programs offered by IGNOU are internationally recognized.

Learning management systems are used for e-learning needs. An LMS is software that helps create, manage, organize, and deliver online learning materials to the classroom [19–21]. The main use of an LMS is to create content and track students' progress. Statistics from the initial closing period (March to June 2020) during the COVID-19 pandemic show that the number of applications and student activity for learning with LMS has nearly doubled from pre-pandemic levels of use of the software.

The popularity of online education is understandably increasing dramatically beginning in 2020. The year-over-year growth rate for 2020 is estimated at nearly 20%, as confirmed by Technavio's statistics. Not only that, experts are predicting the e-learning market to grow 21% through 2024, which will further increase the market capacity by more than $14 billion.

The needs of education in India are constantly being monitored by the government. The Indian telecommunication satellite EDUSAT was created in 2004 by the Indian Space Research Organisation exclusively for the needs of education [22–24]. The main purpose of the project is to build satellite technology and an interactive distance learning system based on it. The project reflects India's commitment to using space technology for its development, especially for the education of citizens from the most remote, rural areas.

3 E-Learning Benefits

In this regard, the prospects and key benefits of e-education are very clear. Advances in information and communication technologies have opened up great opportunities for the transfer and acquisition of knowledge at universities, both on campus and remotely [25–27]. Students with disabilities who are unable to study at a traditional university and anyone wishing to update their knowledge can successfully use the new means of communication [20]. In addition, new technologies offer greater access to sources of information and facilitate communication between scholars and professors, as well as building a network of academic institutions [28].

Thanks to open universities and distance learning, students are systematically improving their qualifications and developing their competencies [29]. The needs of citizens for continuing, lifelong education are steadily deepening. Institutions of higher education offer programs to meet this demand. Direct access and flexibility are key elements of these initiatives. The introduction of open universities is designed to increase the chances of higher education, make the higher education system more democratic, and guarantee lifelong learning. The growing popularity of e-education is a definite answer to the "elitism" of classical university education. This is confirmed by statistics, as the main criterion for the majority of students in India who choose e-education is its accessibility [30] (see Fig. 1).

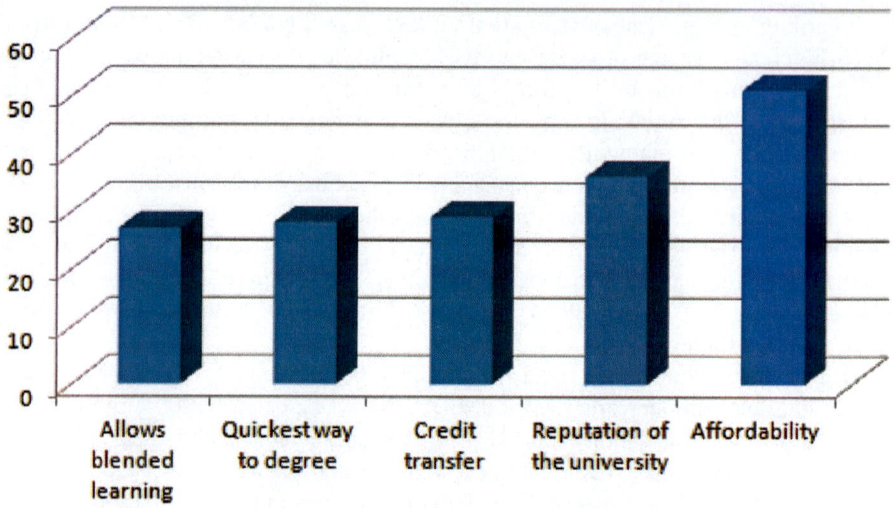

Fig. 1. Selection Criteria for Online Courses in India.

4 COVID-19 and the Impact of the Pandemic on Distance Education

The COVID-19 pandemic and subsequent restrictive measures affected India's educational system [31]. As people locked themselves in their homes for months, schoolchildren and students had no way to continue their studies in a traditional way [32, 33]. What they thought was an unexpected, pleasant weekend and what they enjoyed turned out to be a catastrophic mess. Many were deprived of the opportunity to study, so the government needed to find an alternative, and that alternative was a distributed educational environment [34, 35].

Online education is a growing area of the education system in India [36]. Since the new National Education Policy (NEP) was passed in 2020, there have been many changes to the education system that also included online learning, but with the new set of rules there have been many drawbacks.

According to UNESCO, since the COVID-19 upsurge, 1.37 billion students in 138 countries have been affected by school and university closures. Nearly 60.2 million school teachers and university professors were not in attendance.

With the entire country in lockdown, electronic education proved to be the best and only option left [33, 37]. University departments create accounts on online video conferencing platforms like Zoom, Skype, Google Classroom, Meet and others to interact with students [38, 39].

These new media provide the ability to access content on demand at any time or on any digital platform, but it has become a challenge for administration and students alike [40]. Digital media today is a mixture of the traditional way of learning with books and notebooks, and digital software with e-books and pdf files.

5 Main Disadvantages of Online Education

5.1 Socioeconomic Disconnect

India is a diverse country, and there is a wide variety of cultures and beliefs within that diversity, and India suffers from a huge socio-economic divide - the division into castes such as upper, lower, and middle, and this is a major disadvantage [41, 42]. Even a third of the population does not receive an online education.

In rural areas, the main problems are lack of Internet connection, to a lesser extent lack of electricity, and inability or inaccessibility to purchase appropriate devices [43]. Many teachers have complained about the inaccessibility of learning materials for elementary school students because they cannot attend online classes due to the lack of a network.

As an alternative to this, many classes are held by sending pre-recorded videos via WhatsApp or YouTube so that children can learn at their own convenience, but even here there are difficulties, such as a lack of understanding of the lessons. As a result, students are still deprived of the knowledge they need.

5.2 Gender Differentiation

Another disadvantage is the differentiation between girls and boys in this country [44, 45]. According to a recent survey conducted in Bihar public schools, out of 733 children, only 28% of girls had smartphones, compared to 36% of boys.

Half of these families could not afford the Internet, and their children had to rely on lessons that were televised. But in these cases, girls were found to spend more time doing homework than boys, which often coincided with TV broadcast times.

Another factor was that in most of these cases parents refused to talk to the girls and offered to take the survey on their behalf. In some of the cases where teachers were able to reach girls, there were no conversations because they had to communicate with a male family member, which made them realize how difficult it was for girls to learn through online resources.

5.3 Disadvantages of the Internet

Using the Internet as a source of online education is a big problem. Google, Firefox, and other web browsers are good sources of information, but using these platforms for online learning is not easy. For teachers who have always worked in schools and universities, it is completely unfair to expect them to be up-to-date on digital content creation and its effective delivery online, and for children to adapt easily to it.

Two major factors such as body language and eye contact are difficult to grasp in the online classroom. Many questions arise as to whether students are paying attention in class, whether they understand the material in class, and whether they have time to absorb the teacher's speech. These issues are also relevant for traditional classes, but it is even more difficult to answer them in relation to online lessons.

5.4 Mental Health of Students

It is a mental strain for young children (5–10 years old) because their period of concentration is short and they need physical activity to maintain their attention. But it's not just for younger students; students in higher education complain about the same thing.

Exact science needs more hands-on lab work, thesis projects, and hands-on activities to supplement the theoretical aspects, which are severely limited and nearly impossible in online education.

5.5 Excessive Screen Time

Excessive screen time also becomes a problem for students who take online classes. With online learning going away, students will be consumed by their smartphone/laptop screens, and this is bad for their mental health. It overburdens their eyesight and brain. After relaxing the restrictions and with more online classes, the government has limited classes to 60 min for students and 45 min classes for school children.

6 Conclusion

The COVID-19 pandemic and the restrictive measures that followed it have caused serious problems in India's education system. With the closure of educational institutions for an extended period, education systems are increasingly leaning towards digital learning platforms. With a new set of regulations formed by the Indian government in 2020, we can state that digital ways of learning will become the new norm. But this will come with many challenges, as mentioned above.

At least 50% of India's population lives in rural areas and lacks basic needs such as electricity, water and shelter. In order to educate this section of society, we need to first meet their basic needs. The government needs to take steps to provide them with proper housing and 24-h access to electricity, only then will the new principles of national education have any impact on this sector of Indian society, otherwise only the upper class will benefit and the rest of the population will continue to live in darkness and illiteracy.

In the near future online education will become a part of everyone's life, and we have to be ready for the change. It is no longer just an alternative, but a necessity. With the help of new technologies it is possible to completely rebuild the education system, but it is a long way to go without shocks to society.

References

1. Singh, S.: An analysis of factors responsible for the rapid growth of e-learning in India. PalArch's J. Archaeol. Egypt/Egyptol. **17**(12), 1454–1466 (2020)
2. Nagaratinam, D.: Satellite based education for distance education through EDUSAT. Lang. India **15**(10), 149–166 (2015)
3. Phutela, N., Dwivedi, S.: A qualitative study of students' perspective on e-learning adoption in India. J. Appl. Res. High. Educ. **12**(4), 545–559 (2020). https://doi.org/10.1108/JARHE-02-2019-0041

4. Joseph, G.V., Thomas, K.A.: Integration of technology initiatives with educational neuroscience and its impact on technology readiness to technology adoption by HSS Teachers, Kerala. In: Thomas, K.A., Kureethara, J.V., Bhattacharyya, S. (eds.) Neuro-Systemic Applications in Learning, pp. 423–444. Springer, Cham (2021). https://doi.org/10.1007/978-3-030-72400-9_21
5. Raman, R., Venkatasubramanian, S., Achuthan, K., Nedungadi, P.: Computer science (CS) education in indian schools: situation analysis using Darmstadt model. ACM Trans. Comput. Educ. (TOCE) **15**(2), 1–36 (2015)
6. Vanitha, P.S., Alathur, S.: Factors influencing E-learning adoption in India: learners' perspective. Educ. Inf. Technol. **26**(5), 5199–5236 (2021)
7. Malik, S., Rana, A.: E-Learning: role, advantages, and disadvantages of its implementation in higher education. Int. J. Inf. Commun. Comput. Technol. **8**(1), 403–408 (2020)
8. Agrawal, V., Agrawal, A., Agarwal, S.: Assessment of factors for e-learning: an empirical investigation. Ind. Commer. Train. **48**(8), 409–415 (2016). https://doi.org/10.1108/ICT-03-2016-0015
9. Abdollahi, D.: A study of pedagogical aspects of a virtual university. Int. J. Educ. Psychol. Res. **4**(1), 12 (2018)
10. Natarajan, M.: Impact of e-Learning in virtual university environment. J. Libr. Inf. Commun. Technol. **4**(1–2), 65–72 (2015)
11. Lakshmi, Y.V., Das, J., Majid, I.: Assessment of e-Learning readiness of academic staff & students of higher education institutions in Gujarat, India. Indian J. Educ. Technol. **2**(1), 31 (2020)
12. Taso, K., Chakrabarty, A.: E-learning in higher education in India: experiences and challenges—an exploratory Study. In: Peng, S.-L., Son, L.H., Suseendran, G., Balaganesh, D. (eds.) Intelligent Computing and Innovation on Data Science. LNNS, vol. 118, pp. 715–723. Springer, Singapore (2020). https://doi.org/10.1007/978-981-15-3284-9_77
13. Majhi, M., Meher, V.: A comparative study of motivation among the learners of Indira Gandhi national open university and Odisha State open university. Online J. Distance Educ. e-Learning **10**(2), 169–179 (2022)
14. Ghosh, C.K.: Electronic media learning materials of Indira Gandhi national open university, India: an analytical study. Turkish Online J. Distance Educ. **14**(4), 210–221 (2013)
15. Subrahmanyam, V.V.: Use of Technology for Learner Support Services: A Case Study of IGNOU. In: Anjana (eds.) Technology for Efficient Learner Support Services in Distance Education, pp. 113–132. Springer, Singapore (2018). https://doi.org/10.1007/978-981-13-2300-3_6
16. Arifulina, R.U., Katushenko, O.A.: Analysis of domestic and foreign trends of individualization of the educational process in the university: analytical report. Vestnik Minin Univ. **9**(4), 2 (2021). https://doi.org/10.26795/2307-1281-2021-9-4-2
17. Jena, P.K.: Academic counselling services in Indira Gandhi national open university (IGNOU). Int. J. Adv. Res. (IJAR) **6**(4), 441–448 (2018)
18. Kaur, S., Gothwal, P.: E-Learning in India: Some dynamic initiatives in higher education. ACADEMICIA: Int. Multi. Res. J. **11**(4) 1–10 (2021)
19. Aljarbouh, A. et al.: Application of the k-medians clustering algorithm for test analysis in e-learning. In: Silhavy, R., Silhavy, P., Prokopova, Z. (eds.) Software Engineering Application in Systems Design. CoMeSySo 2022. Lecture Notes in Networks and Systems, vol. 596, pp. 249–256. Springer, Cham (2023). https://doi.org/10.1007/978-3-031-21435-6_21
20. Deepika, V., Soundariya, K., Karthikeyan, K., Kalaiselvan, G.: 'Learning from home': role of e-learning methodologies and tools during novel coronavirus pandemic outbreak. Postgrad. Med. J. **97**(1151), 590–597 (2021)
21. Singh, G., Bhatnagar, V., Gupta, R., Kumar, G.: Exploration of e-learning vs traditional learning in India. Human. Soc. Sci. Rev. **8**(2), 69–76 (2020)

22. Gothwal, P.: An overview of e-learning initiatives in Indian higher education institutions. Know. Manag. High. Educ. Inst. **1**, 176–182 (2021)
23. Mahajan, N., Garg, S.: Strive to Lead in Quality Education: A Study on Online Education During the Pandemic in India. In: Technology Training for Educators From Past to Present, pp. 276–292. Springer, Singapore (2022)
24. Tunga, S.K.: Role of academic libraries in e-learning process in India. KIIT J. Libr. Inf. Manag. **8**(2), 89–95 (2021)
25. Chahal, J., Rani, N.: Exploring the acceptance for e-learning among higher education students in India: combining technology acceptance model with external variables. J. Comput. High. Educ. **34**, 844–867 (2022)
26. Gavinolla, M.R., Swain, S.K., Livina, A.: Research contribution to the progress of digital learning in India. In: Mishra, D., Chuang, Y. (eds.) Learning How to Learn Using Multimedia. LNET, pp. 103–122. Springer, Singapore (2021). https://doi.org/10.1007/978-981-16-1784-3_8
27. Ahmed, S.U., Ahmad, T., Ahmad, N., Ahamad, M.: Current trends and future perspectives of e-learning in India. In: Proceedings 2021 Sustainable Leadership and Academic Excellence International Conference (SLAE), pp. 1–6. IEEE, Manama, Bahrain (2021)
28. Hassouni, B.E., et al.: Realization of an educational tool dedicated to teaching the fundamental principles of photovoltaic systems. J. Phys: Conf. Ser. **1399**(2), 022044 (2019). https://doi.org/10.1088/1742-6596/1399/2/022044
29. Nikolaeva, I., Sleptsov, Y., Gogoleva, I., Mirzagitova, A., Bystrova, N., Tsarev, R.: Statistical hypothesis testing as an instrument of pedagogical experiment. In: AIP Conference Proceedings, vol. 2647, p. 020037 (2022). https://doi.org/10.1063/5.0104059
30. Statista. https://www.statista.com. Accessed 17 Jan 2023
31. Sharma, D., Singh, A.: E-Learning In India duringcovid-19: challenges and opportunities. Eur. J. Mol. Clin. Med. **7**(7), 2020 (2021)
32. van Cappelle, F., Chopra, V., Ackers, L., Gochyyev, P.: An analysis of the reach and effectiveness of distance learning in India during school closures due to COVID-19. Int. J. Educ. Dev. **85**, 102439 (2021). https://doi.org/10.1016/j.ijedudev.2021.102439
33. Mathivanan, S.K., et al.: Adoption of e-learning during lockdown in India. Int. J. Syst. Assur. Eng. Manag. **1**, 1–10 (2021)
34. Karattuthodi, M.S., Thorakkattil, S.A., Chandrasekhar, D., Poonkuzhi, N.P., Ageeli, M.M.A., Madathil, H.: Pharmacy Student's challenges in virtual learning system during the second COVID 19 wave in southern India. Soc. Sci. Human. Open **5**(1), 100241 (2022). https://doi.org/10.1016/j.ssaho.2021.100241
35. Singh, H.K., et al.: A survey of E-learning methods in nursing and medical education during COVID-19 pandemic in India. Nurse Educ. Today **99**, 104796 (2021). https://doi.org/10.1016/j.nedt.2021.104796
36. Radha, R., Mahalakshmi, K., Kumar, V.S., Saravanakumar, A.R.: E-Learning during lockdown of Covid-19 pandemic: a global perspective. Int. J. Control Autom. **13**(4), 1088–1099 (2020)
37. Khan, M.A., Nabi, M.K., Khojah, M., Tahir, M.: Students' perception towards e-learning during COVID-19 pandemic in India: an empirical study. Sustainability **13**(1), 57 (2020)
38. Tsarev, R.Y., et al.: An approach to developing adaptive electronic educational course. In: Silhavy, R. (ed.) CSOC 2019. AISC, vol. 986, pp. 332–341. Springer, Cham (2019). https://doi.org/10.1007/978-3-030-19813-8_34
39. Vedavathi, N., Bharadwaj, R.S.: Deep flamingo search and reinforcement learning based recommendation system for e-learning platform using social media. Procedia Comput. Sci. **215**, 192–201 (2022). https://doi.org/10.1016/j.procs.2022.12.022

40. Pal, D., Vanijja, V.: Perceived usability evaluation of Microsoft Teams as an online learning platform during COVID-19 using system usability scale and technology acceptance model in India. Child Youth Serv. Rev. **119**, 105535 (2020). https://doi.org/10.1016/j.childyouth.2020.105535
41. Dash, G., Chakraborty, D.: Transition to e-learning: by choice or by force–a cross-cultural and transnational assessment. Prabandhan: Indian J. Manag. **14**(3), 8–23 (2021)
42. Karabuschenko, N.B., Pilishvili, T.S., Chkhikvadze, T.V., Sorokoumova, S.N., Dyukareva, Y.: Comparative analysis of the intellectual foundations of emotion recognition among Asian and Russian students. Vestnik Minin Univ. **9**(2), 7 (2021). https://doi.org/10.26795/2307-1281-2021-9-2-7
43. Ganie, M.A., Molvi, S.A., Ali, M., Hussein, A.H.: Impact of e-learning on child education and development in rural areas of India. In: Proceedings of the International Conference on E-Business, e-Commerce, e-Management, e-Learning and e-Governance, pp. 77–82. University of Greenwich, London, UK (2020). https://doi.org/10.978.819252/12223
44. Rutskiy, V. et al.: The relationship between gender inequality and the performance of large corporations. In: Silhavy, R., Silhavy, P., Prokopova, Z. (eds.) Data Science and Algorithms in Systems. CoMeSySo 2022. Lecture Notes in Networks and Systems, vol. 597, pp. 576–586. Springer, Cham (2023). https://doi.org/10.1007/978-3-031-21438-7_46
45. Rakshit, S., Sahoo, S.: Biased teachers and gender gap in learning outcomes: evidence from India. J. Dev. Econ. **161**, 103041 (2023). https://doi.org/10.1016/j.jdeveco.2022.103041

Evolution of Information Security in Banking: From Incident Monitoring to Business Continuity

Vyacheslav V. Yashkin[1], Kirill V. Pitelinskiy[2], Sergei A. Kesel[2](✉), Andrey S. Boyar-Sozonovitch[3], and Yuriy N. Philippovich[2]

[1] OJSC Ak Bars Holding, 1, Dekabristov Street, Kazan 420066, Russia
[2] Moscow Polytechnic University, 38, Bol'shaya Semonovskaya Street, Moscow 107023, Russia
sakesel161@gmail.com
[3] Moscow Aviation Institute (National Research University), 4, Volokolamskoe Highway, Moscow 125993, Russia

Abstract. According to the concept of corporate risk management the basic types of risks (operational, credit, market and economic) are studied, and their role in providing economic and information security (EIS) of organizations of credit and financial (banking) sphere is discussed. As a research methodology, the tools of simulation modeling (SM), which allows obtaining the required characteristics of the dynamics of business processes, are applied. The way to describe the activities of the organization is a model in the form of a system of dynamic contour flows (information, financial, material, energy and personnel). Operational and business risks (often arising in the activities of economic entities) and explicitly affecting their EIS are considered in sufficient detail. It is noted that the risks associated with the disruption of the continuity of business processes can be included in the classification systems of risks grouped according to various criteria. In particular, they can include an extensive list of risks caused, for example, by errors of internal business (so-called process risks, legal risks, risks of using an unsuccessful business management model and risks of lack of liquidity to ensure uninterrupted funding of business processes). #CSOC1120.

Keywords: Simulation Modeling · Operational Risk · Optimization · Decision-Making · Credit and Financial Sphere · Information Security · Economic Security · Quality Management

1 Introduction

The current socio-economic situation in the Russian Federation requires an increase in investment in the real sectors of the economy through the support of promising projects and high-tech enterprises. Business entities in carrying out their activities are based on the criterion of maximizing profits in conditions of uncertainty and, therefore, risk.

The success of the decision maker (DM) is determined by the understanding and attitude to risk. The risk of commercial and entrepreneurial activity is an important

R. Silhavy and P. Silhavy (Eds.): CSOC 2023, LNNS 723, pp. 518–537, 2023.
https://doi.org/10.1007/978-3-031-35317-8_47

element of management theory and practice. Investment risks are random risks, where positive and negative results are possible (defined by the probability of damage due to risky operations). In addition to the risks of business and commercial activities (mainly financial), there are also non-financial risks: operational risks. Operational risks are important because it is the continuity of an organization's business processes that depends on them.

Operational risks include EIS risks, personnel risks, information system risks, project risks (due to operational risks), risks of lack of internal control and other types of operational risks. Credit and financial institutions (banks) are not outside of these processes, but instead find themselves in these problems, exacerbated by threats in the information sphere and cybersecurity. Below we will consider financial (investment risks) and non-financial risks (EIB risks).

2 Theoretical Basis

Classification of risks allows arranging them in homogeneous clusters, and then applying to them already concrete methods of risk management for their analysis and estimation. Specification of a possible type of risk for concrete activity of the subject of economy is possible through use of classification (see Fig. 1, Fig. 2) [1].

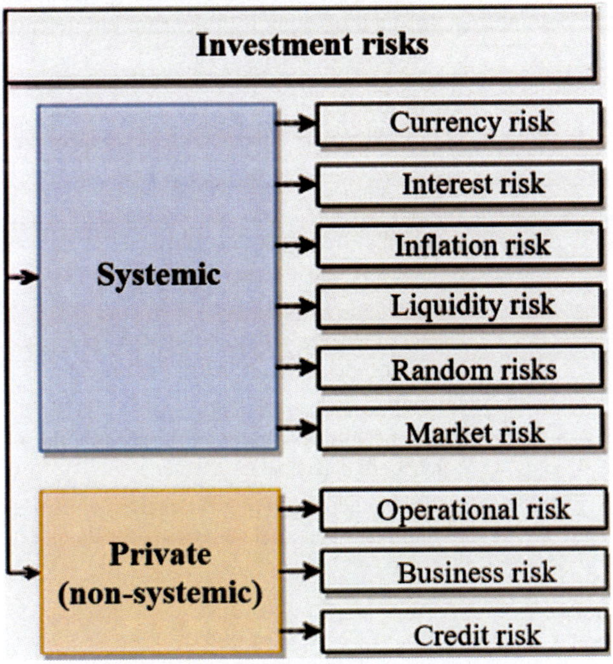

Fig. 1. Classification of investment risks.

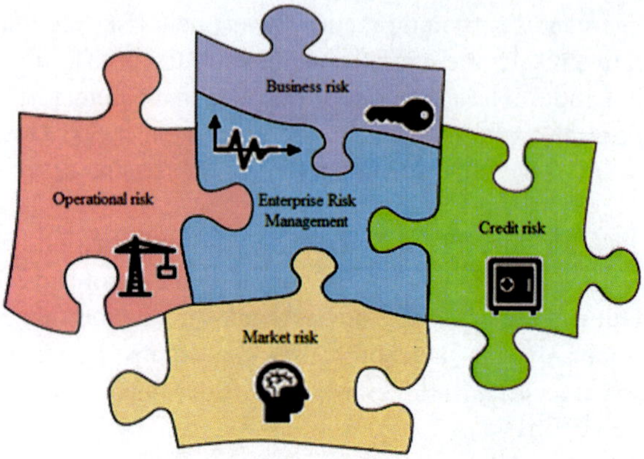

Fig. 2. Possible classification risks by ERisk.

Corporate risk management in business contains methods and processes of risk management and assessment of opportunities to achieve the objectives set by DM. According to Enterprise Risk Management (ERM) concept of corporate risk management the following types of risks are distinguished: operational, credit, market and business risks (Table 1). They include an extensive list of risks caused, for example, by errors in internal business (so-called process risks, legal risks, risks of using an unsuccessful business management model and risks of lack of liquidity to ensure uninterrupted financing of business processes).

Table 1. Classification of risks in corporate risk management.

Type of risk	Composition of risk
Economic	Caused by errors in business activities or in making management decisions
Market	Deterioration of financial position from changes in market dynamics (interest rates, foreign capital participation and/or share prices)
Operational	(Non-)direct losses from inadequate realization of internal business processes: from wrongful behavior of personnel or external events
Credit	Losses from inability or unwillingness of partners to fulfill their financial obligations

For 2020–2021, the following trends in the dynamics of the priorities of socio-economic risks were observed (see Figs. 3, 4, 5, 6 and 7) [2, 3].

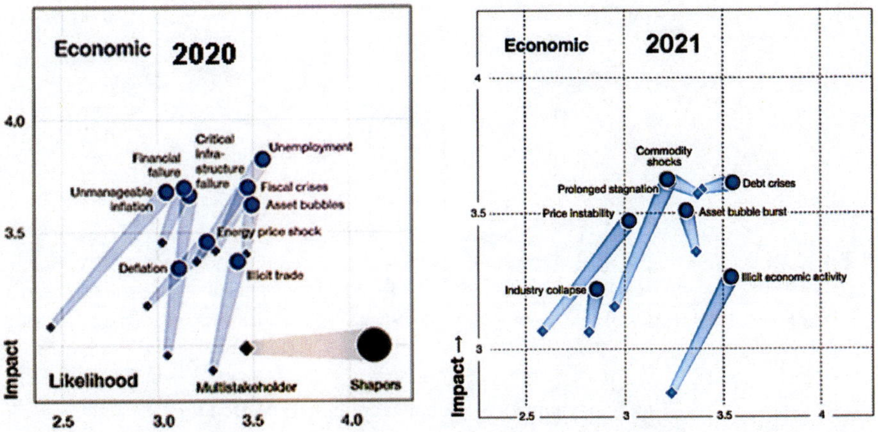

(a) (b)

Fig. 3. Dynamics of global economic risk trends.

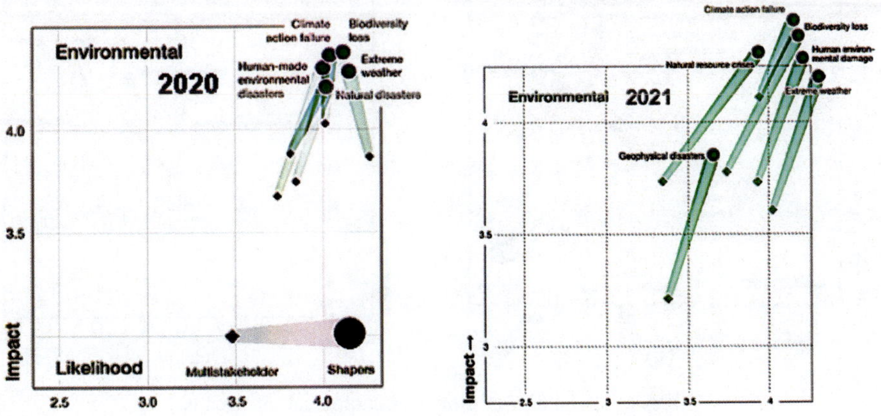

(a) (b)

Fig. 4. Dynamics of global environmental risk trends.

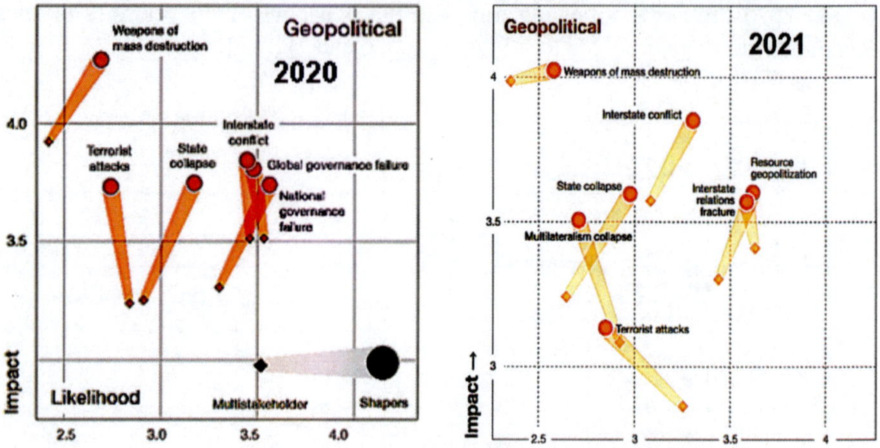

(a) (b)

Fig. 5. Dynamics of global geopolitical risk trends.

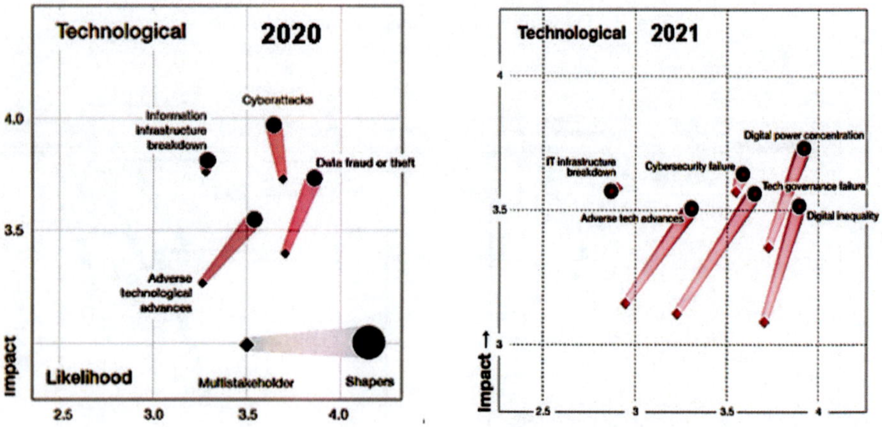

(a) (b)

Fig. 6. Dynamics of global technological risk trends.

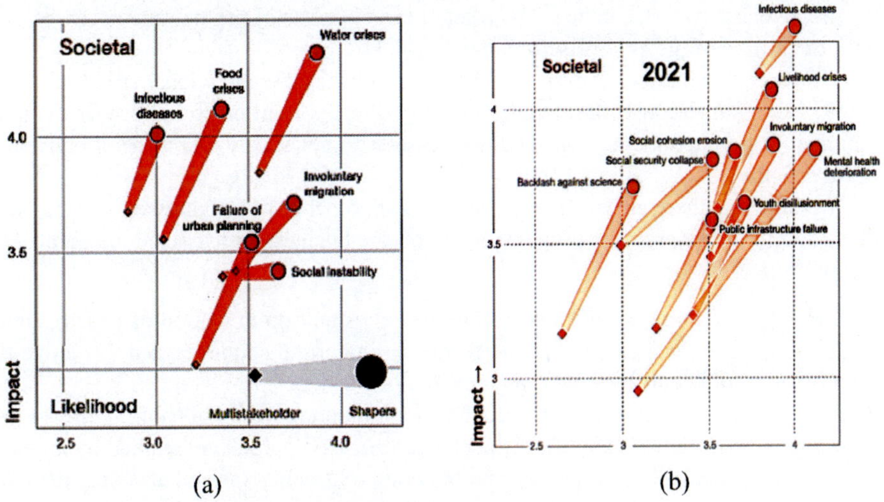

Fig. 7. Dynamics of global societal risk trends.

3 Methodology

To identify and systematize the risks of economic entities when building internal risk management systems, financiers usually apply a cluster approach. According to the authors, the most stringent here are the requirements relating to the activities of financial institutions (including in the field of EIS). In particular, having studied Basel II Accord methodology of risk classification, we can understand that it is based on the principle of multidimensional risk classification, i.e. volumetric geometric model approach. All of the above risks have a significant impact on the continuity of business processes implemented by an economic entity, which affects their EIS (because it affects the competitiveness and business value).

GOST R ISO 22301-2014 sets the requirements for effective Business Continuity Management System (BCMS) to protect against incidents, reduce the likelihood of their occurrence, response and recovery from incidents. The activity of the enterprise can be described as a synergistic interaction of dynamic contour flows (information, financial, material, energy and personnel). In [4] it is shown that to reduce the risk of incidents and its structural optimization it makes sense to apply SM.

BCMS is an element of an organization's management system, including activity planning, policies, organizational structure, responsibility allocation, procedures, processes and resources.

Thus, an effective BCMS should (according to the authors):

- ensure the implementation of business processes in planning the continuity of business operations, as well as in the context of incidents and disruptions generated by them;
- take into account the impact of uncertainty and risk of incidents.

It is necessary to clearly understand what organizational and resource conditions are necessary and sufficient to ensure the operation of the enterprise at the proper level of its effectiveness (which can be set by a set of target values of key parameters associated with the profile of the enterprise/organization) [1].

However, in practice, there is always an established organizational structure of the enterprise/organization, in the activities of which there is a certain probability of incidents that distort the implemented business processes and negatively affect the effectiveness of its activities.

We note that there are two methodological approaches to the study of possible incidents and to the assessment of their impact on the business processes implemented by the enterprise:

1) Carrying out a number of interrupting measures and optimization of their resource support (here the measures are performed in a certain sequence and minimize the possibility of occurrence of an incident);
2) Representation of an incident occurring with a certain probability in the form of a multidimensional random value, the characteristics of which are studied by methods of statistical modeling (through directed computational experiment) to estimate the severity of its possible consequences and the frequency of occurrence.

Risk is a probabilistic category, therefore for its assessment mathematical apparatuses of the probability theory and econometrics are used, and at their algorithmic realization are the methods of statistical modeling (here application of methods of mathematical programming and the game theory is also possible).

Active use of economic and mathematical methods allows conducting a quantitative and qualitative analysis of economic phenomena, give a quantitative assessment of the values of risk and uncertainty and choose the best solution for a particular economic situation. At the same time, formal methods and SM allow describing various economic situations and assessing the impact of management decisions on them (without conducting resource-intensive natural experiments).

In the presence of risk and associated moral, financial and other losses, there is a need for risk management. Here there is an algorithm of cyclic risk management process in a complex economic system (Table 2).

Table 2. Algorithm of cyclic risk management.

№	Composition of action
Step 1	Identification (qualitative assessment) of system risk and its quantification
Step 2	Selection of a method to minimize risk in the considering (sub)system
Step 3	Selection of a management decision to minimize risk
Step 4	Implementation of managing impact on risk
Step 5	Monitoring the results of managing impact on risk
Step 6	Based on the results of monitoring, return to Steps 2, 3, 4 or 7
Step 7	Completion of the management cycle

Simulation is a method often used to solve practical economic problems, useful for clarifying the relation between qualitative and quantitative variables (in the presence of uncertainty and risk factors, under conditions of turbulence and when the structure of the object/process under study is not fully known). Applying simulation, we proceed from the realities of the object to its SM, then to the results of the simulation, and then to the numerical solution. Combination of simulation with decision-support tools leads to increase of efficiency of the decision-making process (increase of quality and reduction of time for realization).

Of course, to analyze large arrays of heterogeneous information and to draw rare phenomena for which no statistically significant sample is accumulated, we have to actively use software to search for optimal solutions using neural networks and genetic algorithms. Decision support systems specialized in their subject area and based on artificial intelligence are becoming increasingly popular: production models, semantic networks, frame structures, etc. The use of the method of hierarchy analysis (T. Saaty method) and models of fuzzy expert evaluations of key performance indicators of economic systems is seen as promising.

The practice of socio-economic relations formed in the XXI century dictates strict requirements for information security (IS) as the most important asset of any modern enterprise/organization. High standards of information protection form the effective security policy of the enterprise and constant improvement of IS methods and technologies.

The share of intangible assets in the value of enterprises is growing, the effective management of which becomes significant due to the increasing role of innovation, intellectual resources and IT technologies and other aspects that form competitive business processes (see Fig. 8).

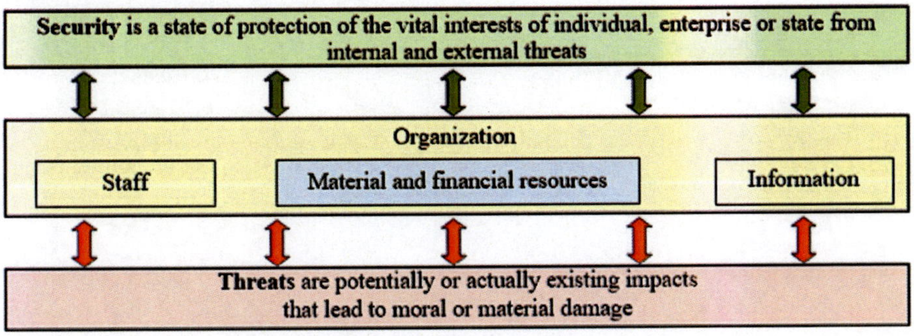

Fig. 8. Factors affecting effective operation of the enterprise/organization.

Before the implementation of IS systems, it is necessary to understand what the subject of the economy will have to be protected against, with each enterprise/organization and each specific case being assessed for its own priority IS risks, associated with the features of the protected organizational structures implementing business processes [1,

5–7]. Therefore, in addition to risks, it is necessary to identify the factors that affect effective operation of the enterprise (see Fig. 9):

1) logistical system failure (social engineering, cyberattacks, increased traffic, etc.);
2) information risks (denial of authorship, receipt/transmission of erroneous or outdated information, etc.)
3) information disruption (business process stoppage);
4) breach of confidentiality (industrial espionage, competitive intelligence, cyberattacks, etc.).

The consequences are the following: decrease in competitiveness, monetary and image losses, etc. Any of these consequences can lead to the termination of further economic activity.

Data on technologies and their implementation should be protected from all kinds of leaks and delays at all stages of the life cycle. There is always the issue of the quality of the personnel involved in the enterprise/organization. Unresponsive personnel and/or their unprofessionalism increase the likelihood of confidentiality breaches and failure of key system nodes. Through social engineering, cyberattacks and other disruptive influences, the dynamic flow (see below) can be stopped or disconnected without the possibility of recovery.

To improve the efficiency and effectiveness of management within systems approach we require the use of scientific methods. For technical and industrial-technical systems they have long been known and widely used, but for complex socio-economic systems management issues they are still poorly developed (because the number of possible variants of organizational structures is infinite, but viable are only those of them which have a well-defined configuration).

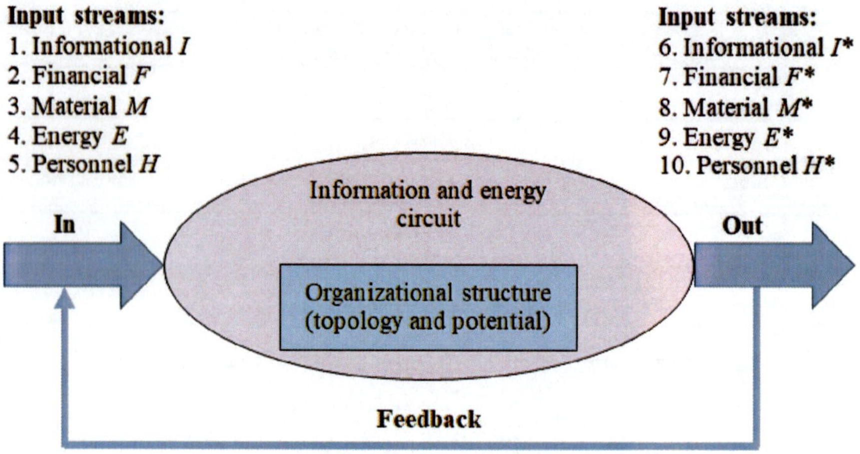

Fig. 9. Contour flows of the enterprise/organization.

Similar organizational structures are built on similar principles (see Fig. 9): they differ only in topology (i.e. location in space and links between their elements) and their potentials. Such connections can be described as a system of dynamic contour flows, reflecting the actual business processes of the organization [4].

A contour flow is a flow of resources through a closed chain of links of organizational structure, in which the subjects and objects of management are combined by means of direct and reverse communication. We describe the composition and structure of dynamic contour flows circulating in the enterprise.

The input of the enterprise/organization is fed by the flows:

1) information (data of different nature and on different media);
2) financial resources (securities, borrowed/equity funds, etc.);
3) tangible assets (raw materials, semi-finished products, fixed assets, components);
4) energy resources (electricity, fuel, hot water, etc.);
5) labor resources (with their competences, skills and abilities).

The potentials of the organizational structure are human and technological potential, and the output will be a product or a service [4].

Each l-th stream can generate stream j (according to the accepted numbering and with the reproduction coefficient $®_{lj}$, which can be either negative or positive, depending on whether there is decrease or increase in the generated flow) (see Fig. 10). Anyone can be decomposed into separate flows from different sources.

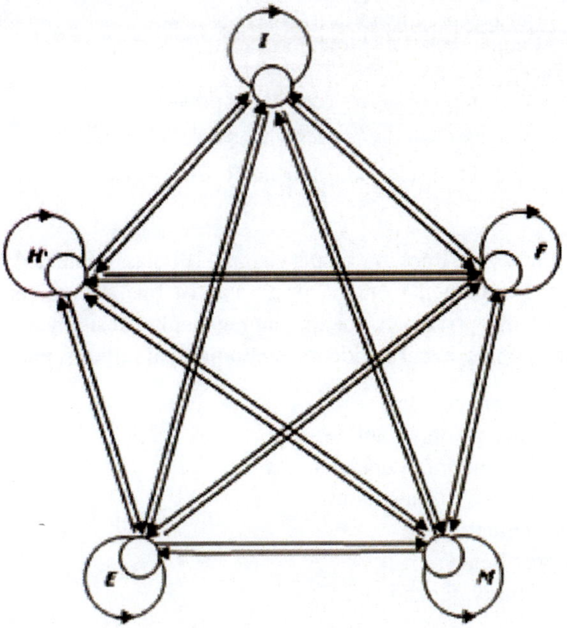

Fig. 10. Dynamic contour graph of the organizational structure.

Within the framework of solution of abovementioned problem, it is necessary to reveal mismatch of design and actual metrics of organizational structure (establishment, based on needs of a specific organization, its structure and functioning schemes) to solve optimization problem for dynamic contour flows (information, financial, material, energy and personnel) of the enterprise/organization:

$$e = |r_k - p_k| \rightarrow min, \tag{1}$$

where k is the hierarchical level of the system in question; p_k, r_k is the actual design and measured metrics at the k-th level of the hierarchy [8].

3.1 Operational Resilience

Operational resilience is the bank's ability to perform critical operations under abnormal conditions, due to systemic activities to identify threats and potential failures (including those related to cyberattacks), protect against them, detect, respond and minimize the impact. Figure 11 shows an enlarged algorithm for restructuring the Russian financial market [9].

Step 1: Analysis of the current state of the financial market	**Technology development**	**Increased attention of regulators to the development of competitive environment**
	Changes in the expectations of consumers of financial services	**Globalization and geopolitical risks**
	Changing in the approaches to supervision of financial institutions	**Macroeconomic trends**
Step 2: Challenges and opportunities. Analysis of prospects		
Step 3: Implementation of a set of measures for the development of the financial market		

Fig. 11. Factors affecting the Russian finance market.

Basel Committee on Banking Supervision [10] has published for consideration COVID-19 related documents on the principles of operational risk and operational resilience. We list the principles for enhancing banks' resilience to operational risk emergencies (pandemics, cyber incidents, technological failures, natural disasters, etc.):

- operational risk management;
- business continuity planning and testing;
- mapping relations and interdependencies;
- operational resilience management;
- incident management;
- cybersecurity resiliency;
- managing third-party dependencies.

Earlier, Central Bank of the Russian Federation issued regulatory requirements for operational resilience:

- Regulation of the Bank of the Russian Federation of April 08, 2020 No. 716-P "On requirements for the operational risk management system in a credit institution and banking group";

- Regulation of the Bank of the Russian Federation of December 07, 2020 No. 744-P "On the procedure for calculating the amount of operational risk ("Basel III") and supervision of its observance by the Bank of the Russian Federation".

Ensuring operational resilience of the business is an important component of the operational resilience of the bank, requiring synergistic interaction of the following areas of the bank's activities (see Fig. 12).

Fig. 12. Elements of the target concept of bank business sustainability.

3.2 Relation Between Cybersecurity, Data Quality and Business Resilience

We consider achieving operational sustainability of the bank through ensuring the sustainability of the following areas:

- cybersecurity;
- quality and availability of information resources and the speed of their processing;

- business processes being implemented.

Ensuring the sustainability and continuity of the bank's business processes and achieving an appropriate level of cybersecurity is impossible without the accumulation of a large body of data that allows for retrospective analysis of incidents and events. Thus, it is possible to build an integrated system of monitoring and counteracting already known vulnerabilities. It should be noted that zero-day vulnerabilities and Nation-State attacks are difficult to predict and require a specialized amount of data that only a narrow range of organizations have.

Another success factor in building the business processes implemented by the bank is the presence of: synergy in knowledge management, strategy and methodological development apparatus, advanced technologies, maturity of built business processes and qualified personnel. In the bank, achieving a sustainable cyber security system starts with the formation of a three-year strategy. After that, people with knowledge and competencies emerge, building business processes (including communication between departments). The "strategy-people-process" triad is regular in E. Deming's cycle and J. Boyd's OODA cycle (see Fig. 13, Fig. 14).

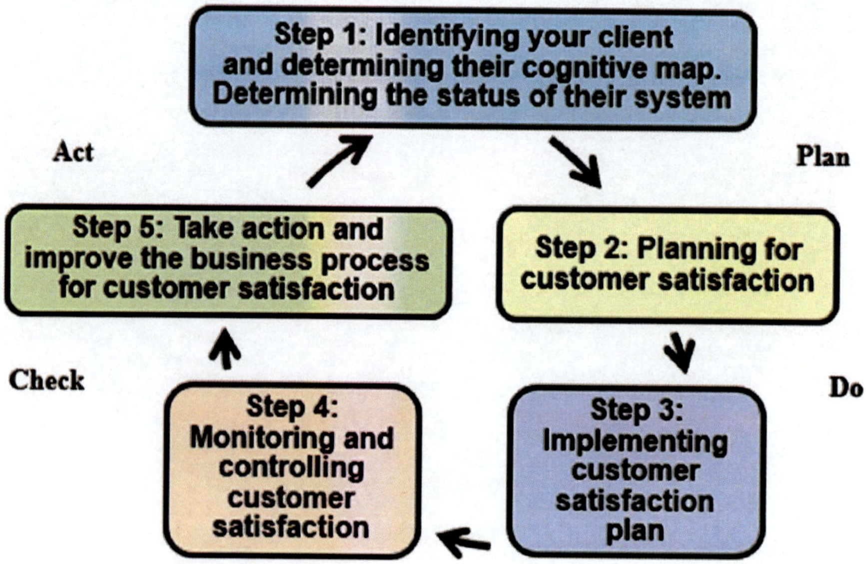

Fig. 13. Quality management cycle within the framework of W.E. Deming's PDCA concept.

In the opinion of the authors, in the future achievement of sustainability of bank activity will be defined within the direction of Enterprise Architecture: integral methodology and practice of management of carrying out changes in the organization with a complex structure (see Fig. 15).

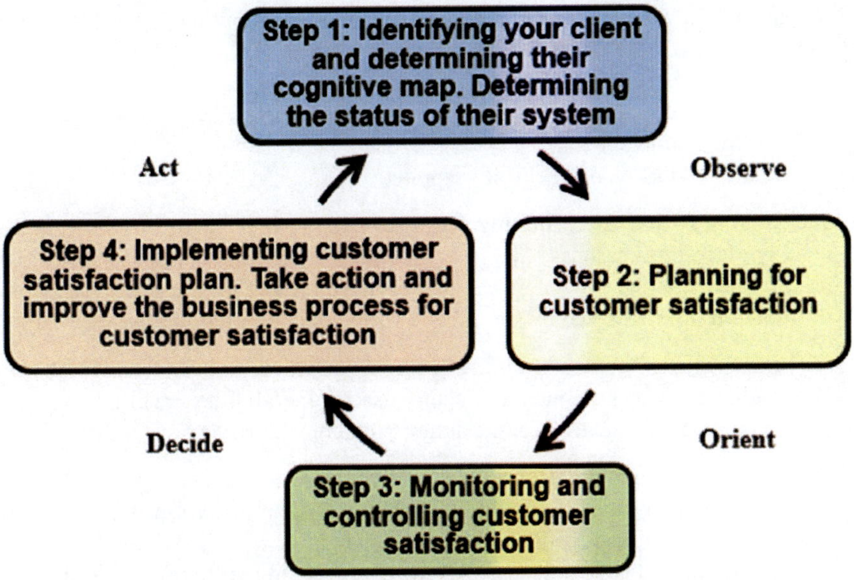

Fig. 14. Quality management cycle in the framework of J. Boyd's OODA concept.

Strategy	Processes	Materials	Roles	Geography	Events
Why are we doing this?	How are we going to do that?	What do we need?	Who do we need?	Where do we do it?	When is it necessary to do this?

Fig. 15. Enterprise architecture scheme.

3.3 Topical IS Risks

The main IS risks are computer attacks on information resources (including critical information infrastructure facilities). In February 2022, their number increased at least 4 times compared to February 2021 [11].

We list the possible types of cyberattacks:

- DDoS attack: the main type of attack to bring a computing system to the point of denial of service. The result is complete or partial inaccessibility of a corporate resource on the Internet lasting up to several hours;
- content spoofing: hacking one of the pages, replacing its content with some provocative message on it (with the aim of propaganda and reputational damage to the site owner). The content is most often replaced with unreliable information;
- phishing is a type of Internet scam aimed at gaining access to confidential user data by sending out mass emails on behalf of agencies, popular brands and companies.

The strength of DDoS attacks is in the mass nature of the attackers. We can create communities in any available messenger with instructions on how to organize this type of attack. Communities of hackers publish instructions on how to set up their PC and coordinate the direction and timing of attacks to destabilize target information resources. Now anyone can become a participant in a cyberincident using their PC or smartphone.

It's enough to click on a special link in browser and the computing power of device will be directed to attack the target.

Introduction testing utilities are used for content spoofing attacks, which can be used to identify vulnerabilities in a web resource and, if compromised, can be used to spoof its content, for example.

To counter DDoS-type attacks, it is applied:

- geo-filtering by provider: limiting access to content is applied for political, social, legal or marketing reasons [12];
- use of traffic filtering protectors and tools is necessary to limit user access to certain information (implemented by specialized software and/or hardware [13].

To protect against "content spoofing" attacks, we use current versions of web servers and their add-ons, where the most frequently used vulnerabilities are eliminated.

Based on practical experience, the authors recommend paying special attention to:

- constant monitoring of all external services;
- phishing technology as a vector of intrusion into the enterprise/organization (constant work with and training of staff);
- creating backups of key systems and equipment (backing up configurations), creating a set of spare tools and accessories in case of hardware failures, searching for analogues of functioning equipment, rebalancing resources from less important to key objects;
- control of updates and the perimeter of external communications.

Performing backups is mandatory, it is a requirement of the Central Bank of the Russian Federation, which obliges banks to maintain backups for all critical processes (Regulation of the Central Bank of the Russian Federation of April 08, 2020 No. 716-P "On requirements for operational risk management system in a credit institution and banking group"). Here, it is important to identify IT systems that support the execution of critical processes. They require backing up data at least once a day (Recovery Time Objective 24 h).

The best solution to the whole set of problems related to cybersecurity of banks is the implementation and improvement of modern Security Governance, Risk, Compliance (SGRC) platforms [14].

3.4 SGRC Platforms in Banking Security

To establish an integrated information security management system in the banking sector, SGRC solutions have been developed. These solutions facilitate prompt management decision-making and enable the objective assessment and treatment of risks, ensuring compliance with documents and information security standards [22]. The SGRC platform is integrated into the organization's infrastructure and populated with assets via integration mechanisms. Through integration with AD systems and information security systems, as well as manual input of asset information and platform enrichment, the automation of multiple operations is possible. The participation of necessary stakeholders enables task management implementation, which generates a single field for the functioning of the entire information security organization [22].

The functionality of SGRC depends on its developmental direction and commonly includes asset management, vulnerabilities, risks and audit management, and compliance. Furthermore, there is a module for visualization and reporting, while additional modules are available for raising awareness and interacting with service providers. Each module is linked to the processes that GRC can support, with several blocks closing several processes simultaneously.

Sharing SGRC with other IT and information security solutions enhances the IT management function and its integration into corporate processes. The presence of analytics and visualization functionalities within the system provides managers with the ability to manage both internal and external checks for compliance with requirements and legislative norms.

The implementation of SGRC systems, along with the adoption of international continuity management principles and best global practices, enables banking organizations to sustain and improve their businesses. Certification of the business continuity management system completes this process and creates the basis for effective management during crisis situations [23].

4 Results and Discussion

The following possible steps to minimize the negative effects of incidents on the implemented business processes of the enterprise/organization are seen:

- study of theoretical sources on the problems of management activities under conditions of risk factors and uncertainty;
- organizing, summarizing and processing theoretical and practical managerial experience for further practical application, staff training, etc.;
- revealing the resource and theoretical-methodological basis for optimal managerial decisions of the enterprise/organization under specific conditions;
- development/correction of the existing organizational structure of management, forms and methods and composition of implemented business processes.

To conduct competent preparation and further verification of business processes, it is necessary to qualitatively approach the modeling of effective structure of audit power in the organization [15] (see Fig. 16).

To choose a way to improve the quality of the implemented business processes (because of the variability of ways to improve the efficiency of the enterprise/organization, and given the important role of their management systems in the implementation of the mechanism of production efficiency), it makes sense to use a decision support system [16]. In the works of this author the system of information and analytical support of the mechanism of production efficiency improvement, correlating with the concept of dynamic contour flow management of the enterprise/organization [17] is studied.

The application of the system approach in the management of contour flows (in logistics management) "...provides end-to-end management of flow processes, increases the quality of management and contributes to the emergence of synergistic effect. In the conditions of globalization, economic crisis, sanctions aspects, increasing role of logistics

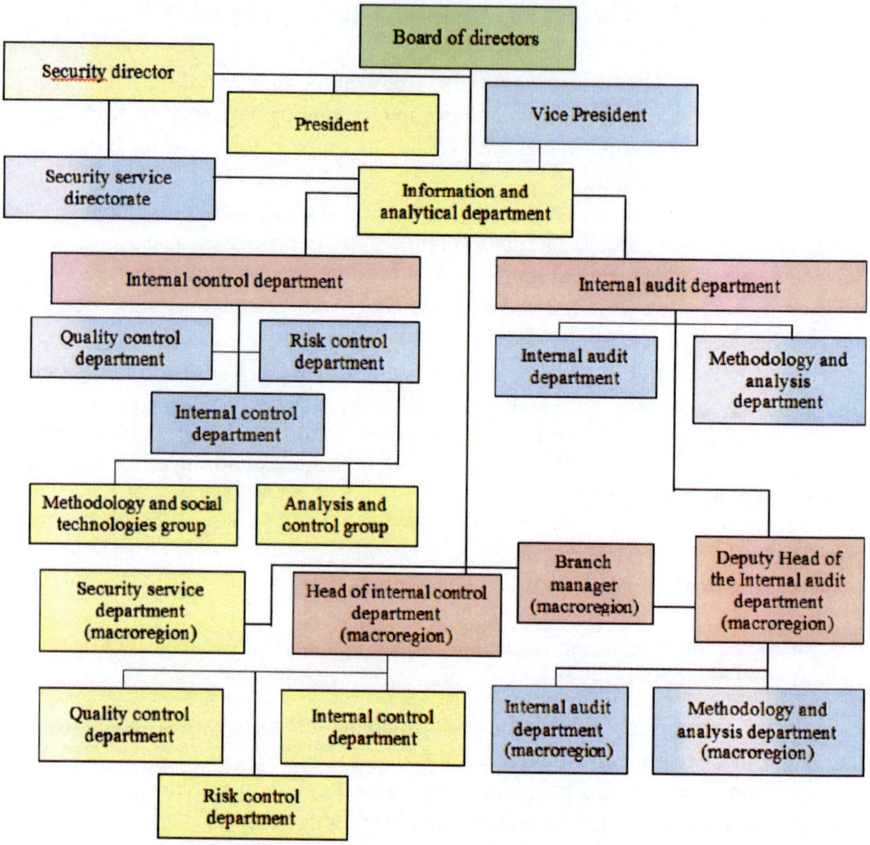

Fig. 16. Structure of the enterprise/organization that has branches.

costs the formation of logistics systems on the methodological basis of logistics theory and within a system paradigm using a combination of process and object approaches is an important element of improving the management efficiency of a modern dynamically developing enterprise" [18].

Solving the problems of improving the reliability of production and organizational systems based on the application of logistic principles allows (through effective management and organization) to provide them with optimal control actions to improve efficiency, reliability, competitiveness and adaptability, which has a positive impact on their EIS as economic entities.

Most of the real complex systems exist under uncertainty, so they cannot be described by deterministic models, but through the use of numerical simulation, DM can optimize them, track the dynamics of functioning and obtain possible models of the system. Since determining the metrics of the quality of the organization is quite difficult, it requires the use of computers. Analytical solution of this problem is complicated by random fluctuations of its parameters, so it is better to apply methods of IS.

IS is a method of research, which consists in the simulation of the process of functioning of the system or its elements. The essence of IS is the implementation of algorithms imitating the behavior (its characteristics and properties) of the real system is in the necessary for the study of the system (in the opinion of DM) composition, volume and area of change of its parameters. IS for real systems is easier than directed experiments on the organization, because it is often possible to observe situations when experiments on real systems are unacceptable. IS allows taking into account the presence in the system of continuous and discrete elements, their nonlinearities, random perturbations and other factors that complicate the conduct of analytical studies. IS is an effective method (and sometimes the only one, especially in design) to study large and poorly structured systems.

Every DM needs to know the classification and basics of development of managerial decisions, because the effectiveness of companies depends on their efficiency and quality. IS gives the number of alternatives of managerial decisions (as a set of instances of models optimal in some sense) needed by DM. The main advantage of IS is the possibility in conditions of uncertainty to build a model, not having complete and accurate data, and find out "what will happen if…?".

For this purpose, DM has to specify:

- methods for solving a number of standard management problems, with the choice of appropriate tools being determined by the laws of development of objects of this or that class;
- input parameters of the decision-making system or types of situations studied through IS;
- system of functional parameters at the output of IS, allowing the adoption of alternative management decisions to avoid or eliminate arising problems in the activity of the enterprise/organization.

For example, in [19] IS is studied as a tool for the functioning of the organization to obtain near-optimal results of management decisions by identifying their shortcomings even at the stage of goal prediction and with the implementation of the management goal with the identified problems. In this way it is possible to improve the effectiveness of the management process in general and DM management decisions in particular.

The algorithm implementing the model reproduces the process of functioning of the system in time, and its constituent phenomena, their temporal dynamics and logical structure are simulated. This gives the possibility to DM to determine the parameters of the state of the process at a given point in time by the initial conditions. Here, IS in the form of mass service networks are widely used [20].

In order to minimize the mismatch of control parameters, automatic control theory (ACT) can also be applied. Within the framework of the conceptual apparatus of ACT [21] the concept of modeling of decision-making process of DM is given, which will allow developing the theory of decision-making in terms of studying and classifying the specifics of human thought processes depending on external conditions. It was found that the main difference between the rational and intuitive-heuristic models of decision-making is the phenomenon of "enrichment" of incoming information by individual personality traits of DM, which are described within the methodological apparatuses of ACT and IS. These authors obtained a taxonomy of typical elements of ACT and studied their correspondence to standard models of decision-making (as specificity of realization of the process of making managerial decisions and behavior of DM for a certain period of their professional activity). At the same time, the choice of a specific strategy of management of the enterprise/organization by DM is greatly influenced by the existing regulatory framework of economic activity.

5 Conclusion

As a summary of the article the authors can make the following conclusions.

1. Factors affecting the effective operation of the enterprise/organization are discussed. The concept of ERM examines the role of risk in ensuring the continuity of business processes implemented by the enterprise/organization.
2. Quality management of business processes realized by the enterprise/organization is a continuous cyclic procedure (such as E. Deming's PDCA and J. Boyd's OODA cycles), related to the optimization of the dynamics of the circulating contour flows in the organization.
3. Achievement of organization business continuity at the level of requirements of GOST R ISO 22301-2014 and GOST R 53647.3-2015 standards is possible with functioning of complete SGRC-system and its continuous improvement.
4. Uninterrupted operation of a company's business in any industry is the result of synergy between business, IT and EIS. Detection of critical processes (business impact analysis), those, which must be protected in the first place, is the basis for creation of business continuity system. Carrying out of qualitative analysis of business critical processes creates business transparency for a company management, helps it better understand its business and forms a basis for stable development/functioning of a business.

Use of methodological apparatuses of mathematical and IS allows DM to significantly improve the quality of managerial decisions, to standardize (unify) them and reduce the time for their preparation for implementation.

References

1. Boyar-Sozonovitch, A.S., Buikin, A.Y., Pitelinskiy, K.V.: Features of enterprise risk management associated with operational risks. Amazonia Investiga **10**(46), 9–19 (2021)
2. World Economic Forum. The Global Risks Report (2020). http://www3.weforum.org/docs/WEF_Global_Risk_Report_2020.pdf. Accessed 23 Mar 2023
3. World Economic Forum. The Global Risks Report (2021). https://www3.weforum.org/docs/WEF_The_Global_Risks_Report_2021.pdf. Accessed 23 Mar 2023
4. Pitelinskiy, K.V.: Modeling of dynamic loop flows as a method of business continuity management. Qual. Manag. Methods **11**, 16–21 (2018)
5. Boyar-Sozonovitch, A.S., Pitelinskiy, K.V., Ermolatiy, D.A.: Innovation economy: aspects of economic and information security in logistics innovation. Amazonia Investiga **8**(21), 6–13 (2019)
6. Pitelinskiy, K.V., Khachatryan, A.Yu., Bereza, A.O.: Violations of the requirements of PBU 14/2007 on the rights to intangible assets: consequences and minimization of risks. Account. Constr. Organ. **2**, 69–76 (2015)

7. Pitelinskiy, K.V., Antipenkova, A.A., Titov, V.R.: Risks associated with intangible assets of an economic entity: identification and management. Bull. Moscow Int. Acad. **1**, 116–123 (2016)
8. Pitelinskiy, K.V.: On the prospects of using fractal modeling for the study of technical and social systems. Interbranch Inf. Serv. **4**(141), 12–26 (2007)
9. Central Bank of the Russian Federation: The Central Bank of the Russian Federation: the main directions for the development of the financial market of the Russian Federation for the period 2019–2021 (2019). https://rulaws.ru/acts/Osnovnye-napravleniya-razvitiya-finansovogo-rynka-Rossiyskoy-Federatsii-na-period-2019---2021-godov-(ra/. Accessed 23 Mar 2023
10. BIS.org. Basel Committee releases consultative documents on principles for operational risk and operational resilience (2020). https://www.bis.org/press/p200806.htm. Accessed 23 Mar 2023
11. TASS.ru. The number of cyberattacks on the Russian Federation since the beginning of the special operation in Ukraine has increased at least four times (2022). https://tass.ru/ekonomika/14384505. Accessed 23 Mar 2023
12. Shkuro, A.: Geo-blocking: why are we limited access to information on the Internet? (2020). https://nag.ru/material/38847. Accessed 23 Mar 2023
13. Carbonsoft.ru. Network traffic filtering: types and features (2018). https://www.carbonsoft.ru/network-traffic-filtering/. Accessed 23 Mar 2023
14. Cisoclub.ru. Comparison of SGRC systems (Security Governance, Risk, Compliance) (2020). https://cisoclub.ru/sgrc. Accessed 23 Mar 2023
15. Goncharenko, G.Yu., Ermakov, I.K., Ermolatiy, D.A., Pitelinskiy, K.V.: Computer psychology or a universal approach to the vulnerabilities of confidential information. Inf. Secur. Issues **4**, 62–68 (2018)
16. Shendrikova, O.O.: Simulation modeling in the process of supporting and making managerial decisions. Prod. Organ. **1**(56), 99–103 (2013)
17. Pitelinsky, K.V.: Organization as a system of dynamic planimetric flows. Sustain. Innov. Dev. Des. Manag. **3**, 17–29 (2009). www.rypravlenie.ru/?p=411. Accessed 23 Mar 2023
18. Sokolova, O.G., Karkh, D.A.: Logistic hierarchy as the basis of enterprise system management. Journal of New Economy **4**(66), 99–107 (2016)
19. Bogdanova, E.A., Sherstyankina, A.A.: Simulation modeling as a decision-making tool. NovaUm. Ru **6**, 25–28 (2017)
20. Walrand, J.: An Introduction to Queueing Netwoks. Prentice Hall, New Jersey (1988)
21. Rizun, N., Taranenko, Y.: Simulation models of human decision-making processes. Manag. Dyn. Knowl. Econ. **2**(2), 241–264 (2014)
22. Yashkin, V.V., Kesel, S.A., Makovey, S.O., Domnikov, A.S.: SGRC system as a basis for building business processes and measuring the digital sustainability of a business. Lect. Notes Netw. Syst. **597**, 933–951 (2023)
23. Ak Bars Bank was the first in Russia to confirm compliance with international and national requirements for business continuity management systems. https://www.akbars.ru/news/ak-bars-bank-pervim-v-rossii-podtverdil-sootvetstvie-trebovaniyam-neprerivnosti-biznesa/. Accessed 23 Mar 2023

Information Systems Development Planning Based on Concepts Proactive Control and Management of Viability

Valerii Zakharov[1](✉), Boris Sokolov[1], and Igor Kimyaev[2]

[1] St. Petersburg Federal Research Center of the RAS, Saint Petersburg, Russia
valeriov@yandex.ru
[2] LLC "Nornikel Sputnik", Moscow, Russia

Abstract. This paper addresses the problems of improving the efficiency of ground-based aviation transport and logistics subsystems of modern airports and updating the corresponding integrated information and control modules based on the development of programs of their modernization proactive control. To increase the viability level of these objects, the paper proposes to base on the concepts and principles of targeted complication of the components of the internal structure of information systems. The article presents a formal problem statement, proposes new ways of developing elements of model and algorithmic support for planning the evolutionary replacement of operators is including decision makers, by functionally equivalent information-management software and hardware complexes, which can be used for automation and intellectualization of control processes of existing, upgraded and prospective ground-based aircraft transport and logistic systems operation.

At the same time, due to the structures heterogeneity (topological, technical, technological, informational, etc.) of these objects it is impossible to exclude the decision makers from the corresponding control loop and conduct the synthesis of integrated information and control systems, which are based only on formal and mathematical approaches.

Keywords: Complexity management · integrated management system · EIS update · transport and logistics system

1 Introduction

The ground-based aviation transport and logistics system (GBATLS) of a modern airport is a complex object (CO) [1, 2]. It includes a set of interconnected controlled objects (ContO) and information management subsystems (IMS), which are represented as production and storage units, technological and transport and logistics lines, active mobile objects (autolifts), stationary and onboard computing modules that form a dynamic computing network of data, information and knowledge processing and transmission and other cyber-physical systems (CPS) (see Fig. 1, Fig. 2) [2, 3].

R. Silhavy and P. Silhavy (Eds.): CSOC 2023, LNNS 723, pp. 538–545, 2023.
https://doi.org/10.1007/978-3-031-35317-8_48

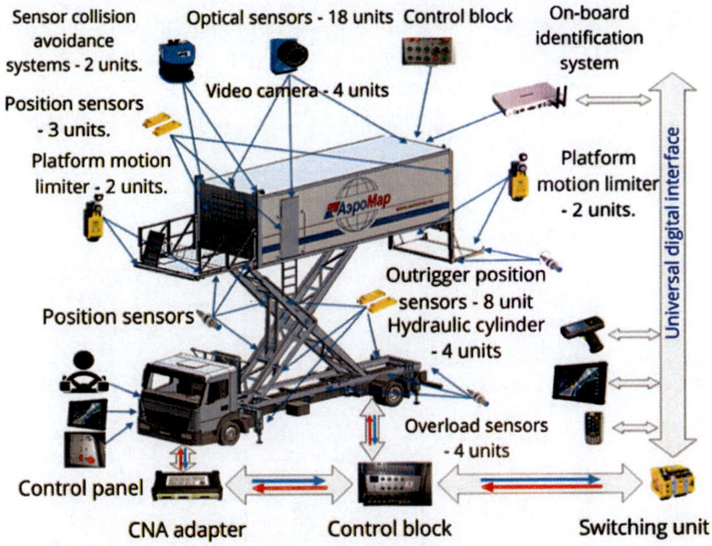

Fig. 1. Autolift elements.

Due to heterogeneity of the object data (topological, technical, technological, information etc.) it is impossible to exclude decision makers (DM) from the corresponding control loop and to synthesize the integrated information and control systems (IICS), which are based exclusively on formal and mathematical approaches. In addition, we note the need to attract to the solution of key management tasks in the structure of CO creative and cognitive abilities of DM with the appropriate qualifications and skills [4].

For correct and timely processing of information flows that circulate in the control loop CO to assist DM, creates software and mathematical support, fully or partially automating their activities. For this purpose, first of all quantitative and qualitative multidimensional models of these objects and systems are developed («digital twins» (DT)) [5, 6].

Meanwhile, the tasks of rational redistribution of control functions between DM and IT systems, as well as proactive planning for the development (modernization) of the IIMS are currently not solved [7]. Therefore, the study topic is new and relevant.

2 Methods of Information Management Systems Modernization Proactive Control

The most complete participation of the DM in the work of the system in the GBATLS structure is taken in those areas and for those subsystems, which are characterized by significant uncertainty of the current state and multi-criteria in the choice of objectives and management and control methods (Fig. 2). This approach is caused primarily by an insufficient level of subject area formalization (as there are no adequate and reliable models), associated with the specific GBATLS elements functioning, incomplete/deficient and low quality of input data, the excess of alternatives for decision-making, as well as the general methodological and technological unpreparedness of the DM [8]. In this case, the competent specialist-operator of the subject area is forming in his head an individual knowledge base, can solve in the operational mode to solve complex tasks of assessing the state and situation related to the CO functioning, as well as predicting and managing and control them.

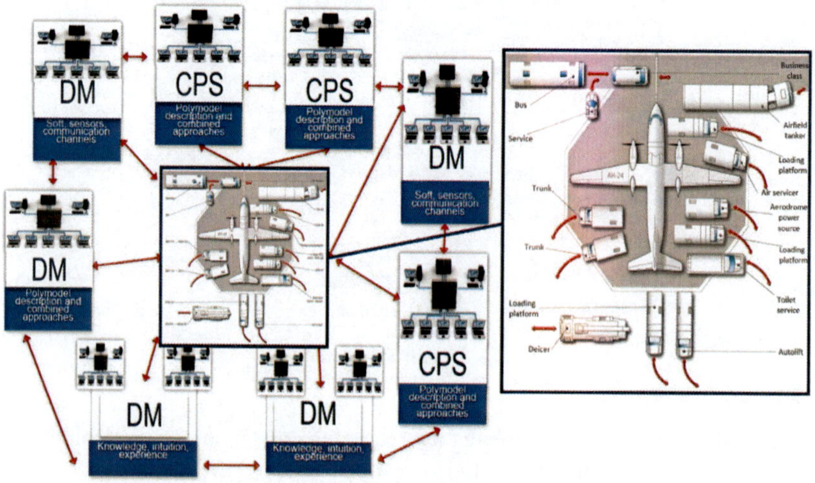

Fig. 2. GBATLS elements interaction scheme.

In fact, the GBATLS IIMS development processes consist in the creation and DT update. The results of these processes are comparable to human intellectual activity. Based on the service-oriented approach, the implemented elements of an integrated information system become an internal service in relation to business processes.

Note that when we are choosing the place and IT-components role (functions) in the IIMS structure at the modernization stage should be based on the following principles: first, the quality of information flows processing should be improved through the creation and use of adequate DT and control models based on them; second, DM should be spared the functions of the most saturated information flows processing routine and regulation of many minor auxiliary subsystems and technological indexes.

Along with this, the synthesizing programs tasks for the GBATLS operation and IIMS modernization are interrelated and must be solved in parallel and proactively.

As a successful operation integral property for the three-component man-machine structure DM $\leftrightarrow$ IIMS $\leftrightarrow$ GBATLS for sustainable functioning (in terms of goals) of GBATLS it is proposed to define its viability [9], that must be ensured at all stages of its life cycle (LC) regardless of the impact of internal and external influences [10]. Unfortunately, in this time the viability of any CO has no generally accepted formalized description.

However, according to the article goals, we define it as the ability of GBATLS and the corresponding IIMS at a given time interval (stage of modernization) to maintain in the ranges required for efficient business activities, indexes are characterizing the business processes efficiency and the interoperability level of (compatibility) GBATLS elements.

3 Formal Problem Statement

The problem synthesis of technologies and programs of processes management and control of IIMS and GBATLS coordinated functioning and modernization is including the replacement of the DM to IT systems, can be formulated as a task of finding the optimal control actions that translate the considered CO from the specified to the required structural state, which characterizes both the current state of objects included in a given type of structure, and the state of relations between them [14].

Therefore, it is necessary to develop principles, approaches, models, methods, algorithms, that allow to find such values of control and state of existing and control system of GBATLS $< U^t, S_\delta^{*tf} >$, at which

$$J_\theta(J_\iota^t, X_\chi^t, \Gamma_\chi^t, Z_\chi^t, F_{<\chi,\chi'}^t, \prod_{(\delta,\tilde{\delta})}^t, t \in (t_0, t_f]) \to \underset{<U^t, S_\delta^{*tf}> \in \Delta_1^{(g)}}{\text{extr}}$$

$$\Delta_1^{(g)} = \{< U^t, S_\delta^{tf} > \left| R_\beta\left(X_\chi^t, \Gamma_\chi^t, Z_\chi^t, F_{<\chi,\chi'}^t, \prod_{(\tilde{\delta},\tilde{\tilde{\delta}})}^t\right) \le \tilde{R}_g; \right.$$

$$U^t = \prod_{<\delta_1,\delta_2>}^{t_1} \circ \prod_{<\delta_2,\delta_3>}^{t_2} \circ \ldots \circ \prod_{<\delta,\delta>}^{t_f}; \beta \in \mathbf{B}\},$$

where

J_θ - composite index of viability;

$\theta \in \Theta$ – set of indexes numbers;

χ– set of indexes corresponding to GBATLS structures (topological, technical, technological, information etc.)

$T = (t_0, t_f]$– time interval at which GBATLS operates and upgrades, i.e. there is the decision maker is replaced by the IT components;

$X_\chi^t = \left\{x_{\chi l}^t, l \in L_\chi\right\}$– set of elements included in the structure dynamic alternative system graph (DASG);

G_χ^t – set of DASG nodes;

$\Gamma_\chi^t = \left\{\gamma_{<\chi,l,l'>}^t, l, l' \in L_\chi\right\}$– set of DASG edges of type G_χ^t, it's showing the relationships between its elements at time t;

$Z_\chi^t = \left\{z_{<\chi,l,l'>}^t, l, l' \in L_\chi\right\}$– set of parameter values;

$F_{<\chi,\chi'>}^t$– the mappings of GBATLS different structures into each other at time t,

$\Delta_1^{(g)}$– set of dynamic alternatives (a set of structure and parameters of GBATLS business process of inherited and modernized IIMS (replacement of the operator by IT), as well as the set of programs of their functioning);

U^t – controlling actions, that allow synthesizing the structures of the modernized and implemented IIMS GBATLS;

$\tilde{R}_g$– given values (constants);

«o» – mapping composition operation;

B – set of spatial and temporal numbers, technical and technological constraints that determine the processes of implementation of GBATLS modernization programs.

The problem of synthesizing technologies and programs for controlling the interaction of GBATLS elements is providing the highest level of IIMS elements interoperability under various conditions of internal and external disturbing influences should be solved.

$$J_\iota^{<t,y>}(C_\chi^{<t,y>}, N_\chi^{<t,y>}, D_\chi^{<t,y>}, W_\chi^{<t,y>}, V^{<t,y>}, t \in (t_0, t_f]) \to \underset{<V^{<t,y>}, S_\delta^{*t_f}> \in \Delta_2^{(g)}}{extr}$$

$$\Delta_2^{(g)} = \{< V^t, S_\delta^{t_f} > \Big| R_o\left(C_\chi^{<t,y>}, N_\chi^{<t,y>}, D_\chi^{<t,y>}, W_\chi^{<t,y>}, V^{<t,y>}, t \in (t_0, t_f]\right) \le \tilde{R}_o^{(g)};$$
$$V^{<t,y>} = U^{<t,y>} \times \Xi^{<t>}, o \in \mathbf{O}\},$$

where

$J_\iota^{<t,y>}$ – composite index, that characterizes the level of interoperability of NATLS elements;

$C_\chi^{<t,y>} = \left\{c_{<\chi,l,l'>}^{<t,y>}, l, l' \in L_\chi\right\}$ – set of parameter values, that characterizes the relationship between GBATLS elements;

$N_\chi^{<t,y>}$ – set of nodes included in the GBATLS control system including the DM;

$D_\chi^{<t,y>} = \left\{d_{<\chi,\rho,\varepsilon,\varepsilon'>}^{<t,y>}, \rho, \varepsilon, \varepsilon' \in L_\chi\right\}$ – set of parameter values, that characterizes the types of detectable data, information and knowledge formats circulating in the GBATLS IIMS;

$W_\chi^{<t,y>} = \left\{w_{<\chi,\nu,\kappa,\kappa'>}^{<t,y>}, \nu, \kappa, \kappa' \in L_\chi\right\}$ – set of parameter values, that characterizes the ways of input, processing, transfer of information in the GBATLS control system;

$V^{<t,y>}$ – set of input actions formed on the Cartesian product of allowed controlling actions $U^{<t,y>}$ and the set of perturbation actions;

$\Xi^{<t>}$ – set of perturbation actions;

$\Delta_2^{(g)}$ – set of dynamic alternatives i.e., a set of variants of collection, processing, transfer of data, information and knowledge (physical by means of exchange of text files, through API technology, etc.), as well as a set of programs to manage the GBATLS elements interaction;

$R_o\left(C_\chi^{<t,y>}, N_\chi^{<t,y>}, D_\chi^{<t,y>}, W_\chi^{<t,y>}, V^{<t,y>}, t \in (t_0, t_f]\right) \le \tilde{R}_o^{(g)}$ – set of restrictions, that imposed on the technologies and programs synthesis processes to manage the interaction of GBATLS elements and subsystems;

$\tilde{R}_o^{(g)}$ – given values (constants);

o – current restriction number;

The presented problems class is characterized by large dimensionality, non-stationarity and the need to take into account uncertainty factors and multi-criteria aims of various subsystems [13]. One possible approach to the development of coordinated management programs for LC at the IIMS GBATLS modernization stage is to rely on the integrated modeling principles and involvement of logic-dynamic models to describe a given subject area [14]. The attraction of this mathematical construct allows at the planning stage for IIMS GBATLS functioning and modernization to use control and management models of basic, auxiliary and supporting material and information operations, as well as to describe the consumption of different types of resources used in a single time scale [7, 15].

The proposed approach allows to formally present the management and control processes of transport and logistics and information processes in GBATLS in a single language of description and at the same time connect the interoperability index with the viability index based on the analytical-simulation models application and the original fuzzy-probability convolution [16–18].

Due to the limited volume of the article, a detailed description of methods and algorithms for the synthesis of optimal programs for solving network planning problems would be redundant. We invite readers to pay attention to these publications in order to get an exhaustive idea of the subject of the article [14, 19].

4 Results

As a result of the joint solution of the above tasks, programs will be developed to manage the GBATLS IIMS modernization (comprehensive or integrated plan) and the elements interaction using different processing and transfer channels of data, information and knowledge [20]. At the formal level of description, the interoperability index J_l^t can be related to the GBATLS viability index, which could not be done before.

The use of optimal control theory methods to solve the problems of IIMS GBATLS modernization integrated planning allow to synthesize the «reference» transition programs to a new structural state, which will ensure the maintenance of key indexes of CO viability at the required target level within a given time interval.

The difference from the familiar results is that based on the proposed approach, it is possible to link the viability index with the interoperability of GBATLS functioning processes indexes, obtained in the new business processes implementation, with the infor-mation services volume provided by the operator and/or DT.

We were able to conduct a multi-criteria assessment of the costs caused by the operation and modernization of IIMS elements, which are the basis of control and management in modern GBATLS.

5 Discussion

The paper presents a new approach to complexity management as applied to solve the problems of evolutionary creation and implementation (modernization) of equivalent information management software and hardware complexes used in existing and prospective transport and logistics systems. As a generalized performance index is proposed to use the viability index, which at the formal level of description was able to link with the GBATLS interoperability index. In this work theoretical-multiple problem statement has been carried out. Appropriate models, methods and solution algorithms have been proposed.

The further direction of research is the development of the existing model-algorithmic support in order to analyze the influence of internal and external environment factors on the processes of DT structural and parametric adaptation at the stage of GBATLS modernization [21].

Additional information about the results of our research is available on the laboratory's website [22].

Funding. The study was carried out at the expense of a grant Russian Science Foundation No. 22–79-00301.

References

1. Okhtilev, M., Sokolov, B., Yusupov, R.: Intelligent technologies for monitoring and controlling the structural dynamics of complex technical objects (2006)
2. Sokolov, B., Zakharov, V., Baranov, A.: Combined models and algorithms on modern proactive intellectual scheduling under industry 4.0 environment. IFAC-PapersOnLine **55**, 1331–1336 (2022). https://doi.org/10.1016/j.ifacol.2022.09.575
3. Sampigethaya, K., Poovendran, R.: Aviation cyber-physical systems: foundations for future aircraft and air transport. Proc. IEEE **101**, 1834 (2013). https://doi.org/10.1109/JPROC.2012.2235131
4. Ignatiev, M.B., Marley, E., v, Mikhailov V v, Spesivtsev A v,: Modeling Weakly Formalized Systems Based on Explicit and Implicit Expert Knowledge, p. 501. Politeh-Press, St Petersburg (2018)
5. Hajjem, E., Benderbal, H.H., Hamani, N., Dolgui, A.: Digital twin framework for reconfigurable manufacturing systems: challenges and requirements (2021)
6. Luo, D., Thevenin, S., Dolgui, A.: A digital twin-driven methodology for material resource planning under uncertainties (2021)
7. Zakharov, V.: Combined optimization algorithm of complex technical object functioning and its information system modernization. In: Vasiliev, Y.S., Pankratova, N.D., Volkova, V.N., et al. (eds.) System Analysis in Engineering and Control, pp. 487–497. Springer International Publishing, Cham (2022)
8. Mikoni, S.V., Sokolov, B.V., Yusupov, R.M.: Qualimetry of models and multiple-model complexes (2018)
9. Beer, S.: Cybernetics and Management. Wiley, Hoboken (1959)
10. Sokolov, B., Pavlov, A., Potriasaev, S., Zakharov, V.: Methodology and technologies of the complex objects proactive intellectual situational management and control in emergencies (2020)
11. Pavlov, A., Ivanov, D., Pavlov, D., Slinko, A.: Optimization of network redundancy and contingency planning in sustainable and resilient supply chain resource management under conditions of structural dynamics. Ann. Oper. Res. **5**, 1–30 (2019). https://doi.org/10.1007/s10479-019-03182-6
12. Ivanov, D., Sokolov, B., Kaeschel, J.: A multi-structural framework for adaptive supply chain planning and operations control with structure dynamics considerations. Eur. J. Oper. Res. **200**, 409–420 (2010). https://doi.org/10.1016/j.ejor.2009.01.002
13. Ivanov, D., Dolgui, A., Sokolov, B., Ivanova, M.: Integrated simulation-optimization modeling framework of resilient design and planning of supply chain networks. IFAC-PapersOnLine **55**, 2713–2718 (2022). https://doi.org/10.1016/j.ifacol.2022.10.121
14. Ivanov, D., Sokolov, B., Chen, W., et al.: A control approach to scheduling flexibly configurable jobs with dynamic structural-logical constraints. IISE Trans. **53**, 21–38 (2021). https://doi.org/10.1080/24725854.2020.1739787
15. Sokolov, B.V, Kalinin, V.N., Zakharov, V.V.: Integrated planning and scheduling of enterprise information system modernization. In: CEUR Workshop Proceedings, pp 3–12 (2020)
16. Sokolov, B., Zakharov, V., Krylov, A.: Complex scheduling of measurement and calculation systems functioning. In: CEUR Workshop Proceedings, pp 58–63 (2021)

17. Popov, V.D., Spesivtsev, A.V., Sukhoparov, A.I., Spesivtsev, V.A.: Convolution of multi-criteria expert estimates in a context of uncertainty. In: Proceedings of 2017 20th IEEE International Conference on Soft Computing and Measurements, SCM 2017, pp 203–206 (2017)
18. Kimyaev, I.T., Spesivtsev, A., v,: Ontological and fuzzy-possibility approach to the synthesis of the DM functional equivalent for management of hierarchical systems. In: Silhavy, R. (ed.) Artificial Intelligence Trends in Systems, pp. 575–584. Springer International Publishing, Cham (2022)
19. Dolgui, A., Ivanov, D., Sethi, S.P., Sokolov, B.: Scheduling in production, supply chain and Industry 4.0 systems by optimal control: fundamentals, state-of-the-art and applications. Int. J. Prod. Res. **57**, 411–432 (2019)
20. Bakhmut, A.D., et al.: Methods of conceptual modeling of intelligent decision support systems for managing complex objects at all stages of its life cycle. In: Abraham, A., Kovalev, S., Tarassov, V., Snasel, V., Sukhanov, A. (eds) Proceedings of the Third International Scientific Conference "Intelligent Information Technologies for Industry" (IITI2018). IITI'18 2018. Advances in Intelligent Systems and Computing, vol. 875, pp. 171–180. Springer, Cham (2019). https://doi.org/10.1007/978-3-030-01821-4_18
21. Ushakov, V.: Approximation a reachability area in the state space for a discrete task. In: Silhavy, R. (ed.) CSOC 2020. AISC, vol. 1226, pp. 617–624. Springer, Cham (2020). https://doi.org/10.1007/978-3-030-51974-2_57
22. Litsam Homepage. http://www.litsam.ru. Accessed 13 Mar 2023

Increasing of Throughput of a Steganographic Communication Channel for Secured Telecom Networks of Railway Transport Based on Multialphabet Coding Method

Jurii Ryabinin[1], Oleg Finko[1(✉)], Alexander Kurakin[2], and Nikolay Kramskoi[2]

[1] North-Caucasus Federal University, 2 Kulakova Street, 355029 Stavropol, Russia
ofinko@yandex.ru

[2] The Special Technology Center Inc., 10, Karavaevskaya Street, 197377 St. Petersburg, Russia

Abstract. The method of increasing the throughput of a steganographic channel for data transmission by implementing the various format containers system (for instance, jpeg, bmp, mp3, mpeg4, avi etc.) with preseted level of steganographic resistance and data error security (random or intentional) is proposed. This is achieved by fragmenting data into polynomial symbols of an extended algebraic representation of data based on a multialphabet modular code. A model for covert transmission of error-protected multithreaded data by steganography methods is proposed; the results of comparison of multiformat and singleformat modes of the steganographic system are presented.

Keywords: Steganographic system · throughput of the steganographic channel for data transfer · multinomial modular arithmetic · steganographic resistance

1 Introduction

The safety of railway transport traffic could be achieved in various ways, including the applying of real-time Close-circuit television (CCTV, hereinafter) systems, Digital video recording (DVR, hereinafter) systems combined with cybersecurity methods [1, 2].

Video monitoring of railway facilities is carried out with smart CCTV systems for terminals, train crossings, high risk railway areas etc. The system includes the data transfer channels (CCTV cameras, communication lines and communication devices) for the traffic control center and train operator's cab [2].

DVR systems for locomotives are designed for record and store of audio and video data from the operator's cab, video data from CCTV all along the train etc. [2].

The key safety requirement is to prevent emergencies during the railway traffic movement. Therefore, It is essential to provide data security for DVR.

In order to make proper data security in telecommunication networks with large amount of multimedia traffic, the following could be applied:

1) the high-speed real-time cryptographic security solutions [3, 4] for large amount of multimedia data (100–200 Gb/sec or faster);

R. Silhavy and P. Silhavy (Eds.): CSOC 2023, LNNS 723, pp. 546–556, 2023.
https://doi.org/10.1007/978-3-031-35317-8_49

2) the Quantum key distribution (QKD, hereinafter) systems [5, 6] which provide the high efficiency of multimedia data transfer systems and resistance for key compromise;
3) the digital steganographic methods [7, 8] that provide data transfer and storage of identification and authentication information for the entities that can be telemetry [9] and other service data (the device data, location and time of registration, speed and acceleration of the movement of the locomotive etc.). Digital steganography methods can also ensure the security of local user group data by access restriction.

The well-known disadvantage of steganographic data transmission channel (SDTC, hereinafter) is the low throughput in comparison with throughput of communication channel, which the SDTC is based on (with the preseted level of steganographic resistance). Thus, the SDTC throughput is the highest possible (potentially possible) amount of data that steganographic system (stegosystem) is capable of transmitting over given channel per unit of time. Steganographic resistance is the ability of stegosystem to hide from detected intruder the secret data transmission; the ability to resist the intruder's attempts to destroy, edit (distort) or delete the secret data.

The data in stegosystems is stored in stegocontainers (containers) – a file or a part of multimedia file with the corresponding format (the certain specification, describing the file structure and it's interpretation in accordance with file structure; data-flow transmission via network and storage). It is natural to assume that it is possible to increase the throughput of SDTC by extending the amount of stegocontainers that provide the transmission of stegomessages (data embedded in a container). The capacity of stegomessage that can be embedded in a container without reducing the level of steganographic resistance is the main characteristics of stegocontainer.

During the transmission of stegomessages via SDTC various errors may occur. Some authors, for instance [7], in order to detect and fix the errors the methods of redundant monoalphabet code are applied (ex. Reed–Solomon codes). The disadvantage of the proposed methods is the assumption that the capacity of stegocontainers are equal, which may cause a reduction of throughput of the SDTC (since the stegocontainers of the wrong capacity will be discarded).

In [10–12] the authors proposed to implement multialphabet coding methods based on the system of residual classes [13–15] to prepare stegomessages for embedding in stegocontainers of various capacity. Further in the text this method is generalized for the case of using stegocontainers for various formats and an assessment of throughput of the SDTC is made in comparison with the classic case of embedding in single format container.

2 Methods

The working principle of the multiformat mode of stegomessages transfer via SDTC is showed on Fig. 1. An indicator allowing to evaluate the effectiveness of multiformat data transmission is the probability that the message $M_{u,\,w}$ (or the parts of the message $m_{u,\,1}|m_{u,\,1}|\cdots|m_{u,\,s} = M_{u,\,w}$; "|" is the concatenation symbol), is to be transmitted in preseted time frame.

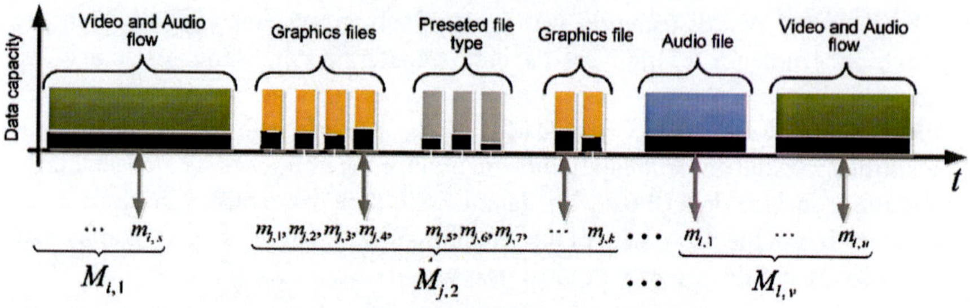

Fig. 1. The generalized time diagram of the sequence of the files in SDTC (dashes indicate the target data (parts of stegomessages), embedded in stegocontainers).

We will take into account the assumption that some errors may occur during the transmission of stegomesagges in SDTC due to the influence of random factors or malicious actions. Therefore, when we had chosen the method of redundant coding of data, which is to be transmited via SDTC, we gave our preference to modular redundant codes.

Let the $\mathbb{F}[x]$ be the set of all polynominals with coefficients from $\mathbb{F}$ field (ex. GF_2) and indefinite x. Consider a set of polynomials $p_1(x),\ p_2(x),\ \ldots,\ p_k(x) \in \mathbb{F}$ that do not have common divisors in pairs which is called modules (polynomial bases) and the $P(x) = \prod_{j=1}^{k} p_j(x)$. It is commonly known [13–15], that any polynominal $M(x)$, which degree is less than $P(x)$ (written as $\deg(M(x)) < \deg(P(x))$) can be uniquely represented by a vector of polynomial residues $\mathbf{M} = (m_1(x),\ m_2(x),\ \ldots,\ m_k(x))$ which we are calling a polynomial word of an algebraic representation (AR, hereinafter) of data based on a polynomial modular code, where $m_j(x) = M(x) \mod p_j(x); j = 1,\ 2,\ ...,\ k$; $m_j(x)$ are polynomials, which we are calling polynomial symbols (or just symbols) of AR. As the polynominal bases $p_1(x),\ p_2(x),\ \ldots,\ p_k(x)$ are the symbols of different alphabets, so the polynomial modular code belongs to the class of multialphabet codes.

3 Results

Let the $M(x)$ be the message of $\mathbf{M}$, which is to be transmited via SDTC in polynominal form. Then the $M(x)$ can be represented as follows:

$$M(x) = \underbrace{w_{j_k-1}x^{j_k-1} + \cdots + w_{j_{k-1}}x^{j_{k-1}}}_{m_1(x)x^{\deg(p_1(x))-1}} + \underbrace{w_{j_2-1}x^{j_2-1} + \cdots + w_{j_1}x^{j_1}}_{m_2(x)x^{\deg(p_2(x))-1}} + \cdots$$

$$\cdots + \underbrace{w_{j_1-1}x^{j_1-1} + \cdots + w_0}_{m_k(x)x^0}$$

for $\sum_{s=0}^{k-1} j_s$ and numbers of $w_{j_s} \in \{0,\ 1\}$. That is:

$$M(x) = m_1(x)x^{\deg(p_1(x))-1} + m_2(x)x^{\deg(p_2(x))-1} + \cdots + m_k(x)x^0.$$

In common case, the $\deg(m_u(x)) \neq \deg\big(m_y(x)\big)$ and the $u \neq y; u, y = 1,\ 2,\ \ldots,\ k$. Then the $M(x)$ can be considered as a vector of coefficients $M(x)$:

$$\mathbf{M} = \left[m_1(x)\ m_2(x) \cdots m_k(x)\right].$$

which can be interpreted as the AR symbols of some a priori unknown $Z(x)$, that is isomorphic to $M(x)$.

Let us take a set of polynomial bases such as $p_1(x),\ p_2(x),\ \ldots,\ p_k(x) \in \mathbb{F}[x]$, that $\deg(M(x)) < \deg(P(x))$ and $\deg\big(m_j(x)\big) < \deg\big(p_j(x)\big)$. Therefore, $Z(x) = (m_1(x),\ m_2(x),\ \ldots,\ m_k(x))$, where $m_j(x) = Z(x) \mod p_j(x); j = 1,\ 2,\ \ldots,\ k$.

Then the system is as follows:

$$\begin{cases} Z(x) = m_1(x) \mod p_1(x); \\ Z(x) = m_2(x) \mod p_2(x); \\ \cdots\cdots\cdots\cdots \\ Z(x) = m_k(x) \mod p_k(x), \end{cases}$$

If the conditions mentioned above are met, it has a unique solution $Z(x)$:

$$Z(x) = \left(\sum_{j=1}^{k} P_j(x)\, y_j(x)\, m_j(x)\right) \mod P(x), \tag{1}$$

where $P_j(x) = \frac{P(x)}{p_j(x)}$, $y_j(x) = P_j^{-1}(x) \mod p_j(x)$, $P(x) = \prod_{j=1}^{k} p_j(x)$. $Z(x)$ can be unequivocally restored [13–15].

The AR symbol expansion procedure is implemented as follows. Introduce the r redundant polynomial bases $p_{k+1}(x),\ \ldots,\ p_{k+r}(x)$ and get the r of test polynomial symbols: $m_{k+1}(x) = Z(x) \mod p_{k+1}(x),\ \ldots,\ m_{k+r}(x) = Z(x) \mod p_{k+r}(x)$, assuming that the

$$\deg(p_1(x)),\ \ldots,\ \deg(p_k(x)) > \deg(p_{k+1}(x)) > \cdots > \deg(p_{k+r}(x)), \tag{2}$$

where $\gcd(p_a(x),\ p_b(x)) = 1, a \neq b, a$ and $b = 1,\ 2,\ \ldots,\ k + r$.

This is how the polynomial word $m_1(x),\ m_2(x),\ \ldots,\ m_{k+r}(x)$ of the extended AR (ExAR, hereinafter) is formed. The use of the data AR expansion procedure allows us to detect distortions and perform the correction of distorted AR symbols [13–15]. In this case, the choice of polynomial bases $p_1(x),\ \ldots,\ p_{k+r}(x)$ on the transmitting side has to be implemented taking into account (2) and on the basis of a preliminary analysis of the available containers in accordance with rule as follows [13–15]:

$$\deg(p_i(x)) > \lceil \text{ant}\, log\ L_i \rceil,$$

$i = 1,\ 2,\ \ldots,\ k + r$, L_i is the maximum amount of bit that can be embedded in corresponding container with preseted steganographic resistance, $\lceil a \rceil$ – smallest integer of a.

Thus, the analysis of the container allows us to determine the size of useful data that is to be embedded in the container in each case, as the $\deg\big(m_j(x)\big) < \deg\big(p_j(x)\big)$. Therefore, for the receiving side, along with the useful data it is necessary to transmit the service data in a form of a number of a module from the table (assuming the fact that the receiving and the transmitting side already have the module tables); along with the module number, it is also presented a polynomial symbol ExAR which is to be embaddedd in the container, so the $L_i = \tilde{L}_i + N_i^{(\text{service data})}$, where $\tilde{L}_i$ is the actual number of bit that can be embedded in a corresponding container with a preseted steganographic resistance, $N_i^{(\text{service data})}$ is an amount of bit (fixed) that is required for embed the service data.

Criterion:

$$\deg(Z(x)) < \deg(P(x)), \tag{3}$$

determines the presence or absence of possible errors in the received sequence [13–15].

The metric of the ExAR is the minimal distance $D_{\min}$ is the weight of difference between two ExAR words. The ability of the ExAR to reliably detect and/or fix q-fold errors is determined accordingly: $q_{\text{обнар}} \le D_{\min} - 1$ and $q_{\text{испр}} \le 2^{-1}(D_{\min} - 1)$, the q-fold error is the random distortion q of ExAR symbols.

Thus, the sequence $m_1(x),\ m_2(x),\ \ldots,\ m_{k+r}(x)$ that was formed is subject to transmission via SDTC. The presented model is not restricted by the formats of containers that the resulting polynomial ExAR symbols will be embedded; it also ensures the embedding the useful data in various container formats with the necessary steganographic resistance, with the multiformat mode of steganographic system.

It is known that the container use factor (or efficiency factor $K_{\text{effic. coeff.}}$) is determined as follows: $K_{\text{effic. coeff.}} = \frac{|M_{\max}(T)|}{|T|}$, where M is a corresponding hiden message, T is a container, $|M|$ it the size of message, $|T|$ is the capacity of a container, $|M_{\max}(T)|$ is the highest possible size of the message that can be hidden in a container. It can be seen that $|M| \le |M_{\max}(T)|$ the preseted level $K_{\text{effic. coeff.}}$ of steganographic system is reached. Therefore, if the $m_1(x),\ m_2(x),\ \ldots,\ m_{k+r}(x)$ of the ExAR data will satisfy the condition: $\deg(m_i(x)) \le |M_{\max}(T)|$, then the embedded data will not exceed the preseted detection threshold or stegoresistance, which imposes a limit on the size of the message that is to be hide.

On the receiving side, possibly distorted ExAR polynomials are extracted from the set of containers $m_1^*(x),\ m_2^*(x),\ \ldots,\ m_{k+r}^*(x)$ and calculating the $Z(x)$ in accordance with the rule (1) and performing the comparison (3). If the (3) is failed to perform, the occurred error is noted and the procedure of searching and correction of distorted symbols is started. The symbols "*" and "*"hereinafter mean the probability of error occurrence in the received sequence of polynomial ExAR; and reciving the incorrect ExAR symbol during the correction procedure. Performing the (3) means that received ExAR symbols are correct. When the redundant polynomial symbols have been discarded $m_{k+1}^{**}(x),\ \ldots,\ m_{k+r}^{**}(x)$, the procedure of adding information polynomials is performed:

$$M^{**}(x) = m_1^{**}(x)x^{\deg(p_{i,1}(x))-1} + m_2^{**}(x)x^{\deg(p_{i,2}(x))-1} + \cdots + m_k^{**}(x),$$

which in vector form can be represented as the result of the concatenation of the coefficients $m_1^{**}(x),\ m_2^{**}(x),\ \ldots,\ m_k^{**}(x)$, i.e.:

$$\mathbf{M}^{**} = \mathbf{m}_1^{**} |\ \mathbf{m}_2^{**} | \cdots |\ \mathbf{m}_k^{**},$$

where $\mathbf{m}_j^{**} = \left[w_{N-1}^{**}\ w_{N-2}^{**} \cdots w_{i,0}^{**} \right]$; $j = 1,\ 2,\ \ldots,\ k$; N is the vector length. Then:

$$\mathbf{M} = w_{j_k-1}^{**},\ \ldots,\ w_{j_{k-1}}^{**} |\ w_{j_2-1}^{**},\ \ldots w_{j_1}^{**} | \cdots | w_{j_1-1}^{**},\ \ldots,\ w_0^{**},$$

for $\sum_{s=0}^{k-1} j_s$ and numbers of $w_{j_s} \in \{0, 1\}$.

Thus, the ExAR of the data to be transmitted via SDTC allows us to restore the parts of corrupted and/or missing ExAR symbols (no more than r symbols, if $r \geq 2^{-1}(D_{\min} - 1)$), which increases the level of data protection to prevent the attack of intruder, leading to degradation of the data transmitted via the SDTC.

4 Discussions

The absence of a container of the required format at corresponding moment of time can cause delays of the target data transfer. Therefore, when we expand the stegosystem's ability to embed the data of various formats, the probability $R(i)$ of the event occurrence is increasing; the event occurance means that there is one or more $i \geq 1$ that is suitable for embedding in the container.

During the event the system receives 1, 2, 3 etc. suitable for embedding containers. Let's assume that the stationarity of the stegocontainers receiving flow is for low amount of time. The reciving flows of the each format containers are the simplest (or stationary Poission) with intensities λ_k ($k = 1, 2, 3 \ldots$). Based on the additivity property N of Poisson flows with intensities λ_k is forming a Poisson flow with intensity $\lambda_\Sigma = \sum_{k=1}^{N} \lambda_k$ (Fig. 2).

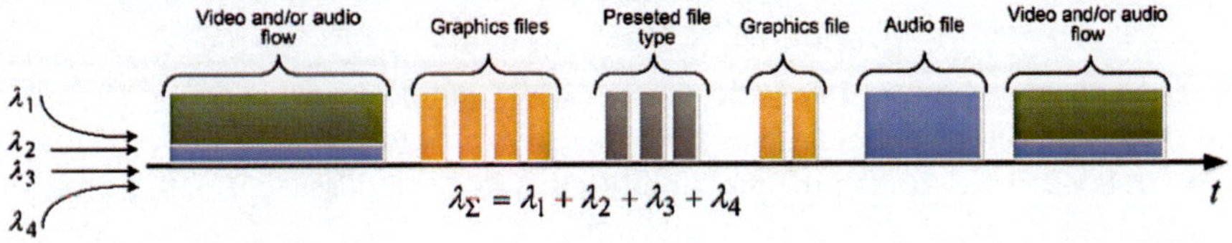

Fig. 2. The time diagram of the total Poisson flow of the various format containers.

Then the probability of receiving i of the containers during the period of time t is determined by the classic expression $P(i) = \frac{(\lambda t)^i}{i!} e^{-\lambda t}$, where $i \in I$. Figure 3 representing the results of simulation modeling (1000 random iterations, which are independent for each flow), the relative frequency of occurrence for different number of containers ($i = \overline{1,\ 5}$) $W(i)$ for the corresponding length of the period of time for three single-format flows for the period of time t equal to 1 s.

Following the Fig. 3, we can see that the total Poisson flow has a higher relative frequency $W(i)$ of receiving the stegocontainers, which are suitable for message embedding. Therefore, the multiformat channel SDTC, described by total flow, has significant advantages over singleformat in terms of the probability of embedding a message in a fixed amount of time.

Let's determine the probability $R(i)$ of receiving at least one container suitable for embedding the target data, which will allow us to evaluate the ability of the system to embed and transmit stegodata within the corresponding time limit.

The probability of appearance no less than k of stegocontainers is equal: $R(k) = P(I \geq k) = 1 - (P(0) + P(1) + \cdots + P(k-1))$, where $P(i) = \frac{a^i}{i!} e^{-a}$, $\lambda t = a$.

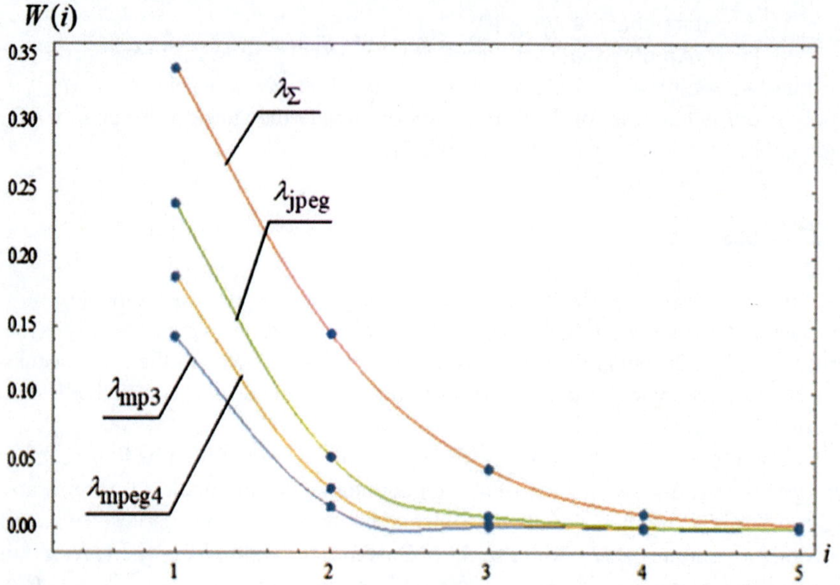

Fig. 3. The dependence of the relative frequency $W(i)$ and occurrence of a given number i of containers.

Example. Let the flow intensities: $\lambda_{jpeg} = 20$ (unit per min.), $\lambda_{mp3} = 10$ (unit per min.), $\lambda_{mpeg4} = 5$ (unit per min.), and the intensity of the total flow is: $\lambda_{\Sigma} = 35$ (unit per min.). It is necessary to find the probabilities that in 1 s. at least 1, 2, 3, 4, 5 containers will be received.

The mathematical expectations of the flows are equal: $a_{jpeg} = 1/3$ (unit per min.), $a_{mp3} = 1/6$ (unit per min.), $a_{mpeg4} = 1/12$ (unit per min.), $a_{\Sigma} = 7/12$ (unit per min.). The calculation of the dependencies of the probabilities $R(i)$ are presented in Fig. 4.

Let's increase the intensity: $\lambda_{jpeg} = 60$ (unit per min.), $\lambda_{mp3} = 20$ (unit per min.), $\lambda_{mpeg4} = 15$ (unit per min.) and, therefore: $\lambda_{\Sigma} = 95$ (unit per min.) (Fig. 5).

Let's increase the observation time interval to 5 (sec.). We will get the dependencies shown in Fig. 6.

Thus, if the system is ready to embed a message in a container within 5 s, the multiformat mode of the system provides a probability of 0.95 of embedding a message at relatively low rates of singleformat flows.

For an observation period equal to 1 min, the range of dependences of the probability of occurrence no less than i containers is shown in Fig. 7.

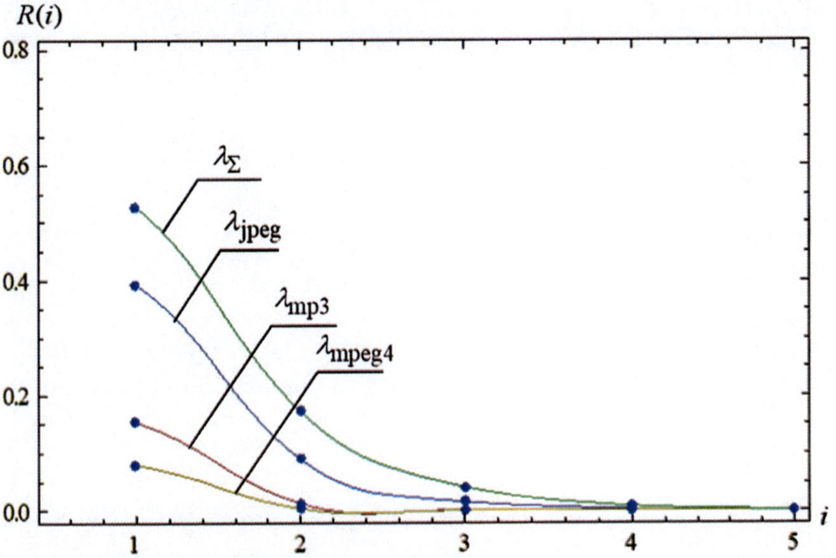

Fig. 4. An example of estimating the probability $R(i)$ of receiving at least one container suitable for embedding the target data (i is the number of containers).

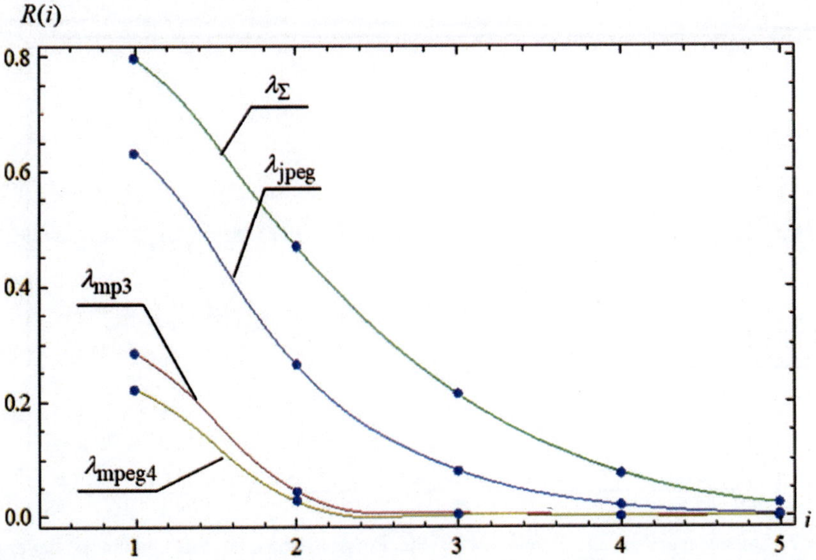

Fig. 5. Dependences of the intensities of the Poisson flows of the containers under given conditions (time interval – 1 sec.; i is the number of containers).

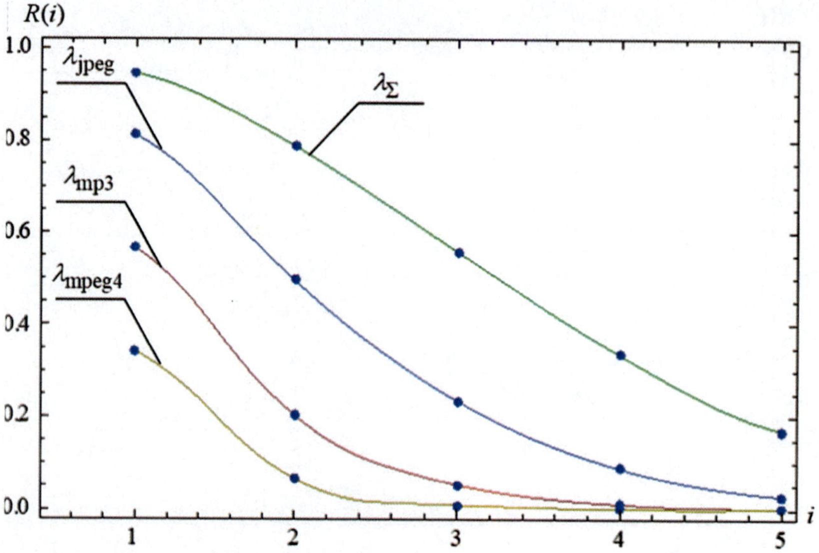

Fig. 6. Dependences of the intensities of the Poisson flows of the containers under given conditions (time interval – 5 sec.; *i* is the number of containers).

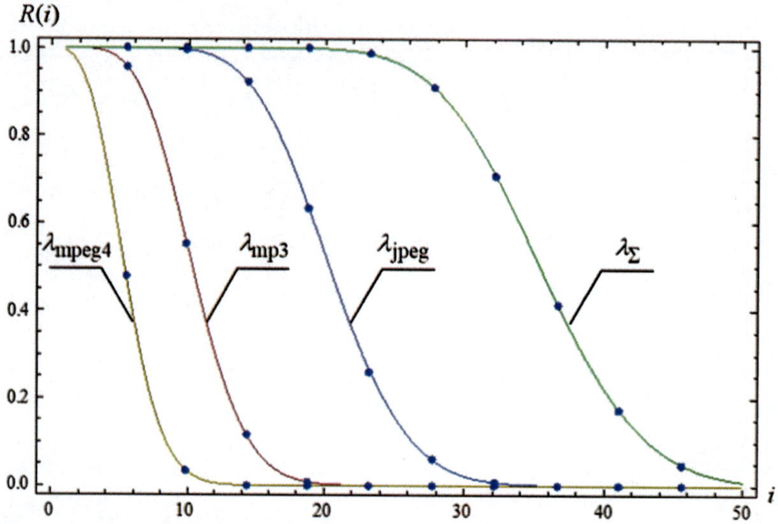

Fig. 7. Dependences of the intensities of the Poisson flows of the containers under given conditions.

5 Conclusion

The method that allows to increase throughput of the SDTC at preseted levels of steganographic resistance and data protection from errors by transmitting them in a multiformat mode is proposed. Before the transmission of target data via SDTC, the fragmentation into polynomial residues of the extended algebraic representation is performed. That method allows us to solve two issues: firstly, it can increase the usage factor of various format containers (with preseted level of steganographic resistance), that is, to increase their efficient use and, consequently, increase the throughput of the SDTC; and, secondly, it can increase the transmission speed of target information under the conditions of the occurrence of probable errors in the SDTC (the same as usage of error-correcting codes in a communication channel).

Acknowledgments. The study was carried out with the financial support of OJSC "RZD".

References

1. Plekhanov, P.A.: Ensuring integrated safety in railway transport in the context of strategic development. In: Proceedings of Petersburg State Transport University, vol. 17, issue 4, 552–565 (2020). https://doi.org/10.20295/1815-588X-2020-4-552-565
2. Szmigiel, P., Kasprzyk, Z.: Assessment of availability of railway video surveillance system. In: Siergiejczyk, M., Krzykowska, K. (eds.) ISCT21 2019. AISC, vol. 1032, pp. 430–439. Springer, Cham (2020). https://doi.org/10.1007/978-3-030-27687-4_43
3. Xiao, C., Zhang, L., Xie, Y., Liu, W., Liu, D.: Hardware/Software adaptive cryptographic acceleration for big data processing. Secur. Commun. Netw. **2018**, 1–24 (2018). https://doi.org/10.1155/2018/7631342
4. Malina, L., Smekal, D., Ricci, S., Hajny, J., Cíbik, P., Hrabovsky, J.: Hardware-accelerated cryptography for software-defined networks with P4. In: Maimut, D., Oprina, A.-G., Sauveron, D. (eds.) SecITC 2020. LNCS, vol. 12596, pp. 271–287. Springer, Cham (2021). https://doi.org/10.1007/978-3-030-69255-1_18
5. Wolf, R.: Quantum key distribution. Lecture Notes in Physics, vol. 988. Springer, Cham (2021). ISBN 978-3-030-73990-4. https://doi.org/10.1007/978-3-030-73991-1
6. Mehic, M., Rass, S., Fazio, P., Voznak, M.: Fundamentals of quantum key distribution. In: Mehic, M., Rass, S., Fazio, P., Voznak, M. (eds.) Quantum Key Distribution Networks: A Quality of Service Perspective, pp. 1–28. Springer International Publishing, Cham (2022). https://doi.org/10.1007/978-3-031-06608-5_1
7. Fridrich, J.: Steganography in Digital Media: Principles, Algorithms, and Applications. Cambridge University Press (2009). https://doi.org/10.1017/CBO9781139192903
8. Yahya, A.: Steganography Techniques for Digital Images. Springer International Publishing, Cham (2019). https://doi.org/10.1007/978-3-319-78597-4
9. Patent 2765811 Russian Federation, H04L 9/00 (2021.08) G06F 11/08 (2021.08) H04N 7/08 (2021.08). In: Ryabinin, J.E., Finko, O.A., Kurakin, A.S. (eds.) Method for Hidden Secure Transmission of Telemetry Data in Robotic Complexes. Patent holder – KVVU (2021)
10. Ryabinin, J., Finko, O.A.: Steganographic system resistant to attack in the extended modular arithmetic Izvestiya SFedU. Eng. Sci. **2**(151), 167–174 (2014)
11. Ryabinin, J.E., Finko, O.A.: Mathematical Model of Hidden, Noseproof Information Transfer Provide in the Modular Code. Science. Innovations. Technologies, vol. 2, 53–62. North-Caucasus Federal University (2017)
12. Ryabinin, J., Finko, O.: Some principles of building steganographic communication networks. In: Choraś, M., Choraś, R.S., Kurzyński, M., Trajdos, P., Pejaś, J., Hyla, T. (eds.) CORES/IP&C/ACS -2021. LNNS, vol. 255, pp. 240–251. Springer, Cham (2022). https://doi.org/10.1007/978-3-030-81523-3_24

13. Mandelbaum, D.M.: On Efficient Burst Correcting Residue PolynomialCodes/Information and Control, pp. 319–330 (1970). https://doi.org/10.1016/S0019-9958(70)90168-3
14. Molahosseini, A.S., Sousa, L.: Introduction to residue number system: structure and teaching methodology. In: Molahosseini, A.S., de Sousa, L.S., Chang, C.-H. (eds.) Embedded Systems Design with Special Arithmetic and Number Systems, pp. 3–17. Springer, Cham (2017). https://doi.org/10.1007/978-3-319-49742-6_1
15. Ananda Mohan, P.V.: Residue Number Systems: Theory and Applications, Birkh¨auser, Basel (2016). ISBN 978-3-319-41385-3, x+351 pp. https://doi.org/10.1007/978-3-319-41385-3

Stabilization of Second-Order Systems Using PI Controllers with Delayed Proportional and Integral Parts

Martin Strmiska[1(✉)], Libor Pekař[1], and José Mario Araújo[2]

[1] Tomas Bata University, Zlín, Czechia
strmiska@utb.cz
[2] Federal Institute of Education, Science and Technology of Bahia, Salvador, Brazil

Abstract. The present paper deals with the stabilization, optimization, and control of second-order systems using a PI delayed controller. This study analyzes the control of linear-time invariant (LTI) systems with time delay in the integral and proportional parts of the PI delayed controller and obtains the region of global stability in the two-dimensional plane. The proposed stabilization method is verified through software simulations with two benchmark examples: the control of an inverted pendulum and the control of the stick-slip vibration. The results show the effectiveness and practical applicability of the proposed method in all analyses and control. Additionally, the optimization of LTI systems with time delay in the integral and proportional parts of the PI delayed controller is discussed, and selected heuristic methods are compared. Future research aims to implement modern optimization algorithms for solving non-smooth, non-convex, and non-Lipschitz optimization problems in this field.

Keywords: Stability · linear-time invariant (LTI) systems · second-order systems · PI delayed controller · heuristic optimization methods

1 Introduction

Stabilization is a technique to achieve asymptotic stability in a controlled system. The foundation of stability analysis of dynamical systems, including nonlinear and linear time-invariant systems, can be traced back to Lyapunov's works and systems theory texts [1–3]. Various methods, such as the Routh-Hurwitz criterion, can determine stability, provided that the control system does not involve time delay [4]. However, since for systems with some time delay, analytical methods are often insufficient to decide on the system's stability, since the application of numerical mathematical techniques or frequency domain methods, such as the Nyquist criterion [5, 6].

Real-world scenarios frequently involve time delays, as seen in the transportation of materials [7], signal transmission [8], control of epidemics [9], traffic systems [10], network control systems [11, 12], control of RNA messenger in human cells [13], and among others [12]. Time delays are common in control systems and can have adverse

R. Silhavy and P. Silhavy (Eds.): CSOC 2023, LNNS 723, pp. 557–572, 2023.
https://doi.org/10.1007/978-3-031-35317-8_50

effects such as closed-loop instability and noise amplification. However, time delays can be added to the controller structure artificially and, in some cases, to achieve specific benefits while keeping the design as simple as possible. For example, the designer may introduce output feedback with a delay; or add a delay to the proportional, integration, or both parts of the PI controller [14–18, 24]. An undesired consequence of implementing delays in controller design is that the infinite-dimensional spectrum reveals because of the quasi-polynomial properties of the characteristic equation [19, 20]. Ensuring that stable roots of the system dominate can be a complex task for the designer. Stability charts that vary according to system parameters, such as the delay and controller gains, often provide valuable insights for the stability analysis or design of the system [17, 20, 21, 21, 23].

This paper focuses on the stabilization, optimization, and control of second-order systems using a PI delayed controller with time delay in the integral and proportional parts. The goal is to obtain the controller parameters in the region of global stability in a two-dimensional plane and verify the proposed method using MATLAB Simulink computations of two benchmark examples. The results demonstrate the effectiveness and practical applicability of the proposed method in all analyses and control. The paper also discusses the optimization of LTI systems with time delay and compares selected heuristic methods.

The article consists of two main sections: analytical and numerical. The analytical section englobes the principle of the design method, the determination of the stabilization pair for the delayed PI controller, and heuristic methods. The proposed approach is verified in the numerical section through two examples, where mathematical formulas and stability regions are derived and described. The article concludes with a brief discussion of all the results and future directions in the conclusion section.

2 Analytical Section

2.1 Principle of the Design Method

Consider a closed-loop system [17, 18] in Fig. 1 with PI delayed controller $C(s)$ and a controlled system $G(s)$.

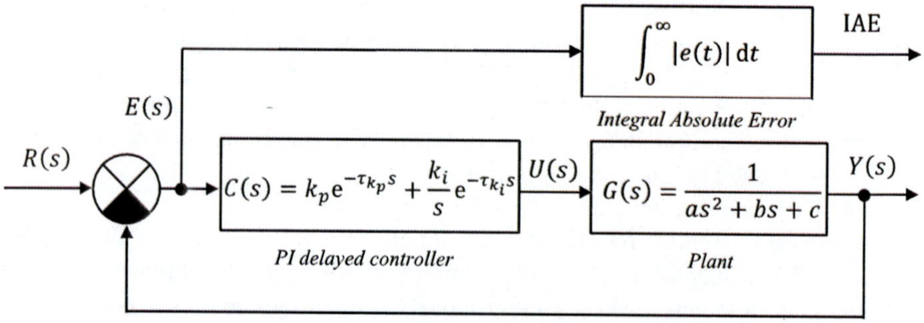

Fig. 1. The closed-loop control design scheme.

The closed-loop control system regulates the system plant to keep it at the desired level. Within the framework of the closed-loop control, the input reference signal is represented by $R(s)$, the control error by $E(s)$, the controller by $C(s)$, the control action by $U(s)$, the controlled plant by $G(s)$, the output signal by $Y(s)$, and the integral absolute error by IAE (more information about IAE is in the Subsect. 2.3).

The controller $C(s)$ and the controlled plant $G(s)$ are defined by transfer functions

$$C(s) = k_p \mathrm{e}^{-\tau_{k_p} s} + \frac{k_i}{s} \mathrm{e}^{-\tau_{k_i} s}, \quad G(s) = \frac{1}{as^2 + bs + c}, \tag{1}$$

where k_p represents a proportional gain, k_i an integral gain, $\tau_{k_p} \in \mathbb{R}_+$ is a delay connected to proportional gain, $\tau_{k_i} \in \mathbb{R}_+$ is a delay connected to integral gain, s is a complex variable, and a, b, c are real coefficients.

2.2 Determination of the Stabilization Pair (k_p, k_i) for PI Delayed Controller

Now, we will find a stabilizing pair of values (k_p, k_i) that will allow us to identify the stable region of the closed-loop system. Let use the output signal $Y(s)$ [1, 2] of the closed-loop control in Fig. 1 defined as

$$Y(s) = \frac{G(s)C(s)}{1 + G(s)C(s)} R(s). \tag{2}$$

The numerator of the $1 + G(s)C(s)$ is a characteristic quasi-polynomial $\Delta(s)$ [1, 4] of the closed-loop system

$$\Delta(s) = \mathrm{num}[1 + G(s)C(s)]. \tag{3}$$

An asymptotically stable system is characterized by having all roots of its characteristic equation located in the left half-plane of the complex plane [1–3]. To obtain the characteristic quasi-polynomial Δ(s), we substitute the expressions for $G(s)$ and $C(s)$, as defined in (1), into Eq. (3), resulting in

$$\Delta(s) = as^3 + bs^2 + \left(c + k_p \mathrm{e}^{-\tau_{k_p} s}\right)s + k_i \mathrm{e}^{-\tau_{k_i} s}. \tag{4}$$

After substituting $s = o + \omega i$ into the characteristic quasipolynomial (4), we get an expression with a real and an imaginary part. By setting the real $\Re[\Delta(o + \omega i)] = 0$ and imaginary $\Im[\Delta(o + \omega i)] = 0$ parts to zero, we get a system of two equations with two unknowns variables k_p, k_i.

After solving the system, we achieve the following equations:

$$\begin{aligned} k_p &= \frac{ao\left(o^2 - 3\omega^2\right) + b\left(o^2 - \omega^2\right) + co + \omega\left[\left[a\left(3o^2 - \omega^2\right) + 2bo + c\right]\cot\omega\tau_{ki}\right]}{o\sin\omega\left(\tau_{kp} - \tau_{ki}\right) - \omega\cos\omega\left(\tau_{kp} - \tau_{ki}\right)} \mathrm{e}^{\tau_{kp} o} \sin\omega\tau_{ki}, \\ k_i &= \frac{\omega(2ao + b)\cos\omega\tau_{k_p} + \left[a\left(o^2 - w^2\right) + bo + c\right]\sin\omega\tau_{k_p}}{\omega\cos\omega\left(\tau_{kp} - \tau_{ki}\right) - o\sin\omega\left(\tau_{kp} - \tau_{ki}\right)} \left(o^2 + \omega^2\right) \mathrm{e}^{\tau_{ki} o}. \end{aligned} \tag{5}$$

It is noteworthy that the characteristic quasi-polynomial (4) exhibits at least one real root. By substituting $s = o$ into the characteristic quasi-polynomial, we derive

$$\forall k_p \in \mathbb{R} : k_i = -o\left(ao^2 + bo + c + k_p e^{-\tau_{k_p} o}\right) \mathrm{e}^{\tau_{k_i} o}. \tag{6}$$

The stability regions [8, 9, 11, 12] can be determined by utilizing the Eqs. (5) and (6). These equations provide a mathematical framework for analyzing the stability of a given system, and they are used to predict how the system will respond to various inputs.

2.3 Closed-Loop System Optimization

Once a point within the stable region is chosen, the control of the circuit becomes stable, but there is a possibility of oscillation or a longer stabilization delay. Thus, optimizing the system is essential. To achieve this, we utilize heuristic optimization techniques, specifically pattern search, Simulated annealing, and genetic algorithm, as they have demonstrated remarkable effectiveness in control-loop optimization.

The integral error is a significant metric in evaluating constant reference tracking performance [30]. Once a set of controller parameters is selected with a guarantee of closed-loop stability, the designer can create a solution for fast reference tracking. This is somewhat challenging as lower gains in the set produce slow transient responses, while higher gains result in excessively oscillatory responses. Integrals of the error, whether absolute or quadratic, with or without time weighting, are considered excellent cost functions due to their ability to contain stationary solutions. In this research, the chosen error metric is the integral of the absolute error, also known as IAE

$$\mathrm{IAE} = \int_0^{\infty} |e(t)|\mathrm{d}t. \tag{7}$$

As the IAE is dependent on the controller gains, the optimal IAE design can be achieved by solving the following optimization problem

$$\begin{aligned} &\min_{k_p, k_i} \mathrm{IAE} \\ &s.t.(k_p, k_i) \in B \end{aligned}. \tag{8}$$

The optimization problem is to locate a suitable box region B that intersects the stability region in charts in the (k_p, k_i) plane. As a result, the most cost-effective method of calculating the solution is through heuristic methods that do not require derivatives.

The paper's authors explored three heuristic approaches to solve the optimization problem presented in Eq. (8). They then compared the results obtained with each approach to illustrate their capabilities. Namely, the three methods were simulated annealing, genetic algorithm, and pattern search. The reader is referred to [25] for more detail.

The first approach, simulated annealing (SA) [25, 26], is a numerical heuristic optimization technique that is based on the principles of thermodynamics. The technique's name comes from the equation that calculates the probability, which corresponds to material-controlled cooling during annealing. The algorithm works by combining elements of local search with the Metropolis algorithm, which is the algorithm for accepting a worse solution. At each iteration of the simulated annealing algorithm, a new point is randomly generated, and if the generated candidate shows a better value of the criterion function, it replaces the original solution. Otherwise, the original solution is replaced by the generated candidate with only a certain probability.

The second approach is the genetic algorithm (GA) [25, 27], a widely used optimization technique based on the principles of genetics and mechanisms of natural selection. The algorithm modifies a population of individual solutions at each iteration. At each iteration, individuals from the current population are randomly selected and used as parents to produce children for the next generation. Over subsequent generations, the population converges toward an optimal solution.

The third approach is the pattern search algorithm (PS) [25, 28], which is based on searching the neighborhood of the criterion function at discrete points that are calculated according to a given pattern. The template is a set of directional vectors defined over the space of optimized parameters, which is the definition domain of the criterion function. At each iteration of the pattern search algorithm, the pattern either moves to the point that best minimizes its objective function or shrinks in size if no point is better than the current point until the expected accuracy has been achieved or the algorithm reaches a maximum number of predetermined iterations.

3 Numerical Section

This section uses two examples: controlling an inverted pendulum [29] and controlling stick-slip vibration [30]. These examples serve as benchmarks to verify the proposed approach. Two tables presenting the optimization results are provided below and will be referred to in the following subsections.

3.1 Numerical Example I

Consider a fixed delay $\tau_{k_i} = 0.5$s of the following system that describes the inverted pendulum

$$G(s) = \frac{1}{s^2 - 2}, \tag{9}$$

which has two real poles $s_1 = \sqrt{2}$ and $s_2 = -\sqrt{2}$. Since one root lies in a right half-plane, it is an unstable open-loop plant system; i.e., uncontrolled system is asymptotically unstable.

Now, considering the open-loop transfer function (9) and the controller (1), the characteristic quasi-polynomial $\Delta(s)$ of the closed-loop system can be computed as

$$\Delta(s) = s^3 + \left(k_p \mathrm{e}^{-\tau_{k_p} s} - 2\right)s + k_i \mathrm{e}^{-0.5s}. \tag{10}$$

Using Eqs. (5), we obtain the stabilization pair $\left(k_p, k_i\right)$:

$$\begin{aligned} k_p &= \frac{o\left(o^2 - 3\omega^2\right) - 2o + \omega\left[\left(3o^2 - \omega^2 - 2\right)\cot 0.5\,\omega\right]}{o\sin\omega\left(\tau_{kp} - 0.5\right) - \omega\cos\omega\left(\tau_{kp}, - 0.5\right)} \mathrm{e}^{\tau_{kp} o} \sin 0.5\omega, \\ k_i &= \frac{2o\omega\cos\omega\tau_{k_p} + \left(o^2 - w^2 - 2\right)\sin\omega\tau_{k_p}}{\omega\cos\omega\left(\tau_{kp} - 0.5\right) - o\sin\omega\left(\tau_{kp} - 0.5\right)} \left(o^2 + \omega^2\right) \mathrm{e}^{0.5o}. \end{aligned} \tag{11}$$

Using the stability region equations, we can graph the stability regions in the (k_p, k_i) plane.. Selecting a random point from stability region α may result in oscillatory response or eventual stabilization, which may be deemed unacceptable for industrial applications. Thus, optimizing the stable region is crucial for improving controller adjustment.

Now, we are going to investigate the stability region of the system for a fixed delay $\tau_{ki} = 0.5$ and gradually change the value of the delay parameter $\tau_{kp} \in \{0, 0.005, 0.01\}$.

Partial Evaluation of Regulation and Its Optimization

The results of the controller settings after optimizing the control loop can be seen in Table 1. Figures 2, 4, 6 show the area for optimization (yellow highlighted rectangle), a randomly selected point *P* used to compare optimized and non-optimized results, and finally, optimized point *O*. For reading the labels of the curves, we can read directly from the figure legend. Graphs 3, 5 and 7 display the closed-loop response at the initial point *P* before optimization and the optimized responses obtained via the simulated annealing, genetic algorithm, and pattern search optimization methods. Through simulations, it has been demonstrated that the selected heuristic optimization methods are suitable for stabilizing the unstable controlled system, and the PI delayed controller effectively achieves stabilization.

Table 1. Computed solutions with the heuristic methods – inverted pendulum.

Controller configuration	Start point & domain	Optimal solution		
		Method	k_p	k_i
$\tau_{k_p} = 0$ $\tau_{k_i} = 0.5$	$P[375, 450]$	*Simulated annealing*	354.5	772.2
	$k_p \in \langle 300, 450 \rangle$	*Genetic algorithm*	354.5	772.2
	$k_i \in \langle 100, 800 \rangle$	*Pattern search*	354.5	772.1
$\tau_{k_p} = 0.005$ $\tau_{k_i} = 0.5$	$P[350, 825]$	*Simulated annealing*	355.2	949.9
	$k_p \in \langle 340, 360 \rangle$	*Genetic algorithm*	355.0	928.9
	$k_i \in \langle 700, 950 \rangle$	*Pattern search*	355.1	950.0
$\tau_{k_p} = 0.01$ $\tau_{k_i} = 0.5$	$P[45, 65]$	*Simulated annealing*	40.0	65.1
	$k_p \in \langle 40, 50 \rangle$	*Genetic algorithm*	41.2	69.5
	$k_i \in \langle 50, 80 \rangle$	*Pattern search*	40.0	65.1

Option I $(\tau_{ki} = 0.5, \tau_{kp} = 0)$.

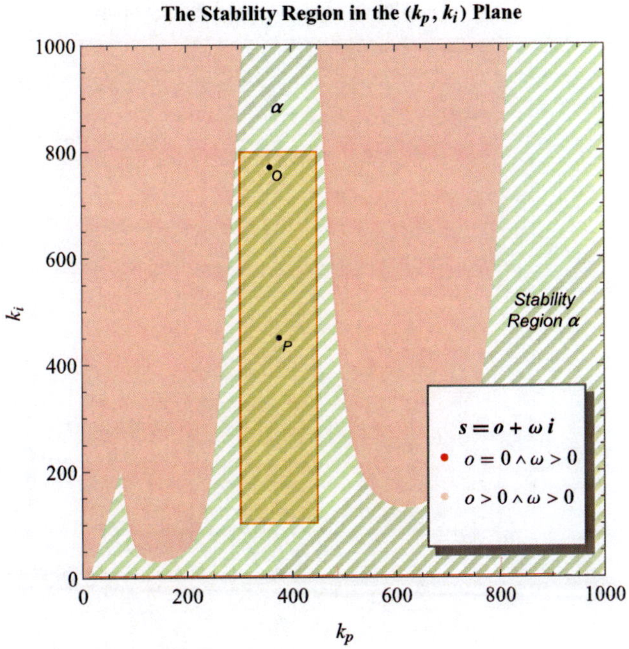

Fig. 2. The stability region of G for $\tau_{ki} = 0.5$ and $\tau_{ki} = 0$.

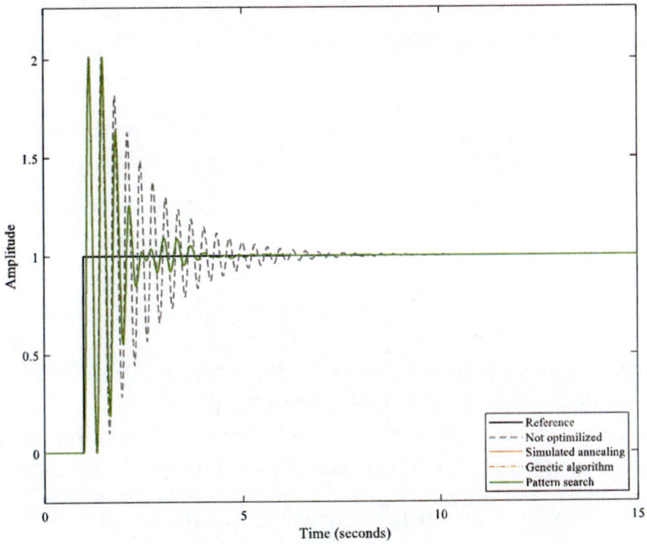

Fig. 3. The unit step response of the closed-loop system.

Option II $\left(\tau_{ki} = 0.5, \tau_{kp} = 0.005\right)$.

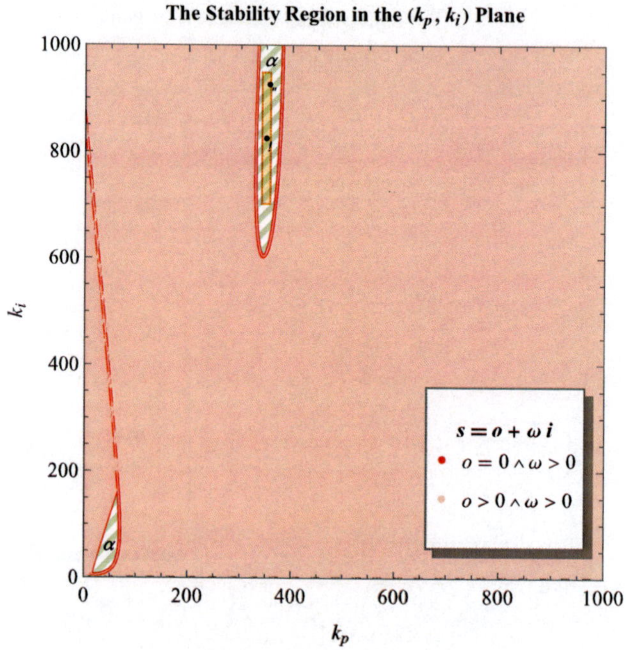

Fig. 4. The stability region of G for $\tau_{ki} = 0.5$ and $\tau_{ki} = 0.005$.

Option III $\left(\tau_{ki} = 0.5, \tau_{kp} = 0.01\right)$.

3.2 Numerical Example II

Consider a fixed delay $\tau_{k_i} = 0.25$s of the following system that describes the stick-slip vibration problem

$$G(s) = \frac{1}{s^2 - 0.1s + 10}, \tag{12}$$

which has two imaginary poles $s_1 = 0.05 - 3.16i$ and $s_2 = 0.05 + 3.16i$. Since one root lies in a right half-plane, it is an unstable open-loop plant.

Now, considering the open-loop transfer function (12) and the controller (1), the characteristic quasi-polynomial $\Delta(s)$ of the closed-loop system can be computed as

$$\Delta(s) = s^3 - 0.1s^2 + \left(k_p e^{-\tau_{k_p} s} + 10\right)s + k_i e^{-0.25s} \tag{13}$$

With Eqs. (5), we obtain the stabilization pair $\left(k_p, k_i\right)$:

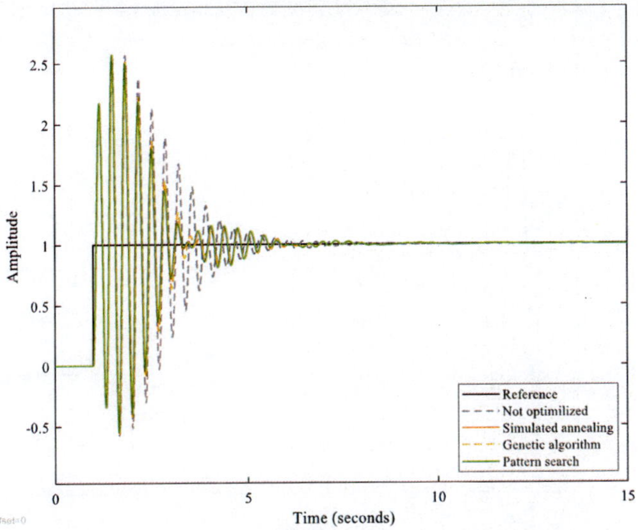

Fig. 5. The unit step response of the closed-loop system .

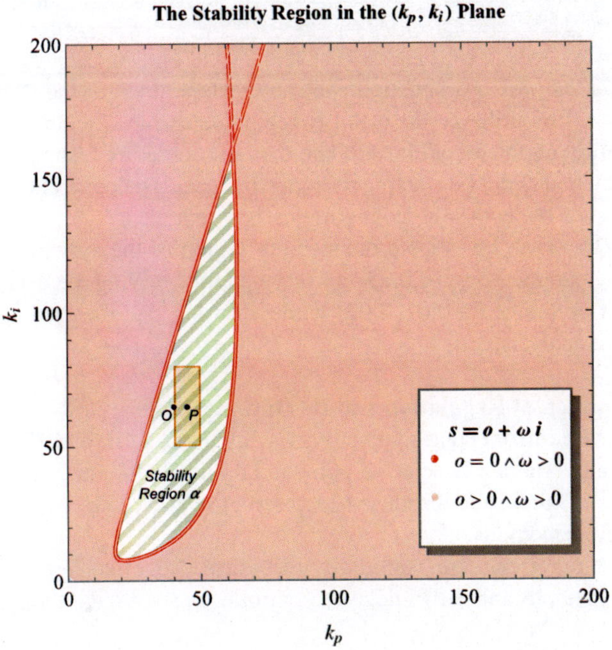

Fig. 6. The stability region of G for $\tau_{ki} = 0.5$ and $\tau_{ki} = 0.01$.

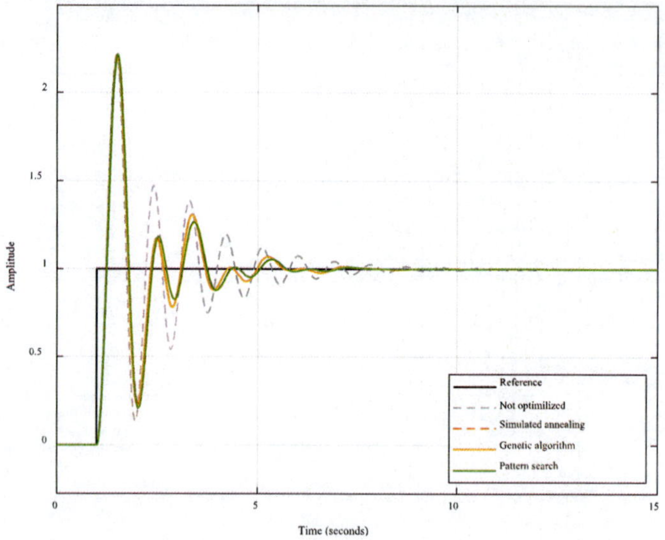

Fig. 7. The unit step response of the closed-loop system .

$$\begin{aligned} k_p &= \frac{10o - 0.1o^2 + o^3 + 0.1\omega^2 - 3o\omega^2 + \omega\left[(10 - 0.2o + 3o^2 - \omega^2)\cot 0.25\omega\right]}{o\sin\omega(\tau_{kp} - 0.25) - \omega\cos\omega(\tau_{kp} - 0.25)e^{\tau_{kp}o}\sin 0.25\omega}, \\ k_i &= \frac{\omega(2o - 0.1)\cos\omega\tau_{k_p} + (10 - 0.1o + o^2 - \omega^2)\sin\omega\tau_{k_p}}{\omega\cos\omega(\tau_{kp} - 0.25) - o\sin\omega(\tau_{kp} - 0.25)}\left(o^2 + \omega^2\right)e^{0.25o}. \end{aligned} \quad (14)$$

Just like in the previous situation, we can create a graph of the stability regions in the plane (k_p, k_i) by utilizing the stability region equations. Picking a point randomly from the stability region α could result in either an oscillatory response or eventual stabilization, which may not be suitable for industrial uses. Hence, optimizing the stability region to enhance the controller configuration is essential.

Our objective is to investigate the stability region of the system by maintaining a constant delay parameter τ_{ki} of 0.25 and subsequently varying the delay parameter τ_{kp} within the set of values {0, 0.01, 0.02} while ensuring that the system remains stable throughout the analysis.

Partial Evaluation of Regulation and Its Optimization

Table 2 shows optimized controller settings, while Figs. 8, 10 and 12 display the optimization area, the randomly selected point and optimized point. The labeled curves in Figs. 9, 11 and 13 represent the closed-loop response at the start and after optimization through heuristic methods.

As in the previous case, simulations confirm the effectiveness of the chosen optimization methods and the PI delayed controller in stabilizing the system.

Option I $(\tau_{ki} = 0.25, \tau_{kp} = 0)$

Table 2. Computed solutions with the heuristic methods – stick-slip vibration.

Controller configuration	Start point & domain	Optimal solution		
		Method	k_p	k_i
$\tau_{k_p} = 0$ $\tau_{k_i} = 0.25$	$P[175, 325]$ $k_p \in \langle 150, 200\rangle$ $k_i \in \langle 250, 400\rangle$	*Simulated annealing*	150.0	399.9
		Genetic algorithm	152.7	399.9
		Pattern search	150.0	400.0
$\tau_{k_p} = 0.01$ $\tau_{k_i} = 0.25$	$P[37.5, 475]$ $k_p \in \langle 125, 150\rangle$ $k_i \in \langle 400, 550\rangle$	*Simulated annealing*	127.9	481.0
		Genetic algorithm	128.0	480.9
		Pattern search	127.8	479.9
$\tau_{k_p} = 0.02$ $\tau_{k_i} = 0.25$	$P[136, 525]$ $k_p \in \langle 130, 142\rangle$ $k_i \in \langle 450, 600\rangle$	*Simulated annealing*	131.6	550.6
		Genetic algorithm	131.1	547.3
		Pattern search	132.3	555.9

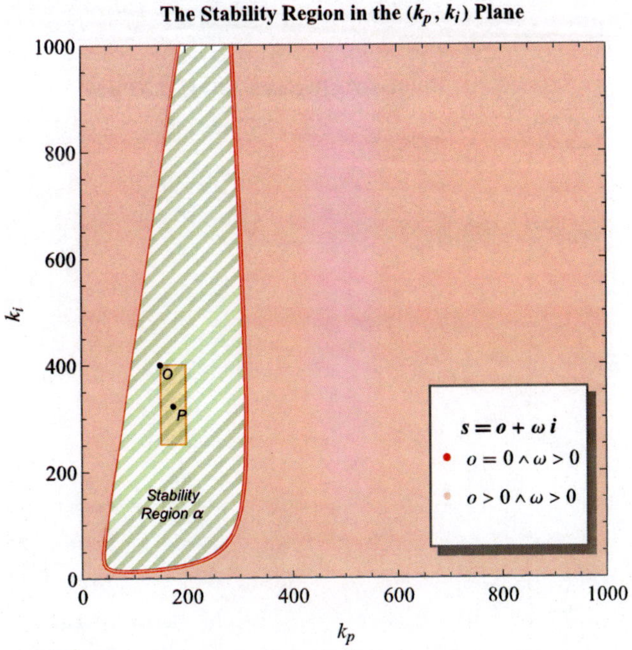

Fig. 8. The stability region of G for $\tau_{ki} = 0.25$ and $\tau_{ki} = 0$.

Option II $\left(\tau_{ki} = 0.25, \tau_{kp} = 0.01\right)$

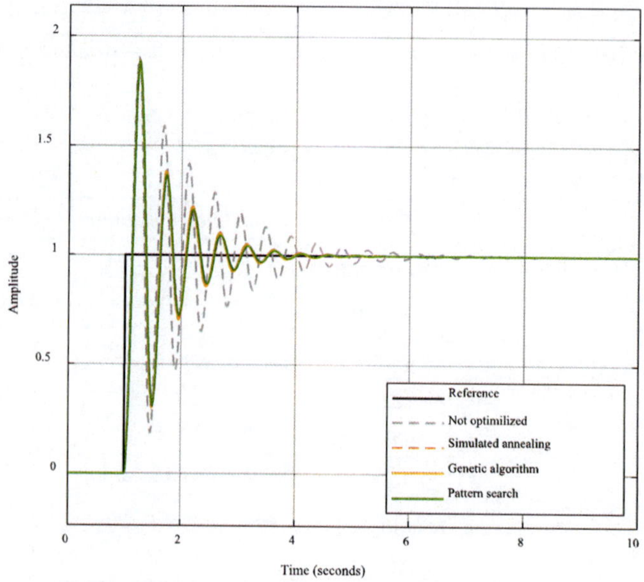

Fig. 9. The unit step response of the closed-loop system .

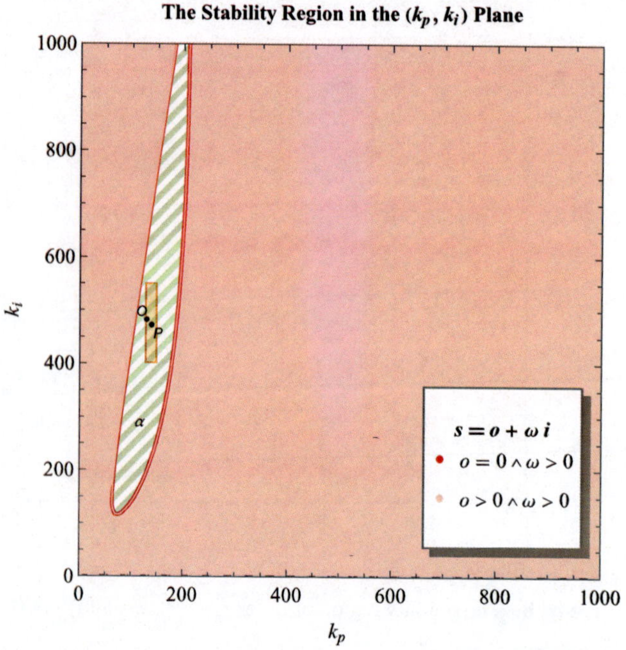

Fig. 10. The stability region of G for $\tau_{ki} = 0.25$ and $\tau_{ki} = 0.01$.

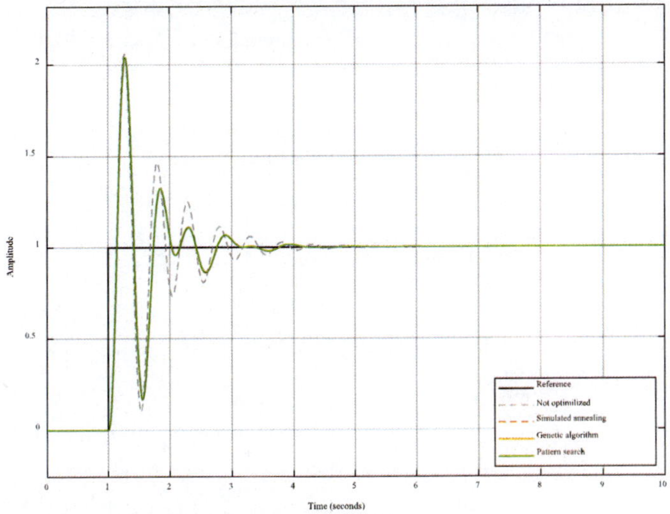

Fig. 11. The unit step response of the closed-loop system..

Option III $\left(\tau_{ki} = 0.25, \tau_{kp} = 0.02\right)$

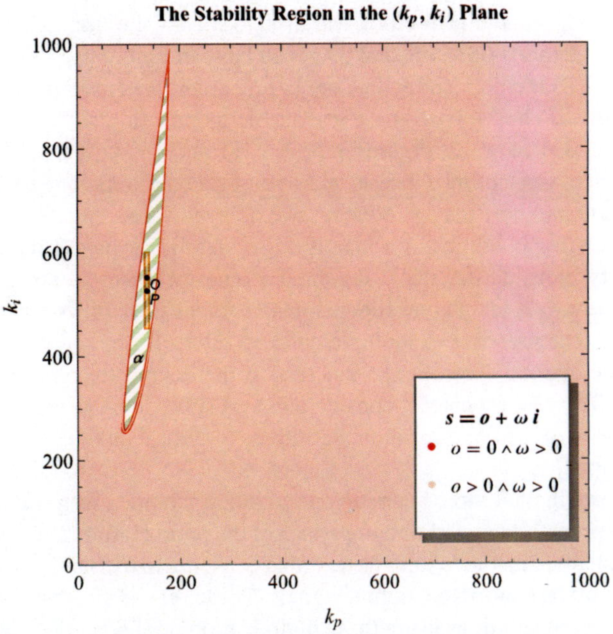

Fig. 12. The stability region of G for $\tau_{ki} = 0.25$ and $\tau_{ki} = 0.02$.

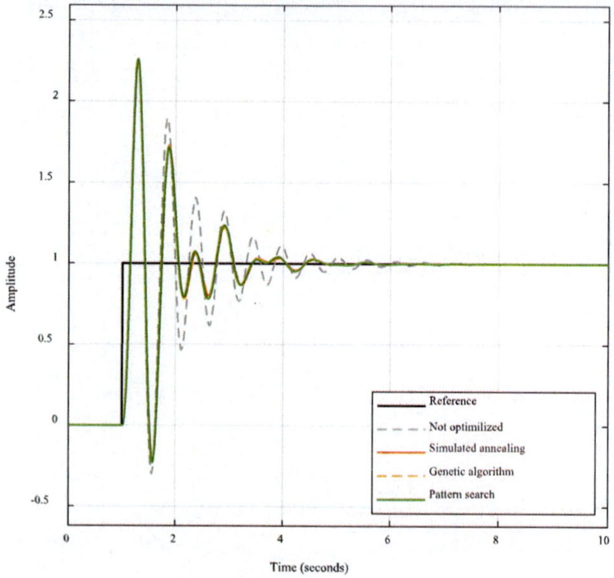

Fig. 13. The unit step response of the closed-loop system.

3.3 Results Discussion

The MATLAB Optimization Toolbox was used to solve the optimization problem using three heuristic methods (GA, PS, and SA) with built-in functions. A personal computer Acer Spin 3 (8 GB memory and Intel Core i3 processor) was used for computation. The box domain *B* was chosen based on a reasonable intersection with the stable regions α in Figs. 2, 4, 6, 8, 10 and 12. Figures 3, 5, 7, 9, 11 and 13 show the unit step response for the reference, non-optimized case, and optimized solutions for each method.

It is important to note that introducing two delayed values into the controller results in more challenging conditions for identifying a stabilizing pair. This can be observed in the stability plots, as the stable regions become increasingly smaller. As a result, finding an appropriate set of controller parameters that can stabilize the system becomes more difficult. The presence of multiple delayed values further complicates the task of identifying the optimal controller parameters, as the delay can significantly impact the system's stability. Hence, careful consideration must be given to the selection of delay values and the choice of optimization algorithms to ensure an effective control strategy can be developed.

We must say that all three heuristic optimization methods were initialized from the same starting point to ensure a fair comparison of the optimization. Despite their different approaches, all three methods were observed to converge to the same results, which were characterized by fast and stable output signals. These results suggest a high probability of obtaining global solutions using these optimization methods.

4 Conclusion

The use of PI controllers is a fascinating topic with various applications in different fields. This article aims to determine the stability regions for second-order systems, which is a complicated task if the closed-loop system involves one or more delays. For such cases, numerical mathematical methods are used to find the stability regions. These regions help us identify the suitable controller parameters and determine whether the system is asymptotically stable, marginally stable, or unstable. In our study, we have focused on second-order plant systems for stability regions analysis. The stabilization pair that corresponds to the graph in the global stability region simplifies the control requirements for the designer. We can extend this study to stabilization using higher-order polynomial plant systems or optimization methods where optimal parameters can be identified for setting the controller.

Acknowledgement. This study received financial support from the Czech Science Foundation through grant number GAČR 21-45465L.

References

1. Lyapunov, A.M.: The General Problem of the Stability of Motion, pp. 1–251. Kharkiv Mathematical Society, Kharkiv (1892)
2. Zadeh, L.A., Deoser, C.A.: Linear System theory: The State Space Approach. Dover Publications (2008)
3. Khalil, H.K.: Nonlinear Systems. Prentice Hall (2014)
4. Barnett, S.: Polynomials and linear control systems. Dekker (1983)
5. Hristuvarsakelis, D., Levine, W.S.: Handbook of Networked and Embedded Control Systems Birkhäuser, Boston (2005)
6. Xia, Y., Fu, M., Shi, P.: Analysis and Synthesis of Dynamical Systems with Time-Delays. Springer Berlin Heidelberg, Berlin, Heidelberg (2009)
7. Mahmoud, M.S.: An overview of time-delay control systems. In: Control Strategy for Time-Delay Systems, pp. 1–82. Elsevier (2021). https://doi.org/10.1016/B978-0-12-820599-0.00006-9
8. Wong, K., Lee, Y.E., Lee, H.J.: Optimal transmission of messages in computer networks – an optimal control problem involving control-dependent time-delayed arguments. J. Inequal. Appl. **2022**(1) (2022). https://doi.org/10.1186/s13660-022-02823-y
9. COVID-19 model via Caputo type fractional derivative. Mathematical Methods in the Applied Sciences (2020)
10. Muhammad, N., Mehdi, K., Guilhem, S., Sami, K., Mahdi, Z., Fouad, H.S.: Estimation of Time Delay caused by Point Geometry in Public Transport (2018)
11. Yu, D., Wang, G., Ding, Q., Li, T., Ya, J.: Effects of bounded noise and time delay on signal transmission in excitable neural networks. Chaos Solitons Fractals **157**, 111929 (2022)
12. Mirkin, L., Palmor, Z.J.: Control issues in systems with loop delays. In: Hristu-Varsakelis, D., Levine, W.S. (eds.) Handbook of Networked and Embedded Control Systems. Control Engineering. Birkhäuser Boston (2005)
13. Kennell, D., Talkad, V.: Messenger RNA potential and the delay before exponential decay of messages. J. Molecul. Biol. (1976)

14. Peng, J., Xiang, M., Wang, L., Xie, W., Sun, H., Yu, J.: Nonlinear primary resonance in vibration control of cable-stayed beam with time delay feedback. Mech. Syst. Signal Process. **137**, 106488 (2020)
15. Jnifene, A.: Active vibration control of flexible structures using delayed position feedback. Syst. Control Lett. **56**(3), 215–222 (2007)
16. Elmadssia, S., Saadaoui, K., Zaguia, A., Ezzedine, T., Wang, Q.-G.: Stabilization domains for second order delay systems. IEEE Access **9**, 53518–53529 (2021)
17. Strmiska, M., Pekař, L., Araújo, J.M.: Stabilization of second-order systems using a P/I-delayed controller. In: Silhavy, R., Silhavy, P., Prokopova, Z. (eds.) Software Engineering Application in Systems Design: Proceedings of 6th Computational Methods in Systems and Software 2022, Volume 1, pp. 531–539. Springer International Publishing, Cham (2023). https://doi.org/10.1007/978-3-031-21435-6_46
18. Strmiska, M., Araújo, J.M., Pekař, L.: Comparison of stabilization with P/I-delayed controllers for second-order systems using built-in MATLAB heuristic optimization methods. In: Silhavy, R., Silhavy, P., Prokopova, Z. (eds.) Software Engineering Application in Systems Design: Proceedings of 6th Computational Methods in Systems and Software 2022, Volume 1, pp. 652–662. Springer International Publishing, Cham (2023). https://doi.org/10.1007/978-3-031-21435-6_55
19. Busłowicz, M.: Sufficient conditions for instability of delay differential systems. Int. J. Control **37**(6), 1311–1321 (1983)
20. Pekař, L., Strmiska, M., Song, M., Dostálek, P.: Numerical gridding stability charts estimation using quasi-polynomial approximation for TDS, In: Proceedings of the 23rd International Conference on Process Control (PC), pp. 290–295 (2021)
21. Insperger, T., Stépán, G.: Stability charts for fundamental delay-differential equations. In: Insperger, T., Stépán, G. (eds.) Semi-Discretization for Time-Delay Systems: Stability and Engineering Applications, pp. 13–37. Springer New York, New York, NY (2011). https://doi.org/10.1007/978-1-4614-0335-7_2
22. Szaksz, B., Stepan, G.: Stability charts of a delayed model of vehicle towing. IFAC-PapersOnLine **54**(18), 64–69 (2021)
23. Lv, Z.-H., Zhang, J.-F., Ouyang, H.: Receptance-based computation of stability crossing curves for single-input-multiple-output second-order linear systems with two time-delays. Int. J. Struct. Stabil. Dyn. **22**(01) (2022). https://doi.org/10.1142/s021945542250002x
24. Vyhlídal, T., Olgac, N., Kučera, V.: Delayed resonator with acceleration feedback – complete stability analysis by spectral methods and vibration absorber design. J. Sound Vibrat. **333**(25), 6781–6795 (2014). https://doi.org/10.1016/j.jsv.2014.08.002
25. Floudas, C., Pardalos, P.: Encyclopedia of Optimization, 2nd edn. Springer, USA (2009)
26. What Is Simulated Annealing? MATLAB [cit. 2023-03-01]. https://www.mathworks.com/help/gads/what-is-simulated-annealing.html
27. Create generic algorithm options structure. MATLAB [cit. 2023–03–01]. https://www.mathworks.com/help/gads/psoptimset.html
28. Create pattern search options structure. MATLAB [cit. 2023–03–03]. https://www.mathworks.com/help/gads/psoptimset.html
29. Mirza, A., Hussain, S.: Robust controller for nonlinear & unstable system: inverted pendulum. AMSE J. Control Design Simul. **55**(3), 449–460 (2000)
30. Liu, W., Yang, F., Zhu, X., Chen, W.: Stick-slip vibration behaviors of BHA and its control method in highly-deviated wells. Alexandria Eng. J. **61**(12), 9757–9767 (2022). ISSN 1110-0168

The Problem of Large Local Fluctuations Appearance

Roman I. Dzerjinsky[1(✉)], Sergey V. Sidorov[2], and Timur E. Anosov[1]

[1] MIREA – Russian Technological University, 78, Vernadsky Pros, Moscow 119454, Russia
9015111295@mail.ru

[2] Peoples' Friendship University of Russia, Moscow, Russia

Abstract. The question of the appearance of fluctuations of physical quantities is considered on the basis of a hydrodynamic model. A possible mechanism for the formation of large fluctuations is discussed in the framework of a hydrodynamic model based on the Burgers equation. Solutions of the Burgers equation in a complex-valued space are considered. A transformation to a system of differential equations of the first order is carried out with a further study of the behavior of the system's solutions on the entire set of fixed points. An analysis is made of two different cases of system behavior for different values of the divergence of the vector field of the original system. The case of the occurrence of turbulence in a system with perturbations having a different type of solution is considered in more detail - in the form of a solitary wave, in the form of a switching wave, and in the form of a wave packet with many oscillations. A bifurcation mechanism for the appearance of turbulence is considered, based on two hypotheses - Landau's hypothesis and the hypothesis of strange attractors. It is shown that there are critical values of physical quantities at which their small fluctuations can generate large local fluctuations, which can be another mechanism for the appearance of dynamic chaos. #CSOC1120.

Keywords: Fluctuation Formation · Burger's Equation · Turbulence · Dynamic Chaos

1 Introduction

Fluctuations or random deviations of physical quantities from their average values occur in any quantities that depend on random factors. Although the fluctuations are negligible, the problem of fluctuations is not trivial. In particular, the small value of fluctuations does not exclude the possibility of significant local fluctuations in small regions of the system considered. Such large local fluctuations play a key role in explaining many important physical phenomena, such as the scattering of light, which determines, in particular, the color of the sky, the excitation of plasma oscillations, and so on. It is possible that large local fluctuations may be the reason of turbulence appearance.

Fluctuations are studied using a wide range of methods from statistical physics to quantum mechanics [1–7]. In particular, in [7], based on the quantum mechanical

R. Silhavy and P. Silhavy (Eds.): CSOC 2023, LNNS 723, pp. 573–580, 2023.
https://doi.org/10.1007/978-3-031-35317-8_51

model in the hydrodynamic approximation, the dynamics of fluctuation relaxation was studied. In this work, using the Sukhanov equation, obtained in the framework of the hydrodynamic description of the quantum mechanical model of the environment, the dynamics of individual fluctuations was studied. The authors have shown that each individual fluctuation behaves like a perturbation which decays with time in space. In other words, a local fluctuation can be considered as a dissipative damped wave. In this paper, we discuss a possible mechanism of large fluctuations formation considering a hydrodynamic model based on the Burgers equation.

2 Methods

Consider the solutions of the Burgers equation [8]

$$\frac{\partial w(x,t)}{\partial t} + w(x,t)\frac{\partial w(x,t)}{\partial x} = \mu\frac{\partial^2 w(x,t)}{\partial x^2} \tag{1}$$

with a complex-valued function

$$w(x,t) = u(x,t) + iv(x,t)$$

and with complex local viscosity $\mu = \alpha + i\beta$. Due to the fact that a fluctuation, which is a local perturbation, behaves like a wave, we consider the solution of Eq. (1) in the autowave approximation, that is, in the form of a solution to a running flat wave:

$$w(x,t) = \tilde{w}(x-ct) = \tilde{w}(\xi)$$

where $\xi = x - ct$, c is the perturbation propagation velocity in the environment. In this approximationwe have

$$\frac{\partial w(x,t)}{\partial x} = \frac{d\tilde{w}(x-ct)}{dx} = \frac{d\tilde{w}(\xi)}{d\xi}, \frac{\partial^2 w(x,t)}{\partial x^2} = \frac{d^2\tilde{w}(\xi)}{d\xi^2}, \frac{\partial w(x,t)}{\partial x} = -c\frac{d\tilde{w}(\xi)}{d\xi}$$

In this case, Eq. (1) can be transformed into a second-order ordinary differential equation

$$\mu w''(\xi) = (\tilde{w}(\xi) - c)\cdot \tilde{w}'(\xi) \tag{2}$$

where $\tilde{w}(\xi) = \tilde{u}(\xi) + i\tilde{v}(\xi)$. We represent the complex-valued Eq. (2) as a system of two real equations

$$\begin{aligned} \alpha\tilde{u}'' - \beta\tilde{v}'' &= (\tilde{u} - c)\tilde{u}' - \tilde{v}\tilde{v}', \\ \beta\tilde{u}'' + \alpha\tilde{v}'' &= (\tilde{u} - c)\tilde{v}' + \tilde{v}\tilde{u}'. \end{aligned} \tag{3}$$

For further research, it is convenient to write the system of two second-order Eqs. (3) as a system of first-order equations. To do this, we transform the system of Eqs. (3) to the following form

$$\begin{aligned} |\mu|^2\tilde{u}'' &= (\tilde{u} - c)\big(\alpha\tilde{u}' + \beta\tilde{v}'\big) + \tilde{v}\big(\beta\tilde{u}' - \alpha\tilde{v}'\big), \\ |\mu|^2\tilde{v}'' &= (\tilde{u} - c)\big(\alpha\tilde{v}' - \beta\tilde{u}'\big) + \tilde{v}\big(\alpha\tilde{u}' + \beta\tilde{v}'\big). \end{aligned}$$

and change the variables

$$\tilde{u} = u_1, \tilde{u}' = u_2, \tilde{\upsilon} = u_3, \tilde{\upsilon}' = u_4.$$

Then we obtain a system of four ordinary differential equations of the first order

$$\begin{aligned} u_1' &= u_2, \\ |\mu|^2 u_2' &= (u_1 - c)(\alpha u_2 + \beta u_4) + u_3(\beta u_2 - \alpha u_4), \\ u_3' &= u_4, \\ |\mu|^2 u_4' &= (u_1 - c)(\alpha u_4 - \beta u_2) + u_3(\alpha u_2 + \beta u_4). \end{aligned} \tag{4}$$

A feature of system (4) is that the fixed points are the set $\{u_1, 0, u_3, 0\}$–that is, the fixed points fill the entire plane (u_1, u_3). Let us estimate the nature of the stability of these fixed points. Note that the divergence of the vector field of the system (4)

$$divF(u) = 2\frac{\alpha(u_1 - c) + \beta u_3}{|\mu|^2}$$

depends on phase variables, and the plane

$$\alpha u_1 + \beta u_3 = \alpha c \tag{5}$$

divides the entire four-dimensional phase space of system (4) into two subspaces: into the area with positive divergence, where all fixed points are unstable, and the area with negative divergence - the dissipative area. According to Lyapunov stability, the nature of the fixed points in these regions is determined in the first approximation by the eigenvalues of the Jacobi matrix $J(u^*)$

$$J(u^*) = \begin{pmatrix} 0 & 1 & 0 & 0 \\ \frac{\alpha u_2^* + \beta u_4^*}{|\mu|^2} & \frac{\alpha(u_1^* - c) + \beta u_3^*}{|\mu|^2} & \frac{\beta u_2^* - \alpha u_4^*}{|\mu|^2} & \frac{\beta(u_1^* - c) - \alpha u_3^*}{|\mu|^2} \\ 0 & 0 & 0 & 1 \\ \frac{\alpha u_4^* - \beta u_2^*}{|\mu|^2} & \frac{\alpha u_3^* - \beta(u_1^* - c)}{|\mu|^2} & \frac{\alpha u_2^* + \beta u_4^*}{|\mu|^2} & \frac{\alpha(u_1^* - c) + \beta u_3^*}{|\mu|^2} \end{pmatrix}$$

in these points. Solving the equation

$$|J(u^*) - \lambda E| = 0$$

or

$$\begin{vmatrix} -\lambda & 1 & 0 & 0 \\ 0 & \frac{\alpha(u_1^* - c) + \beta u_3^*}{|\mu|^2} - \lambda & 0 & \frac{\beta(u_1^* - c) - \alpha u_3^*}{|\mu|^2} \\ 0 & 0 & -\lambda & 0 \\ 0 & \frac{\alpha u_3^* - \beta(u_1^* - c)}{|\mu|^2} & 0 & \frac{\alpha(u_1^* - c) + \beta u_3^*}{|\mu|^2} - \lambda \end{vmatrix} = 0$$

we get the characteristic equation

$$\lambda^2\left(\left(\lambda - \frac{\alpha(u_1^* - c) + \beta u_3^*}{|\mu|^2}\right)^2 + \frac{\left(\beta(u_1^* - c) - \alpha u_3^*\right)^2}{|\mu|^4}\right) = 0 \tag{6}$$

which determines the λ eigenvalues of theJacobi matrix at each point $u^* = \left(u_1^*, 0, u_3^*, 0\right)$ of four-dimensional space. So, the two eigenvaluesare equal to zero, and the other two are determined by the equation

$$\left(\lambda - \frac{\alpha(u_1^* - c) + \beta u_3^*}{|\mu|^2}\right)^2 + \frac{\left(\beta\left(u_1^* - c\right) - \alpha u_3^*\right)^2}{|\mu|^4} = 0 \tag{7}$$

from which we have

$$\lambda = \frac{\alpha(u_1^* - c) + \beta u_3^*}{|\mu|^2} \pm i \frac{\beta\left(u_1^* - c\right) - \alpha u_3^*}{|\mu|^2} \tag{8}$$

The eigenvalues in Eq. (8) are complex conjugate, except for those that lie in the plane

$$\beta u_1 - \alpha u_3 = \beta c$$

which is perpendicular to the plane (5). The real part of the eigenvalues in (8) is determined by the divergence $divF(u)$ of the vector field, and the sign of the real part coincides with the sign of the divergence.

3 Results

Let us study the behavior of solutions to system (4) on the entire set of fixed points. Let us pay attention to the following circumstance. If the coordinates of a fixed point are $\left(u_1^*, 0, u_3^*, 0\right)$, then the fluctuation Δu of the coordinate u_1 or u_3 with $u_2^* = u_4^* = 0$ leads to the fact that from the original fixed point $\left(u_1^*, 0, u_3^*, 0\right)$ we move to a fixed point $\left(u_1^* + \Delta u, 0, u_3^*, 0\right)$ [or a fixed point $\left(u_1^*, 0, u_3^* + \Delta u, 0\right)$]. As the fluctuation decays, the phase trajectory returns to the initial fixed point. At the same time, the nature of the flow movement remains unchanged, moreover, laminar. This case is of no interest.

The behavior of the solutions looks quite different if the fluctuation affects the coordinates u_2^* or u_4^*, which are derivatives of coordinates u_1^*, u_3^*. There are two possible cases.

1. The fluctuation of the coordinates u_2 (or u_4) occurs when the values of the variables u_1, u_3 lie in the dissipative region, where the solutions of system (4) are stable. In this case, the perturbations of the variables u_1 and u_3 are comparable with the magnitude of the fluctuation. The dynamics of these perturbations is described in [7] – immediately after they appear, they begin to decay, propagating in space in the form of a wave, and the system returns to its original state. If the variables u_1 and u_3 lie in the dissipative region, then small fluctuations of any coordinates do not affect the nature of the flow.
2. The fluctuation of the coordinates u_2 (or u_4) occurs when the values of the variables u_1, u_3 lie in a region with positive divergence of the vector field. It is numerically shown that in this case any small fluctuation of the variables u_2 (or u_4) leads to the appearance of huge perturbations of the variables u_1 and u_3, in comparison with the

magnitude of the initial fluctuation, which means that the small fluctuation leads to the appearance of large local fluctuations in the system.

The perturbation of variables u_1, u_3, which has arisen in a region with positive divergence, develops until the trajectory does not appear in the dissipative region, where it converges to stable fixed point. Trajectory has a spiral form due to the fact that the eigenvalues of the Jacobi matrix of system (4) are complex conjugate. In this case, the amplitude of perturbations is comparable with the value of the initial values of the variables u_1, u_3, but can significantly exceed them. The shape of the spatial perturbation (in fact, the wave) depends on the initial conditions, on the viscosity, on the ratio of the real and imaginary α and β parts of local coefficient of viscosity. It can have a different form, for example, a solitary wave, a soliton (Fig. 1), a switching wave (Fig. 2) or a wave packet with many oscillations (Fig. 3).

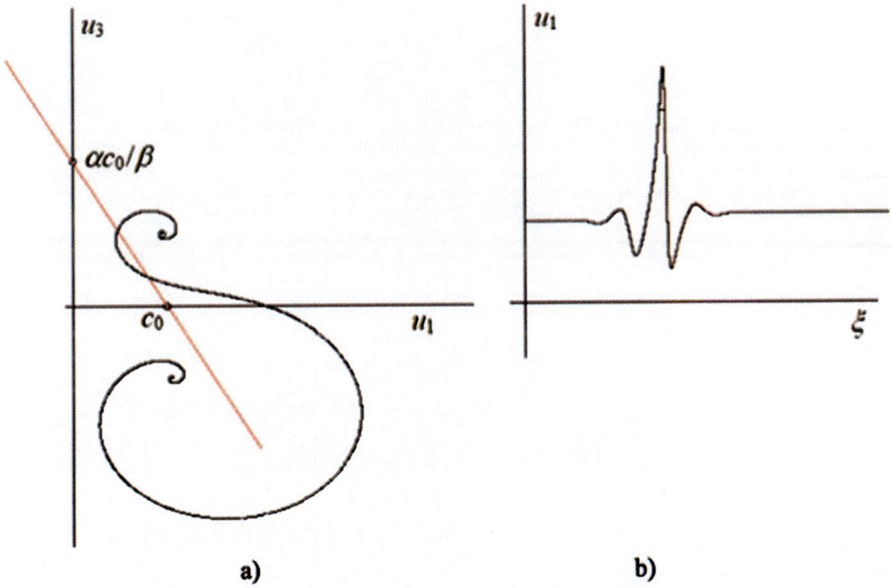

Fig. 1. Phase portrait (a) and solution in a form of solitary wave (b).

Small fluctuations of physical quantities in the flow (for example, velocity, pressure, density, temperature, and so on), which always take place, at certain low speeds do not have a noticeable effect on the nature of the flow as a whole. However, there are critical values of these quantities, in particular flow rates, at which small fluctuations can give rise to large local fluctuations of these quantities.

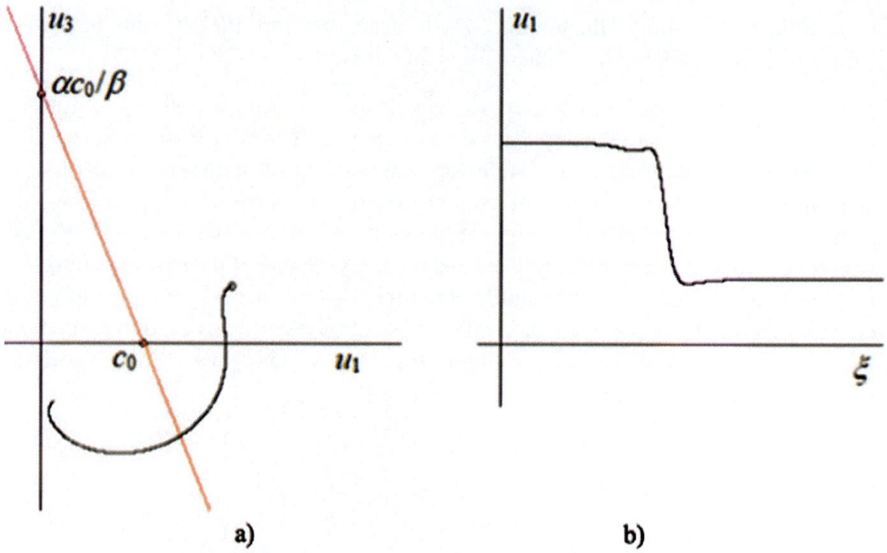

Fig. 2. Phase portrait (a) and solution in a form of switching wave (b).

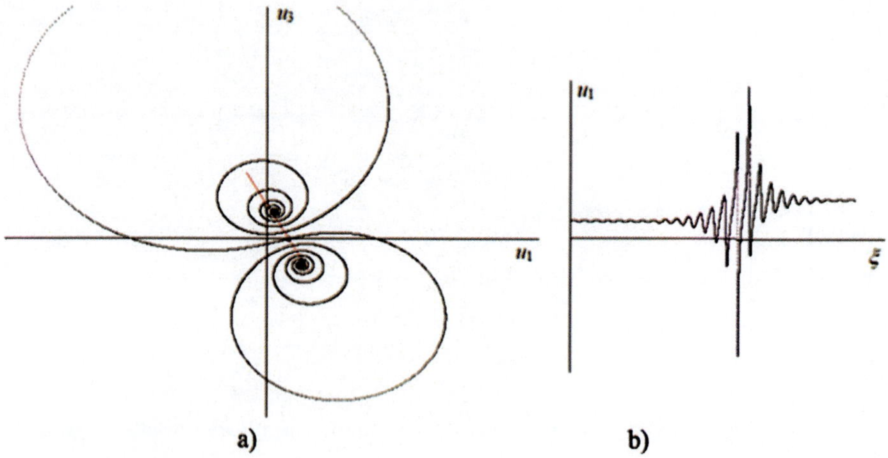

Fig. 3. Phase portrait (a) and solution in a form of wave packet (b).

4 Discussions

We note that the results of this work allow us to take a fresh look at the problem of turbulence. Until now, the bifurcation mechanism of turbulence appearance has been considered the most proven. This mechanism includes two hypotheses – the Landau hypothesis and the hypothesis based on the formation of strange attractors. According to Landau's hypothesis, the basic scenario of transition to turbulence is the birth bifurcation of the limit cycle - the Andronov-Hopf bifurcation. The solution complication occurs as a result of repeated bifurcations with a certain parameter change. First, after the birth of the limit cycle, a two-dimensional invariant torus is born, then at the next bifurcation, a three-dimensional invariant torus, then a four-dimensional torus, and so on until the birth of an n-dimensional invariant torus having n different independent harmonics, which leads at $n \to \infty$ to the chaotic behavior of the dynamic system.

The hypothesis of strange attractors is based on cascades of bifurcations of double-period limit cycle, as well as cascades of bifurcations with cycles having topological periods according to the Sharkovskii order [9]. In the works [10, 11] it was shown that these cascades take place not only for limit cycles, but also for two-dimensional invariant tori. In any case, the bifurcation mechanism creates an ultrahigh density of trajectories in a narrow region of the phase space, which leads to chaotic behavior of solutions in dynamical systems and, in particular, to turbulence.

5 Conclusion

We studied the solutions of the Burgers equation with a complex-valued function of fluid flow velocity and analyzed two different cases of system behavior for different values of the divergence of the vector field of the original system. Two main cases were considered - the case of damped oscillations, when small fluctuations of any coordinates do not affect the nature of the flow movement, and the case of the appearance of huge disturbances, in comparison with the magnitude of the initial fluctuation, that is, the appearance of large local fluctuations in the system. As a result, small fluctuations of physical quantities in the flow (for example, velocity, pressure, density, temperature, and so on), which always take place, at certain low speeds do not have a noticeable effect on the nature of the flow as a whole. However, there are critical values of these quantities, in particular flow rates, at which small fluctuations can give rise to large local fluctuations of these quantities.

References

1. Chattopadhyay, A., Dutta, S., Mukherjee, D.: Neetu, Quantum mechanics of Plancherel growth. Nucl. Phys. B **966**, 115368 (2021). https://doi.org/10.1016/j.nuclphysb.2021.115368
2. Diep, H.T., Desgranges, G.: Dynamics of the price behavior in stock markets: a statistical physics approach. Physica A **570**, 125813 (2021). https://doi.org/10.1016/j.physa.2021.125813
3. Gritsay, A.A.: Mathesis. Odessa, Ukraine (1924)
4. Jia, L., Ke, J., Wang, J.: Fluctuation behavior analysis of stochastic exclusion financial dynamics with random jump. Phys. A **542**, 123503 (2020). https://doi.org/10.1016/j.physa.2019.123503
5. Mandelshtam, L., Tamm, I.: The uncertainty relation between energy and time in nonrelativistic quantum mechanics. Izv. Akad. Nauk SSSR (ser. Fiz.) **9**, 122–128 (1945)
6. Melia, F.: Classicalization of quantum fluctuations at the Planck scale in the Rh = ct universe. Phys. Lett. B **818**, 136362 (2021). https://doi.org/10.1016/j.physletb.2021.136362
7. Rudoi, Y., Sukhanov, A.D.: Thermodynamic fluctuations within the Gibbs and Einstein approaches. Usp. Fiz. Nauk **170**(12), 1265–1296 (2000)

8. Burgers, J.M.: A mathematical model illustrating the theory of turbulence. Adv. Appl. Mech. **1**, 171–199 (1948)
9. Sharkovskii, A.N.: Coexistence of cycles of a continuous map of the line to itself. Int. J. Bifurcation Chaos **5**(5), 1263–1273 (1995)
10. Akhromeeva, T.S., Kurdyumov, S.P., Malineckiyj, G.G., Samarskiyj, A.A.: Nonstationary dissipative structures and diffusion-induced chaos in nonlinear media. Phys. Rep. **176**(5–6), 189–370 (1989)
11. Sidorov, S.V.: On the stability of the numerical simulation of periodic solutions of nonlinear differential equations. dynamics of heterogeneous systems. Proc. ISA RAS **11**, 78–84 (2007)

A Way to Visualize Disjunctive Disorders in Coal Seams Based on Seismic Data

Ivan Stepanov, Danila Shabanov(✉), and Leonid Burmin

Kemerovo State University, 6, Krasnaya Street, Kemerovo 650043, Russian Federation
alvarsondar@gmail.com

Abstract. Within the framework of this article, the relevance of the development of an information system for detecting disjunctive disorders of rocks as preventive work before mining coal underground is substantiated, and the process of developing a module designed to visualize disjunctive disorders of rocks is described. The article describes the problems of mining operations in conditions of complex occurrence of layers, analyzes the negative consequences from the point of view of economics and safety of the possible occurrence of faults in coal-bearing rocks when they are extracted from the bowels of the earth. There is a review of the main provisions of seismic exploration as a source of geological exploration data, the results of which are used as initial data, as they allow to clarify the structure of the formation and identify the location of disorders. An algorithm is proposed for working with the initial data of seismic exploration obtained in sgy format using an instrumental component, which is a software library of the Python programming language. The mathematical component of the algorithm of linear interpolation of spatial data using triangulation is also considered. This algorithm is used to visualize coal seams in three-dimensional space, which is a key element of the functionality of the module being disassembled. #CSOC1120.

Keywords: Information system · Disjunctive Disorders · Visualization

1 Introduction

Mining operations involve a significant number of safety and profitability hazards. An important negative factor affecting underground mining operations is the occurrence of disjunctive rock disorders. Disjunctive disorders are characterized by a discontinuity in the solid structure of rock bodies. Thus, during the formation of geological faults there is a displacement of rocks on the surface of the occurrence of a fault in the vertical or horizontal direction.

This displacement can cause significant difficulties in the mining process. In particular, there will be an urgent need to modify mining equipment settings to adapt to a non-continuous formation. This can lead to a long downtime for the equipment during the selection of necessary changes in its configuration, which will have a negative impact on the economic condition of the mining project, can lead to serious losses. These problems are of particular relevance when conducting mining operations in the Kuznetsk

R. Silhavy and P. Silhavy (Eds.): CSOC 2023, LNNS 723, pp. 581–586, 2023.
https://doi.org/10.1007/978-3-031-35317-8_52

coal basin, which is characterized by difficult conditions, expressed in discontinuities of rock structures, having a complexly folded shape [1].

If the fault characteristics are incorrectly determined and, as a consequence, improper setting of mining equipment, the efficiency of the mining process may fall, because some of the minerals may remain in the ground, unaffected by the equipment, as well as there may be damage to the equipment itself, which will lead to difficulties in its operation.

Based on the above, it can be noted that the presence of mining faults is a significant problem in the mining industry, which requires methods of preventive indication of disjunctive disorders. There are various ways of detecting disjunctive disorders. In this work we use the results of seismic exploration methods.

2 Method

Seismic exploration is based on the assumption that different rocks, due to their properties, have different velocities of elastic waves. In addition, the different properties of rocks lead to the fact that when the waves hit the interface between the two layers of the rock mass, they are reflected or refracted. Thus, by recording the time of arrival of the reflected waves, it is possible to determine their speed, and then compare it with the speed characteristic of a particular rock and on this basis to compose the structure of the Earth in a given area.

One of the main methods of seismic exploration is the method of reflected waves. The essence of the reflected wave method is to study the characteristics of waves reflected from the interface of media with different acoustic stiffness, calculated as the product of the density of the medium on the velocity of the elastic wave propagation in it [2] (Fig. 1).

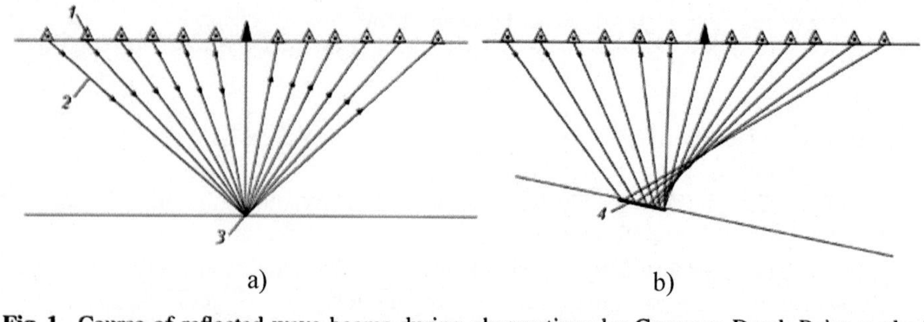

Fig. 1. Course of reflected wave beams during observations by Common Depth Point method [3]: a - horizontal boundary; b - sloping boundary; 1 - receiver; 2 - wave beam incoming from the source; 3 - common depth point; 4 - common reflective plate.

Registration of information in the framework of this method is carried out along the lines of geophysical profiles, sections of the surface accommodating an ordered interval-distributed system of seismic receivers. There is a configuration both with connection by wires and wireless variant. As wires, a seismic wire is used, a stranded wire, providing for the possibility of connecting a given number of seismic receivers, or geophones.

3 Results

An important tool accompanying the methods of detecting disjunctive disorders can be tools for visualization of the mining allotment in the process of seismic exploration, as well as mining strata and arisen geological faults. Within the framework of measures on foreign software import substitution it was decided to develop the domestic information system, allowing in particular to visualize the carbonaceous massif on the territory of the allotment on the basis of seismic data. The relevance of such a system is due to the fact that a very common format for working with vector data used in the mining industry is dwg-format (or its open analogue dxf). This format was developed by Autodesk for the AutoCAD system, but access to it was blocked for users from the Russian Federation.

The objects of visualization in this case will be seismic profiles, which are the lines of location of sensors receiving seismic values and rock strata with different variants of disjunctive disorders.

As input files for the visualization module can be fragments of an image of the mining site on the map or a schematic image in vector format of the mining allotment without reference to specific geospatial coordinates. The second option may be preferable because geospatial data may be subject to information security and represent an economic secret that is inadmissible for distribution.

The Cartesian coordinate system is used for orientation in space. On it, the user has to set the coordinates of the zero point, from which the coordinates of various elements to be applied to the vector image will later be deferred. Thus, the image elements can be transposed from the mining allotment scheme to the map with referencing to geospatial coordinates. Such a point may be, for example, the point of the beginning of mining operations on the surface of the earth or the point of the beginning of seismological research work. To map a point to a given coordinate, the user can either enter the coordinates from the zero point or, for a point on the same plane, specify the distance and the angle of the point. In this case, the coordinates of the desired point will be calculated as follows:

$$x = \cos(\alpha) \cdot d, \ y = \sin(\alpha) \cdot d$$

where α is the angle value, d is the distance to the point.

Seismic profile lines may already be plotted on the mine allotment plan, or may be plotted using appropriate module methods that use as input the coordinate of the beginning of the seismic profile line relative to the zero point, the length of the line, and the angle of incidence relative to the abscissa axis passing through the zero point. In both cases, it is required to schematically display the sensors on the profile line, in the specified coordinates. Sensor coordinates must be stored in two formats: coordinates in two-dimensional space relative to the zero point, and coordinates in one-dimensional space relative to the origin point of the seismic profile line. In this way, it will be possible to transfer to the image the data from the seismic sensors obtained as a result of seismic exploration.

The files obtained as a result of seismic operations are in a special SEG-Y format. The SEG-Y format provides storage of signal values obtained as a result of seismic exploration, as well as data on the nature of seismic operations in the format of binary headers common to all traces and separate for each trace [4]. As a result, it is possible to extract from the SEG-Y file the data about the number of sensors on the profile, receiving and recording signals, their distance from the point of wave generation.

Working with SEG-Y format data can be done, in particular, using the program library obspy of the Python programming language, which allows to read data from headers and values of seismic signals. It is also possible to open files in vector format, in particular in dxf format, and to edit them by applying various graphical primitives using the Python language, in particular using the ezdxf program library. Thus, Python and the above-mentioned libraries were chosen as a toolkit for implementing the visualization module.

By taking data from the seismic exploration files, it is possible to draw lines of seismic profiles, and on these lines to put sensors whose coordinates will be calculated based on the values of the above-mentioned headers. This will help visualize the seismic exploration process in detail.

Coal seams with associated disjunctive disorders are mapped using data on the depth of the coal seam relative to the seismic profile sensors. Analysis of the seismic exploration results helps to match the sensor coordinate and the depth of coal seam occurrence at that coordinate. Thus, a three-dimensional figure consisting of a set of polyhedrons is built, each of which represents an area of coal seam occurrence on the segment between two sensors on the seismic profile. The data from the seismic profiles also allows to determine in the area of which sensors the fault was detected. Visualization can reflect potentially dangerous areas of disjunctive disorders detected under certain sensors.

However, for detailed quality visualization it is not enough to display data from only one seismic profile, it is necessary to combine the information obtained from different profiles. In addition, due to various negative factors, the seismic exploration may not provide detailed depth data for all of the sensors in a seismic profile. In such a case, it is necessary to use interpolation methods to determine the depth of the coal seam at intermediate points where seismic profiling has not been conducted.

In this work, the triangulation method with linear interpolation was used to interpolate depth values at points in the three-dimensional space. This method is based on the division of interpolation space into a set of triangles, the vertices of which are used as given points with a known value of the desired value. This partition, called Delaunay triangulation, must obey two conditions: the minimum polygon covering all triangles must be convex, and the circle circumscribed around any constructed triangle must not fall inside any of the given points of triangulation [5].

Thus, to find out the value of the variable Z at the point (X_0, Y_0), which fell into the triangle formed by interpolation, with known coordinates of vertices, the following formula can be used [6]:

$$Z = \frac{-a * X_0 - b * Y_0 - d}{c},$$

$$a = Y_1(Z_2 - Z_3) + Y_2(Z_3 - Z_1) + Y_3(Z_1 - Z_2),$$

$$b = Z_1(X_2 - X_3) + Z_2(X_3 - X_1) + Z_3(X_1 - X_2),$$

$$c = X_1(Y_2 - Y_3) + X_2(Y_3 - Y_1) + X_3(Y_1 - Y_2),$$

$$d = X_1(Y_2Z_3 - Y_3Z_2) + X_2(Y_3Z_1 - Y_1Z_3) + X_3(Y_1Z_2 - Y_2Z_1),$$

where (X_1, Y_1), (X_2, Y_2), (X_3, Y_3) are coordinates of points of the triangle, which contains the point with the interpolated value; Z_1, Z_2, Z_3 are values of Z variable in the points of triangle.

It should be noted that this method is better to apply when the data is evenly distributed in the plane. In our case data are concentrated along profile lines, usually at a fixed distance from each other, so triangulation with linear interpolation seems to be the most optimal.

4 Discussion

So we obtained data on the depth of the coal seam beneath the sensors located on the seismic profile lines in the manner shown in Fig. 2, as well as the potential location of the faults relative to the sensors.

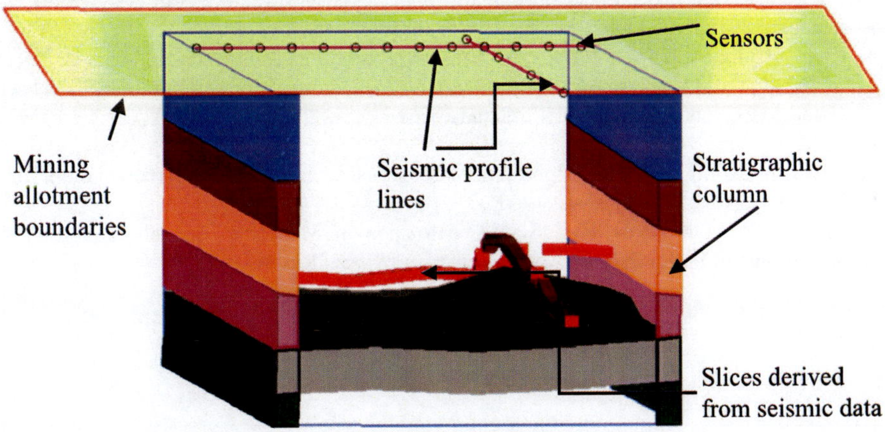

Fig. 2. Visualized coal seam.

To perform interpolation, a grid was built within the space bounded on the abscissa and ordinate axes by the lengths of the sensor locations on the seismic profiles. Then, using methods of the program library matpotlib of the Python language, triangulation was carried out for points with known values of the depth of occurrence. After that, using the same library, the depth values of the layer were calculated in the nodes of the grid, and the layer was displayed as a set of polygons with vertices in the nodes of the grid. In addition, potential fault zones were mapped under the given sensors. The results of interpolation make it clear that in these areas there is a violation of the monotonic emplacement of the seam, from which we can conclude that these zones may indeed be zones of disjunctive disorders.

5 Conclusion

In this work, we have demonstrated that a well executed disjunctive disorder identification can be an important factor in the success of mining operations, as the presence of faults can have a direct impact on the efficiency of the mining process. Visualization of the results will allow to visualize more clearly the situation on the site of underground mining works, which should have a positive impact on the efficiency of the mining planning process, taking into account the identified irregularities. In this regard, we proposed an algorithm for visualizing disjunctive disorders, which includes working with sgy format files using the functionality provided by the Python ezdfx software library, as well as a method of triangulation with linear interpolation, which allows to restore the formation structure in the area of mining operations.

References

1. Zhukov, E.M., Kropotov, Y., Luginin, I.A., Poloshkov, S.I.: Relevance of seismic works for prediction of anomalous zones in the coal massif during detailed exploration of coal deposits. Young Sci. **3**(107), 95–98 (2016)
2. Kozlovskiy, E.A.: Rock Encyclopedia. Sov. Encyclopedia, Moscow, USSR (1984–1991)
3. Khmelevsky, V.K.: Geophysics. "KDU", "Dobrosvet", Moscow, Russia (2018)
4. Rune, H., Levin, S.A.: SEG-Y_r2.0: SEG-Y revision 2.0 Data Exchange format. Society of Exploration Geophysicists, Tulsa, OK, US (2017). https://seg.org/Portals/0/SEG/News%20and%20Resources/Technical%20Standards/seg_y_rev2_0-mar2017.pdf. Accessed 0 March 2022
5. Skvortsov, A.V., Mirza, N.S.: Algorithms of Construction and Analysis of Triangulation. Tomsk University Press, Tomsk, Russia (2006)
6. Maltsev, K.A., Mukharamova, S.S.: Construction of Models of Spatial Variables (With Application of Surfer Packet). Kazan University, Kazan, Russia (2014)

Applying Ant Colony Optimisation When Choosing an Individual Learning Trajectory

Rukiya Deetjen-Ruiz[1(✉)], Oleg Ikonnikov[2,3], Shahzool Hazimin Azizam[4], Darío Salguero García[5], Juan Carlos Orosco Gavilán[6], Ivan Otcheskiy[7], and Roman Tsarev[8,9]

[1] Zayed University, Abu Dhabi, United Arab Emirates
officialrukiyadeetjen@gmail.com
[2] Siberian Federal University, Krasnoyarsk, Russia
[3] Reshetnev Siberian State University of Science and Technology, Krasnoyarsk, Russia
[4] Universiti Sultan Zainal Abidin, Terengganu, Malaysia
[5] Almería University, Almería, Spain
[6] Universidad Privada del Norte, Lima, Perú
[7] Tyumen State University, Tyumen, Russia
[8] MIREA - Russian Technological University (RTU MIREA), Moscow, Russia
[9] Bauman Moscow State Technical University, Moscow, Russia

Abstract. The effectiveness of learning depends in many ways on the organization of the educational process. These days, the educational environment is becoming more flexible and responsive to the needs of students. The traditional form of learning is expanding through the use of new approaches and teaching methods, learning systems, and information technology. Individual learning trajectory allows the learner to regulate the order of studying course modules and the pace of mastering the subject material. Decision-making in choosing the learning trajectory can be supported by specialized methods and tools. This paper proposes the use of ant colony optimisation to support decision making on the choice of an individual learning trajectory. #CSOC1120.

Keywords: Ant Colony Optimisation · Individual Learning Trajectory · Algorithm · Graph · Path · E-learning

1 Introduction

Modern education is characterized by rapid growth of informatization, active use of information and communication technologies in teaching [1–5]. Digitalization of education is based on the use of technologies that promote individualization of learning, and thus effectively improve the quality of students' knowledge [6–8]. The problem of individualization of learning to this day is a burning issue in e-learning and pedagogy in general [6, 9].

In order to implement the principle of individualization of educational activities, a teacher needs to organize students' independent activities, build individual learning

R. Silhavy and P. Silhavy (Eds.): CSOC 2023, LNNS 723, pp. 587–594, 2023.
https://doi.org/10.1007/978-3-031-35317-8_53

trajectories using modern information and communication technologies based on the functioning of learning management systems, and educational organizational support services [10–12]. The creation of digital educational tools with modular courses as independent learning is an effective way to organize individual education [10, 13, 14]. Often such learning takes place in a mode that allows students to design their own educational path within the framework of the digital educational environment.

Researchers have different interpretations of the concept of individual learning trajectories [15–17]. The definition of an individual educational trajectory provides for an individual educational path, i.e. the sequence of mastering the content and its components, and the developed way of its implementation, i.e. the technology of organizing the educational process.

In addition, researchers consider the abilities and interests of the individual, the way they perceive information, learning by individual trajectories. Thus, the ability to construct individual learning trajectories is due to differences in cognitive styles, that is, memory, thinking, and perception of information. That is why it is necessary to remember the person-centered approach to learning when constructing individual learning trajectories (focus on a certain student with his/her abilities and peculiarities of perception and learning of information).

Thus, an individual learning trajectory, as a form of educational activity organization, is a personal way for a particular student to achieve a set educational objective [18]. Let us consider the functions of a student in the process of learning on an individual trajectory in details:

- the learner's awareness of the benefits of learning the material on the individual trajectory, the desire for self-education, and the need for creativity;
- familiarization with the expected learning outcomes;
- determination of the aspects of the studied topics that are of interest to students;
- choice of forms and methods of learning, the pace of material learning;
- correction of the chosen individual learning trajectory, control of learning achievements, reflection.

Despite the high degree of independence and self-education of students, the role and functions of the teacher in the process of constructing an individual learning trajectory should not be neglected. Thus, the functions of the teacher include:

- motivating the student in constructing individual learning trajectories;
- familiarizing the student with the expected learning outcomes;
- providing the choice of content, methods, forms and pace of learning within the curriculum, methods of control;
- implementation of monitoring and reflection.

As the students gain experience in choosing an individual trajectory, the teacher's role decreases but does not disappear.

Thus, individual learning trajectories may vary in the degree of independence, level of complexity, timing, and pace, but to create an optimal effective course with the student's choice of an individual learning trajectory, the teacher also needs to have a number of professional skills:

- determine the initial level of students' knowledge, their characteristics, and degree of interest in order to create an individual trajectory in accordance with the learning objective;
- create an information and education environment, design and develop modular courses, ensure management of digital resources and their automation, in which the individual learning trajectories of students are implemented;
- create diagnostic and controlling materials (e-tests) to evaluate the results of learning on the selected individual learning trajectories, its improvement and, as a consequence, the provision of counseling and reflection activities [19–21];
- construct individual learning trajectories for both talented students and students with special needs, to ensure that everyone has a choice based on their individual abilities, interests, and needs;
- introduce information and communication technologies as part of learning with the construction of an individual learning trajectory;
- form research and creative skills with the help of visualization tools for students in the work on individual learning trajectories.

An individual trajectory is constructed with the help of an additional set of methodological elements: the choice of methods, forms, and tools for learning content. To support the decision-making about the choice of an individual learning trajectory we propose to use the ant colony optimisation algorithm.

2 Method

Ant colony optimisation has proven itself in solving a number of problems presented on the graph [22–26]. This technique is based on the behavior of an ant colony when searching for food and delivering it to the nest [27–30]. The path an ant takes is marked with a special substance, pheromone, which allows the ants following it to determine the direction of its movement. The presence of a pheromone on a particular path does not uniquely determine the direction of the next ant, but increases the probability that it chooses this direction. This approach makes the ant colony optimisation flexible, robust, and allows to find the optimal solution. The probability of moving in a certain direction, namely from vertex i to vertex j, is defined as follows [27]:

$$p_{ij} = \frac{\eta_{ij}^{\beta} \tau_{ij}^{\alpha}}{\sum\limits_{\substack{k \in \text{ available} \\ \text{edges}}} \eta_{ik}^{\beta} \tau_{ik}^{\alpha}},$$

where $\eta_{ij} = \frac{1}{L_{ij}}$, L_{ij} is the distance, η_{ij} is the visibility. The smaller the distance the higher the visibility;

τ_{ij} is the concentration of pheromone on the edge (i, j);

α is the significance of pheromone;

β is the significance of visibility;

α and β are constants.

After an ant has passed over the graph vertices, the concentration of pheromone changes. First, the ant left the pheromone behind on the path it took, and the pheromone concentration increased:

$$\tau_{ij} = \tau_{ij}^{\text{previous}} + \Delta\tau,$$

where Δτ is the change in pheromone concentration: $\Delta\tau = \frac{Q}{L_i}$, Q is a constant; L_i is the length of the path taken by the ant (sum of weights of edges).

In addition, as in nature, with time the concentration of pheromone on each path decreases due to evaporation. This makes it possible to find the optimal path due to the fact that the importance of non-optimal path, determined by the concentration of pheromone, decreases. Pheromone evaporation changes the concentration of pheromone as follows:

$$\tau_{ij} = (1 - \rho)\tau_{ij}^{\text{previous}},$$

where ρ is the rate of pheromone evaporation.

The ant colony optimisation algorithm is executed iteratively. The number of iterations, as well as the values of constants, is set by the researcher.

In this paper the algorithm is used to choose individual learning trajectory. The vertices of the graph are modules of an educational course, the relationships between which are established on the basis of the choice of course modules by previous students. For each student with the help of an ant colony optimisation algorithm an individual learning trajectory is built, which is a sequence of course modules (the optimal path in the graph). To implement the ant algorithm, a vertex attendance factor is introduced, representing students' interest in a particular module. The pheromone concentration on each edge of the graph is updated on the basis of the intensity of visiting vertices and passing through edges adjacent to these vertices.

3 Results

When selecting the individual learning trajectory, the ant colony optimisation is based on the importance of vertices, determined by the concentration of pheromone on adjacent edges. Initially, the concentration of pheromone on all edges is taken equal to one $\tau_{ij} = 1$ for all (i, j). Later, the concentration of pheromone changes and when the next vertex is selected, the pheromone concentration changed due to the students who previously studied the course affects the decision. The higher the pheromone concentration associated with the next vertex, the more likely that vertex will be selected.

Figure 1 shows the individual learning trajectory generated for one of the students. The course contains 16 modules, and from the Fig. 1, one can see that, based on the student's abilities and interests, he is offered a certain sequence of modules, different from their linear succession from the first module to the sixteenth.

The construction of an individual learning trajectory has a number of advantages:

- each learner can choose an individual, appropriate to his abilities, solution to the set learning tasks, and achievement of the goal;

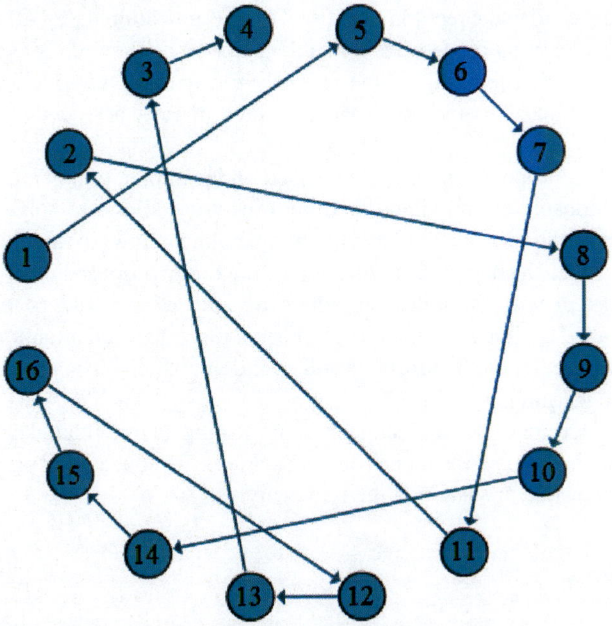

Fig. 1. Individual learning trajectory of a student.

- the opportunity to choose the level of complexity of the material studied, the implementation of their potential for students with different levels of progress;
- choice of deadlines and a suitable pace for mastering the material;
- organization of self-control, reflection, and correction of the individual learning trajectory to achieve the required result;
- contribution to the maximum involvement of the student in the study of the material, and increases the motivation for learning in general.

Thus, the results of applying the ant colony optimisation algorithm when choosing an individual learning trajectory, presented by us, guarantee the success of mastering the course. Self-management of the process and visibility contribute to improving the quality of education and individual results of students.

4 Conclusion

Nowadays, information technology is increasingly impacting the educational environment. It is already impossible to imagine modern education without Learning Management Systems, Massive Open Online Courses, various methods and tools designed to organize learning and support the educational process. One of the advances in education is the teaching of students on an individual trajectory, which has now become available to a wide range of learners. Determining an individualized learning trajectory is an important objective in terms of learning effectiveness and requires a particular focus. The use of methods and tools that can support decision making in choosing an individualized learning trajectory is an essential issue. These methods and tools can automate the selection process and make recommendations for the educational process in terms of creating individual learning trajectories for students based on their abilities and needs. The important point is that learning on the properly formed individual trajectory increases the involvement and motivation of students, which directly affects the effectiveness of learning.

The approach proposed in this paper, based on the application of ant colony optimisation when choosing an individual learning trajectory, allows to solve the mentioned problems. The experience accumulated by the ant colony allows to find the optimal learning trajectory for each student. Studying according to the proposed individual trajectory is more consistent with the individual characteristics of the student than the general approach for all. This allows the student to keep a more focused attention on the course being studied, to feel more comfortable and confident, which ultimately leads to a better mastery of the learning material.

This approach can be developed in terms of forming an individual learning trajectory not only within the framework of one course, but the entire educational program, covering the period of the student's study at the university.

References

1. Neffati, O.S., et al.: An educational tool for enhanced mobile e-Learning for technical higher education using mobile devices for augmented reality. Microprocess. Microsyst. **83**, 104030 (2021). https://doi.org/10.1016/j.micpro.2021.104030
2. Prasad, V.N., Muthappa, A.K.K.: An efficient framework for the similarity prediction with query recommendation in E-learning system. Concurr. Comput. Pract. Exper. **34**(22), e7145 (2022). https://doi.org/10.1002/cpe.7145
3. Suri, P.A., Syahputra, M.E., Amany, A.S.H., Djafar, A.: Systematic literature review: the use of virtual reality as a learning media. Procedia Comput. Sci. **216**, 245–251 (2023). https://doi.org/10.1016/j.procs.2022.12.133
4. Tsarev, R.: Application of Kahoot! in education gamification. Int. J. Adv. Stud. **7**(1), 9–17 (2017). https://doi.org/10.12731/2227-930X-2017-1-9-17
5. Ung, L.-L., Labadin, J., Mohamad, F.S.: Computational thinking for teachers: development of a localised E-learning system. Comput. Educ. **177**, 104379 (2022). https://doi.org/10.1016/j.compedu.2021.104379
6. Dietrich, J., Greiner, F., Weber-Liel, D., Berweger, B., Kämpfe, N., Kracke, B.: Does an individualized learning design improve university student online learning? A randomized field experiment. Comput. Hum. Behav. **122**, 106819 (2021). https://doi.org/10.1016/j.chb.2021.106819
7. Kotsyuba, I., Shikov, A., Karpik, K., Galperin, M., Kudriashov, A., Silko, J.: Recommendation web service for choosing an individual educational path in the field of transportation systems' programming. Transp. Res. Procedia **63**, 591–599 (2022). https://doi.org/10.1016/j.trpro.2022.06.052
8. Meirbekov, A., Maslova, I., Gallyamova, Z.: Digital education tools for critical thinking development. Thinking Skills Creat. **44**, 101023 (2022). https://doi.org/10.1016/j.tsc.2022.101023
9. Tsarev, R.Y., et al.: An approach to developing adaptive electronic educational course. In: Silhavy, R. (ed.) CSOC 2019. AISC, vol. 986, pp. 332–341. Springer, Cham (2019). https://doi.org/10.1007/978-3-030-19813-8_34

10. Hassouni, B.E., et al.: Realization of an educational tool dedicated to teaching the fundamental principles of photovoltaic systems. J. Phys. Conf. Ser. **1399**(2), 022044 (2019). https://doi.org/10.1088/1742-6596/1399/2/022044
11. Ibrahim, T.S., Saleh, A.I., Elgaml, N., Abdelsalam, M.M.: A fog based recommendation system for promoting the performance of E-Learning environments. Comput. Electr. Eng. **87**, 106791 (2020). https://doi.org/10.1016/j.compeleceng.2020.106791
12. Nikolaeva, I., Sleptsov, Y., Gogoleva, I., Mirzagitova, A., Bystrova, N., Tsarev, R.: Statistical hypothesis testing as an instrument of pedagogical experiment. AIP Conf. Proc. **2647**, 020037 (2022). https://doi.org/10.1063/5.0104059
13. Joy, J., Pillai, R.V.G.: Review and classification of content recommenders in E-learning environment. J. King Saud Univ. Comput. Inform. Sci. **34**(9), 7670–7685 (2022). https://doi.org/10.1016/j.jksuci.2021.06.009
14. Membrive, A., Silva, N., Rochera, M.J., Merino, I.: Advancing the conceptualization of learning trajectories: a review of learning across contexts. Learn. Cult. Soc. Interact. **37**, 100658 (2022). https://doi.org/10.1016/j.lcsi.2022.100658
15. Arias, A.M., Davis, E.A.: Supporting children to construct evidence-based claims in science: Individual learning trajectories in a practice-based program. Teach. Teach. Educ. **66**, 204–218 (2017). https://doi.org/10.1016/j.tate.2017.04.011
16. Cuevas-Vallejo, A., Orozco-Santiago, J., Paz-Rodríguez, S.: A learning trajectory for university students regarding the concept of vector. J. Math. Behav. **70**, 101044 (2023). https://doi.org/10.1016/j.jmathb.2023.101044
17. Poell, R.F., Van der Krogt, F.J.: An empirical typology of hospital nurses' individual learning paths. Nurse Educ. Today **34**(3), 428–433 (2014). https://doi.org/10.1016/j.nedt.2013.04.005
18. Sysoyev, P.V.: Individual education path. Sci. Educ. **1**, 121–131 (2014)
19. Aljarbouh, A., et al.: Application of the K-medians clustering algorithm for test analysis in E-learning. Lect. Notes Networks Syst. **596**, 249–256 (2023). https://doi.org/10.1007/978-3-031-21435-6_21
20. Moravec, T., Štěpánek, P., Valenta, P.: The influence of using e-learning tools on the results of students at the tests. Procedia. Soc. Behav. Sci. **176**, 81–86 (2015). https://doi.org/10.1016/j.sbspro.2015.01.446
21. Pavlenko, D., Barykin, L., Nemeshaev, S., Bezverhny, E.: Individual approach to knowledge control in learning management system. Procedia Comput. Sci. **169**, 259–263 (2020). https://doi.org/10.1016/j.procs.2020.02.162
22. Benabdellah, N.C., Gharbi, M.: Ant colont algorithm and new pheromone to adapt units sequence to learners' profiles. Int. J. Comput. Sci. Appl. **12**(1), 26–39 (2015)
23. Chernigovskiy, A.S., Kapulin, D.V., Noskova, E.E., Yamskikh, T.N., Tsarev, R.Y.: Production scheduling with ant colony optimization. IOP Conf. Ser. Earth Environ. Sci. **87**(6), 062002 (2017). https://doi.org/10.1088/1755-1315/87/6/062002
24. Bernard, J., Popescu, E., Graf, S.: Improving online education through automatic learning style identification using a multi-step architecture with ant colony system and artificial neural networks. Appl. Soft Comput. **131**, 109779 (2022). https://doi.org/10.1016/j.asoc.2022.109779
25. Wu, Y., Gong, M., Ma, W., Wang, S.: High-order graph matching based on ant colony optimization. Neurocomputing **328**, 97–104 (2019). https://doi.org/10.1016/j.neucom.2018.02.104
26. Wu, L., Huang, X., Cui, J., Liu, C., Xiao, W.: Modified adaptive ant colony optimization algorithm and its application for solving path planning of mobile robot. Expert Syst. Appl. **215**, 119410 (2023). https://doi.org/10.1016/j.eswa.2022.119410
27. Dorigo, M.: Optimization, Learning and Natural Algorithms. Politecnico di Milano, Milan, Italy (1992)

28. Dorigo, M., Stützle, T.: Ant Colony Optimization. MIT Press, Cambridge, US (2004)
29. Mohan, B.C., Baskaran, R.: A survey: ant Colony Optimization based recent research and implementation on several engineering domain. Expert Syst. Appl. **39**(4), 4618–4627 (2012). https://doi.org/10.1016/j.eswa.2011.09.076
30. Neto, N.R.F., Filho, M.G.: Literature review regarding Ant Colony Optimization applied to scheduling problems: guidelines for implementation and directions for future research. Eng. Appl. Artif. Intell. **26**(1), 150–161 (2013). https://doi.org/10.1016/j.engappai.2012.03.011

DAO Tokens: The Role for the Web 3.0 Industry and Pricing Factors

Vladislav Rutskiy[1,2], Iskandar Muda[3], Fadoua Joudar[4], Filippov Ilia[1,2], Svetlana Lyubaya[5], Alexandra Kuzmina[6], and Roman Tsarev[7,8](✉)

[1] Siberian Federal University, Krasnoyarsk, Russia
[2] Saint-Petersburg National Research University of Information Technologies, Mechanics and Optics (ITMO University), Saint-Petersburg, Russia
[3] Universitas Sumatera Utara, Medan, Indonesia
[4] Faculty of Economics and Management, Hassan First University of Settat, Settat, Morocco
[5] Stavropol State Agrarian University, Stavropol, Russia
[6] National University of Science and Technology MISIS, Moscow, Russia
[7] MIREA - Russian Technological University (RTU MIREA), Moscow, Russia
tsarev.sfu@mail.ru
[8] Bauman Moscow State Technical University, Moscow, Russia

Abstract. The paper discusses the factors influencing the pricing of cryptocurrency tokens for decentralized autonomous organization (DAO) projects. The formation and development of DAOs are among the promising applications of blockchain technology. DAO Project participants who own governance tokens can influence important decisions regarding project development through a voting function. The authors demonstrate the prospects for using DAO tokens as a tool for developing the Web 3.0 industry and conduct econometric modeling of the factors influencing the pricing of DAO tokens on panel data of the top 10 DAO projects in the industry. The analysis results show that the trading dynamics of DAO tokens depend more on the internal factors of development of this sector of the blockchain industry and only partially on the price quotes of traditional financial assets. However, the modeling results during the period of growth of the blockchain industry and DAO segment in 2021 differ somewhat from the modeling results for the period of blockchain industry stagnation in 2022. In 2022, DAO projects show greater dependence on traditional financial markets, demonstrating a more mature state of the DAO segment in the blockchain industry.

Keywords: Decentralized Autonomous Organization · DAO · Blockchain · Cryptocurrency · Web 3.0 · Pricing Factors

1 Introduction

The openness and transparency of blockchain technology have helped it gain immense popularity and widespread use in many industries over the past 10 years. Blockchain, or distributed ledger technology, has enabled the creation of an entirely new market with a current market capitalization of over a trillion dollars. Among the most developed

R. Silhavy and P. Silhavy (Eds.): CSOC 2023, LNNS 723, pp. 595–604, 2023.
https://doi.org/10.1007/978-3-031-35317-8_54

segments of the market are cryptocurrencies (such as Bitcoin and Ethereum), decentralized finance (Avalanche, Dai), and non-fungible tokens (CryptoPunks, Bored Ape Yacht Club). The popularity of this technology and its widespread use is a response to the established traditional structure of financial institutions, where all decisions are made centrally and by a small group of people.

One of the most relevant directions for the development of blockchain at the moment is decentralized autonomous organizations (DAO) projects. They are based on decentralized management by the project community and the absence of a hierarchy of power when making important decisions for the project. DAOs help people around the world collaborate on projects in real-time and create value for everyone, using cryptocurrencies and tokens as investment assets and pools of shared resources [1]. These projects can become the basis for a new standard of the Web 3.0, which is about equal participation of each user in creating and managing value.

In this work, the authors aimed to identify the factors that have a significant impact on the pricing of DAO project tokens. First, we described the features of DAO technology and examined examples of the most popular DAO projects. Then, we selected the factors that could affect the price of DAO tokens and identified the most significant factors using econometric modeling.

2 DAO in the Blockchain Industry: Technology Features and Examples of DAOs

The technology of distributed ledger (blockchain) was created in 2009 and represents a digital database without central management [2]. Satoshi Nakamoto was the first to use this technology, creating the first cryptocurrency bitcoin based on blockchain [3]. Blockchain is used in Bitcoin as a database that stores all transactions made from one account of a decentralized network to another. Blockchain introduced mechanisms that allow for decentralized management and avoid supervision and control by third parties [4]. Due to this, the rapid development of the technology has led to the emergence of various applications of blockchain technology, in particular, smart contracts and Decentralized Autonomous Organizations (DAOs) formed on their basis, which can operate autonomously, without the control or intervention of third parties.

The history of DAOs began with a major failure. The first DAO project represented an experiment in the form of a venture fund with a new organizational structure known as "Decentralized Autonomous Organizations (DAOs)" [5]. The founders' goal of the first DAO was to create organizations that did not require managers and hierarchies, which would be replaced by automated tasks based on smart contracts in the blockchain protocol [6]. The first DAO was based on the cryptocurrency Ethereum (ETH), and it was planned that investors would select projects for financing through voting. Unfortunately, the first DAO failed: due to vulnerabilities in the code, hackers were able to steal 11.5 million ETH (at the time worth $50 million), nearly a third of the funds raised through crowdfunding, and the project was closed. Fortunately, most of the vulnerabilities have been addressed, and DAO projects are only growing in number today.

Researchers interpret the definition of the term Decentralized Autonomous Organizations (DAO) differently. But they all agree that DAO is an organization working on the blockchain, the management of which is decentralized based on computer code in the form of smart contracts and algorithms [7]. The feature of DAO technology is based on several principles: tokenization, decentralization, automation, autonomy, and transparency.

Tokenization in DAO involves the use of blockchain-based tokens to provide voting rights. This principle allows only the cryptocurrency token holders to participate in making decisions on important issues for the community project.

Decentralization in DAO helps members of digital communities collaborate to achieve common goals without any formal hierarchy, control, or intervention [8]. In DAO, a joint peer-to-peer (P2P) community is created, where participants take on different roles and automate tasks necessary for the organization to function [9]. Unlike hierarchical organizations, DAOs do not have executive boards or CEOs who make decisions for everyone else. Instead, the P2P community itself handles these responsibilities by voting, proposing ideas, and making decisions. Thus, relationships in DAO follow the principles of equality, voluntariness, and mutual benefit.

There are several roles in DAO necessary for its normal functioning. One of them is founders, who create DAO and establish its goals and rules in a White Paper [10]. Their main goal is to attract investors and organize strategy of the project development. Typically, the White Paper presents a decentralized philosophy, concepts, and goals, as well as technical, organizational, and investment information about DAO's roadmap. Founders often open source of the blockchain code and smart contracts to facilitate joint software development with other developers interested in their project [11].

If investors like the project, they invest in DAO and become its members. By purchasing DAO cryptocurrency tokens, they gain the right to perform certain functions in the community (voting, exchange, idea proposal). Thus, with tokens, they participate in community management.

Automation in DAO is ensured through blockchain-encoded smart contracts. Smart contracts are programmable agreements that allow automating relationships between two or more parties regarding property rights and transactions in the community. Smart contracts in DAO help solve these issues without involving a centralized authority [12]. These contracts promote clear interaction, transparency, and equal trust between DAO participants. Thanks to them, procedures for coordination, confirmation, verification, approval, and decision-making in DAO are standardized and automated [13].

Autonomy of work in DAO is achieved by combining machine and human management [14]. The first consists of ensuring the operation of the network through algorithms and smart contracts written in the project code. Human management is demonstrated outside the network, for example, when community members suggest and discuss new ideas on social networks or forums and make decisions. Open and transparent discussion of important ideas of the project development is especially relevant in management of environmental projects [15]. Each DAO project creates its own tokens, which serve as proof of the rights and interests of each investor in the DAO. For example, they give investors the right to vote "for" or "against" cooperation with a new project. They are also used to incentivize community members to participate more actively in decision-making.

Transparency of DAO work is achieved through the use of blockchain principles, thanks to which each DAO participant can see all transactions, and smart contracts ensure interaction between parties without intermediaries.

The above principles create an advantage for DAO in the form of a significant reduction in internal and external transaction costs, including management costs. Since DAO usually operates according to cooperation models defined by all interested parties, it is easier to achieve consensus and trust within the organization. Thus, this reduces transaction and communication costs.

The DAO structure has gained widespread adoption among blockchain projects. According to the analytical agency Cointelegraph Research [16], there were 4,000 DAOs in mid-2022. The majority of them (40%) were in the decentralized finance (DeFi) sector, with the most well-known projects being the anonymous payment system Dash (the first functioning DAO), decentralized exchanges (DEX) Uniswap and PancakeSwap, and lending and borrowing platforms Compound and Aave. Other sectors with DAOs include NFT projects (17%), with popular examples like Pleasr DAO, an investment fund for NFT collections. Additionally, DAOs are applied in various other areas, such as KyberDAO for decision-making, and DAOstack, DAOhaus, and Aragon for facilitating DAO creation and management: creation of decentralized applications (dApp), cryptoprotocols, DAOs and their management [17].

There are also DAOs in fashion, such as Saint Fame, and in providing a decentralized system for hiring professionals, such as Raid Guild. Audius allows musicians to control the distribution of their work and earn revenue directly from their listeners. The Internet Computer project aims to realize the idea of Web 3.0, which we will discuss later. Ethereum Name Service enables domain registration on the Ethereum blockchain. DAOs are also heavily used in Metaverses, such as Decentraland, a platform with its own virtual world where participants can create their own spaces and communities.

To operate a DAO, the development of special governance tokens is required. These tokens implement the community governance logic: based on a majority voting system, token holders express their opinions regarding the development of the project protocol [18]. For example, this can include questions about transaction fees, calculating rewards for network participants, or selecting best practices for development. By investing in a promising project, investors receive these tokens and gain voting rights in the community according to their token holdings [5]. Governance tokens usually follow the principle of "one token, one vote," providing the means for decentralized governance through equal distribution of voting rights among the community. The token price is usually determined by supply and demand on cryptocurrency exchanges.

The use of the DAO mechanism provides opportunities for the development of various projects in the Web3.0 industry. The main distinctive feature of the Web3.0 Internet model from previous versions 1.0 and 2.0 is the decentralization of business models [19]. Currently, we are in the Web 2.0 stage, where a centralized model prevails: user data (personal data, transaction history, and credit ratings) is aggregated and stored by technology giants such as Google, Amazon, Apple, and others. This model creates certain risks and distrust. For example, we do not know how securely our data is stored, what data is collected, and what is not. In addition, these companies can resell this data. Also, the power over collected personal data is in the hands of a smaller number of managers and directors.

These drawbacks have become the basis for the emergence of the Web 3.0 Internet, where there is no centralized authority and everything is transparent. In this regard, the DAO management structure is ideal for Web 3.0. Smart contracts can perform the function of automatic data processing, which allows avoiding intermediaries. Another advantage is that smart contracts can work forever. Through owning DAO tokens, community members can participate in DAO management and express their opinion. Thus, based on DAO and its tokens, a decentralized Web 3.0 can be created and developed.

3 Econometric Modeling of Factors Affecting DAO Token Pricing

As previously mentioned, the price of DAO tokens is determined by free trade between buyers and sellers on cryptocurrency exchanges. Therefore, the goal of this study is to examine the factors affecting DAO token pricing, considering the factors affecting cryptocurrency assets identified in earlier studies.

We should mention that this kind of research study was not performed before in the academic literature. Most of the studies on DAO projects either explain technological features and business opportunities of the DAO phenomenon or focus on technological peculiarities of the chosen DAO projects. Our task is different. Previously some studies on modelling pricing factors such as token returns showed that DeFi and NFT being innovative sectors of blockchain industry showed some independence from the rest of the blockchain industry and possibilities for risk diversification for investors in blockchain assets and traditions financial assets [20]. We are going to reveal possible independence of DAO sector on the rest of the blockchain industry and traditional financial markets using econometric modelling approaches of relevant academic studies. Thus, through performing current study we could formulate better understanding of the recent technological and business trends of the Web 3.0 industry.

To do so, firstly, we should limit the amount of data to the most relevant projects and crucial indicators to obtain more reliable conclusions. We have selected 10 leading DAO projects as the object of our study because these projects have the highest impact on the DAO sector and help us to form a portrait of an average DAO project of the sector. When selecting DAO projects, we considered the following 5 criteria, which are normally key criteria for analyzing the dynamics and state of DAO projects by the industry experts and academic researchers [16]:

1) Market capitalization of the project's tokens.
2) Number of active project participants.
3) Industry affiliation of the project (the structure of the DAO market was taken into account: DeFi and investments – 45%, Infrastructure and tools – 14%, Metaverse, GameFi and NFT – 26%, Social and political projects – 15%).
4) Number of proposals in DAO voting.
5) Number of governance token holders.

As a result, we selected the following projects for analysis: Uniswap, Lido DAO, Decentraland, Aave, Curve DAO, PancakeSwap, Maker, Synthetix, Kusama and Balancer.

Our goal was to assess the influence of blockchain industry factors (using indicators of key cryptocurrencies as an example) and traditional financial markets (using stock indices as an example) on the pricing of tokens of an average DAO project.

Thus, in the quantitative part of this study, we conducted an econometric analysis of the relationship between the pricing of tokens of the 10 leading DAO projects (indicators of DAO token prices or trading volume of DAO tokens as the dependent variable in the regression model) and several possible factors affecting DAO token pricing (independent variables).

The main goal of this study was to test the hypothesis of how independent the DAO token pricing factors are, or if they depend greatly on factors of the blockchain industry and traditional financial markets.

To test our hypothesis, firstly we conducted a correlation analysis of the variables in the model. During the analysis, we found it more reasonable to use the trading volume of DAO tokens (weekly data) as the dependent variable rather than the price of DAO tokens. After the initial selection of significant variables, we then built a regression model based on panel data for the 10 DAO projects using the pooled least squares method. All calculations were performed using the applied econometric software package Gretl 1.9.92.

The final set of studied variables is presented in Table 1. The panel data values for the variables were obtained from databases stored by Coinmarketcap, the key blockchain industry tracker [21] and by Bloomberg, the global financial information agency [22].

We tested daily and weekly data of panel data series and finally chose the weekly data as it showed much better results when modelling. Academic studies mention that weekly data could be more useful than daily data for modelling cryptocurrencies pricing dynamics [20, 23]. We chose weekly values of data to build models for two periods: the period from January 1, 2021, to December 31, 2021, and the period from January 1, 2022, to December 31, 2022. The choice of these periods for modelling is connected with the need to study the trends of the blockchain industry and DAO sector. So, we have built two models as we needed to study the difference between the relevant token pricing factors during periods of growth (2021) and decline (2022) in the cryptocurrency markets. In 2021, the cryptocurrency market capitalization grew fourfold to $2.2 trillion, but by the next year, the market began to stagnate, and by the end of 2022, the capitalization volume amounted to $795 million according to the industry tracker Coinmarketcap.

Based on the analysis of data for 2021, we found that due to weak correlation, several factors should be excluded. After removing statistically insignificant variables from the model, we obtained the following regression equation Y_{2021} with average quality (R-squared = 0.27), demonstrating the dependence of DAO token trading volume on DAO Market Capitalization, Ethereum Trading Volume, and Binance Coin Trading Volume:

$$Y_{2021} = -1,71 + 0,2969 * \text{DAO\_token\_MC} + 0,0032 * \text{ETH\_TV} + 0,0254 * \text{BNB\_TV}.$$

Next, we built a model based on data for 2022. Similarly to the 2021 data model, we excluded several factors that showed weak correlation with the dependent variable and low statistical significance. As a result, we obtained the following regression equation Y_{2022} with average quality (R-squared = 0.36), demonstrating the dependence of DAO

Table 1. Description of endogenous and exogenous variables of regression model.

Endogenous (dependent) variables			
Indicator	*Unit*	*Period*	*Comments*
DAO token Trading Volume (TV)	USD	For 1st model: 01.01.2021 – 31.12.2021 For 2nd model: 01.01.2022 – 31.12.2022	As the dependent variable, weekly trading volumes of tokens of 10 DAO projects were taken for analysis (Uniswap, Lido DAO, Decentraland, Aave, Curve DAO, PancakeSwap, Maker, Synthetix, Kusama, Balancer). The weekly token volume is calculated as the product of its market price and the volume of token trading for the week.
Exogenous (explanatory) variables			
Indicator	*Unit*	*Period*	*Comments*
DAO token price (P)	USD	For 1st model: 01.01.2021 – 31.12.2021; For 2nd model: 01.01.2022 – 31.12.2022	Market value of token at the end of the week for 10 DAO projects selected for analysis.
DAO Market Capitalization (MC)	USD		Market capitalization of token at the end of the week for 10 DAO projects selected for analysis.
Bitcoin Price (P)	USD		An innovative private payment network and a new kind of decentralized (private) "digital" money. Bitcoin serves as a key indicator of the cryptocurrency assets market. The first cryptocurrency in terms of capitalization.
Bitcoin Market Capitalization (MC)	USD		
Bitcoin Trading Volume (TV)	USD		
Ethereum Price (P)	USD		Cryptocurrency and the most popular platform for creating decentralized autonomous organizations based on smart contracts. The second-largest cryptocurrency in terms of capitalization. Most DAO projects operate on the Ethereum technology.
Ethereum Market Capitalization (MC)	USD		
Ethereum Trading Volume (TV)	USD		
Binance Coin Price (P)	USD		A cryptocurrency on the Binance Smart Chain launched by Binance, one of the world's largest centralized cryptocurrency exchanges. A number of DAO projects operate on the Binance Smart Chain.
Binance Market Capitalization (MC)	USD		
Binance Coin Trading Volume (TV)	USD		
S&P 500 Index (P)	USD		The S&P 500 Index is built on the quotes of the 500 largest-cap companies traded on the US stock exchanges. It is one of the key indicators of traditional financial markets.
NASDAQ Index (P)	USD		The NASDAQ-100 Index is built on the stock quotes of the 100 largest non-financial corporations on the NASDAQ platform, reflecting the dynamics of the value of companies in the sectors of industry, technology, retail, telecommunications, biotechnology, healthcare, transportation, media and services. The index reflects the conjuncture in the US high-tech markets and the reaction of the business community to political and economic events in the country that affect the high-tech markets. The USA is one of the countries with the most massive investments in cryptocurrencies.
Russell 2000 Index (P)	USD		The Russell 2000 Index was launched in 1984 by the Frank Russell Company. It is a U.S. index and is managed by FTSE Russell, which is a subsidiary of the London Stock Exchange (LSE) Group. The index is comprised of 2,000 small-cap companies.

token trading volume on DAO Market Capitalization, Ethereum Trading Volume, and S&P 500 Index:

$$Y_{2022} = -1, 16 + 0, 2591 * DAO\_token\_MC + 0, 0051 * ETH\_TV + 250718 * SP500.$$

As we can see, both models have two recurring factors – DAO project market capitalization (DAO_token_MC) and weekly Ethereum cryptocurrency trading volume (ETH_TV). This means that the importance of Ethereum and DAO project capitalization factors was present in both the growth and decline periods of the cryptocurrency market in 2021 and 2022. The significance of these factors means greater importance of internal development factors in the DAO sector of the blockchain industry for pricing of DAO tokens.

However, there are also differences in the models. For example, in 2021, the Binance Coin cryptocurrency trading volume factor (BNB_TV) was found significant because it was a year of rapid growth for both DAO projects and the Binance ecosystem, which was also used to create DAOs. In 2022, when the DAO market became more mature and a prolonged market correction occurred in the blockchain industry as a whole, traditional financial markets (S&P 500), which also experienced a correction, were found to be more significant factors. Moreover, the significance of the model increased.

We need to note that in both models, the R-squared value did not exceed 0.4, indicating the need for further research in this area. We could perform this, for example, through expansion of DAO projects dataset and introduction of more intra-industry indicators such as dynamics of number of users holding DAO tokens (number of crypto wallets).

4 Conclusions

DAO is a logical continuation of the development of the blockchain industry, based on decentralization and transparency. In our paper, we have examined in detail the concepts and principles of DAO, as well as built a regression model to identify the factors influencing the token pricing for 10 leading DAO projects. The main hypothesis was to study whether the prices of DAO tokens are the result of larger market forces, such as changes in the quotes of key cryptocurrencies (Bitcoin, Ethereum) and traditional financial assets (S&P 500, NASDAQ, Russell 2000 indices), or whether the value of tokens is mainly determined by internal industrial factors.

The modeling results showed that trading volumes of DAO tokens are more dependent on internal development factors and only partially dependent on the price quotes of traditional financial assets. This can be explained by the fact that the main function of DAO tokens is the right to participate in community management, rather than speculation, so their value can change in accordance with the project's development. It can also be associated with the purpose and objectives of the DAO project, with the possibilities of implementing the rights of token owners using smart contracts.

The relationship between DAO tokens and traditional financial assets, identified for the period of 2022, can be explained by the fact that the DAO market is becoming more mature and stable during this period, while at the same time, there is a correction on both the cryptocurrency market and traditional financial markets.

Thus, it can be stated that DAO projects are a new direction in the blockchain industry, promising for the development of the global Web 3.0 ecosystem. DAO technology can contribute to faster development of Web 3.0 thanks to its decentralization, transparency, reliability, and resistance to hacking. The prospects for the application of blockchain and DAO technology for Web 3.0 also depend on regulatory measures taken by legislators, which can create obstacles or stimulate industry development. In general, the opportunities and limitations of DAO dissemination in the digital world and other areas require more detailed quantitative and qualitative research.

References

1. Van Rijmenam, M.: Sociomateriality in the Age of Emerging Information Technologies: How Bid Data Analytics, Blockchain and Artificial Intelligence Affect Organisations PhD Dissertation University of Technology Sydney Business School (2019)
2. Rutskiy, V., et al.: The price determinants of bitcoin as a new digital form of money. In: Silhavy, R., Silhavy, P., Prokopova, Z. (eds.) CoMeSySo 2020. AISC, vol. 1294, pp. 784–792. Springer, Cham (2020). https://doi.org/10.1007/978-3-030-63322-6_66
3. Casino, F., Dasaklis, T.K., Patsakis, C.: A systematic literature review of blockchain-based applications: Current status, classification and open issues. Telemat. Inform. **36**, 55–81 (2019). https://doi.org/10.1016/j.tele.2018.11.006
4. Wang, S., Ding, W., Li, J., Yuan, Y., Ouyang, L., Wang, F.-Y.: Decentralized autonomous organizations: concept, model, and applications. IEEE Trans. Comput. Soc. Syst. **6**(5), 870–878 (2019). https://doi.org/10.1109/TCSS.2019.2938190
5. Hsieh, Y.-Y., Vergne, J.-P.: Bitcoin and the rise of decentralized autonomous organizations. J. Organ. Des. **7**(1), 1–16 (2018). https://doi.org/10.1186/s41469-018-0038-1
6. Buterin, V.: DAOs, DACs, DAs and more: an incomplete terminology guide. Ethereum Blog **6**, 2014 (2014)
7. Santana, C., Albareda, L.: Blockchain and the emergence of Decentralized Autonomous Organizations (DAOs): an integrative model and research agenda. Technol. Forecast. Soc. Change **182** (2022). https://doi.org/10.1016/j.techfore.2022.121806
8. Blockchain technology for decentralized autonomous organizations Adv. Comput. **115**, 115–140 (2019). https://doi.org/10.1016/bs.adcom.2019.06.001
9. De Filippi, P., Hassan, S.: Decentralized autonomous organizations. Internet Policy Rev. **10** (2), 1–10 (2021). https://doi.org/10.14763/2021.2.1556
10. DuPont, Q.: Cryptocurrencies and Blockchains. Wiley, Cambridge (2019)
11. Blockchain: the birth of decentralized governance, Economics and Business, Working Paper Series, 1608, Pompeu Fabra University (2018)
12. Lumineau, F., Wang, W., Schilke, O.: Blockchain governance – a new way of organizing collaboration Organ. Sci. **32**(2), 500–521 (2021). https://doi.org/10.1287/orsc.2020.1379
13. Kutsyk, P., Redchenko, K., Voronko, R.: Management control and modern decentralized technologies. Baltic J. Econ. Stud. **6**(4), 98–102 (2020). https://doi.org/10.30525/2256-0742/2020-6-4-98-102
14. De Filippi, P., Loveluck, B.: The invisible politics of bitcoin: governance crisis of a decentralised infrastructure. Internet Policy Rev. **5**(4) (2016). https://doi.org/10.14763/2016.3
15. Eliseeva, E.: Environmental management as an important element of the concept of sustainable development of the organization. In: Proceedings of the 19th International Multidisciplinary Scientific GeoConference SGEM, vol. 19(5.3), pp. 299–306. Sofia, Bulgaria (2019). https://doi.org/10.5593/sgem2019/5.3/S21.038

16. DAO: The Evolution of Organization, Cointelegraph. https://research-backend.cointelegraph.com/uploads/attachments/cl9ctvb5i07wzr4qnhgz2eu7v-cointelegraph-research-dao-report-october.pdf
17. Kronovet, D.: Aragon, DAOstack, Colony, Moloch. http://kronosapiens.github.io/blog/2019/06/16/aragon-daostack-colony-moloch.html
18. Jensen, J.R.: Empirical evidence from four governance token distributions (2020). https://doi.org/10.48550/arXiv.2102.10096
19. McKinsey&Company: Web3 beyond the hype. September 26, 2022. https://www.mckinsey.com/industries/financial-services/our-insights/web3-beyond-the-hype
20. Qiao, X., Zhu, H., Tang, Y., Peng, C.: Time-frequency extreme risk spillover network of cryptocurrency coins, DeFi tokens and NFTs. Financ. Res. Lett. **51**, 103489 (2023). https://doi.org/10.1016/j.frl.2022.103489
21. Coinmarketcap. https://coinmarketcap.com/
22. Bloomberg. https://bloomberg.com
23. Rutskiy, V., et al.: Non-fungible tokens: value for digital arts and pricing factors. Lect. Notes Networks Syst. **596**, 805–816 (2023). https://doi.org/10.1007/978-3-031-21435-6_70

Modeling the Well-Being of the Population and Its Factors Using the Well-Being Index

Vladislav Rutskiy[1], Darío Salguero García[2], Elena Denisova[3], Fukalyak Alina[1], Nikolay Okashev[4], Igor Devederkin[4], Natalia Bystrova[5], Evgeniya Eliseeva[6,7], and Roman Tsarev[8,9](✉)

[1] Siberian Federal University, Krasnoyarsk, Russia
[2] Almería University, Almería, Spain
[3] Kaluga Branch of the Financial University Under the Government of the Russian Federation, Kaluga, Russia
[4] Stavropol State Agrarian University, Stavropol, Russia
[5] Minin State Pedagogical University of Nizhny Novgorod, Nizhny Novgorod, Russia
[6] National University of Science and Technology MISIS, Moscow, Russia
[7] Financial University Under the Government of the Russian Federation, Moscow, Russia
[8] MIREA - Russian Technological University (RTU MIREA), Moscow, Russia
tsarev.sfu@mail.ru
[9] Bauman Moscow State Technical University, Moscow, Russia

Abstract. The paper examines and evaluates the factors of population well-being based on the author's approach to modeling the well-being index. Increasing population well-being is one of the important tasks of economic policy in various countries. Therefore, the development of approaches to the analysis and modeling of population well-being as a multi-faceted phenomenon is relevant. The authors of the paper propose a composite approach to assessing well-being considering more subjective areas and assessments of population well-being, as well as more objective areas and assessments of economic activity and state policies to support the population. An author's methodology for measuring the integral well-being index of the population in the regional economy was developed and tested by the authors using Russian regions as an example. An econometric modeling of factors of the integral well-being index was also conducted. The modeling results showed that economic activity in regions is more significant for integral population well-being than the quality of human capital. This speaks to the importance of prioritizing the creation of conditions for economic development and improving the material well-being of the population compared to socio-economic policy in other areas of population well-being.

Keywords: Population Well-Being · Well-Being Index · Well-Being Factors · Regional Economy · Economic Development

1 Introduction

An important direction of economic policy in various countries is the increase of population well-being, which directly affects the economic development of a country. In academic literature and practical socio-economic policy, there are different approaches

R. Silhavy and P. Silhavy (Eds.): CSOC 2023, LNNS 723, pp. 605–614, 2023.
https://doi.org/10.1007/978-3-031-35317-8_55

to defining and measuring population well-being. There are various specific indicators and indexes that allow analyzing the level of well-being [1]. However, the formation of composite indicators of well-being, considering various social and economic indicators, as well as the measurement of factors of integrated well-being, are still insufficiently studied research questions.

Such research tasks are especially relevant for developing and transitional economies, in particular, for the economy of many Russian regions, which are characterized by a high level of differentiation both in terms of economic development and population well-being [2].

The authors of this study aimed to develop and evaluate a type of composite approach to measuring the well-being of the population. To achieve this goal, different approaches to studying the well-being of the population were compared first. Then, considering various approaches, a methodology for assessing the integrated index of well-being of the population was developed. The proposed methodology was tested on data from Russian regions. The factors that could affect the integrated well-being of the population were also selected, and econometric modeling was carried out to identify the most important factors influencing the integrated well-being index.

2 Approaches to Defining and Measuring the Well-Being of the Population

Economic development means development in many areas of society, such as technologies, economics, and social well-being. At the same time, the difference in the living standards between different population groups and among countries and regions is becoming more noticeable, and the risks of achieving and maintaining the level of prosperity are growing. Therefore, the tasks of forming mechanisms for regulating economic development and ensuring social stability and a decent standard of living for the population are becoming increasingly relevant, which is also related to the tasks of correctly defining and measuring population well-being.

To reliably assess the level of population well-being both at the international and regional levels, it is necessary to consider both objective and subjective indicators of well-being. The results of the analysis could be influenced by subjective perception, self-evaluation, self-awareness, and it is important to understand that assessing population well-being is a complex task in terms of selecting reliable factors and constructing a more objective assessment.

First of all, it is necessary to distinguish between two concepts: "welfare" and "well-being" [3, 4]. They could have two very different meanings in economics. Welfare and well-being are the general state in which special attention is paid to satisfaction, financial and material standard of living, happiness. Welfare is more often related to the assessment of the state of the population of the entire country, including certain parameters of its economy and the effectiveness of socio-economic policies, and researchers largely refer well-being to people's subjective assessments – their mental state, happiness, and so on.

Often in international practice, the measurement of well-being occurs through the use of composite indices that take into account a set of objective and subjective indicators (see Table 1) [4–14]. Welfare indices are built on the basis of grouping basic indicators that include various aspects of considering this term. In the practice of different countries, three main groups of indices can be distinguished: objective, subjective, and combined (synthesis of the first two groups). The objective is based on statistical data, the subjective is based on questionnaires and surveys, and the combined includes both approaches.

Table 1. Classification of population well-being indices.

Indicators	Objective						Subjective				Synthesis	
	MEW	ISEW	GPI	IEWB	PQLI	HDI	GNH	GNW	ULF	HPI	BLI	LPI
GDP/GDP per capita	+	+	+	+		+	+	+	+	+	+	+
Environmental damage	+	+	+	+			+	+		+	+	
Education/Literacy			+		+	+	+	+	+		+	+
Healthcare							+	+	+		+	+
Leisure time	+	+	+	+				+	+		+	
Distribution of expenditure			+	+								
Safety		+	+	+					+		+	+
Government management							+	+				+
Employment									+		+	+
Housing availability									+		+	
Life expectancy				+	+	+				+		
Unpaid work	+	+	+									
Civil rights											+	+
Cultural values							+	+				
Infant mortality					+							
Volunteerism				+								
Social energy							+	+				
Satisfaction											+	
Standard of living							+	+				
Religious values							+					

3 Development and Testing of an Integrated Index of Population Well-Being for Regional Economies

To construct a comparable composite index of well-being for different regional economies based on theoretical and applied research, we propose to consider both objective and subjective assessments of well-being, following a number of previous studies [15–17].

As a methodological basis for constructing the index, we used the approach of the Human Development Index (HDI), developed by the United Nations in 1990. This is one of the methods most suitable for aggregating quantitative indicators of well-being

[18, 19]. The calculation of this index is determined as the mean of three indicators: life expectancy, education level, and real per capita GDP.

In our methodology, we propose using 5 key indicators as quantitative criteria for well-being in the most important areas of people's lives in the regional economy:

1. "Life expectancy".
2. "Ratio of average per capita income to the subsistence minimum".
3. "Employment rate for people aged 15–72".
4. "Total living area per person".
5. "Share of domestic expenditures on scientific research and development in the Gross Regional Product (GRP)".

Each indicator for the well-being index will be calculated using formulas of the following form:

$$I = \frac{D_f - D_{min}}{D_{max} - D_{min}}, \quad (1)$$

where I is the index of the given indicator,
D_f is the actual value of the indicator,
D_min is the minimum value of the indicator,
D_max is the maximum value of the indicator.

We have also established the permissible boundaries of the selected indicators for Russian regions based on the analysis of well-being using a wide range of indicators:

1. Life expectancy: maximum value is 76 years, minimum value is 55 years.
2. Ratio of average per capita income to the subsistence minimum: maximum value is 4.27, minimum value is 1.59.
3. Employment rate for population aged 15–72 years: maximum value is 78.5 years, minimum value is 22.5 years.
4. Total area of residential premises per person on average: maximum value is 33.4 sq. m., minimum value is 12 sq. m.
5. Share of internal expenditures on scientific research and development in GRP: maximum value is 5.94%, minimum value is 0.01%.

Next, we tested the methodology for constructing the index with the aim of tracking the dynamics of the well-being of the population of 30 Russian regions between the periods of 2005, 2010, and 2015, where it was possible to consider subjective and objective factors of well-being in spatial data analysis. The database used for analyzing the subjective and objective factors of well-being was formed within the framework of the Russian Longitudinal Monitoring Survey by the National Research University Higher School of Economics.

The integrated assessment of well-being for regions was conducted in 3 groups:

- Regions with an index below 0.5 – low level of well-being.
- Regions with an index from 0.5 to 0.8 – average level of well-being.
- Regions with an index from 0.8 to 1 – high level of well-being.

The results of the analysis show that the well-being index is increasing in all regions of Russia between the observation periods of 2005, 2010, and 2015.

In 2005, all Russian regions were characterized by a low level of well-being, except for the Moscow region (0.549), the Republic of Komi (0.531), and the Republic of Tatarstan (0.558), where an average level of the well-being index was observed due to the indicators.

In 2010, all regions were already characterized by an average level of well-being, except for Altai Krai (an increase in the well-being index from 0.410 to 0.481), Amur Oblast (an increase in the well-being index from 0.312 to 0.443), and Kabardino-Balkar Republic (an increase in the well-being index from 0.350 to 0.489).

In 2015, all Russian regions reached an average level of well-being. The maximum values were achieved in the Moscow region (0.865) and the Republic of Tatarstan (0.819).

4 Econometric Modeling of Factors of Integrated Well-Being of the Population

In the next stage of this research, we formulated and tested a number of hypotheses about the influence of subjective and objective socio-economic factors on the composite well-being index through econometric modeling. We examined 29 key indicators of possible well-being factors. All factors underwent a normalization procedure similar to the normalization of the components of the composite well-being index.

After creating indices for all factors, we identified hypotheses and tested them for a sample of 30 regions using spatial data for 2005, 2010, and 2015.

Table 2 highlights the factors that are significant in different regression models for 2005, 2010, and 2015 (the number of asterisks corresponds to the level of significance, "0" corresponds to an insignificant factor).

The results of the modeling indicate that all models have sufficiently high levels of statistical significance (for example, the coefficient of determination R2 exceeds 0.9). Additionally, a number of interesting substantive conclusions can be drawn from the modeling.

In the intermediate regression models of 2005, it can be seen that there are indicators that have an inverse relationship with the index of well-being, such as the number of people below the subsistence level, the number of registered unemployed, the number of theater-goers per 1000 population, and the proportion of total area equipped with hot water supply.

In the intermediate regression models of 2010, all indicators have a direct relationship with the index of well-being, except for the subsistence minimum and the proportion of total area equipped with water supply, which have an inverse relationship with the index of well-being.

In the intermediate regression models of 2015, there are three factors that have a negative correlation with the well-being index: the size of the subsistence minimum; the number of registered crimes per 100,000 population; the share of total area equipped with a water supply system.

In further regression analysis, insignificant variables were excluded from the models to increase the reliability of the modeling results. The resulting regression equations showing the factors of well-being for Russian regions for 2005, 2010, and 2015 are

Table 2. Importance of well-being factors in Russian regions in 2005, 2010, and 2015 according to the results of regression analysis.

	Indicators	2005			2010		2015	
		Models			Models		Models	
		1.1	1.2	1.3	2.1	2.2	3.1	3.2
x1	Per capita income			***	***	**	***	***
x2	Number of people below the subsistence minimum	***	***			*		
x3	Number of registered unemployed	***	***					
x7	Average size of assigned pensions	***	***		***	**	***	***
x8	Life expectancy	***	***					
x9	Subsistence minimum			***	***	***	***	***
x13	Share of expenses on goods and services		**	*	**	**	0	
x14	Share of expenses on real estate					***		
x15	Number of registered crimes per 100,000 population							***
x16	Number of public buses per 100,000 population			**				
x17	Number of private cars per 1,000 population			**	***		**	**
x18	Density of public roads				***		***	***
x19	Passenger traffic			*				
x20	Number of theater-goers per 1,000 population	*	**					

(*continued*)

Table 2. (*continued*)

	Indicators	2005			2010		2015	
		Models			Models		Models	
		1.1	1.2	1.3	2.1	2.2	3.1	3.2
x21	Total amount of subsidies for housing and utilities	***	**					
x22	Emissions of pollutants into the air from stationary sources	*						
x23	Discharge of polluted wastewater into surface water bodies	***	***					
x24	Share of total area equipped with water supply			**	**		**	***
x26	Share of total area equipped with heating		**	**	**		**	*
x27	Share of total area equipped with baths			*				
x28	Share of total area equipped with hot water supply	***	***	*				
x29	Share of population's expenditures on food in total consumer spending			**				
R^2 value in regression models		0,93	0,94	0,87	0,92	0,90	0,92	0,94

presented below:

$$Index_{2005} = 2,40508 * x1 - 1,70749 * x9 + 0,179593 * x14 + 0,361658 \quad (2)$$

$$Index_{2010} = 1,47611 * x1 - 1,11084 * x9 + 0,108575 * x14 + 0,523087 \quad (3)$$

$$Index_{2015} = 0,988068 * x - 0,682600 * x9 + 0,254069 * x14 + 0,591155 \quad (4)$$

As a result of the study, we obtained a number of unexpected conclusions. For example, the life expectancy index is not a significant factor in the final regression models. It can be concluded that economic activity factors are much more important for the population than factors of human capital, such as health and life expectancy. This conclusion also confirms the importance of the factor of purchasing real estate. Based on this, it can be argued that Russia has a rather average than high level of economic development in terms of the achieved level of well-being of the population.

Also, based on the results of the study, it can be concluded that the implementation of a number of priorities to increase the well-being of the population in the regions is expedient in the Russian socio-economic policy:

1. Increase of the orientation of production on consumer preferences.
2. Increase of the competitiveness and investment attractiveness of private enterprises and entrepreneurs, increase of the global competitiveness of business, and reduction of administrative barriers.
3. Increase of the scale of financing of social obligations of the state: healthcare, public safety, culture, science, education, targeted social support for the population, and effective demographic policy.
4. Improving the quality of the professional education system at all its levels.

5 Conclusions

Assessing the well-being of the population is an important task for both economic research and economic practice. Measuring the integral well-being of the population in economic research and practical socio-economic policy of different countries is possible by considering both more subjective (evaluative) and more objective (parametric) aspects of well-being.

The authors of this study compared different approaches and proposed a composite approach to well-being. This approach suggests considering both more subjective spheres and evaluations of population well-being, as well as more objective spheres and evaluations related to the objective state of the population, its economic activity, and corresponding policies ("welfare" and "well-being"). Based on this approach and using the index method and the model of the human development index, we developed an original methodology for measuring the integral index of population well-being in the regional economy. We tested this methodology on the example of Russian regions for three time periods (2005, 2010, 2015) to track trends in the dynamics of well-being.

We also conducted econometric modeling of possible factors of integral well-being of the population of Russian regions with an assessment of the relationship between the well-being index and its various factors in the economy of Russian regions based on a spatial selection of statistical data from Rosstat and from the Russian monitoring of the economic situation and health of the population run by Higher School of Economics. The same time intervals of 2005, 2010, and 2015 were used as research periods to draw conclusions about the most important factors of the dynamics of the well-being index.

The modeling results showed that economic activity in Russian regions is more significant for the integral well-being of the population than the quality of human capital, in particular, the level of population health. Some of the most significant factors were such indicators as per capita income, subsistence minimum, and the proportion of expenditures on real estate purchases. The obtained results indicate the importance of prioritizing the creation of conditions for the development of the economy of Russian regions and improving the material well-being of the population compared to socio-economic policies in other spheres of population well-being. These conclusions indicate the importance of further research on the topic.

References

1. Van der Slycken, J., Bleys, B.: A Conceptual exploration and critical inquiry into the theoretical foundation(s) of economic welfare measures. Ecol. Econ. **176**, 106753 (2020). https://doi.org/10.1016/j.ecolecon.2020.106753
2. Moroshkina, M.V.: Interregional differentiation of Russian regions: trends and prospects for rapprochement. Theor. Appl. Econ. **3**, 48–60 (2018)
3. Fleche, S., Smith, C., Sorsa, P.: Exploring Determinants of Subjective Well-being in OECD Countries: Evidence from the World Value Survey. OECD Economics Department Working Papers, vol. 921 (2011)
4. Andreeva, O.N.: System of indicators of the level and quality of life of the population. Soc. Demograph. Struct. **1**, 75–77 (2013)
5. Becker, G., Philipson, T., Soares, R.: The Quantity and Quality of Life and the Evolution of World Inequality. NBER Working Paper, vol. 9765 (2003)
6. Human Development Index. http://www.hdr.undp.org. Accessed 4 April 2023
7. ULF Systems. http://www.ulf-systems.de. Accessed 4 April 2023
8. Gross National Happiness. http://www.grossnationalhappiness.com. Accessed 4 April 2023
9. Happy Planet Index. http://www.happyplanetindex.org. Accessed 4 April 2023
10. Legatum Prosperity Index. http://www.prosperity.com. Accessed 4 April 2023
11. Karabuschenko, N.B., Pilishvili, T.S., Chkhikvadze, T.V., Sorokoumova, S.N., Dyukareva, Y.: Comparative analysis of the intellectual foundations of emotion recognition among asian and russian students. Vestnik Minin Univ. **9**(2), 7 (2021). https://doi.org/10.26795/2307-1281-2021-9-2-7
12. Rutskiy, V., et al.: An empirical study on price determinants of the primary real estate market. Lect. Notes Networks Syst. **597**, 587–599 (2023). https://doi.org/10.1007/978-3-031-21438-7_47
13. Rutskiy, V., et al.: The relationship between gender inequality and the performance of large corporations. Lect. Notes Networks Syst. **597**, 576–586 (2023). https://doi.org/10.1007/978-3-031-21438-7_46
14. Slepenkova, E.A., Aksyonov, S.I.: Formation and development of pedagogical education in Russia: XVIII - XX centuries. Vestnik Minin Univ. **9**(1), 7 (2021). https://doi.org/10.26795/2307-1281-2021-9-1-7
15. Pesta, B.J., McDaniel, M.A., Bertsch, S.: Toward an index of well-being for the fifty U.S. states. Intelligence **38**(1), 160–168 (2010). https://doi.org/10.1016/j.intell.2009.09.006
16. Greco, S., Ishizaka, A., Resce, G., Torrisi, G.: Measuring well-being by a multidimensional spatial model in OECD Better Life Index framework. Socio-Econ. Plan. Sci. **70** (2019). https://doi.org/10.1016/j.seps.2019.01.006

17. Yee, S.H., Paulukonis, E., Buck, K.D.: Downscaling a human well-being index for environmental management and environmental justice applications in Puerto Rico. Appl. Geogr. **123**, 102231 (2020). https://doi.org/10.1016/j.apgeog.2020.102231
18. Baganov, V.Y.: Human development index as an indicator of sustainable development of the region. Baikal Res. J. **4**, 9–15 (2013)
19. Efimova, L.A.: Human development index in Russia: analysis and prospects. Region. Econ. Theory Pract. **22**(397), 53–62 (2015)

Automatic Programming Assessment Systems

Janet Liebenberg(✉)

North-West University, Potchefstroom, South Africa
janet.liebenberg@nwu.ac.za

Abstract. COVID-19 has led to unparalleled disruptions in education and particularly assessment. Automatic Programming Assessment (APA) allows instant feedback which can support real-time learning. This paper aimed to better understand students' experiences with an APA system. Qualitative research was conducted to investigate incremental changes and interpret the reaction of students as a response to the feedback they received from an APA system. The study of exceptional cases revealed that there are individuals that struggle to learn only through the feedback of the APA system. It was observed that there were cases where students experienced the feedback as confusing or misleading which is not conducive to learning. APA systems may effectively support learning and APA systems are valuable as a formative assessment tool, provided that the feedback is informative. This detailed study of how students reacted to the system's actions is useful in pointing the way to more intelligent and flexible systems.

Keywords: Automatic assessment systems · Introductory programming · Automatic programming assessment

1 Introduction

COVID-19 has led to unparalleled disruptions in education and particularly assessment. There is an urgent need to ensure the continued sustainability and equity of education systems. What makes automatic programming assessment (APA) very appealing is the fact that it allows instant feedback in order to support real-time learning [1–3]. This research sought to better understand how different students learn through the use of an APA system (APAS) with formative feedback, allowing students to resubmit their work and improve their solutions as they go along. [4] investigated quantitatively what steps students took toward solutions for programming assignments by considering the incremental changes students make and the correlating score between sequential submissions, measured by various metrics such as cyclomatic complexity, state-space, and the six Halstead measures of complexity of the program. In this research, a more qualitative approach was adhered to in order to understand how different students experienced the use of an APAS. Incremental changes were investigated and the reaction of students as a response to the feedback they received from the APAS was interpreted to receive direction to more intelligent and flexible systems.

R. Silhavy and P. Silhavy (Eds.): CSOC 2023, LNNS 723, pp. 615–628, 2023.
https://doi.org/10.1007/978-3-031-35317-8_56

2 Related Work

2.1 Automatic Programming Assessment

Automatic program assessment systems have been used for more than 50 years [5]. In a review of APA systems by [6], developed in the period 2006 to 2010, it was observed that APAs are mainly used in programming contests and introductory programming courses. A tremendous number of systems and systems for APA have been developed [2]. [7] and [8] remark that APA systems are rarely used outside the institutions in which they are developed and cite several systems which are not available or have not been updated in a long time.

An automated assessment system is suitable for monitoring the performance of students and improving their programming ability over time [9]. Automated assessment can reduce educators' effort, besides benefiting students with immediate feedback. Good and immediate feedback is one of the key components of learning programming [10].

[11] point to the complexity of being able to provide sensible feedback as it is nearly impossible to anticipate all errors that can occur in novice programs and to have test cases to identify each of the anticipated errors. They propose similarity clustering to improve the accuracy of feedback. [12] report that the design of automatically assessed exercise tests was one of the most difficult challenges they faced when applying Moodle plug-in VPL for automatic assessment of programming assignments.

[13] saw how insufficient feedback, such as complicated compiler error messages or mere pass/fail feedback can hinder student progress. However, it is possible to induce positive behaviour with formative feedback.

[14] found that it takes students more submission attempts in the APA when they are given questions that contain some starter code, than when they have to write their solution from scratch. However, when writing code from scratch, the students' code quality can be impaired because the students are not required to actually understand the concept being tested and might be able to find a way to bypass or hack the tests of the APA.

[15] suggest that there is no agreement on what constitutes valid measures of student learning in computing. Researchers speculate that students' poor performance may be indicative of inaccurate measures of their ability and knowledge [15, 16].

[17] state that traditional code-writing exam questions seem to require in addition to a mastery of several concepts, the ability to design with or synthesize those concepts. As a result, students obtain marks for peripheral code in addition to code that satisfies the aim of the question and the mark awarded may therefore not alert students to core misconceptions. [17] propose that single-concept questions - questions targeting one concept, or adding one concept over a previous concept question - are more effective formative feedback systems.

2.2 Benefits and Drawbacks of APAS

Many benefits of applying automatic assessment of programming assignments have been reported. Automatic assessment is bound to be consistent and objective [18, 19], enables rapid feedback [2, 18, 20], and allows students to submit multiple improved versions of the programs they have written [19, 21]. It can play a motivational role to engage students in the educational process [22, 23]. The most appealing benefit seems to be the possibility of saving time [2]. This comes as no surprise as it has been reported that assessment is one of the most often mentioned tasks that lecturers find burdensome [24]. [21] report how they effectively saved time when using a system, which can automatically identify correct code, reducing manual assessment to involve only code, which contains errors.

[23] describe a number of challenges associated with applying automatic assessment of programming tasks. [2] mentions the problem that many APAS suffer from inflexibility and unfair grading. An important challenge, which is often overlooked, is that considerable time and effort need to be devoted to the implementation of resources for automated assessment [3, 25–27]. Another problem is that the development of new exercises often requires considerable technical skills beyond the scope of the content being assessed [26, 28]. To address this problem [29] and [2] call for standardisation.

3 Methodology

3.1 Scenario

The study was conducted at a university in South Africa. The participants were a convenience sample of 187 first-year students registered for a Java programming course in the second semester.

In the seven weekly practical lab sessions prior to the study's session, the students completed an assignment while they could ask for help and feedback from the tutors and lecturer. Once a student had completed the assignment, the program was submitted and manually assessed by a tutor.

3.2 Data Collection

The data used in this research originates from an assignment that was automatically assessed using an APA system (APAS) called Fitchfork which applies dynamic testing-oriented assessment. The system executes the student's solution with several test inputs and each time verifies the output of the student's solution against a regular expression specifying the expected output for the given test input. This approach has two major drawbacks, namely that it requires a functional program and that the choice of test cases for evaluation may hamper the quality of the assessment.

Students were expected to complete seven tasks. The code the students had written was uploaded to the APAS during the session. Although they could complete the tasks in any order, it was assumed that they should complete them in the given order which resembles a step-by-step guide to write a complete program which takes a user-specified number of integers, determines the minimum and maximum of these values and do some calculations involving the minimum and maximum.

Task1: User Input
The first task required that the students write a simple program to prompt the user for an integer value and echo the value in an output statement. The task was assessed

with a single test case. The feedback for anticipated output lines produced by the student programs is shown in Table 1. The presence of a colon in the prompt and not in the output line was emphasised. This is because the system determines the number of iterations of the loop by counting the number of colons in the output. This caused frustration for the students since they did not understand why this formatting issue was deemed so important (Fig. 1).

```
Enter an integer: 9
You entered 9
That's all folks!
```

Fig. 1. Expected output for Task 1

Table 1. Assessment of Task 1 with test value = 22

Line	Possible output	Message	Mark (max)
1	Enter an integer:	PASS prompt line	1 (1)
	Other	FAIL The prompt should end with a colon and a space	0 (1)
2&3	You entered 22 That's all folks!	PASS output line	3 (3)
	You entered: 22 That's all folks!	FAIL Program output line 2 should NOT contain a colon	2 (3)
	22	FAIL The input should be on the same line as the prompt	1 (3)
		FAIL The acknowledgement line is missing	1 (3)
	Other	FAIL User input	0 (3)

Task 2: Loops
The second task instructed the students to add code to the program completed in the first task to prompt the user for the number of integers specified through the first prompt. Each prompt should have the number of the prompt. There should be a line of 20 dashes after the last prompt.

The evaluation was done by counting the total number of prompts and using a distinct output line of dashes at the end to determine if the number of prompts is in accordance with the value entered in the first prompt. Table 2 shows the error messages that were received by students for typical incorrect student programs. The numeric value which should be included in each prompt is also evaluated for correctness. The last error message was given if the student program would produce at least one prompt after the first prompt, in case the input of the first prompt was 0.

Task 3: Minimum and Maximum
For this task, the students had to change the program so that it establishes the value of the

Table 2. Loop errors

Error message (Max mark = 9.5)
Too few prompts
Too many prompts
Some prompts may be on the wrong line when counted from the top
Prompt values not shown
Prompt values off-by-one (too low)
Prompt values off-by-one (too high)
Program does not behave correctly when no values should be entered

smallest as well as the largest input value. The program must print the outcome below the line of dashes after the prompts. Use full sentences to display the results, each on its own line. The first output line should contain the minimum value and the second line the maximum. There should be a line of at least 20 dashes after the output.

If the student's solution to the previous task is incorrect, the assessment of the task to determine maximum and minimum values among the input values is practically impossible. The assessment thus starts by verifying that distinct lines of dashes are found where expected when counted from the top of the output lines. If they are not found there, the error message Some output may be on the wrong line when counted from the top is given. In this case, the task cannot be assessed any further.

Table 3 shows the messages that students received when assessing the identification of the minimum and maximum values in different situations.

Tasks 4 to 6: Methods for Different Calculations

The students were expected to write three different independent methods, each to perform a different calculation. The code for the definition of each of these methods had to be uploaded to each own upload slot for assessment. These methods are respectively called getSum, getDif and getQuo.

The getSum method must take two integer parameters and return the sum of its parameters. The getDiff method must take two integer parameters and return the difference between its parameters by subtracting the first parameter from the second parameter. The getQuo must take two integer parameters and return a double. The method should calculate the quotient of its parameters by dividing the second parameter by the first parameter and return the result.

Since the APAS merely assesses expected output for known inputs, the data type of the input parameters could not be evaluated. It could, however, be established if the return type of the method is correct by evaluating the format of the output when printing. For example, if the return type of getQuo is not a double one can observe whether the return value is truncated to an integer or not. Students were awarded 2 out of a possible 4 marks if the return value produced was of the expected type but determined by dividing the first parameter by the second parameter instead of the other way around. Students who applied implicit casting by having the parameters to be of type double were not penalized.

Table 3. Test cases used to assess finding the minimum and maximum values

Item	Test values	Possible output	Message	Mark (max)
Minimum	−8, −12, −6. −3, −7	−12	PASS Smallest (input random and all negative)	2 (2)
		other	FAIL Smallest value not identified correctly (input random and all negative)	0 (2)
	−3, −1, 0, 5, 7	−3	PASS Smallest (input ascending)	2 (2)
		other	FAIL Smallest value not identified correctly (input ascending)	0 (2)
	8, 6, 5, 4, 3	3	PASS Smallest (input descending and all positive)	2 (2)
		other	FAIL Smallest value not identified correctly (input descending and all positive)	0 (2)
Maximum	−8, −12, −6. −3, −7	−3	PASS Largest (input random and all negative)	2 (2)
		other	FAIL Largest value not identified correctly (input random and all negative)	0 (2)
	−3, −1, −0, 5, 7	7	PASS Largest (input ascending)	2 (2)
		other	FAIL Largest value not identified correctly (input ascending)	0 (2)
	8, 6, 5, 4, 3	8	PASS Largest (input descending and all positive)	2 (2)
		other	FAIL Largest value not identified correctly (input descending and all positive)	0 (2)

Table 4 shows the error messages that were displayed to students when assessing the functionality of these methods.

Task 7: Main

To test their final main method, we configured Fitchfork to compile their main method along with bogus implementations of the three required methods. When calling the

Table 4. Test cases used to evaluate the methods

Method	Test values	Possible output	Message	Mark (max)
getSum	1, 3	4	PASS Addition	2 (2)
		4.0	FAIL addition: Result should be an integer	1 (2)
		other	FAIL addition	0 (2)
getDiff	1, 3	2	PASS subtraction	3 (3)
		2.0	FAIL subtraction: Result should be an integer	2 (3)
		– 2 or – 2.0	FAIL subtraction: Values subtracted in the wrong order	1 (3)
		other	FAIL subtraction	0 (3)
getQuo	1, 3 and	3.0 and 3.1428	PASS division	4 (4)
	7, 22	3.0 and 3.0	PASS division but answer should be a real number	2 (4)
		0.33333 and 0.31818	FAIL Division in wrong order yet functionally correct for real answers	2 (4)
		other	FAIL division	0 (4)

methods from their main method, it should then display the values returned by our bogus methods instead of the correct values. If they would upload a file containing more than only their main method the student would receive a compiler error referring to a Duplicate Method for each of the required methods. In this case, they will likely get three messages, one for each method definition that was included in the source that should not have been included.

Students were awarded 1 mark each for each method call, 1 mark for the perfect spacing of the output and two marks for formatting the result of the quotient correctly with 2 decimal digits as prescribed.

Table 5 shows the error messages that were displayed for typical expected incorrect student programs apart from the usual compiler errors which will result if their code contains such errors.

3.3 Analysis

The uploaded files were downloaded after the students completed the assignment. The uploaded files as well as the marks and feedback given to students were analysed quantitatively and qualitatively. The quantitative analysis included metrics such as marks per task, final marks of students, number of uploads per task, and total number of uploads per student. These metrics led to the identification of exceptional cases.

Table 5. Main method errors

Message (Max mark = 6)	Diagnosis
getSum was not called correctly	
getDiff was not called correctly	Bogus values do not appear in student output
getQuo was not called correctly	
Incorrect formatting of answer of quotient	Value not shown with two decimal digits
Bad spacing. Consider putting spaces before and after operators in your output	The parser is unable to parse the output lines correctly
Output seems out of sync	The number of input prompts does not match the required number and/or the input value is not on the same line as the prompt for one or more of the prompts

The qualitative analysis included an investigation of these exceptional cases to deepen our understanding of these individual students' experiences with the APAS and to provide direction to an accurate and accommodating system.

4 Results and Discussion

There were seven tasks and at least one upload per task was required to complete the assignment. There were 18 students who achieved low marks and had 10 or less uploads. All but one did not complete all the tasks. The student who did complete all the tasks is discussed as Exception 1 in more detail in Sect. 4.1.

There are 74 students in the category of students who used more than 16 uploads during the assignment. Three of them uploaded more than 30 times and one of these students is discussed as Exception 2 in more detail in Sect. 4.2.

4.1 Exception 1

An exceptional student in the category of bottom eighteen students is one who submitted solutions to all the tasks, whereas the rest of the students in this category omitted many tasks. This student's final mark is 18.5%. The number of uploads and marks rewarded to this student is shown in Table 6.

He/she started well by getting full marks for task one on his/her first attempt. However, from the second task forward, it went from bad to worse.

In the second task, the student had newlines in the prompts. This resulted in the student receiving only the generic feedback that *some prompts may be on the wrong line when counted from the top*, and failed to assess the solution further. In the second try the student made substantial improvements, yet failed to remove the superfluous newlines and got the same mark of 2.5 for the first prompt, and the same feedback as before and gave up.

Table 6. Number of uploads per task

Task Nr	Task Name	Number of uploads	Mark	Max
1	Input	1	4	4
2	Loop	2	2.5	9.5
3	MinMax	2	1	12
4	getSum	2	0	2
5	getDiff	1	0	3
6	getQuo	1	0	4
7	Main	1	0	6

Continuing with the third task, the student extended the same flawed program resulting in the same unsatisfactory results. It is worth mentioning that the student's solution was partially correct and again improved with his/her second try even though Fitchfork was unable to evaluate it or provide proper feedback. It is likely that he/she ignored the system and relied on feedback from a tutor or a peer.

For the fourth, fifth and sixth tasks, the student misunderstood the instructions. In each case, he/she wrote a whole program which prompted the user for two values and then performs the required operation, instead of writing and uploading only the required method. This resulted in compiler errors and zero marks.

In the final task the student, again, extended the flawed program he/she was using in the second and third uploads. The solution was otherwise also wrong as he/she failed to call the prescribed methods. Instead, the student calculated the values in place. He/she, however, did not get good feedback because he/she made a filename error and uploaded the whole program instead of only the main method. The APAS failed this student since his/her code was never actually assessed. The system was unable to provide proper feedback and the student did not grasp the quirks of the system. He/she fell victim to typical requirements for Fitchfork to work properly i.e. he/she included newlines which were not required and did not upload only prescribed code segments for assessment. If the student's code was assessed manually, he/she probably would have received much higher marks.

4.2 Exception 2

Table 7 shows the number of uploads for each of the tasks for a particular student. This student made multiple uploads in all tasks and ended up with full marks.

When investigating the student's progress in each of the tasks, we observed that the student struggled with naming the file and class according to the specifications in the first two tasks. In the first task, he/she first got a compiler error message. Instead of realising that the class or file name could be wrong, he/she changed the output line in his/her program to better match the specification example for this line. When receiving the same error message with his/her second upload, he/she figured out how to fix the problem and then got full marks for his/her third upload of the second task.

Table 7. a Number of uploads per task for Exception 2

Task	Number of uploads	Mark	Max
Input	3	4	4
Loop	2	9.5	9.5
MinMax	2	12	12
getSum	3	2	2
getDiff	4	3	3
getQuo	3	4	4
Main	16	6	6

In the third task where the minimum and maximum value had to be determined, he/she initialized the variables as follows: max = min = 0;. With the first upload, he/she received error messages stating that the smallest and largest values were not identified correctly in the case of testing respectively with only positive input values and only negative input values. He/she learned from his/her mistakes and corrected them to: max = -100000000; min = 100000000; and got full marks for the second upload. Although it worked for these big numbers, the automatic assessment did not penalize the students for not programming more robustly by stating max = MIN_INT; min = MAX_INT;

In the task where students had to write the GetSum method his/her initial solution was flawless, but he/she received a compiler error message indicating that the method should be static in this context. Instead of realizing the core problem, he/she changed the program instructions, which did not make any functional difference, and received the same compiler error. Finally, he/she registered that the method should be static, resulting in full marks for this task.

When writing the GetDiff method, the student needed four uploads to obtain full marks. With the first upload, he/she received compiler errors for a misspelled method name and the use of an undeclared variable. The student corrected the undeclared variable problem for his/her second upload, but again got a compiler error message for the misspelled method name. For his/her third upload, he/she corrected the spelling of the method name and received an error message indicating that he/she is subtracting in the wrong order, which was corrected in the final upload.

In the GetQuo task, it was specified that they have to write a method which takes two integer parameters and returns a double. The student's first upload did not return a double and he, therefore, got the message: *PASS division but answer should be a real number*. He/she consequently changed the return type to double but received the same error message. We assume he/she realized that integer division caused the error message. The student corrected the problem by changing the input variable type to double, and got full marks. It is evident that the student would not have gotten full marks if it was assessed manually, as the task expects the students to do explicit type casting to address this problem whereas this student applied implicit type casting to resolve the issue.

In the final task, this student used 16 uploads to complete the task. He/she uploaded three times before realizing the compiler errors were caused by him not uploading only the main method as instructed. In his/her fourth upload, he/she received yet another compiler error indicating a mismatch of data types of the return type of the getQuo method and the variable he/she uses to receive this value. With the next upload (5th) he/she received a message *Incorrect formatting of quotient output line*. He/she used four uploads to correct the error. Two uploads produced compiler errors because of syntax errors in the formatting instruction and in one he/she reverted to his/her previous solution before fixing the problem in his/her ninth upload. Unfortunately, this output still did not receive full marks owing to the order in which the values should be passed to the getQuo function. The student resolved the problem by trial and error using seven more uploads in which he/she tested all possible orders of both the output statement and the call statement to finally hit the correct one in his/her sixteenth upload.

5 Conclusion

Although this study was conducted pre-pandemic, the sudden move to remote emergency teaching and learning brought about an accelerated shift to automatic assessment systems and consequently in the computing fields to APASs.

The best feature of APASs is probably the fact that they can provide instant feedback. In contrast with waiting several minutes for a tutor to attend to his/her problem, or sometimes not getting feedback at all as was the case during the lockdown, all students who used the APAS in this lab session got feedback on each submission. Students can respond to feedback by correcting their programs and in doing so, learn as they go along.

In our experience, students are laid-back in following instructions, and the APAS requires careful attention to instructions. Therefore, the APAS provides better opportunities for students to develop the skill of paying attention to instructions and to detail. This skill has a particular significance in IT.

As with any automatic processing of code (such as compilers and APASs), the error messages produced by APASs may be inaccurate and sometimes misleading. For example, Exception 1 as discussed in Sect. 4.1 where the student passed the first task in a single upload, receiving a simple pass result. As this task was aimed at emphasising the need for proper alignment of the output. Consequently, he/she failed to understand the message that his/her output lines are unaligned in subsequent tasks. This confirms the experience reported by [13] who observed that mere pass messages can hinder student progress.

Some students do not benefit from the instant feedback produced by the APAS. It seems that they do not read or understand the feedback. They tend to continue blindly completing the assignment as if there is no guidance.

In automatic assessment, it is inevitable that the student solutions should comply with precise specifications. This observation corroborates the finding of [4] that students who had rated APASs negatively almost unanimously disliked the level of precision required from their programs. Furthermore, this required precision puts a damper on the creativity of the students.

Because of the way in which Fitchfork assesses the output of the program, it is important that the input required by the student's program should be on the same line as the prompt, otherwise the extra line occupied for input would push the expected output a line down and it would not be assessed as such. Exception 1 discussed in Sect. 4.1 is an illustrative example of this problem.

It is extremely taxing and difficult to set a memo [3]. It is also error-prone [30]. Our investigation of how the students learned while doing this assignment also revealed how errors in our memo might have confused them. Exception 2 discussed in Sect. 4.2 is an illustrative example of this problem.

Since this was the students' first encounter with an APAS, we suspect that some of the problems mentioned above will get better over time as the students get acquainted with the system.

In this study, the majority of our students managed to complete small programming tasks, but a deep investigation of exceptional cases revealed that there are individuals that struggle to learn only through the feedback of the APAS.

In light of the above-mentioned consequences of APASs in the context of post-pandemic pedagogy, we conclude that APASs may effectively support learning. It is, however, essential that the feedback is as informative as possible. Ideally, automatic assessment should be accompanied by feedback from people such as tutors and peers in a lab situation to merge the best of both worlds. This detailed study of how individual students reacted to the system's actions is very valuable in pointing the way to more intelligent and accommodating systems.

References

1. Liebenberg, J., Pieterse, V.: Investigating the feasibility of automatic assessment of programming tasks. J. Inf. Technol. Educ. Innov. Pract. **17**, 201–223 (2018)
2. Ullah, Z., et al.: The effect of automatic assessment on novice programming: strengths and limitations of existing systems. Comput. Appl. Eng. Educ. **26**(6), 2328–2341 (2018)
3. Pieterse, V., Liebenberg, J.: Automatic vs manual assessment of programming tasks. In: Proceedings of the 17th Koli Calling International Conference on Computing Education Research (2017)
4. Pettit, R., et al.: An empirical study of iterative improvement in programming assignments. In: Proceedings of the 46th ACM Technical Symposium on Computer Science Education, pp. 410–415 (2015)
5. Douce, C., Livingstone, D., Orwell, J.: Automatic test-based assessment of programming: a review. J. Educ. Resour. Comput. **5**(3), 4 (2005)
6. Ihantola, P., et al.: Review of recent systems for automatic assessment of programming assignments. In: Proceedings of the 10th Koli Calling international Conference on Computing Education Research, pp. 86–93 (2010)
7. Mekterović, I., et al.: Building a comprehensive automated programming assessment system. IEEE Access **8**, 81154–81172 (2020)
8. Cipriano, B.P., Fachada, N., Alves, P.: Drop project: an automatic assessment tool for programming assignments. SoftwareX **18**, 101079 (2022)
9. Ho, S.B., et al.: Engineering meaningful computing education: programming learning experience model. In: 2019 IEEE International Conference on Industrial Engineering and Engineering Management (IEEM). IEEE (2019)
10. Lokar, M.: Project Tomo: automated feedback service in teaching programming in Slovenian high schools. In: Proceedings of the 8th Computer Science Education Research Conference (2019)

11. Combéfis, S., Schils, A.: Automatic programming error class identification with code plagiarism-based clustering. In: Proceedings of the 2nd International Code Hunt Workshop on Educational Software Engineering, pp. 1–6 (2016)
12. Lepp, M., et al.: Self-and Automated Assessment in Programming MOOCs. In: Proceedings of the International Computer Assisted Assessment Conference, pp. 72–85 (2016)
13. Tirronen, V., Tirronen. M.: A framework for evaluating student interaction with automatically assessed exercises. In: Proceedings of the 16th Koli Calling International Conference on Computing Education Research (2016)
14. Battestilli, L., Korkes, S.: Writing effective autograded exercises using Bloom's Taxonomy. In 2020 ASEE Virtual Conference, American Society for Engineering Education (2020)
15. Tew, A.E., Guzdial, M.: Developing a validated assessment of fundamental CS1 concepts. In: Proceedings of the 41st ACM Technical Symposium on Computer Science Education, pp. 97–101 (2010)
16. Lister, R.: Computing education research - Geek genes and bimodal grades. ACM Inroads **1**(3), 16–17 (2010)
17. Zingaro, D., Petersen, A., Craig. M.: Stepping up to integrative questions on CS1 exams. In: Proceedings of the 43rd ACM Technical Symposium on Computer Science Education (2012)
18. Arifi, S.M., et al., Automatic program assessment using static and dynamic analysis. In: Proceedings of the 2015 Third World Conference on Complex Systems (WCCS), pp. 1–6 (2015)
19. Staubitz, T., et al., CodeOcean-A versatile platform for practical programming excercises in online environments. In: Proceedings of the 2016 IEEE Global Engineering Education Conference (EDUCON), pp. 314–323 (2016)
20. Liu, L., et al., Metis: A multi-faceted hybrid book learning platform. In: Proceedings of the 2016 ACM Symposium on Document Engineering, pp. 31–34 (2016)
21. Del Fatto, V., Dodero, G., Gennari, R., Gruber, B., Helmer, S., Raimato, G.: Automating assessment of exercises as means to decrease MOOC teachers' efforts. In: Mealha, Ó., Divitini, M., Rehm, M. (eds) Citizen, Territory and Technologies: Smart Learning Contexts and Practices. SLERD 2017. Smart Innovation, Systems and Technologies, vol. 80. Springer, Cham (2018). https://doi.org/10.1007/978-3-319-61322-2_20
22. Šťastná, J., et al.: Security measures in automated assessment system for programming courses. Acta Informat. Pragensia **4**(3), 226–241 (2015)
23. Staubitz, T., et al., Towards practical programming exercises and automated assessment in Massive Open Online Courses. In: Proceedings of the 2015 IEEE International Conference onTeaching, Assessment, and Learning for Engineering (TALE), pp. 23–30 (2015)
24. Pieterse, V., Sonnekus, I.P.:Why are we doing IT to ourselves? In: Proceedings of the 33rd annual conference of the Southern African Computer Lecturers' Association (SACLA), p. Paper 9 (2009)
25. Ala-Mutka, K.: A survey of automated assessment approaches for programming assignments. Comput. Sci. Educ. **15**(2), 83–102 (2005)
26. Pieterse, V.: Automated assessment of programming assignments. In: Proceedings of the 3rd Computer Science Education Research Conference (CSERC 2013), pp. 45–56 (2013)
27. Watanobe, Y., Rahman, M.M., Rage, U.K., Penugonda, R.: Online automatic assessment system for program code: architecture and experiences. In: Fujita, H., Selamat, A., Lin, J.CW., Ali, M. (eds.) Advances and Trends in Artificial Intelligence. From Theory to Practice. IEA/AIE 2021. LNCS, vol .12799. Springer, Cham (2021). https://doi.org/10.1007/978-3-030-79463-7_23
28. Korhonen, A., Malmi, L.: Algorithm simulation with automatic assessment. ACM SIGCSE Bull. **32**(3), 160–163 (2000)

29. Solms, F., Pieterse, V.: Towards a generic DSL for automated marking systems. In: Gruner, S. (ed.) ICT Education. SACLA 2016. Communications in Computer and Information Science, vol. 642. Springer, Cham (2016). https://doi.org/10.1007/978-3-319-47680-3_6
30. Pieterse, V., Janse van Vuuren, H.: Experience in the formulation of memoranda for an automarker of simple programming tasks. In: Proceedings of the 44th annual Southern African Computer Lecturers' Association (SACLA), pp. 210–214 (2015)

On the Importance of Signal-to-Noise Estimation While Testing the Communication Channel Quality and Detecting the Phase-Shift Keyed Signals

Oleg Chernoyarov[1,2(✉)], Alexey Glushkov[3], Baktybek Karimov[4], Vladimir Litvinenko[5], and Alexandra Salnikova[1]

[1] National Research University "MPEI", Krasnokazarmennaya Str. 14, 111250 Moscow, Russia
chernoyarovov@mpei.ru

[2] National Research Tomsk State University, Lenin Avenue 36, 634050 Tomsk, Russia

[3] Zhukovsky-Gagarin Air Force Academy, Marshal Nedelin Str. 133, 394052 Voronezh, Russia

[4] Kyrgyz State Technical University Named After I. Razzakov, Manas Avenue 66, 720044 Bishkek, Kyrgyzstan

[5] Voronezh State Technical University, Moscow Avenue 14, 394026 Voronezh, Russia

Abstract. The most important characteristic of a working communication channel is the current signal-to-noise ratio. In the paper, to estimate it, it is proposed to use the variance of the logarithm of the ratio of the amplitudes of the adjacent information symbols. In this case, for the Gaussian channel (that is the channel in which the interferences can be approximated by Gaussian noise), there is no need to determine the signal and interference levels, and no intervention is required in the operation of the communication system. The resulting estimate can be used to control channel selection and the change in the symbol rate, as well as in a number of other tasks. The necessary calculation relationships are determined and the statistical simulation is carried out for the case of receiving a phase-shift keyed signal. #CSOC1120.

Keywords: Radio signal · Constant amplitude · Unknown signal and noise levels · Decision determining statistics · Signal-to-noise ratio estimate · Signal detection · Statistical simulation

1 Introduction

The amplitude of the symbol received by the phase-shift keyed (PSK) signals does not depend on the transmitted information, and at the end of symbol reception (the clock synchronization), its value can be generated at the demodulator output [1, 2]. In the absence of interferences, the ratio of the amplitudes of the adjacent symbols is equal to unity. The presence of Gaussian noise leads to Rayleigh oscillations in the amplitude response of the demodulator [3, 4]. The amplitudes of the symbols received by the amplitude phase-shift keyed (APSK) signals depend on the transmitted information, but

R. Silhavy and P. Silhavy (Eds.): CSOC 2023, LNNS 723, pp. 629–638, 2023.
https://doi.org/10.1007/978-3-031-35317-8_57

their ratio a is fixed, for example, its value can be 3 or 1/3 [1, 2], or it is 2.5, 1.4 and their reciprocals [5–7]. That is, in this case, the ratio of the symbol amplitudes can be normalized to unity.

Estimating the signal-to-noise ratio (SNR) is closely related to the problem of signal detection [8, 9]. In this case, the aim is to minimize the required amount of a priori information about the signal and interference, and it is desirable to ensure that the false alarm probability is independent of the noise level. Below, it is shown that the random variable which is the logarithm of the ratio of the amplitudes of the adjacent information symbols can be effectively used to solve such tasks.

The variance of the specified variable characterizes how greatly the range of values of the samples of the logarithm of the ratio of the amplitudes of the adjacent information symbols differ from the mean value (which is zero). In the absence of a signal, this variance is large enough and does not depend on the noise level, while in the presence of a signal it decreases with the increase in the SNR that allows you to make the necessary decisions, control the quality of a working communication channel without interfering with the transmitted signals, and estimate the SNR.

2 Signal and Noise Models

The discrete signals of a particular (but widely used) type are considered, they have an amplitude that does not depend on the transmitted information. For example, they can be PSK signals that take the form

$$s(t) = S\cos\left[2\pi f_0 t + \psi(t)\right],$$

where S is the constant amplitude, f_0 is the carrier frequency and $\psi(t)$ is the current phase. Based on the responses of the quadrature signal processing channels [1, 2] of the demodulator, at the moment of the end of the symbol, a sample can be formed that is proportional to the amplitude ξ_k of the k-th received symbol. The presence of the Gaussian noise $n(t)$ in the input realization of the observable data $\xi(t)$ leads to random fluctuations of the sample ξ_k the value of which obeys the Rayleigh law of distribution with the probability density

$$w_n(x) = x\exp\left(-x^2\Big/2\sigma_n^2\right)\Big/\sigma_n^2, \tag{1}$$

if $\xi(t) = n(t)$, or the Rice law of distribution [3, 4, 10, 11] with the probability density

$$w_s(x) = x\exp\left[-\left(x^2+S^2\right)\Big/2\sigma_n^2\right] I_0\left(xS\big/\sigma_n\right)\Big/\sigma_n^2, \tag{2}$$

if $\xi(t) = s(t) + n(t)$ [3, 4, 10, 11]. Here σ_n^2 is the Gaussian noise sample dispersion at the receiver input and $I_0(x)$ is the zero-order modified Bessel function.

3 Logarithm of the Symbol Amplitude Ratio

The logarithm of the amplitude ratio

$$Y_k = \xi_k\big/\xi_{k-1} \tag{3}$$

of the adjacent symbols is equal to

$$G_k = \ln\left(\xi_k \big/ \xi_{k-1}\right). \tag{4}$$

The sampling G_k is of the size N, and $k = \overline{1, N}$. Then the G_k variance is estimated and, finally, the decision on the SNR value can be made, based on this estimate.

Following [12], the probability density $u(y)$ of the ratio (3), where ξ_k and ξ_{k-1} are the random variables with the same probability density $w(x)$, is calculated as

$$u(y) = \int_0^\infty w(x)\, w(yx)\, |x|\, \mathrm{d}x, \tag{5}$$

and then, correspondingly, the probability density $f(g)$ of the values G_k (4) can be obtained as

$$f(g) = \exp(g)\, u(\exp(g)). \tag{6}$$

By substituting $w_n(x)$ (1) into (5), after calculating the integral, the probability density $u_n(x)$ can be found for the case when only the noise is present at the receiver input:

$$u_n(y) = 2y \Big/ \left(1 + y^2\right)^2 \tag{7}$$

and then, from (6) and (7), one gets the expression for the probability density of the random variable (4), if $\xi(t) = n(t)$:

$$f_n(g) = 2\exp(2g) \Big/ \left[1 + \exp(2g)\right]^2. \tag{8}$$

As it follows from (7), (8), the probability densities $u_n(y)$ and $f_n(g)$ do not depend on the noise dispersion, their diagrams are presented in Fig. 1. There, the curve $f_n(g)$ (8) is symmetrical to the y-axis.

The mean value and the mean square of the ratio Y_k (3) are determined in the following way:

$$Y_{mean} = \int_0^\infty y u_n(y)\, \mathrm{d}y = \frac{\pi}{2},\ Y_{mean}^2 = \int_0^\infty y^2 u_n(y)\, \mathrm{d}y = \int_0^\infty \frac{z\, \mathrm{d}z}{(1+z)^2}$$
$$= \left[\frac{1}{1+z} + \ln(1+z)\right]\Bigg|_0^\infty \to \infty.$$

Thus, the variance of the random variable (3) is unlimited.

As for the logarithm of the ratio G_k (4), it has a zero mean value and its variance is finite and equal to

$$\sigma_G^2 = \int_{-\infty}^\infty g^2 f_n(g)\, \mathrm{d}g = \pi^2 \big/ 12 = 0.822. \tag{9}$$

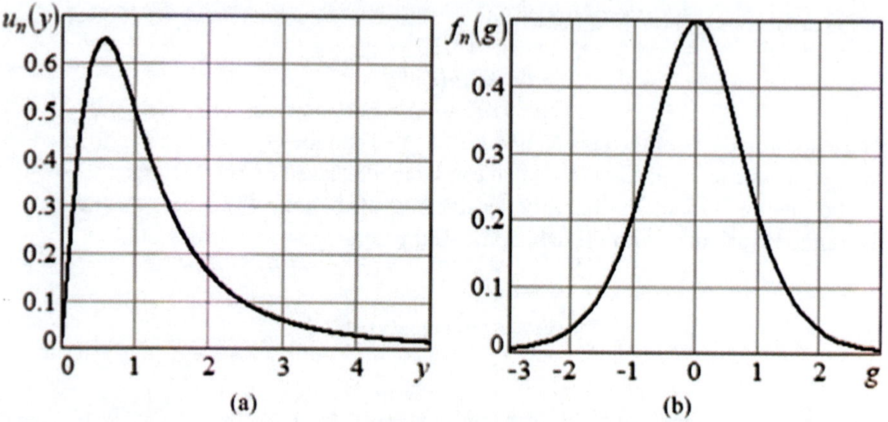

Fig. 1. The probability densities of the amplitude ratio of the adjacent symbols (a) and its logarithm (b)

Therefore, the random variable (4) is more feasible to use than (3).

In the presence of the signal, the probability density of the samples ξ_k of the realization of observable data $\xi(t)$ is described by the Rice formula (2), and the probability density $u_s(y)$ of the ratio (3) is

$$u_s(y) = \int_{-\infty}^{\infty} yz^3 \exp\left[-\left(z^2 + (yz)^2 + 4h^2\right)/2\right] I_0\left(zh\sqrt{2}\right) I_0\left(yzh\sqrt{2}\right) \mathrm{d}z. \tag{10}$$

Here

$$h^2 = S^2 / 2\sigma_n^2 \tag{11}$$

is the input power SNR.

Accounting for (6), the probability density $f_s(g)$ of the logarithm of the ratio of the adjacent symbols takes the form

$$f_s(g) = \exp(g)\, u_s(\exp(g)). \tag{12}$$

The integral (10) can be calculated numerically only. The dependences $f_s(g)$ (12) for the different values of h^2 (11) are shown in Fig. 2. The curve 1 corresponds to $h^2 = 10$, the curve 2 – $h^2 = 40$, the curve 3 – $h^2 = 100$. As one can see from Fig. 2, when a signal appears, the probability density $f_s(g)$ (12) becomes significantly "narrower", that is, the variance of the values of the logarithm of the amplitude ratio decreases sharply with the increase in the SNR, especially compared to the case demonstrated in Fig. 1b. And the curves describing the probability density $f_s(g)$ (12) are almost symmetrical with respect to the y-axis.

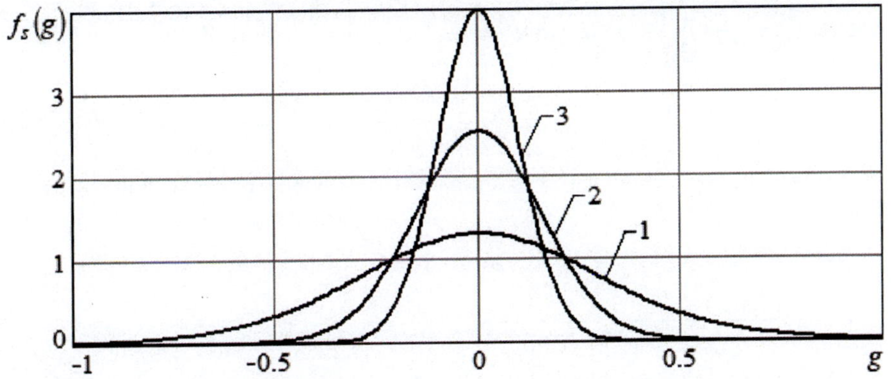

Fig. 2. The probability density of the logarithm of the ratio of the adjacent symbols in the presence of the information signal

4 The Estimate of the Variance of the Values of the Logarithm of the Symbol Amplitude Ratio

The variance of the logarithm of the ratio of the amplitudes of the adjacent symbols (4) in the absence of a signal is equal to σ_G^2 (9) and does not depend on the noise level. In the presence of a signal, according to the results of numerical calculations, the mean value of the random variable (4) can practically be taken equal to zero, then its variance is determined by the expression

$$\sigma_G^2 = \int_{-\infty}^{\infty} g^2 f_s(g)\, dg. \tag{13}$$

The dependence of σ_G^2 (13) on SNR h^2 (expressed in dB) is shown by the solid line in Fig. 3, with the dashed line demonstrating the value of the variance σ_G^2 (9) in the absence of a signal being 0.822.

An increase in the signal level leads to a decrease in the variance of the logarithm of the adjacent symbol amplitude ratio (4) and it can be used to detect a signal or to estimate the SNR ratio in a working channel with the help of the curve in Fig. 3.

The variance σ_G^2 of the sample sequence G_k of the size N where $k = \overline{1, N}$ is estimated in the case of the zero mean by the value [13]

$$\hat{\sigma}_G^2 = \frac{1}{N-1} \sum_{k=1}^{N} G_k^2. \tag{14}$$

The estimate $\hat{\sigma}_G^2$ (14) is the random one, its mean value being σ_G^2 and the variance is

$$D\left(\hat{\sigma}_G^2\right) = \frac{1}{N}\left(\mu_4 - \frac{N-3}{N-1}\sigma_G^4\right), \tag{15}$$

where $\mu_4 = \frac{1}{N-1}\sum_{k=1}^{N} G_k^4$. The value

$$\Delta = \sqrt{D(\hat{\sigma}_G^2)} \quad (16)$$

determines the error of the variance estimate (14) which decreases with the increase in the sampling size N.

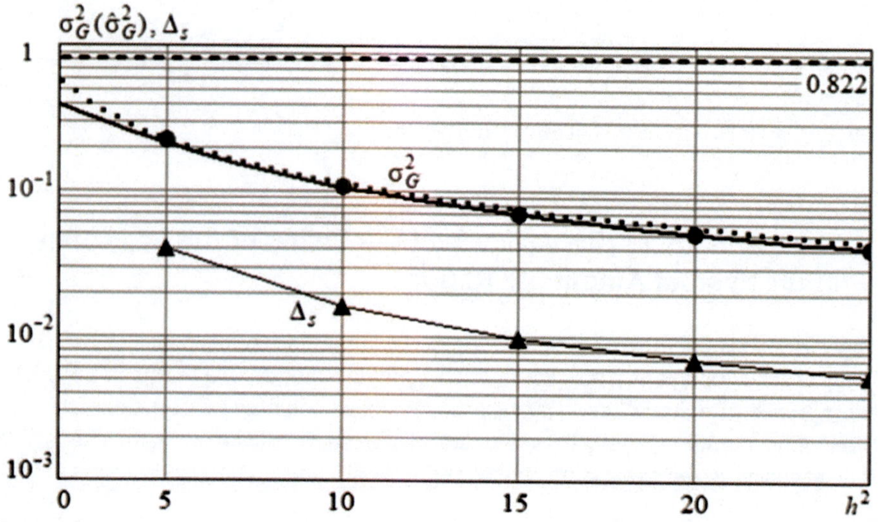

Fig. 3. Theoretical and experimental dependences of the variance of the logarithm of the ratio of the adjacent symbols and its mean square deviation

Statistical simulation of the operation of the digital demodulator of differential binary phase-shift keyed (DBPSK) signals [14] has been carried out with an estimation of the logarithm of the adjacent symbol amplitude ratio.

In Fig. 4, the histogram $P(g)$ of the probabilities of the values of the logarithm of the adjacent symbol amplitude ratio (4) is presented in the absence of a signal ($S = 0$), the sampling size being $N = 128$ and the number of simulation cycles – 8192 (the dashed line shows the calculation curve from Fig. 1b). According to this histogram, the mean value of G_k lies within $-(1.5 \div 4) \cdot 10^{-4}$ (it is much less than the standard deviation, approximately zero) and the average variance is 0.822 which agrees well with (9).

In Fig. 5, the histogram $P(\hat{\sigma}_G^2)$ of the estimates of the probabilities of occurrence of the values $\hat{\sigma}_G^2$ (14) under the same conditions is plotted. Here the mean value of the variance estimate $\hat{\sigma}_G^2$ is 0.823 which is in good agreement with the theoretical value (9) of 0.822. It can be seen that the histogram is asymmetric and noticeably differs from the Gaussian one.

In Fig. 6, by dots, the dependence of the standard deviation (error) Δ (16) of the variance estimate $\hat{\sigma}_G^2$ (14) upon the sampling size N is shown. It is obtained by means of simulation in the case when $h^2 = 20$. By the dashed line, demonstrating the empirically fitted dependence $\Delta(N)$ is demonstrated taking in the form

$$\Delta(N) = \Delta(32)\sqrt{32/N}, \tag{17}$$

where the value $\Delta(32)$ depends on the SNR (11) and, for example, is taken equal to 0.255 at $h^2 = 20$.

As it follows from Fig. 6 and other simulation results, the estimate (17) is in good agreement with the theoretical results (15), (16) and it can be used to determine the error in estimating the variance.

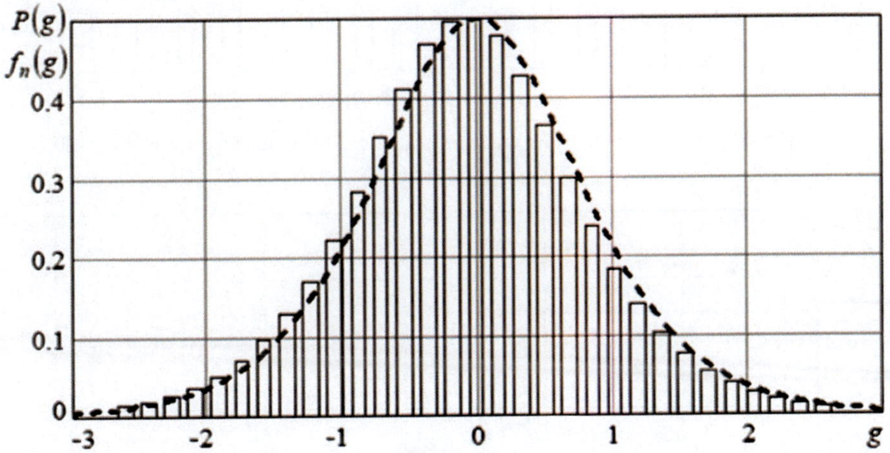

Fig. 4. The histogram of the values of the logarithm of the adjacent symbol amplitude ratio in the absence of a signal

In the presence of a signal, the variance estimate $\hat{\sigma}_G^2$ (14) decreases with the increase in the SNR h^2 (11), as it is shown in Fig. 3. There, the dots mark the results of the statistical simulation of the estimate (14), and the dotted line presents the empirical approximation $\sigma_G^2(h^2)$ of the experimental dependence $\hat{\sigma}_G^2(h^2)$ obtained by fitting:

$$\sigma_G^2\left(h^2\right) = 1.12\Big/h^2. \tag{18}$$

Thus, the variance σ_G^2 of the logarithm of the adjacent information symbol amplitude ratio is inversely proportional to the SNR (11).

The line with the triangles in Fig. 3 shows the values of the standard deviation Δ_s of the variance estimate $\hat{\sigma}_G^2\left(h^2\right)$ obtained as a simulation result in the presence of a signal and with the sampling size being $N = 128$. As it can be seen, the error Δ_s is approximately 7 times less than the estimated variance or it is 14%. In the case of the increasing N, it decreases according to the law $1\Big/\sqrt{N}$, as in the absence of a signal (17), and can be approximately estimated by the empirical expression

$$\Delta_s(N)\Big/\sigma_G^2 = \sqrt{128}\Big/7\sqrt{N} = 1.616\Big/\sqrt{N}. \tag{19}$$

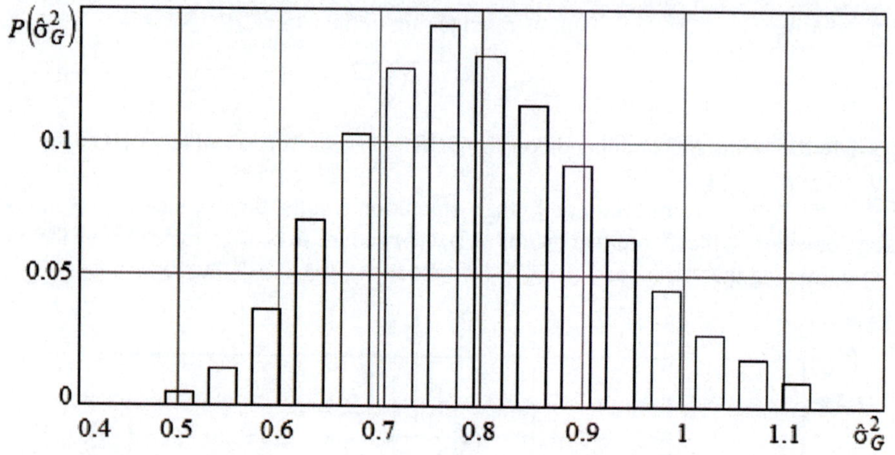

Fig. 5. The histogram of the variance estimates (14) in the absence of a signal

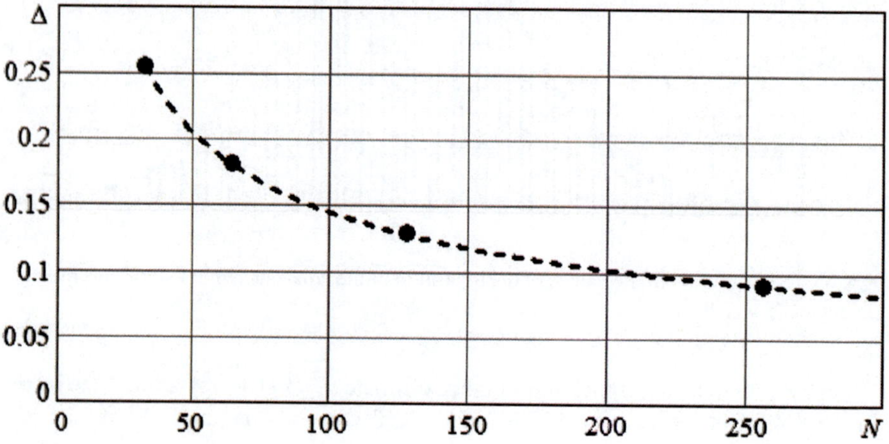

Fig. 6. The error of the variance estimate (14) in the absence of a signal

5 The Signal-To-Noise Estimate

Within the framework of the signal and noise models under consideration, the variance σ_G^2 of the logarithm of the adjacent information symbol amplitude ratio depends only on the SNR h^2 (according to (13) and Fig. 3) and the estimate (18) obtained from the results of statistical simulation. At the moments of clock synchronization at the output of the demodulator, based on the sampling G_k where $k = \overline{1, N}$, the variance estimate $\hat{\sigma}_G^2$ (14) is found, according to which the SNR is approximately determined from the expression (18) in the form

$$h^2 = 1.12 \Big/ \hat{\sigma}_G^2.$$

The error Δ_s (19) of the variance estimate determining the SNR estimation error depends on the sampling size N and becomes less than 3% at $N \geq 2^{12} = 4096$.

With a small sampling size, for example, $N = 64$, it is possible to make evaluative decisions about the quality of the communication channel (for example 3 gradations: "bad", "satisfactory", "good") or about the presence of a signal in it in the working channel without the need to estimate the noise and signal levels.

6 Conclusion

An effective algorithm is proposed for determining the signal-to-noise ratio for the information signal with a constant amplitude by means of a statistical estimation of the variance of the logarithm of the adjacent information symbol amplitude ratio in a working communication channel without the need for a priori information about the levels of noise and signal. It is shown that in the absence of a signal, this variance does not depend on the noise level and is equal to 0.822, and when a signal appears, it decreases according to the law that depends on the signal-to-noise ratio at the receiver input only. And to determine the specified signal-to-noise ratio, an experimental estimate of individual values of the signal and noise levels is not required. The measurement error decreases inversely to the input signal-to-noise ratio and the square root of the sample size. The proposed solution can be used to control the quality of the communication channel and the presence of an information signal in it.

Acknowledgements. The work was supported by the Ministry of Education and Science of the Russian Federation (research project No. FSWF-2023-0012).

References

1. Sklar, B., Harris, F.J.: Digital Communications: Fundamentals and Applications, 3rd edn. Pearson, London (2020)
2. Proakis, J.G., Salehi, M.: Digital Communications, 5th edn. McGraw-Hill Higher Education, New York (2007)
3. Levin, B.R.: Theoretical Principles of Radioengineering Statistics. Defense Technical Information Center, Virginia (1968)
4. Van Trees, H.L., Bell, K.L., Tian, Z.: Detection, Estimation, and Modulation Theory, Part I. Detection, Estimation, and Filtering Theory, 2nd edn.. Wiley, New York (2013)
5. Chernoyarov, O., Litvinenko, V., Kaung Myat San, Glushkov, A., Litvinenko, Y.: On the demodulation of the ADPSK signals. In: 14th International Conference "ELEKTRO2022", pp. 1–5. IEEE, Krakow (2022)
6. Liang, D., Ng, S.X., Hanzo, L.: Near-capacity turbo coded soft-decision aided DAPSK/Star-QAM for amplify-and-forward based cooperative communications. IEEE Trans. Commun. **61**(3), 1080–1087 (2013)
7. Genç, F., Yengel, E., Savaşçihabeş, A., Ertuğ, Ö.: On the optimum ring ratio determination for 16-DAPSK modulation in OFDM systems. In: 2014 22nd Signal Processing and Communications Applications Conference, SIU, pp. 742–745. IEEE, Trabzon (2014)
8. Urkowitz, H.: Energy detection of unknown deterministic signals. Proc. IEEE **55**(4), 523–531 (1967)

9. Litvinenko, V.P., Litvinenko, Y.V., Matveev, B.V., Chernoyarov, O.V., Makarov, A.A.: The new detection technique of phase-shift keyed signals. In: 2016 International Conference on Applied Mathematics. Simulation and Modelling, AMSM, pp. 355–358. Atlantis Press, Zhengzhou (2016)
10. Ibe, O.: Fundamentals of Applied Probability and Random Processes, 2nd edn. Academic Press, New York (2014)
11. Forbes, C., Evans, M., Hastings, N., Peacock, B.: Statistical Distributions, 4th edn. Wiley, New York (2010)
12. Feller, W.: An introduction to Probability Theory and Its Applications, vol. 1, 3rd edn. Wiley, New York (1991)
13. Korn, G.A., Korn, T.M.: Mathematical Handbook for Scientists and Engineers: Definitions, Theorems, and Formulas for Reference and Review. Dover Publications, Mineola (2000)
14. Chernoyarov, O.V., Glushkov, A.N., Litvinenko, V.P., Litvinenko, Y.V., Matveev, B.V.: Fast digital algorithms for the noncoherent demodulation of the differential phase-shift keyed binary signals. Int. Rev. Electr. Eng. **13**(4), 334–341 (2018)

High-Order Non-uniform Grid Scheme for Numerical Analysis of Shortest Queue Control Problem with a Small Parameter

Sergey A. Vasilyev(✉), Mohamed A. Bouatta, Shahmurad K. Kanzitdinov, and Galina O. Tsareva

Peoples' Friendship University of Russia (RUDN University), 6 Miklukho-Maklaya Street, Moscow 117198, Russian Federation
vasilyev-sa@rudn.ru
http://www.rudn.ru

Abstract. The shortest queue problem (SQP) is widely applied for load-balancing procedure in telecommunication systems. Now SQP is one of the most intensively studied and many investigators give full consideration to the SQP research for large-scale queueing systems (LSQS) with a huge number of identical devices because of the integration of 5G/6G networks is carried out at a very high speed. In this work we apply the high-order non-uniform grid numerical scheme for the investigation of the LSQS dynamics with identical single-service devices. We assume that identical services provide by devices with exponentially distributed service time and there is a Poisson incoming flow of requests for LSQS services with finite intensity. LSQS implements the service discipline so that each input request provides a random selection from any m-set devices such device that has the s-th shortest (or equivalently, the $(m-s)$-th longest) queue size. The evolution dynamics LSQS can be considered using the solutions analysis, where the solutions can be obtained by solving a system of differential equations of infinite degree which can be found with the use of the Markov chains method. We consider the control problem for LSQS, which is formulated in the form of the boundary value problem (BVP) for this system of differential equations with a small parameter. We use the truncation procedure for this singularly perturbed BVP and investigate BVP for the finite order system of differential equations. We apply the high-order non-uniform grid scheme for numerical solving of the truncated BVP. The grid scheme demonstrates good convergence of solutions of the singularly perturbed BVP when a small parameter tend to zero. The results of the LSQS dynamics simulation demonstrate that this LSQS with the control capable of serving a huge number of input requests.

Keywords: Infinite systems of differential equations · Control problem with a small parameter · Singular perturbed boundary value problem ·

S. A. Vasilyev—This paper has been supported by the RUDN University Strategic Academic Leadership Program.

R. Silhavy and P. Silhavy (Eds.): CSOC 2023, LNNS 723, pp. 639–653, 2023.
https://doi.org/10.1007/978-3-031-35317-8_58

Shortest queue problem · Numerical analysis of the boundary value problem · Layer-adapted piecewise uniform Shishkin-type meshes

1 Introduction

Modern researches in the field of telecommunication technologies, 5G/6G networks, Internet of Things (IoT) and large-scale queueing systems (LSQS) is carried out using mathematical methods [8,16,19] and simulation approaches [2,15,17].

There are many considerations of LSQS with shortest queues [4,10,20] and optimal control problems for LSQS [6,7].

The problem of the queueing systems scaling is of particular interest due to the significance of 5G/6G networks stability analysis [1,3]. Scaling methods demonstrate high efficiency with the use of asymptotics approaches for LSQS modeling [5,11,12].

Recently, there are some high-efficiency layer-adapted methods for numerical analysis of control problems with singularly perturbed systems differential equations which are modifications of non-uniform mesh methods [9,13,14].

In papers [18] we consider the evolution of LSQS dynamics that can be found by solving a system of differential equations of infinite degree. We formulate the singularly perturbed Cauchy problem for this system of differential equations with a small parameter. We use the truncation procedure for this singularly perturbed Cauchy problem and consider the finite order system of differential equations. We apply a high-order non-uniform grid scheme for numerical solving of the truncated Cauchy problem. The grid scheme demonstrates good convergence of solutions of the singularly perturbed Cauchy problem when a small parameter tend to zero.

In this paper we use our approach for LSQS from [18] and we apply numerical methods for analysis of the evolution of LSQS dynamics with $n \to \infty$ identical single-service devices, each with its own exponentially distributed service times of mean $\bar{t} = 1/\mu$, where $\mu > 0$ is a service intensity. We consider LSQS for which there is a Poisson incoming flow of requests with the intensity $n\lambda > 0$. LSQS implements the service discipline so that each input request provides a randomly selection from any m-set devices such device that has the s-th shortest (or equivalently, the $(m-s)$-th longest) queue size, $1 \le s \le m$. The evolution dynamics LSQS can be investigated through behavior analysis of the functions $x_k^{s,m}(t)$ $(k = 0,1,2,\ldots; s,m = 1,2,\ldots; 1 \le s \le m)$, which can be found by solving the boundary value problem (BVP) for the system of differential equations infinite degree. We formulate the singularly perturbed control problem for this LSQS with a small parameter. We apply the truncation procedure for this singularly perturbed control problem and formulate the control problem in the form of BVP for the finite degree system of the differential equations. We apply the high-order non-uniform grid scheme of the Shishkin-type for numerical solving of this BVP. We use a small parameter for time-scaling processes analysis for LSQS-control. The grid scheme demonstrates good convergence of solutions of the singularly perturbed BVP when a small parameter $\varepsilon \to 0$. The results of the numerical simulation show that this LSQS with the control can hold with a high incoming flow of requests.

2 Shortest Queue Systems and a Small Parameter

In the work [18] we study the shortest queue problem for LSQS with FCFS identical single-service device with its own exponentially distributed service times, where the value of the average service time is $\bar{t} = 1/\mu$ ($\mu > 0$ is a service intensity). We propose the use of the Poisson arrival rate of requests with $n\lambda > 0$ in the model. We suppose that it is possible to select m identical devices for each request upon it arrival randomly and immediately. The next procedure with each request is a choice of one single device among m selected devices that has the s-th shortest ($(m-s)$-th longest) queue length in the choice moment. If there is more than one device with the s-th shortest queue size, the choice between them is made randomly and the request is sent to the chosen device after selection immediately.

Let $\mathbf{x}^{s,m}(t) = \{x_k^{s,m}(t)\}_{k=0}^{\infty}$ be an infinite dimensional vector, where the functions $x_k^{s,m}(t)$ are the shares of the identical devices that have the queue lengths not less than the value k. Thus, there are the next formulas:

$$x_{k-1}^{s,m}(t) \geq x_k^{s,m}(t),\ x_0^{s,m}(t) = 1,\ \sum_{k=1}^{\infty} x_k^{s,m}(t) < \infty,\ t \geq 0,$$

$$x_k^{s,m}(t) = \zeta_k^{s,m}/n,\ 0 \leq \zeta_k^{s,m} \leq n,$$

$$k \in \mathcal{Z}_+,\ \mathcal{Z}_+ = \{0,1,2,\ldots\},\ s,m \in \mathcal{N},\ \mathcal{N} = \{1,2,\ldots\}.$$

The functions $x_k^{s,m}(t)$ become deterministic in the infinite limit $n \to \infty$ and we can find $x_k^{s,m}(t)$ by solving the Cauchy problem for the infinite system of differential equations with small parameter $\varepsilon > 0$ in the such form [18]:

$$\begin{cases} \varepsilon^{b_k} \dot{x}_k^{s,m}(t) = \mu\left(x_{k+1}^{s,m}(t) - x_k^{s,m}(t)\right) - \lambda \Delta h_{s,m}(x_k^{s,m}(t)), \\ x_0^{s,m}(t) = 1,\ x_k^{s,m}(0) = g_k \geq 0,\ g_k \geq g_{k+1},\ k \geq 1,\ t \geq 0, \end{cases} \tag{1}$$

where $\mathbf{b} = \{b_k\}_{k=1}^{\infty}$, $(b_k \geq 0,\ b_k \in \mathbf{R})$ is a numerical sequence (we use the time transformation in this form $\bar{t}_k = \varepsilon^{-b_k} t$) and $\mathbf{g} = \{g_k\}_{k=0}^{\infty}$ $(g_0 = 1, g_k \geq g_{k+1}, g_k \geq 0, g_k \in \mathbf{R})$ is a non-increasing sequence. The function $\Delta h_{s,m}(x_k^{s,m}(t))$ has the form for $1 \leq s \leq m$ $(s,m \in \mathcal{N})$:

$$\Delta h_{s,m}(x_k^{s,m}(t)) = \left(h_{s,m}(x_k^{s,m}(t)) - h_{s,m}(x_{k-1}^{s,m}(t))\right),\ h_{s,m}(x_k^{s,m}(t))$$

$$= \sum_{l=0}^{s-1} C_m^l (1 - x_k^{s,m}(t))^l (x_k^{s,m}(t))^{m-l} = \sum_{l=0}^{s-1} \sum_{p=0}^{l} \frac{(-1)^{l-p} m!}{p!(m-l)!(l-p)!} (x_k^{s,m}(t))^{m-p},$$

$$\Delta h_{s,m}(x_k^{s,m}(t)) = \sum_{l=0}^{s-1} \sum_{p=0}^{l} (-1)^{l-p} C_m^l C_l^p [(x_k^{s,m}(t))^{m-p} - (x_{k-1}^{s,m}(t))^{m-p}]$$

$$= \sum_{l=0}^{s-1}\sum_{p=0}^{l} \frac{(-1)^{l-p}m!}{p!(m-l)!(l-p)!}[(x_k^{s,m}(t))^{m-p} - (x_{k-1}^{s,m}(t))^{m-p}].$$

We can rewrite the Cauchy problem (1) in this way

$$\begin{cases} \dot{\mathbf{x}}^{s,m} = \mathbf{F}(\mathbf{x}^{s,m}, \mu, \lambda, \varepsilon, \mathbf{b}, t), \\ \mathbf{x}^{s,m}(0) = \mathbf{g}, \end{cases} \tag{2}$$

$$\mathbf{x}^{s,m}(t) = (x_1^{s,m}(t), \ldots, x_n^{s,m}(t), \ldots), \mathbf{F}(t) = (F_1(t), \ldots, F_n(t), \ldots),$$
$$F_k(t) = \varepsilon^{-b_k}[\mu[x_{k+1}^{s,m}(t) - x_k^{s,m}(t)] - \lambda \Delta h_{s,m}(x_k^{s,m}(t))], k \geq 1,$$
$$\mathbf{x}^{s,m}(0) = (x_1^{s,m}(0), x_2^{s,m}(0), \ldots, x_n^{s,m}(0), \ldots), x_0^{s,m}(t) = 1,$$
$$\mathbf{b} = (b_1, b_2, \ldots, b_n, \ldots), \mathbf{g} = (g_1, g_2, \ldots, g_n, \ldots),\ g_k \geq 0,\ g_0 = 1,\ g_k \geq g_{k+1},\ k \geq 1.$$

If we allow $b_k = 0$, $k = 1, 2, \ldots, l$, $b_k > 0$, $k = l+1, l+2, \ldots$ $(l \geq 2)$, then there is a Tikhonov problem [18].

3 Large-Scale Queueing Systems and the Optimal Execution Problem with a Small Parameter

We formulate the optimal execution problem on the segment $[0, T_0]$ with a small parameter ε in such a way that among permissible controls $\mathbf{u}^{s,m}(t)$ find an optimal control $\mathbf{u}^{s,m*}(t)$ under the influence of which the LSQS evolves in such a way that there is a minimum integral

$$J(\mathbf{x}^{s,m}(t), \mathbf{u}^{s,m}(t)) = \frac{1}{2}\sum_{k=1}^{\infty}\int_0^{T_0} \left((x_k^{s,m}(t))^2 + (u_k^{s,m}(t))^2\right) dt,$$

$$\begin{cases} J(\mathbf{x}^{s,m}(t), \mathbf{u}^{s,m}(t)) \to \min, \\ \dot{\mathbf{x}}^{s,m} = \mathbf{G}(\mathbf{x}^{s,m}(t), \mu, \lambda, \varepsilon, \mathbf{b}, t; \mathbf{u}^{s,m}(t)), \\ \mathbf{x}^{s,m}(0) = \mathbf{g}^0,\ \mathbf{u}^{s,m}(T_0) = \mathbf{g}^f, \end{cases} \tag{3}$$

$$\mathbf{x}^{s,m}(t) = (x_1^{s,m}(t), \ldots, x_n^{s,m}(t), \ldots),\ \mathbf{G}(t) = (G_1(t), \ldots, G_n(t), \ldots),$$
$$G_k(t) = \varepsilon^{-b_k}[\mu\left(x_{k+1}^{s,m}(t) - x_k^{s,m}(t)\right) - \lambda \Delta h_{s,m}(x_k^{s,m}(t)) + u_k^{s,m}(t)]$$
$$= \varepsilon^{-b_k}[F_k(t) + u_k^{s,m}(t)], k \geq 1,\ \mathbf{b} = (b_1, b_2, \ldots, b_n, \ldots),$$
$$\mathbf{x}^{s,m}(0) = (x_1^{s,m}(0), x_2^{s,m}(0), \ldots, x_n^{s,m}(0), \ldots), x_0^{s,m}(t) = 1,$$
$$\mathbf{u}^{s,m}(T_0) = (u_1^{s,m}(T_0), u_2^{s,m}(T_0), \ldots, u_n^{s,m}(T_0), \ldots),$$
$$\mathbf{g}^0 = (g_1^0, g_2^0, \ldots, g_n^0, \ldots),\ \mathbf{g}^f = (g_1^f, g_2^f, \ldots, g_n^f, \ldots),$$

where T_0 is a positive and fixed value, $\mathbf{u}^{s,m}(t) = (u_1^{s,m}(t), \ldots, u_n^{s,m}(t), \ldots)$ is a control for LSQS (3), $\{u_k^{s,m}(t)\}_{k=1}^{\infty}$ are control variables, $\varepsilon > 0$ is a small parameter, $\mathbf{b} = \{b_k\}_{k=1}^{\infty}$, $(b_k \geq 0)$ is a numerical sequence of positive real numbers,

$\mathbf{g}^0 = \{g_k^0\}_{k=1}^{\infty}$ $(g_k^0 \geq 0,\ g_k^0 \geq g_{k+1}^0,\ k \geq 1)$ is a non-increasing non-negative sequence of real numbers and $\mathbf{g}^f = \left\{g_k^f\right\}_{k=1}^{\infty}$ is a sequence of real numbers.

Thus, the solution of the problem (3) allows us to find a minimum of the functional:

$$I[T_0] = \sum_{k=1}^{\infty} \alpha_k\, x_k^{s,m}(t) \to \min, \tag{4}$$

where α_k $(k = 1, 2, \ldots)$ are fixed numbers.

We can consider a Hamilton function of the form:

$$H\left(\mathbf{\Psi}, \mathbf{x}^{s,m}, \mu, \lambda, \varepsilon, \mathbf{b}, t; \mathbf{u}^{s,m}\right) = \sum_{k=1}^{\infty} \Psi_k(t)\, G_k(t), \tag{5}$$

where $\mathbf{\Psi} = (\Psi_1(t), \Psi_2(t), \ldots)$ $(\{\Psi_k(t)\}_{k=1}^{\infty})$ are conjugate functions which can be found as a solution to the following Cauchy problem:

$$\begin{cases} \dot{\Psi}_k = -\partial H\left(\mathbf{\Psi}, \mathbf{x}^{s,m}, \mu, \lambda, \varepsilon, \mathbf{b}, t; \mathbf{u}^{s,m}\right)/\partial x_k^{s,m},\ k = 1, 2, \ldots, \\ \Psi_k(T_0) = -\alpha_k,\ k = 1, 2, \ldots \end{cases} \tag{6}$$

So we can rewrite the differential equations from the system (3)

$$\dot{\mathbf{x}}^{s,m} = \mathbf{G}(\mathbf{x}^{s,m}(t), \mu, \lambda, \varepsilon, \mathbf{b}, t; \mathbf{u}^{s,m}(t)),$$

in the form:

$$\dot{x}_k^{s,m} = \partial H\left(\mathbf{\Psi}, \mathbf{x}^{s,m}, \mu, \lambda, \varepsilon, \mathbf{b}, t; \mathbf{u}^{s,m}\right)/\partial \Psi_k,\ k = 1, 2, \ldots$$

Thus, if piecewise continuous control $\mathbf{u}^{s,m*}(t)$ $(t \in [0, T_0])$ provides minimal value to the functionality $I[T_0]$ (4) for the LSQS evolution, which is described by a system of differential equations (3), then there is a non-zero vector function $\mathbf{\Psi}^*(t)$, which satisfies the conditions (6), that provide maximum of the Hamilton function H together with the function $\mathbf{u}^{s,m*}$ for any $t \in [0, T_0]$

$$H(\mathbf{\Psi}^*, \mathbf{x}_{opt}^{s,m}, \mu, \lambda, \varepsilon, \mathbf{b}, t; \mathbf{u}^{s,m*}) = \max_{\mathbf{u}^{s,m*}, t \in [0,T_0]} H(\mathbf{\Psi}^*, \mathbf{x}^{s,m}, \mu, \lambda, \varepsilon, \mathbf{b}, t; \mathbf{u}^{s,m*})$$

where $\mathbf{x}_{opt}^{s,m}(t)$ is an optimal evolution of LSQS on the segment $[0, T_0]$.

4 Truncated Optimal Execution Problem with a Small Parameter

We can write the optimal execution problem (3) in a truncated form:

$$\begin{cases} \tilde{J}(\tilde{\mathbf{x}}^{s,m}, \tilde{\mathbf{u}}^{s,m}(t)) \to \min, \\ \dot{\tilde{\mathbf{x}}}^{s,m} = \tilde{\mathbf{G}}(\tilde{\mathbf{x}}^{s,m}(t), \mu, \lambda, \varepsilon, \mathbf{b}, t; \tilde{\mathbf{u}}^{s,m}(t)), \\ \tilde{\mathbf{x}}^{s,m}(0) = \tilde{\mathbf{g}}^0, \tilde{\mathbf{u}}^{s,m}(T_0) = \tilde{\mathbf{g}}^f, \end{cases} \tag{7}$$

$$\tilde{J}(\tilde{\mathbf{x}}^{s,m},\tilde{\mathbf{u}}^{s,m}(t))=\frac{1}{2}\sum_{k=1}^{n}\int_0^{T_0}\left((x_k^{s,m}(t))^2+(u_k^{s,m}(t))^2\right)dt,$$

$$\tilde{\mathbf{x}}^{s,m}(t)=(x_1^{s,m}(t),\ldots,x_{n+1}^{s,m}(t)), x_0(t)=1, \tilde{\mathbf{G}}(t)=(\tilde{G}_1(t),\ldots,\tilde{G}_{n+1}(n)),$$

$$\tilde{G}_k(t)=\varepsilon^{-b_k}\left[\mu\left(x_{k+1}^{s,m}(t)-x_k^{s,m}(t)\right)-\lambda\Delta h_{s,m}(x_k^{s,m}(t))+u_k^{s,m}(t)\right]$$

$$=\varepsilon^{-b_k}[F_k(t)+u_k^{s,m}(t)],\ k=\overline{1,n-1},$$

$$\tilde{G}_n(t)=\varepsilon^{-b_n}[-\mu x_n^{s,m}(t)-\lambda\Delta h_{s,m}(x_n^{s,m}(t))+u_n^{s,m}(t)]$$

$$=\varepsilon^{-b_n}[\tilde{F}_n(t)+u_n^{s,m}(t)],\ \tilde{F}_n=-\mu x_n^{s,m}(t)-\lambda\Delta h_{s,m}(x_k^{s,m}(t)),$$

$$\tilde{G}_{n+1}(t)=\frac{1}{2}\sum_{k=1}^{n}\left((x_k^{s,m}(t))^2+(u_k^{s,m}(t))^2\right),$$

$$\tilde{\mathbf{x}}^{s,m}(0)=(x_1^{s,m}(0),x_2^{s,m}(0),\ldots,x_{n+1}^{s,m}(0)),\ x_0^{s,m}(0)=1,$$

$$\tilde{\mathbf{u}}^{s,m}(T_0)=(u_1^{s,m}(T_0),u_2^{s,m}(T_0),\ldots,u_n^{s,m}(T_0)),$$

$$\tilde{\mathbf{g}}^0=(g_1^0,g_2^0,\ldots,g_{n+1}^0),\ g_k^0\geq 0,g_k^0\geq g_{k+1}^0,k=\overline{1,n},\ g_{n+1}^0=0,$$

$$\tilde{\mathbf{g}}^f=(g_1^f,g_2^f,\ldots,g_n^f),$$

where $\tilde{\mathbf{u}}^{s,m}(t)=(u_1^{s,m}(t),\ldots,u_n^{s,m}(t))$ is a control for LSQS, $\{u_k^{s,m}(t)\}_{k=1}^n$ are control variables, $\varepsilon>0$ is a small parameter, $\mathbf{b}=\{b_k\}_{k=1}^n$ $(b_k\geq 0)$ are positive real numbers, $\tilde{\mathbf{g}}^0=\{g_k^0\}_{k=1}^{n+1}$ are non-increasing non-negative real numbers and $\tilde{\mathbf{g}}^f=\left\{g_k^f\right\}_{k=1}^n$ are real numbers.

We can use an additional variable in the system (7)

$$x_{n+1}^{s,m}(t)=\frac{1}{2}\sum_{k=1}^{n}\int_0^{T_0}\left((x_k^{s,m}(t))^2+(u_k^{s,m}(t))^2\right)dt,$$

where we have the differential equation and the initial condition for this variable in this case:

$$\dot{x}_{n+1}^{s,m}(t)=\frac{1}{2}\sum_{k=1}^{n}\left((x_k^{s,m}(t))^2+(u_k^{s,m}(t))^2\right),\ x_{n+1}^{s,m}(0)=0.$$

Thus, the solution of the problem (7) allows us to find a minimum of the functional:

$$\tilde{I}[T_0]=\sum_{k=1}^{n+1}\alpha_k\,x_k^{s,m}(t)\to\min, \tag{8}$$

where α_k $(k=1,2,\ldots,n+1)$ are fixed numbers.

Let this functional has the form: $\tilde{I}[T_0]=x_{n+1}^{s,m}(t)\to\min$, where $\alpha_k=0$ $(k=1,2,\ldots,n)$ and $\alpha_{n+1}=1$.

Hence, we can write a Hamilton function in this way:

$$\tilde{H}\left(\tilde{\mathbf{\Psi}}, \tilde{\mathbf{x}}^{s,m}, \mu, \lambda, \varepsilon, \mathbf{b}, t; \tilde{\mathbf{u}}^{s,m}\right) = \sum_{k=1}^{n+1} \Psi_k(t)\, \tilde{G}_k(t) = \sum_{k=1}^{n-1} \varepsilon^{-b_k} \Psi_k [F_k(t) + u_k^{s,m}(t)]$$

$$+ \varepsilon^{-b_n} \Psi_n(t) [\tilde{F}_n(t) + u_n^{s,m}(t)] + \frac{1}{2} \Psi_{n+1}(t) \sum_{k=1}^{n} \left((x_k^{s,m}(t))^2 + (u_k^{s,m}(t))^2 \right),$$

where $\tilde{\mathbf{\Psi}} = (\Psi_1(t), \Psi_2(t), \ldots, \Psi_{n+1}(t))$ ($\{\Psi_k(t)\}_{k=1}^{n+1}$) are conjugate functions which can be found as a solution to the following Cauchy problem:

$$\begin{cases} \dot{\tilde{\Psi}}_k = -\partial \tilde{H}\left(\tilde{\mathbf{\Psi}}, \tilde{\mathbf{x}}^{s,m}, \mu, \lambda, \varepsilon, \mathbf{b}, t; \tilde{\mathbf{u}}^{s,m}\right) / \partial x_k^{s,m}, \; k = \overline{1, n+1}, \\ \Psi_k(T_0) = 0, \; k = \overline{1, n}, \\ \Psi_{n+1}(T_0) = -1, \end{cases} \tag{9}$$

where

$$\dot{\tilde{\Psi}}_k = -\sum_{j=1}^{n+1} \Psi_j(t) \frac{\partial \tilde{G}_j(t)}{\partial x_k^{s,m}} = -\left[\sum_{j=1}^{3} \Psi_{j-2}(t) \frac{\partial \tilde{G}_{j-2}(t)}{\partial x_k^{s,m}} \right] - \Psi_{n+1}(t) x_k^{s,m}(t)$$

$$= -\varepsilon^{-b_{k-1}} \mu \Psi_{k-1} + \varepsilon^{-b_k} \Psi_k \left[\mu + \lambda h_{s,m}^*(x_k^{s,m}(t)) \right] - \varepsilon^{-b_{k+1}} \lambda \Psi_{k+1} h_{s,m}^*(x_k^{s,m}(t))$$
$$- \Psi_{n+1} x_k^{s,m}(t), \; k = \overline{1, n-1},$$

$$\dot{\tilde{\Psi}}_n = -\varepsilon^{-b_{n-1}} \mu \Psi_{n-1} + \varepsilon^{-b_n} \Psi_n \left[\mu + \lambda h_{s,m}^*(x_n^{s,m}(t)) \right] - \Psi_{n+1} x_n^{s,m}(t),$$

$$h_{s,m}^*(x_k^{s,m}(t)) = \frac{\partial h_{s,m}(x_k^{s,m}(t))}{\partial x_k^{s,m}} = \sum_{l=0}^{s-1} \sum_{p=0}^{l} \frac{(-1)^{l-p} (m-p) m!}{p!(m-l)!(l-p)!} (x_k^{s,m}(t))^{m-p-1}.$$

The function $\tilde{H}(\tilde{\mathbf{\Psi}}, \tilde{\mathbf{x}}^{s,m}, \mu, \lambda, \varepsilon, \mathbf{b}, t; \tilde{\mathbf{u}}^{s,m})$ is maximized if there are the following conditions $\partial \tilde{H}(\tilde{\mathbf{\Psi}}, \tilde{\mathbf{x}}^{s,m}, \mu, \lambda, \varepsilon, \mathbf{b}, t; \tilde{\mathbf{u}}^{s,m}) / \partial u_k^{s,m} = 0$, $k = \overline{0, n}$, thus, we get the equalities $\varepsilon^{-b_k} \Psi_k(t) + \Psi_{n+1}(t) u_k^{s,m}(t) = 0$, $k = \overline{1, n}$. If we use the condition $\Psi_{n+1} = -1$ (9), we have the following equality $\Psi_k(t) = \varepsilon^{b_k} u_k^{s,m}(t)$ $k = \overline{1, n}$. Hence, we obtain the following system of differential equations:

$$\dot{u}_k^{s,m} = \varepsilon^{-b_k} [\mu(u_k^{s,m} - u_{k-1}^{s,m}) + \lambda(u_k^{s,m} - u_{k+1}^{s,m}) h_{s,m}^*(x_k^{s,m}(t)) + x_k^{s,m}(t)], \; k = \overline{1, n-1},$$

$$\dot{u}_n^{s,m} = \varepsilon^{-b_n} [\mu(u_n^{s,m} - u_{n-1}^{s,m}) + \lambda u_n^{s,m} h_{s,m}^*(x_n^{s,m}(t)) + x_n^{s,m}(t)].$$

We can also obtain the following system of differential equations

$$\dot{\tilde{x}}_k^{s,m} = \partial \tilde{H}\left(\tilde{\mathbf{\Psi}}, \tilde{\mathbf{x}}^{s,m}, \mu, \lambda, \varepsilon, \mathbf{b}, t; \tilde{\mathbf{u}}^{s,m}\right) / \partial \tilde{\Psi}_k = \tilde{G}_k(t), \; k = \overline{1, n},$$

Thus, we can consider the boundary value problem for the system of the differential equations to find solutions $\mathbf{w}^{s,m}(t)$ and $\mathbf{u}^{s,m}(t)$ in the form:

$$\begin{cases} \dot{\mathbf{w}}^{s,m} + \mathbf{M}(t) = \mathbf{N}(t), \\ \mathbf{w}_1^{s,m}(0) = \tilde{\mathbf{g}}^0, \mathbf{w}_2^{s,m}(T_0) = \tilde{\mathbf{g}}^f, \end{cases} \tag{10}$$

$$\mathbf{w}^{s,m}(t) = (\mathbf{w}_1^{s,m}(t), \mathbf{w}_2^{s,m}(t)) = (x_1^{s,m}(t), \dots, x_n^{s,m}(t), u_1^{s,m}(t), \dots, u_n^{s,m}(t)),$$

$$w_0^{s,m}(t) = 0,\ \mathbf{w}_1^{s,m}(t) = (x_1^{s,m}(t), \dots, x_n^{s,m}(t)),\ \mathbf{w}_2^{s,m}(t) = (u_1^{s,m}(t), \dots, u_n^{s,m}(t)),$$

$$\mathbf{M}(t) = (\mathbf{M}_1(t), \mathbf{M}_2(t)),\ \mathbf{N}(t) = (\mathbf{N}_1(t), \mathbf{N}_2(t)),$$

$$\mathbf{M}_1(t) = (M_{11}(t), \dots, M_{1n}(t)),\ \mathbf{M}_2(t) = (M_{21}(t), \dots, M_{2n}(t)),$$

$$\mathbf{N}_1(t) = (N_{11}(t), \dots, N_{1n}(t)),\ \mathbf{N}_2(t) = (N_{21}(t), \dots, N_{2n}(t)),$$

$$M_{1k}(t) = \varepsilon^{-b_k}[\mu(x_k^{s,m}(t) - x_{k+1}^{s,m}(t)) - u_k^{s,m}(t)]$$

$$= \varepsilon^{-b_k}[\mu(w_k^{s,m}(t) - w_{k+1}^{s,m}(t)) - w_{n+k}(t)],\ k = \overline{1, n-1},$$

$$M_{1n}(t) = \varepsilon^{-b_n}[\mu x_n^{s,m}(t) - u_n^{s,m}(t)] = \varepsilon^{-b_n}[\mu w_n^{s,m}(t) - w_{n+k}(t)],$$

$$M_{2k}(t) = \varepsilon^{-b_k}[\mu(u_{k-1}^{s,m}(t) - u_k^{s,m}(t)) - x_k^{s,m}(t)]$$

$$= \varepsilon^{-b_k}[\mu(w_{n+k-1}(t) - w_{n+k}(t)) - w_k^{s,m}(t)],\ k = \overline{1, n-1},$$

$$M_{2n}(t) = \varepsilon^{-b_n}[\mu(u_{n-1}^{s,m}(t) - u_n^{s,m}(t)) - x_n^{s,m}(t)]$$

$$= \varepsilon^{-b_n}[\mu(w_{2n-1}(t) - w_{2n}(t)) - w_n^{s,m}(t)],$$

$$N_{1k}(t) = -\varepsilon^{-b_k}\lambda\Delta h_{s,m}(x_k^{s,m}(t)) = -\varepsilon^{-b_k}\lambda\Delta h_{s,m}(w_k^{s,m}(t)),\ k = \overline{1, n-1},$$

$$N_{1n}(t) = -\varepsilon^{-b_n}\lambda\Delta h_{s,m}(x_n^{s,m}(t)) = -\varepsilon^{-b_n}\lambda\Delta h_{s,m}(w_n^{s,m}(t)),$$

$$N_{2k}(t) = \varepsilon^{-b_k}\lambda[u_k^{s,m}(t) - u_{k+1}^{s,m}(t)]h_{s,m}^*(x_k^{s,m}(t))$$

$$= \varepsilon^{-b_k}\lambda[w_{n+k}(t) - u_{n+k+1}^{s,m}(t)]h_{s,m}^*(w_k^{s,m}(t)),\ k = \overline{1, n-1},$$

$$N_{2n}(t) = \varepsilon^{-b_n}\lambda u_n^{s,m} h_{s,m}^*(x_n^{s,m}(t)) = \varepsilon^{-b_n}\lambda w_{2n} h_{s,m}^*(w_n^{s,m}(t)).$$

$$\mathbf{w}_1^{s,m}(0) = (x_1^{s,m}(0), \dots, x_n^{s,m}(0)),\ \mathbf{w}_2^{s,m}(T_0) = (u_1^{s,m}(T_0), \dots, u_n^{s,m}(T_0)),$$

$$\tilde{\mathbf{g}}^0 = (g_1^0, g_1^0, \dots, g_n^0),\ g_k^0 \geq 0,\ g_k^0 \geq g_{k+1}^0, k = \overline{1, n-1},$$

$$\tilde{\mathbf{g}}^f = (g_1^f, g_2^f, \dots, g_n^f),$$

where $\tilde{\mathbf{g}}^0 = \{g_k^0\}_{k=1}^n$ are non-increasing non-negative real numbers and $\tilde{\mathbf{g}}^f = \left\{g_k^f\right\}_{k=1}^n$ are real numbers.

5 High-Order Non-uniform Grid Scheme for Numerical Analysis of the Optimal Execution Problem with a Small Parameter

We apply a piecewise-uniform grid $\bar{\Omega}_t$ ($t_0 = 0$; $t_{i-1} < t_i$, $i = \overline{1, N}$; $t_N = T_0$) for numerical analysis of the problem (10)

$$\bar{\Omega}_t = (t_i | t_i = i\tau_1,\ i = \overline{0, K};\ t_i = t_K + (i - K)\tau_2,\ i = \overline{K+1, N}),$$

$$\tau_1 = \delta/K,\ \tau_2 = (T_0 - \delta)/(N - K),\ \delta = \bar{C}\varepsilon\, ln(\varepsilon^{-1}),$$

where a value $\bar{C}$ is determined analytically by asymptotic estimates of solutions of the problem (10). Thus, this piecewise-uniform grid $\bar{\Omega}_t$ has K small steps τ_1 and $(N - K)$ big steps τ_2 on the segment $[0, T_0]$.

We can consider a finite-difference approximation of the problem (10) in the following form:

$$\begin{cases} (L_h \mathbf{w}^{s,m})_i + \mathbf{M}_i(\mathbf{w}^{s,m}) = \mathbf{N}_i(\mathbf{w}^{s,m}),\ i = \overline{0, N}, \\ \mathbf{w}_{10}^{s,m} = \tilde{\mathbf{g}}^0,\ \mathbf{w}_{2N}^{s,m} = \tilde{\mathbf{g}}^f, \end{cases} \tag{11}$$

$$\hbar_i = 0.5(t_{i+1} - t_{i-1}),\ h_i = t_i - t_{i-1};\ 0 \le t_i \le T_0,\ i = \overline{0, N};$$

$$(L_h \mathbf{w}^{s,m})_i = \frac{1}{2\hbar_i}\left(\frac{h_i}{h_{i+1}}(\mathbf{w}_{i+1}^{s,m} - \mathbf{w}_i^{s,m}) + \frac{h_{i+1}}{h_i}(\mathbf{w}_i^{s,m} - \mathbf{w}_{i-1}^{s,m})\right),$$

$$\mathbf{M}_i(\mathbf{w}^{s,m}) = \mathbf{M}(\mathbf{w}^{s,m}(t_i)),\ \mathbf{N}_i(\mathbf{w}^{s,m}) = \mathbf{N}(\mathbf{w}^{s,m}(t_i)),$$

$$\mathbf{w}_i = \mathbf{w}^{s,m}(t_i).$$

We can obtain a system of linear algebraic equations:

$$\begin{cases} w_{k0} = g_k^0,\ k = \overline{1, n},\ w_{0i} = 0,\ i = \overline{0, N} \\ A_{ki}w_{ki-1} + B_{ki}w_{ki} + C_{ki}w_{ki+1} + Q_k w_{k+1i} + R_k w_{n+ki} = N_{ki}, \\ k = \overline{1, n},\ i = \overline{1, N-1}, \\ A_{ki}w_{ki-1} + B_{ki}w_{ki} + C_{ki}w_{ki+1} + S_k w_{k-ni} + T_k w_{n+k-1i} = N_{ki}, \\ k = \overline{n+1, 2n},\ i = \overline{1, N-1}, \\ w_{kN} = g_k^f,\ k = \overline{n+1, 2n}, \end{cases} \tag{12}$$

where

$$A_{ki} = -\frac{h_{i+1}}{2\hbar_i h_i},\ C_{ki} = \frac{h_i}{2\hbar_i h_{i+1}}\ i = \overline{1, N-1},\ k = \overline{1, 2n},$$

$$B_{ki} = \frac{1}{2\hbar_i}\left(\frac{h_{i+1}}{h_i} - \frac{h_i}{h_{i+1}}\right) + \varepsilon^{-b_k}\mu,\ i = \overline{1, N-1},\ k = \overline{1, n},$$

$$B_{ki} = \frac{1}{2\hbar_i}\left(\frac{h_{i+1}}{h_i} - \frac{h_i}{h_{i+1}}\right) - \varepsilon^{-b_{k-n}}\mu,\ i = \overline{1, N-1},\ k = \overline{n+1, 2n},$$

$$Q_k = -\varepsilon^{-b_k}\mu,\ k = \overline{1, n-1}, Q_n = 0,\ R_n = -1,\ k = \overline{1, n},$$

$$S_k = -\varepsilon^{-b_{k-n}}\mu,\ k = \overline{n+1, 2n}, T_{n+1} = 0, T_k = \varepsilon^{-b_{k-n}}\mu,\ k = \overline{n+2, 2n},$$

$$N_{ki} = -\varepsilon^{-b_k}\lambda\Delta h_{s,m}(w_{ki}),\ i = \overline{1, N-1},\ k = \overline{1, n},$$

$$N_{ki} = \varepsilon^{-b_{k-n}}\lambda[w_{ki} - w_{k+1i}]h^*(w_{ki}),\ i = \overline{1, N-1},\ k = \overline{n+1, 2n-1},$$

$$N_{ki} = \varepsilon^{-b_{k-n}}\lambda w_{ki}h^*(w_{ki}),\ i = \overline{1, N-1},\ k = 2n,$$

$$h^*(w_{ki}) = h^*_{s,m}(w_k^{s,m}(t_i)).$$

We use a modification of the Gauss method with the choice of the main element (the block matrix sweep method) for solving the systems of linear equations with a block-tape matrix (12). The idea of the algorithm is to implement the Gauss method on a packed array in which nonzero elements of matrices are placed. The solution of nonlinear systems of Eqs. (12) is obtained as a result of the iterative process. We solve the homogeneous boundary value problem (11) with $\mathbf{N}_i = 0$, $i = \overline{0, N}$ and find the solution $\mathbf{w}_i^{s,m,0}$ at the first stage of calculations. We solve the boundary value problem (11) with $\mathbf{N}_i(\mathbf{w}^{s,m,0})\, i = \overline{0, N}$ at the second stage of calculations. These iterations are repeated until convergence. The value

$$R_{kj} = 0.2 \max_{i=\overline{0,N}} \left| w_{ki+1}^{s,m,j+1} - w_{ki+1}^{s,m,j} \right|$$

is estimated for each $k = \overline{1, 2n}$ and iteration $j = 0, 1, \ldots$. If this value R_{kj} is greater than a permissible error δ, then we make the next iteration.

The numerical analysis of the solutions $x_k^{s,m}(t), u_k^{s,m}(t)\ (k = 1, 5, 10, 15, 20;$ $m = 2, 3;\ 1 \le s \le m)$ is presented in the figures (see Fig. 1, 2, 3, 4 and 5).

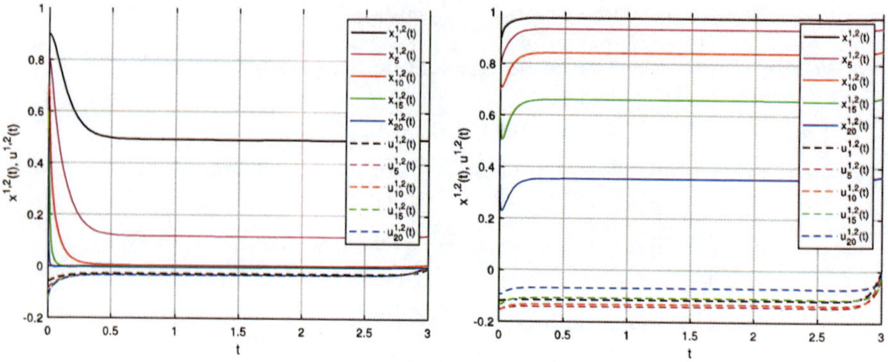

Fig. 1. The solutions $x_k^{1,2}(t)$, $u_k^{1,2}(t)$ for the parameters $s = 1$, $m = 2$, the small parameter $\varepsilon = 0.001$ and the values $\mu = 5$, $\lambda = 4$ (the left graph), $\lambda = 6$ (the right graph).

Thus, we present results from numerical experiments. We used the following parameters for numerical simulation: the arrival request rates are $\lambda = 4$ (the low incoming mode on the left graphs) and $\lambda = 6$ (the high incoming mode on

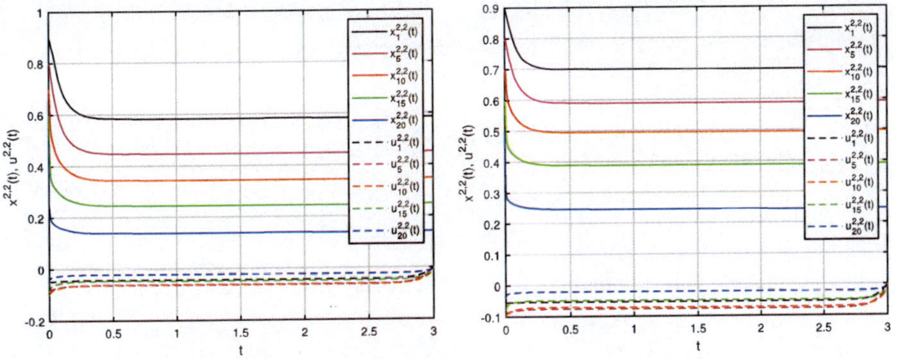

Fig. 2. The solutions $x_k^{2,2}(t)$, $u_k^{2,2}(t)$ for the parameters $s = 2$, $m = 2$, the small parameter $\varepsilon = 0.001$ and the values $\mu = 5$, $\lambda = 4$ (the left graph), $\lambda = 6$ (the right graph).

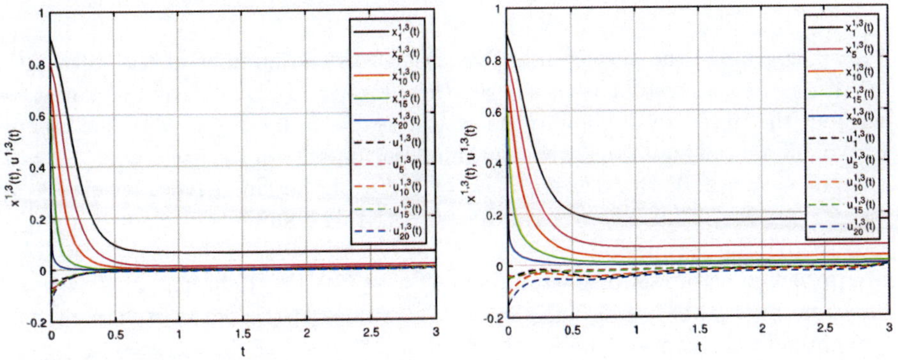

Fig. 3. The solutions $x_k^{1,3}(t)$, $u_k^{1,3}(t)$ for the parameters $s = 1$, $m = 3$, the small parameter $\varepsilon = 0.001$ and the values $\mu = 5$, $\lambda = 4$ (the left graph), $\lambda = 6$ (the right graph).

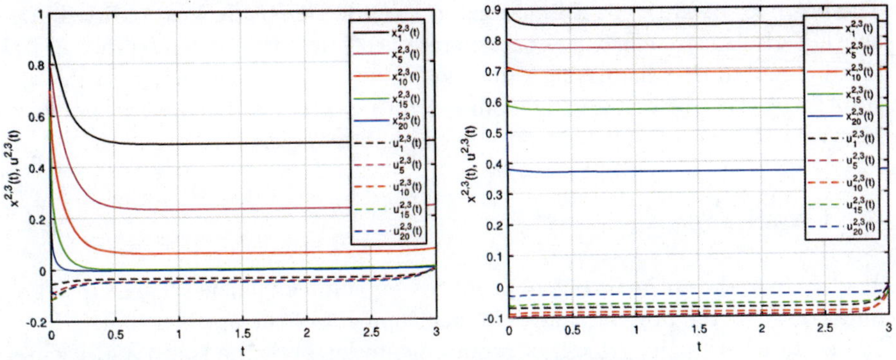

Fig. 4. The solutions $x_k^{2,3}(t)$, $u_k^{2,3}(t)$ for the parameters $s = 2$, $m = 3$, the small parameter $\varepsilon = 0.001$ and the values $\mu = 5$, $\lambda = 4$ (the left graph), $\lambda = 6$ (the right graph).

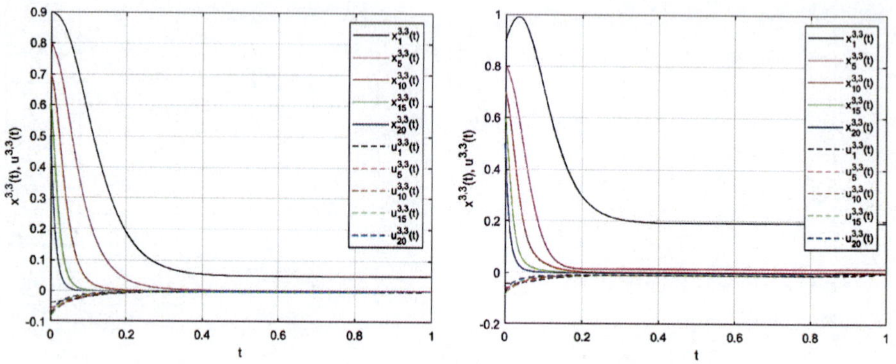

Fig. 5. The solutions $x_k^{3,3}(t)$, $u_k^{3,3}(t)$ for the parameters $s = 3$, $m = 3$, the small parameter $\varepsilon = 0.001$ and the values $\mu = 5$, $\lambda = 4$ (the left graph), $\lambda = 6$ (the right graph).

the right graph), the service intensity is $\mu = 5$, the number of the differential equations in the system is $n = 20$, the number of the differential equations without the small parameters in the system is $l = 9$ (i.e. $1 \leq k \leq 9$), the number of the differential equations with the small parameters in the system is $n - l = 11$ (i.e. $10 \leq k \leq 20$), the degrees $b_k (k = \overline{1, 20})$ of the small parameters ε^{b_k} are $b_k = 0\,(1 \leq k \leq 9)$ and $b_k = 1/k\,(10 \leq k \leq 20)$, the number of steps of the grid is $N = 10^4$, the permissible error is $\delta = 10^{-6}$. The values of the initial conditions are presented as the set of the numbers $g_k^0 = (175 - 4k)/190$, $g_k^f = 0$, $(k = \overline{1, 20})$, the time interval is $t \in [0, 3]$ $(T_0 = 3)$. The values of the parameters s, m, ε are presented in the captions under the figures (see Fig. 1, 2, 3, 4 and 5).

In the Fig. 1, 2, 3, 4 and 5, we plot the results of the numerical stimulation of the solutions $x_k^{s,m}(t)$, $u_k^{s,m}(t)(s = 1, 2, 3; m = 2, 3; s \leq m)$. It is shown that LSQS has an stable service mode in case the low load $(\lambda = 4, \mu = 5)$ and the high load $(\lambda = 6, \mu = 5)$ conditions. The solutions have the left transition layers. The behavior of the control functions $u_k^{s,m}(t)$ demonstrates load redistribution, which leads to more efficient processing requests over the time interval $t \in [0, 3]$. It is illustrated that increasing values s and m $(s \leq m)$ leads LSQS to the more stable execution of services, incoming requests processed more efficiently and solutions $x_k^{s,m}(t)$, $u_k^{s,m}(t)$ have only left boundary layers.

6 Conclusion

The study of the optimal control problems for complex queue systems is of great importance at the present time. The development of the optimal control theory is associated with the growth of requirements for both the speed and accuracy of control systems and the transition to new telecommunications technologies. A further increase in performance of complex systems is possible only with the correct allocation of limited distribution resources. Thus, the construction of high-precision control systems require to take into account the interaction of individual parts (channels) of the system in the synthesis of regulators. The synthesis of such complex multidimensional (multi-connected) systems is of great complexity in the modern theory of optimal control. The implementation of 5G/6G technology requires special attention of developers of new products in the field of wireless communication, as these technologies will be able to provide higher data transfer speeds, shorter latency, as well as greater energy efficiency compared to 4G technologies currently used. Developers will inevitably face a number of technical problems when designing the 5G/6G architecture, which must cope with a more complex multi-user environment and the use of channels at higher frequencies. The modern studies of theoretical LSQS models makes it possible to understand the problems which will be solved for the implementation of 5G/6G technology.

In this pare we use the high-order non-uniform grid numerical scheme for investigation of the LSQS dynamics with identical single-service devices. We consider the LSQS, which provides identical services by the huge number of devices. We assume that each device have exponentially distributed service time and there is a Poisson incoming flow of requests for LSQS services with the finite intensity. We investigated the case when LSQS implements the service discipline so that each input request provides a random selection from any m-set devices such device that has the s-th shortest (or equivalently, the $(m-s)$-th longest) queue size. We used the fact that the evolution dynamics LSQS can be described using the analysis of the behavior of solutions, which can be obtained by solving a system of differential equations of infinite degree. This system can be found if we use the Markov chains method. We formulate the control problem for LSQS and study the boundary value problem (BVP) for this system of differential equations with a small parameter. We apply the truncation procedure for this singularly perturbed BVP and consider BVP for the finite order system of differential equations. We use a high-order non-uniform grid scheme for numerical solving of the truncated BVP. We take into account different sets of parameters for time-scaling processes analysis for LSQS. The grid scheme demonstrates good convergence of solutions of the singularly perturbed BVP when a small parameter tend to zero. The results of the LSQS dynamics simulation demonstrate that this LSQS with the control capable of serving a huge number of input requests. We suppose that it is possible to investigate an optimal control problem for such LSQS and successfully solve it. Our ability to solve such problems is very important because it will save technical and financial resources for 5G/6G networks implementation.

This paper has been supported by the RUDN University Strategic Academic Leadership Program (recipient S.A. Vasilyev, mathematical model development, simulation model development, numerical analysis).

References

1. Banerjee, S., Budhiraja, A., Puha, A.L.: Heavy traffic scaling limits for shortest remaining processing time queues with heavy tailed processing time distributions. Ann. Appl. Probab. **32**(4), 2587–2651 (2022). https://doi.org/10.1214/21-AAP1741
2. Beliakov, G., Divakov, D.V.: Aggregation with dependencies: capacities and fuzzy integrals. Fuzzy Sets Syst. **446**, 222–232 (2022). https://doi.org/10.1016/j.fss.2021.03.009
3. der Boor, M.V., Borst, S.C., Van Leeuwaarden, J.S., Johan, S.H., Mukherjee, D.: Scalable load balancing in networked systems: a survey of recent advances. SIAM Rev. **64**(3), 554–622 (2022). https://doi.org/10.1137/20M1323746
4. Bu, Q.-H., Liu, L.-W., Tang, J.-S., Zhao, Y.-Q.Q.: Approximations for a queueing game model with join-the-shortest-queue strategy. J. Oper. Res. Soc. China 1–16 (2022). https://doi.org/10.1007/s40305-021-00382-5
5. Bushkova, T., Moiseeva, S., Moiseev, A., Sztrik, J., Lisovskaya, E., Pankratova, E.: Using infinite-server resource queue with splitting of requests for modeling two-channel data transmission. Methodol. Comput. Appl. Probab. **24**, 1753–1772 (2022). https://doi.org/10.1007/s11009-021-09890-6
6. Chen, H., Wang, Z.: Optimal control for parallel queues with a single batch server. Oper. Res. Lett. **50**(4), 377–383 (2022). https://doi.org/10.1016/j.orl.2022.04.004
7. Haviv, M., Ravner, L.: A survey of queueing systems with strategic timing of arrivals. Queueing Syst. **99**, 163–198 (2021). https://doi.org/10.1007/s11134-021-09717-8
8. Kondratyeva, A., et al.: Characterization of dynamic blockage probability in industrial millimeter wave 5G deployments. Future Internet **14**(7), 193 (2022). https://doi.org/10.3390/fi14070193
9. Kumar, S., Sumit, S., Vigo-Aguiar, J.: A high order convergent numerical method for singularly perturbed time dependent problems using mesh equidistribution. Math. Comput. Simul. **199**, 287–306 (2022). https://doi.org/10.1016/j.matcom.2022.03.025
10. Liu, X., Gong, K., Ying, L.: Steady-state analysis of load balancing with Coxian-2 distributed service times. Naval Res. Logist. **69**(1), 57–75 (2022). https://doi.org/10.1002/nav.21986
11. Nazarov, A., Dudin, A., Moiseev, A.: Pseudo steady-state period in non-stationary infinite-server queue with state dependent arrival intensity. Mathematics **10**(15), 2661 (2022). https://doi.org/10.3390/math10152661
12. Nazarov, A., Phung-Duc, T., Izmailova, Y.: Asymptotic-diffusion analysis of multiserver retrial queueing system with priority customers. In: Dudin, A., Nazarov, A., Moiseev, A. (eds.) ITMM 2020. CCIS, vol. 1391, pp. 236–250. Springer, Cham (2021). https://doi.org/10.1007/978-3-030-72247-0_18
13. Negero, N.T.: A uniformly convergent numerical scheme for two parameters singularly perturbed parabolic convection-diffusion problems with a large temporal lag. Results Appl. Math. **16**, 100338 (2022). https://doi.org/10.1016/j.rinam.2022.100338
14. Ranjan, K.R., Gowrisankar, S.: Uniformly convergent NIPG method for singularly perturbed convection diffusion problem on Shishkin type meshes. Appl. Numer. Math. **179**, 125–148 (2022). https://doi.org/10.1016/j.apnum.2022.04.015

15. Sevastianov, L.A., Lovetskiy, K.P., Kulyabov, D.S.: Multistage collocation pseudo-spectral method for the solution of the first order linear ODE. In: Proceedings 8th International Conference on Information Technology and Nanotechnology, ITNT 2022 (2022). https://doi.org/10.1109/itnt55410.2022.9848731
16. Sidorov, S., Faizliev, A., Balash, V., Balash, O., Krylova, M., Fomenko, A.: Extended innovation diffusion models and their empirical performance on real propagation data. J. Mark. Anal. **9**, 99–110 (2021). https://doi.org/10.1057/s41270-021-00106-x
17. Tiutiunnik, A.A., Divakov, D.V., Lovetskiy, K.P., Sevastianov, A.L.: Single-mode propagation of adiabatic guided mode in a smooth transition between planar waveguides. In: Proceedings Laser Physics, Photonic Technologies, and Molecular Modeling. Event: XXV Annual Conference Saratov Fall Meeting 2021; and IX Symposium on Optics and Biophotonics 2021, Saratov, Russian Federation, vol. 12193, p. 1219305 (2022). https://doi.org/10.1117/12.2626384
18. Vasilyev, S.A., Bouatta, M.A., Tsareva, G.O.: High-order non-uniform grid scheme for numerical analysis of queueing system with a small parameter. In: Silhavy, R., Silhavy, P., Prokopova, Z. (eds.) CoMeSySo 2022. LNCS, vol. 597, pp. 785–797. Springer, Cham (2023). https://doi.org/10.1007/978-3-031-21438-7_66
19. Yarkina, N., Correia, L.M., Moltchanov, D., Gaidamaka, Y., Samouylov, K.: Multi-tenant resource sharing with equitable-priority-based performance isolation of slices for 5G cellular systems. Comput. Commun. **188**, 39–51 (2022). https://doi.org/10.1016/j.comcom.2022.02.019
20. Zhou, X., Shroff, N., Wierman, A.: Asymptotically optimal load balancing in large-scale heterogeneous systems with multiple dispatchers. Perform. Eval. **145**, 102146 (2021). https://doi.org/10.1016/j.peva.2020.102146

Low Cost IoT-Based Automated Locust Monitoring System, Kazungula, Zambia

Brian Halubanza[1](✉), Jackson Phiri[1], Mayumbo Nyirenda[1], Phillip O. Y. Nkunika[2], and Douglas Kunda[3]

[1] Computer Science Department, University of Zambia, Lusaka, Zambia
bhalubanza@gmail.com, jackson.phiri@unza.zm, mayumbo.nyirenda@cs.unza.zm
[2] Biology Department, University of Zambia, Lusaka, Zambia
pnkunika@unza.zm
[3] ZCAS University, Lusaka, Zambia
douglas.kunda@zcasu.edu.zm

Keywords: World Population · Food Production · Technology · Agriculture · Climate Change · Pests · Locust Infestations · Internet of Things (IoT) · Sensors · Data Collection · Predictive Analytics · Pest Control · Deep Learning · Design Science Research Methodology

1 Introduction

The world's population is expected to reach 9.7 Billion by 2050 and the demand for food will increase. The use of technology to find better ways of enhancing food production is being encouraged so as to meet the anticipated consumption demand. Agriculture is however not free of other factors that threaten its existence [1]. Climate change and pests threaten the world's food security [2]. Locust infestations can cause significant damage to agricultural crops, and managing these pests can be challenging [3, 4]. One potential solution is to use Deep Learning (DL) embedded low-cost Internet of Things (IoT) devices to monitor and control locust populations. This approach involves deploying IoT sensors in locust breeding areas to collect data on locust presence and environmental conditions such as temperature and humidity. This information can then be used to predict and monitor locust swarms, enabling farmers to take proactive measures to prevent infestations. There are few trained agricultural extension officers in countries that have a history of locust invasion. This is partly due to the long periods of locust absence and lack of accurate predictive systems in Africa. A 2021 article in the journal "Agriculture" examined the role of extension services in managing locust invasions in Ethiopia. The article noted that the country has a decentralized extension system, with extension workers based in each kebele (local administrative unit). However, the authors suggested that the extension system may not be well-equipped to address locust invasions, as there is a lack of trained personnel and resources [5]. A 2020 report by the World Bank on locusts in East Africa noted that many countries in the region have limited capacity for

R. Silhavy and P. Silhavy (Eds.): CSOC 2023, LNNS 723, pp. 654–666, 2023.
https://doi.org/10.1007/978-3-031-35317-8_59

monitoring and responding to locust outbreaks. The report stated that "most of the countries affected by the desert locust invasion have weak agricultural extension systems" [6]. Furthermore, the Food and Agriculture Organization (FAO) of the United Nations has reported on the importance of early warning systems and response mechanisms for locust invasions. In a 2019 publication, the FAO noted that effective response to locust outbreaks requires trained personnel, including "extension workers, scouts, and monitors" [7]. There is hence a need to find better ways of monitoring the areas that are prone to locust invasion amidst trained extension officers' challenges. There is great usability, for instance, in video surveillance and monitoring than never before. Internet of Things (IoT) hardware and software are not only convenient to handle but also easy to install. The cost of microcomputers and other related hardware to process data has drastically reduced and are affordable to an ordinally user.

1.1 IoT Technology

Internet of Things (IoT) refers to the interconnection of physical devices, vehicles, buildings, and other items embedded with electronics, software, sensors, and network connectivity that enable these objects to collect and exchange data [8]. Some of the low cost IoT options include a Raspberry Pi which is a series of low-cost, single-board computers that can be used for a variety of IoT applications [9]. It has built-in Wi-Fi and Bluetooth connectivity, and can be programmed using Python or other programming languages.

Temperature and humidity sensors are widely used in many applications, such as in monitoring the temperature and humidity in indoor environments, greenhouses, and warehouses. The DHT22 sensor is a popular choice for measuring temperature and humidity and can be used with a Raspberry Pi to collect data [10]. Another example is the AM2320 sensor, which is a high-precision sensor that can measure temperature and humidity with a higher accuracy [11] (Fig. 1).

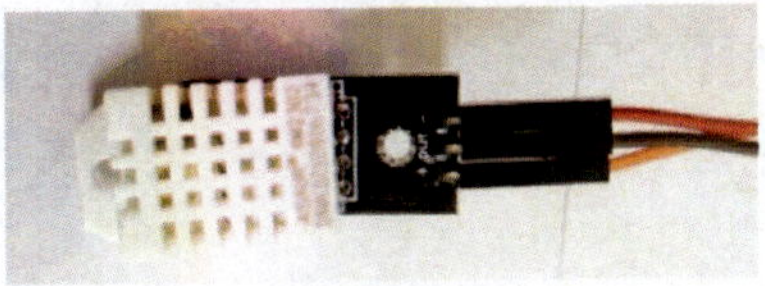

Fig. 1. DHT22 Sensor

Pi cameras are commonly used in various applications such as surveillance systems, wildlife monitoring, and video conferencing. The Raspberry Pi camera module is a popular choice, which can be controlled with Python code and can capture still images or video with a resolution of up to 1080p [12]. The camera can be used in various research applications, such as monitoring plant growth or animal behavior [13] (Fig. 2).

A GPRS (General Packet Radio Service) module is a wireless communication device that enables data transmission over the mobile network. When connected to a Raspberry Pi 4B, it can provide a way for the Pi to communicate with other devices or servers remotely. One popular GPRS module for use with the Raspberry Pi is the SIM800L

Fig. 2. Pi Camera

module [14]. This module can be connected to the Pi via the GPIO pins and allows for data transmission over 2G networks. Another popular option is the SIM7600 module, which supports 4G networks and can provide faster data transmission speeds [15]. This module can also be connected to the Pi via the GPIO pins or a USB interface. GPRS modules are commonly used in IoT applications, such as remote monitoring and control systems, as they provide a reliable and cost-effective means of data transmission over long distances [16] (Fig. 3).

Fig. 3. GPRS Module

1.2 Deep Learning in Locust Management

Deep learning is a subfield of machine learning that uses artificial neural networks with multiple layers to extract high-level features from raw data and make predictions or decisions based on that data [17]. It has become a powerful tool in a wide range of applications, including computer vision, natural language processing, and speech recognition. Deep learning algorithms use a hierarchical approach to learn representations of the data, with each layer of the neural network learning increasingly complex features [18]. This enables deep learning models to automatically extract features from raw data, making them highly effective at tasks such as image recognition and speech rccognition. The most popular type of neural network used in deep learning is the convolutional neural network (CNN), which is particularly effective at processing image and video data [19]. CNNs use convolutional layers to extract spatial features from the input data and pooling layers to reduce the dimensionality of the output. Other types of neural networks used in deep learning include recurrent neural networks (RNNs), which are used for sequence data such as natural language processing, and generative adversarial networks (GANs), which can be used for tasks such as image generation [17].

Deep learning has shown a promise in helping to manage locust swarms. A team of researchers used a deep learning algorithm to detect and track locust swarms in real-time using satellite imagery. They were able to accurately predict the direction and speed of locust swarms, which could help with early detection and prevention of crop damage [20]. Another study by [21] used deep learning to identify different species of locusts based on their physical characteristics. This could be useful in developing targeted control strategies for specific locust species. [22] also used deep learning to predict locust outbreaks by analyzing weather patterns and other environmental factors. Furthermore, [23] used deep learning to detect locust swarms from aerial imagery. Their method was able to accurately detect and track locust swarms in real-time.

There is potential to use IoT technology and deep learning to monitor locust populations in order to identify potential risks and manage locust populations more effectively. Monitoring locust populations through deep learning embedded IoT technology could provide early warning signs of locust infestation and allow for more effective and timely management measures.

1.3 Problem Identification and Motivation

The area under study is a vast area which spans more than 19000 Km^2 and monitoring the area using a few locust experts is practically impossible. The available technology requires the presence of the locust contact personnel to be physically in the field if they are to report on any traces of invasion to the Locust Control Center. According to [24], the surveyed area is managed by a single Agricultural Camp Officer and efforts were being put in place to train community leaders in the use of eLocust3 system, a measure which will help to beef up the monitoring aspect. This measure will however come with its own challenges such as false alarms and lack of power in the remote areas to keep the tablet charged as often as possible. The researchers were hence motivated to provide an innovative solution that could automatically capture and record locust data without human intervention.

The objectives of this study were organized as follows;

i. To design a low cost IoT system based on Deep Learning to automatically capture locusts.
ii. To save the captured images and related environmental variable to the web server for use by stakeholders.

2 Related Works

Artificial intelligence (AI) can be embedded into Internet of Things (IoT) systems for more advanced and efficient monitoring of locust populations. [25] describes an intelligent monitoring system for locust detection based on IoT and deep learning. The system consists of a network of IoT devices with sensors for collecting environmental data such as temperature, humidity, and sound, which are then processed by a deep learning algorithm for locust detection and classification. The authors report experimental results demonstrating the effectiveness of the system in detecting locusts in real-world environments. The authors evaluated their system in a field experiment and report an accuracy of over 90% in detecting locusts. The proposed system has the potential to provide early warning of locust outbreaks and help farmers take timely preventive measures to minimize crop damage. The system could also facilitate the monitoring and tracking of locust populations, which could aid in the development of effective control strategies.

[26] designed a system for detecting locusts based on IoT and machine learning technologies. The system utilizes an IoT device equipped with sensors to collect environmental data such as temperature, humidity, and wind speed, which are then processed by a machine learning algorithm to predict the presence of locusts. The authors report experimental results demonstrating the effectiveness of the system in detecting locusts under various environmental conditions and reported an accuracy of over 95% in detecting locusts. The proposed system has the potential to provide early warning of locust outbreaks and help farmers take timely preventive measures to minimize crop damage.

[27] presents a system for real-time monitoring of locusts using AI and IoT technologies. The system consists of an IoT device with sensors for collecting locust-related data and a deep learning-based algorithm for locust detection and classification. The system is designed to continuously monitor locust activity and provide real-time alerts to farmers when locusts are detected. The authors also report experimental results demonstrating the effectiveness of the system in detecting locusts in real-world environments with an accuracy of 93% in detecting locusts. Overall, the system has the potential to improve early detection and control of locust outbreaks, reducing the negative impact of locusts on agriculture.

[28] presents a real-time object detection system that utilizes deep learning algorithms on a Raspberry Pi board. The authors use a pre-trained deep learning model for object detection called You Only Look Once (YOLO), which is known for its speed and accuracy. The YOLO model is trained on the COCO (Common Objects in Context) dataset, which contains 80 different object categories. The authors modify the YOLO algorithm to run on the Raspberry Pi board by optimizing the code for the ARM architecture and reducing the resolution of the input images. They evaluate the performance of their system by testing it on a variety of objects, including people, vehicles, and animals.

The experimental results show that the system can achieve real-time object detection with a frame rate of 5–6 frames per second on the Raspberry Pi 3 Model B board. The accuracy of the system is comparable to the YOLO model running on a desktop computer. The authors conclude that their system can be used for a variety of applications, such as surveillance, robotics, and smart homes. The study demonstrates the potential of using AI and Raspberry Pi for real-time object detection in resource-constrained environments. It could potentially provide an affordable and effective solution for early detection and control of locust outbreaks too.

[29] provides a detailed account of the development and evaluation of a smart pest control system using the Raspberry Pi platform for detecting and controlling locust swarms. The authors note that locust swarms have become a major concern for agriculture and food security, and that the use of intelligent pest control systems can help to minimize their impact. They furthermore highlight the various techniques and technologies that have been used for pest control, including chemical pesticides, biological control, and integrated pest management. The authors note that these methods have limitations, including potential harm to the environment and non-target organisms, as well as potential resistance to pesticides by pest populations. The article then presents the design and development of a smart pest control system using the Raspberry Pi platform, which includes a monitoring module, a wireless communication module, and a pest control module. The authors describe the various components of the system, including the hardware and software design, and present the results of several experiments that evaluated the performance of the system in detecting and controlling locust swarms.

3 Methodology

The research used Design Science Research (DSR) Methodology. DSR is a methodology for creating and evaluating innovative artifacts or solutions to address practical problems or challenges in various domains. [30] describe the DSR methodology as a process that involves the following stages:

1. Problem identification and motivation: This stage involves identifying a practical problem or challenge in a specific domain that requires a solution or improvement. The motivation for solving the problem is also identified.
2. Define objectives for a solution: In this stage, the researcher defines the objectives or goals for the solution to be developed. These objectives should be aligned with the problem and the motivation for solving it.
3. Design and development: This stage involves designing and developing the solution or artifact to address the problem. The solution should be innovative, practical, and effective in addressing the identified problem.
4. Demonstration and evaluation: In this stage, the solution is demonstrated and evaluated in a real-world setting to determine its effectiveness in addressing the problem. The evaluation should be based on predefined criteria and metrics.
5. Communication and knowledge transfer: This stage involves communicating the results of the DSR process to stakeholders and transferring knowledge about the solution to ensure its adoption and use (Fig. 4).

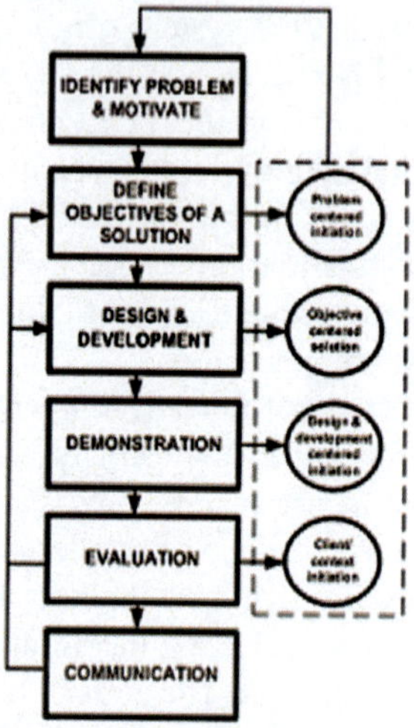

Fig. 4. Design Science Research Methodology [31]

The DSR methodology is iterative and involves continuous feedback and improvement throughout the process. It is an action-oriented approach that emphasizes the creation of practical solutions to real-world problems.

The IoT Deep learning embedded model was designed using Raspberry Pi 4 model B and connected to it was a Pi Camera and Temperature/Humidity sensor (DHT22) coupled with GPRS module for internet access. Embedded with a fast processor, the raspberry Pi 4 Model B is a brilliant microcomputer with two 4K HDMI ports as outputs. It also has a fast Gigabyte Ethernet port, two USB 3.0 ports and a slot for fast microSD card [32]. DHT22 sensor collects both temperature and humid and data collected through this sensor will be of value in predicting the favourable conditions that favours the movement of locusts [4]. Fitted with a Simcard, GPRS enables access to the internet for easy data collection through the web browser. It also facilitates the remote access to the remote model.

The model Block diagram was connected as shown in Fig. 5.

The IoT components were powered up using a 30 W solar panel and 12 V battery. A solar charge controller electronic device, also known as a solar regulator regulated the voltage and current from a solar panel to a battery. It prevents overcharging and ensures that the battery is charged efficiently. To ensure that the Raspberry Pi was receiving a stable and clean power supply, a Buck DC to DC converter was added to the setup. A Buck DC to DC converter is a type of voltage regulator that steps down the voltage from the solar charge controller to a stable and regulated 5 V required by the Raspberry Pi [33–35]. The converter will also filter out any noise or fluctuations in the input voltage, ensuring that the output voltage is stable and within the required range (Fig. 6).

The low cost IoT-AI embedded design was first implemented using the video that was captured in the study area during the African Migratory Locust outbreak in December 2020. It was further demonstrated in real time using both the AML and Red Locust insects captured in the study area. Only images that fit the criteria of 90% identification

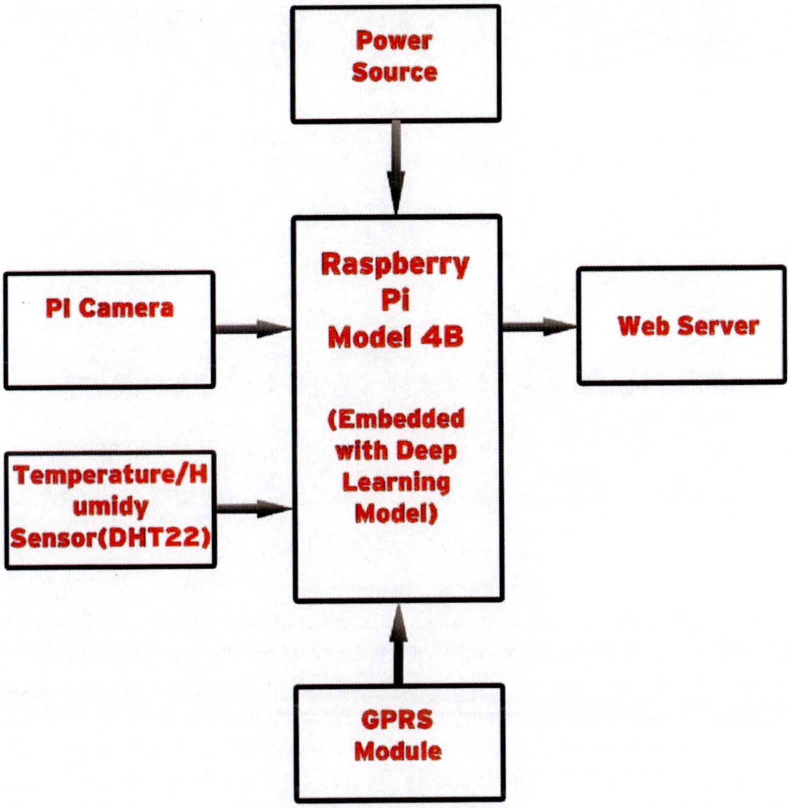

Fig. 5. IoT Block Diagram

Fig. 6. Solar Power and Buck DC to DC Converter

threshold were uploaded to the remote database for easy access through the browser. The GPRS/GPS provided a platform for data accessibility through the internet by use of Airtel mobile network Sim card. The Daily Temperature and Humidity levels at the time of capture of locust images were made possible through the use of DHT22 sensor (Fig. 7).

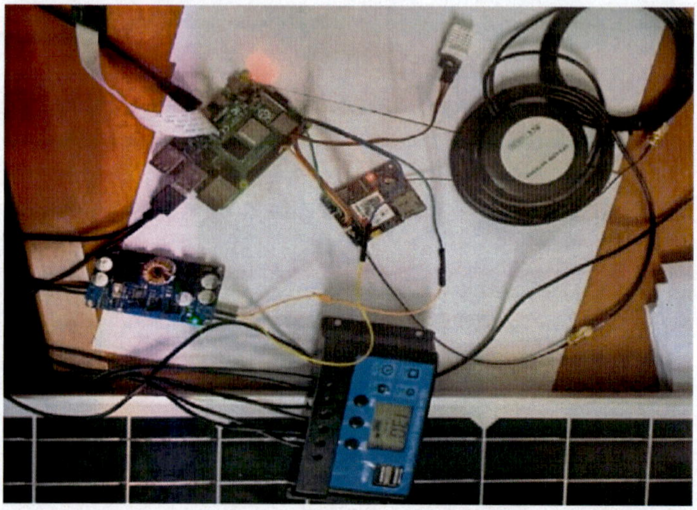

Fig. 7. Low Cost IoT

This study employed the use of low-cost Internet of things (IoT) and Deep Learning technologies to monitor locust invasion at an early stage. The deep learning model and dataset aspects of the study were adopted from a recent research which looked at the "detection of *Locusta migratoria* and *Nomadacris septemfasciata*" in the study area [2]. The low cost IoT employed a Pi Camera serially connected to a deep learning embedded Raspberry Pi 4 model B to automatically capture locust images, the temperature and humidity sensors captured the environmental variables [1] that favours the proliferation of locust presence while the GPRS module provided the internet access.

4 Results

The AI embedded low cost IoT model was tested using a single component at a time. Firstly, a General Packet Radio Service (GPRS) module, a device that enables wireless data communication over the cellular network using the GPRS technology, enabled packet-based data transmission, allowing the microcomputer to send and receive data from the DHT22 Sensor and Pi Camera over the internet via a cellular connection. Connected via a serial interface, SIM800, a popular GPRS module manufactured by SIMCom, included a cellular modem, antenna, and communication protocols for establishing a data connection with an Airtel cellular network.

The second aspect involved the Raspberry Pi Camera's capability of capturing the locust images in real time and the AI model's ability to detect the type of locust. The Raspberry Pi Camera module with 5-megapixel sensor was able to capture high-quality images of locusts. Locusts that were captured from the study area during data collection were instrumental in testing the object detection phase. Output from this stage showed that only images that fit the classification accuracy of 90% were considered for upload to the web server as shown in Fig. 8 (Fig. 9).

Fig. 8. Locust detection [2]

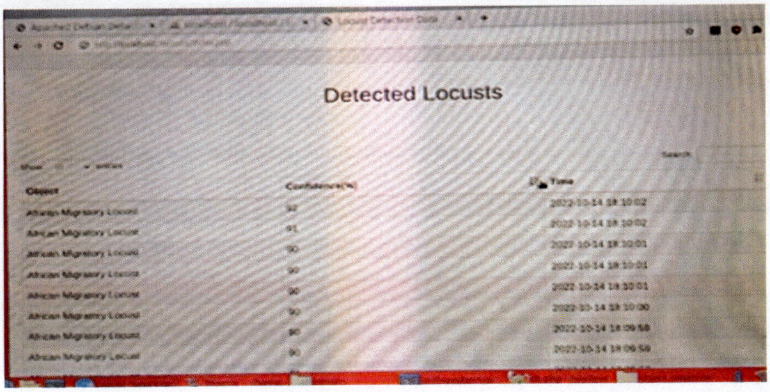

Fig. 9. Detection Results

The DHT22 sensor was connected to the Raspberry Pi using the following pinout: DHT22 VCC to Raspberry Pi 5V, DHT22 GND to Raspberry Pi GND and DHT22 DATA to Raspberry Pi GPIO 4. The temperature and humidity levels at the image capturing time were also recorded. The output from each of the IoT components where later organized and displayed in the web browser and this was in line with the research objectives.

The evaluation of the research looked at the importance of the identified problem and the feasibility of it being solved. The problem as identified by [24] was crucial to ultimately reduce the locust experts staff challenges in the study area. A novelty idea was hence needed to achieve the desired objective of having a well-designed system to monitor locust invasion with minimal effort. Combining DHT22 temperature and humidity results with locust detection accuracy percentages provide useful insights into the relationship between environmental factors and locust presence. Furthermore, it is now possible to identify patterns or correlations between temperature, humidity, and locust activity, which could be used to develop more effective locust detection and control strategies. It's a simple solution full of clarity from the user perspective and there is consistency in the collected data. It is also easy to use with a possibility of expanding it hence making it a robust innovation. The designed model provides efficiency and effectiveness in the real time monitoring of locust invasion without much human effort. Data from the model is effortlessly disseminated to the users.

5 Conclusion

In conclusion, the implementation of a low-cost IoT-based automated Locust monitoring system in Kazungula, Zambia, would bring numerous benefits to the local farmers and agricultural sector. The system would provide real-time monitoring and early detection of locust swarms, allowing for timely and effective control measures to be taken to prevent crop damage and losses. Additionally, the low-cost nature of the system would make it accessible to small-scale farmers who may not have the resources to invest in expensive monitoring technologies.

Furthermore, the use of IoT technology would enable remote monitoring and data collection, reducing the need for farmers to physically monitor their fields, which would save time and resources. The system would also provide valuable data insights that could be used for future planning and decision-making in the agricultural sector.

Overall, a low-cost IoT-based automated Locust monitoring system would significantly benefit the agricultural sector in Kazungula, Zambia, and could serve as a model for other developing countries facing similar challenges with locust swarms.

Acknowledgments. The authors hereby acknowledge the immense help received from Scholarship funders - International Development Research Centre (IDRC) and Swedish International Development Cooperation Agency (SIDA); Scholarship Programme-Artificial Intelligence for Development (AI4D) Africa Scholarship Fund Manager-Africa Center for Technology Studies (ACTS), University of Zambia; and lastly but not the least Mulungushi University for staff grant assistance.

References

1. Klein, I., Uereyen, S., Eisfelder, C., Pankov, V., Oppelt, N., Kuenzer, C.: Application of geospatial and remote sensing data to support locust management. Int. J. Appl. Earth Obs. Geoinf. **117**, 103212 (2023)
2. Halubanza, B., Phiri, J., Nyirenda, M., Nkunika, P.O.Y., Kunda, D.: Detection of Locusta migratoria and Nomadacris septemfasciata (Orthoptera: Acrididae) using MobileNet V2 quantized convolution neural network, Kazungula, Zambia. In: Silhavy, R. (ed.) CSOC 2022, vol. 503, pp. 490–501. Springer, Cham (2022). https://doi.org/10.1007/978-3-031-09073-8_43

3. Salim, S.A., Amin, M.R., Rahman, M.S., Arafat, M.Y., Khan, R.: An IoT-based smart agriculture system with locust prevention and data prediction. In: 2021 8th International Conference on Information Technology, Computer and Electrical Engineering (ICITACEE), pp. 201–206. IEEE, September 2021
4. Ali, B., Khalid, H., Rao, M.: Self-sustainable non-toxic locust control using electric field & real-time remote monitoring using android application. J. Independent Stud. Res. Comput. **20**(1), 31–38 (2022)
5. Basha, H., Legesse, G., Beyene, A.: Role of extension services in managing desert locust (Schistocerca gregaria Forskål) invasion in Ethiopia: challenges and opportunities. Agriculture **11**(1), 36 (2021)
6. World Bank: East Africa locust crisis: impacts, responses, and challenges (2020). https://openknowledge.worldbank.org/handle/10986/34586
7. FAO: Desert locust crisis appeal: FAO's response (2019). http://www.fao.org/3/ca5185en/ca5185en.pdf
8. Shen, W., Wang, X., Li, X.: The Internet of Things: a survey. J. Internet Technol. **16**(6), 919–926 (2015)
9. Upton, E., Halfacree, G.: Raspberry Pi User Guide. Wiley, Hoboken (2014)
10. Adafruit: DHT22 temperature-humidity sensor + extras (2022). https://www.adafruit.com/product/385
11. Aosong: AM2320 digital temperature and humidity sensor (2022). https://www.aosong.com/en/products/humidity/11.html
12. Raspberry Pi: Camera module (2022). https://www.raspberrypi.org/products/camera-module-v2/
13. Cao, Y., Li, H., Wang, L., Hu, X.: A low-cost and effective monitoring system based on Raspberry Pi for animal behavior observation. J. Sens. **2021**, 1–9 (2021). https://doi.org/10.1155/2021/6675353
14. Wu, Y.: Design of smart home system based on Raspberry Pi and IoT. J. Phys. Conf. Ser. **1199**(4), 042041 (2019)
15. Budagavi, M.: Implementing an IoT-based system for remote monitoring and control of smart devices. In: Advances in Intelligent Systems and Computing, vol. 1091, pp. 315–325. Springer (2020)
16. Akinwande, O.O., Sanusi, T.A., Ajayi, O.A.: Design and implementation of IoT-based smart farming system for crop management. J. Eng. Res. Rep. **15**(2), 1–14 (2020)
17. Goodfellow, I., Bengio, Y., Courville, A.: Deep Learning. MIT Press, Cambridge (2016)
18. LeCun, Y., Bengio, Y., Hinton, G.: Deep learning. Nature **521**(7553), 436–444 (2015)
19. Schmidhuber, J.: Deep learning in neural networks: an overview. Neural Netw. **61**, 85–117 (2015)
20. Wang, S., Xia, Y., Zhang, Y.: Deep learning for real-time detection and tracking of desert locusts. J. Appl. Remote Sens. **15**(4), 044514 (2021)
21. Cholleti, H., Wehner, R., Swaminathan, S.: A deep learning approach to locust species identification. Front. Robot. AI **7**, 83 (2020)
22. Liu, J., Xu, W., Yu, L.: A deep learning approach to predicting desert locust outbreaks. Remote Sens. **13**(11), 2225 (2021)
23. Das, A., et al.: Detection and tracking of locust swarms in aerial imagery using deep learning. Int. J. Remote Sens. **42**(23), 9288–9310 (2021)
24. Halubanza, B., Phiri, J., Nkunika, P.O.Y., Nyirenda, M., Kunda, D.: Toward locust management: challenges and technological opportunities, Sikaunzwe, Zambia. Zambia ICT J. **6**(1), 61–65 (2022). https://doi.org/10.33260/zictjournal.v6i1.152
25. Wu, J., Zou, Y., Sun, X., Yang, Y., Zhou, J., Liu, J.: An intelligent monitoring system for locust detection based on IoT and deep learning. Comput. Electron. Agric. **185** (2021). https://doi.org/10.1016/j.compag.2021.106126

26. Zheng, J., Zhang, Y., Gao, L., Wang, Y.: Locust detection based on IoT and machine learning. In: IEEE International Conference on Computer and Communications (ICCC), pp. 1411–1416 (2019). https://doi.org/10.1109/ICCC.2019.8793388
27. Sun, H., Zhang, Y., Liu, W., Luo, S.: Real-time monitoring of locusts based on AI and IoT technologies. J. Intell. Fuzzy Syst. **40**(4), 6537–6548 (2021). https://doi.org/10.3233/JIFS-201660
28. Ghosh, S., Pathak, S.: Real-time object detection on Raspberry Pi using deep learning. Procedia Comput. Sci. **165**, 621–630 (2019). https://doi.org/10.1016/j.procs.2019.12.220
29. Liu, R., Li, G., Li, Y., Wang, Y., Li, C., Li, H.: Smart pest control system for locusts based on Raspberry Pi. Int. J. Distrib. Sens. Netw. **16**(8), 1550147720947432 (2020)
30. Brocke, J., Budde, L., Schenk, T.: Design science research. In: Durand, T.M., Grodal, J.J. (eds.) The Palgrave Encyclopedia of Strategic Management. Palgrave Macmillan (2020). https://doi.org/10.1007/978-3-030-02006-0_84-1
31. Azasoo, J.Q., Boateng, K.O.: A retrofit design science methodology for smart metering design in developing countries. In: 2015 15th International Conference on Computational Science and Its Applications, pp. 1–7. IEEE, June 2015
32. Pajankar, A.: Introduction to Raspberry Pi. In: Practical Linux with Raspberry Pi OS, pp. 1–34. Apress, Berkeley (2021)
33. Sahin, M., Ozdemir, E., Kose, E.: Design and implementation of a solar-powered Raspberry Pi system for IoT applications. J. Ambient. Intell. Humaniz. Comput. **11**(10), 4283–4293 (2020). https://doi.org/10.1007/s12652-019-01432-8
34. Deka, D., Baruah, P., Sarma, K.B.: Design of a standalone solar powered Raspberry Pi based monitoring system for agriculture. In: 2019 IEEE International Conference on Sustainable Energy Technologies and Systems (ICSETS), pp. 548–552. IEEE, December 2019. https://doi.org/10.1109/ICSETS48194.2019.9069360
35. Senthilkumar, N.N., Sankaralingam, S.: Design and implementation of solar-powered smart irrigation system using Raspberry Pi. In: 2018 International Conference on Innovations in Information, Embedded and Communication Systems (ICIIECS), pp. 1–6. IEEE, March 2018. https://doi.org/10.1109/ICIIECS.2018.8690334

Monitor and Analyze Sensor Data from a Connected Vehicle Thanks to Cloud Edge Computing

Olfa Souki[1,2,3,4](✉), Raoudha Ben Djemaa[1,2,3,4], Ikram Amous[1,2,3,4], and Florence Sedes[1,2,3,4]

[1] Faculty of Sciences Economics and Management of Sfax, Sfax, Tunisia
ikram.amous@enetcom.usf.tn, florence.sedes@irit.fr

[2] Higher Institute of Computer Science and Communication Technology Hammam Sousse, Sousse, Tunisia
soukiolfa92@gmail.com

[3] National School of Electronics and Telecommunications of Sfax, Sfax, Tunisia

[4] Toulouse Institute of Computer Science Research Paul Sabatier Toulouse, Toulouse, France

Abstract. Progress on innovative wireless sensors and connected vehicles on intelligent transport systems (ITS) is a recent research topic. It involves the collection and processing of large quantities of sensor data. Many types of sensors embedded in today's vehicles like cameras, lidar, and radar are useful for monitoring vehicle data. Reliable, low-cost detection, connectivity, computational capability, and powerful analytics usher in a new era of vehicle context detection and vehicle network detection (VANET). However, given the concerns about latency, bandwidth, cost, security, and privacy, as well as the growing capabilities of edge computing devices, it is necessary to examine edge and cloud computing together in order to make informed judgments based on context and performance. So, in this paper, we have proposed three algorithms focused on an intelligent transportation system (ITS) based on both Cloud and Edge computing for monitoring and analyzing vehicle sensor data and its environment. Our proposal is capable of receiving sensor data from the vehicle, analyzing it, and giving effective plans in real-time to the driver suited to his context. These plans can be helpful in reducing traffic congestion and educating the driver on the road situation. Thanks to our proposal, we have contributed to the vehicle, road safety, and traffic efficiency by providing real-time information to the operator. Finally, thanks to the measurements of recall, accuracy, F1 score, and evaluation of execution time, we have approved the performance of our proposal.

Keywords: Intelligent transportation system (ITS) · vehicular traffic · Monitoring as a service · Analyzing as a service · CloudIoT · Edge Computing · connected vehicle

R. Silhavy and P. Silhavy (Eds.): CSOC 2023, LNNS 723, pp. 667–683, 2023.
https://doi.org/10.1007/978-3-031-35317-8_60

1 Introduction

Advances in technology present an opportunity for road administrations around the world to transform the way they manage and operate their road and highway systems. As a result of these advances, an intelligent traffic system called ITS is being developed [20].

Intelligent Transport Systems (ITS) play a key role in transport and traffic management systems all over the world. They are a combination of advanced information and communication technologies used in transportation and traffic management systems to enhance the safety, efficiency, and sustainability of road transportation systems, reduce congestion and improve drivers' experiences. There is no end of possibilities. Through technological developments in a variety of fields, the perception of the state of the transport system has changed dramatically with the adoption of further progress, such as computing equipment, positioning systems, sensing technologies, telecommunications, data processing networks, virtual operations, and planning methods.

Connected objects such as vehicles, traffic signals, various motion sensors, traffic sensors, and congestion sensors make up a complex, dynamic, and heterogeneous vehicle network. Within that network, every object produces a lot of data, called Big Data. Big Data can not only be used for the automation of vehicles, but also for the development of new technologies. The intelligent transport system consists of several related concepts, such as VANET, Vehicle-to-Vehicle (V2V), Li-Fi technology, RFID, and LORAWAN technology [1]. These concepts support the development of leading-edge technologies and services. Automated operations generally require human involvement. As well, the performance of the road network can be monitored and adjusted in real-time.

On the other hand, the data that used to be collected by costly infrastructure is now being provided by prolific new sources. Intelligent systems are currently providing real-time historic data analysis.

Road users' choices, formerly dictated solely by road signs, can now be influenced by a wide range of distribution channels, such as mobile devices and vehicle systems. These opportunities will only materialize if road authorities have a clear and well-thought-out strategy to consolidate disparate systems, services, and business approaches over time.

The most commonly used way to define a context is to list the list of information that is considered an application context [18]. On the basis of these applications, a context can therefore have different meanings: the duration of an event, the localization of an object [12], an interaction history [6], personal characteristics, knowledge, and interests [16].

The rest of this document is structured in the following way: Sect. 2 examines the best available literature. Section 3 describes how we contribute. Next, Sect. 4 presents preliminary findings from our experimental assessment. Our findings are discussed in Sect. 4. Finally, Sect. 5 summarises the document and outlines the broad directions for future research.

2 Related Works

In the last few years, intelligent transportation systems attracted the attention of various researchers. These systems play a prominent role in discovering new technologies and approaches to offer the best transportation services [4,7,9,11,13].

Researchers in [4] assessed how to use VANETs for collecting vehicular traffic measurements. They provided two VANET sampling protocols, named SAME and TOME. They also designed and implemented an application for one of them, to perform real-time incident detection. A distributed autonomous FCD collection protocol was defined based on VANET. In addition, they showed accurate real-time speed estimation based on FCD sampling through VANET. Moreover, they ran traffic simulations based on real vehicular GPS traces. Finally, they suggested an incident detection algorithm based on standard accuracy GPS positions.

In the [7], the authors survey the scope and evolution of a basic Bluetooth sensor system with system attributes and design decisions illustrated via a reference design. A basic Bluetooth sensor system for traffic monitoring consists of a Bluetooth probe device(s) that scans for another Bluetooth-enabled device (s) within its radio proximity and then stores the data for future analysis and use.

However, the authors of [13] proposed a self-powered wireless smart sensor, powered by the vibration energy of trains via a maglev porous nanogenerator (MPNG) when the train is running. A wireless smart sensor can be sustainably powered by MPNG arrays, transmitting real-time data to a cellphone, demonstrating the capability for the Internet of Things, especially in the train monitoring system.

Concerning [11], integrity monitoring is an essential step in the positioning service, which needs to be fulfilled for integrated vehicular navigation systems used in intelligent transportation systems to o bind the positioning errors within a pre-defined integrity risk. The knowledge of many relevant aspects including the structure, positioning methodology, and different errors affecting the positioning solution of the individual and integrated systems is required to Develop such innovative integrity monitoring techniques. Moreover, knowledge is needed for the current mitigation techniques of these errors, for possible fault detection and exclusion algorithms, and for the computation of protection levels.

The authors of [9] discuss the impact and implications of 5G on Intelligent Transporting Systems (ITS) from various dimensions. They present an overview of the technological context and the economic benefits of the 5G and how key vertical industries will be affected in a smart city, i.e., energy, healthcare, manufacturing, entertainment, and automotive and public transport.

This review paper [10] provided an exhaustive study of existing incident detection techniques. Current techniques have been studied based on traffic monitoring and incident detection. This paper presented a relative study on roadways incident detection methods in terms of working with the pros and cons. Also, proprietary incident detection techniques have been explored to determine applications for incident detection and management. The authors in [10] presented a comprehensive overview of the field-related research. They compared all available accident detection techniques and presented a comparative study of strengths and weaknesses related to available incident detection techniques.

Another study conducted by [2] qualitatively reviewed Cloud computing in VANETs (CC-V) with an emphasis on layered architecture, network components, taxonomy, and future challenges. Specifically, a four-layered architecture for CC-V was suggested including perception, coordination, artificial intelligence, and smart application layers. Three network components of CC-V, namely, vehicle, connection, and computation were explored with their cooperative roles. The taxonomy of CCV was presented considering the major themes of research including design of architecture, data dissemination, security, and applications. Related literature on each theme was critically investigated with a comparative assessment of recent advances. Finally, some open research challenges were identified as future issues. The challenges were the outcome of the critical and qualitative assessment of the literature on CC-V.

Researchers in [3] discussed the application of VANET technology to road traffic monitoring. They also set forth a distributed communication protocol for vehicle traffic monitoring. Real-world simulations of the major highways in Rome and Italy were calibrated using a large sample of floating car data. These simulations allowed us to determine the operational limits of communication protocols and assess the accuracy of traffic state estimates.

In [5], the authors defined and evaluated two traffic monitoring approaches that can be used to leverage the potential of VANETs. Protocols were based upon message exchange using a multi-hop path formed by links between vehicles equipped with DSRC devices. To show the potential of these approaches, the authors simulated the main expressway in Rome and Italy, using a road traffic model calibrated by exploiting an extensive sample of floating car data. They also carried out joint road traffic network and communication network simulations. In fact, they pointed out that the proposed protocols can collect quite accurate average speed estimates for different road sections with very little effort.

In [8], the authors provided a comprehensive survey on information dissemination in vehicular networks by focusing on some critical issues such as security and privacy, adaptability, scalability, and so on. For this, they classified the existing information dissemination solutions into eight new classes based on their design and optimization objectives and give a qualitative overview of each class based on their objectives, forwarding strategy, features, and advantages. The authors presented open research issues to improve the efficiency of information dissemination protocols as much as possible.

From Table 1, firstly most authors have a static approach that does not take into account any real-time changes. They propose some ITS VANET based and deployed into Cloud computing (CC), Vehicular Cloud (VC), Vehicular Cloud computing (VCC), and Vehicle to broadband Cloud (V2B). Although some works have used the Vehicular Cloud, the latter suffers from several problems. It admits a low bandwidth and high latency. Also, most papers exposed in Table 1,

Table 1. Monitoring and analyzing approach on Intelligent transportation system

Author	Cloud services	Sensor Data	Analyze method	Receive Data	Technical method	Real-Time	Context-aware
[10]	X	location	micro-controller (MCU)	GPS, GSM/GPRS	VANET-based ITS	X	X
[2]	Vehicular cloud	computing power, Internet, and storage	X	Vehicle sensor network	artificial intelligence	Yes	X
[3]	X	Traffic measure-ment	trace driven traffic	dissemination protocol	floating car data in Rome, GPS	X	Location
[5]	road traffic monitoring	vehicle sensor devices	X	Dedicated Short Range Communica-tion protocol	realistic simulation expressway in Rome	X	geo-Location
[8]	vehicle-to-broadband cloud (V2B)	road traffic	Analytical scheme (SMF)	dissemination protocol	X	X	X

didn't take into account the context of the vehicle in terms of energy, state, status, location, etc. All papers presented didn't use Edge or Fog Computing as an intermediary between Cloud Computing and Vehicle to reduce latency and increase the bandwidth of Cloud services.

In this paper, we will fix all these problems to give an adaptive and autonomic system that reacts to any change in vehicle context in real time through Cloud Edge Computing.

3 Contribution

The contribution of this paper focuses especially on monitoring connected vehicles and analyzing their contextual data and environmental sensor data to prepare for future self-adaptive intelligent transportation systems (ITS).

Based on the literature review, the current Cloud-based ITS offered three traditional cloud models, IaaS, PaaS, and SaaS. Although it provided various services with good quality of service (QoS), this Cloud- system provided low bandwidth and high latency. A huge bulk of research proposed applying edge-cloud computing to deal with these issues and fix them.

To this end, in this paper, we propose to deploy our proposed algorithm in Cloud Edge environments which can be a solution to reduce latency and increase the bandwidth of basic Cloud computing environments. This one is called ITS Cloud-Edge of things (ITS-CEoT).

This hybridization approach used not only appropriate vehicular cloud computing tools for immediately using close-by vehicles, but also Edge cloud that provides more substantive and persistent applications. Although this approach reduces the cost of cloud services, it requires not only the ability to distribute applications across tiers (local cloud, vehicular cloud, edge cloud, and core cloud) and context-aware applications, but also the ability to determine the best division of an application across multiple tiers and nodes and manage that division as resources changed. The changes in resources can be at the vehicle tier to save battery, at the vehicular cloud tier as vehicles move in and out of range, and at the edge cloud tier as services are migrated to new servers with different resources, such as compute, memory, storage, etc. To realize future Intelligent transportation systems, it is crucial to deploy distributed applications across tiers, manage real-time deployment, together and migrate services between edge cloud nodes without a noticeable reduction in operations. When we talk about Edge Computing in ITS, we can refer mainly to three vital components, namely Real-time data processing and basic analytics, Data catching, and Machine to machine process as shown in Fig. 1

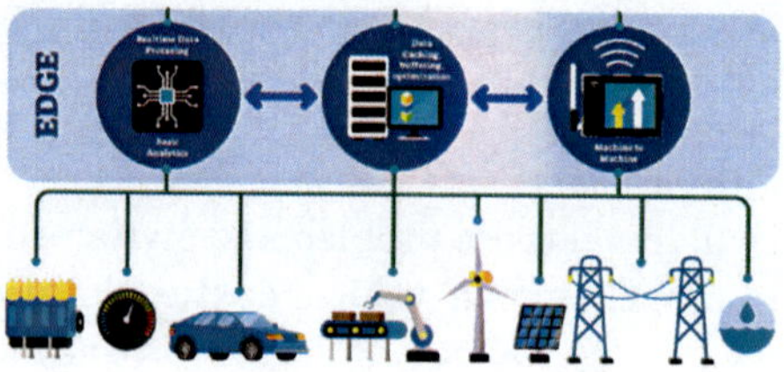

Fig. 1. The Edge Components

Our proposition emerge due to the necessity of supporting different types of IoT applications where the ITS cloud Edge computing works on the distinct roles mentioned. Different models are also needed to provide flexible and on-demand cloud Edge services based on users' needs.

Our suggestion contains different steps divided into two principal phases: The monitoring phase and the analyzing phase as shown in Fig. 2. To benefit from these services, the user must subscribe to these services to receive all alerts and suggested results.

The Monitoring phase plays a key role in receiving and previous treatment of data. It is discussed in detail in Sect. 3.1.

Moreover, the Analyzing phase, we distinguish two types of analysis which are analyzing Event-triggered and Time-triggered. These types are detailed in Sect. 3.3.

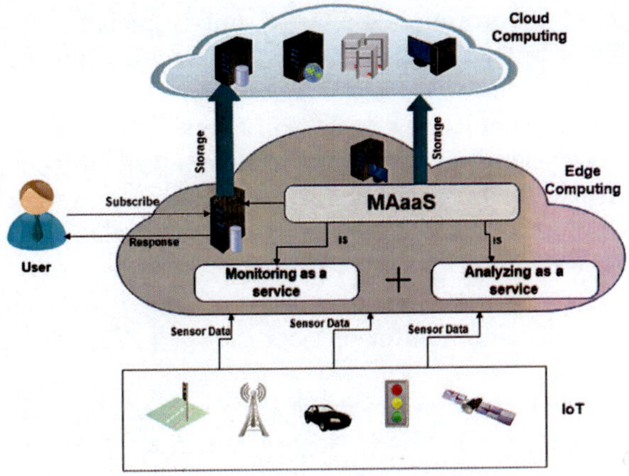

Fig. 2. Monitoring and analyzing in general

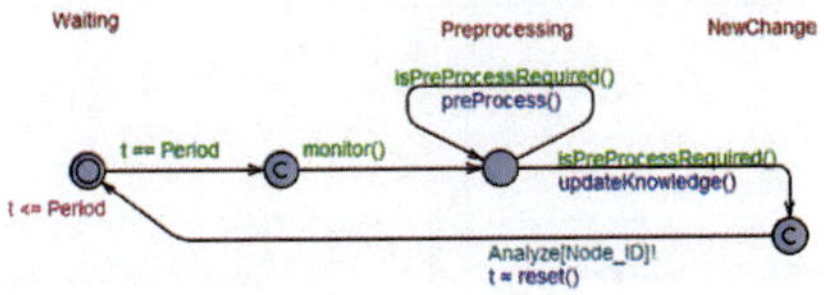

Fig. 3. Time-triggered monitoring

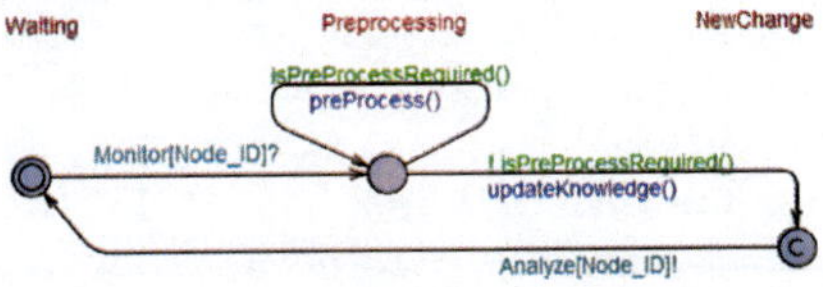

Fig. 4. Event-triggered monitoring

Cloud Edge provides data storage, data management, data analysis, and presentation to the IoT system. In this model, the IoT system generates and aggregates a large amount of monitoring and sensing data for cloud Edge applications. In this vein, we can list some examples, such as a weather monitoring system, transportation, and logistics surveillance in a smart city [17].

3.1 Monitoring as a Service

During the vehicle monitoring phase, we collect the data from these different sensors in CSV files that we will then analyze it. We can distinguish two types of monitoring which are: time-triggered monitoring and event-triggered monitoring. We started with time-triggered monitoring as shown in Fig. 3. We have fixed a time pipeline which is t = period = 15 ms as a time of waiting to start monitoring each time.

During monitoring, and before detecting a changing in any feature a preprocessing step is tested each time to know if it is required or not and detect a new change to start the analyzing process. Each monitor object is defined by its ID named Node_ID. When time-triggered monitoring process detects a new change and before starting analyzing a knowledge update will be effected to constitute the new context of the monitoring object in our case it is vehicle context and its environment.

Event-triggered monitoring is similar to time-triggered monitoring. However, the only difference is that the steps are triggered by the detection of a new change or event.

Then both of the monitoring types will check if the data are pre-processed correctly to update knowledge and launch the analysis process phase.

To model our monitoring services, their constraints, and their inputs/outputs, we consider a set of services $M_s = \{M_{s1}, M_{s2}, ..., M_{sj}\}$ when j designs the number of monitoring services (Fig. 4).

From the generated set of monitor services, we distinguish a subset that corresponds to the efficient monitor services $EM_s = \{EM_{s1}, EA_{s2}, ..., EA_{sv}\}$ with $v \leq j$.

The set of relevant monitor services RM_s is part of the overall set of analysis services M_s and can be linked by the expression $RM_s \subset M_s$.

Algorithm 1: Monitoring of connected object algorithm

```
Data: Connected object
Result: Sensor_ Dataset
begin
for Node_ID from dataset do
    for each_Features_from_List_Feature do
        while Feature_IS_Changed do
            Pre_Process()
            Analyze_Process()
        end
    end
    Monitoring()
end
end
```

3.2 Cleaning Data

Before starting the analysis, and during the monitoring phase we should have some previous steps to execute. These steps can be summarized into three processes which are: Cleaning data, aggregation data, and building a database as shown in Fig. 5.

The different features was shown in Fig. 7 which are:

- date: which corresponds to the date and time of record.
- position: geolocation of the vehicle composed by latitude and longitude.
- mesure: This one is the speed of the vehicle at the moment of recording.

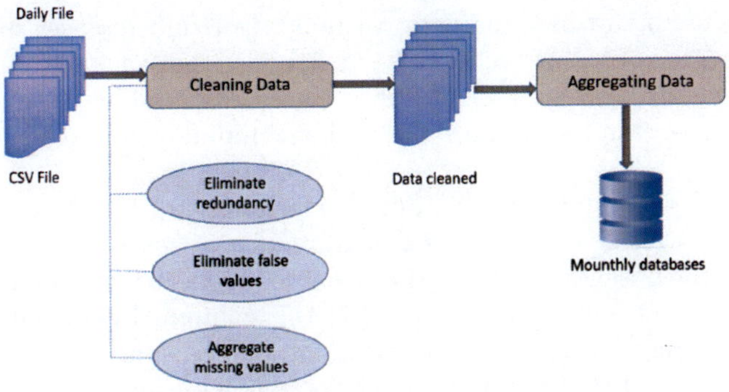

Fig. 5. Pre-processing steps

- limite: represents the maximum value of speed not to be exceeded and if it is exceeded an alert should be generated to the driver to reduce his speed.
- Name: represents the name of the nearest parking available.
- Type: represents the type of parking.
- Category: represents the category of the parking.
- Status: represents the status of the parking.
- Latitude: represents the latitude of the parking.
- Longitude: represents the longitude of the parking.
- location: represents geolocation as a couple (latitude, longitude) of the parking. Regarding the step of cleaning data, we have developed a data-cleaning algorithm described as follows in Algorithm 2.

Algorithm 2: Data-cleaning algorithm

```
Data: Sensor Data
Result: Cleaned data-set
begin ;                                                    /* read data */
missing_values=["n/a", "na", "–"," ","Null","null","0"]
line< –file.read()
while (line!=EOF) do
    t< –Verif(Features.values IN missing_values)
    if (t=TRUE) then
        Avg(Features.values) ;                  /* a missing value is find */
        ;                                  /* an average value will be putted */
    else
        nextLine()
    end
end
end
```

As an example of how the data cleaning algorithm replaces missing values with the average of feature values, we can consider that we have found an empty speed record for a given vehicle, the replacement is done by calculating the average between the maximum speed and the minimum speed recorded for this vehicle.

Each time we find a missing value that belongs to the list of missing values mentioned in the algorithm, the calculation of average is performed except in cases where the missing field is a string type, the replacement is done by inserting the most frequent value that appeared in this column. In general, in our case, we are only interested in numerical data that can be consistent and relevant in the construction of the contextual profile of a given vehicle.

3.3 Analyzing as a Service

The analysis phase deals with the input data provided by the previous Monitor phase. After executing a set of algorithms, the analysis phase can generate a semantic request which will be passed to the Plan phase if there is a need for change. Indeed, during this phase, complex reasoning about the symptoms provided by the monitor process is performed. This phase is also influenced by the stored knowledge data, if available.

Analyzing Event-triggered processes is generally preceded by the monitoring process which allows to control of connected objects, detect context variations and collect the data to be analyzed. The Event-triggered Analyzing process is used to detect a new event that should be analyzed (Fig. 6).

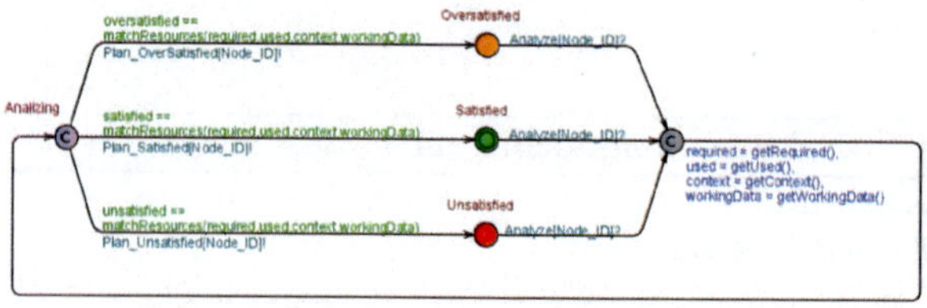

Fig. 6. Analyzing process Event-triggered

To model our analysis services, their constraints, and their inputs/outputs, we consider a set of services $A_s = \{A_{s1}, A_{s2}, ..., A_{sn}\}$ when n designs the number of analyzing services.

From the generated set of analyses services, we distinguish a subset that corresponds to the efficient analyses services $EA_s = \{EA_{s1}, EA_{s2}, ..., EA_{sk}\}$ with $k \leq n$.

The set of relevant analysis services RA_s is part of the overall set of analysis services A_s and can be linked by the expression $RA_s \subset A_s$.

Concerning the features, the analyzing algorithm receives a set of efficient features $EF = \{EF_1, EF_2, ..., EF_i\}$ which i designs the number of efficient features. The set of EF is a subset of the total features $F = \{F_1, F_2, ..., F_k\}$ stored in our dataset.

Algorithm 3 shows the different steps followed to realize the first analysis method of data Event-Triggered systems.

Algorithm 3: Analyzing Event-Triggered algorithm

```
Data: Efficient_Features from Monitoring
Result: Adaptable Plans
begin ;                                              /* check context data */
while (Efficient_Features.read()!=NULL) do
    ;                                       /* Verify each value of features */
    read(Efficient_Features.value);
    if (Efficient_Features.value.Haschange()=TRUE) then
        Changing(Current_Context, New_Context)
        ;                                                /* Update Context */
    else
        Analyze(Efficient_Features);
        ;                                                   /* Re_analyze */
    end
end
end
```

In this Event-Triggered algorithm, after detecting a new event, the algorithm starts to check the different important features that constitute the context of the vehicle which are energy level (diesel, gas, battery), geolocation (longitude and latitude), and speed. Our algorithm will compare the current vehicle context and the newly detected context with the aim of choosing the right plan for this context. Much pertinent information will be given to the driver of the vehicle such as the number of nearest cars, the state of the next traffic light, and the availability of parking at arrival time.

4 Experimentation Results and Discussion

In this section, we will show the experimentation results of proposed algorithms, and give one of the generated plans that have been displayed to the user according to his context as shown in Fig. 10. Also, we will evaluate the different results obtained from the following evaluation metrics, namely Recall and Precision. This section will end up with a discussion reflecting on the performance of our approach.

4.1 Monitoring Evaluation

To monitor any object we will need a different sensor to collect data about the object and its environment. In our research, we utilized real data captures using IoT sensors installed in the streets of Chicago US and saved them in a dataset downloadable from this link [19]. The different data which is streamed daily from many heterogeneous vehicle sensors, roads, and parking, is shown in Fig. 7. This figure shows examples of missing values, displayed as a red circle.

In intelligent transportation systems, a connected vehicle for instance has several contextual data, such as location or geolocation, diesel level (energy), the state after diagnosis, connectivity, etc. These features are sufficient to constitute the context of a connected vehicle. The missing data may be troublesome, especially those from the features that constitute the context of a vehicle or its environment. Therefore, to solve this issue, we proposed a data-cleaning algorithm which will be evaluated in the next subsection. To evaluate the performance of the proposed monitoring algorithm, we calculate the precision, the recall and the F1_score as following in these Eqs. (1), (2) and (3).

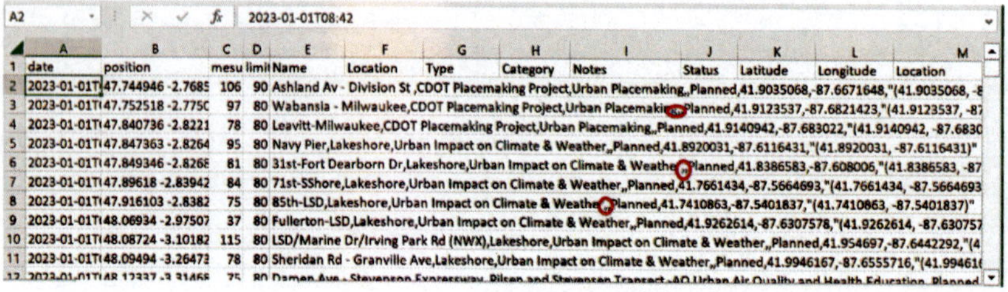

A2 2023-01-01T08:42

	A	B	C	D	E F G H I J K L M
1	date	position	mesu	limit	Name Location Type Category Notes Status Latitude Longitude Location
2	2023-01-01T	47.744946 -2.7685	106	90	Ashland Av - Division St ,CDOT Placemaking Project,Urban Placemaking,,Planned,41.9035068,-87.6671648,"(41.9035068, -8
3	2023-01-01T(	47.752518 -2.775C	97	80	Wabansia - Milwaukee,CDOT Placemaking Project,Urban Placemaki,,Planned,41.9123537,-87.6821423,"(41.9123537, -87
4	2023-01-01T(	47.840736 -2.8221	78	80	Leavitt-Milwaukee,CDOT Placemaking Project,Urban Placemaking,,Planned,41.9140942,-87.683022,"(41.9140942, -87.6830
5	2023-01-01T(	47.847363 -2.8264	95	80	Navy Pier,Lakeshore,Urban Impact on Climate & Weather,,Planned,41.8920031,-87.6116431,"(41.8920031, -87.6116431)"
6	2023-01-01T(	47.849346 -2.8268	81	80	31st-Fort Dearborn Dr,Lakeshore,Urban Impact on Climate & Weathe,,Planned,41.8386583,-87.608006,"(41.8386583, -87
7	2023-01-01T(	47.89618 -2.83942	84	80	71st-SShore,Lakeshore,Urban Impact on Climate & Weather,,Planned,41.7661434,-87.5664693,"(41.7661434, -87.5664693
8	2023-01-01T(	47.916103 -2.8382	75	80	85th-LSD,Lakeshore,Urban Impact on Climate & Weathe,,Planned,41.7410863,-87.5401837,"(41.7410863, -87.5401837)"
9	2023-01-01T(	48.06934 -2.97507	37	80	Fullerton-LSD,Lakeshore,Urban Impact on Climate & Weather,,Planned,41.9262614,-87.6307578,"(41.9262614, -87.630757
10	2023-01-01T(	48.08724 -3.10182	115	80	LSD/Marine Dr/Irving Park Rd (NWX),Lakeshore,Urban Impact on Climate & Weather,,Planned,41.954697,-87.6442292,"(4
11	2023-01-01T(	48.09494 -3.26473	78	80	Sheridan Rd - Granville Ave,Lakeshore,Urban Impact on Climate & Weather,,Planned,41.9946167,-87.6555716,"(41.99461(

Fig. 7. Sensor data in CSV file

$$Precision(1) = \frac{RMS}{TMS} \quad (1) \qquad Recall(1) = \frac{RMS}{TMS - RMS} \quad (2)$$

With:

RMS: Relevant Monitoring Service.
TMS: Total Monitoring Service.

The precision is defined by the number of relevant (pertinent) monitoring services returned by the proposed algorithm and divided by the total count of monitoring services.

The recall is defined count of relevant monitoring services retrieved, divided by the subtraction between the total number of monitoring services and the relevant monitoring services retrieved.

Regarding the F1-Score equation of the monitoring algorithm, it is as follows:

$$F1 - Score(1) = \frac{2 * recall * precision}{recall + precision} \quad (3)$$

F-Measure is the weighted harmonic mean of recall and precision. We combine the precision and recall scores into a single quality metric using a standard F-measure.

We started by illustrating the different precision, recall, and F1 Score values of the monitoring algorithm as reported in Table 2.

Table 2. Precision, Recall and F1_Score results of monitoring algorithm

Number of monitoring service	Precision 1	Recall 1	F1_Score 1
5	0.9	0.92	0.91
10	0.92	0.98	0.95
20	0.96	1.0	0.98
30	0.8	0.99	0.89

After further testing, we could denote that the value of precision has an average of 0.89 which indicates the performance of our proposed monitoring of the connected vehicle algorithm.

Regarding the recall, we have obtained an average of 0.97 which is very high and shows our proposed algorithm's performance.

4.2 Cleaning Data Algorithm Evaluation

In this subsection, we will illustrate the input and output of the data-cleaning algorithm. Then, to evaluate our proposed algorithm, we calculate the precision, the recall, and the F1_Score.

For the data-cleaning algorithm, the precision equation was calculated as Eq. 4.

$$Precision(2) = \frac{RRMV}{TRMV} \quad (4) \qquad Recall(2) = \frac{RRMV}{TRMV - RRMV} \quad (5)$$

With:

RRMV: Relevant Replacement of Missing Values.
TRMV: Total Replacement of Missing Values.

The precision value is defined by the number of pertinent replacements of missing values returned by our proposed algorithm and divided by the count of missing value replacements. The recall is defined by the count of relevant missing values replacements divided by the subtraction between the count of missing values replacements and the count of relevant missing values replacements.

$$F1 - Score(2) = \frac{2 * recall * precision}{recall + precision} \quad (6)$$

The precision, recall and F1-score values of our proposed data-cleaning algorithm are shown in the Table 3.

Table 3. Precision, Recall and F1_Score results of data-cleaning algorithm

Number of features	Precision 2	Recall 2	F1_Score 2
2	0.98	0.96	0.97
5	0.96	0.94	0.95
7	0.98	0.99	0.99
10	0.99	1.0	0.99

We illustrated the precision, recall, and F1-Score values of the data cleaning algorithm in Fig. 8.

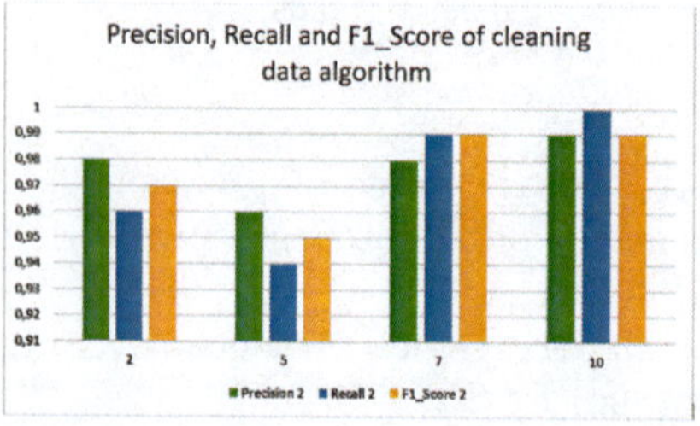

Fig. 8. Evaluation metrics of cleaning data algorithm

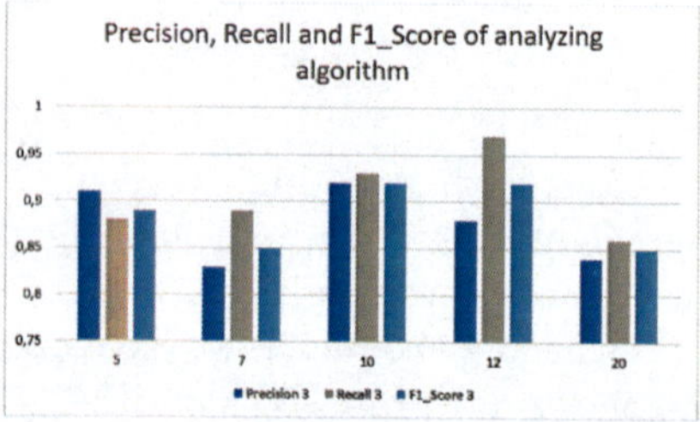

Fig. 9. Evaluation metrics of analyzing algorithm

From the different results obtained, we have an average of 0.97 for the precision and the recall values which is high and shows the performance of our proposed algorithm.

4.3 Analyzing Evaluation

In this subsection, we evaluate our proposed analyzing algorithm by calculating the precision and the recall values. Because of the importance of precision and recall results, we should calculate the F1-score as shown in Eq. 9.

$$Precision(3) = \frac{RAS}{TAS} \quad (7) \qquad Recall(3) = \frac{RAS}{TAS - RAS} \quad (8)$$

With:

RAS: Relevant Analyzing service.
TAS: Total Analyzing service.

Regarding the F1-Score equation of our proposed analyzing algorithm which combines the precision and the recall value as follows:

$$F1 - Score(3) = \frac{2 * recall * precision}{recall + precision} \tag{9}$$

Figure 9 displays the different values obtained by calculating the previous equations and based on how many features are analyzed. From these values, we obtained an average precision of 0.87, an average recall of 0.9, and 0.88 for the F1-score, which showcases the performance of our algorithm.

The analyzing algorithm outputs a set of plans generated in real-time according to the vehicle context. Figure 10 illustrates one of the plans at a given time.

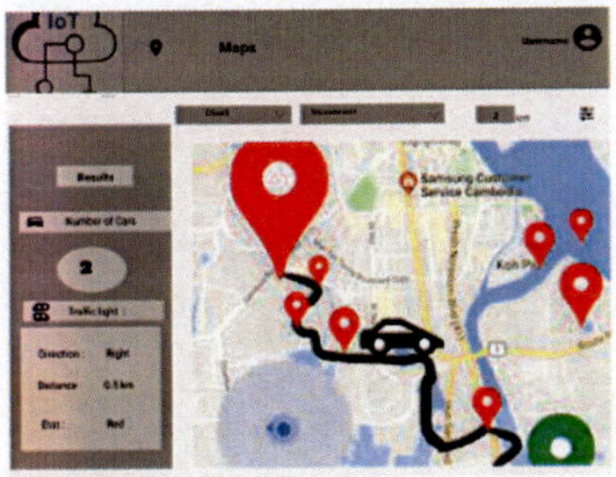

Fig. 10. A proposed plan to show real-time information of a corresponding vehicle context

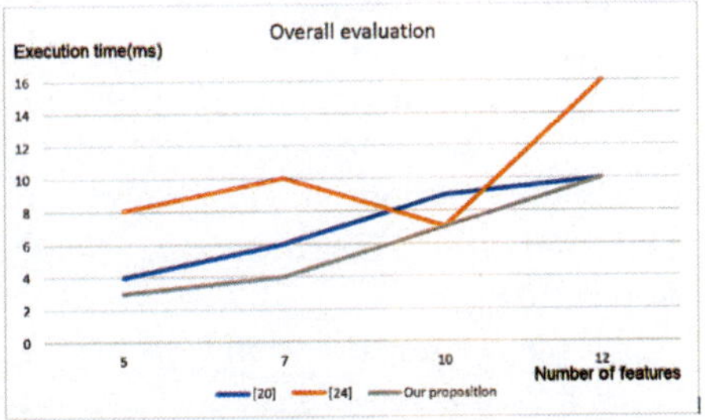

Fig. 11. Overall evaluation

For instance, to monitor vehicles' flow and calculate their number, we use our proposed algorithms which take data from an integrated camera in the sensors of each traffic light. In addition, traffic data is sent to our cloud Edge process. After the analysis of this flow of data, our cloud Edge gives the user the total number of vehicles closest to his location to know the situation of road congestion, as shown in Fig. 10.

Using different evaluation metrics which were the precision, the recall, and the F1 score, we obtained significant results that attest to the performance and the efficiency of our proposed algorithm. Also, we compared our proposition with other works in literature like [15] and [14].

In the intelligent transportation field, the authors in [14] recommended an agent-based information dissemination model for VANETs. The proposed model used the cognitive agent concept for realizing intelligent information dissemination. Some analysis performance parameters can be suggested, such as bandwidth, push latency and push/pull decision latency [14].

In Fig. 11, we have plotted the evaluation curve of the execution time as a function of the number of features.

Comparing different results to [15] results which are Cloud assisted, and after obtaining our metric scores which were precision, recall, F1-score, and the execution time as well, we can conclude that our proposal is more performing than those proposed by [15] and [14].

5 Conclusion and Future Work

In this paper, we have proposed three algorithms in order to monitor and analyze heterogeneous vehicle sensor data and road sensors. It focuses on an intelligent transportation system (ITS) using Edge Cloud computing called (ECC). The latter is used to interconnect IoT edge devices and the conventional cloud computing system.

Our proposition is a context-aware connected vehicle. It is able to monitor vehicles with different sensors and road contexts with the purpose of analyzing real-time data and providing users with an efficient plan suitable for any new context.

Then, to evaluate our proposition, we calculated the Recall, precision, and F1 Score values of each proposed algorithm.

Finally, we compared our proposition with other works in terms of execution time, which further demonstrated the efficiency and effectiveness of our proposition.

As for future directions, we will explore the output plans of the analyzing algorithm to achieve dynamic and automatic learning.

Acknowledgment. This work was funded by the "PHC Utique" program of the French Ministry of Foreign Affairs and Ministry of higher education and research and the Tunisian Ministry of higher education and scientific research in the CMCU project number 18G1431.

References

1. Agarwal, V., Sharma, S., Agarwal, P.: IoT based smart transport management and vehicle-to-vehicle communication system. In: Pandian, A.P., Fernando, X., Islam, S.M.S. (eds.) Computer Networks, Big Data and IoT. LNDECT, vol. 66, pp. 709–716. Springer, Singapore (2021). https://doi.org/10.1007/978-981-16-0965-7_55
2. Aliyu, A., et al.: Cloud computing in VANETs: architecture, taxonomy, and challenges. IETE Tech. Rev. **35**(5), 523–547 (2018)
3. Baiocchi, A., Colombaroni, C., Cuomo, F., De Felice, M., Fusco, G.: Vehicular traffic monitoring through VANETs: simulation and analysis in a real case study, pp. 1–11. University of Roma La Sapienza (2013)
4. Baiocchi, A., Cuomo, F., De Felice, M., Fusco, G.: Vehicular ad-hoc networks sampling protocols for traffic monitoring and incident detection in intelligent transportation systems. Transp. Res. Part C Emerg. Technol. **56**, 177–194 (2015)
5. De Felice, M., Baiocchi, A., Cuomo, F., Fusco, G., Colombaroni, C.: Traffic monitoring and incident detection through VANETs. In: 2014 11th Annual Conference on Wireless On-Demand Network Systems and Services (WONS), pp. 122–129. IEEE (2014)
6. Dourish, P.: Seeking a foundation for context-aware computing. Hum.-Comput. Interact. **16**(2–4), 229–241 (2001)

7. Friesen, M.R., McLeod, R.D.: Bluetooth in intelligent transportation systems: a survey. Int. J. Intell. Transp. Syst. Res. **13**(3), 143–153 (2015). https://doi.org/10.1007/s13177-014-0092-1
8. Ghebleh, R.: A comparative classification of information dissemination approaches in vehicular ad hoc networks from distinctive viewpoints: a survey. Comput. Netw. **131**, 15–37 (2018)
9. Gohar, A., Nencioni, G.: The role of 5G technologies in a smart city: the case for intelligent transportation system. Sustainability **13**(9), 5188 (2021)
10. Hamdi, M.M., Audah, L., Rashid, S.A., Alani, S.: VANET-based traffic monitoring and incident detection system: a review. Int. J. Electr. Comput. Eng. **11**(4), 3193–3200 (2021). ISSN: 2088-8708
11. Hassan, T., El-Mowafy, A., Wang, K.: A review of system integration and current integrity monitoring methods for positioning in intelligent transport systems. IET Intell. Transp. Syst. **15**(1), 43–60 (2021)
12. Horvitz, E., Dumais, S., Koch, P.: Learning predictive models of memory landmarks. In: Proceedings of the Annual Meeting of the Cognitive Science Society, vol. 26 (2004)
13. Jin, L., et al.: Self-powered wireless smart sensor based on maglev porous nanogenerator for train monitoring system. Nano Energy **38**, 185–192 (2017)
14. Kakkasageri, M., Manvi, S.: Intelligent information dissemination in vehicular ad hoc networks. Int. J. Ad Hoc Sens. Ubiquit. Comput. **2**, 112–123 (2011)
15. Liu, C., Ke, L.: Cloud assisted Internet of Things intelligent transportation system and the traffic control system in the smart city. J. Control Decis. **10**(2), 174–187 (2022)
16. Oppermann, R., Specht, M., Jaceniak, I.: Hippie: a nomadic information system. In: Gellersen, H.-W. (ed.) HUC 1999. LNCS, vol. 1707, pp. 330–333. Springer, Heidelberg (1999). https://doi.org/10.1007/3-540-48157-5_37
17. Othman, M.M., El-Mousa, A.: Internet of Things & cloud computing Internet of Things as a service approach. In: 2020 11th International Conference on Information and Communication Systems (ICICS), pp. 318–323. IEEE (2020)
18. Souki, O., Djemaa, R.B., Amous, I., Sèdes, F.: A Survey of Middlewares for self-adaptation and context-aware in Cloud of Things environment. Procedia Comput. Sci. **207**, 2804–2813 (2022)
19. Data.World: There are 990 open data datasets available on data world, June 2022. https://data.world/cityofchicago/array-of-things-locations
20. WSP: Systèmes de transport intelligents (STI), May 2022. https://www.wsp.com/fr-GL/services/systemes-de-transport-intelligents-sti

User-Oriented Process Analysis of Using the DIZU-EVG Instrument for Educational Video Games

Yavor Dankov(✉)

Faculty of Mathematics and Informatics, Sofia University "St. Kliment Ohdridski", Sofia, Bulgaria
yavor.dankov@fmi.uni-sofia.bg

Abstract. The paper focuses on educational video games and the users who play and learn through these games. More specifically, on the ability to visually analyze the results achieved by users of educational maze video games created using the APOGEE platform. This paper presents a **user-oriented process analysis of using the software tool DIZU-EVG (Data visualIZation instrUment for Educational Video Games)** focusing on analyzing the behavior of the players and learners. The paper furthers the development of the DIZU-EVG instrument designed by the author. The paper presents **an overview context model of the DIZU-EVG instrument in the platform**, explaining the primary communication and data exchange between the DIZU-EVG tool and the primary systems external to the instrument within the APOGEE platform. The creation and integration of the DIZU-EVG instrument into the APOGEE platform will enrich it with a new, original, and valuable tool that is not available on the platform. This tool will enable users to view their results, thanks to specialized dashboards personalized for each unique user, depending on their profile on the platform. The possibility of visual analysis of the results achieved by players and learners in a personalized way, according to the user's profile, will contribute to the overall understanding of the results achieved and an incentive to improve these results. Thus, users will be able to independently take timely individual actions to correct their failures in the game and increase the efficiency and effectiveness of their perception of the educational material integrated into the educational video game.

Keywords: Educational video games · Data visualization · Serious games · Game-based learning · User-oriented analysis · DIZU-EVG instrument

1 Introduction

Nowadays, more and more video game users need interactive and innovative ways to impact them, their emotions ultimately, and their game experiences. As a variety of so-called serious games, educational video games are gaining popularity and attracting the attention of diverse users and numerous developers, educators, and practitioners [1–4]. Educational video games are video games that combine the ability to meet consumer

R. Silhavy and P. Silhavy (Eds.): CSOC 2023, LNNS 723, pp. 684–693, 2023.
https://doi.org/10.1007/978-3-031-35317-8_61

needs for entertainment, immersion in a game environment, and interactivity, along with the educational and gaming components integrated into these games [5]. Educational video games train users of knowledge and skills on a particular topic and educate morals and values in consumers.

Educational video games result from the work of many specialists who undergo the design, development, testing, and integration of games into interactive, captivating architectural [6, 7] and aesthetic environments [8–10] to be accessible to users (learners and players), which is intended. Educational video games are applied and used in the learning process. Various educational and learning goals and strategies are integrated into these games, harmonizing with the relevant education system and program [11–14]. For these reasons, educational video games educate and upbrought specialized groups of learners who play these games. The learners perceive knowledge and skills on a specific educational topic interactively and entertainingly, thanks to the pedagogical and educational components integrated into the game and the integrated gaming components. Therefore, designers and creators of video games must design these games depending on the defined group of learners for which this game is intended. This is so that this game can be as helpful and practical as possible for these users, and they have the opportunity to perceive the material on which they are trained [15]. Therefore, among the critical factors in designing, creating (and improving the design of the already made games) educational video games, it is possible to determine users' characteristics, requirements and needs or define a group of trainees. The personalization of the content of educational video games and the adaptation of the game content to the dynamic behavior of the player or the learner in the game are possible precisely according to the characteristics of users and their behavior during gameplay [16–19].

Often, in the initial phases of educational game design, these characteristics of users (static and dynamic) are used to achieve and realize a practical and valuable game designed for the respective learners and players. Once the educational game has already been created, it is played by the players and learners. The users who played the game then evaluate it (through a particular way to collect feedback from players and trainees, such as a poll, interview, questionnaire, etc.). A complete analysis and evaluation of the game by its creators are made based on the obtained results. Depending on the overall assessment, designers may decide to take steps towards improving the game's design and enhancing the degree of satisfaction and learning and playing user experience in the game. An example of the successful implementation of the described process is the APOGEE platform for creating educational video games [20]. With the help of this platform, educational maze video games are designed and developed, which is realized with the help of numerous integrated assistive and analytics instruments supporting these processes [21, 22]. The design of educational games on the platform is predominantly oriented toward the users, or the so-called User-Oriented Design is applied in the APOGEE platform [23]. Through the collected characteristics for the learners and players, unique user models for learners and players for each unique user are created, distinguishing the user's playing style and learning style, preferences, etc. [24]. Every individual user has a platform account created, which stores all data on it, as well as the relevant delegated rights to use the various tools integrated into the APOGEE platform. With the help of this platform, many successful educational video games have been created, in which the author of this paper has made a high contribution to their creation and practical studies [25, 26].

The paper focuses on educational video games and the users who play and learn through these games. More specifically, on the ability to visually analyze the results achieved by users of educational maze video games designed and created using the APOGEE platform. The platform is described and developed in detail in numerous co-authored publications by the author of this paper [22, 27]. The platform is enriched with various tools supporting the game design and development processes [21, 22]. It is also improved with integrated analytics tools that help the evaluation process of the designed, created, and played games [22, 28].

This paper presents a **user-oriented process analysis of using the software tool DIZU-EVG (Data visualIZation instrUment for Educational Video Games)** [29], focusing on analyzing the behavior of the users (players and learners) who will use the tool. This paper furthers the author's research on the DIZU-EVG instrument designed by the author of this paper. The basic concept of the tool, as well as the detailed description and models of the instrument's functionalities, were presented in the previous publications of the author of this paper [29, 30] in his role as a PostDoc and Recognized researcher (R2) and implemented within the project "Young Scientists and Postdoctoral Students - 2", funded by the Bulgarian Ministry of Education and Science. The paper presents **an overview context model of the DIZU-EVG instrument in the platform**, which explains the primary communication and data exchange between the DIZU-EVG tool and the primary systems external to the instrument located in the APOGEE platform.

The creation and integration of the DIZU-EVG into the APOGEE platform will enrich it with a new and valuable tool that was not developed and available to the platform. This tool will enable educational maze video game users to view their results, thanks to specialized dashboards personalized for each unique user, depending on their profile on the platform. The possibility of visual analysis of the results achieved by users (players and learners) in a personalized way, according to the user's profile, will contribute to the overall understanding of the results achieved and an incentive to improve these results. Thus, users will be able to independently (without the intervention of educators and pedagogues) take timely individual actions to correct their failures in the game and increase the efficiency and effectiveness of their perception of the educational material integrated into the educational video game.

The paper continues as follows: Sect. 2 presents an overview context model of the communication of the DIZU -EVG instrument in the APOGEE platform. The model describes the primary communication and data exchange between the DIZU-EVG instrument and the other instruments within the APOGEE platform. Section 3 presents the proposed user-oriented process model of using the DIZU-EVG instrument functionalities. The model focuses on analyzing the instrument functionalities used by players and learners. Section 4 introduces the paper's conclusive words with the tool's contribution and its benefits to the instrument's users - players and learners.

2 Overview of the DIZU-EVG Communication in the Platform

The DIZU-EVG tool will enable the APOGE platform to have unique and valuable functionalities designed and presented in the author's previous publications [29, 30]. The original instrument provides all three categories of users with an easily accessible, customizable (personalization), according to the respective user's profile, way to visualize the game and learning results of educational video games.

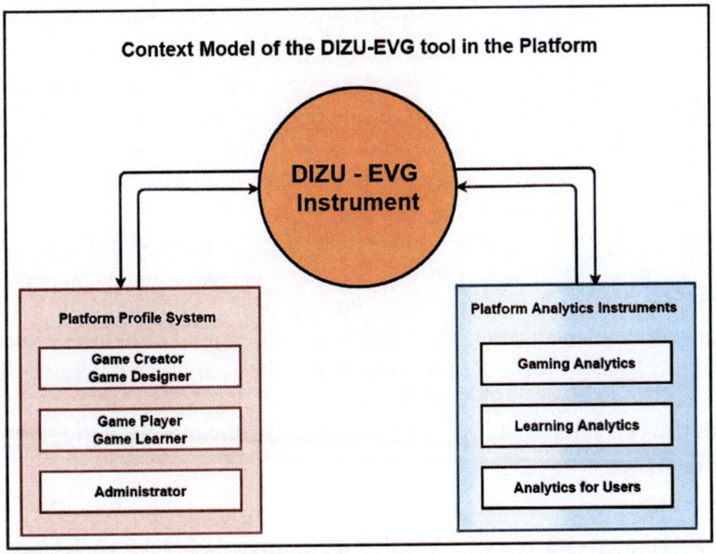

Fig. 1. Overview context model of the DIZU-EVG instrument in the platform

For the successful operation of the instrument, it is necessary to implement an analysis of the tool's communication with the available integrated software instruments in an APOGEE platform for the creation of educational maze video games, where the conduct of experimental and practical research will take place. For this purpose, **an overview context model of the communication of the DIZU -EVG instrument was developed.** The model is illustrated in Fig. 1.

The model presents **the primary communication and data exchange between the DIZU-EVG instrument and the primary systems external to the instrument located in the APOGEE platform**. Initially, the DIZU-EVG tool communicated with the profile system of the platform for each user to be recognized as a registered user and logged in to their profile on the software platform. This will enable the instrument to retrieve user profile data and grant the appropriate usage rights to the tool for that user, depending on the profile used. A necessary and preliminary condition is that the user has logged into his profile (*Game Designer / Game Creator; Game Player / Game Learner; Administrator* [29, 30]) in the platform to provide access to the relevant functionalities of the DIZU tool.

The model illustrated in Fig. 1 also presents the exchange of data and information between the DIZU-EVG instrument and the integrated analytics tools in the APOGEE platform. One of the most essential and fundamental functionalities of the DIZU tool is the ability to visualize gaming and learning results from educational video games. This is done by synchronizing the DIZU-EVG tool with the other integrated tools in the APOGEE platform and extracting the profiled result data from the analytics tools. In this regard, the analytics tools in the APOGEE platform are designed and created precisely to collect, process, and analyze the data generated by the game sessions and the behavior of the users on the platform. The analytics tools are responsible for collecting, processing, analyzing, and synthesizing this data and distributing it to the individual user profiles. These are the gaming and learning results profiled for individual users that are stored in the database of the analytics tools on the APOGE platform. The DIZU-EVG tool, through the *Gather Gameplay Data* functionality and its sub-functionalities (detailed in [29]), retrieves this profiled data to use it to create profiled dashboards for each unique user according to the profile. These profiled dashboards are created and stored in the instrument's database and are ready to be visualized to the users through the *Visualize Data* functionality of the tool.

The model (Fig. 1) can also be applied to the analysis of the communication and data exchange of the DIZU-EVG tool during its integration into other software platforms for creating educational video games. As a limiting condition, this platform must have a profile system that associates a given user with a unique profile to that user and also have integrated analytical tools responsible for analyzing and processing game data related to each user's profile. This will ensure the tool's ability to visualize customized dashboards for the relevant users, proof of the originality and capabilities of the DIZU-EVG instrument.

3 User-Oriented Process Model of Using the DIZU-EVG Instrument

This paper presents the **user-oriented process model of using the DIZU-EVG instrument**, described in Fig. 2. The model describes the analysis of the instrument functionalities used by users (players and learners) logged in to their relevant player and learner profiles in the platform.

The model **describes in detail the processes of using the primary functionalities of the instrument available to these users**. Learners or players intend to start using the tool. The process begins by accessing the instrument through interface and going through the *Access DIZU-EVG Instrument* process. As described in Sect. 2 and illustrated in Fig. 1, a prerequisite for using the functionalities of the DIZU-EVG instrument is that each user must have registration in the platform, and each user must be logged in to their profile, providing certain delegated rights to the user. Therefore, the initial step is to check that this user is already logged in to the APOGEE platform. If the user is not logged in to the platform, it goes to the *Log in to Platform Profile* process. Therefore, the system checks that the user already has a registered account, and if such a user is not registered, the user goes through the *Register New User* process The player or learner's individual data is filled in within this process, and its unique profile is created. In the second possibility, where a user is already registered, this user is redirected to the *Enter User Information* process, where the user must enter their username and password.

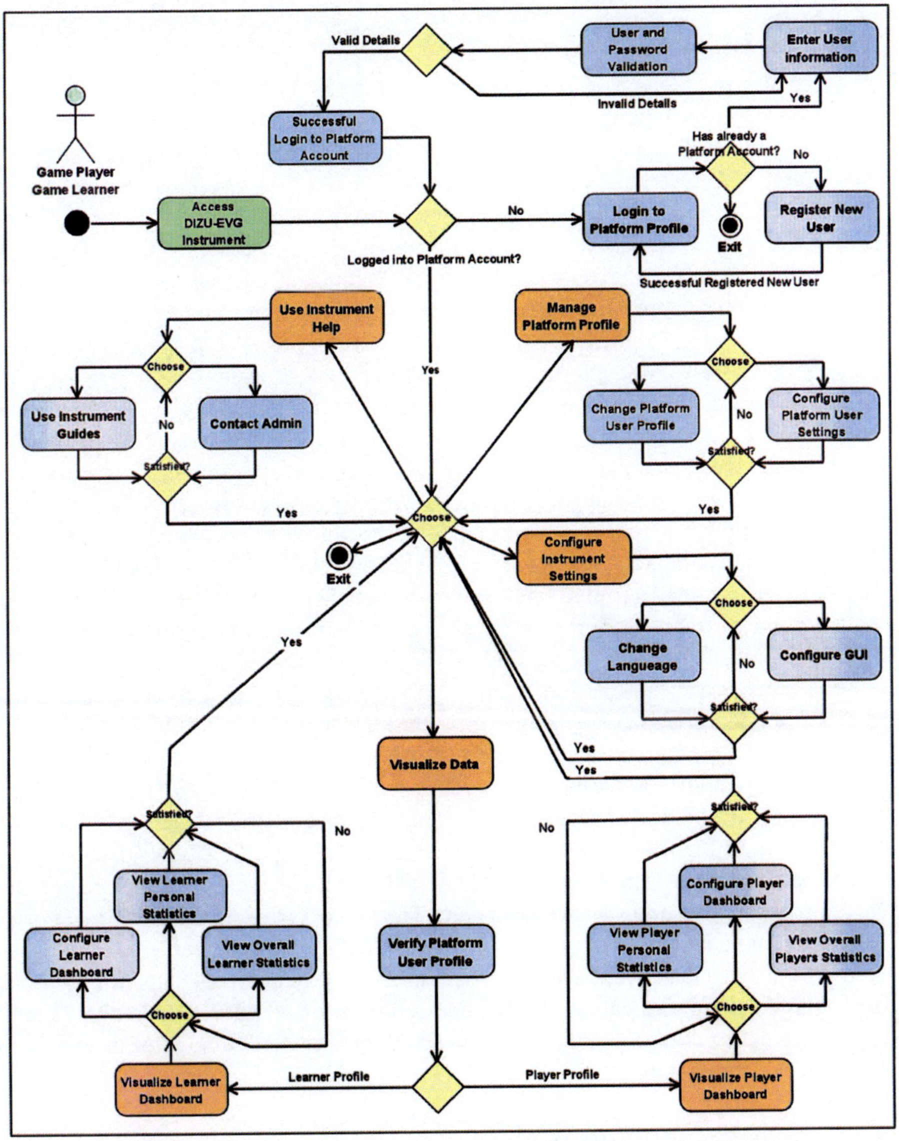

Fig. 2. User-oriented Process Model of Using the DIZU-EVG instrument

These data go through the data validation process, and again there are two possible options: 1) In one case, the data entered by the user does not successfully pass through the process of their validation, and accordingly, these data are not valid. In this case, the user must return to the *Enter User Information* process and re-enter their username and password; and 2) the second case where the user has entered their username and password, and this data successfully pass through the process of their validation - respectively the data entered are valid. In this case, the user goes through the *Successful Login to Platform Account* process, and therefore, the user can use the functionalities of the DIZU-EVG instrument.

Once players or learners have logged successfully into their profiles, they can choose from which primary functionality (accessible to players and learners) of the DIZU-EVG instrument to use. The model presents an analysis of the process of using the functionalities *Manage Platform Profile, Configure Instrument Settings, Use Instrument Help*, and *Visualize Data*. Suppose the user uses the functionality for the *Manage Platform Profile*. In that case, the user can decide to go through the selection processes to set up an account and change the profile if this user is registered as a player and learner and has learner and player profiles. Once the user is satisfied with the settings and changes made, the user ends the work, and the process goes to the starting position of choice between the primary functionality of the tool. The described process of using the *Configure Instrument Settings* functionality in the model is similar. In this process, players and learners can change the language of the tool, as well as change the characteristics of the graphic user interface and customize it according to their preferences. Using the *Use Instrument Help* functionality, users can contact an administrator (if there is a specific problem or inquiry) and utilize the functionality to support users through the instrument user guides. Therefore, upon completing the work with the relevant functionality, the system sends a request to the users to determine whether they are satisfied with the job. After completing the work, the instrument returns the users to the initial selection position for using the primary functionalities and the possibility of exit from the instrument.

The process of using the functionality to visualize data from games and learning results from educational video games goes through the verification of the user's profile process. Depending on whether the user has entered the platform as a player or a learner, the DIZU-EVG instrument provides access to the *Visualize Player Dashboard* functionality and the *Visualize Learner Dashboard* functionality. Players go through the *Visualize Player Dashboard* process, where users can view profiled player data as statistics on personal success in the game and general statistics according to other players and customize specific settings on the dashboard. The process of using the *Visualize Learner Dashboard* functionality is only available to users with profiles of learners. They can set up a personalized dashboard and review summaries and individual statistics presented on the dashboard for the learners. Again, upon completion of the work, users are referred to the original position for selecting between the primary functionalities or to exit the DIZU-EVG instrument.

4 Conclusion and Future Work

The creation and integration of the DIZU-EVG instrument into the APOGEE platform will enrich it with a new, original, and valuable tool that is not available on the platform. The paper presented an overview context model of the communication of the DIZU -EVG instrument that was developed. The model explained in detail the primary communication and data exchange between the DIZU-EVG tool and the primary systems external to the instrument located in the APOGEE platform. The user-oriented process model of using the DIZU-EVG instrument described the processes of using the primary functionalities of the instrument available to these users, logged in to their relevant player and learner profiles in the platform.

This tool will enable educational maze video game users to view their results, thanks to specialized dashboards personalized for each unique user, depending on their profile on the platform. The possibility of visual analysis of the results achieved by users (players and learners) in a personalized way, according to the user's profile, will contribute to the overall understanding of the results achieved and an incentive to improve these results. Thus, users will be able to independently (without the intervention of educators and pedagogues) take timely individual actions to correct their failures in the game and increase the efficiency and effectiveness of their perception of the educational material integrated into the educational video game.

Future work is planned to develop the instrument further and conduct experiments and surveys related to the use of the instrument by all users and stakeholders, including designers of educational video games. The design of the software architecture as well as the refinement of the instrument, are planned as future work.

Acknowledgements. This research is supported by the Bulgarian Ministry of Education and Science under the National Program "Young Scientists and Postdoctoral Students – 2".

References

1. Boyle, E., et al.: An update to the systematic literature review of empirical evidence of the impacts and outcomes of computer games and serious games. Comput. Educ. **94**, 178–192 (2016). https://doi.org/10.1016/j.compedu.2015.11.003
2. Riopel, M., et al.: Impact of serious games on science learning achievement compared with more conventional instruction: an overview and a meta-analysis. Stud. Sci. Educ. **55**(2), 169–214 (2019). https://doi.org/10.1080/03057267.2019.1722420
3. Kara, N.: A systematic review of the use of serious games in science education. Contemp. Educ. Technol. **13**(2), ep295 (2021). https://doi.org/10.30935/cedtech/9608
4. Ullah, M., et al.: Serious games in science education. A systematic literature review. Virtual Reality Intell. Hardware **4**(3), 189–209 (2022). https://doi.org/10.1016/j.vrih.2022.02.001
5. Dyulicheva, Y., Glazieva, A.: Game based learning with artificial intelligence and immersive technologies: an overview. In: Proceedings of CS&SE@SW 2021: 4th Workshop for Young Scientists in Computer Science & Software Engineering, 18 December 2021, Kryvyi Rih, Ukraine, CEUR Workshop Proceedings, vol. 3077 (2022)
6. Deykov, Y., Andreeva, A.: Current aspects of the virtual design of expo-environment - gallery, museum, church. In: V International Conference Modern Technologies in Cultural Heritage, vol. 5, pp. 17–22. Technical University of Sofia (TU-Sofia), ISSN: 2367–6523 (2017)
7. Andreeva, A.: Conceptual requirements for non-traditional exhibition spaces. Architect. Des. Art Aspects. Bul. J. Eng. Des. (40), 53–58, Mechanical Engineering Faculty, Technical University of Sofia (TU-Sofia). ISSN:1313–7530 (2019)
8. Deykov, Y.: Museum of museums. Digitized art in a virtual museum environment. In: Ivanova, M. (ed.) First International Conference Modern Technologies in Cultural Heritage, vol. 1, pp. 45–48, Technical University of Sofia (TU-Sofia). ISSN: 2367–6523 (2013)
9. Andreeva, A.: Design of experimental exhibition space. Bul. J. Eng. Des. (37), 27–33, Mechanical Engineering Faculty, Technical University of Sofia (TU-Sofia). ISSN:1313–7530 (2018)
10. Andreeva, A.: Colorful and general-artistic aspects of architecture and design. Viewpoints. In: Aesthetic Achievements of the Exhibition Activities of Technical University—Sofia 2009–2019, vol. 1, pp. 78–96. Technical University of Sofia (TU-Sofia) (2019)

11. Aleksieva-Petrova, A., Dorothee, A., Petrov, M.: A survey for policies and strategies for ICT implementation in the learning process. In: INTED2018 Proceedings, pp. 192–197 (2018)
12. Adame, E.A., et al.: Serious games and growth mindsets: an experimental investigation of a serious gaming intervention. Int. J. Game-Based Lear. (IJGBL) **12**(1) (2022). https://doi.org/10.4018/IJGBL.308787
13. Pozo, J., Cabellos, B., Sánchez, D.: Do teachers believe that video games can improve learning? Heliyon **8**(6) (2022). https://doi.org/10.1016/j.heliyon.2022.e09798
14. Alshar'e, M., Albadi, A., Jawarneh, M., Tahir, N., Amri, M.: Usability evaluation of educational games: an analysis of culture as a factor affecting children's educational attainment. Adv. Hum.-Comput. Interact. **2022** (2022). Article ID 9427405. https://doi.org/10.1155/2022/9427405
15. Ibrahim, R., Azami, H., Masrom, S., Magalingam, P., Yaacob, S., Aljuaid, N.: Towards an integrated educational games evaluation model: a review of study. In: 2022 International Congress on Human-Computer Interaction, Optimization and Robotic Applications (HORA), Ankara, Turkey (2022). https://doi.org/10.1109/HORA55278.2022.9799961
16. Göbel, S., Wendel, V.: Personalization and adaptation. In: Dörner, R., Göbel, S., Effelsberg, W., Wiemeyer, J. (eds.) Serious Games, pp. 161–210. Springer, Cham (2016). https://doi.org/10.1007/978-3-319-40612-1_7
17. Terzieva, V., Paunova-Hubenova, E., Bontchev, B.: Personalization of educational video games in APOGEE. In: Brooks, A., Brooks, E.I. (eds.) ArtsIT/DLI -2019. LNICSSITE, vol. 328, pp. 477–487. Springer, Cham (2020). https://doi.org/10.1007/978-3-030-53294-9_34
18. Bontchev, B., Naydenov, I., Adamov, I.: Controlling adaptation in affective serious games. In: 2021 International Conference Automatics and Informatics (ICAI), Varna, Bulgaria, pp. 159-162 (2021). https://doi.org/10.1109/ICAI52893.2021.9639656
19. Bontchev, B.P., Terzieva, V., Paunova-Hubenova, E.: Personalization of serious games for learning. Interact. Technol. Smart Educ. **18**(1), 50–68 (2021). https://doi.org/10.1108/ITSE-05-2020-0069
20. APOGEE Project site. https://apogee.online/. Accessed 3 Feb 2023
21. Dankov, Y., Bontchev, B., Terzieva, V.: Design and creation of educational video games using assistive software instruments. In: Ahram, T.Z., Karwowski, W., Kalra, J. (eds.) AHFE 2021. LNNS, vol. 271, pp. 341–349. Springer, Cham (2021). https://doi.org/10.1007/978-3-030-80624-8_42
22. Dankov, Y., Bontchev, B.: Towards a taxonomy of instruments for facilitated design and evaluation of video games for education. In: Proceedings of the 21st International Conference on Computer Systems and Technologies (CompSys-Tech2020), pp. 285–292. ACM, New York (2020). https://doi.org/10.1145/3407982.3408010
23. Dankov, Y., Antonova, A., Bontchev, B.: Adopting user-centered design to identify assessment metrics for adaptive video games for education. In: Ahram, T., Taiar, R. (eds.) IHIET 2021. LNNS, vol. 319, pp. 289–297. Springer, Cham (2022). https://doi.org/10.1007/978-3-030-85540-6_37
24. Terzieva, V., Bontchev, B., Dankov, Y., Paunova-Hubenova, E.: How to tailor educational maze games: the student's preferences. Sustainability **14**, 6794 (2022). https://doi.org/10.3390/su14116794
25. Paunova-Hubenova, E., Dankov, Y., Terzieva, V., Vassileva, D., Bontchev, B., Antonova, A.: Ready to play - a comparison of four educational maze games. In: Krouska, A., Troussas, C., Caro, J. (eds.) Novel & Intelligent Digital Systems: Proceedings of the 2nd International Conference (NiDS 2022). NiDS 2022. Lecture Notes in Networks and Systems, vol. 556, pp. 84–94. Springer, Cham (2023). https://doi.org/10.1007/978-3-031-17601-2_9
26. Bontchev, B., Antonova, A., Terzieva, V., Dankov, Y.: "Let us save Venice"—an educational online maze game for climate resilience. Sustainability **14**, 7 (2022). https://doi.org/10.3390/su14010007

27. Bontchev, B.; Vassileva, D.; Dankov, Y.: The APOGEE software platform for construction of rich maze video games for education. In: Proceedings of the 14th International Conference on Software Technologies (ICSOFT 2019). SCITEPRESS - Science and Technology Publications, Lda, Setubal, PRT, pp. 491–498 (2019). https://doi.org/10.5220/0007930404910498
28. Dankov, Y., Birov, D.: General architectural framework for business visual analytics. In: Shishkov, B. (ed.) BMSD 2018. LNBIP, vol. 319, pp. 280–288. Springer, Cham (2018). https://doi.org/10.1007/978-3-319-94214-8_19
29. Dankov, Y.: DIZU-EVG – an instrument for visualization of data from educational video games. In: Silhavy, R. (eds.) Software Engineering Perspectives in Systems. CSOC 2023. Lecture Notes in Networks and Systems, Springer, Cham (2022)
30. Dankov, Y.: Conceptual model of a data visualization instrument for educational video games. In: Abraham, A., Pllana, S., Casalino, G., Ma, K., Bajaj, A. (eds.) Intelligent Systems Design and Applications. ISDA 2023. Lecture Notes in Networks and Systems. Springer, Cham (2023)

Simulation Analysis of the Oligopoly Game in Telecommunications Industry and the Dynamic Pricing for 5G/6G Services

Sergey A. Vasilyev(✉), Daniel P. Acosta, Mohamed A. Bouatta, Igor V. Levichev, and Kanzitdinov S. Kanzitdinovich

Peoples' Friendship University of Russia (RUDN University), 6 Miklukho-Maklaya Street, Moscow 117198, Russian Federation
vasilyev-sa@rudn.ru
http://www.rudn.ru

Abstract. In this paper we consider a model of dynamic pricing in the 5G/6G telecommunications market with incomplete competition and use the simulation analysis of dynamic pricing for 5G/6G services in case when there is the oligopoly game in telecommunications industry. We consider network services pricing in 5G/6G telecommunications with multi-period technology diffusion model. The time path of 5G/6G telecommunication market demand, the price dynamics, the social surplus and the company profits, when the telecommunication companies implement new technologies, are studied. The relationship between users, the telecommunication companies and the telecommunication services costs for new 5G/6G technologies are investigated. The telecommunication industry equilibrium is considered with free telecommunication market entry and exit. We study the optimal multi-period dynamics of the telecommunication companies' profits for the case when the companies implements 5G/6G innovations using research and development (R&D). We analyze R&D expenditure on the services and the innovation process of 5G/6G technology, assuming a different number of the telecommunication companies undertaking R&D. We study the multi-period diffusion model of 5G/6G technology in the telecommunications market and also investigate the stability of this process depending on the number of companies that are present on the market.

Keywords: Access dynamic pricing · Oligopoly game in telecommunications industry · 5G/6G telecommunications · Research and Development (R&D) · Multi-period diffusion model · Simulation analysis

---

S. Vasilyev—This paper has been supported by the RUDN University Strategic Academic Leadership Program.

R. Silhavy and P. Silhavy (Eds.): CSOC 2023, LNNS 723, pp. 694–704, 2023.
https://doi.org/10.1007/978-3-031-35317-8_62

1 Introduction

Research and development (R&D) in telecommunications play an important role in industrial policy. 5G/6G mobile networks have become a necessary element of cities and suburbans infrastructures. 5G/6G technology requires high flexibility and intelligence in the following aspects: spectrum sharing, millimeter-wave (mmWave) communications, integrated Internet of Things (IoT) access, massive Multiple Input — Multiple Output (MIMO), smart antennas, Big Data, Cloud computing, and many other disruptive technologies [2,4,21,25,26]. One of the most efficient instrument in the modelling of telecommunication systems, service systems, traffic systems is queuing models [1,9,24].

5G/6G will see initial deployments in urban areas and this is perhaps fortuitous from an IoT perspective, since many new applications of IoT will support Smart Cities, Smart Campuses and Smart Buildings. Bandwidth demand for a number of Smart City applications is the main driver for enhanced mobile broadband (eMBB)-based 5G/6G services in general, and new-generation 5G/6G IoT applications, in particular. Recent economics models and cost analysis of 5G/6G network infrastructure have explored within of a single country infrastructure 5G/6G operators [7,8,11–14,16,18].

Over the last two decade, the deregulation of telecommunications industry plays an important role within government policy and in economic literature [10,13,14,17,18].

On the basis of an telecommunications directive on access and interconnection, the telecommunication services have already been defined as a relevant market in most countries of the world, since each telecommunication company has a monopoly power in case new technology service providing. The technological shifts of the telecommunications markets give companies opportunity for new technologies implementation and as a consequence resulted in asymmetries among companies. Same companies implement new technologies early (market technology leaders) and same companies late (market technology followers). Technology leaders have the opportunity to introduce a recent telecommunication technology which implies a lower cost of new services. But a leader who implements a new technology on the telecommunication market earlier has an extra market advantage and a market power because of reliability or reputation for quality, realizes in the form of extra utility for those users who wishes to subscribe for the new technology services (brand loyalty).

Thus, new technology asymmetries of telecommunication market, together with their rapid development, allow us to reconsider the role of the regulatory mechanism. In case new technology implementation the following questions arise: is the cost-based new service pricing reasonable, or do other ways of asymmetric or symmetric price regulation seem rational? What is the effect of new service price regulation on companies' profit, on consumer surplus, on social welfare? These questions are extremely important in connection with the implementation of the 5G/6G technology that is related to processes the generation of new technology (Research and Development, R&D).

In the literature on the technological changes it has become almost a convention to analyze and the diffusion of new technology, as separate components in an overall process [15,19]. In the R&D field, P. Dasgupta, J. Stiglitz [5], look at the determinants of R&D spending, and assume that any new technology generated is immediately used or diffused and B. Godin, G. Gaglio, D. Vinck [6], P. Stoneman, E. Bartoloni, M. Baussola [20] consider models of the spread of new technologies under the assumption that technology is provided regardless of the expected economic effect or the resulting distribution path.

This work is a continuation of the studies started in [22,23] where it is considered the telecommunication market model with optimal pricing scheme of 5G/6G services and mobile services pricing with the multiple access for 5G/6G technology.

In this paper we consider network services pricing in 5G/6G telecommunications with multi-period technology diffusion model whereby the time path of ownership of the telecommunication technology is determined by the interaction and the characteristics of the users, the telecommunication companies and the time profile of telecommunication services costs of the new 5G/6G technology.

In this paper we argue that there is a link between R&D spending and 5G/6G multi-period technology diffusion; that it is important to realize the existence of this link and that once this link is formally integrated into our modelling we gain a number of new insights into both the determination of R&D and the 5G/6G technology diffusion path. To illustrate our argument, before attempting the formal modelling, we can summarize the basic elements of our approach. We consider a 5G/6G technological advance that is embodied in 5G/6G network implementation that may be in great demand either users and/or manufacturers of devices and equipments. There is a set of users across which differing characteristics lead to demand for new telecommunication services being derived as a function of its price trajectory and its technological characteristics. New telecommunication services is provided by the telecommunication companies and, given the costs of a new technology implementation and services technological characteristics, the interaction of supply and demand determine a time profile of prices for new services, the companies profit and social surplus dynamics.

2 Innovation, Diffusion and 5G/6G Networks

In order to highlight and concentrate upon the linkages between the separate stages of 5G/6G technological change process we proceed by using a particularly simple multi-period diffusion model as opposed to a more complex multi-period model. Let 5G/6G services have a life cycle of at least n periods $(t = 1, 2, \ldots, n)$.

We assume that there are N_t $(N_1 < N_2,\, t = 1, 2, ..., n)$ numbers of potential users (NoPU) of 5G/6G services at the beginning of periods $t = 1, 2, \ldots, n$. The number and characteristics distribution of potential users are assumed sufficiently large. We may note that we assume that for any given user, the benefits he obtains are independent of the number of other users. This may suggest that the model is most appropriate to the case of a user stimulate innovation.

We suppose that there is a total market demand for 5G/6G services that is shown by the users. Let the function $D_t(p_t)$ be an individual demand of each user for 5G/6G services where $p_t \geq 0$ is a price of services (for example a monthly payment for using 5G/6G services in Monetary Units, MU).

We assume that dependence of price on demand has the form:

$$D_t(p_t) = \alpha_t - \beta_t p_t,\ t = 1, 2, \ldots, n,$$

where $\alpha_t > 0$ (Units of Services, UoS) and $\beta_t > 0$ (Units of Services per Monetary Units, UoSpMU) are positive parameters.

We admit that a total telecommunication market demand for 5G/6G services can be represented as

$$Q_t(p_t) = N_t D_t(p_t) = N_t(\alpha_t - \beta_t p_t) = A_t - B_t p_t,$$

where $A_t = N_t \alpha_t$, $B_t = N_t \beta_t$.

We can write the inverted total market demand in this way:

$$p_t(Q_t) = a_t - b_t Q_t, \tag{1}$$

$$a_t = A_t / B_t = \alpha_t / \beta_t > 0,\ b_t = 1/B_t = (N_t \beta_t)^{-1} > 0.$$

We suppose that users of 5G/6G technology perfect foresight on prices p_t (Monetary Units, MU) at the beginning of periods $t = 1, 2, \ldots, n$. The market diffusion path of 5G/6G services displays the result from the interaction of the period valuation functions $(p_t(Q_t), p_{t+1}(Q_{t+1}))$ or $(Q_t(p_t), Q_{t+1}(p_{t+1}))$ in the assumption of perfect foresight and effective investment in new technologies by companies.

Let M_t be the number of companies until periods $t = 1, 2, \ldots, n$ that provide a homogeneous service (the market 5G/6G service suppliers). We consider an oligopoly market model where a small group of large telecommunication companies (oligopolists) compete by choosing quantities of 5G/6G services.

We assume that each company has marginal costs c_t (where $c_t > c_{t+1}$), total fixed costs f_t, expenditures R_t on R&D during periods $t = 1, 2, \ldots, n$ and let a typical company sales $q_t (t = 1, 2, \ldots, n)$ units of services have the forms:

$$q_1 = \frac{Q_1}{M_1},\ q_{t+1} = q_t + \frac{Q_{t+1} - Q_t}{M_{t+1}},\ t > 1,$$

then we can write the total profit of leader $TP_{Lt,t+1}$ and follower $TP_{Ft,t+1}$ companies in this form:

$$TP_{Lt,t+1} = p_t q_t + p_{t+1} q_{t+1} - (c_t + c_{t+1}) q_t - c_{t+1} q_{t+1} - f_t - f_{t+1} - R_t,\ t \geq 1, \tag{2}$$

$$TP_{Ft,t+1} = p_{t+1} q_{t+1} - c_{t+1} q_{t+1} - f_{t+1},\ t \geq 1, \tag{3}$$

where follower companies enters the market at the next moment and have not expenditures R_t on R&D.

Under Cournot-Nash conjectures that each leader's company considers other competitor' behaviour regarding sales level as independent of its own, maximisation of yields

$$\frac{\partial TP_{Lt,t+1}}{\partial q_t} = p_t + q_t\frac{\partial p_t}{\partial q_t} + (p_{t+1} - c_{t+1})\frac{\partial q_{t+1}}{\partial q_t} - (c_t + c_{t+1}) = 0, \tag{4}$$

$$\frac{\partial TP_{Ft,t+1}}{\partial q_{t+1}} = p_{t+1} - c_{t+1} = 0, \tag{5}$$

where the equilibrium stability conditions have the form:

$$\frac{\partial^2 TP_{Lt,t+1}}{\partial q_t^2} \leq 0, \tag{6}$$

$$\frac{\partial^2 TP_{Ft,t+1}}{\partial q_{t+1}^2} \leq 0. \tag{7}$$

The equilibrium solution (q_t^*, q_{t+1}^*) of this system is obtained in the form:

$$q_t^* = r_t^{-1}[(a_{t+1} - c_{t+1})k_{t+1} - c_t], \tag{8}$$

$$q_{t+1}^* = [b_{t+1}M_{t+1}r_t]^{-1}[(b_{t+1}k_{t+1}^2M_{t+1}(k_{t+1}^2 - 1) - 2b_tM_t)c_{t+1} + \tag{9}$$
$$+(2b_tM_{t+1} - b_{t+1}k_{t+1}^2M_{t+1})a_{t+1}],$$

where $k_t = 1 - M_t/M_{t+1}$, $r_t = 2(b_tM_t - b_{t+1}k_{t+1}^2M_{t+1})$, and the equilibrium prices (p_t^*, p_{t+1}^*) have form:

$$p_t^* = a_t - b_tM_tq_t^*,\ p_{t+1}^* = a_{t+1} - b_{t+1}M_{t+1}q_{t+1}^*,$$

We can calculate the equilibrium industry demands until periods $t = 1, 2, \ldots, n$ form:

$$Q_t^*(p_t^*) = N_t(\alpha_t - \beta_tp_t^*),\ Q_{t+1}^*(p_{t+1}^*) = N_{t+1}(\alpha_{t+1} - \beta_{t+1}p_{t+1}^*).$$

The equilibrium total profit for each company if it provides 5G/6G services during all periods $t = 1, 2, \ldots, n-1$ (the market technology leader) have form:

$$TP_{L,t,t+1}^* = (p_t^* - c_t)q_t^* + (p_{t+1}^* - c_{t+1})(q_{t+1}^* - q_t^*) - f_t - f_{t+1} - R_t,$$

and the equilibrium total profit for each company if it provides 5G/6G services during only periods $t = 2, \ldots, n$ (the market technology follower) have form:

$$TP_{F,t,t+1}^* = (p_{t+1}^* - c_{t+1})q_{t+1}^* - f_{t+1},$$

where the equilibrium total industry profits during periods $t = 1, 2, \ldots, n$ can be obtained form:

$$TP_{t,t+1}^* = M_t\,TP_{L,t,t+1}^* + (M_{t+1} - M_t)\,TP_{F,t,t+1}^*.$$

The social surplus (SS^*) during periods $t = 1, 2, \ldots, n-1$ is the sum of the equilibrium total industry profits, TP^*, and the equilibrium total consumer surplus (CS^*) during all periods where the equilibrium consumer surplus during periods $t = 1, 2, \ldots, n$ are

$$CS_t^* = 0.5[a_t - p_t^*]Q_t^*(p_t^*), CS_{t+1}^* = 0.5[a_{t+1} - p_{t+1}^*]Q_{t+1}^*(p_{t+1}^*),$$

$$CS_{t,t+1}^* = CS_t^* + CS_{t+1}^*, \; SS_{t,t+1}^* = TP_{t,t+1}^* + CS_{t,t+1}^*.$$

The equilibrium stability condition gives the following inequalities:

$$\left|\frac{b_{t+1}}{b_t}\left(\frac{M_t}{M_{t+1}} - 1\right) - 1\right| < \sqrt{1 - 2\frac{b_{t+1}}{b_t}}, \; b_{t+1} < 0.5b_t.$$

For the diffusion path to 5G/6G services sales in both periods it is necessary that $Q_{t+1} > Q_t > 0$. This requires that, even with the perfect foresight assumed, some sales are made in next period, and also that it is profitable to make sales in previous period. We assume that these conditions hold and thus that a diffusion process over time does in fact take place. We may argue that a social surplus optimal diffusion would imply that all users are supplied by 5G/6G services in previous period for whom the premium from having the services in period is greater than or equal to the cost savings from delaying provide of 5G/6G services until next period.

Thus given perfect foresight, optimal diffusion will only be produced if the supplying industry is competitive. If a monopoly or oligopoly supplier can perfectly discriminate pricing a separate price to each buyer (a necessary condition for which was that buyers had myopic price expectations rather than perfect foresight), then the resulting diffusion path would be the same the social surplus optimal path. However under these conditions the monopoly supplier appropriates all the surplus as profits, whereas under perfect foresight with a large number of suppliers, although the diffusion path is social surplus optimal the suppliers have zero surplus, the users obtaining the benefits.

Thus, although the same diffusion path may result from different expectations and structural assumptions, the distribution cf the surplus will generally be very different. It is however precisely this surplus that is the incentive to the development of 5G/6G technology, and thus one may expect such development to be sensitive to both the expectations regime of 5G/6G technology buyers and the structure of the supplying industry.

3 Simulation Analysis of Oligopoly Game

For example, we can consider the 4-period model when there are only two companies in the oligopoly 5G/6G telecommunication market and each company offers for each user who owned devices with 5G/6G services. We assume that the number of the telecommunication companies is two ($M_1 = 2$, the duopoly) at the initial moment $t = 1$ and more than two ($M_t > 2$, the oligopoly) at the initial moments $t = 2, 3, 4$. Thus, the parameter $k = 1 - M_1/M_2$ changes within $k \in (0, 1)$ if $M_t \to \infty$ $(t = 2, 3, 4)$. The parameters of the demand function α_t, β_t, number of users N_t, marginal costs c_t, fixed costs f_t and expenditures in on R&D R_t have the values from Table 1.

Table 1. Model parameters

Parameters	$t=1$	$t=2$	$t=3$	$t=4$
α_t, UoS	5	10	12	35
β_t, UoSpMU	3,1	1,41	0,7	0,3
N_t, million NoPU	10	20	50	100
c_t, MU	0.005	0,002	0,001	0,001
f_t, million MU	0,5	0,2	0,1	0,1
R_t, million MU	10	0	0	0

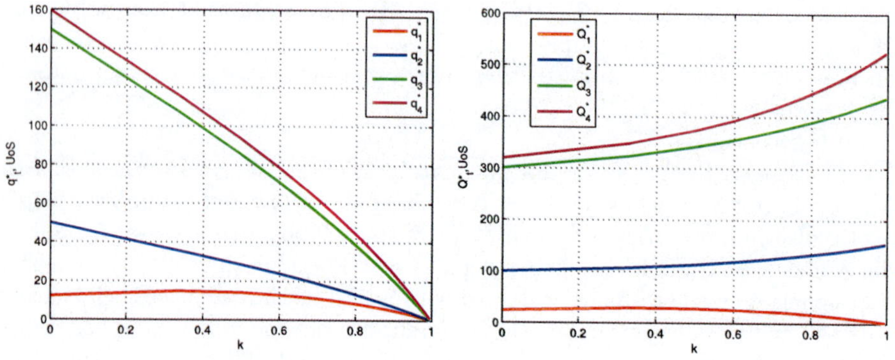

Fig. 1. The optimal company dynamics $q_t^*(k)$ (the left graph), the total market demand $Q_t^*(k)$ (the right graph).

Figure 1 illustrates the optimal dynamics of services sales $q_t^*(k)$ and the optimal industry dynamics of the total market demand $Q_t^*(k)$ for the 5G/6G services during the periods $t = \overline{1, 4}$. The optimal company sales of 5G/6G services $q_1^*(k)$ increase in the interval $k \in [0, 0.4]$, but then $q_1^*(k)$ decrease in the interval $k \in [0.4, 1]$ during the first period $t = 1$, and $q_t^*(k)$ decrease during each period $t = 2, 3, 4$, if following companies are expected to enter on the 5G/6G services market at the beginning of the next periods for $k \to 1$. We can see that the total market demand $Q_t^*(k)$ increase in the interval $k \in [0, 0.4]$, but then $q_1^*(k)$ decrease in the interval $k \in [0.4, 1]$ during the first period $t = 1$, and $q_t^*(k)$ increase during each period $t = 2, 3, 4$, when the parameter increase $k \to 1$. Figure 2 demonstrates the optimal industry prices dynamics $p_t^*(k)$ and the optimal total profits companies dynamics $TP_{L,1,2}^*(k)$, $TP_{F,1,2}^*(k)$ for the 5G/6G services during the

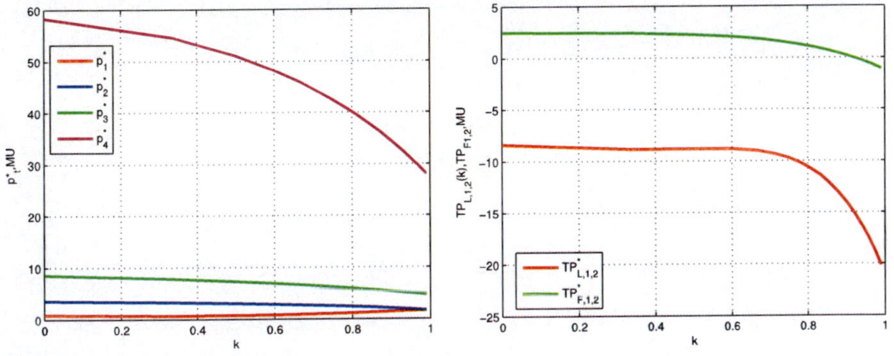

Fig. 2. The optimal prices dynamics $p_t^*(k)$ (the left graph), the optimal total profits companies dynamics $TP_{L,1,2}^*(k), TP_{F,1,2}^*(k)$ (the right graph).

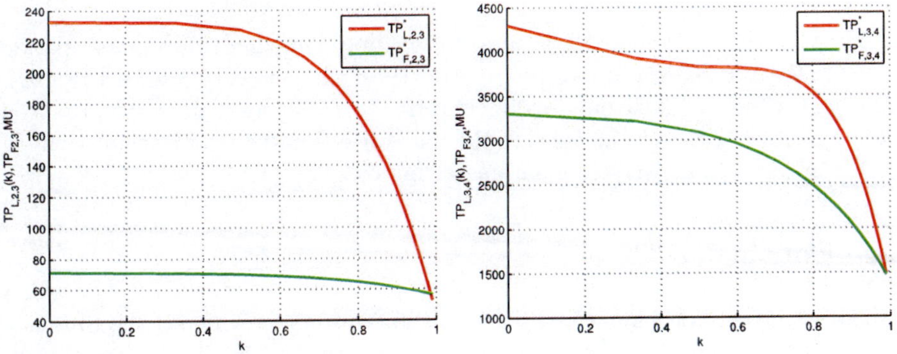

Fig. 3. The optimal total profits companies dynamics $TP_{L,2,3}^*(k), TP_{F,2,3}^*(k)$ (the left graph), the optimal total profits companies dynamics $TP_{L,3,4}^*(k), TP_{F,3,4}^*(k)$ (the right graph).

periods $t = \overline{1,4}$. The optimal 5G/6G services price dynamics $p_1^*(k)$ demonstrates increasing trend during the period $t = 1$. If a limited number of following companies grows on the 5G/6G services market at the beginning of periods $t = 2, 3, 4$, the optimal prices dynamics $p_t^*(k)$ demonstrates decreasing trend. Thus, we can see 5G/6G services prices are lower, if a large number following companies plan to enter on this market at the beginning of periods $t = 2, 3, 4$. Thus, if industry competition increases, companies profits decrease what causes the company to reduce 5G/6G services prices. New users are interested in attractive 5G/6G services prices and industry demand increases.

Figures 2, 3 and 4 show optimal the optimal industry total profits companies dynamics $TP_{L,t,t+1}^*(k)$, $TP_{F,t,t+1}^*(k)$ and the total industry profits $TP_{t,t+1}^*(k)$ dynamics during the periods $t = \overline{1,4}$, which decrease in all periods, when the parameter increase $k \to 1$. Optimal total profits of leading companies $TP_{L,1,2}^*(k)$ is less than optimal total profits of following companies $TP_{F,1,2}^*(k)$, what is

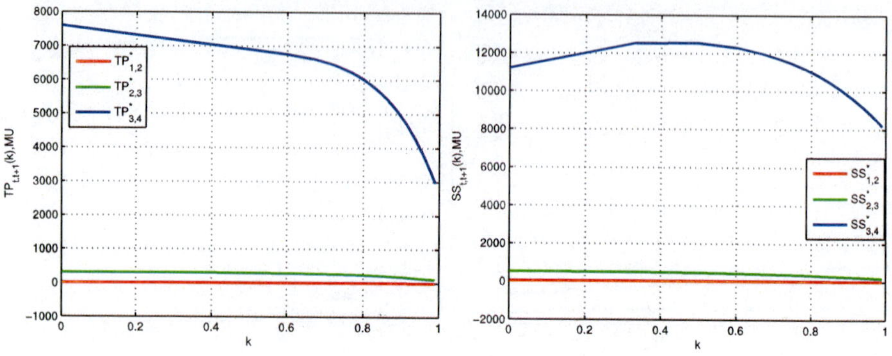

Fig. 4. The total industry profits $TP^*_{t,t+1}(k)$ dynamics (the left graph), the social surplus $SS^*_{t,t+1}(k)$ dynamics (the right graph).

related to the additional R&D costs of leading companies during the periods $t = 1$. We can see the opposite trend in the next periods $t = 2, 3, 4$. Figure 4 illustrates the social surplus $SS^*_{t,t+1}(k)$ dynamics, where it decrease in periods $t = 1, 2, 3$, but $SS^*_{3,4}(k)$ increase in the interval $k \in [0, 0.4]$ and it decrease in the interval $k \in [0.4, 1]$, when the parameter increase $k \to 1$.

4 Conclusion

In this paper we develop a multi-period diffusion model whereby the time path of ownership of the 5G/6G technology is determined by the interaction and the characteristics of the users and the telecommunication industry and the time profile of implementation 5G/6G technology costs of new services. The nature of profit companies stream provides the incentive to 5G/6G technology R&D. By specifying the competitive game in R&D and the R&D production relationship one can proceed to find R&D spending, the costs of implementation new telecommunication services and/or its technological characteristics. This closes the circle enabling one to consider as endogenously determined R&D, the extent of telecommunication technological advance and/or the production cost of the new equipments, the market structure of the telecommunication industry, the time profile of prices for new services and its diffusion path.

We can use economical these methods that may be useful for of 5G/6G, Internet of Things (IoT), big data and Smart Cities investment analysis such as the payback period of the investment, the index of profitability of investments, the net present value, the internal rate of return, the average weighted period of the life cycle of a new technology investment projects. The deployment of 5G/6G technologies is a very difficult economic task and requires the involvement of all participants in the telecommunications market. The governments of many countries support the concept of creating a single infrastructure operator for the development of 5G/6G networks. Our model can be useful for analyzing the dynamics of the telecommunications industry in the case of the presence of a single infrastructure operator on the 5G/6G market.

This paper has been supported by the RUDN University Strategic Academic Leadership Program (recipient S.A. Vasilyev, mathematical model development, simulation model development, numerical analysis).

References

1. Adou, Y., Markova, E., Gaidamaka, Y.: Modeling and analyzing preemption-based service prioritization in 5G networks slicing framework. Future Internet **14**(10), 299 (2022). https://doi.org/10.3390/fi14100299
2. Attanasio, B., La Corte, A., Scata, M.: Evolutionary dynamics of MEC's organization in a 6G scenario through EGT and temporal multiplex social network. ICT Express **7**(2), 138–142 (2021). https://doi.org/10.1016/j.icte.2021.05.006
3. Bouras, C., Kokkalis, S., Kollia, A.: Techno-economic comparison of MIMO and DAS cost models in 5G networks. Wireless Netw. **26**, 1–15 (2020). https://doi.org/10.1007/S11276-018-1780-6
4. Cuomo, S., Somma, V.D., Piccialli, F.: Pricing estimation of a barrier option in an IoT scenario. Future Gener. Comput. Syst. **110**, 407–412 (2020). https://doi.org/10.1016/J.FUTURE.2018.01.027
5. Dasgupta, P., Stiglitz, J.: Industrial structure and the nature of innovative activity. Econ. J. **90**, 266–293 (1980). https://doi.org/10.7916/D8BZ6H08
6. Godin, B., Gaglio, G., Vinck, D.: Handbook on Alternative Theories of Innovation, Edward Elgar Publishing, Cheltenham (2021)
7. Hui, H., Ding, Y., Shi, Q., Li, F., Song, Y., Yan, J.: 5G network-based internet of things for demand response in smart grid: a survey on application potential. Appl. Energy **257**, 113972 (2020). https://doi.org/10.1016/j.apenergy.2019.113972
8. Knieps, G., Bauer, J.M.: Internet of things and the economics of 5G-based local industrial networks. Telecommun. Policy **46**(4), 102261 (2022). https://doi.org/10.1016/j.telpol.2021.102261
9. Kondratyeva, A., et al.: Characterization of dynamic blockage probability in industrial millimeter wave 5G deployments. Future Internet **14**(7), 193 (2022). https://doi.org/10.3390/fi14070193
10. Li, X., Nosheen, S., Haq, N.U., Gao, X.: Value creation during fourth industrial revolution: use of intellectual capital by most innovative companies of the world. Technol. Forecast. Soc. Chang. **163**, 120479 (2021). https://doi.org/10.1016/j.techfore.2020.120479
11. Maeng, K., Kim, J., Shin, J.: Demand forecasting for the 5G service market considering consumer preference and purchase delay behavior. Telematics Inform. **47**, 101327 (2020). https://doi.org/10.1016/j.tele.2019.101327
12. Moussaoui, M., Bertin, E., Crespi, N.: Telecom business models for beyond 5G and 6G networks: towards disaggregation? In: 1st International Conference on 6G Networking (6GNet), Paris, France, vol. 2022, pp. 1–8 (2022). https://doi.org/10.1109/6GNet54646.2022.9830514
13. Oughton, E.J., Lehr, W.: Surveying 5G techno-economic research to inform the evaluation of 6G wireless technologies. IEEE Access **10**, 25237–25257 (2022). https://doi.org/10.1109/ACCESS.2022.3153046

14. Oughton, E.J., Jha, A.: Supportive 5G infrastructure policies are essential for universal 6G: assessment using an open-source techno-economic simulation model utilizing remote sensing. IEEE Access **9**, 101924–101945 (2021). https://doi.org/10.1109/ACCESS.2021.3097627
15. Palm, A.: Innovation systems for technology diffusion: an analytical framework and two case studies. Technol. Forecast. Soc. Chang. **182**, 121821 (2022). https://doi.org/10.1016/j.techfore.2022.121821
16. Parcu, P.L., Innocenti, N., Carrozza, C.: Ubiquitous technologies and 5G development. Who is leading the race?, Telecommun. Policy **46**(4), 102277 (2022). https://doi.org/10.1016/j.telpol.2021.102277
17. Peng, Y., Li, J., Hai, H., Jiang, X.-Q., Al-Hazemi, F., Park, S.: Cost optimization of distributed data centers via computing workload distribution for next generation network systems. Phys. Commun. **46**, 101340 (2021). https://doi.org/10.1016/j.phycom.2021.101340
18. Rendon Schneir, J., et al.: Cost assessment of multi-tenancy for a 5G broadband network in a dense urban area. Digital Policy Regul. Gov. **22**(2), 53–70 (2020). https://doi.org/10.1108/DPRG-10-2019-0086
19. Sidorov, S., Faizliev, A., Balash, V., Balash, O., Krylova, M., Fomenko, A.: Extended innovation diffusion models and their empirical performance on real propagation data. J. Mark. Anal. **9**, 99–110 (2021). https://doi.org/10.1057/s41270-021-00106-x
20. Stoneman, P., Bartoloni, E., Baussola, M.: The Microeconomics of Product Innovation. Published to Oxford Scholarship (2018). https://doi.org/10.1093/oso/9780198816676.001.0001
21. Strinati, E.C., Barbarossa, S.: 6G networks: beyond shannon towards semantic and goal-oriented communications. Comput. Netw. **190**, 107930 (2021). https://doi.org/10.1016/j.comnet.2021.107930
22. Vasilyev, S., Levichev, I., Acosta, D. P.: Network asymmetries and pricing in 5G telecommunications. In: 12th International Congress on Ultra Modern Telecommunications and Control Systems and Workshops (ICUMT), Brno, Czech Republic, vol. 2020, pp. 173–178 (2020). https://doi.org/10.1109/ICUMT51630.2020.9222416
23. Vasilyev, S., Levichev, I., Acosta, D. P.: Dynamic pricing model for 5G/6G telecommunications. In: 13th International Congress on Ultra Modern Telecommunications and Control Systems and Workshops (ICUMT), Brno, Czech Republic, vol. 2021, pp. 85–90 (2021). https://doi.org/10.1109/ICUMT54235.2021.9631647
24. Yarkina, N., Correia, L.M., Moltchanov, D., Gaidamaka, Y., Samouylov, K.: Multi-tenant resource sharing with equitable-priority-based performance isolation of slices for 5G cellular systems. Comput. Commun. **188**, 39–51 (2022). https://doi.org/10.1016/j.comcom.2022.02.019
25. Ye, N., Yu, J., Wang, A., Zhang, R.: Help from space: grant-free massive access for satellite-based IoT in the 6G era. Digital Commun. Netw. (2021). https://doi.org/10.1016/j.dcan.2021.07.008
26. Zheng, Z., Wang, L., Zhu, F., Liu, L.: Potential technologies and applications based on deep learning in the 6G networks. Comput. Electr. Eng. **95**, 107373 (2021). https://doi.org/10.1016/j.compeleceng.2021.107373

Blockchain: A Background for the Sake of Understanding

Mahyuddin K. M. Nasution[1(✉)], F. Rizal Batubara[2], Marischa Elveny[2,3], Arif Ridha Lubis[2], and Rima Aprilia[4,5]

[1] Data Science and Computational Intelligence Research Group, Universitas Sumatera Utara, Medan 20155, Sumatera Utara, Indonesia
mahyuddin@usu.ac.id

[2] Postgraduate Student of Computer Science, Fasilkom-TI, Universitas Sumatera Utara, Medan 20155, Sumatera Utara, Indonesia

[3] Data Science and Computational Intelligence Research Group, Excellent Center of Innovation and New Science, Universitas Medan Area, Medan 20223, Indonesia

[4] Postgraduate Student of Mathematics, FMIPA, Universitas Sumatera Utara, Medan 20155, Sumatera Utara, Indonesia

[5] Mathematics Program Study, FST, Universitas Islam Sumatera Utara, Medan 20223, Indonesia

Abstract. Technology is always present later to answer the problems faced by humans. One of them is about security and ease of transaction. The concept is as conflicting affairs in both their answers and their resolutions. An answer provides insight into getting the appropriate technology. Technology has the characteristics of connecting, which is enough to help special human activities and avoid various behaviors from human interests. From an ever-evolving concept, blockchain is the answer, but it requires understanding to solve related issues, where it requires a background from a different perspective. This paper reveals the essence of human interest in interacting and transacting based on several conditions of human attitude towards this technology.

Keywords: Bitcoin · computer network · organization · technology · cryptography · social network

1 Introduction

Human interests are the background for the emergence of technology. Since the discovery of computers that frame reasoning as computer science, computer machines have brought about many changes in all aspects of life, especially the shift from manual to automatic contexts in all business affairs of any organization [1,2]. Humans develop an organization to merge interests and build partnerships to make them stronger. Organizations build their business with rules where they administrate all activities as interactions or transactions into documents [3]. These documents not only require protection at various levels,

R. Silhavy and P. Silhavy (Eds.): CSOC 2023, LNNS 723, pp. 705–717, 2023.
https://doi.org/10.1007/978-3-031-35317-8_63

but these documents are also assets for the organization [4]. Then the organization organizes it into a system that makes it easy to carry out transactions and interactions. After that, now an organization connects to other organizations through computer networks [5]. Following the principles of the social network, the computer network not only broadens organizational interactions but makes transactions easier to take an appropriate position [6,7]. However, the easier the transaction takes place, the transactions require increasingly better protection.

Organizations interact with one another, dealing among themselves in a pair or more based on their interests. Thus, the protection of organizations indirectly connects and separates these organizations. Organizations separate from each other along with their documents to form blocks, but organizations relate to each other with different levels of regulation to realize an order, something that is interrelated [8,9]. Therefore, using interrelation words - such as organization, document, and model - inspires the use of the word "block" [10,11], while words - such as interaction, transaction, and relationship - accumulate to the meaning of the word "chain" [12–14]. Chain and block provide new technological concepts that have the potential to improve human well-being, namely two words and their meanings which form the background of the importance of technology. This paper aims to elaborate on that concept - called a blockchain - in which understanding it is of interest today.

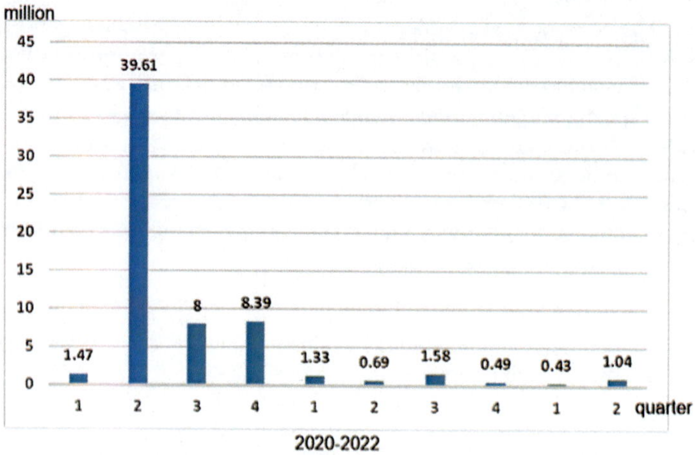

Fig. 1. Account hacking in Indonesia - Surfshark

2 Phenomenon and Paradigm

Behind convenience, security is the background for the need for the presence of technology. The emergence of the internet and the web has supported computer networks to bridge the gap between places, times, or cultures [15]. Disconnecting between the past and the present will potentially lose the future as distance also creates different cultures. The internet and the web as technology are not strong enough to change human attitudes towards their interests, especially those related to self-interest [16, 17]. In other words, while computer network technology has the potential to bridge various gaps, it cannot connect current customs and cultures without rules. Habits always have the potential to be at odds with natural reasoning or common sense [18,19]. Among that habits form behavior in which some humans betray nature for profit, and not a few people also abuse their neighbors and even their country for personal gain [20,21]. Therefore, there is no guarantee of the security of personal secrets, organizations to countries where documents or databases and systems play a role in constantly facing destructive tests: Theft of data and secrets or illegal transactions potentially from inside or outside the organization [22]. For example, Fig. 1 shows the number of accounts in Indonesia that were hacked in the stated time by the cybersecurity company: Surfshark. Globally, the same cybersecurity company has recorded that hackers do hacking 2.3 billion accounts since the beginning of 2020. Therefore, technology is needed to protect them, which is always vulnerable to breaches, and the answer may be based on community management or through organizations [23].

The human interest in welfare is the background for the interests of every human being in technology. Humans - when dealing with the level of welfare accomplishment - cause phenomena of human attitudes in dealing with this gap [24]. Data or information security always follows the challenges of that human attitude [25,26]. Its security has been the subject of discussion for a long time and has flowed with the flow of the times until now. Among these streams, some phenomena challenge changes in attitude and produce rules or provisions as agreements which generally refer to as paradigms, for example, cryptography in its particularity [27,28].

2.1 Phenomena for Consideration

For a long time, humans have fought over technology to make their lives easier. One of the technology is algorithms [29]. When various algorithms appear to improve computer performance, algorithms have become the most key asset in an organization, and until today algorithms have always been protected as intellectual property [30]. However, those algorithms do not significant if the real core of the technology presence is missing, namely data. Finally, people, organizations, or governments inevitably fight over data to dominate the world [31].

Phenomena are facts - and when the computer encodes them - the facts become data [32]. Thus, a phenomenon is a fact with a position: A fact that has a distance from its average more than other facts, where the experts usually describe it as a graphically remote point [33]. However, being present follows the flow of the same type of fact. Therefore, the tendency of interaction between any two organizations describes the bonds that grow and develop between the organizations to form a chain between them [34]. The flow of information - as a modality in interactions - is facts or data from one organization to another, and it may be that its implementation violates the agreed terms [35]. A disagreement causes disturbance in the interaction until dissolving the agreement [36]. So, every interaction requires data protection, protection that forms a block, which is something that comes from and becomes a paradigm [37]. Therefore, the phenomenon becomes the background for the interest in technology. Blockchain is the call of technology.

2.2 Paradigm for Solution

Based on the interests of every human being, the interests of each organization also vary according to their business. Each organization develops rules where the rules are attached to itself, which sometimes intersect with conditions from different organizations, which administratively, the organization implements into any system, in this case, an information system as a representation of that organization [38]. Any system will consist of several components: One component interrelates with another, namely structured and unstructured databases, software, hardware, and another. Organizations design, model, and build databases according to their interests and conditions. Also, the organization adapts to suit customers' interests [39], minimizes risks, and maximizes benefits based on theory or formulation of database models [40]. Likewise, databases by organizations implement appropriate data protection feasibility, use reliable software and robust database systems, and implement capable hardware against environmental challenges. Therefore, the system interface with the user must be familiar, at least automatically guiding its use or indirectly conveying messages to the user and his environment [41]. In this case, the hardware responds to the requirements of the system and its users by complementing the interface of any assistive software [42].

Many systems apply encryption and decryption techniques to store data based on a system administrator's key, but appropriate users may interpret it according to their level of authority [43,44]. However, there are not a few systems that only use keywords to access data in the database management system. Information systems generally contain private or public data. The information systems also consist of data that only organizations can access it. The guarantee of privacy data protection - by the organization and the system - is the main problem [40,41]. Failure to protect that data or track an error is to become their responsibility [45]. Even so, if accidentally thrown out of the system by the user for his private data, the system must be able to protect the data or at least have an early warning against any destruction. Even though public data may be freely accessed by anyone, to prevent misuse, the system always protects data [46]. Therefore, any transaction, whether in data or financial, must follow an agreement between organizations, where they put it in documents that mutually protect the system [47,48].

All protection by the system from the organization will chain the organization to other organizations. Thus, transactions between two organizations are transparent only between them and are autonomous but separate from the interactions between the organizations. The importance of the presence of a technology such as a blockchain - is the background for integrating several supporting technologies as a solution to various problems of technology.

3 Discussion

The presence of blockchain as a technology has a background from human awareness of having tools to achieve prosperity. They broaden human interests to the blockchain with their study interests as background to achieve that prosperity. These efforts are to understand the early literature on the emergence of blockchain.

An understanding of the ins and outs of data protection accumulates to a concept of technology known as the blockchain. Blockchain is a technology related to transactions between organizations [49–53]. Blockchain is a decentralized ledger for doing transactions safely [54–59], a built-in social network of organizations involving computer networks [60]. A transaction may involve digital currency [54,61,62], but it does not rule out that the transaction involves data [51,63,64], provided that the data is following its interests and origin has special protection. Thus, the ledger is a distributed database [54] technically decentralized [49,50], transparent [53], secured way [65,66], fairness [67] and openness [68], or electronic database [69] is also transactional [70,71] and public [56,71].

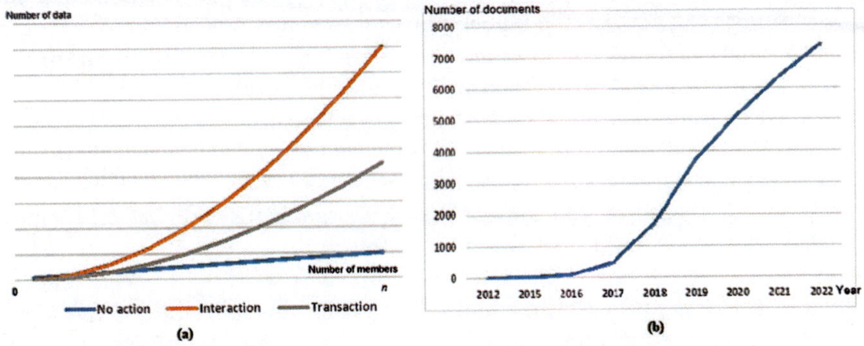

Fig. 2. (a) Predictive picture of data volume, (b) Number of documents based on title "Blockchain" in Scopus database

Each organization becomes a member of the network, wherein a social network, the members become the vertices of the graph, or $v \in V$, while transactions as directed edges from one organization to another, $e_{ij} = v_i v_j \neq v_j v_i = e_{ji}$, $e_{ij}, e_{ji} \in E$, and $v_i, v_j \in V$, and graph $G(E, V)$ as ledgers. Each $e \in E$ has its weight and direction. A transaction is a vector or block data structure of a distributed database where each block is timestamped and related to the previous block via any information. Therefore, blockchain is a distributed and immutable digital record system where sharing among many parties freely by following a consensus [49].

Considering the blockchain as a distributed database, which puts all financial transactions and not into it, naturally, the database has a structure according to the data model of various data structures. This distributed database has a variety of structures that have the potential to record data. As a ledger, of course, this database - together with software: database management systems and information systems - is a polyglot that resides on a server to serve the transactions and interactions of all members. The purpose of the database is to immortalize the block data structure as evidence of the transactions of individual members and their interactions [54]. Therefore, n members have $n(n-1)$ traffic of information as proof of their interactions and a database that records all this activity. So, there are $n(n-1)$ records, where those records grow in an ever-increasing accumulation according to the number of members. In other words, the blockchain is a distributed collection of digital event records [64]. In addition, there are potential $\frac{1}{2}n(n-1)$ transactions between members. The transaction is the distributed computation delivery of bitcoins or other cryptocurrencies [72]. Thus, the blockchain provides the underlying technology for bitcoin. In this case, blockchain is a keyword in the field of Financial Technology (Fintech) [73]. With a distributed consensus method over the internet infrastructure, a blockchain involves a distributed resource management framework and renders any single entity unable to subvert it [57]. With its ease of distribution, blockchain has attracted the attention of many governments/countries, financial institutions, high-tech companies, capital markets, or large and small companies [72], so some are competing to launch blockchain-enabled products [73]. It causes the block data structure grows following the graphic in Fig. 2(a): An overview of the complexity of data in interactions and transactions as the background of interest in the study of blockchain from a literature standpoint. Meanwhile, Fig. 2(b) explains human effort in understanding blockchain as a technology.

Therefore, a decentralized blockchain consists of databases with connections to one another based on the consensus of the organizations that own them, the polyglot. Such a structure causes each database to be distributed and shared between block structures. One database connects itself peer to peer, especially between investors and entrepreneurs [56,58]. The centralized database aims to support deals creation and transactions [54], procuring its infrastructure [49], and providing crypto-ledger platforms independent of others [50].

Blockchain - as a technology - is a radical innovation that impacts transactions, where payments, investments, and shares require security and legal processing [54]. Members - who exchange digital money - require secure transactions. For example, in transactions between members as customers and dealer goods, blockchain is a bridge that reconciles members' weaknesses and dealers' constancy [74]. With blockchain, members or organizations may own, control, and share their data with that security without limiting each other's privacy [56]. Because it is safe in transactions, this technology is efficient and low-cost [58].

Blockchain makes it possible to store transactions transparently so that it disrupts the monopoly rigidity of information [53] and ensures the transmission of information to the destination properly - especially regarding virtual money [75].

Transparency also causes any transaction information to be immutable retaining any information from transactions: including errors [76]. Thus, blockchain indirectly simplifies transactions by transferring equity [58].

The spirit of the internet has encouraged blockchain to contribute to transaction fairness [67], where data management principles apply in terms of measurement and authentication, coordination, and financing [68]. Thus, through the peer-to-peer transaction - fairness - there is compliance with regulations and security in data management and funds [58].

Internet capabilities also introduce openness to the blockchain, which makes the technology consistent in executing transactions [68], where there are limits to protecting privacy due to its exposure to public issues [77]. In this case, the blockchain guarantees every consideration of all members decide and carry out joint governance based on agreement [58].

Blockchain is technology as a human endeavor to increase convenience and provide services to policymakers. Before making a transaction, the members and policymakers know the ongoing or previous conditions, manage transactions according to regulations, and fight fraud and money laundering [58]. Perhaps, blockchain as a technology will encourage good law enforcement - avoiding unfair ways of upholding the law in Indonesia [78], for example - which results in ethically damaging the behavior of officials from the central level to the regions involved in cases about money laundering [79,80], which recently uncovered case (early 2023).

4 Conclusion

Understanding technology starts from the concept of the technology itself. Blockchain is a technology that has the potential to help people transact safely without worrying about the destructive attitude of human interests. Therefore, through understanding it, there are several characteristics that this technology must fulfill. The nature of it is some concept of understanding about it. Furthermore, research on this matter will bring up several implementations based on the understanding of all the studies in the literature as a background for the importance of this paper.

Acknowledgment. This paper is the result of doctoral dissertation research (Penelitian Disertasi Doktor, PDD) 2022 with the support of the Ministry of Education and Culture of the Republic of Indonesia.

References

1. Bohleber, P., Roman, M., Barbante, C., Vascon, S., Siddiqi, K., Pelillo, M.: Ice core science meets computer vision: challenges and perspectives. Front. Comput. Sci. **317**, 690276 (2021). https://doi.org/10.3389/fcomp.2021.690276
2. Nasution, M.K.M., Hidayat, R., Syah, R.: Computer science. Int. J. Adv. Sci. Eng. Inf. Technol. **12**(3), 11421159 (2022). https://doi.org/10.18517/ijaseit.12.3.14832

3. Kox, M., Staring, R.: If you don't have documents or a legal procedure, you are out! making humanitarian organizations partner in migration control. Eur. J. Criminol. **19**(5), 974–993 (2022). https://doi.org/10.1177/1477370820932079
4. Lugo, H., Segura, D., Michopoulos, V., Conway, P., West, A.: Secure document and asset tracking. J. Commun. Softw. Syst. **9**(1), 24–34 (2013). https://doi.org/10.24138/jcomss.v9i1.155
5. Zhou, Y., Chen, W., Li, L., Gong, L., Zhang, T.: Heterogeneous multi-agent system for brain-computer interaction in routing and forwarding with Memristive neuron networks. IEEE Trans. Parallel Distrib. Syst. **33**(11), 3233–3248 (2022). https://doi.org/10.1109/TPDS.2021.3137837
6. Nasution, M.K.M.: Social network extraction based on web: 1. Related superficial methods. In: IOP Conference Series: Materials Science and Engineering, vol. 300, no. 1 (2018). https://doi.org/10.1088/1757-899X/300/1/012056
7. Nasution, M.K.M.: Semantic interpretation of search engine resultant. In: IOP Conference Series: Materials Science and Engineering, vol. 300, no. 1 (2018). https://doi.org/10.1088/1757-899X/300/1/012053
8. Stockton, D.A., Fowler, C., Debono, D., Travaglia, J.: World Health Organization building blocks in rural community health services: an integrative review. Health Sci. Reports **4**(2) (2021). https://doi.org/10.1002/hsr2.254
9. Rapp, T.L., Davis, W.D., Gilson, L.L.: The 2022 conceptual issue: highlighting the individual, team, and organizational building blocks of effective organizations. Group Org. Manag. **47**(2), 143–147 (2022). https://doi.org/10.1177/10596011221089718
10. Ailem, M., Role, F., Nadif, M.: Sparse poisson latent block model for document clustering. IEEE Trans. Knowl. Data Eng. **29**(7), 1563–1576 (2017). https://doi.org/10.1109/TKDE.2017.2681669
11. Chabert-Liddell, S.-C., Barbillon, P., Donnet, S., Lazega, E.: A stochastic block model approach for the analysis of multilevel networks: an application to the sociology of organizations. Comput. Stat. Data Anal. **158** (2021). https://doi.org/10.1016/j.csda.2021.107179
12. Reimers, K., Li, M.: Interaction of technological and institutionai change in the development of an electronic commerce system in china's pharmaceutical distribution chain a transaction cost perspective. In: HEALTHINF 2008–1st International Conference on Health Informatics, Proceedings. 1st International Conference on Health Informatics, HEALTHINF 2008, 28 January 2008 through 31 January 2008, vol. 2, pp. 3–12 (2008). ISBN 978-989811116-6
13. Kobayashi, R., Hato, E.: Modeling of microscopic interactions among land transactions and recursive trip chains. In: Proceedings of the 25th International Conference of Hong Kong Society for Transportation Studies, HKSTS 2021: Sustainable Mobility. 25th International Conference of Hong Kong Society for Transportation Studies: Sustainable Mobility, HKSTS 2021 Hong Kong, 9 December 2021 through 10 December, pp. 443–451 (2021). ISBN 978-988158149-5
14. Syah, R., Nasution, M.M., Shol, V.V., Kireeva, N., Jalil, A.T., Chen, T.-C., Aravindhan, S., Abood, E.S., Alkaim, A.F.: Designing a green supply chain transportation system for an automotive company based on bi-objective optimization. Found. Comput. Decis. Sci. **47**(2), 193207 (2022)
15. Nasution, M.K.M., Nuradi, T.E., Syah, R.: SumutSiana: a framework for applying ICT to preserve the cultural heritage of Sumatera Utara Indonesia. J. Telecommun., Electr. Comput. Eng. **9**(2–4), 123128 (2017). ISSN 21801843

16. Oram, D.: The internet age - ethics, culture, time, place and requirements: the ethical triangle revisited. In: Proceedings of the 2nd International Conference on Internet Technologies and Applications ITA. 2nd International Conference on Internet Technologies and Applications, ITA 074 September 2007 through 7 September 2007, vol. 07, pp. 558–561 (2007). ISBN 978-094688154-3
17. Batubara, F.R., Ubacht, J., Janssen, M.: Challenges of blockchain technology adoption for e-government: a systematic literature review. In: ACM International Conference Proceeding Series (2018). https://doi.org/10.1145/3209281.3209317
18. Lubis, A.R., Nasution, M.K.M., Sitompul, O.S., Zamzami, E.M.: Obtaining value from the constraints in finding user habitual words. In: 2020 International Conference on Advancement in Data Science, E-Learning and Information Systems, ICADEIS 2020 (2020). https://doi.org/10.1109/ICADEIS49811.2020.9277443
19. Lubis, A.R., Nasution, M.K.M., Sitompul, O.S., Zamzami, E.M.: A new approach to achieve the users habitual opportunities on social media. IAES Int. J. Artif. Intell. **12**(1), 4147 (2023). https://doi.org/10.11591/ijai.v12.i1.pp41-47
20. Eckey, R., Agarwal, D.P., Goedde, H.W.: Genetic factors in alcohol metabolism: implications in alcohol drinking habits, alcohol abuse and alcoholism — [Genetisch bedingte Variabilites Alkoholstoffwechsels und ihr Einfluuf Trinkverhalten und Neigung zum Alkoholismus]. *Zeitschrift fr Rechtsmedizin* **103**(3), 169-190 (1990). https://doi.org/10.1007/BF00207339
21. Timsinha, S., Kar, S.M., Agrawal, P.: Drug abuse and alcohol consumption as a social habit in Nepal. Med.-Leg. Update **11**(2), 42–43 (2011). ISSN 0971720X
22. Nerurkar, P.: Illegal activity detection on bitcoin transaction using deep learning. Soft Comput. Article Press (2023). https://doi.org/10.1007/s00500-022-07779-1
23. Peng, Y., Huang, W.: Using blockchain technology and sharing culture to promote sustainable forest management in tribal communities. J. Environ. Public Health (2022). https://doi.org/10.1155/2022/1529407
24. Gottfried, R.R.: On gardening and human welfare, or, the role of attitudes and natural capital in sustainable welfare. Agric. Hum. Values **9**(4), 36–47 (1992). https://doi.org/10.1007/BF02217963
25. Hadlington, L., Binder, J., Stanulewicz, N.: Exploring role of moral disengagement and counterproductive work behaviours in information security awareness. Comput. Hum. Behav. **114** (2021). https://doi.org/10.1016/j.chb.2020.106557
26. Nasution, M.K.M.: World on data perspective. World **3**(3), 736–752. https://doi.org/10.3390/world3030041
27. Gudes, E., Koch, H.S., Stahl, F.A.: The application of cryptography for data base security. In: AFIPS Conference Proceedings - 1976 National Computer Conference AFIPS 1976, pp. 97–107 (1976). https://doi.org/10.1145/1499799.1499814
28. Gadde, S., Amutharaj, J., Usha, S.: A security model to protect the isolation of medical data in the cloud using hybrid cryptography. J. Inf. Secur. Appl. **73** (2023). https://doi.org/10.1016/j.jisa.2022.103412
29. Zhao, N.: Improvement of cloud computing medical data protection technology based on symmetric encryption algorithm. J. Test. Eval. **52**(1) (2024). https://doi.org/10.1520/JTE2021-0456
30. Cao, J.: Mode optimization and rule management of intellectual property rights protection of educational resource data based on machine learning algorithm. Complexity (2021). https://doi.org/10.1155/2021/1909518
31. Joamets, K.: Plagiarism as a legal phenomenon and algorithm-based decision making. TalTech J. Eur. Stud. **12**(2), 146–164 (2022). https://doi.org/10.2478/bjes-2022-0015

32. Prez Garrido, B., Sebrek, S.S., Semenova, V., Bal, D., Gbor, M.: Addressing the phenomenon of overtourism in budapest from multiple angles using unconventional methodologies and data. Sustainability (Switzerland) **14**(4) (2022). https://doi.org/10.3390/su14042268
33. Alieva, D.: Weber's contribution to the sociology of everydayness (Weberov prnos k sociolgii kadodennosti). Sociologia (Slovakia) **44**(5), 513–547 (2012). ISSN 00491225
34. Naudts, B., Tavernier, W., Verbrugge, S., Colle, D., Pickavet, M.: Deploying SDN and NFV at the speed of innovation: toward a new bond between standards development organizations, industry fora, and open-source software projects. IEEE Commun. Mag. **54**(3), 46–53 (2016). https://doi.org/10.1109/MCOM.2016.7432171
35. Nasution, M.K.M., Hardi, M., Sitepu, R.: Using social networks to assess forensic of negative issues. In: Proceedings of 2016 4th International Conference on Cyber and IT Service Management, CITSM 2016 (2016). https://doi.org/10.1109/CITSM.2016.7577513
36. Chambers, L.M.R.: Mind the gap: the consideration of financial technologies and blockchain in the reform of the vertical agreements block exemption regulation. Competition Law J. **18**(3), 116–121 (2019). https://doi.org/10.4337/clj.2019.03.04
37. Treiblmaier, H., Petrozhitskaya, E.: Is it time for marketing to reappraise B2C relationship management? The emergence of a new loyalty paradigm through blockchain technology. J. Bus. Res. **159** (2023). https://doi.org/10.1016/j.jbusres.2023.113725
38. Dolan, L., Kavanaugh, B., Korinek, K., Sandler, B.: Off the chain: blockchain technology—an information organization system. Tech. Serv. Q. **36**(3), 281–295 (2019). https://doi.org/10.1080/07317131.2019.1621571
39. Elveny, M., Syah, R., Nasution, M.K.M.: Taxonomy business model to customer behaviour in E-metric ecosystem. J. Adv. Res. Dy. Control Syst. **12**(7), 99104 (2020)
40. Nasution, M.K.M., Elveny, M.: Data modeling as emerging problems of data science. Data Sci. Semant. Technol.: Theor. Pract. Appl., 71–90 (2022). https://doi.org/10.1002/9781119865339.ch3
41. Nasution, M. K.M., Syah, R.: Data management as emerging problems of data science. Data Sci. Semant. Technol.: Theor. Pract. Appl., 91–104 (2022). https://doi.org/10.1002/9781119865339.ch4
42. Doan, T.T.H., Thiemann, P.: A typed programmatic interface to contracts on the blockchain. In: Oh, H. (ed.) APLAS 2021. LNCS, vol. 13008, pp. 222–240. Springer, Cham (2021). https://doi.org/10.1007/978-3-030-89051-3_13
43. Pillai, B.G., Lal, N.D.: Blockchain-based asymmetric searchable encryption: a comprehensive survey. Int. J. Eng. Trends Technol. **70**(7), 355–365 (2022). https://doi.org/10.14445/22315381/IJETT-V70I7P237
44. Kim, Y., Park, B., Kim, S.-Y.: A selective encryption/decryption method of sensitive music usage history information on theme, background and signal music blockchain network. J. Web Eng. **21**(4), 1265–1282 (2022). https://doi.org/10.13052/jwe1540-9589.21411
45. Giacalone, M., Santarcangelo, V., Donvito, V., Schiavone, O., Massa, E.: Big data for corporate social responsibility: blockchain use in Gioia del Colle DOP. Qual. Quant. **55**(6), 1945–1971 (2021). https://doi.org/10.1007/s11135-021-01095-w
46. Rainero, C., Modarelli, G.: Food tracking and blockchain-induced knowledge: a corporate social responsibility tool for sustainable decision-making. Br. Food J. **123**(12), 4284–4308 (2021). https://doi.org/10.1108/BFJ-10-2020-0921

47. Martnez-Ros, F.O., Marmolejo-Saucedo, J.A., Abascal-Olascoaga, G.: A new protocol based on blockchain technology for transparent operation of corporate social responsibility. In: Strategy, Power and CSR: Practices and Challenges in Organizational Management, pp. 205–233 (2020). https://doi.org/10.1108/978-1-83867-973-620201012
48. Ezzi, F., Jarboui, A., Mouakhar, K.: Exploring the relationship between blockchain technology and corporate social responsibility performance: empirical evidence from European firms. J. Knowl. Econ. Arti. Press (2022). https://doi.org/10.1007/s13132-022-00946-7
49. Avital, M., Beck, R., King, J.L., Rossi, M., Teigland, R.: Jumping on the blockchain bandwagon: Lessons of the past and outlook to the future. In: 2016 International Conference on Information Systems, ICIS 2016 (2016)
50. Pilkington, M.: Blockchain technology: principles and applications. In: Research Handbook on Digital Transformations, pp. 225–253 (2016). https://doi.org/10.4337/9781784717766.00019
51. Weber, I., Xu, X., Riveret, R., Governatori, G., Ponomarev, A., Mendling, J.: Untrusted business process monitoring and execution using blockchain. Lecture Notes in Computer Science (including subseries Lecture Notes in Artificial Intelligence and Lecture Notes in Bioinformatics) 9850 LNCS, pp. 329–347 (2016). https://doi.org/10.1007/978-3-319-45348-4_19
52. Xu, X., et al.: The blockchain as a software connector. In: Proceedings - 2016 13th Working IEEE/IFIP Conference on Software Architecture WICSA 2016, pp. 182–191 (2016). https://doi.org/10.1109/WICSA.2016.21
53. Petersen, M., Hackius, N., Kersten, W.: Blockchain for manufacturing and logistics - basics, benefits, and use cases. ZWF Zeitschrift fuer Wirtschaftlichen Fabrikbetrieb **111**(10), 626–629 (2016). https://doi.org/10.3139/104.111603
54. Singh, S., Singh, N.: Blockchain: future of financial and cyber security. In: Proceedings of the 2016 2nd International Conference on Contemporary Computing and Informatics IC3I 2016, pp. 463–467 (2016). https://doi.org/10.1109/IC3I.2016.7918009
55. Fanning, K., Centers, D.P.: Blockchain and its coming impact on financial services. J. Corp. Account. Finan. **27**(5), 53–57 (2016). https://doi.org/10.1002/jcaf.22179
56. Yue, X., Wang, H., Jin, D., Li, M., Jiang, W.: Healthcare data gateways: found healthcare intelligence on blockchain with novel privacy risk control. J. Med. Syst. **40**(10), 1–8 (2016). https://doi.org/10.1007/s10916-016-0574-6
57. Hari, A., Lakshman, T.V.: The Internet blockchain: A distributed, tamper-resistant transaction framework for the internet. In: HotNets 2016 - Proceedings of the 15th ACM Workshop on Hot Topics in Networks, pp. 204–210 (2016). https://doi.org/10.1145/3005745.3005771
58. Zhu, H., Zhou, Z.Z.: Analysis and outlook of applications of blockchain technology to equity crowdfunding in China. Financ. Innovat. **2**(1), 1–11 (2016). https://doi.org/10.1186/s40854-016-0044-7
59. Sundara, T.A., Gaputra, I., Aulia, S.: Study on blockchain visualization. Int. J. Inf. Vis. **1**(3), 76–82 (2017). https://doi.org/10.30630/joiv.1.3.23
60. Buccafurri, F., Lax, G., Nicolazzo, S., Nocera, A.: Tweetchain: an alternative to blockchain for crowd-based applications. In: Cabot, J., De Virgilio, R., Torlone, R. (eds.) ICWE 2017. LNCS, vol. 10360, pp. 386–393. Springer, Cham (2017). https://doi.org/10.1007/978-3-319-60131-1_24
61. Zhao, J.L., Fan, S., Yan, J.: Overview of business innovations and research opportunities in blockchain and introduction to the special issue. Financ. Innovat. **2**(1), 1–7 (2016). https://doi.org/10.1186/s40854-016-0049-2

62. Cao, S., Cao, Y., Wang, X., Lu, Y.: A review of researches on blockchain. In: 16th Wuhan International Conference on E-Business, WHICEB 2017, pp. 108–117 (2017)
63. Kim, K.J., Hong, S.P.: Study on rule-based data protection system using blockchain in P2P distributed networks. Int. J. Secur. Appl. **10**(11), 201–210 (2016). https://doi.org/10.14257/ijsia.2016.10.11.18
64. Sharples, M., Domingue, J.: The blockchain and kudos: a distributed system for educational record, reputation and reward. In: Verbert, K., Sharples, M., Klobučar, T. (eds.) EC-TEL 2016. LNCS, vol. 9891, pp. 490–496. Springer, Cham (2016). https://doi.org/10.1007/978-3-319-45153-4_48
65. Konashevych, O.: The concept of the blockchain-based governing: current issues and general vision. In: Proceedings of the European Conference on e-Government, ECEG Part F129463, pp. 79–85 (2017)
66. Ahram, T., Sargolzaei, A., Sargolzaei, S., Daniels, J., Amaba, B.: Blockchain technology innovations.: IEEE Technology and Engineering Management Society Conference. In: TEMSCON 2017, pp. 137–141 (2017)
67. Cai, J., Li, S., Fan, B., Tang, L.: Blockchain based energy trading in energy internet. Dianli Jianshe/Electr. Power Constr. **38**(9), 24–31 (2017). https://doi.org/10.3969/j.issn.1000-7229.2017.09.004
68. Zhang, N., Wang, Y., Kang, C., Cheng, J., He, D.: Blockchain technique in the Energy Internet: preliminary research framework and typical applications. Zhongguo Dianji Gongcheng Xuebao/Proc. Chin. Soc. Electr. Eng. **36**(15), 4011–4022 (2016). https://doi.org/10.13334/j.0258-8013.pcsee.161311
69. Lou, A.T.F., Li, E.Y.: Integrating innovation diffusion theory and the technology acceptance model: the adoption of blockchain technology from business managers' perspective. In: Proceedings of the International Conference on Electronic Business (ICEB), December 2017, pp. 299–302 (2017)
70. Kawase, Y., Kasahara, S.: Transaction-confirmation time for bitcoin: a queueing analytical approach to blockchain mechanism. In: Yue, W., Li, Q.-L., Jin, S., Ma, Z. (eds.) QTNA 2017. LNCS, vol. 10591, pp. 75–88. Springer, Cham (2017). https://doi.org/10.1007/978-3-319-68520-5_5
71. Ellervee, A., Matulevicius, R., Mayer, N.: A comprehensive reference model for blockchain-based distributed ledger technology. In: CEUR Workshop Proceedings 1979, pp. 320–333 (2017)
72. Yuan, Y., Wang, F.-Y.: Towards blockchain-based intelligent transportation systems. In: IEEE Conference on Intelligent Transportation Systems, Proceedings, ITSC, pp. 2663–2668 (2016). https://doi.org/10.1109/ITSC.2016.7795984
73. Yammanuru, P.R., Jain, A., Vinayakaram, H.: Enabling enterprise blockchain AppDev teams. In: Proceedings - 23rd Annual Conference on Advanced Computing and Communications, ADCOM 2017, pp. 34–39 (2017). https://doi.org/10.1109/ADCOM.2017.00012
74. Zhu, Y., Guo, R., Gan, G., Tsai, W.-T.: Interactive incontestable signature for transactions confirmation in bitcoin blockchain. In: Proceedings - International Computer Software and Applications Conference, vol. 1, pp. 443–448. https://doi.org/10.1109/COMPSAC.2016.142
75. Jemel, M., Serhrouchni, A.: Decentralized access control mechanism with temporal dimension based on blockchain. In: Proceedings - 14th IEEE International Conference on E-Business Engineering, ICEBE 2017 - Including 13th Workshop on Service-Oriented Applications, Integration and Collaboration, SOAIC 2017, pp. 177–182 (2017)

76. Sankar, L.S., Sindhu, M., Sethumadhavan, M.: Survey of consensus protocols on blockchain applications. In: 2017 4th International Conference on Advanced Computing and Communication Systems, ICACCS 2017 (2017). https://doi.org/10.1109/ICACCS.2017.8014672
77. Xu, L., et al..: Enabling the Sharing Economy: Privacy Respecting Contract based on Public Blockchain. In: BCC 2017 - Proceedings of the ACM Workshop on Blockchain, Cryptocurrencies and Contracts, co-located with ASIA CCS 2017, pp. 15–21 (2017). https://doi.org/10.1145/3055518.3055527
78. Sukardi, S., Purnama, H.R.: Restorative justice principles in law enforcement and democracy in Indonesia. J. Indonesian Legal Stud. **7**(1), 155–190 (2022). https://doi.org/10.15294/jils.v7i1.53057
79. Ali, M., Nurhidayat, S., Prabowo, M.S., Muhammad, R.: Corruption, asset origin and the criminal case of money laundering in Indonesian law. J. Money Laund. Control **25**(2), 455–466 (2022). https://doi.org/10.1108/JMLC-03-2021-0022
80. Rifai, E., Tisnanta, H.S.: Role of law enforcement to prevent cyber laundering and asset recovery from overseas. Int. J. Cyber Criminol. **16**(1), 110–122 (2022). https://doi.org/10.5281/zenodo.4766559

Application of Micropython to Design Industry 4.0 Solutions

Gabriel Gaspar[1,2(✉)], Juraj Dudak[1,2], Ivan Sladek[1], Vojtech Simak[3], Stefan Sedivy[4], and Lubos Halvon[5]

[1] Research Centre, University of Zilina, Univerzitna 8215/1, 010 26 Zilina, Slovakia
{gabriel.gaspar,juraj.dudak,ivan.sladek}@uniza.sk

[2] Faculty of Materials Science and Technology in Trnava, Slovak University of Technology in Bratislava, Jana Bottu 2781/25, 917 24 Trnava, Slovakia
{gabriel.gaspar,juraj.dudak}@stuba.sk

[3] Faculty of Electrical Engineering and Information Technology, University of Zilina, Univerzitna 8215/1, 010 26 Zilina, Slovakia
vojtech.simak@uniza.sk

[4] TNtech, S.R.O., Lucna 1014/9, 014 01 Bytca, Slovakia
ssedivy@tntech.eu

[5] National Forest Centre, T. G. Masaryka 2175/22, 960 01 Zvolen, Slovakia
lubos.halvon@nlcsk.org

Abstract. Microcontrollers are reaching into our lives from all sides and, over time, have replaced mechanically and electrically complex solutions, rendering them obsolete. They find their place in solutions ranging from the simplest handheld devices to advanced solutions for homes and professionals to highly demanding industry, science, and technology applications. They also have an indispensable place in automated control, from the most straightforward sensors, through actuators and actuators, to programmable logic controllers and control panels. Microcontrollers can be programmed using several familiar techniques. Assembler is suitable for handling the lowest registers, peripherals, and routines. For standard designs, the C language is used. For rapid prototyping, several solutions are available such as Micropython, Circuitpython, and similar. For this publication, Micropython has been chosen as an implementation of the higher-level Python language for selected microcontrollers. A problem with a PID controller was chosen as an example of rapid device prototyping. It is broadly used in industrial automation, and its control algorithm is one of the most used in industry. The purpose of the PID controller in this example is to control the output value according to the desired setpoint so that the system's output is achieved as accurately as possible. When selecting the type of controller, it is necessary to determine the transfer function of the controlled system and take them into account.

Keywords: Microcontroller · Micropython · PID controller

R. Silhavy and P. Silhavy (Eds.): CSOC 2023, LNNS 723, pp. 718–727, 2023.
https://doi.org/10.1007/978-3-031-35317-8_64

1 Introduction

Microcontrollers are reaching into our lives from all sides, and over time, they have replaced mechanically and electrically complex solutions that have become obsolete. They find their place in solutions ranging from the simplest handheld devices through advanced solutions for homes and professionals to highly demanding industry, science, and technology applications. In the same way, they have an indispensable place in technical means of automated control, from the most straightforward sensors, through actuators, motors, and grippers [1], to programmable logic controllers and control panels. Examples of applications are:

- IoT solutions, which already cover a broad application area. These are devices with applications in the field of personal consumption (also as wearable electronics), medical use (IoMT - Internet of Medical Things), and finally, in the industry (IIoT - Industrial Internet of Things).
- Single-purpose control boards and general-purpose logic controllers are used for deployment areas ranging from consumer electronics products (home appliances, telephones, TVs, and others) to industrial solutions such as actuators, sensors, controls, and more.
- Displays with added functionality - configurable HMI touch panels, e-ink displays, and various all-in-one solutions for industrial automation.

Microcontrollers can be programmed using several familiar techniques. An assembler is suitable for handling the lowest registers, peripherals, and routines. For standard designs, the C language is used. Several solutions are available for rapid application prototyping, such as Micropython, Cpython, and the like. For this publication, Micropython has been chosen as the implementation of the higher-level Python language for selected microcontrollers. Thanks to its widely available, proven libraries and programming approach, it has become a standard tool for rapid application development [2].

2 Materials and Methods

In this section, we discuss several aspects of the prototyped device design. Based on our previous research [3], we describe the hardware design using a microcontroller from the ESP32 family. We further discuss the system identification of the controlled system using MATLAB toolboxes. Finally, we discuss the implementation of the PID controller using the library for Micropython.

2.1 Hardware Design

For our design, we chose the NodeMCU-32S board [4] with the ESP-32S microcontroller in Fig. 1, which is designed to be scalable and adaptive. It contains two cores that can be individually controlled, and the clock frequency of the CPU is adjustable from 80 MHz to 240 MHz. The user can also turn the CPU off and use the low-power coprocessor to monitor peripherals for exceeding thresholds. The ESP32 includes a broad set of peripherals ranging from capacitive touch sensors, a Hall sensor SD card interface, Ethernet, high-speed SPI, UART, I2S, and I2C.

Fig. 1. NodeMCU-32S board.

An example of a simple controlled system we will consider is the circuit of a power resistor and a thermistor, according to Fig. 2, where the power resistor realizes the controlled heating.

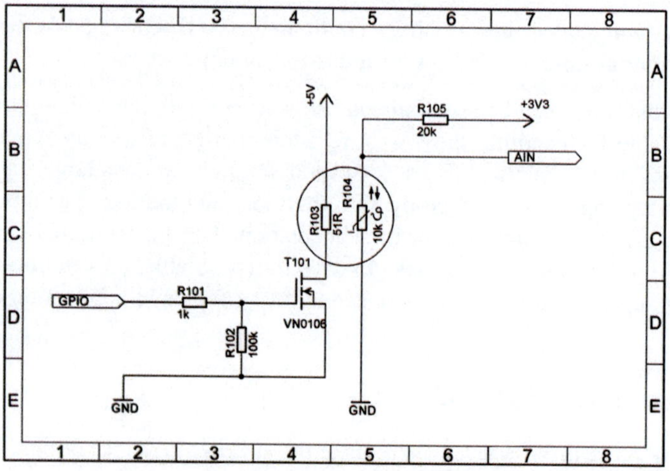

Fig. 2. The wiring of the regulated system for PID control.

In the above system, it is necessary to ensure good thermal contact, to eliminate the influence of ambient factors that could introduce a critical error in the control circuit. The transistor represents the actuator through which the heat (controlled variable) generated on the power resistor is manipulated. A change in the output voltage at the photoresistor proportional to the heat indirectly provides information about the controlled quantity.

2.2 System Identification

We started with measuring the regulated system's step response for the system identification task. We applied a PWM-modulated power supply to resistor R103 to control its heating. We connected thermistor R104 to a voltage divider in series with the resistor. The AD converter of the microcontroller realized the reading of the changing voltage values. The graphical representation of the transient characteristics of the regulated system is shown in Fig. 3.

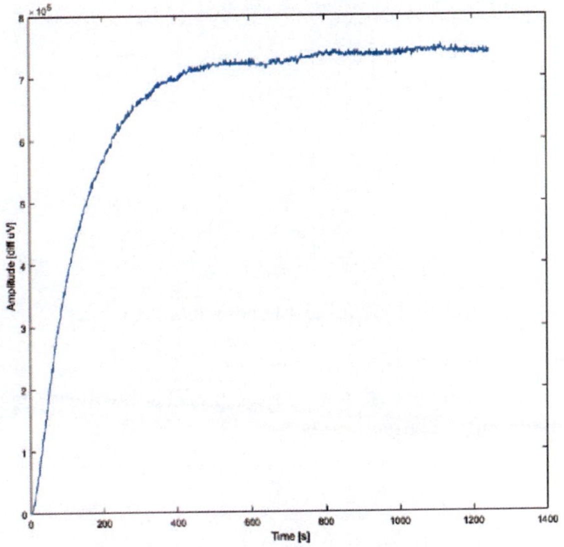

Fig. 3. Step response of the regulated system.

We then used MATLAB's *systemIdentification* toolbox to identify the system as a second-order system. A graphical comparison of the measured and calculated data is shown in Fig. 4.

The identified transfer function is shown in Eq. (1):

$$TF = \frac{141.1}{s^2 + 0.13s + 0.0009579} \quad (1)$$

2.3 PID Controller Tuning

After identifying the controlled system, we must tune the PID controller appropriately. There are several ways of achieving it, from which we have chosen tuning using a MATLAB toolbox. We inserted the identified transfer function into the workspace as follows:

```
sys=tf(141.1,[1 0.13 0.0009579])
```

For tuning the PID controller, we have chosen the *pidTuner* toolbox. We imported the transfer function and selected the controller type. Since the controlled system was

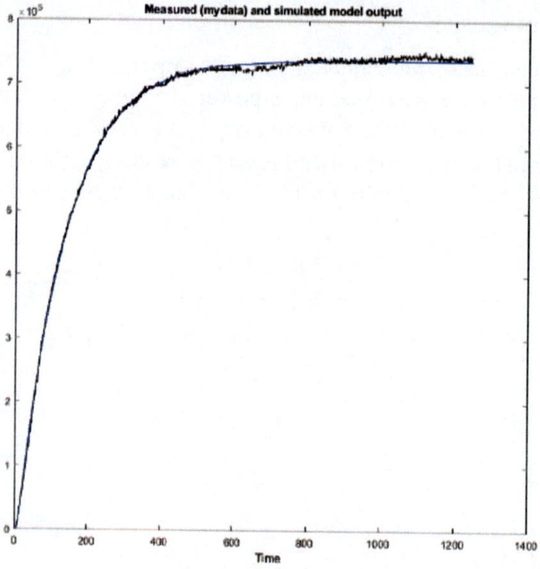

Fig. 4. SystemIdentification toolbox output

identified as a second-order system, we selected the PI controller type. The toolbox then offered the optimal control course in Fig. 5.

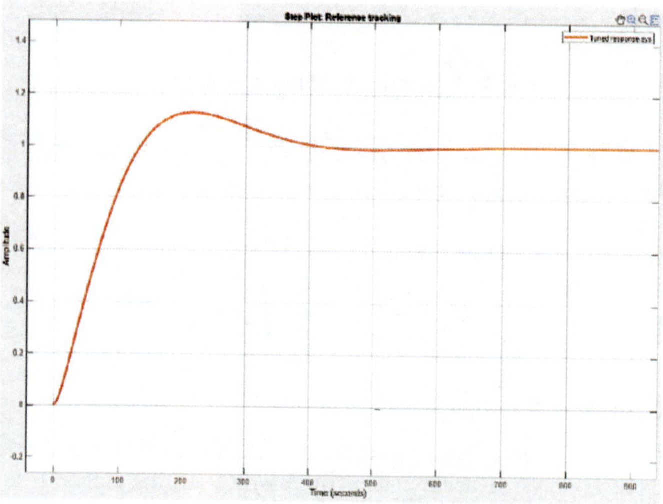

Fig. 5. PidTuner step plot reference tracking.

Figure 6 shows the optimized PI controller parameter, including other estimated parameters.

Controller Parameters

	Tuned
Kp	8.6749e-06
Ki	1.5922e-07
Kd	n/a
Tf	n/a

Performance and Robustness

	Tuned
Rise time	95.9 seconds
Settling time	373 seconds
Overshoot	12.7 %
Peak	1.13
Gain margin	Inf dB @ Inf rad/s
Phase margin	60 deg @ 0.0142 rad/s
Closed-loop stability	Stable

Fig. 6. PidTuner controller parameters.

We used the parameters to set the PID controller in the microcontroller program as follows: Kp = 2.3464 and Ki = 0.0385.

2.4 PID Controller Implementation

Proportional-integral-derivative (PID) controller is often used in industrial automation, and its control algorithm is one of the most used in industry. The PID algorithm consists of three essential components: proportional, integral, and derivative, which can be modified to achieve the optimum system response. The purpose of the PID controller is to control the output value according to the desired setpoint so that the system's output is achieved as accurately as possible. When selecting the type of controller, the transfer characteristics of the system to be controlled must be determined and considered. In Micropython, the implementation of the PID controller is provided in the software. There are several freely available modules, of which the *simple-pid* module [5], which houses the PID class, has been chosen for this article. The constructor of the *PID* class is as follows:

```
PID (Kp, Ki, Kd, setpoint=1, sample_time=None, scale='us')
```

Creates and returns a new PID object using the following parameters:

- Kp - proportional component (gain) of the controller,
- Ki - integral component of the controller,
- Kd - derivative component of the controller,
- setpoint - the desired value of the controlled variable,
- sample_time - sampling period, depends on the setting of the scale parameter,
- scale - sampling time in seconds' s', milliseconds' ms', microseconds' us', or nanoseconds' ns'.

After setting the above parameters of the PID controller, it can be used in a program loop. The principle of operation of the module is then as follows:

1. The current value of the regulated variable is inserted into the variable *val*:

```
val = controlled_system.update(0)
```

2. Subsequently, a new value of the control variable is calculated based on the current state of the system *val* and stored in the variable *u*, which is then written into the controlled system:

```
u = PID(val)
val=controlled_system.update(u)
```

The example software solution uses the methods of the *time* and *PID* (simple-pid) modules, the *PWM*, *Pin,* and *ADC* classes of the *machine* module. First, the parameters of the peripherals used need to be set. The PWM output is implemented on GPIO13 with a switching frequency of 1000 Hz and an initial duty cycle of 0%. The AD converter is implemented on GPIO36 with a 10-bit resolution and 11dB attenuation. The PID controller settings are set Kp = 0.1, Ki = 1, Kd = 0, setpoint = 2400000, with resolution in milliseconds and a sampling period of 50 ms. Consequently, in the infinite loop implemented by the while loop in Listing 1, occurs cyclically:

- measuring the value on the ADC (regulated variable),
- printing the value of the regulated variable to the terminal,
- calculating the value of the intervention (control variable) using the PWM output,